T0367553

Regression Modeling with Actuarial and Financial Applications

Statistical techniques can be used to address new situations. This is important in a rapidly evolving risk management and financial world. Analysts with a strong statistical background understand that a large data set can represent a treasure trove of information to be mined and can yield a strong competitive advantage.

This book provides budding actuaries and financial analysts with a foundation in multiple regression and time series. Readers will learn about these statistical techniques using data on the demand for insurance, lottery sales, foreign exchange rates, and other applications. Although no specific knowledge of risk management or finance is presumed, the approach introduces applications in which statistical techniques can be used to analyze real data of interest. In addition to the fundamentals, this book describes several advanced statistical topics that are particularly relevant to actuarial and financial practice, including the analysis of longitudinal, two-part (frequency/severity), and fat-tailed data.

Datasets with detailed descriptions, sample statistical software scripts in R and SAS, and tips on writing a statistical report, including sample projects, can be found on the book's Web site: http://research.bus.wisc.edu/RegActuaries.

INTERNATIONAL SERIES ON ACTUARIAL SCIENCE

The International Series on Actuarial Science, published by Cambridge University Press in conjunction with the Institute of Actuaries and the Faculty of Actuaries, will contain textbooks for students taking courses in or related to actuarial science, as well as more advanced works designed for continuing professional development or for describing and synthesizing research. The series will be a vehicle for publishing books that reflect changes and developments in the curriculum, that encourage the introduction of courses on actuarial science in universities, and that show how actuarial science can be used in all areas in which there is long-term financial risk.

There is an old saying, attributed to Sir Issac Newton:

"If I have seen far, it is by standing on the shoulders of giants."

I dedicate this book to the memory of two giants who helped me, and everyone who knew them, see farther and live better lives:

James C. Hickman
and
Joseph P. Sullivan.

Regression Modeling with Actuarial and Financial Applications

EDWARD W. FREES
University of Wisconsin, Madison

CAMBRIDGE
UNIVERSITY PRESS

University Printing House, Cambridge CB2 8BS, United Kingdom

One Liberty Plaza, 20th Floor, New York, NY 10006, USA

477 Williamstown Road, Port Melbourne, VIC 3207, Australia

4843/24, 2nd Floor, Ansari Road, Daryaganj, Delhi - 110002, India

79 Anson Road, #06-04/06, Singapore 079906

Cambridge University Press is part of the University of Cambridge.

It furthers the University's mission by disseminating knowledge in the pursuit of education, learning and research at the highest international levels of excellence.

www.cambridge.org
Information on this title: www.cambridge.org/9780521135962

First published 2010

A catalogue record for this publication is available from the British Library

Library of Congress Cataloging in Publication data
Frees, Edward W.
Regression modeling with actuarial and financial applications / Edward W. (Jed) Frees.
p. cm.
Includes index.
ISBN 978-0-521-76011-9 (hardback : alk. paper) 1. Insurance – Statistical methods. 2. Finance – Statistical methods. 3. Regression analysis. I. Title.
HG8781.F67 2010
519.5'36–dc22 2009032791

ISBN 978-0-521-13596-2 Paperback

Additional resources for this publication at http://research.bus.wisc.edu/RegActuaries

Contents

Preface

Actuaries and other financial analysts quantify situations using data – we are "numbers" people. Many of our approaches and models are stylized, based on years of experience and investigations performed by legions of analysts. However, the financial and risk management world evolves rapidly. Many analysts are confronted with new situations in which tried-and-true methods simply do not work. This is where a toolkit like regression analysis comes in.

Regression is the study of relationships among variables. It is a generic statistics discipline that is not restricted to the financial world – it has applications in the fields of social, biological, and physical sciences. You can use regression techniques to investigate large and complex data sets. To familiarize you with regression, this book explores many examples and data sets based on actuarial and financial applications. This is not to say that you will not encounter applications outside of the financial world (e.g., an actuary may need to understand the latest scientific evidence on genetic testing for underwriting purposes). However, as you become acquainted with this toolkit, you will see how regression can be applied in many (and sometimes new) situations.

Who Is This Book For?

This book is written for financial analysts who face uncertain events and wish to quantify the events using empirical information. No industry knowledge is assumed, although readers will find the reading much easier if they have an interest in the applications discussed here! This book is designed for students who are just being introduced to the field as well as industry analysts who would like to brush up on old techniques and (for the later chapters) get an introduction to new developments.

To read this book, I assume knowledge comparable to a one-semester introduction to probability and statistics – Appendix A1 provides a brief review to brush up if you are rusty. Actuarial students in North America will have a one-year introduction to probability and statistics – this type of introduction will help readers grasp concepts more quickly than a one-semester background. Finally, readers will find matrix, or linear, algebra helpful though not a prerequisite for reading this text.

Different readers are interested in understanding statistics at different levels. This book is written to accommodate the "armchair actuary," that is, one who passively reads and does not get involved by attempting the exercises in the text. Consider an analogy to football or any other game. Just like the armchair quarterback of football, there is a great deal that you can learn about the game just by watching. However, if you want to sharpen your skills, you have to go out and play the game. If you do the exercises or reproduce the statistical analyses in the text, you will become a better player. Still, this text interweaves examples with the basic principles. Thus, even the armchair actuary can obtain a solid understanding of regression techniques through this text.

What Is This Book About?

The table of contents provides an overview of the topics covered, which are organized into four parts. The first part introduces linear regression. This is the core material of the book, with refreshers on mathematical statistics, distributions, and matrix algebra woven in as needed.

The second part is devoted to topics in time series. Why integrate time series topics into a regression book? The reasons are simple, yet compelling: most accounting, financial, and economic data become available over time. Although cross-sectional inferences are useful, business decisions need to be made in real time with currently available data. Chapters 7–10 introduce time series techniques that can be readily accomplished using regression tools (and there are many).

Nonlinear regression is the subject of the third part. Many modern-day predictive modeling tools are based on nonlinear regression – these are the workhorses of statistical shops in the financial and risk management industry.

The fourth part concerns actuarial applications, topics that I have found relevant in my research and consulting work in financial risk management. The first four chapters of this part consist of variations of regression models that are particularly useful in risk management. The last two chapters focus on communications, specifically report writing and designing graphs. Communicating information is an important aspect of every technical discipline, and statistics is certainly no exception.

How Does This Book Deliver Its Message?

Chapter Development

Each chapter has several examples interwoven with theory. In chapters in which a model is introduced, I begin with an example and discuss the data analysis without regard to the theory. This analysis is presented at an intuitive level, without reference to a specific model. This is straightforward, because it amounts to little more than curve fitting. The goal is to have students summarize data sensibly without having the notion of a model obscure good data analysis. Then, an introduction to the theory is provided in the context of the introductory example. One or more additional examples follow that reinforce the theory already introduced and provide a context for explaining additional theory. In Chapters 5 and 6, which do introduce not models but rather techniques for analysis, I begin with an introduction of the technique. This introduction is then followed by an example that reinforces the explanation. In this way, the data analysis can be easily omitted without loss of continuity, if time is a concern.

Real Data

Many of the exercises ask the reader to work with real data. The need for working with real data is well documented; for example, see Hogg (1972) or Singer and Willett (1990). Some criteria of Singer and Willett for judging a good data set include authenticity, availability of background information, interest and relevance to substantive learning, and availability of elements with which readers can identify. Of course, there are some important disadvantages to working with real data. Data sets can quickly become outdated. Further, the ideal data set for illustrating a specific statistical issue is difficult to find. This is because with real data, almost by definition, several issues occur simultaneously. This makes it difficult to isolate a specific aspect. I particularly enjoy working with large data sets. The larger the data set, the greater the need for statistics to summarize the information content.

The larger the data set, the greater the need for statistics to summarize the information content.

Statistical Software and Data

My goal in writing this text is to reach a broad group of students and industry analysts. Thus, to avoid excluding large segments, I chose not to integrate any specific statistical software package into the text. Nonetheless, because of the applications orientation, it is critical that the methodology presented be easily accomplished using readily available packages. For the course taught at the University of Wisconsin, I use the statistical packages SAS and R. On the book's Web site, at

http://research.bus.wisc.edu/RegActuaries,

users will find scripts written in SAS and R for the analyses presented in the text. The data are available in text format, allowing readers to employ any statistical packages that they wish. When you see a display such as this in the margin, you will also be able to find this data set (*TermLife*) on the book's Web site.

Ⓡ **EMPIRICAL**
Filename is "TermLife"

Technical Supplements

The technical supplements reinforce and extend the results in the main body of the text by giving a more formal, mathematical treatment of the material. This treatment is, in fact, a supplement because the applications and examples are described in the main body of the text. For readers with sufficient mathematical background, the supplements provide additional material that is useful in communicating to technical audiences. The technical supplements provide a deeper, and broader, coverage of applied regression analysis.

I believe that analysts should have an idea of "what is going on under the hood," or "how the engine works." Most of these topics will be omitted from the first reading of the material. However, as you work with regression, you will be confronted with questions such as, "Why?" and you will need to get into the details to see exactly how a certain technique works. Further, the technical supplements provide a menu of optional items that an instructor may wish to cover.

Suggested Courses

There is a wide variety of topics that can go into a regression course. Here are some suggested courses. The course that I teach at the University of Wisconsin is the first on the list in the following table.

Audience	Nature of Course	Suggested Chapters
One-year background in probability and statistics	Survey of regression and time series models	Chapters 1–8, 11–13, 20–21, main body of text only
One-year background in probability and statistics	Regression and time series models	Chapters 1–8, 20–21, selected portions of technical supplements
One-year background in probability and statistics	Regression modeling	Chapters 1–6, 11–13, 20–21, selected portions of technical supplements
Background in statistics and linear regression	Actuarial regression models	Chapters 10–21, selected portions of technical supplements

In addition to the previously suggested courses, this book is designed as supplemental reading for a time series course as well as a reference book for industry analysts. My hope is that college students who use the beginning parts of the book in their university courses will find the later chapters helpful in their industry positions. In this way I hope to promote lifelong learning!

Acknowledgments

It is appropriate to begin the acknowledgment section by thanking the students in the actuarial program here at the University of Wisconsin; students are important partners in the knowledge creation and dissemination business at universities. Through their questions and feedback, I have learned a tremendous amount over the years. I have also benefited from excellent assistance from those who have helped me pull together all the pieces for this book, specifically Missy Pinney, Peng Shi, Yunjie (Winnie) Sun, and Ziyan Xie.

I have enjoyed working with several former students and colleagues on regression problems in recent years, including Katrien Antonio, Jie Gao, Paul Johnson, Margie Rosenberg, Jiafeng Sun, Emil Valdez, and Ping Wang. Their contributions are reflected indirectly throughout the text. Because of my long association with the University of Wisconsin–Madison, I am reluctant to go back further in time and provide a longer list, for fear of missing important individuals. I have also been fortunate to have a more recent association with the Insurance Services Office (ISO). Colleagues at ISO have provided me with important insights into applications.

Through this text that features applications of regression into actuarial and financial industry problems, I hope to encourage the fostering of additional partnerships between academia and industry.

I am pleased to acknowledge detailed reviews that I received from my colleagues Tim Welnetz and Margie Rosenberg. I also wish to thank Bob Miller for permission to include our joint work on designing effective graphs in Chapter 21. Bob has taught me a lot about regression over the years.

Moreover, I am happy to acknowledge financial support through the Assurant Health Professorship in Actuarial Science at the University of Wisconsin–Madison.

Saving the most important for last, I thank my family for their support. Ten thousand thanks to my mother, Mary; my brothers Randy, Guy, and Joe; my wife, Deirdre; and our sons, Nathan and Adam.

1

Regression and the Normal Distribution

Chapter Preview. Regression analysis is a statistical method that is widely used in many fields of study, with actuarial science being no exception. This chapter provides an introduction to the role of the normal distribution in regression, the use of logarithmic transformations in specifying regression relationships, and the sampling basis that is critical for inferring regression results to broad populations of interest.

1.1 What Is Regression Analysis?

Statistics is about data. As a discipline, it is about the collection, summarization, and analysis of data to make statements about the real world. When analysts collect data, they are really collecting information that is quantified, that is, transformed to a numerical scale. There are easy, well-understood rules for reducing the data, through either numerical or graphical summary measures. These summary measures can then be linked to a theoretical representation, or model, of the data. With a model that is calibrated by data, statements about the world can be made.

Statistics is about the collection, summarization, and analysis of data to make statements about the real world.

Statistical methods have had a major impact on several fields of study:

- In the area of data collection, the careful design of *sample surveys* is crucial to market research groups and to the auditing procedures of accounting firms.
- *Experimental design* is a subdiscipline devoted to data collection. The focus of experimental design is on constructing methods of data collection that will extract information in the most efficient way possible. This is especially important in fields such as agriculture and engineering where each observation is expensive, possibly costing millions of dollars.
- Other applied statistical methods focus on managing and predicting data. *Process control* deals with monitoring a process over time and deciding when intervention is most fruitful. Process control helps manage the quality of goods produced by manufacturers.
- *Forecasting* is about extrapolating a process into the future, whether it be sales of a product or movements of an interest rate.

Regression analysis is a statistical method used to analyze data. As we will see, the distinguishing feature of this method is the ability to make statements about

Table 1.1 Galton's 1885 Regression Data

Height of Adult Child in Inches	Parents' Height											
	<64.0	64.5	65.5	66.5	67.5	68.5	69.5	70.5	71.5	72.5	>73.0	Total
>73.7	—	—	—	—	—	—	5	3	2	4	—	14
73.2	—	—	—	—	—	3	4	3	2	2	3	17
72.2	—	—	1	—	4	4	11	4	9	7	1	41
71.2	—	—	2	—	11	18	20	7	4	2	—	64
70.2	—	—	5	4	19	21	25	14	10	1	—	99
69.2	1	2	7	13	38	48	33	18	5	2	—	167
68.2	1	—	7	14	28	34	20	12	3	1	—	120
67.2	2	5	11	17	38	31	27	3	4	—	—	138
66.2	2	5	11	17	36	25	17	1	3	—	—	117
65.2	1	1	7	2	15	16	4	1	1	—	—	48
64.2	4	4	5	5	14	11	16	—	—	—	—	59
63.2	2	4	9	3	5	7	1	1	—	—	—	32
62.2	—	1	—	3	3	—	—	—	—	—	—	7
<61.2	1	1	1	—	—	1	—	1	—	—	—	5
Total	14	23	66	78	211	219	183	68	43	19	4	928

Source: Stigler (1986).

variables after having controlled for values of known explanatory variables. As important as other methods are, it is regression analysis that has been the most influential one. To illustrate, an index of business journals, ABI/INFORM, lists more than 24,000 articles using regression techniques over the thirty-year period 1978–2007. And these are only the applications that were considered innovative enough to be published in scholarly reviews!

Regression analysis of data is so pervasive in modern business that it is easy to overlook the fact that the methodology is barely more than 120 years old. Scholars attribute the birth of regression to the 1885 presidential address of Sir Francis Galton to the anthropological section of the British Association of the Advancement of Sciences. In that address, described in Stigler (1986), Galton provided a description of regression and linked it to *normal curve* theory. His discovery arose from his studies of properties of natural selection and inheritance.

® **Empirical**
Filename is "Galton"

To illustrate a dataset that can be analyzed using regression methods, Table 1.1 displays some data included in Galton's 1885 paper. The table displays the heights of 928 adult children, classified by an index of their parents' height. Here, all female heights were multiplied by 1.08, and the index was created by taking the average of the father's height and rescaled mother's height. Galton was aware that the parents' and the adult child's height could each be adequately approximated by a normal curve. In developing regression analysis, he provided a single model for the joint distribution of heights.

Regression analysis is a method to quantify the relationship between a variable of interest and explanatory variables.

Table 1.1 shows that much of the information concerning the height of an adult child can be attributed to, or "explained," in terms of the parents' height. Thus, we use the term *explanatory variable* for measurements that provide information

Figure 1.1 Ten deutsche mark – German currency featuring the scientist Gauss and the normal curve.

about a variable of interest. Regression analysis is a method to quantify the relationship between a variable of interest and explanatory variables. The methodology used to study the data in Table 1.1 can also be used to study actuarial and other risk management problems, the thesis of this book.

1.2 Fitting Data to a Normal Distribution

Historically, the normal distribution had a pivotal role in the development of regression analysis. It continues to play an important role, although we will be interested in extending regression ideas to highly "nonnormal" data.

Formally, the normal curve is defined by the function

$$\mathrm{f}(y) = \frac{1}{\sigma\sqrt{2\pi}} \exp\left(-\frac{1}{2\sigma^2}(y-\mu)^2\right). \tag{1.1}$$

This curve is a probability density function with the whole real line as its domain. From equation (1.1), we see that the curve is symmetric about μ (the mean and median). The degree of peakedness is controlled by the parameter σ^2. These two parameters, μ and σ^2, are known as the *location* and *scale parameters*, respectively. Appendix A3.1 provides additional details about this curve, including a graph and tables of its cumulative distribution that we will use throughout the text.

Appendix A3.1 provides additional details about the normal curve, including a graph and distribution table.

The normal curve is also depicted in Figure 1.1, a display of an out-of-date German currency note, the ten Deutsche Mark. This note contains the image of the German Carl Gauss, an eminent mathematician whose name is often linked with the normal curve (it is sometimes referred to as the *Gaussian curve*). Gauss developed the normal curve in connection with the theory of least squares for fitting curves to data in 1809, about the same time as related work by the French scientist Pierre LaPlace. According to Stigler (1986), there was quite a bit of acrimony between these two scientists about the priority of discovery! The normal curve was first used as an approximation to histograms of data around 1835 by Adolph Quetelet, a Belgian mathematician and social scientist. As with many good things, the normal curve had been around for some time, since about 1720, when Abraham de Moivre derived it for his work on modeling games of

Table 1.2 Summary Statistics of Massachusetts Automobile Bodily Injury Claims

Variable	Number	Mean	Median	Standard Deviation	Minimum	Maximum	25th Percentile	75th Percentile
Claims	272	0.481	0.793	1.101	−3.101	3.912	−0.114	1.168

Note: Data are in logs of thousands of dollars.

chance. The normal curve is popular because it is easy to use and has proved successful in many applications.

Ⓡ EMPIRICAL
Filename is "MassBodilyInjury"

Example: Massachusetts Bodily Injury Claims. For our first look at fitting the normal curve to a set of data, we consider data from Rempala and Derrig (2005). They considered claims arising from automobile bodily injury insurance coverages. These are amounts incurred for outpatient medical treatments that arise from automobile accidents, typically sprains, broken collarbones, and the like. The data consist of a sample of 272 claims from Massachusetts that were closed in 2001 (by "closed," we mean that the claim is settled and no additional liabilities can arise from the same accident). Rempala and Derrig were interested in developing procedures for handling mixtures of "typical" claims and others from providers who reported claims fraudulently. For this sample, we consider only those typical claims, ignoring the potentially fraudulent ones.

Table 1.2 provides several statistics that summarize different aspects of the distribution. Claim amounts are in units of logarithms of thousands of dollars. The average logarithmic claim is 0.481, corresponding to $1,617.77 (=1000 exp(0.481)). The smallest and largest claims are −3.101 ($45) and 3.912 ($50,000), respectively.

For completeness, here are a few definitions. The *sample* is the set of data available for analysis, denoted by $y_1, \ldots, y_n$. Here, n is the number of observations, y_1 represents the first observation, y_2 the second, and so on up to y_n for the nth observation. Here are a few important summary statistics.

Basic Summary Statistics

(i) The *mean* is the average of observations, that is, the sum of the observations divided by the number of units. Using algebraic notation, the mean is

$$\overline{y} = \frac{1}{n}(y_1 + \cdots + y_n) = \frac{1}{n}\sum_{i=1}^{n} y_i .$$

(ii) The *median* is the middle observation when the observations are ordered by size. That is, it is the observation at which 50% are below it (and 50% are above it).

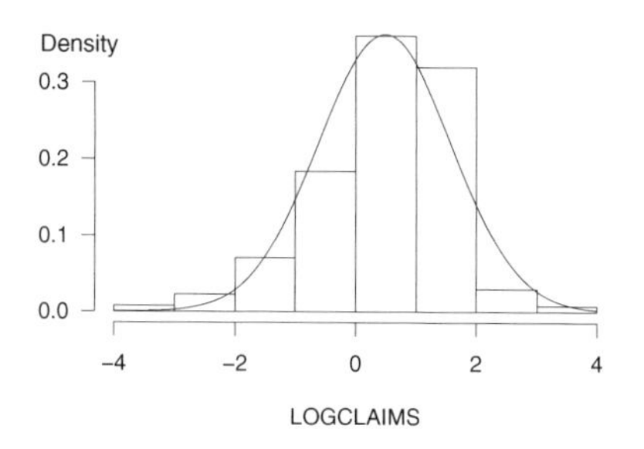

Figure 1.2 Bodily injury relative frequency with normal curve superimposed.

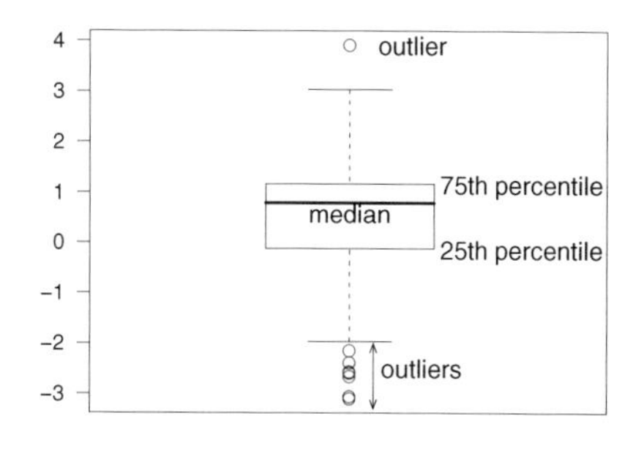

Figure 1.3 Box plot of bodily injury claims.

(iii) The *standard deviation* is a measure of the spread, or scale, of the distribution. It is computed as

$$s_y = \sqrt{\frac{1}{n-1}\sum_{i=1}^{n}(y_i - \overline{y})^2}.$$

(iv) A *percentile* is a number at which a specified fraction of the observations is below it, when the observations are ordered by size. For example, the 25th percentile is the number so that 25% of observations are below it.

To help visualize the distribution, Figure 1.2 displays a *histogram* of the data. Here, the height of the each rectangle shows the relative frequency of observations that fall within the range given by its base. The histogram provides a quick visual impression of the distribution; it shows that the range of the data is approximately (−4,4), that the central tendency is slightly greater than zero, and that the distribution is roughly symmetric.

Normal Curve Approximation

Figure 1.2 also shows a normal curve superimposed, using $\overline{y}$ for μ and s_y^2 for σ^2. With the normal curve, only two quantities (μ and σ^2) are required to summarize the entire distribution. For example, Table 1.2 shows that 1.168 is the 75th percentile, which is approximately the 204th ($= .75 \times 272$) largest observation from the entire sample. From the equation (1.1) normal distribution, we see that $z = (y - \mu)/\sigma$ is a standard normal, of which 0.675 is the 75th percentile. Thus, $\overline{y} + 0.675 s_y = 0.481 + 0.675 \times 1.101 = 1.224$ is the 75th percentile using the normal curve approximation.

Box Plot

A quick visual inspection of a variable's distribution can reveal some surprising features that are hidden by statistics: numerical summary measures. The *box plot*, also known as a box-and-whiskers plot, is one such graphical device. Figure 1.3 illustrates a box plot for the bodily injury claims. Here, the box captures the

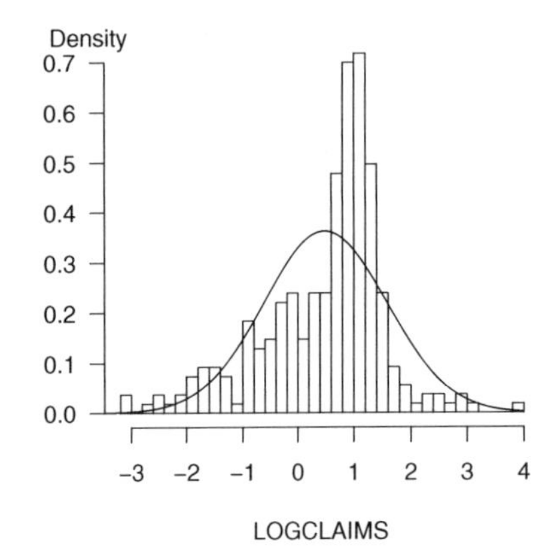

Figure 1.4 Redrawing of Figure 1.2 with an increased number of rectangles.

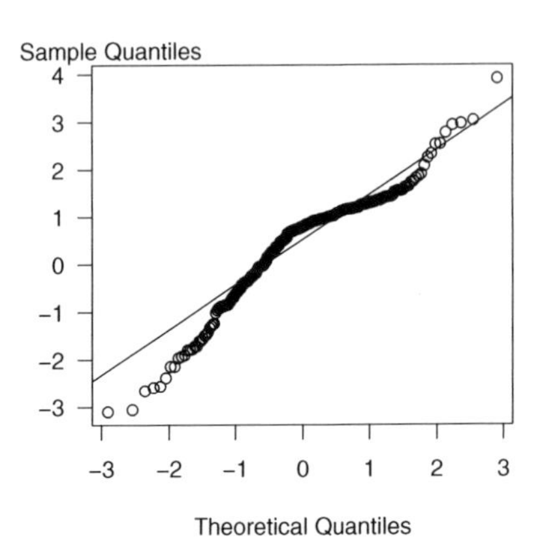

Figure 1.5 A *qq* plot of bodily injury claims, using a normal reference distribution.

middle 50% of the data, with the three horizontal lines corresponding to the 75th, 50th, and 25th percentiles, reading from top to bottom. The horizontal lines above and below the box are the "whiskers." The upper whisker is 1.5 times the *interquartile range* (the difference between the 75th and 25th percentiles) above the 75th percentile. Similarly, the lower whisker is 1.5 times the interquartile range below the 25th percentile. Individual observations outside the whiskers are denoted by small circular plotting symbols and are referred to as "outliers."

Graphs are powerful tools; they allow analysts to readily visualize nonlinear relationships that are hard to comprehend when expressed verbally or by mathematical formula. However, by their very flexibility, graphs can also readily deceive the analyst. Chapter 21 will underscore this point. For example, Figure 1.4 is a redrawing of Figure 1.2; the difference is that Figure 1.4 uses more, and finer, rectangles. This finer analysis reveals the asymmetric nature of the sample distribution that was not evident in Figure 1.2.

Quantile-Quantile Plots

Increasing the number of rectangles can unmask features that were not previously apparent; however, there are, in general, fewer observations per rectangle, meaning that the uncertainty of the relative frequency estimate increases. This represents a trade-off. Instead of forcing the analyst to make an arbitrary decision about the number of rectangles, an alternative is to use a graphical device for comparing a distribution to another known as a *quantile-quantile*, or *qq*, plot.

Points in a qq *plot close to a straight line suggest agreement between the sample and the reference distributions.*

Figure 1.5 illustrates a *qq* plot for the bodily injury data using the normal curve as a reference distribution. For each point, the vertical axis gives the quantile using the sample distribution. The horizontal axis gives the corresponding quantity using the normal curve. For example, earlier we considered the 75th percentile point. This point appears as (1.168, 0.675) on the graph. To interpret a *qq* plot, if the quantile points lie along the superimposed line, then the sample and the normal reference distribution have the same shape. (This line is defined by connecting the 75th and 25th percentiles.)

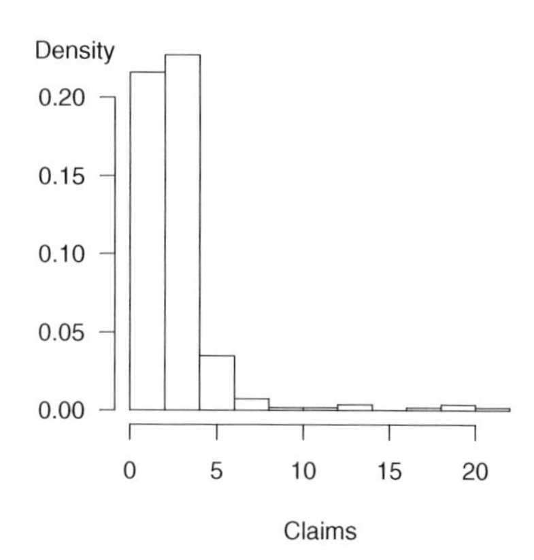

Figure 1.6 Distribution of bodily injury claims. Observations are in (thousands of) dollars, with the largest observation omitted.

In Figure 1.5, the small sample percentiles are consistently smaller than the corresponding values from the standard normal, indicating that the distribution is skewed to the left. The difference in values at the ends of the distribution are due to the outliers noted earlier that can also be interpreted as the sample distribution having larger tails than the normal reference distribution.

1.3 Power Transforms

In the Section 1.2 example, we considered claims without justifying the use of the logarithmic scaling. When analyzing variables such as assets of firms, wages of individuals, and housing prices of households in business and economic applications, it is common to consider logarithmic units instead of the original units. A log transform retains the original ordering (e.g., large wages remain large on the log wage scale) but serves to pull in extreme values of the distribution.

To illustrate, Figure 1.6 shows the bodily injury claims distribution in (thousands of) dollars. To graph the data meaningfully, the largest observation ($50,000) was removed prior to making this plot. Even with this observation removed, Figure 1.6 shows that the distribution is heavily lopsided to the right, with several large values of claims appearing.

A right-skewed distribution has long tails on the right and a concentration of mass on the left. Many insurance claims distributions are right skewed.

Distributions that are lopsided in one direction or the other are known as *skewed*. Figure 1.6 is an example of a distribution skewed to the right, or positively skewed. Here, the tail of the distribution on the right is longer, and there is a greater concentration of mass to the left. In contrast, a left-skewed, or negatively skewed, distribution has a longer tail on the left and a greater concentration of mass to the right. Many insurance claims distributions are right skewed (see Klugman, Panjer, and Willmot, 2008, for extensive discussions). As we saw in Figures 1.4 and 1.5, a logarithmic transformation yields a distribution that is only mildly skewed to the left.

Logarithmic transformations are used extensively in applied statistics work. One advantage is that they serve to symmetrize distributions that are skewed. More generally, we consider *power transforms*, also known as the *Box-Cox family of*

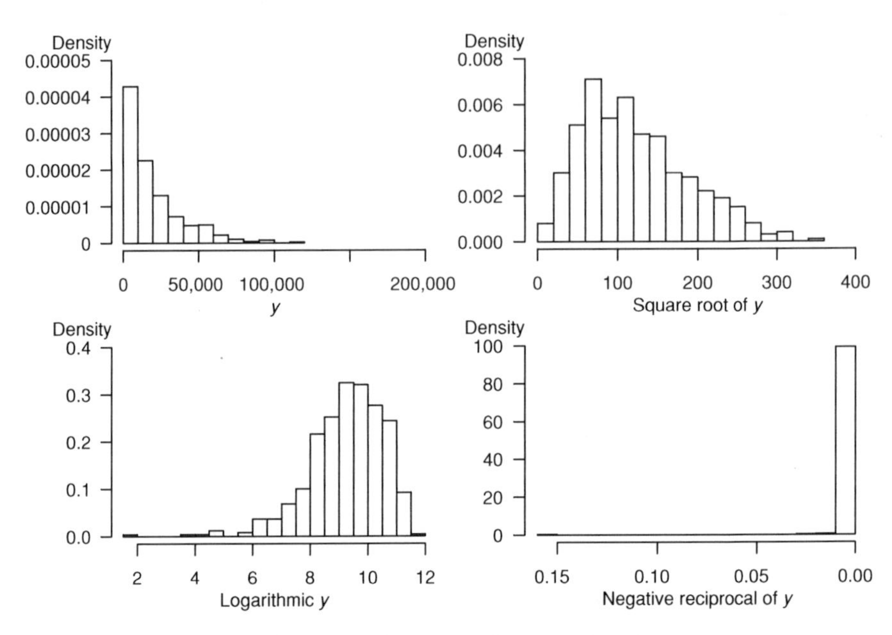

Figure 1.7 500 simulated observations from a chi-square distribution. The upper-left panel is based on the original distribution. The upper-right panel corresponds to the square root transform, the lower left to the log transform, and the lower right to the negative reciprocal transform.

transforms. In this family of transforms, in lieu of using the response y, we use a transformed, or rescaled version, y^{λ}. Here, the power λ (lambda, a Greek letter "el") is a number that may be user specified. Typical values of λ that are used in practice are $\lambda = 1, 1/2, 0$, or -1. When we use $\lambda = 0$, we mean $\ln(y)$, that is, the natural logarithmic transform. More formally, the Box-Cox family can be expressed as

$$y^{(\lambda)} = \begin{cases} \frac{y^{\lambda}-1}{\lambda} & \lambda \neq 0 \\ \ln(y) & \lambda = 0 \end{cases}.$$

As we will see, because regression estimates are not affected by location and scale shifts, in practice, we do not need to subtract 1 or divide by λ when rescaling the response. The advantage of the foregoing expression is that, if we let λ approach 0, then $y^{(\lambda)}$ approaches $\ln(y)$, from some straightforward calculus arguments.

To illustrate the usefulness of transformations, we simulated 500 observations from a chi-square distribution with two degrees of freedom. Appendix A3.2 introduces this distribution (which we will encounter again in studying the behavior of test statistics). The upper-left panel of Figure 1.7 shows that the original distribution is heavily skewed to the right. The other panels in Figure 1.7 show the data rescaled using the square root, logarithmic, and negative reciprocal transformations. The logarithmic transformation, in the lower-left panel, provides the best approximation to symmetry for this example. The negative reciprocal transformation is based on $\lambda = -1$, and then multiplying the rescaled observations by -1, so that large observations remain large.

1.4 Sampling and the Role of Normality

A *statistic* is a summary measure of data, such as a mean, median, or percentile. Collections of statistics are very useful for analysts, decision makers, and everyday consumers for understanding massive amounts of data that represent complex situations. To this point, our focus has been on introducing sensible techniques to summarize variables; techniques that will be used repeatedly thought this text. However, the true usefulness of the discipline of statistics is its ability to say something about the unknown, not merely to summarize information already available. To this end, we need to make some fairly formal assumptions about the manner in which the data are observed. As a science, a strong feature of the discipline of statistics is the ability to critique these assumptions and offer improved alternatives in specific situations.

It is customary to assume that the data are drawn from a larger population that we are interested in describing. The process of drawing the data is known as the *sampling*, or *data generating*, *process*. We denote this sample as $\{y_1, \ldots, y_n\}$. So that we may critique, and modify, these sampling assumptions, we list them here in detail:

A statistic is a summary measure of a sample. Statistics, as a discipline, can be used to infer behavior about a larger population from a sample.

Basic Sampling Assumptions
1. E $y_i = \mu$.
2. Var $y_i = \sigma^2$.
3. $\{y_i\}$ are independent.
4. $\{y_i\}$ are normally distributed.

In this basic setup, μ and σ^2 serve as *parameters* that describe the location and scale of the parent population. The goal is to infer something sensible about them on the basis of statistics such as $\overline{y}$ and s_y^2. For the third assumption, we assume independence among the draws. In a sampling scheme, this may be guaranteed by taking a simple random sample from a population. The fourth assumption is not required for many statistical inference procedures because central limit theorems provide approximate normality for many statistics of interest. However, a formal justification of some statistics, such as t-statistics, requires this additional assumption.

Assumption 4 is not required for many statistical inference procedures because central limit theorems provide approximate normality for many statistics of interest.

Section 1.9 provides an explicit statement of one version of the central limit theorem, giving conditions in which $\overline{y}$ is approximately normally distributed. This section also discusses a related result, known as an *Edgeworth approximation*, that shows that the quality of the normal approximation is better for symmetric parent populations when compared to skewed distributions.

Linear regression is the study of weighted averages.

How does this discussion apply to the study of regression analysis? After all, so far we have focused only on the simple arithmetic average $\overline{y}$. In subsequent chapters, we will emphasize that linear regression is the study of weighted averages; specifically, many regression coefficients can be expressed as weighted averages with appropriately chosen weights. Central limit and Edgeworth approximation theorems are available for weighted averages – these results will ensure approximate normality of regression coefficients. To use normal curve approximations in a regression context, we will often transform variables to achieve approximate symmetry.

Table 1.3 Terminology for Regression Variables

y-Variable	x-Variable
Outcome of interest	Explanatory variable
Dependent variable	Independent variable
Endogenous variable	Exogenous variable
Response	Treatment
Regressand	Regressor
Left-hand-side variable	Right-hand-side variable
Explained variable	Predictor variable
Output	Input

1.5 Regression and Sampling Designs

We often transform variables to achieve approximate symmetry to use normal curve approximations in a regression context.

Approximating normality is an important issue in practical applications of linear regression. Parts I and II of this book focus on linear regression, where we will learn basic regression concepts and sampling design. Part III will focus on *nonlinear* regression, involving binary, count, and fat-tailed responses, where the normal is not the most helpful reference distribution. Ideas concerning basic concepts and design are also used in the nonlinear setting.

In regression analysis, we focus on one measurement of interest: the *dependent variable*. Other measurements are used as *explanatory variables*. A goal is to compare differences in the dependent variable in terms of differences in the explanatory variables. As noted in Section 1.1, regression is used extensively in many scientific fields. Table 1.3 lists alternative terms that you may encounter as you read regression applications.

In the latter part of the nineteenth century and early part of the twentieth century, statistics was beginning to make an important impact on the development of experimental science. Experimental sciences often use *designed studies*, where the data are under the control of an analyst. Designed studies are performed in laboratory settings, where there are tight physical restrictions on every variable that a researcher thinks may be important. Designed studies also occur in larger field experiments, where the mechanisms for control are different than in laboratory settings. Agriculture and medicine use designed studies. Data from a designed study are said to be *experimental data*.

In designed studies, the data are under the control of an analyst. Data from a designed study are said to be experimental data.

To illustrate, a classic example is to consider the yield of a crop such as corn, where each of several parcels of land (the observations) are assigned various levels of fertilizer. The goal is ascertain the effect of fertilizer (the explanatory variable) on the corn yield (the response variable). Although researchers attempt to make parcels of land as much alike as possible, differences inevitably arise. Agricultural researchers use *randomization techniques* to assign different levels of fertilizer to each parcel of land. In this way, analysts can explain the variation in corn yields in terms of the variation of fertilizer levels. Through the use of randomization techniques, researchers using designed studies can infer that the treatment has a *causal effect* on the response. Chapter 6 discusses causality further.

Example: Rand Health Insurance Experiment. How are medical-care expenditures related to the demand for insurance? Many studies have established a positive relation between the amount spent on medical care and the demand for health insurance. Those in poor health anticipate using more medical services than similarly positioned people in good or fair health and seek higher levels of health insurance to compensate for the anticipated expenditures. They obtain this additional health insurance by (i) selecting a more generous health insurance plan from an employer, (ii) choosing an employer with a more generous health insurance plan, or (iii) paying more for individual health insurance. Thus, it is difficult to disentangle the cause-and-effect relationship of medical-care expenditures and the availability of health insurance.

A study reported by Manning et al. (1987) sought to answer this question using a carefully designed experiment. In this study, enrolled households from six cities, between November 1974 and February 1977, were *randomly assigned* to one of 14 different insurance plans. These plans varied by the cost-sharing elements, the co-insurance rate (the percentage paid on out-of-pocket expenditures, which varied by 0%, 25%, 50%, and 95%) as well as the deductible (5%, 10%, or 15% of family income, up to a maximum of $1,000). Thus, there was a random assignment to levels of the treatment, the amount of health insurance. The study found that more favorable plans resulted in greater total expenditures, even after controlling for participants' health status.

For actuarial science and other social sciences, designed studies are the exception rather than the rule. For example, if we want to study the effects of smoking on mortality, it is highly unlikely that we could get study participants to agree to be randomly assigned to smoker and nonsmoker groups for several years just so that we could observe their mortality patterns! As with the Section 1.1 Galton study, social science researchers generally work with *observational data*. Observational data are not under control of the analyst.

With observational data, we cannot infer causal relationships, but we can readily introduce measures of *association*. To illustrate, in the Galton data, it is apparent that "tall" parents are likely to have "tall" children, and conversely "short" parents are likely to have "short" children. Chapter 2 will introduce a correlation and other measures of association. However, we cannot infer causality from the data. For example, there may be another variable, such as family diet, that is related to both variables. Good diet in the family could be associated with tall heights of parents and adult children, whereas poor diet might stifle, growth. If this were the case, we would call family diet a *confounding variable*.

With statistical control, we seek to compare a y and an x, "controlling for" the effects of other explanatory variables.

In designed experiments such as the Rand Health Insurance Experiment, we can control for the effects of variables such as health status through random assignment methods. In observational studies, we use *statistical control* rather than experimental control. To illustrate, for the Galton data, we might split our observations into two groups, one for good family diet and one for poor family diet, and examine the relationship between parents' and children's height for each subgroup. This is the essence of the regression method, to compare a y and an x, "controlling for" the effects of other explanatory variables.

Of course, to use statistical control and regression methods, one must record family diet and any other measures of height that may confound the effects of parents' height on the height of their adult child. The difficulty in designing studies is trying to imagine all of the variables that could possibly affect a response variable, an impossible task in most social science problems of interest. To give some guidance on when enough is enough, Chapter 6 discusses measures of an explanatory variable's importance and its impact on model selection.

1.6 Actuarial Applications of Regression

This book introduces the statistical method of regression analysis. The introduction is organized around the traditional triad of statistical inference:

- Hypothesis testing
- Estimation
- Prediction

Further, this book shows how this methodology can be used in applications that are likely to be of interest to actuaries and to other risk analysts. As such, it is helpful to begin with the three traditional areas of actuarial applications:

- Pricing
- Reserving
- Solvency testing

Pricing and Adverse Selection

Regression analysis can be used to determine insurance prices for many lines of business. For example, in private passenger automobile insurance, expected claims vary by the insured's sex, age, location (city versus rural), vehicle purpose (work or pleasure), and a host of other explanatory variables. Regression can be used to identify the variables that are important determinants of expected claims.

In competitive markets, insurance companies do not use the same price for all insureds. If they did, "good risks," those with lower-than-average expected claims, would overpay and leave the company. In contrast, "bad risks," those with higher-than-average expected claims, would remain with the company. If the company continued this flat-rate pricing policy, premiums would rise (to compensate for claims by the increasing share of bad risks) and market share would dwindle as the company loses good risks. This problem is known as *adverse selection*. Using an appropriate set of explanatory variables, classification systems can be developed so that each insured pays his or her fair share.

Reserving and Solvency Testing

Both reserving and solvency testing are concerned with predicting whether liabilities associated with a group of policies will exceed the capital devoted to meeting obligations arising from the policies. Reserving involves determining the appropriate amount of capital to meet these obligations. Solvency testing is about assessing the adequacy of capital to fund the obligations for a block of business. In some practice areas, regression can be used to forecast future obligations to help determine reserves (see, e.g., Chapter 19). Regression can also be used to compare characteristics of healthy and financially distressed firms for solvency testing (see, e.g., Chapter 14).

Other Risk Management Applications

Regression analysis is a quantitative tool that can be applied in a broad variety of business problems, not just in the traditional areas of pricing, reserving, and solvency testing. By becoming familiar with regression analysis, actuaries will have another quantitative skill that can be brought to bear on general problems involving the financial security of people, companies, and governmental organizations. To help you develop insights, this book provides many examples of potential nonactuarial applications through featured vignettes labeled as "examples" and illustrative datasets.

To help understand potential regression applications, start by reviewing the several datasets featured in the Chapter 1 Exercises. Even if you do not complete the exercises to strengthen your data summary skills (which requires the use of a computer), a review of the problem descriptions will help you become more familiar with types of applications in which an actuary might use regression techniques.

1.7 Further Reading and References

This book introduces regression and time series tools that are most relevant to actuaries and other financial risk analysts. Fortunately, there are other sources that provide excellent introductions to these statistical topics (although not from a risk management viewpoint). Particularly for analysts that intend to specialize in statistics, it is helpful to get another perspective. For regression, I recommend Weisburg (2005) and Faraway (2005). For time series, Diebold (2004) is a good source. Moreover, Klugman, Panjer, and Willmot (2008) provide a good introduction to actuarial applications of statistics; this book is intended to complement the Klugman et al. book by focusing on regression and time series methods.

Chapter References

Beard, Robert E., Teivo Pentikäinen, and Erkki Pesonen (1984). *Risk Theory: The Stochastic Basis of Insurance*, 3rd ed. Chapman & Hall, New York.

Diebold, Francis X. (2004). *Elements of Forecasting*, 3rd ed. Thomson, South-Western, Mason, Ohio.

Faraway, Julian J. (2005). *Linear Models in R*. Chapman & Hall/CRC, New York.

Hogg, Robert V. (1972). On statistical education. *American Statistician* 26, 8–11.

Klugman, Stuart A., Harry H. Panjer, and Gordon E. Willmot (2008). *Loss Models: From Data to Decisions*. John Wiley & Sons, Hoboken, New Jersey.

Manning, Willard G., Joseph P. Newhouse, Naihua Duan, Emmett B. Keeler, Arleen Leibowitz, and M. Susan Marquis (1987). Health insurance and the demand for medical care: Evidence from a randomized experiment. *American Economic Review* 77, no. 3, 251–277.

Rempala, Grzegorz A., and Richard A. Derrig (2005). Modeling hidden exposures in claim severity via the EM algorithm. *North American Actuarial Journal* 9, no. 2, 108–128.

Rosenberg, Marjorie A., Edward W. Frees, Jia Feng Sun, Paul Johnson Jr., and James M. Robinson (2007). Predictive modeling with longitudinal data: A case study of Wisconsin nursing homes. *North American Actuarial Journal* 11, no. 3, 54–69.

Singer, Judith D., and J. B. Willett (1990). Improving the teaching of applied statistics: Putting the data back into data analysis. *American Statistician* 44, 223–230.

Stigler, Steven M. (1986). *The History of Statistics: The Measurement of Uncertainty before 1900*. Belknap Press, Harvard University Press, Cambridge, Massachusetts.

Weisberg, Sanford (2005). *Applied Linear Regression*, 3rd ed. John Wiley & Sons, New York.

1.8 Exercises

Ⓡ **Empirical**
Filename is "HealthExpend"

1.1. **MEPS Health Expenditures.** This exercise considers data from the Medical Expenditure Panel Survey (MEPS), conducted by the U.S. Agency of Health Research and Quality. MEPS is a probability survey that provides nationally representative estimates of health-care use, expenditures, sources of payment, and insurance coverage for the U.S. civilian population. This survey collects detailed information on individuals of each medical-care episode by type of services, including physician office visits, hospital emergency room visits, hospital outpatient visits, hospital inpatient stays, all other medical provider visits, and use of prescribed medicines. This detailed information allows one to develop models of health-care use to predict future expenditures. You can learn more about MEPS at http://www.meps.ahrq.gov/mepsweb/.

We consider MEPS data from panels 7 and 8 of 2003 that consists of 18,735 individuals between ages 18 and 65. From this sample, we took a random sample of 2,000 individuals that appear in the file "HealthExpend." From this sample, there are 157 individuals that had positive inpatient expenditures. There are also 1,352 that had positive outpatient expenditures. We will analyze these two samples separately.

Our dependent variables consist of amounts of expenditures for inpatient (EXPENDIP) and outpatient (EXPENDOP) visits. For MEPS, outpatient events include hospital outpatient department visits, office-based provider visits, and emergency room visits excluding dental services. (Dental services, compared to other types of health-care services, are more predictable and occur on a more regular basis.) Hospital stays with the same date of admission and discharge, known as zero-night stays, were included in outpatient counts and expenditures. (Payments associated with emergency room visits that immediately preceded an inpatient stay were included in the inpatient expenditures. Prescribed medicines that can be linked to hospital admissions were included in inpatient expenditures, not in outpatient utilization.)

Part 1: Use only the 157 individuals who had positive inpatient expenditures and do the following analysis:

a. Compute descriptive statistics for inpatient (EXPENDIP) expenditures.
   a(i) What is the typical (mean and median) expenditure?
   a(ii) How does the standard deviation compare to the mean? Do the data appear to be skewed?

b. Compute a box plot, histogram, and a (normal) qq plot for EXPENDIP. Comment on the shape of the distribution.

c. Transformations.
   c(i) Take a square root transform of inpatient expenditures. Summarize the resulting distribution using a histogram and a qq plot. Does it appear to be approximately normally distributed?
   c(ii) Take a (natural) logarithmic transformation of inpatient expenditures. Summarize the resulting distribution using a histogram and a qq plot. Does it appear to be approximately normally distributed?

Part 2: Use only the 1,352 individuals who had positive outpatient expenditures.

d. Repeat part (a) and compute histograms for expenditures and logarithmic expenditures. Comment on the approximate normality for each histogram.

Ⓡ **Empirical**
Filename is "WiscNursingHome"

1.2. **Nursing Home Utilization.** This exercise considers nursing home data provided by the Wisconsin Department of Health and Family Services (DHFS). The State of Wisconsin Medicaid program funds nursing home care for individuals qualifying on the basis of need and financial status. As part of the conditions for participation, Medicaid-certified nursing homes must file an annual cost report to DHFS, summarizing the volume and cost of care provided to all of its residents, Medicaid funded and otherwise. These cost reports are audited by DHFS staff and form the basis for facility-specific Medicaid daily payment rates for subsequent periods. The data are publicly available; see http://dhfs.wisconsin.gov/provider/prev-yrs-reports-nh.htm for more information.

The DHFS is interested in predictive techniques that provide reliable utilization forecasts to update its Medicaid-funding rate schedule of nursing facilities. In this assignment, we consider the data in the file "WiscNursingHome" in cost-report years 2000 and 2001. There are 362 facilities in 2000 and 355 facilities in 2001. Typically, utilization of nursing home care is measured in patient days (i.e., the number of days each patient was in the facility, summed over all patients). For this exercise, we define the outcome variable to be total patient years (TPY), the number of total patient days in the cost-reporting period divided by number of facility operating days in the cost-reporting period (see Rosenberg et al., 2007, Appendix 1, for further discussion of this choice). The number of beds (NUMBED) and square footage (SQRFOOT) of the nursing home both measure the size of the facility. Not surprisingly, these continuous variables will be important predictors of TPY.

Part 1: Use cost-report year 2000 data, and do the following analysis:

a. Compute descriptive statistics for TPY, NUMBED, and SQRFOOT.
b. Summarize the distribution of TPY using a histogram and a *qq* plot. Does it appear to be approximately normally distributed?
c. Transformations. Take a (natural) logarithmic transformation of TPY (LOGTPY). Summarize the resulting distribution using a histogram and a *qq* plot. Does it appear to be approximately normally distributed?

Part 2: Use cost-report year 2001 data and repeat parts (a) and (c).

Ⓡ **Empirical**
Filename is "AutoClaims"

1.3. **Automobile Insurance Claims.** As an actuarial analyst, you are working with a large insurance company to help it understand claims distribution for private passenger automobile policies. You have available claims data for a recent year, consisting of

- STATE CODE: codes 01 through 17 used, with each code randomly assigned to an actual individual state
- CLASS: rating class of operator, based on age, sex, marital status, and use of vehicle
- GENDER: operator sex AGE: operator age
- PAID: amount paid to settle and close a claim.

You are focusing on older drivers, 50 and older, for which there are $n = 6{,}773$ claims available.

Examine the histogram of the amount PAID and comment on the symmetry. Create a new variable, the (natural) logarithmic claims paid, LNPAID. Create a histogram and a *qq* plot of LNPAID. Comment on the symmetry of this variable.

Ⓡ **Empirical**
Filename is "HospitalCosts"

1.4. **Hospital Costs.** Suppose that you are an employee benefits actuary working with a medium-sized company in Wisconsin. This company is considering offering, for the first time in its industry, hospital insurance coverage to dependent children of employees. You have access to company records and so have available the number, age, and sex of the dependent children but have no other information about hospital costs from the company. In particular, no firm in this industry has offered this coverage, and so you have little historical industry experience on which you can forecast expected claims.

You gather data from the Nationwide Inpatient Sample of the Healthcare Cost and Utilization Project (NIS-HCUP), a nationwide survey of hospital costs conducted by the U.S. Agency for Healthcare Research and Quality (AHRQ). You restrict consideration to Wisconsin hospitals and analyze a random sample of $n = 500$ claims from 2003 data. Although the data come from hospital records, they are organized by individual discharge, and so you have information about the age and sex of the patient discharged. Specifically, you consider patients aged 0–17 years. In a separate project, you will consider the frequency of hospitalization. For this project, the goal is to model the severity of hospital charges, by age and sex.

a. Examine the distribution of the dependent variable, TOTCHG. Do this by making a histogram and then a *qq* plot, comparing the empirical to a normal distribution.

b. Take a natural log transformation and call the new variable LNTOTCHG. Examine the distribution of this transformed variable. To visualize the logarithmic relationship, plot LNTOTCHG versus TOTCHG.

Ⓡ EMPIRICAL
Filename is "AutoBI"

1.5. **Automobile Injury Insurance Claims.** We consider automobile injury claims data using data from the Insurance Research Council (IRC), a division of the American Institute for Chartered Property Casualty Underwriters and the Insurance Institute of America. The data, collected in 2002, contain information on demographic information about the claimant, attorney involvement, and economic loss (LOSS, in thousands), among other variables. We consider here a sample of $n = 1{,}340$ losses from a single state. The full 2002 study contains more than 70,000 closed claims based on data from 32 insurers. The IRC conducted similar studies in 1977, 1987, 1992, and 1997.

a. Compute descriptive statistics for the total economic loss (LOSS). What is the typical loss?

b. Compute a histogram and (normal) *qq* plot for LOSS. Comment on the shape of the distribution.

c. Partition the dataset into two subsamples, one corresponding to those claims that involved an ATTORNEY (=1) and the other to those in which an ATTORNEY was not involved (=2).

c(i) For each subsample, compute the typical loss. Does there appear to be a difference in the typical losses by attorney involvement?

c(ii) To compare the distributions, compute a box plot by level of attorney involvement.

c(iii) For each subsample, compute a histogram and *qq* plot. Compare the two distributions.

Ⓡ EMPIRICAL
Filename is "NAICExpense"

1.6. **Insurance Company Expenses.** Like other businesses, insurance companies seek to minimize expenses associated with doing business to enhance profitability. To study expenses, this exercise examines a random sample of 500 insurance companies from the National Association of Insurance Commissioners' (NAIC) database of more than 3,000 companies. The NAIC maintains one of the world's largest insurance regulatory databases; we consider here data that are based on 2005 annual reports for all the property and casualty insurance companies in the United States. The annual reports are financial statements that use statutory accounting principles.

Specifically, our dependent variable is EXPENSES, the nonclaim expenses for a company. Although not needed for this exercise, nonclaim expenses are based on three components: unallocated loss adjustment, underwriting, and investment expenses. The unallocated loss adjustment expense is the expense not directly attributable to a claim but indirectly associated with settling claims; it includes items such as the salaries of claims adjusters, legal fees, court costs, expert witnesses, and investigation costs. Underwriting expenses consist of policy acquisition costs, such as commissions, as well as the portion of administrative, general, and other expenses attributable to underwriting operations. Investment expense are those expenses related to investment activities of the insurer.

a. Examine the distribution of the dependent variable, EXPENSES. Do this by making a histogram and then a qq plot, comparing the empirical to a normal distribution.
b. Take a natural log transformation and examine the distribution of this transformed variable. Has the transformation helped to symmetrize the distribution?

Ⓡ **EMPIRICAL**
Filename is "UNLifeExpectancy"

1.7. **National Life Expectancies.** Who is doing health care right? Health-care decisions are made at the individual, corporate, and government levels. Virtually every person, corporation, and government has a different perspective on health care; these result in a wide variety of systems for managing health care. Comparing different health-care systems help us learn about approaches other than our own, which in turn help us make better decisions in designing improved systems.

Here, we consider health-care systems from $n = 185$ countries throughout the world. As a measure of the quality of care, we use LIFEEXP, the life expectancy at birth. This dependent variable and several explanatory variables are listed in Table 1.4. From this table, you will note that although there are 185 countries consider in this study, not all countries provided information for each variable. Data not available are noted under the column "Num Miss." The data are from the UN Human Development Report.

a. Examine the distribution of the dependent variable, LIFEEXP. Do this by making a histogram and then a qq plot, comparing the empirical to a normal distribution.
b. Take a natural log transformation and examine the distribution of this transformed variable. Has the transformation helped to symmetrize the distribution?

1.9 Technical Supplement – Central Limit Theorem

Central limit theorems form the basis for much of the statistical inference used in regression analysis. Thus, it is helpful to provide an explicit statement of one version of the central limit theorem.

Central Limit Theorem. Suppose that $y_1, \ldots, y_n$ are independently distributed with mean μ, finite variance σ^2 and $\mathrm{E}|y|^3$ is finite. Then,

$$\lim_{n\to\infty} \Pr\left(\frac{\sqrt{n}}{\sigma}(\bar{y} - \mu) \leq x\right) = \Phi(x)$$

for each x, where $\Phi(.)$ is the standard normal distribution function.

Table 1.4 Life Expectancy, Economic, and Demographic Characteristics of 185 Countries

Variable	Description	Num Miss	Mean	Median	Standard Deviation	Minimum	Maximum
BIRTH ATTEND	Births attended by skilled health personnel (%)	7	78.25	92.00	26.42	6.00	100.00
FEMALE BOSS	Legislators, senior officials, and managers, % female	87	29.07	30.00	11.71	2.00	58.00
FERTILITY	Total fertility rate, births per woman	4	3.19	2.70	1.71	0.90	7.50
GDP	Gross domestic product, in billions of USD	7	247.55	14.20	1,055.69	0.10	12,416.50
HEALTH EXPEND	2004 health expenditure per capita, PPP in USD	5	718.01	297.50	1,037.01	15.00	6,096.00
ILLITERATE	Adult illiteracy rate, % aged 15 and older	14	17.69	10.10	19.86	0.20	76.40
PHYSICIAN	Physicians, per 100,000 people	3	146.08	107.50	138.55	2.00	591.00
POP	2005 population, in millions	1	35.36	7.80	131.70	0.10	1,313.00
PRIVATE HEALTH	2004 private expenditure on health, % of GDP	1	2.52	2.40	1.33	0.30	8.50
PUBLIC EDUCATION	Public expenditure on education, % of GDP	28	4.69	4.60	2.05	0.60	13.40
RESEAR CHERS	Researchers in R&D, per million people	95	2,034.66	848.00	4,942.93	15.00	45,454.00
SMOKING	Prevalence of smoking, (male) % of adults	88	35.09	32.00	14.40	6.00	68.00
LIFEEXP	Life expectancy at birth, in years		67.05	71.00	11.08	40.50	82.30

Source: UN Human Development Report, available at http://hdr.undp.org/en/.
Note: PPP = Purchasing Power Parity.

Under the assumptions of this theorem, the rescaled distribution of $\overline{y}$ approaches a standard normal as the sample size, n, increases. We interpret this as meaning that, for large sample sizes, the distribution of $\overline{y}$ may be approximated by a normal distribution. Empirical investigations have shown that sample sizes of $n = 25$ to 50 provide adequate approximations for most purposes.

When does the central limit theorem not work well? Some insights are provided by another result from mathematical statistics.

Edgeworth Approximation. Suppose that $y_1, \ldots, y_n$ are identically and independently distributed with mean μ, finite variance σ^2 and $\mathrm{E}|y|^3$ is finite. Then,

$$\Pr\left(\frac{\sqrt{n}}{\sigma}(\overline{y} - \mu) \leq x\right) = \Phi(x) + \frac{1}{6}\frac{1}{\sqrt{2\pi}}e^{-x^2/2}\frac{\mathrm{E}(y-\mu)^3}{\sigma^3\sqrt{n}} + \frac{h_n}{\sqrt{n}}$$

for each x, where $h_n \to 0$ as $n \to \infty$.

This result suggests that the distribution of $\overline{y}$ becomes closer to a normal distribution as the skewness, $\mathrm{E}(\overline{y} - \mu)^3$, becomes closer to zero. This is important in insurance applications because many distributions tend to be skewed. Historically, analysts used the second term on the right-hand side of the result to provide a "correction" for the normal curve approximation. See, for example, Beard, Pentikäinen and Pesonen (1984) for further discussion of Edgeworth approximations in actuarial science. An alternative (used in this book) that we saw in Section 1.3 is to transform the data, thus achieving approximate symmetry. As suggested by the Edgeworth approximation theorem, if our parent population is close to symmetric, then the distribution of $\overline{y}$ will be approximately normal.

Part I

Linear Regression

2

Basic Linear Regression

Chapter Preview. This chapter considers regression in the case of only one explanatory variable. Despite this seeming simplicity, most of the deep ideas of regression can be developed in this framework. By limiting ourselves to the one variable case, we are able to express many calculations using simple algebra. This will allow us to develop our intuition about regression techniques by reinforcing it with simple demonstrations. Further, we can illustrate the relationships between two variables graphically because we are working in only two dimensions. Graphical tools prove important for developing a link between the data and a model.

2.1 Correlations and Least Squares

Regression is about relationships. Specifically, we will study how two variables, an x and a y, are related. We want to be able to answer questions such as, If we change the level of x, what will happen to the level of y? If we compare two subjects that appear similar except for the x measurement, how will their y measurements differ? Understanding relationships among variables is critical for quantitative management, particularly in actuarial science, where uncertainty is so prevalent.

It is helpful to work with a specific example to become familiar with key concepts. Analysis of lottery sales has not been part of traditional actuarial practice, but it is a growth area in which actuaries could contribute.

Ⓡ **Empirical** *Filename is "WiscLottery"*

Example: Wisconsin Lottery Sales. State of Wisconsin lottery administrators are interested in assessing factors that affect lottery sales. Sales consists of online lottery tickets that are sold by selected retail establishments in Wisconsin. The tickets are generally priced at \$1.00, so the number of tickets sold equals the lottery revenue. We analyze average lottery sales (SALES) over a 40-week period, April 1998 through January 1999, from fifty randomly selected areas identified by postal (ZIP) code within the state of Wisconsin.

Although many economic and demographic variables might influence sales, our first analysis focuses on population (POP) as a key determinant. Chapter 3 will show how to consider additional explanatory variables. Intuitively, it seems clear that geographic areas with more people have higher sales. So, other things

Table 2.1 Summary Statistics of Each Variable

Variable	Mean	Median	Standard Deviation	Minimum	Maximum
POP	9,311	4,406	11,098	280	39,098
SALES	6,495	2,426	8,103	189	33,181

Source: Frees and Miller (2003).

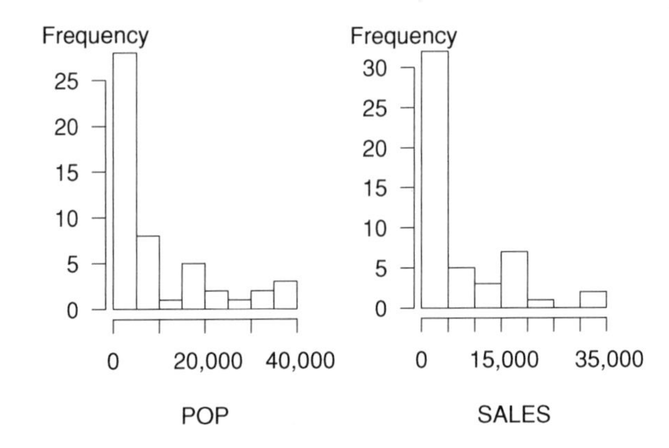

Figure 2.1 Histograms of population and sales. Each distribution is skewed to the right, indicating that there are many small areas compared to a few areas with larger sales and populations.

being equal, a larger $x = POP$ means a larger $y = SALES$. However, the lottery is an important source of revenue for the state, and we want to be as precise as possible.

A little additional notation will be useful subsequently. In this sample, there are fifty geographic areas and we use subscripts to identify each area. For example, $y_1 = 1{,}285.4$ represents sales for the first area in the sample that has population $x_1 = 435$. Call the ordered pair $(x_1, y_1) = (435, 1285.4)$ the first *observation*. Extending this notation, the entire sample containing fifty observations may be represented by $(x_1, y_1), \ldots, (x_{50}, y_{50})$. The ellipses $(\ldots)$ mean that the pattern is continued until the final object is encountered. We will often speak of a generic member of the sample, referring to (x_i, y_i) as the ith observation.

Begin by working with each variable separately.

Datasets can get complicated, so it will help if you begin by working with each variable separately. The two panels in Figure 2.1 show histograms that give a quick visual impression of the distribution of each variable in isolation of the other. Table 2.1 provides corresponding numerical summaries. To illustrate, for the population variable (POP), we see that the area with the smallest number contained 280 people, whereas the largest contained 39,098. The average, over 50 Zip codes, was 9,311.04. For our second variable, sales were as low as \$189 and as high as \$33,181.

As Table 2.1 shows, the basic summary statistics give useful ideas of the structure of key features of the data. After we understand the information in each variable in isolation of the other, we can begin exploring the relationship between the two variables.

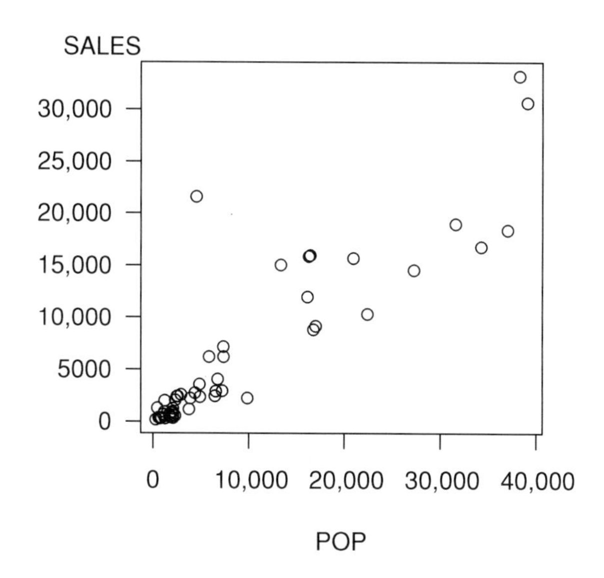

Figure 2.2 A scatter plot of the lottery data. Each of the 50 plotting symbols corresponds to a Zip code in the study. This figure suggests that postal areas with larger populations have greater lottery revenues.

Scatter Plot and Correlation Coefficients – Basic Summary Tools

The basic graphical tool used to investigate the relationship between the two variables is a *scatter plot*, such as in Figure 2.2. Although we may lose the exact values of the observations when graphing data, we gain a visual impression of the relationship between population and sales. From Figure 2.2, we see that areas with larger populations tend to purchase more lottery tickets. How strong is this relationship? Can knowledge of the area's population help us anticipate the revenue from lottery sales? We explore these two questions here.

One way to summarize the strength of the relationship between two variables is through a *correlation* statistic.

> *Definition.* The *ordinary, or Pearson, correlation* coefficient is defined as
>
> $$r = \frac{1}{(n-1)s_x s_y} \sum_{i=1}^{n} (x_i - \overline{x})(y_i - \overline{y}).$$
>
> Here, we use the sample standard deviation $s_y = \sqrt{(n-1)^{-1} \sum_{i=1}^{n} (y_i - \overline{y})^2}$ defined in Section 1.2, with similar notation for s_x.

Although there are other correlation statistics, the correlation coefficient devised by Pearson (1895) has several desirable properties. One important property is that, for any dataset, r is bounded by -1 and 1, that is, $-1 \leq r \leq 1$. (Exercise 2.3 provides steps for you to check this property.) If r is greater than zero, the variables are said to be *(positively) correlated.* If r is less than zero, the variables are said to be *negatively correlated.* The larger the coefficient is in absolute value, the stronger is the relationship. In fact, if $r = 1$, then the variables are perfectly correlated. In this case, all of the data lie on a straight line that goes through the lower-left- and upper-right-hand quadrants. If $r = -1$, then all of the data lie on a line that goes through the upper-left- and lower-right-hand quadrants. The coefficient r is a measure of a *linear* relationship between two variables.

The coefficient r is a measure of a linear relationship between two variables.

The correlation coefficient is said to be *location and scale invariant*. Thus, each variable's center of location does not matter in the calculation of r. For example, if we add \$100 to the sales of each Zip code, each y_i will increase by 100. However, $\overline{y}$, the average purchase price, will also increase by 100, so that the deviation $y_i - \overline{y}$ remains unchanged, or invariant. Further, the scale of each variable does not matter in the calculation of r. For example, suppose we divide each population by 1,000 so that x_i now represents population in thousands. Thus, $\overline{x}$ is also divided by 1,000 and you should check that s_x is also divided by 1,000. Thus, the standardized version of x_i, $(x_i - \overline{x})/s_x$, remains unchanged, or invariant. Many statistical packages compute a standardized version of a variable by subtracting the average and dividing by the standard deviation. Now, let's use $y_{i,std} = (y_i - \overline{y})/s_y$ and $x_{i,std} = (x_i - \overline{x})/s_x$ to be the standardized versions of y_i and x_i, respectively. With this notation, we can express the correlation coefficient as $r = (n-1)^{-1} \sum_{i=1}^{n} x_{i,std} \times y_{i,std}$.

The correlation coefficient is said to be a *dimensionless measure*. This is because we have taken away dollars, and all other units of measures, by considering the standardized variables $x_{i,std}$ and $y_{i,std}$. Because the correlation coefficient does not depend on units of measure, it is a statistic that can readily be compared across different datasets.

The correlation coefficient is location and scale invariant. It is dimensionless.

In the world of business, the term *correlation* is often synonymous with the term "relationship." For the purposes of this text, we use the term *correlation* when referring only to linear relationships. The classic nonlinear relationship is $y = x^2$, a quadratic relationship. Consider this relationship and the fictitious dataset for x, $\{-2, 1, 0, 1, 2\}$. Now, as an exercise (2.2), produce a rough graph of the dataset:

i	1	2	3	4	5
x_i	−2	−1	0	1	2
y_i	4	1	0	1	4

The correlation coefficient for this dataset turns out to be $r = 0$ (check this). Thus, despite the fact that there is a perfect relationship between x and y $(=x^2)$, there is a zero correlation. Recall that location and scale changes are not relevant in correlation discussions, so we could easily change the values of x and y to be more representative of a business dataset.

How strong is the relationship between y and x for the lottery data? Graphically, the response is a scatter plot, as in Figure 2.2. Numerically, the main response is the correlation coefficient, which turns out to be $r = 0.886$ for this dataset. We interpret this statistic by saying that SALES and POP are (positively) correlated. The strength of the relationship is strong because $r = 0.886$ is close to one. In summary, we may describe this relationship by saying that there is a strong correlation between SALES and POP.

Method of Least Squares

Now we begin to explore the question, Can knowledge of population help us understand sales? To respond to this question, we identify sales as the *response* or *dependent variable*. The population variable, which is used to help understand sales, is called the *explanatory* or *independent variable*.

Suppose that we have available the sample data of fifty sales $\{y_1, \ldots, y_{50}\}$ and your job is to predict the sales of a randomly selected Zip code. Without knowledge of the population variable, a sensible predictor is simply $\overline{y} = 6{,}495$, the average of the available sample. Naturally, you anticipate that areas with larger populations will have greater sales. That is, if you also have knowledge of population, then can this estimate be improved? If so, then by how much?

To answer these questions, the first step assumes an approximate linear relationship between x and y. To fit a line to our data set, we use the *method of least squares*. We need a general technique so that, if different analysts agree on the data and agree on the fitting technique, then they will agree on the line. If different analysts fit a dataset using eyeball approximations, in general, they will arrive at different lines, even when using the same dataset.

The method begins with the line $y = b_0^* + b_1^* x$, where the intercept and slope, b_0^* and b_1^*, are merely generic values. For the ith observation, $y_i - \left(b_0^* + b_1^* x_i\right)$ represents the deviation of the observed value y_i from the line at x_i. The quantity

$$SS(b_0^*, b_1^*) = \sum_{i=1}^{n} \left(y_i - \left(b_0^* + b_1^* x_i\right)\right)^2$$

represents the sum of squared deviations for this candidate line. The least squares method consists of determining the values of b_0^* and b_1^* that minimize $SS(b_0^*, b_1^*)$. This is an easy problem that can be solved by calculus, as follows. Taking partial derivatives with respect to each argument yields

$$\frac{\partial}{\partial b_0^*} SS(b_0^*, b_1^*) = \sum_{i=1}^{n} (-2) \left(y_i - \left(b_0^* + b_1^* x_i\right)\right)$$

and

$$\frac{\partial}{\partial b_1^*} SS(b_0^*, b_1^*) = \sum_{i=1}^{n} (-2x_i) \left(y_i - \left(b_0^* + b_1^* x_i\right)\right).$$

The reader is invited to take second partial derivatives to ensure that we are minimizing, not maximizing, this function. Setting these quantities equal to zero and canceling constant terms yields

$$\sum_{i=1}^{n} \left(y_i - \left(b_0^* + b_1^* x_i\right)\right) = 0$$

and

$$\sum_{i=1}^{n} x_i \left(y_i - \left(b_0^* + b_1^* x_i\right)\right) = 0,$$

which are known as the *normal equations*. Solving these equations yields the values of b_0^* and b_1^* that minimize the sum of squares, as follows.

> *Definition.* The *least squares intercept* and *slope estimates* are
>
> $$b_1 = r\frac{s_y}{s_x} \quad \text{and} \quad b_0 = \overline{y} - b_1\overline{x}.$$
>
> The line that they determine, $\widehat{y} = b_0 + b_1 x$, is called the *fitted regression line.*

We have dropped the asterisk, or star, notation because b_0 and b_1 are no longer candidate values.

Does this procedure yield a sensible line for our Wisconsin lottery sales? Earlier, we computed $r = 0.886$. From this and the basic summary statistics in Table 2.1, we have $b_1 = 0.886(8103)/11{,}098 = 0.647$ and $b_0 = 6495 - (0.647)9311 = 469.7$. This yields the fitted regression line

$$\widehat{y} = 469.7 + (0.647)x.$$

The caret, or "hat," on top of the y reminds us that this $\widehat{y}$, or $\widehat{SALES}$, is a fitted value. One application of the regression line is to estimate sales for a specific population, say, $x = 10{,}000$. The estimate is the height of the regression line, which is $469.7 + (0.647)(10{,}000) = 6939.7$.

Example: Summarizing Simulations. Regression analysis is a tool for summarizing complex data. In practical work, actuaries often simulate complicated financial scenarios; it is often overlooked that regression can be used to summarize relationships of interest.

To illustrate, Manistre and Hancock (2005) simulated many realizations of a 10-year European put option and demonstrated the relationship between two actuarial risk measures, the value-at-risk (VaR) and the conditional tail expectation (CTE). For one example, these authors examined lognormally distributed stock returns with an initial stock price of \$100, so that in 10 years the price of the stock would be distributed as

$$S(Z) = 100\exp\left((.08)10 + .15\sqrt{10}Z\right),$$

based on an annual mean return of 10%, standard deviation of 15% and the outcome from a standard normal random variable Z. The put option pays the difference between the strike price, that will be taken to be \$110 for this example, and $S(Z)$. The present value of this option is

$$C(Z) = \mathrm{e}^{-0.06(10)} \max\left(0, 110 - S(Z)\right),$$

based on a 6% discount rate.

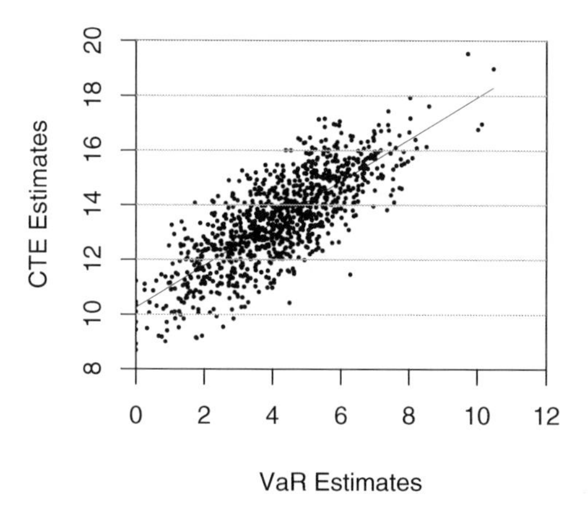

Figure 2.3 Plot of conditional tail expectation (CTE) versus value at risk (VaR). Based on $n = 1{,}000$ simulations from a 10-year European put bond. *Source:* Manistre and Hancock (2005).

To estimate the VaR and CTE, for each i, 1000 i.i.d. standard normal random variables were simulated and used to calculate 1000 present values, $C_{i1}, \ldots, C_{i,1000}$. The 95th percentile of these present values is the estimate of the value at risk, denoted as VaR_i. The average of the highest 50 ($= (1 - .05) \times 1{,}000$) of the present values is the estimate of the conditional tail expectation, denoted as CTE_i. Manistre and Hancock (2005) performed this calculation $i = 1, \ldots, 1{,}000$ times; the result is presented in Figure 2.3. The scatter plot shows a strong but not perfect relationship between the VaR and the CTE, the correlation coefficient turns out to be $r = 0.782$.

2.2 Basic Linear Regression Model

The scatter plot, correlation coefficient, and the fitted regression line are useful devices for summarizing the relationship between two variables for a specific dataset. To infer general relationships, we need models to represent outcomes of broad populations.

This chapter focuses on a basic linear regression model. The "linear regression" part comes from the fact that we fit a line to the data. The "basic" part is because we use only one explanatory variable, x. This model is also known as a "simple" linear regression. This text avoids this language because it gives the false impression that regression ideas and interpretations with one explanatory variable are always straightforward.

We now introduce two sets of assumptions of the basic model, the *observables* and the *error* representations. They are equivalent, but each will help us as we later extend regression models beyond the basics:

Basic Linear Regression Model
Observables Representation Sampling Assumptions

F1. E $y_i = \beta_0 + \beta_1 x_i$.
F2. $\{x_1, \ldots, x_n\}$ are nonstochastic variables.
F3. Var $y_i = \sigma^2$.
F4. $\{y_i\}$ are independent random variables.

The observables representation focuses on variables that we can see (or observe), (x_i, y_i). Inference about the distribution of y is conditional on the observed explanatory variables, so that we may treat $\{x_1, \ldots, x_n\}$ as nonstochastic variables (assumption F2). When considering types of sampling mechanisms for (x_i, y_i), it is convenient to think of a *stratified random sampling* scheme, where values of $\{x_1, \ldots, x_n\}$ are treated as strata, or groups. Under stratified sampling, for each unique value of x_i, we draw a random sample from a population. To illustrate, suppose you are drawing from a database of firms to understand stock return performance (y) and want to stratify on the basis of the size of the firm. If the amount of assets is a continuous variable, then we can imagine drawing a sample of size 1 for each firm. In this way, we hypothesize a distribution of stock returns conditional on firm asset size.

As a digression, you will often see reports that summarize results for the "top 50 managers" or the "best 100 universities," measured by some outcome variable. In regression applications, make sure that you do not select observations based on a dependent variable, such as the highest stock return, because this is stratifying that is based on the y, not the x. Chapter 6 will discuss sampling procedures in greater detail.

Stratified sampling also provides motivation for assumption F4, the independence among responses. One can motivate assumption F1 by thinking of (x_i, y_i) as a draw from a population, where the mean of the conditional distribution of y_i given $\{x_i\}$ is linear in the explanatory variable. Assumption F3 is known as *homoscedasticity*, which we will discuss extensively in Section 5.7. See Goldberger (1991) for additional background on this representation.

A fifth assumption that is often implicitly used is as follows:

F5. $\{y_i\}$ are normally distributed.

This assumption is not required for many statistical inference procedures because central limit theorems provide approximate normality for many statistics of interest. However, formal justification for some, such as t-statistics, do require this additional assumption.

In contrast to the observables representation, an alternative set of assumptions focuses on the deviations, or errors, in the regression, defined as $\varepsilon_i = y_i - (\beta_0 + \beta_1 x_i)$:

Basic Linear Regression Model Error Representation Sampling Assumptions
E1. $y_i = \beta_0 + \beta_1 x_i + \varepsilon_i$.
E2. $\{x_1, \ldots, x_n\}$ are nonstochastic variables.
E3. $\mathrm{E}\, \varepsilon_i = 0$ and $\mathrm{Var}\, \varepsilon_i = \sigma^2$.
E4. $\{\varepsilon_i\}$ are independent random variables.

The *error representation* is based on the Gaussian theory of errors (see Stigler, 1986, for a historical background). Assumption E1 assumes that y is in part due

Table 2.2 Summary Measures of the Population and Sample

Data	Summary Measures	Regression Line		Variance
		Intercept	Slope	
Population	Parameters	β_0	β_1	σ^2
Sample	Statistics	b_0	b_1	s^2

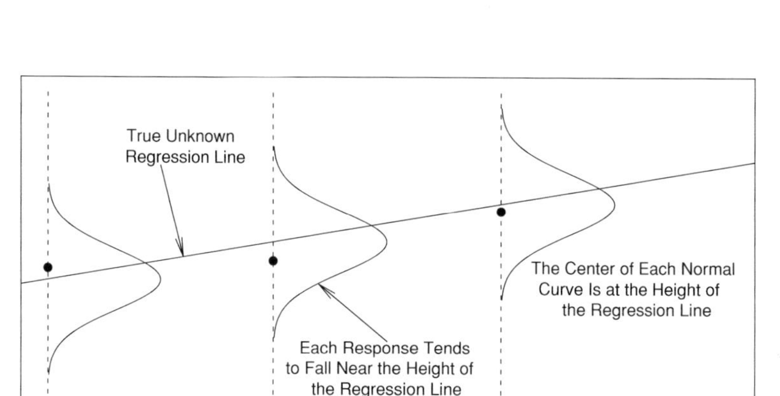

Figure 2.4 The distribution of the response varies by the level of the explanatory variable.

to a linear function of the observed explanatory variable, x. Other unobserved variables that influence the measurement of y are interpreted to be included in the error term, ε_i, which is also known as the *disturbance* term. The independence of errors, E4, can be motivated by assuming that $\{\varepsilon_i\}$ is realized through a simple random sample from an unknown population of errors.

Assumptions E1–E4 are equivalent to F1–F4. The error representation provides a useful springboard for motivating goodness-of-fit measures (Section 2.3). However, a drawback of the error representation is that it draws the attention from the observable quantities (x_i, y_i) to an unobservable quantity, $\{\varepsilon_i\}$. To illustrate, the sampling basis, viewing $\{\varepsilon_i\}$ as a simple random sample, is not directly verifiable because one cannot directly observe the sample $\{\varepsilon_i\}$. Moreover, the assumption of additive errors in E1 will be troublesome when we consider nonlinear regression models.

Figure 2.4 illustrates some of the assumptions of the basic linear regression model. The data (x_1, y_1), (x_2, y_2), and (x_3, y_3) are observed and are represented by the circular opaque plotting symbols. According to the model, these observations should be close to the regression line $\mathrm{E}\, y = \beta_0 + \beta_1 x$. Each deviation from the line is random. We will often assume that the distribution of deviations may be represented by a normal curve, as in Figure 2.4.

The basic linear regression model assumptions describe the underlying population. Table 2.2 highlights the idea that characteristics of this population can be summarized by the parameters β_0, β_1, and σ^2. In Section 2.1, we summarized data from a sample, introducing the statistics b_0 and b_1. Section 2.3 will introduce s^2, the statistic corresponding to the parameter σ^2.

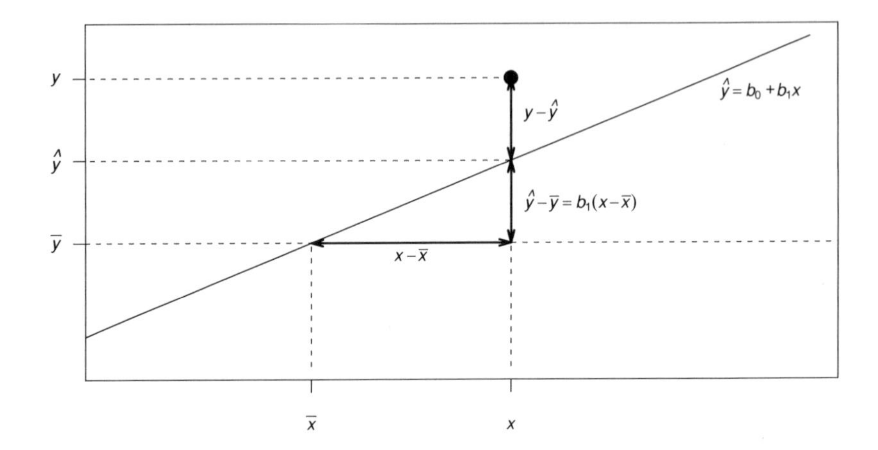

Figure 2.5 Geometric display of the deviation decomposition.

2.3 Is the Model Useful? Some Basic Summary Measures

Although statistics is the science of summarizing data, it is also the art of arguing with data. This section develops some of the basic tools used to justify the basic linear regression model. A scatter plot may provide strong *visual* evidence that x influences y; developing *numerical* evidence will enable us to quantify the strength of the relationship. Further, numerical evidence will be useful when we consider other datasets where the graphical evidence is not compelling.

2.3.1 Partitioning the Variability

The squared deviations, $(y_i - \overline{y})^2$, provide a basis for measuring the spread of the data. If we wish to estimate the ith dependent variable *without* knowledge of x, then $\overline{y}$ is an appropriate estimate and $y_i - \overline{y}$ represents the deviation of the estimate. We use $Total\ SS = \sum_{i=1}^{n}(y_i - \overline{y})^2$, the total sum of squares, to represent the variation in all of the responses.

Suppose now that we also have knowledge of x, an explanatory variable. Using the fitted regression line, for each observation we can compute the corresponding *fitted value*, $\widehat{y}_i = b_0 + b_1 x_i$. The fitted value is our estimate *with* knowledge of the explanatory variable. As before, the difference between the response and the fitted value, $y_i - \widehat{y}_i$, represents the deviation of this estimate. We now have two "estimates" of y_i; these are $\widehat{y}_i$ and $\overline{y}$. Presumably, if the regression line is useful, then $\widehat{y}_i$ is a more accurate measure than $\overline{y}$. To judge this usefulness, we algebraically decompose the total deviation as

$$\begin{array}{ccccc} \underbrace{y_i - \overline{y}} & = & \underbrace{y_i - \widehat{y}_i} & + & \underbrace{\widehat{y}_i - \overline{y}} \\ \text{total} & = & \text{unexplained} & + & \text{explained} \\ \text{deviation} & & \text{deviation} & & \text{deviation} \end{array} \tag{2.1}$$

Interpret this equation as "the deviation without knowledge of x equals the deviation with knowledge of x plus the deviation explained by x." Figure 2.5 is a geometric display of this decomposition. In the figure, an observation above the line was chosen, yielding a positive deviation from the fitted regression line, to make the graph easier to read. A good exercise is to draw a rough sketch corresponding to Figure 2.5 with an observation below the fitted regression line.

Now, from the algebraic decomposition in equation (2.1), square each side of the equation and sum over all observations. After a little algebraic manipulation, this yields

$$\sum_{i=1}^{n} (y_i - \overline{y})^2 = \sum_{i=1}^{n} (y_i - \widehat{y}_i)^2 + \sum_{i=1}^{n} (\widehat{y}_i - \overline{y})^2 . \tag{2.2}$$

We rewrite this as *Total* SS = *Error* SS + *Regression* SS, where SS stands for sum of squares. We interpret

- *Total* SS as the total variation without knowledge of x
- *Error* SS as the total variation remaining after the introduction of x
- *Regression* SS as the difference between *Total* SS and *Error* SS, or the total variation explained through knowledge of x

When squaring the right-hand side of equation (2.1), we have the cross-product term $2(y_i - \widehat{y}_i)(\widehat{y}_i - \overline{y})$. With the algebraic manipulation, one can check that the sum of the cross-products over all observations is zero. This result is not true for all fitted lines but is a special property of the least squares fitted line.

In many instances, the variability decomposition is reported through only a single statistic.

Definition. The *coefficient of determination* is denoted by the symbol R^2, called "*R-square*," and defined as follows:

$$R^2 = \frac{\textit{Regression } SS}{\textit{Total } SS}.$$

We interpret R^2 to be the proportion of variability explained by the regression line. In one extreme case where the regression line fits the data perfectly, we have *Error* $SS = 0$ and $R^2 = 1$. In the other extreme case where the regression line provides no information about the response, we have *Regression* $SS = 0$ and $R^2 = 0$. The coefficient of determination is constrained by the inequalities $0 \leq R^2 \leq 1$, with larger values implying a better fit.

2.3.2 The Size of a Typical Deviation: *s*

In the basic linear regression model, the deviation of the response from the regression line, $y_i - (\beta_0 + \beta_1 x_i)$, is not an observable quantity because the parameters β_0 and β_1 are not observed. However, by using estimators b_0 and b_1, we can approximate this deviation using

$$e_i = y_i - \widehat{y}_i = y_i - (b_0 + b_1 x_i),$$

known as the *residual*.

Residuals will be critical to developing strategies for improving model specification in Section 2.6. We now show how to use the residuals to estimate σ^2. From a first course in statistics, we know that if one could observe the deviations ε_i, then a desirable estimate of σ^2 would be $(n-1)^{-1}\sum_{i=1}^{n}(\varepsilon_i - \bar{\varepsilon})^2$. Because $\{\varepsilon_i\}$ are not observed, we use the following.

Definition. An estimator of σ^2, the *mean square error (MSE)*, is defined as

$$s^2 = \frac{1}{n-2}\sum_{i=1}^{n} e_i^2. \tag{2.3}$$

The positive square root, $s = \sqrt{s^2}$, is called the *residual standard deviation.*

Comparing the definitions of s^2 and $(n-1)^{-1}\sum_{i=1}^{n}(\varepsilon_i - \bar{\varepsilon})^2$, you will see two important differences. First, in defining s^2, we have not subtracted the average residual from each residual before squaring. This is because the average residual is zero, a special property of least squares estimation (see Exercise 2.14). This result can be shown using algebra and is guaranteed for all datasets.

s^2 is an unbiased estimator of σ^2.

Second, in defining s^2 we have divided by $n-2$ instead of $n-1$. Intuitively, dividing by either n or $n-1$ tends to underestimate σ^2. The reason is that, when fitting lines to data, we need at least two observations to determine a line. For example, we must have at least three observations for there to be any variability about a line. How much "freedom" is there for variability about a line? We will say that the error degrees of freedom is the number of observations available, n, minus the number of observations needed to determine a line, 2 (with symbols, $df = n-2$). However, as we saw in the least squares estimation subsection, we do not need to identify two actual observations to determine a line. The idea is that if an analyst knows the line and $n-2$ observations, then the remaining two observations can be determined, without variability. When dividing by $n-2$, it can be shown that s^2 is an unbiased estimator of σ^2.

We can also express s^2 in terms of the sum of squares quantities. That is,

$$s^2 = \frac{1}{n-2}\sum_{i=1}^{n}(y_i - \widehat{y}_i)^2 = \frac{Error\ SS}{n-2} = MSE.$$

This leads us to the *analysis of variance*, or *ANOVA*, table:

ANOVA Table

Source	Sum of Squares	df	Mean Square
Regression	*Regression SS*	1	*Regression MS*
Error	*Error SS*	$n-2$	*MSE*
Total	*Total SS*	$n-1$	

The ANOVA table is merely a bookkeeping device used to keep track of the sources of variability; it routinely appears in statistical software packages as part of the regression output. The mean square column figures are defined to be the sums of square (SS) figures divided by their respective degrees of freedom (df). In particular, the mean square for errors (*MSE*) equals s^2, and the regression sum of squares equals the regression mean square. This latter property is specific to the regression with one variable case; it is not true where we consider more than one explanatory variable.

The error degrees of freedom in the ANOVA table is $n-2$. The total degrees of freedom is $n-1$, reflecting the fact that the total sum of squares is centered about the mean (at least two observations are required for positive variability). The single degree of freedom associated with the regression portion means that the slope, plus one observation, is enough information to determine the line. This is because it takes two observations to determine a line and at least three observations for there to be any variability about the line.

The analysis of variance table for the lottery data is as follows:

ANOVA Table

Source	Sum of Squares	df	Mean Square
Regression	2,527,165,015	1	2,527,165,015
Error	690,116,755	48	14,377,432
Total	3,217,281,770	49	

From this table, you can check that $R^2 = 78.5\%$ and $s = 3{,}792$.

2.4 Properties of Regression Coefficient Estimators

The least squares estimates can be expressed as a weighted sum of the responses. To see this, define the weights

$$w_i = \frac{x_i - \overline{x}}{s_x^2(n-1)}.$$

Because the sum of x-deviations $(x_i - \overline{x})$ is zero, we see that $\sum_{i=1}^n w_i = 0$. Thus, we can express the slope estimate

$$b_1 = r\frac{s_y}{s_x} = \frac{1}{(n-1)s_x^2}\sum_{i=1}^n (x_i - \overline{x})(y_i - \overline{y}) = \sum_{i=1}^n w_i (y_i - \overline{y}) = \sum_{i=1}^n w_i y_i. \tag{2.4}$$

The exercises ask the reader to verify that b_0 can also be expressed as a weighted sum of responses, so our discussion pertains to both regression coefficients. Because regression coefficients are weighted sums of responses, they can be affected dramatically by unusual observations (see Section 2.6).

Regression coefficients are weighted sums of the responses.

Because b_1 is a weighted sum, it is straightforward to derive the expectation and variance of this statistic. By the linearity of expectations and Assumption F1, we have

$$\mathrm{E}\, b_1 = \sum_{i=1}^{n} w_i \,\mathrm{E}\, y_i = \beta_0 \sum_{i=1}^{n} w_i + \beta_1 \sum_{i=1}^{n} w_i x_i = \beta_1 .$$

That is, b_1 is an unbiased estimator of β_1. Here, the sum $\sum_{i=1}^{n} w_i x_i = \left[s_x^2(n-1)\right]^{-1} \sum_{i=1}^{n} (x_i - \overline{x}) x_i = \left[s_x^2(n-1)\right]^{-1} \sum_{i=1}^{n} (x_i - \overline{x})^2 = 1$. From the definition of the weights, some easy algebra also shows that $\sum_{i=1}^{n} w_i^2 = 1/\left(s_x^2(n-1)\right)$. Further, the independence of the responses implies that the variance of the sum is the sum of the variances, and thus we have

$$\mathrm{Var}\; b_1 = \sum_{i=1}^{n} w_i^2 \mathrm{Var}\; y_i = \frac{\sigma^2}{s_x^2(n-1)} .$$

Replacing σ^2 by its estimator s^2 and taking square roots leads to the following.

Definition. The *standard error* of b_1, the estimated standard deviation of b_1, is defined as

$$se(b_1) = \frac{s}{s_x \sqrt{n-1}} . \qquad (2.5)$$

This is our measure of the reliability, or precision, of the slope estimator. Using equation (2.5), we see that $se(b_1)$ is determined by three quantities, n, s, and s_x, as follows:

A standard error is an estimated standard deviation.

- If we have more observations so that n becomes larger, then $se(b_1)$ becomes smaller, other things equal.
- If the observations have a greater tendency to lie closer to the line so that s becomes smaller, then $se(b_1)$ becomes smaller, other things equal.
- If values of the explanatory variable become more spread out so that s_x increases, then $se(b_1)$ becomes smaller, other things equal.

Smaller values of $se(b_1)$ offer a better opportunity to detect relations between y and x. Figure 2.6 illustrates these relationships. Here, the scatter plot in the middle has the smallest value of $se(b_1)$. Compared with the middle plot, the left-hand plot has a larger value of s and thus $se(b_1)$. Compared with the right-hand plot, the middle plot has a larger s_x, and thus smaller value of $se(b_1)$.

Equation (2.4) also implies that the regression coefficient b_1 is normally distributed. That is, recall from mathematical statistics that linear combinations of normal random variables are also normal. Thus, if Assumption F5 holds, then b_1 is normally distributed. Moreover, several versions of central limit theorems

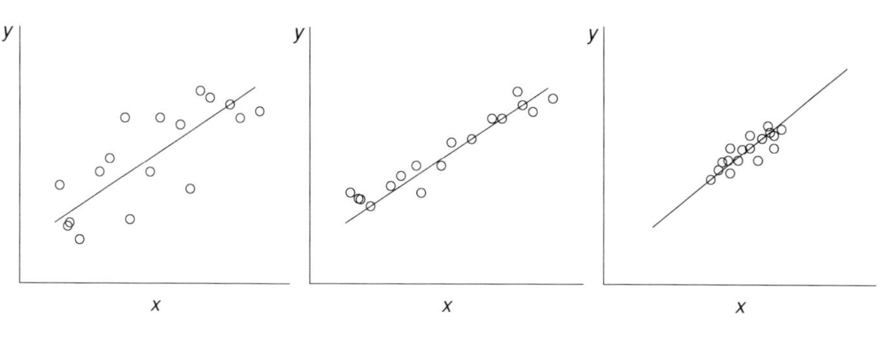

Figure 2.6 These three scatter plots exhibit the same linear relationship between y and x. The plot on the left exhibits greater variability about the line than the plot in the middle. The plot on the right exhibits a smaller standard deviation in x than the plot in the middle.

exists for weighted sums (see, e.g., Serfling, 1980). Thus, as discussed in Section 1.4, if the responses y_i are even approximately normally distributed, then it will be reasonable to use a normal approximation for the sampling distribution of b_1. Using $se(b_1)$ as the estimated standard deviation of b_1, for large values of n we have that $(b_1 - \beta_1)/se(b_1)$ has an approximate standard normal distribution. Although we will not prove it here, under Assumption F5 $(b_1 - \beta_1)/se(b_1)$ follows a t-distribution with degrees of freedom $df = n - 2$.

2.5 Statistical Inference

Having fit a model with a dataset, we can make a number of important statements. Generally, it is useful to think about these statements in three categories: (i) tests of hypothesized ideas, (ii) estimates of model parameters, and (ii) predictions of new outcomes.

2.5.1 *Is the Explanatory Variable Important? The t-Test*

We respond to the question of whether the explanatory variable is important by investigating whether $\beta_1 = 0$. The logic is that if $\beta_1 = 0$, then the basic linear regression model no longer includes an explanatory variable x. Thus, we translate our question of the importance of the explanatory variable into a narrower question that can be answered using the hypothesis-testing framework. This narrower question is, IS $H_0 : \beta_1 = 0$ valid? We respond to this question by looking at the test statistic:

$$t\text{-ratio} = \frac{\text{estimator} - \text{hypothesized value of parameter}}{\text{standard error of the estimator}}.$$

Appendix A3.3 provides additional details about the t-distribution, including a graph and distribution table.

For the case of $H_0 : \beta_1 = 0$, we examine the t-ratio $t(b_1) = b_1/se(b_1)$ because the hypothesized value of β_1 is 0. This is the appropriate standardization because, under the null hypothesis and the model assumptions described in Section 2.4, the sampling distribution of $t(b_1)$ can be shown to be the t-distribution with $df = n - 2$. Thus, to test the null hypothesis H_0 against the alternative $H_a : \beta_1 \neq 0$,

Table 2.3 Decision-Making Procedures for Testing $H_0 : \beta_1 = d$

Alternative Hypothesis (H_a)	Procedure: Reject H_0 in favor of H_a if
$\beta_1 > d$	t-ratio $> t_{n-2,1-\alpha}$.
$\beta_1 < d$	t-ratio $< -t_{n-2,1-\alpha}$.
$\beta_1 \neq d$	$|t\text{-ratio}| > t_{n-2,1-\alpha/2}$.

Note: The significance level is α. Here, $t_{n-2,1-\alpha}$ is the $(1-\alpha)$th percentile from the t-distribution using $df = n-2$ degrees of freedom. The test statistic is t-ratio $= (b_1 - d)/se(b_1)$.

we reject H_0 in favor of H_a if $|t(b_1)|$ exceeds a t-value. Here, this t-value is a percentile from the t-distribution using $df = n-2$. We denote the significance level as α and this t-value as $t_{n-2,1-\alpha/2}$.

Example: Lottery Sales, Continued. For the lottery sales example, the residual standard deviation is $s = 3792$. From Table 2.1, we have $s_x = 11{,}098$. Thus, the standard error of the slope is $se(b_1) = 3792/(11{,}098\sqrt{50-1}) = 0.0488$. From Section 2.1, the slope estimate is $b_1 = 0.647$. Thus, the t-statistic is $t(b_1) = 0.647/0.0488 = 13.4$. We interpret this by saying that the slope is 13.4 standard errors above zero. For the significance level, we use the customary value of $\alpha = 5\%$. The 97.5th percentile from a t-distribution with $df = 50 - 2 = 48$ degrees of freedom is $t_{48,0.975} = 2.011$. Because $|13.4| > 2.011$, we reject the null hypothesis that the slope $\beta_1 = 0$ in favor of the alternative that $\beta_1 \neq 0$.

Making decisions by comparing a t-ratio to a t-value is called a *t-test*. Testing $H_0 : \beta_1 = 0$ versus $H_a : \beta_1 \neq 0$ is just one of many hypothesis tests that can be performed, although it is the most common. Table 2.3 outlines alternative decision-making procedures. These procedures are for testing $H_0 : \beta_1 = d$, where d is a user-prescribed value that may be equal to zero or any other known value. For example, in our Section 2.7 example, we will use $d = 1$ to test financial theories about the stock market.

Alternatively, one can construct probability (p) values and compare these to given significant levels. The p-value is a useful summary statistic for the data analyst to report because it allows the report reader to understand the strength of the deviation from the null hypothesis. Table 2.4 summarizes the procedure for calculating p-values.

Another interesting way of addressing the question of the importance of an explanatory variable is through the correlation coefficient. Remember that the correlation coefficient is a measure of linear relationship between x and y. Let's denote this statistic by $r(y, x)$. This quantity is unaffected by scale changes in either variable. For example, if we multiply the x variable by the number b_1, then the correlation coefficient remains unchanged. Further, correlations are

Alternative Hypothesis (H_a)	$\beta_1 > d$	$\beta_1 < d$	$\beta_1 \neq d$
p-Value	$\Pr(t_{n-2} > t\text{-ratio})$	$\Pr(t_{n-2} < t\text{-ratio})$	$\Pr(\lvert t_{n-2}\rvert > \lvert t\text{-ratio}\rvert)$

Notes: Here, t_{n-2} is a t-distributed random variable with $df = n - 2$ degrees of freedom. The test statistic is t-ratio $= (b_1 - d)/se(b_1)$.

Table 2.4 Probability Values for Testing $H_0 : \beta_1 = d$

unchanged by additive shifts. Thus, if we add a number, say, b_0, to each x variable, then the correlation coefficient remains unchanged. Using a scale change and an additive shift on the x variable can be used to produce the fitted value $\widehat{y} = b_0 + b_1 x$. Thus, using notation, we have $r(y, x) = |r(y, \widehat{y})|$. We may thus interpret the correlation between the responses and the explanatory variable to be equal to the correlation between the responses and the fitted values. This leads, then, to the following interesting algebraic fact, $R^2 = r^2$. That is, the coefficient of determination equals the correlation coefficient squared. This is much easier to interpret if one thinks of r as the correlation between observed and fitted values. See Exercise 2.13 for steps useful in confirming this result.

$R^2 = r^2$.

2.5.2 Confidence Intervals

Investigators often cite the formal hypothesis-testing mechanism to respond to the question, Does the explanatory variable have a real influence on the response? A natural follow-up question is, To what extent does x affect y? To a certain degree, one could respond using the size of the t-ratio or the p-value. However, in many instances, a *confidence interval* for the slope is more useful.

To introduce confidence intervals for the slope, recall that b_1 is our point estimator of the true, unknown slope β_1. Section 2.4 argued that this estimator has standard error $se(b_1)$ and that $(b_1 - \beta_1)/se(b_1)$ follows a t-distribution with $n - 2$ degrees of freedom. Probability statements can be inverted to yield confidence intervals. Using this logic, we have the following confidence interval for the slope β_1:

Definition. A $100(1 - \alpha)\%$ confidence interval for the slope β_1 is

$$b_1 \pm t_{n-2,1-\alpha/2}\, se(b_1). \tag{2.6}$$

As with hypothesis testing, $t_{n-2,1-\alpha/2}$ is the $(1 - \alpha/2)$th percentile from the t-distribution with $df = n - 2$ degrees of freedom. Because of the two-sided nature of confidence intervals, the percentile is $1 - (1 - \text{confidence level})/2$. In this text, for notational simplicity, we generally use a 95% confidence interval, so the percentile is $1 - (1 - .0.95)/2 = 0.975$. The confidence interval provides a range of reliability that measures the usefulness of the estimate.

In Section 2.1, we established that the least squares slope estimate for the lottery sales example is $b_1 = 0.647$. The interpretation is that if a Zip code's population differs by 1,000, then we expect mean lottery sales to differ by \$647. How reliable is this estimate? It turns out that $se(b_1) = 0.0488$ and thus an approximate 95% confidence interval for the slope is

$$0.647 \pm (2.011)(.0488),$$

or (0.549, 0.745). Similarly, if population differs by 1,000, a 95% confidence interval for the expected change in sales is (549, 745). Here, we use the t-value $t_{48,0.975} = 2.011$ because there are 48 ($= n - 2$) degrees of freedom and, for a 95% confidence interval, we need the 97.5th percentile.

2.5.3 Prediction Intervals

In Section 2.1, we showed how to use least squares estimators to predict the lottery sales for a Zip code, outside of our sample, with a population of 10,000. Because prediction is such an important task for actuaries, we formalize the procedure so that it can be used on a regular basis.

To predict an additional observation, we assume that the level of explanatory variable is known and is denoted by x_*. For example, in our previous lottery sales example, we used $x_* = 10{,}000$. We also assume that the additional observation follows the same linear regression model as the observations in the sample.

Using our least square estimators, our point prediction is $\widehat{y}_* = b_0 + b_1 x_*$, the height of the fitted regression line at x_*. We may decompose the prediction error into two parts:

$$\underbrace{y_* - \widehat{y}_*}_{\text{prediction error}} = \underbrace{\beta_0 - b_0 + (\beta_1 - b_1)\, x_*}_{\substack{\text{error in estimating the} \\ \text{regression line at } x_*}} + \underbrace{\varepsilon_*}_{\substack{\text{deviation of the additional} \\ \text{response from its mean}}}.$$

It can be shown that the standard error of the prediction is

$$se(pred) = s\sqrt{1 + \frac{1}{n} + \frac{(x_* - \overline{x})^2}{(n-1)s_x^2}}.$$

As with $se(b_1)$, the terms n^{-1} and $(x_* - \overline{x})^2 / \left[(n-1)s_x^2\right]$ become close to zero as the sample size n becomes large. Thus, for large n, we have that $se(pred) \approx s$, reflecting that the error in estimating the regression line at a point becomes negligible and deviation of the additional response from its mean becomes the entire source of uncertainty.

Definition. A $100(1-\alpha)\%$ prediction interval at x_* is

$$\widehat{y}_* \pm t_{n-2,1-\alpha/2}\, se(pred), \tag{2.7}$$

where the t-value $t_{n-2,1-\alpha/2}$ is the same as used for hypothesis testing and the confidence interval.

For example, the point prediction at $x_* = 10{,}000$ is $\widehat{y}_* = 469.7 + 0.647\,(10{,}000) = 6939.7$. The standard error of this prediction is

$$se(pred) = 3792\sqrt{1 + \frac{1}{50} + \frac{(10{,}000 - 9{,}311)^2}{(50-1)(11{,}098)^2}} = 3829.6.$$

With a t-value equal to 2.011, this yields an approximate 95% prediction interval:

$$6939.7 \pm (2.011)(3829.6) = 6939.7 \pm 7{,}701.3 = (-761.6,\ 14{,}641.0).$$

We interpret these results by first pointing out that our best estimate of lottery sales for a Zip code with a population of 10,000 is $6,939.70. Our 95% prediction interval represents a range of reliability for this prediction. If we could see many Zip codes, each with a population of 10,000, on average we would expect that about 19 out of 20, or 95%, would have lottery sales between $0 and $14,641. It is customary to truncate the lower bound of the prediction interval to zero if negative values of the response are deemed to be inappropriate.

2.6 Building a Better Model: Residual Analysis

Quantitative disciplines calibrate models with data. Statistics takes this one step further, using discrepancies between the assumptions and the data to improve model specification. We will examine the Section 2.2 modeling assumptions in light of the data and use any mismatch to specify a better model; this process is known as *diagnostic checking* (like when you go to a doctor and he or she performs diagnostic routines to check your health).

Statistics uses discrepancies between the assumptions and the data to improve model specification.

We will begin with the Section 2.2 error representation. Under this set of assumptions, the deviations $\{\varepsilon_i\}$ are identically and independently distributed (i.i.d) and, under assumption F5, normally distributed. To assess the validity of these assumptions, one uses (observed) residuals $\{e_i\}$ as approximations for the (unobserved) deviations $\{\varepsilon_i\}$. The basic theme is that if the residuals are related to a variable or display any other recognizable pattern, then we should be able to take advantage of this information and improve our model specification. The residuals should contain little or no information and represent only natural variation from the sampling that cannot be attributed to any specific source. *Residual analysis* is the exercise of checking the residuals for patterns.

If the residuals are related to a variable or display any other recognizable pattern, then we should be able to take advantage of this information and improve our model specification.

There are five types of model discrepancies that analysts commonly look for. If detected, the discrepancies can be corrected with the appropriate adjustments in the model specification.

Model Misspecification Issues

(i) **Lack of Independence**. There may exist relationships among the deviations $\{\varepsilon_i\}$ so that they are not independent.
(ii) **Heteroscedasticity**. Assumption E3 that indicates that all observations have a common (though unknown) variability, known as *homoscedasticity*. *Heteroscedascity* is the term used when the variability varies by observation.
(iii) **Relationships between Model Deviations and Explanatory Variables**. If an explanatory variable has the ability to help explain the deviation ε, then one should be able to use this information to better predict y.
(iv) **Nonnormal Distributions**. If the distribution of the deviation represents a serious departure from normality, then the usual inference procedures are no longer valid.
(v) **Unusual Points**. Individual observations may have a large effect on the regression model fit, meaning that the results may be sensitive to the impact of a single observation.

This list will serve you throughout your study of regression analysis. Of course, with only an introduction to basic models, we have not yet seen alternative models that might be used when we encounter such model discrepancies. In this book's Part II on time series models, we will study lack of independence among data ordered over time. Chapter 5 will consider heteroscedasticity in further detail. The introduction to multiple linear regression in Chapter 3 will be our first look at handling relationships between $\{\varepsilon_i\}$ and additional explanatory variables. We have, however, already had an introduction to the effect of normal distributions, seeing that qq plots can detect nonnormality and that transformations can help induce approximate normality. In this section, we discuss the effects of unusual points.

Much of residual analysis is done by examining a *standardized residual*, a residual divided by its standard error. An approximate standard error of the residual is s; in Chapter 3, we will give a precise mathematical definition. There are two reasons we often examine standardized residuals in lieu of basic residuals. First, if responses are normally distributed, then standardized residuals are approximately realizations from a standard normal distribution. This provides a reference distribution to compare values of standardized residuals. For example, if a standardized residual exceeds two in absolute value, this is considered unusually large and the observation is called an *outlier*. Second, because standardized residuals are dimensionless, we get carryover of experience from one dataset to another. This is true regardless of whether the normal reference distribution is applicable.

Outliers and High Leverage Points. Another important part of residual analysis is the identification of unusual observations in a dataset. Because regression estimates are weighted averages with weights that vary by observation, some

Variables					19 Base Points						A	B	C
x	1.5	1.7	2.0	2.2	2.5	2.5	2.7	2.9	3.0	3.5	3.4	9.5	9.5
y	3.0	2.5	3.5	3.0	3.1	3.6	3.2	3.9	4.0	4.0	8.0	8.0	2.5
x	3.8	4.2	4.3	4.6	4.0	5.1	5.1	5.2	5.5				
y	4.2	4.1	4.8	4.2	5.1	5.1	5.1	4.8	5.3				

Table 2.5 19 Base Points Plus Three Types of Unusual Observations

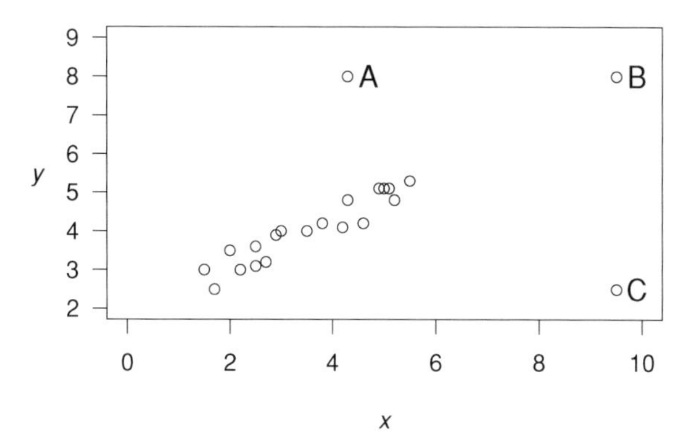

Figure 2.7 Scatter plot of 19 base plus 3 unusual points, labeled A, B and C.

observations are more important than others. This weighting is more important than many users of regression analysis realize. In fact, the example here demonstrates that a single observation can have a dramatic effect in a large dataset.

There are two directions in which a data point can be unusual, the horizontal and vertical directions. By "unusual," I mean that an observation under consideration seems to be far from the majority of the dataset. An observation that is unusual in the vertical direction is called an *outlier*. An observation that is unusual in the horizontal directional is called a *high leverage point*. An observation may be both an outlier and a high leverage point.

Ⓡ **Empirical**
Filename is "OutlierExample"

Example: Outliers and High Leverage Points. Consider the fictitious dataset of 19 points plus three points, labeled A, B, and C, given in Figure 2.7 and Table 2.5. Think of the first nineteen points as "good" observations that represent some type of phenomena. We want to investigate the effect of adding a single aberrant point.

To investigate the effect of each type of aberrant point, Table 2.6 summarizes the results of four separate regressions. The first regression is for the nineteen base points. The other three regressions use the nineteen base points plus each type of unusual observation.

Table 2.6 shows that a regression line provides a good fit for the nineteen base points. The coefficient of determination, R^2, indicates that about 89% of the variability has been explained by the line. The size of the typical error, s, is about 0.29, small compared to the scatter in the y-values. Further, the t-ratio for the slope coefficient is large.

Table 2.6 Results from Four Regressions

Data	b_0	b_1	s	$R^2(\%)$	$t(b_1)$
19 Base Points	1.869	0.611	0.288	89.0	11.71
19 Base Points + A	1.750	0.693	0.846	53.7	4.57
19 Base Points + B	1.775	0.640	0.285	94.7	18.01
19 Base Points + C	3.356	0.155	0.865	10.3	1.44

When the outlier point A is added to the nineteen base points, the situation deteriorates dramatically. The R^2 drops from 89% to 53.7%, and s increases from about 0.29 to about 0.85. The fitted regression line itself does not change that much even though our confidence in the estimates has decreased.

An outlier is unusual in the y-value, but "unusual in the y-value" depends on the x-value. To see this, keep the y-value of point A the same but increase the x-value and call the point B.

When the point B is added to the nineteen base points, the regression line provides a better fit. Point B is close to being on the line of the regression fit generated by the nineteen base points. Thus, the fitted regression line and the size of the typical error, s, do not change much. However, R^2 increases from 89% to nearly 95 percent. If we think of R^2 as $1 - (Error\ SS)/(Total\ SS)$, by adding point B we have increased *Total SS*, the total squared deviations in the y's, even though leaving *Error SS* relatively unchanged. Point B is not an outlier, but it is a high leverage point.

To show how influential this point is, drop the y-value considerably and call this the new point C. When this point is added to the nineteen base points, the situation deteriorates dramatically. The R^2 coefficient drops from 89% to 10%, and the s more than triples, from 0.29 to 0.87. Further, the regression line coefficients change dramatically.

Most users of regression at first do not believe that one point in twenty can have such a dramatic effect on the regression fit. The fit of a regression line can always be improved by removing an outlier. If the point is a high leverage point and not an outlier, it is not clear whether the fit will improve when the point is removed.

Simply because you can dramatically improve a regression fit by omitting an observation does not mean you should always do so! The goal of data analysis is to understand the information in the data. Throughout the text, we will encounter many datasets where the unusual points provide some of the most interesting information about the data. The goal of this subsection is to recognize the effects of unusual points; Chapter 5 will provide options for handling unusual points in your analysis.

All quantitative disciplines, such as accounting, economics, linear programming, and so on, practice the art of *sensitivity analysis*. Sensitivity analysis is a description of the global changes in a system due to a small local change in an

Data	b_0	b_1	s	$R^2(\%)$	$t(b_1)$
With Kenosha	469.7	0.647	3792	78.5	13.26
Without Kenosha	−43.5	0.662	2728	88.3	18.82

Table 2.7 Regression Results with and without Kenosha

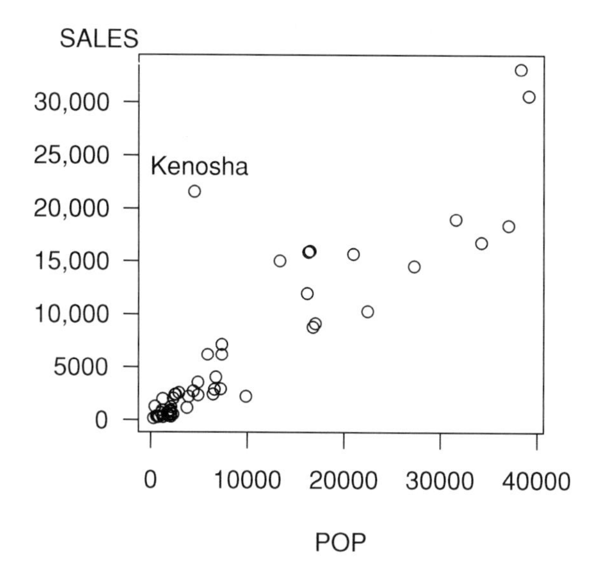

Figure 2.8 Scatter plot of SALES versus POP, with the outlier corresponding to Kenosha marked.

element of the system. Examining the effects of individual observations on the regression fit is a type of sensitivity analysis.

Example: Lottery Sales, Continued. Figure 2.8 exhibits an outlier; the point in the upper-left-hand side of the plot represents a Zip code that includes Kenosha, Wisconsin. Sales for this Zip code are unusually high given its population. Kenosha is close to the Illinois border; residents from Illinois probably participate in the Wisconsin lottery, thus effectively increasing the potential pool of sales in Kenosha. Table 2.7 summarizes the regression fit both with and without this Zip code.

For the purposes of inference about the slope, the presence of Kenosha does not alter the results dramatically. Both slope estimates are qualitatively similar and the corresponding t-statistics are very high, well above cutoffs for statistical significance. However, there are dramatic differences when assessing the quality of the fit. The coefficient of determination, R^2, increased from 78.5% to 88.3% when deleting Kenosha. Moreover, our typical deviation s dropped by more than \$1,000. This is particularly important if we want to tighten our prediction intervals.

To check the accuracy of our assumptions, it is also customary to check the normality assumption. One way of doing this is with the qq plot, introduced

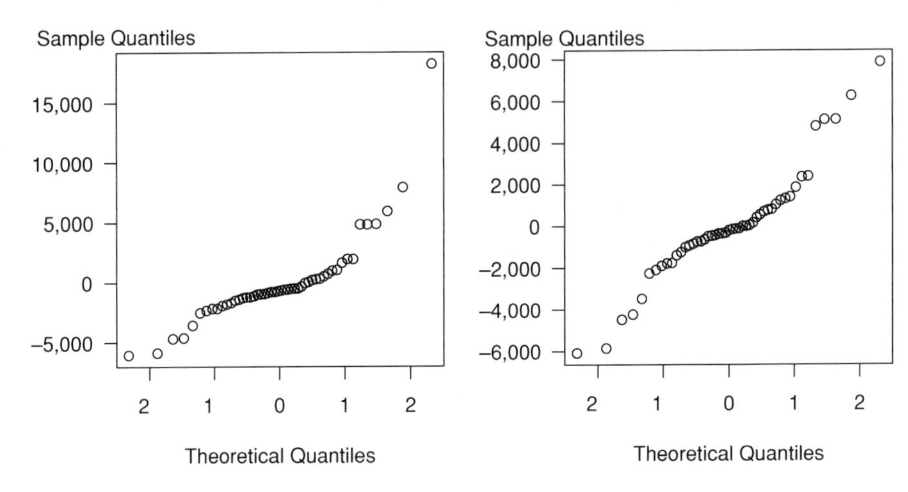

Figure 2.9 The *qq* plots of Wisconsin lottery residuals. The left-hand panel is based on all 50 points. The right-hand panel is based on 49 points, residuals from a regression after removing Kenosha.

in Section 1.2. The two panels in Figures 2.9 are *qq* plots with and without the Kenosha Zip code. Recall that points close to linear indicate approximate normality. In the right-hand panel of Figure 2.9, the sequence does appear to be linear, so residuals are approximately normally distributed. This is not the case in the left-hand panel, where the sequence of points appears to climb dramatically for large quantiles. The interesting thing is that the nonnormality of the distribution is due to a single outlier, not to a pattern of skewness that is common to all the observations.

2.7 Application: Capital Asset Pricing Model

In this section, we study a financial application, the capital asset pricing model (CAPM). The name is something of a misnomer in that the model is really about *returns* based on capital assets, not on the prices themselves. The types of assets that we examine are equity securities that are traded on an active market, such as the New York Stock Exchange (NYSE). For a stock on the exchange, we can relate returns to prices through the following expression:

$$\text{Return} = \frac{\text{price at the end of a period} + \text{dividends} - \text{price at the beginning of a period}}{\text{price at the beginning of a period}}.$$

If we can estimate the returns that a stock generates, then knowledge of the price at the beginning of a generic financial period allows us to estimate the value at the end of the period (ending price plus dividends). Thus, we follow standard practice and model returns of a security.

An intuitively appealing idea, and one of the basic characteristics of the CAPM, is that there should be a relationship between the performance of a security and the market. One rationale is simply that if economic forces are such that the market improves, then those same forces should act on an individual stock, suggesting that it also improve. As noted previously, we measure performance of a security through the return. To measure performance of the market, several market indices exist that summarize the performance of each exchange. We will use the equally weighted index of the Standard & Poor's 500. The S&P500 is the collection of the 500 largest companies traded on the NYSE, where "large" is identified by Standard & Poor's, a financial services rating organization. The equally weighted index is defined by assuming that a portfolio is created by investing \$1.00 in each of the 500 companies.

Another rationale for a relationship between security and market returns comes from financial economics theory. This is the CAPM theory, attributed to Sharpe (1964) and Lintner (1965) and based on the portfolio diversification ideas of Markowitz (1959). Other things equal, investors would like to select a return with a high expected value and a low standard deviation, the latter being a measure of risk. One of the desirable properties of using standard deviations as a measure of riskiness is that it is straightforward to calculate the standard deviation of a portfolio. One needs to know only the standard deviation of each security and the correlations among securities. A notable security is a risk-free one, that is, a security that theoretically has a zero standard deviation. Investors often use a 30-day U.S. Treasury bill as an approximation of a risk-free security, arguing that the probability of default of the U.S. government within 30 days is negligible. Positing the existence of a risk-free asset and some other mild conditions, under the CAPM theory there exists an efficient frontier called the *securities market line*. This frontier specifies the minimum expected return that investors should demand for a specified level of risk. To estimate this line, we can use the equation

$$\mathrm{E}\, r = \beta_0 + \beta_1 r_m,$$

where r is the security return and r_m is the market return. We interpret $\beta_1 r_m$ as a measure of the amount of security return attributed to the behavior of the market.

Testing economic theory, or models arising from any discipline, involves collecting data. The CAPM theory is about ex ante (before-the-fact) returns even though we can only test with ex post (after-the-fact) returns. Before the fact, the returns are unknown and there is an entire distribution of returns. After the fact, there is only a single realization of the security and market return. Because at least two observations are required to determine a line, CAPM models are estimated using security and market data gathered over time. In this way, several observations can be made. For the purposes of our discussions, we follow standard practice in the securities industry and examine monthly prices.

Data

Ⓡ **Empirical** *Filename is "CAPM"*

To illustrate, consider monthly returns over the five-year period from January 1986 to December 1990, inclusive. Specifically, we use the security returns from the Lincoln National Insurance Corporation as the dependent variable (y) and the market returns from the index of the S&P500 as the explanatory variable (x). At the time, the Lincoln was a large, multiline insurance company, with headquarters in the Midwest, specifically in Fort Wayne, Indiana. Because it was

Table 2.8 Summary Statistics of 60 Monthly Observations

	Mean	Median	Standard Deviation	Minimum	Maximum
LINCOLN	0.0051	0.0075	0.0859	−0.2803	0.3147
MARKET	0.0074	0.0142	0.0525	−0.2205	0.1275

Source: Center for Research on Security Prices, University of Chicago.

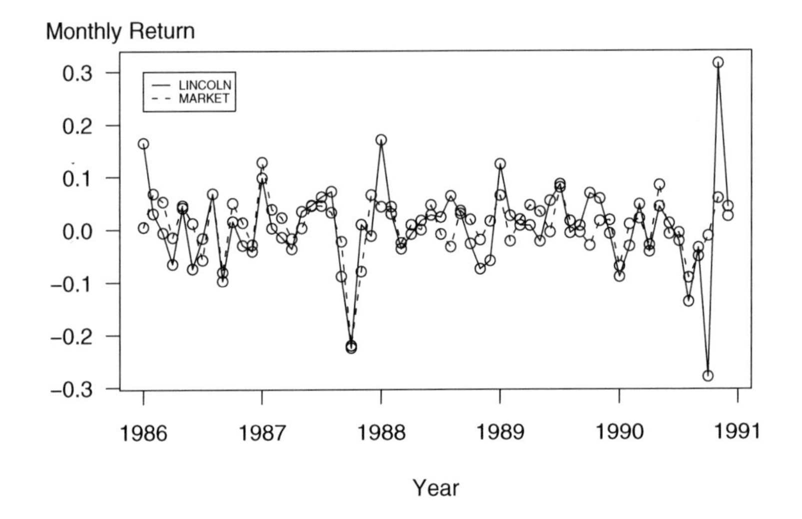

Figure 2.10 Time series plot of returns from Lincoln National Corporation and the market. There are 60 monthly returns over the period January 1986–December 1990.

well known for its prudent management and stability, it is a good company to begin our analysis of the relationship between the market and an individual stock.

We begin by interpreting some of basic summary statistics, in Table 2.8, in terms of financial theory. First, an investor in Lincoln will be concerned that the five-year average return, $\overline{y} = 0.00510$, is less than the return of the market, $\overline{x} = 0.00741$. Students of interest theory recognize that monthly returns can be converted to an annual basis using geometric compounding. For example, the annual return of Lincoln is $(1.0051)^{12} - 1 = 0.062946$, or roughly 6.29%. This is compared to an annual return of 9.26% ($= (100((1.00741)^{12} - 1))$ for the market. A measure of risk, or volatility, that is used in finance is the standard deviation. Thus, interpret $s_y = 0.0859 > 0.05254 = s_x$ to mean that an investment in Lincoln is riskier than the whole market. Another interesting aspect of Table 2.8 is that the smallest market return, −0.22052, is 4.338 standard deviations below its average $((-0.22052 - 0.00741)/0.05254 = -4.338)$. This is highly unusual with respect to a normal distribution.

We next examine the data over time, as is given graphically in Figure 2.10, which shows scatter plots of the returns versus time, called *time series plots*. In Figure 2.10, one can clearly see the smallest market return, and a quick glance at the horizontal axis reveals that this unusual point is in October 1987, the time of the well-known market crash.

The scatter plot in Figure 2.11 graphically summarizes the relationship between Lincoln's return and the return of the market. The market crash is clearly

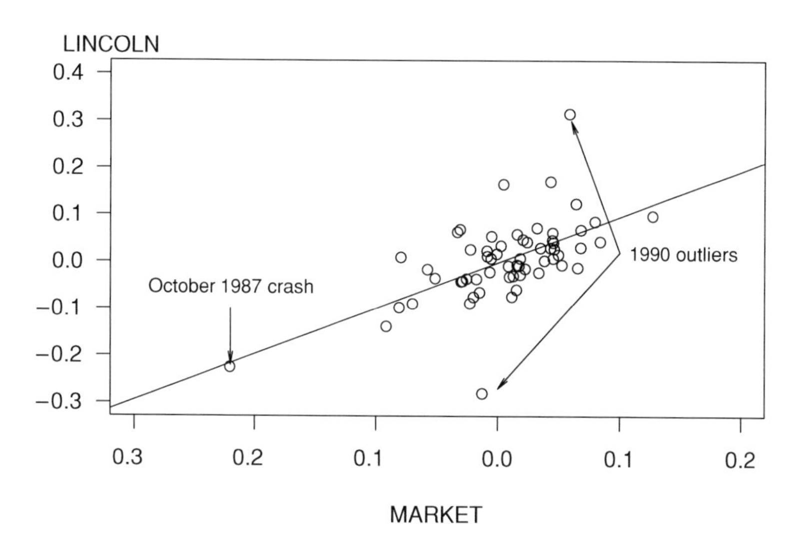

Figure 2.11 Scatter plot of Lincoln's return versus the S&P 500 return. The regression line is superimposed, enabling us to identify the market crash and two outliers.

evident in Figure 2.11 and represents a high leverage point. With the regression line (described subsequently) superimposed, the two outlying points that can be seen in Figure 2.10 are also evident. Despite these anomalies, the plot in Figure 2.11 does suggest that there is a linear relationship between Lincoln and market returns.

Unusual Points

To summarize the relationship between the market and Lincoln's return, a regression model was fit. The fitted regression is

$$\widehat{LINCOLN} = -0.00214 + 0.973 MARKET.$$

The resulting estimated standard error, $s = 0.0696$, is lower than the standard deviation of Lincoln's returns, $s_y = 0.0859$. Thus, the regression model explains some of the variability of Lincoln's returns. Further, the t-statistic associated with the slope b_1 turns out to be $t(b_1) = 5.64$, which is significantly large. One disappointing aspect is that the statistic $R^2 = 35.4\%$ can be interpreted to mean that the market explains only slightly more than a third of the variability. Thus, even though the market is clearly an important determinant, as evidenced by the high t-statistic, it provides only a partial explanation of the performance of the Lincoln's returns.

In the context of the market model, we may interpret the standard deviation of the market, s_x, as *nondiversifiable risk*. Thus, the risk of a security can be decomposed into two components, the diversifiable component and the market component, which is nondiversifiable. The idea is that, by combining several securities, we can create a portfolio of securities that, in most instances, will reduce the riskiness of our holdings when compared with a single security. Again, the rationale for holding a security is that we are compensated through higher expected returns by holding a security with higher riskiness. To quantify the relative riskiness, it is not hard to show that

$$s_y^2 = b_1^2 s_x^2 + s^2 \frac{n-2}{n-1}. \tag{2.8}$$

The riskiness of a security results from the riskiness due to the market plus the riskiness due to a diversifiable component. Note that the riskiness due to the market component, s_x^2, is larger for securities with larger slopes. For this reason, investors think of securities with slopes b_1 greater than one as "aggressive" and slopes less than one as "defensive."

Sensitivity Analysis

The foregoing summary immediately raises two additional issues. First, what is the effect of the October 1987 crash on the fitted regression equation? We know that unusual observations, such as the crash, may potentially influence the fit a great deal. To this end, the regression was rerun without the observation corresponding to the crash. The motivation for this is that the October 1987 crash represents a combination of highly unusual events (the interaction of several automated trading programs operated by the large stock brokerage houses) that we do not want to represent using the same model as our other observations. Deleting this observation, the fitted regression is

$$\widehat{LINCOLN} = -0.00181 + 0.956 MARKET,$$

with $R^2 = 26.4\%$, $t(b_1) = 4.52$, $s = 0.0702$, and $s_y = 0.0811$. We interpret these statistics in the same fashion as the fitted model including the October 1987 crash. It is interesting to note, however, that the proportion of variability explained actually decreases when excluding the influential point. This serves to illustrate an important point. High leverage points are often looked on with dread by data analysts because they are, by definition, unlike other observations in the dataset and require special attention. However, when fitting relationships among variables, they also represent an opportunity because they allow the data analyst to observe the relationship between variables over broader ranges than otherwise possible. The downside is that the relationships may be nonlinear or follow an entirely different pattern when compared to the relationships observed in the main portion of the data.

The second question raised by the regression analysis is what can be said about the unusual circumstances that gave rise to the unusual behavior of Lincoln's returns in October and November 1990. A useful feature of regression analysis is to identify and raise the question; it does not resolve it. Because the analysis clearly pinpoints two highly unusual points, it suggests that the data analyst should go back and ask some specific questions about the sources of the data. In this case, the answer is straightforward. In October 1990, the Travelers' Insurance Company, a competitor, announced that it would take a large write-off in its real estate portfolio because of an unprecedented number of mortgage defaults. The market reacted quickly to this news, and investors assumed that other large publicly traded life insurers would also soon announce large write-offs. Anticipating this news, investors tried to sell their portfolios of, for example, Lincoln's stock, thus causing the price to plummet. However, it turned out that investors overreacted to the news and that Lincoln's portfolio of real estate was indeed sound. Thus, prices quickly returned to their historical levels.

2.8 Illustrative Regression Computer Output

Computers and statistical software packages that perform specialized calculations play a vital role in modern-day statistical analyses. Inexpensive computing capabilities have allowed data analysts to focus on relationships of interest. Specifying models that are attractive merely for their computational simplicity is much less important now than in times before the widespread availability of inexpensive computing. An important theme of this text is to focus on relationships of interest and to rely on widely available statistical software to estimate the models that we specify.

With any computer package, generally the most difficult parts of operating the package are the input, using the commands, and interpreting the output. You will find that most modern statistical software packages accept spreadsheet or text-based files, which makes input of data relatively easy. Personal computer statistical software packages have menu-driven command languages with easily accessible on-line help facilities. Once you decide what to do, finding the right commands is relatively easy.

This section provides guidance in interpreting the output of statistical packages. Most statistical packages generate similar output. Below, three examples of standard statistical software packages, EXCEL, SAS, and R are given. The annotation symbol "[.]" marks a statistical quantity that is described in the legend. Thus, this section provides a link between the notation used in the text and output from some of the standard statistical software packages.

EXCEL Output

```
Regression Statistics
Multiple R                 0.886283[D]
R Square                   0.785497[k]
Adjusted R Square          0.781028[l]
Standard Error             3791.758[j]
Observations                 50[a]

ANOVA
           df              SS              MS                F      Significance F
Regression   1[m]   2527165015 [p]  2527165015 [s] 175.773[u]  1.15757E-17[v]
Residual    48[n]   690116754.8[q]  14377432.39[t]
Total       49[o]   3217281770 [r]

        Coefficients    Standard Error       t Stat          P-value
Intercept     469.7036[b] 702.9061896[d]   0.668230846[f] 0.507187[h]
X Variable 1 0.647095[c] 0.048808085[e]   13.25794257[g] 1.16E-17[i]
```

SAS Output

```
                         The SAS System
                        The REG Procedure
                    Dependent Variable: SALES

                      Analysis of Variance
                            Sum of             Mean
Source                DF       Squares           Square    F Value    Pr > F
Model              1[m]  2527165015[p]  2527165015[s]   175.77[u]  <.0001[v]
Error             48[n]   690116755[q]    14377432[t]
Corrected Total   49[o]  3217281770[r]

        Root MSE              3791.75848[j]    R-Square      0.7855[k]
        Dependent Mean        6494.82900[E]    Adj R-Sq      0.7810[l]
        Coeff Var               58.38119[F]

                        Parameter Estimates
                             Parameter        Standard
Variable     Label        DF     Estimate          Error       t  Value     Pr > |t|
Intercept    Intercept    1   469.70360[b]  702.90619[d]    0.67[f]     0.5072[h]
POP          POP          1     0.64709[c]    0.04881[e]   13.26[g]     <.0001[i]
```

R Output

```
Analysis of Variance Table

Response: SALES
          Df     Sum Sq       Mean Sq        F value         Pr(>F)
POP        1[m] 2527165015[p] 2527165015[s] 175.77304[u] <2.22e-16[v]***
Residuals 48[n]  690116755[q]    14377432[t]
---
Call: lm(formula = SALES ~ POP)

Residuals:
   Min      1Q Median     3Q    Max
 -6047  -1461   -670    486  18229

Coefficients:
             Estimate     Std. Error t value     Pr(>|t|)
(Intercept) 469.7036[b] 702.9062[c]  0.67[f]     0.51      [h]
POP           0.6471[c]   0.0488[e] 13.26[g]    <2e-16 ***[i]
---
Signif. codes:  0 *** 0.001 ** 0.01 * 0.05 . 0.1  1

Residual standard error: 3790[j] on 48[n] degrees of freedom
Multiple R-Squared: 0.785[k],      Adjusted R-squared: 0.781[l]
F-statistic:  176[u] on 1[m] and 48[n] DF,  p-value: <2e-16[v]
```

Legend Annotation Definition, Symbol

[a] Number of observations n.
[b] The estimated intercept b_0.
[c] The estimated slope b_1.
[d] The standard error of the intercept, $se(b_0)$.
[e] The standard error of the intercept, $se(b_1)$.
[f] The t-ratio associated with the intercept, $t(b_0) = b_0/se(b_0)$.
[g] The t-ratio associated with the slope, $t(b_1) = b_1/se(b_1)$.
[h] The p-value associated with the intercept; here, $p - value = Pr(|t_{n-2}| > |t(b_0)|)$, where $t(b_0)$ is the realized value (0.67 here) and t_{n-2} has a t-distribution with $df = n - 2$.
[i] The p-value associated with the slope; here, $p - value = Pr(|t_{n-2}| > |t(b_1)|)$, where $t(b_1)$ is the realized value (13.26 here) and t_{n-2} has a t-distribution with $df = n - 2$.
[j] The residual standard deviation, s.
[k] The coefficient of determination, R^2.
[l] The coefficient of determination adjusted for degrees of freedom, R_a^2. (This term will be defined in Chapter 3.)
[m] Degree of freedom for the regression component: 1, for one explanatory variable.
[n] Degree of freedom for the error component, $n - 2$, for regression with one explanatory variable.
[o] Total degrees of freedoms, $n - 1$.
[p] The regression sum of squares, *Regression SS*.
[q] The error sum of squares, *Error SS*.
[r] The total sum of squares, *Total SS*.
[s] The regression mean square, *Regression MS* = *Regression SS*/1, for one explanatory variable.
[t] The error mean square, $s^2 = Error\ MS = Error\ SS/(n - 2)$, for one explanatory variable.
[u] The F-*ratio* = (*Regression MS*)/(*Error MS*). (This term will be defined in Chapter 3.)
[v] The p-value associated with the F-*ratio*. (This term will be defined in Chapter 3.)
[w] The observation number, i.
[x] The value of the explanatory variable for the ith observation, x_i.
[y] The response for the ith observation, y_i.
[z] The fitted value for the ith observation, $\widehat{y}_i$.
[A] The standard error of the fit, $se(\widehat{y}_i)$.
[B] The residual for the ith observation, e_i.
[C] The standardized residual for the ith observation, $e_i/se(e_i)$. The standard error $se(e_i)$ will be defined in Section 5.3.1.
[D] The multiple correlation coefficient is the square root of the coefficient of determination, $R = \sqrt{R^2}$. This will be defined in Chapter 3.

[E] The average response, $\overline{y}$.
[F] The coefficient of variation of the response is $s_y/\overline{y}$. SAS prints out $100 s_y/\overline{y}$.

2.9 Further Reading and References

Relatively few applications of regression are basic in the sense that they use only one explanatory variable; the purpose of regression analysis is to reduce complex relationships among many variables. Section 2.7 described an important exception to this general rule, the CAPM finance model; see Panjer et al. (1998) for additional actuarial descriptions of this model. Campbell et al. (1997) give a financial econometrics perspective.

Chapter References

Anscombe, Frank (1973). Graphs in statistical analysis. *American Statistician* 27, 17–21.
Campbell, John Y., Andrew W. Lo, and A. Craig MacKinlay (1997). *The Econometrics of Financial Markets*. Princeton University Press, Princeton, New Jersey.
Frees, Edward W., and Tom W. Miller (2003). Sales forecasting using longitudinal data models. *International Journal of Forecasting* 20, 97–111.
Goldberger, Arthur (1991). *A Course in Econometrics*. Harvard University Press, Cambridge, Massachusetts.
Koch, Gary J. (1985). A basic demonstration of the $[-1, 1]$ range for the correlation coefficient. *American Statistician* 39, 201–2.
Lintner, J. (1965). The valuation of risky assets and the selection of risky investments in stock portfolios and capital budgets. *Review of Economics and Statistics* 47, no. 1, 13–37.
Manistre, B. John, and Geoffrey H. Hancock (2005). Variance of the CTE estimator. *North American Actuarial Journal* 9, no. 2, 129–56.
Markowitz, Harry (1959). *Portfolio Selection: Efficient Diversification of Investments*. John Wiley and Sons, New York.
Panjer, Harry H., Phelim P. Boyle, Samuel H. Cox, Daniel Dufresne, Hans U. Gerber, Heinz H. Mueller, Hal W. Pedersen, Stanley R. Pliska, Michael Sherris, Elias S. Shiu, and Ken S. Tan (1998). *Financial Economics: With Applications to Investment, Insurance and Pensions*. Society of Actuaries, Schaumburg, Illinois.
Pearson, Karl (1895). *Royal Society Proceedings* 58, 241.
Serfling, Robert J. (1980). *Approximation Theorems of Mathematical Statistics*. John Wiley and Sons, New York.
Sharpe, William F. (1964). Capital asset prices: A theory of market equilibrium under conditions of risk. *Journal of Finance* 19, no. 3, 425–42.
Stigler, Steven M. (1986). *The History of Statistics: The Measurement of Uncertainty before 1900*. Harvard University Press, Cambridge, Massachusetts.

2.10 Exercises

Sections 2.1–2.2

2.1. Consider the following dataset

i	1	2	3
x_i	2	−6	7.
y_i	3	4	6

Fit a regression line using the method of least squares. Determine r, b_1, and b_0.

2.2. **A Perfect Relationship, yet Zero Correlation.** Consider the quadratic relationship $y = x^2$, with data

i	1	2	3	4	5
x_i	−2	−1	0	1	2.
y_i	4	1	0	1	4

a. Produce a rough graph for this dataset.
b. Check that the correlation coefficient is $r = 0$.

2.3. **Boundedness of the Correlation Coefficient.** Use the following steps to show that r is bounded by -1 and 1 (These steps are due to Koch, 1990).
a. Let a and c be generic constants. Verify

$$0 \leq \frac{1}{n-1}\sum_{i=1}^{n}\left(a\frac{x_i - \overline{x}}{s_x} - c\frac{y_i - \overline{y}}{s_y}\right)^2$$
$$= a^2 + c^2 - 2acr.$$

b. Use the results in part (a) to show $2ac(r-1) \leq (a-c)^2$.
c. By taking $a = c$, use the result in part (b) to show $r \leq 1$.
d. By taking $a = -c$, use the results in part (b) to show $r \geq -1$.
e. Under what conditions is $r = -1$? Under what conditions is $r = 1$?

2.4. **Regression Coefficients Are Weighted Sums.** Show that the intercept term, b_0, can be expressed as a weighted sum of the dependent variables. That is, show that $b_0 = \sum_{i=1}^{n} w_{i,0} y_i$. Further, express the weights in terms of the slope weights, w_i.

2.5. **Another Expression for the Slope as a Weighted Sum.**
a. Using algebra, establish an alternative expression

$$b_1 = \frac{\sum_{i=1}^{n} weight_i \times slope_i}{\sum_{i=1}^{n} weight_i}.$$

Here, $slope_i$ is the slope between (x_i, y_i) and $(\bar{x}, \bar{y})$. Give a precise form for the weight $weight_i$ as a function of the explanatory variable x.
b. Suppose that $\bar{x} = 4$, $\bar{y} = 3$, $x_1 = 2$ and $y_1 = 6$. Determine the slope and weight for the first observation, that is, $slope_1$ and $weight_1$.

2.6. Consider two variables, y and x. Do a regression of y on x to get a slope coefficient that we call $b_{1,x,y}$. Do another regression of x on y to get a slope coefficient that we call $b_{1,y,x}$. Show that the correlation coefficient between x and y is the geometric mean of the two slope coefficients up to sign; that is, show that $|r| = \sqrt{b_{1,x,y} b_{1,y,x}}$.

2.7. **Regression through the Origin.** Consider the model $y_i = \beta_1 x_i + \varepsilon_i$, that is, regression with one explanatory variable without the intercept term. This model is called *regression through the origin* because the true regression line $\mathrm{E}\, y = \beta_1 x$ passes through the origin (the point (0, 0)). For this model, the least squares estimate of β_1 is that number b_1 that minimizes the sum of squares $\mathrm{SS}(b_1^*) = \sum_{i=1}^n \left(y_i - b_1^* x_i\right)^2$.

a. Verify that

$$b_1 = \frac{\sum_{i=1}^n x_i y_i}{\sum_{i=1}^n x_i^2}.$$

b. Consider the model $y_i = \beta_1 z_i^2 + \varepsilon_i$, a quadratic model passing through the origin. Use the result from part (a) to determine the least squares estimate of β_1.

2.8. a. Show that

$$s_y^2 = \frac{1}{n-1}\sum_{i=1}^n (y_i - \overline{y})^2 = \frac{1}{n-1}\left(\sum_{i=1}^n y_i^2 - n\overline{y}^2\right).$$

b. Follow the same steps to show $\sum_{i=1}^n (y_i - \overline{y})(x_i - \overline{x}) = \sum_{i=1}^n x_i y_i - n\overline{x}\,\overline{y}$.

c. Show that

$$b_1 = \frac{\sum_{i=1}^n (y_i - \overline{y})(x_i - \overline{x})}{\sum_{i=1}^n (x_i - \overline{x})^2}.$$

d. Establish the commonly used formula

$$b_1 = \frac{\sum_{i=1}^n x_i y_i - n\overline{x}\,\overline{y}}{\sum_{i=1}^n x_i^2 - n\overline{x}^2}.$$

2.9. **Interpretation of Coefficients Associated with a Binary Explanatory Variable.** Suppose that x_i only takes on the values 0 and 1. Out of the n observations, n_1 take on the value $x = 0$. The n_1 observations have an average y value of $\overline{y}_1$. The remaining $n - n_1$ observations have value $x = 1$ and an average y value of $\overline{y}_2$. Use Exercise 2.8 to show that $b_1 = \overline{y}_2 - \overline{y}_1$.

Ⓡ **Empirical**
Filename is "WiscNursingHome"

2.10. **Nursing Home Utilization.** This exercise considers nursing home data provided by the Wisconsin Department of Health and Family Services (DHFS) and described in Exercise 1.2.

Part 1: Use cost-report year 2000 data, and do the following analysis.

a. Correlations
   - a(i). Calculate the correlation between TPY and LOGTPY. Comment on your result.
   - a(ii). Calculate the correlation among TPY, NUMBED, and SQRFOOT. Do these variables appear highly correlated?
   - a(iii). Calculate the correlation between TPY and NUMBED/10. Comment on your result.

b. Scatter plots. Plot TPY versus NUMBED and TPY versus SQRFOOT. Comment on the plots.

c. Basic linear regression.
   - c(i). Fit a basic linear regression model using TPY as the outcome variable and NUMBED as the explanatory variable. Summarize the fit by quoting the coefficient of determination, R^2, and the t-statistic for NUMBED.

c(ii). Repeat c(i), using SQRFOOT instead of NUMBED. In terms of R^2, which model fits better?

c(iii). Repeat c(i), using LOGTPY for the outcome variable and LOG(NUMBED) as the explanatory variable.

c(iv). Repeat c(iii) using LOGTPY for the outcome variable and LOG(SQRFOOT) as the explanatory variable.

Part 2: Fit the model in Part 1.c(1) using 2001 data. Are the patterns stable over time?

Sections 2.3–2.4

2.11. Suppose that, for a sample size of $n = 3$, you have $e_2 = 24$ and $e_3 = -1$. Determine e_1.

2.12. Suppose that $r = 0$, $n = 15$, and $s_y = 10$. Determine s.

2.13. **The Correlation Coefficient and the Coefficient of Determination.** Use the following steps to establish a relationship between the coefficient of determination and the correlation coefficient.

a. Show that $\widehat{y}_i - \overline{y} = b_1(x_i - \overline{x})$.

b. Use part (a) to show that $Regress\ SS = \sum_{i=1}^{n} (\widehat{y}_i - \overline{y})^2 = b_1^2 s_x^2 (n-1)$.

c. Use part (b) to establish $R^2 = r^2$.

2.14. Show that the average residual is zero, that is, show that $n^{-1} \sum_{i=1}^{n} e_i = 0$.

2.15. **Correlation between Residuals and Explanatory Variables.** Consider a generic sequence of pairs of numbers $(x_1, y_1), \ldots, (x_n, y_n)$ with the correlation coefficient computed as $r(y, x) = \left[(n-1) s_y s_x\right]^{-1} \sum_{i=1}^{n} (y_i - \overline{y}) (x_i - \overline{x})$.

a. Suppose that either $\overline{y} = 0, \overline{x} = 0$ or both $\overline{x}$ and $\overline{y} = 0$. Then, check that $r(y, x) = 0$ implies $\sum_{i=1}^{n} y_i x_i = 0$ and vice versa.

b. Show that the correlation between the residuals and the explanatory variables is zero. Do this by using part (a) of Exercise 2.13 to show that $\sum_{i=1}^{n} x_i e_i = 0$ and then apply part (a).

c. Show that the correlation between the residuals and fitted values is zero. Do this by showing that $\sum_{i=1}^{n} \widehat{y}_i e_i = 0$ and then apply part (a).

2.16. **Correlation and t-Statistics.** Use the following steps to establish a relationship between the correlation coefficient and the t-statistic for the slope.

a. Use algebra to check that

$$R^2 = 1 - \frac{n-2}{n-1} \frac{s^2}{s_y^2}.$$

b. Use part (a) to establish the following quick computational formula for s,

$$s = s_y \sqrt{(1 - r^2) \frac{n-1}{n-2}}.$$

c. Use part (b) to show that

$$t(b_1) = \sqrt{n-2} \frac{r}{\sqrt{1 - r^2}}.$$

Sections 2.6–2.7

2.17. **Effects of an Unusual Point.** You are analyzing a data set of size $n = 100$. You have just performed a regression analysis using one predictor variable and notice that the residual for the 10th observation is unusually large.

a. Suppose that, in fact, it turns out that $e_{10} = 8s$. What percentage of the error sum of squares, *Error SS*, is due to the 10th observation?

b. Suppose that $e_{10} = 4s$. What percentage of the error sum of squares, *Error SS*, is due to the 10th observation?

c. Suppose that you reduce the dataset to size $n = 20$. After running the regression, it turns out that we still have $10 = 4s$. What percentage of the error sum of squares, *Error SS*, is due to the 10th observation?

2.18. Consider a dataset consisting of 20 observations with the following summary statistics: $\overline{x} = 0$, $\overline{y} = 9$, $s_x = 1$, and $s_y = 10$. You run a regression using using one variable and determine that $s = 7$. Determine the standard error of a prediction at $x_* = 1$.

Ⓡ **Empirical**
Filename is "AnscombesData"

2.19. **Summary Statistics Can Hide Important Relationships.** The data in Table 2.9 is due to Anscombe (1973). The purpose of this exercise is to demonstrate how plotting data can reveal important information that is not evident in numerical summary statistics.

a. Compute the averages and standard deviations of each column of data. Check that the averages and standard deviations of each of the x columns are the same, within two decimal places, and similarly for each of the y columns.

b. Run four regressions, (i) y_1 on x_1, (ii) y_2 on x_1, (iii) y_3 on x_1, and (iv) y_4 on x_2. Verify, for each of the four regressions fits, that $b_0 \approx 3.0$, $b_1 \approx 0.5$, $s \approx 1.237$, and $R^2 \approx 0.677$, within two decimal places.

c. Produce scatter plots for each of the four regression models that you fit in part (b).

d. Discuss the fact that the fitted regression models produced in part (b) imply that the four datasets are similar, though the four scatter plots produced in part (c) yield a dramatically different story.

Table 2.9
Anscombe's (1973) Data

Obs num	x_1	y_1	y_2	y_3	x_2	y_4
1	10	8.04	9.14	7.46	8	6.58
2	8	6.95	8.14	6.77	8	5.76
3	13	7.58	8.74	12.74	8	7.71
4	9	8.81	8.77	7.11	8	8.84
5	11	8.33	9.26	7.81	8	8.47
6	14	9.96	8.10	8.84	8	7.04
7	6	7.24	6.13	6.08	8	5.25
8	4	4.26	3.10	5.39	8	5.56
9	12	10.84	9.13	8.15	8	7.91
10	7	4.82	7.26	6.42	8	6.89
11	5	5.68	4.74	5.73	19	12.50

2.20. **Nursing Home Utilization.** This exercise considers nursing home data provided by the Wisconsin Department of Health and Family Services (DHFS) and described in Exercises 1.2 and 2.10.

® **EMPIRICAL**
Filename is "WiscNursingHome"

You decide to examine the relationship between total patient years (LOGTPY) and the number of beds (LOGNUMBED), both in logarithmic units, using cost-report year 2001 data.

a. Summary statistics. Create basic summary statistics for each variable. Summarize the relationship through a correlation statistic and a scatter plot.

b. Fit the basic linear model. Cite the basic summary statistics, include the coefficient of determination, the regression coefficient for LOGNUMBED, and the corresponding t-statistic.

c. Hypothesis testing. Test the following hypotheses at the 5 level of significance using a t-statistic. Also compute the corresponding p-value.
 - c(i). Test $H_0 : \beta_1 = 0$ versus $H_a : \beta_1 \neq 0$.
 - c(ii). Test $H_0 : \beta_1 = 1$ versus $H_a : \beta_1 \neq 1$.
 - c(iii). Test $H_0 : \beta_1 = 1$ versus $H_a : \beta_1 > 1$.
 - c(iv). Test $H_0 : \beta_1 = 1$ versus $H_a : \beta_1 < 1$.

d. You are interested in the effect that a marginal change in LOGNUMBED has on the expected value of LOGTPY.
 - d(i). Suppose that there is a marginal change in LOGNUMBED of 2. Provide a point estimate of the expected change in LOGTPY.
 - d(ii). Provide a 95% confidence interval corresponding to the point estimate in part d(i).
 - d(iii). Provide a 99% confidence interval corresponding to the point estimate in part d(i).

e. At a specified number of beds estimate $x_* = 100$, do these things:
 - e(i). Find the predicted value of LOGTPY.
 - e(ii). Obtain the standard error of the prediction.
 - e(iii). Obtain a 95% prediction interval for your prediction.
 - e(iv). Convert the point prediction in part e(i) and the prediction interval obtained in part e(iii) into total person years (through exponentiation).
 - e(v). Obtain a prediction interval as in part e(iv), corresponding to a 90% level (in lieu of 95%).

2.21. **Initial Public Offerings.** As a financial analyst, you want to convince a client of the merits of investing in firms that have just entered a stock exchange, as an initial public offering (IPO). Thus, you gather data on 116 firms that priced during the six-month time frame of January 1 through June 1, 1998. By looking at this recent historical data, you are able to compute RETURN, the firm's one-year return (as a percentage).

® **EMPIRICAL**
Filename is "IPO"

You are also interested in looking at financial characteristics of the firm that may help you understand (and predict) the return. You initially examine REVENUE, the firm's 1997 revenues in millions of dollars. Unfortunately,

Table 2.10 Initial Public Offering Summary Statistics

	Mean	Median	Standard Deviation	Minimum	Maximum
RETURN	0.106	−0.130	0.824	−0.938	4.333
REV	134.487	39.971	261.881	0.099	1455.761
LnREV	3.686	3.688	1.698	−2.316	7.283
PRICEIPO	13.195	13.000	4.694	4.000	29.000

this variable was not available for six firms. Thus, the statistics here are for the 110 firms that have both REVENUES and RETURNS. In addition, Table 2.10 provides information on the (natural) logarithmic revenues, denoted as LnREV, and the initial price of the stock, denoted as PRICEIPO.

a. You hypothesize that larger firms, as measured by revenues, are more stable and thus should enjoy greater returns. You have determined that the correlation between RETURN and REVENUE is −0.0175.
    - a(i). Calculate the least squares fit using REVENUE to predict RETURN. Determine b_0 and b_1.
    - a(ii). For Hyperion Telecommunications, REVENUE is 95.55 (millions of dollars). Calculate the fitted RETURN using the regression fit in part a(i).

b. Table 2.11 summarizes a regression using logarithmic revenues and returns.
    - b(i). Suppose instead that you use LnREVs to predict RETURN. Calculate the fitted RETURN under this regression model. Is this equal your answer in part a(ii)?
    - b(ii). Do logarithmic revenues significantly affect returns? To this end, provide a formal test of hypothesis. State your null and alternative hypotheses, decision-making criterion, and your decision-making rule. Use a 10% level of significance.
    - b(iii). You conjecture that, other things equal, that firms with greater revenues will be more stable and thus enjoy a larger initial return. Thus, you wish to consider the null hypothesis of no relation between LnREV and RETURN versus the alternative hypothesis that there is a positive relation between LnREV and RETURN. To this end, provide a formal test of hypothesis. State your null and alternative hypotheses, decision-making criterion, and decision-making rule. Use a 10% level of significance.

c. Determine the correlation between LnREV and RETURN. Be sure to state whether the correlation is positive, negative, or zero.

d. You are considering investing in a firm that has LnREV = 2 (so revenues are $e^2 = 7.389$ millions of dollars).
    - d(i). Using the fitted regression model, determine the least squares point prediction.
    - d(ii). Determine the 95% prediction interval corresponding to your prediction in part d(i).

Table 2.11 Regression Results from a Model Fit with Logarithmic Revenues

Variable	Coefficient	Standard Error	t-Statistic
INTERCEPT	0.438	0.186	2.35
LnREV	−0.090	0.046	−1.97

Notes: $s = 0.8136$, and $R^2 = 0.03452$.

e. The R^2 from the fitted regression model is a disappointing 3.5%. Part of the difficulty is due to observation number 59, the Inktomi Corporation. Inktomi sales are 12th smallest of the data set, with LnREV = 1.76 (so revenues are $e^{1.76} = 5.79$ millions of dollars), yet it has the highest first-year return, with RETURN = 433.33.
   e(i). Calculate the residual for this observation.
   e(ii). What proportion of the unexplained variability (error sum of squares) does this observation account for?
   e(iii). Define the idea of a high leverage observation.
   e(iv). Would this observation be considered a high leverage observation? Justify your answer.

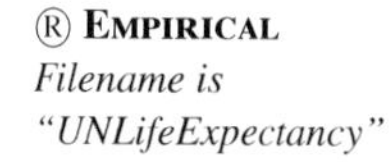

Ⓡ **EMPIRICAL** *Filename is "UNLifeExpectancy"*

2.22. **National Life Expectancies.** We continue the analysis begun in Exercise 1.7 by examining the relation between $y = LIFEEXP$ and $x = FERTILITY$, shown in Figure 2.12. Fit a linear regression model of LIFEEXP using the explanatory variable $x = FERTILITY$.

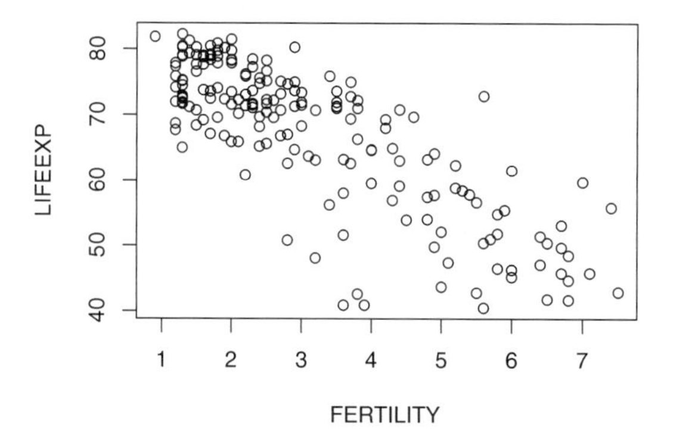

Figure 2.12 Plot of FERTILITY versus LIFEEXP.

a. The United States has a FERTILITY rate of 2.0. Determine the fitted life expectancy.

b. The island nation Dominica did not report a FERTILITY rate and thus was not included in the regression. Suppose that its FERTILITY rate is 2.0. Provide a 95% prediction interval for the life expectancy in Dominica.

c. China has a FERTILITY rate of 1.7 and a life expectancy of 72.5. Determine the residual under the model. How many multiples of s is this residual from zero?

d. Suppose that your prior hypothesis is that the FERTILITY slope is -6.0 and you wish to test the null hypothesis that the slope has increased (i.e., the slope is greater than -6.0). Test this hypothesis at the 5% level of significance. Also compute an approximate p-value.

2.11 Technical Supplement – Elements of Matrix Algebra

Examples are an excellent tool for introducing technical topics such as regression. However, this chapter has also used algebra as well as basic probability and statistics to give you further insights into regression analysis. Going forward, we will be studying multivariate relationships. With many things happening concurrently in several dimensions, algebra is no longer useful for providing insights. Instead, we will need *matrix* algebra. This supplement provides a brief introduction to matrix algebra to allow you to study the linear regression chapters of this text. It re-introduces basic linear regression to give you a feel for things that will be coming up in subsequent chapters when we extend basic linear regression to the multivariate case. Appendix A3 defines additional matrix concepts.

2.11.1 Basic Definitions

A *matrix* is a rectangular array of numbers arranged in rows and columns (the plural of *matrix* is *matrices*). For example, consider the income and age of three people:

$$\mathbf{A} = \begin{matrix} & \textit{Col } 1 & \textit{Col } 2 \\ \textit{Row } 1 \\ \textit{Row } 2 \\ \textit{Row } 3 \end{matrix} \begin{pmatrix} 6{,}000 & 23 \\ 13{,}000 & 47 \\ 11{,}000 & 35 \end{pmatrix}.$$

Here, column 1 represents income, and column 2 represents age. Each row corresponds to an individual. For example, the first individual is 23 years old with an income of \$6,000.

The number of rows and columns is called the *dimension* of the matrix. For example, the dimension of the matrix $\mathbf{A}$ above is 3×2 (read 3 "by" 2). This stands for three rows and two columns. If we were to represent the income and age of 100 people, then the dimension of the matrix would be 100×2.

It is convenient to represent a matrix using the notation

$$\mathbf{A} = \begin{pmatrix} a_{11} & a_{12} \\ a_{21} & a_{22} \\ a_{31} & a_{31} \end{pmatrix}.$$

Here, a_{ij} is the symbol for the number in the ith row and jth column of $\mathbf{A}$. In general, we work with matrices of the form

$$\mathbf{A} = \begin{pmatrix} a_{11} & a_{12} & \cdots & a_{1k} \\ \vdots & \vdots & \ddots & \vdots \\ a_{n1} & a_{n2} & \cdots & a_{nk} \end{pmatrix}.$$

In this case, the matrix $\mathbf{A}$ has dimension $n \times k$.

A *vector* is a special matrix. A *row vector* is a matrix containing only 1 row ($n = 1$). A *column vector* is a matrix containing only one column ($k = 1$). For example,

$$\text{column vector} \rightarrow \begin{pmatrix} 2 \\ 3 \\ 4 \\ 5 \\ 6 \end{pmatrix} \qquad \text{row vector} \rightarrow \begin{pmatrix} 2 & 3 & 4 & 5 & 6 \end{pmatrix}.$$

Notice that the row vector takes much less room on a printed page than the corresponding column vector. A basic operation that relates these two quantities is the *transpose*. The transpose of a matrix $\mathbf{A}$ is defined by interchanging the rows and columns and is denoted by $\mathbf{A}'$ (or $\mathbf{A}^T$). For example,

$$\mathbf{A} = \begin{pmatrix} 6000 & 23 \\ 13{,}000 & 47 \\ 11{,}000 & 35 \end{pmatrix} \qquad \mathbf{A}' = \begin{pmatrix} 6000 & 13{,}000 & 11{,}000 \\ 23 & 47 & 35 \end{pmatrix}.$$

Thus, if $\mathbf{A}$ has dimension $n \times k$, then $\mathbf{A}'$ has dimension $k \times n$.

2.11.2 Some Special Matrices

(i) A *square matrix* is a matrix where the number of rows equals the number of columns; that is, $n = k$.

(ii) The *diagonal numbers* of a square matrix are the numbers of a matrix where the row number equals the column number, for example, a_{11}, a_{22}, and so on. A *diagonal matrix* is a square matrix where all nondiagonal numbers are equal to 0. For example,

$$\mathbf{A} = \begin{pmatrix} -1 & 0 & 0 \\ 0 & 2 & 0 \\ 0 & 0 & 3 \end{pmatrix}.$$

(iii) An *identity matrix* is a diagonal matrix where all the diagonal numbers are equal to 1. This special matrix is often denoted by $\mathbf{I}$.

(iv) A *symmetric matrix* is a square matrix $\mathbf{A}$ such that the matrix remains unchanged if we interchange the roles of the rows and columns. More formally, a matrix $\mathbf{A}$ is symmetric if $\mathbf{A} = \mathbf{A}'$. For example,

$$\mathbf{A} = \begin{pmatrix} 1 & 2 & 3 \\ 2 & 4 & 5 \\ 3 & 5 & 10 \end{pmatrix} = \mathbf{A}'.$$

Note that a diagonal matrix is a symmetric matrix.

2.11.3 Basic Operations

Scalar Multiplication

Let $\mathbf{A}$ be a $n \times k$ matrix and let c be a real number. That is, a real number is a 1×1 matrix and is also called a *scalar*. Multiplying a scalar c by a matrix $\mathbf{A}$ is denoted by $c\mathbf{A}$ and defined by

$$c\mathbf{A} = \begin{pmatrix} ca_{11} & ca_{12} & \cdots & ca_{1k} \\ \vdots & \vdots & \ddots & \vdots \\ ca_{n1} & ca_{n2} & \cdots & ca_{nk} \end{pmatrix}.$$

For example, suppose that $c = 10$ and

$$\mathbf{A} = \begin{pmatrix} 1 & 2 \\ 6 & 8 \end{pmatrix} \quad \text{then} \quad \mathbf{B} = c\mathbf{A} = \begin{pmatrix} 10 & 20 \\ 60 & 80 \end{pmatrix}.$$

Note that $c\mathbf{A} = \mathbf{A}c$.

Addition and Subtraction of Matrices

Let $\mathbf{A}$ and $\mathbf{B}$ be matrices with dimensions $n \times k$. Use a_{ij} and b_{ij} to denote the numbers in the ith row and jth column of $\mathbf{A}$ and $\mathbf{B}$, respectively. Then, the matrix $\mathbf{C} = \mathbf{A} + \mathbf{B}$ is defined to be the matrix with $(a_{ij} + b_{ij})$ in the ith row and jth column. Similarly, the matrix $\mathbf{C} = \mathbf{A} - \mathbf{B}$ is defined to be the matrix with $(a_{ij} - b_{ij})$ in the ith row and jth column. Symbolically, we write this as the following:

$$\text{If} \quad \mathbf{A} = \left(a_{ij}\right)_{ij} \quad \text{and} \quad \mathbf{B} = \left(b_{ij}\right)_{ij}, \text{ then}$$

$$\mathbf{C} = \mathbf{A} + \mathbf{B} = \left(a_{ij} + b_{ij}\right)_{ij} \quad \text{and} \quad \mathbf{C} = \mathbf{A} - \mathbf{B} = \left(a_{ij} - b_{ij}\right)_{ij}.$$

For example, consider

$$\mathbf{A} = \begin{pmatrix} 2 & 5 \\ 4 & 1 \end{pmatrix} \quad \mathbf{B} = \begin{pmatrix} 4 & 6 \\ 8 & 1 \end{pmatrix}.$$

Then

$$\mathbf{A} + \mathbf{B} = \begin{pmatrix} 6 & 11 \\ 12 & 2 \end{pmatrix} \quad \mathbf{A} - \mathbf{B} = \begin{pmatrix} -2 & -1 \\ -4 & 0 \end{pmatrix}.$$

Basic Linear Regression Example of Addition and Subtraction. Now, recall that the basic linear regression model can be written as n equations:

$$y_1 = \beta_0 + \beta_1 x_1 + \varepsilon_1$$
$$\vdots$$
$$y_n = \beta_0 + \beta_1 x_n + \varepsilon_n.$$

We can define

$$\mathbf{y} = \begin{pmatrix} y_1 \\ \vdots \\ y_n \end{pmatrix} \quad \boldsymbol{\varepsilon} = \begin{pmatrix} \varepsilon_1 \\ \vdots \\ \varepsilon_n \end{pmatrix} \quad \text{and} \quad \mathrm{E}\,\mathbf{y} = \begin{pmatrix} \beta_0 + \beta_1 x_1 \\ \vdots \\ \beta_0 + \beta_1 x_n \end{pmatrix}.$$

With this notation, we can express the n equations more compactly as $\mathbf{y} = \mathrm{E}\,\mathbf{y} + \boldsymbol{\varepsilon}$.

Matrix Multiplication
In general, if $\mathbf{A}$ is a matrix of dimension $n \times c$ and $\mathbf{B}$ is a matrix of dimension $c \times k$, then $\mathbf{C} = \mathbf{AB}$ is a matrix of dimension $n \times k$ and is defined by

$$\mathbf{C} = \mathbf{AB} = \left(\sum_{s=1}^{c} a_{is} b_{sj} \right)_{ij} .$$

For example consider the 2×2 matrices

$$\mathbf{A} = \begin{pmatrix} 2 & 5 \\ 4 & 1 \end{pmatrix} \quad \text{and} \quad \mathbf{B} = \begin{pmatrix} 4 & 6 \\ 8 & 1 \end{pmatrix}.$$

The matrix $\mathbf{AB}$ has dimension 2×2. To illustrate the calculation, consider the number in the first row and second column of $\mathbf{AB}$. By the rule presented earlier, with $i = 1$ and $j = 2$, the corresponding element of $\mathbf{AB}$ is $\sum_{s=1}^{2} a_{1s} b_{s2} = a_{11} b_{12} + a_{12} b_{22} = 2(6) + 5(1) = 17$. The other calculations are summarized as

$$\mathbf{AB} = \begin{pmatrix} 2(4) + 5(8) & 2(6) + 5(1) \\ 4(4) + 1(8) & 4(6) + 1(1) \end{pmatrix} = \begin{pmatrix} 48 & 17 \\ 24 & 25 \end{pmatrix}.$$

As another example, suppose

$$\mathbf{A} = \begin{pmatrix} 1 & 2 & 4 \\ 0 & 5 & 8 \end{pmatrix} \quad \mathbf{B} = \begin{pmatrix} 3 \\ 5 \\ 2 \end{pmatrix}.$$

Because $\mathbf{A}$ has dimension 2×3 and $\mathbf{B}$ has dimension 3×1, this means that the product $\mathbf{AB}$ has dimension 2×1. The calculations are summarized as

$$\mathbf{AB} = \begin{pmatrix} 1(3) + 2(5) + 4(2) \\ 0(3) + 5(5) + (2) \end{pmatrix} = \begin{pmatrix} 21 \\ 41 \end{pmatrix}.$$

For some additional examples, we have

$$\begin{pmatrix} 4 & 2 \\ 5 & 8 \end{pmatrix} \begin{pmatrix} a_1 \\ a_2 \end{pmatrix} = \begin{pmatrix} 4a_1 + 2a_2 \\ 5a_1 + 8a_2 \end{pmatrix}.$$

$$\begin{pmatrix} 2 & 3 & 5 \end{pmatrix} \begin{pmatrix} 2 \\ 3 \\ 5 \end{pmatrix} = 2^2 + 3^2 + 5^2 = 38 \qquad \begin{pmatrix} 2 \\ 3 \\ 5 \end{pmatrix} \begin{pmatrix} 2 & 3 & 5 \end{pmatrix} = \begin{pmatrix} 4 & 6 & 10 \\ 6 & 9 & 15 \\ 10 & 15 & 25 \end{pmatrix}.$$

In general, you see that $\mathbf{AB} \neq \mathbf{BA}$ in matrix multiplication, unlike multiplication of scalars (real numbers). Further, we remark that the identity matrix serves the role of "one" in matrix multiplication, in that $\mathbf{AI} = \mathbf{A}$ and $\mathbf{IA} = \mathbf{A}$ for any matrix $\mathbf{A}$, providing that the dimensions are compatible to allow matrix multiplication.

Basic Linear Regression Example of Matrix Multiplication. Define

$$\mathbf{X} = \begin{pmatrix} 1 & x_1 \\ \vdots & \vdots \\ 1 & x_n \end{pmatrix} \text{ and } \boldsymbol{\beta} = \begin{pmatrix} \beta_0 \\ \beta_1 \end{pmatrix}, \text{ to get } \mathbf{X}\boldsymbol{\beta} = \begin{pmatrix} \beta_0 + \beta_1 x_1 \\ \vdots \\ \beta_0 + \beta_1 x_n \end{pmatrix} = \mathrm{E}\,\mathbf{y}.$$

Thus, this yields the familiar matrix expression of the regression model, $\mathbf{y} = \mathbf{X}\boldsymbol{\beta} + \boldsymbol{\varepsilon}$. Other useful quantities include

$$\mathbf{y}'\mathbf{y} = \begin{pmatrix} y_1 & \cdots & y_n \end{pmatrix} \begin{pmatrix} y_1 \\ \vdots \\ y_n \end{pmatrix} = y_1^2 + \cdots + y_n^2 = \sum_{i=1}^{n} y_i^2,$$

$$\mathbf{X}'\mathbf{y} = \begin{pmatrix} 1 & \cdots & 1 \\ x_1 & \cdots & x_n \end{pmatrix} \begin{pmatrix} y_1 \\ \vdots \\ y_n \end{pmatrix} = \begin{pmatrix} \sum_{i=1}^{n} y_i \\ \sum_{i=1}^{n} x_i y_i \end{pmatrix},$$

and

$$\mathbf{X}'\mathbf{X} = \begin{pmatrix} 1 & \cdots & 1 \\ x_1 & \cdots & x_n \end{pmatrix} \begin{pmatrix} 1 & x_1 \\ \vdots & \vdots \\ 1 & x_n \end{pmatrix} = \begin{pmatrix} n & \sum_{i=1}^{n} x_i \\ \sum_{i=1}^{n} x_i & \sum_{i=1}^{n} x_i^2 \end{pmatrix}.$$

Note that $\mathbf{X}'\mathbf{X}$ is a symmetric matrix.

Matrix Inverses

In matrix algebra, there is no concept of "division." Instead, we extend the concept of "reciprocals" of real numbers. To begin, suppose that $\mathbf{A}$ is a square matrix of dimension $k \times k$ and let $\mathbf{I}$ be the $k \times k$ identity matrix. If there exists a $k \times k$ matrix $\mathbf{B}$ such that $\mathbf{AB} = \mathbf{I} = \mathbf{BA}$, then $\mathbf{B}$ is called the *inverse* of $\mathbf{A}$ and is written

$$\mathbf{B} = \mathbf{A}^{-1}.$$

Now, not all square matrices have inverses. Further, even when an inverse exists, it may not be easy to compute by hand. One exception to this rule are diagonal matrices. Suppose that $\mathbf{A}$ is diagonal matrix of the form

$$\mathbf{A} = \begin{pmatrix} a_{11} & \cdots & 0 \\ \vdots & \ddots & \vdots \\ 0 & \cdots & a_{kk} \end{pmatrix}. \quad \text{Then } \mathbf{A}^{-1} = \begin{pmatrix} \frac{1}{a_{11}} & \cdots & 0 \\ \vdots & \ddots & \vdots \\ 0 & \cdots & \frac{1}{a_{kk}} \end{pmatrix}.$$

For example,

$$\underset{\mathbf{A}}{\begin{pmatrix} 2 & 0 \\ 0 & -19 \end{pmatrix}} \underset{\mathbf{A}^{-1}}{\begin{pmatrix} \frac{1}{2} & 0 \\ 0 & -\frac{1}{19} \end{pmatrix}} \;\underset{=}{=}\; \underset{\mathbf{I}}{\begin{pmatrix} 1 & 0 \\ 0 & 1 \end{pmatrix}}.$$

In the case of a matrix of dimension 2×2, the inversion procedure can be accomplished by hand easily even when the matrix is not diagonal. In the 2×2 case, we suppose that if

$$\mathbf{A} = \begin{pmatrix} a & b \\ c & d \end{pmatrix}, \quad \text{then} \quad \mathbf{A}^{-1} = \frac{1}{ad - bc} \begin{pmatrix} d & -b \\ -c & a \end{pmatrix}.$$

Thus, for example, if

$$\mathbf{A} = \begin{pmatrix} 2 & 2 \\ 3 & 4 \end{pmatrix} \quad \text{then} \quad \mathbf{A}^{-1} = \frac{1}{2(4) - 2(3)} \begin{pmatrix} 4 & -2 \\ -3 & 2 \end{pmatrix} = \begin{pmatrix} 2 & -1 \\ -3/2 & 1 \end{pmatrix}.$$

As a check, we have

$$\mathbf{AA}^{-1} = \begin{pmatrix} 2 & 2 \\ 3 & 4 \end{pmatrix} \begin{pmatrix} 2 & -1 \\ -3/2 & 1 \end{pmatrix} = \begin{pmatrix} 2(2) - 2(3/2) & 2(-1) + 2(1) \\ 3(2) - 4(3/2) & 3(-1) + 4(1) \end{pmatrix} = \begin{pmatrix} 1 & 0 \\ 0 & 1 \end{pmatrix} = \mathbf{I}.$$

Basic Linear Regression Example of Matrix Inverses. With

$$\mathbf{X}'\mathbf{X} = \begin{pmatrix} n & \sum_{i=1}^{n} x_i \\ \sum_{i=1}^{n} x_i & \sum_{i=1}^{n} x_i^2 \end{pmatrix},$$

we have

$$(\mathbf{X}'\mathbf{X})^{-1} = \frac{1}{n \sum_{i=1}^{n} x_i^2 - \left(\sum_{i=1}^{n} x_i\right)^2} \begin{pmatrix} \sum_{i=1}^{n} x_i^2 & -\sum_{i=1}^{n} x_i \\ -\sum_{i=1}^{n} x_i & n \end{pmatrix}.$$

To simplify this expression, recall that $\bar{x} = n^{-1} \sum_{i=1}^{n} x_i$. Thus,

$$(\mathbf{X}'\mathbf{X})^{-1} = \frac{1}{\sum_{i=1}^{n} x_i^2 - n\bar{x}^2} \begin{pmatrix} n^{-1} \sum_{i=1}^{n} x_i^2 & -\bar{x} \\ -\bar{x} & 1 \end{pmatrix}. \tag{2.9}$$

Section 3.1 will discuss the relation $\mathbf{b} = (\mathbf{X}'\mathbf{X})^{-1} \mathbf{X}'\mathbf{y}$. To illustrate the calculation, we have

$$\mathbf{b} = (\mathbf{X}'\mathbf{X})^{-1} \mathbf{X}'\mathbf{y} = \frac{1}{\sum_{i=1}^{n} x_i^2 - n\bar{x}^2} \begin{pmatrix} n^{-1} \sum_{i=1}^{n} x_i^2 & -\bar{x} \\ -\bar{x} & 1 \end{pmatrix} \begin{pmatrix} \sum_{i=1}^{n} y_i \\ \sum_{i=1}^{n} x_i y_i \end{pmatrix}$$

$$= \frac{1}{\sum_{i=1}^{n} x_i^2 - n\bar{x}^2} \begin{pmatrix} \sum_{i=1}^{n} \left(\bar{y} x_i^2 - \bar{x} x_i y_i\right) \\ \sum_{i=1}^{n} x_i y_i - n\bar{x}\bar{y} \end{pmatrix} = \begin{pmatrix} b_0 \\ b_1 \end{pmatrix}.$$

From this expression, we may see

$$b_1 = \frac{\sum_{i=1}^{n} x_i y_i - n\overline{x}\,\overline{y}}{\sum_{i=1}^{n} x_i^2 - n\overline{x}^2},$$

and

$$b_0 = \frac{\overline{y}\sum_{i=1}^{n} x_i^2 - \overline{x}\sum_{i=1}^{n} x_i y_i}{\sum_{i=1}^{n} x_i^2 - n\overline{x}^2} = \frac{\overline{y}\left(\sum_{i=1}^{n} x_i^2 - n\overline{x}^2\right) - \overline{x}\left(\sum_{i=1}^{n} x_i y_i - n\overline{x}\,\overline{y}\right)}{\sum_{i=1}^{n} x_i^2 - n\overline{x}^2} = \overline{y} - b_1\overline{x}.$$

These are the usual expressions for the slope b_1 (Exercise 2.8) and intercept b_0.

2.11.4 Random Matrices

Expectations. Consider a matrix of random variables

$$\mathbf{U} = \begin{pmatrix} u_{11} & u_{12} & \cdots & u_{1c} \\ u_{21} & u_{22} & \cdots & u_{2c} \\ \vdots & \vdots & \ddots & \vdots \\ u_{n1} & u_{n2} & \cdots & u_{nc} \end{pmatrix}.$$

When we write the expectation of a matrix, this is shorthand for the matrix of expectations. Specifically, suppose that the joint probability function of u_{11}, $u_{12}, \ldots, u_{1c}, \ldots, u_{n1}, \ldots, u_{nc}$ is available to define the expectation operator. Then we define

$$\mathrm{E}\,\mathbf{U} = \begin{pmatrix} \mathrm{E}u_{11} & \mathrm{E}u_{12} & \cdots & \mathrm{E}u_{1c} \\ \mathrm{E}u_{21} & \mathrm{E}u_{22} & \cdots & \mathrm{E}u_{2c} \\ \vdots & \vdots & \ddots & \vdots \\ \mathrm{E}u_{n1} & \mathrm{E}u_{n2} & \cdots & \mathrm{E}u_{nc} \end{pmatrix}.$$

As an important special case, consider the joint probability function for the random variables $y_1, \ldots, y_n$ and the corresponding expectations operator. Then

$$\mathrm{E}\,\mathbf{y} = \mathrm{E}\begin{pmatrix} y_1 \\ \vdots \\ y_n \end{pmatrix} = \begin{pmatrix} \mathrm{E}y_1 \\ \vdots \\ \mathrm{E}y_n \end{pmatrix}.$$

By the linearity of expectations, for a nonrandom matrix $\mathbf{A}$ and vector $\mathbf{B}$, we have $\mathrm{E}\,(\mathbf{A}\,\mathbf{y} + \mathbf{B}) = \mathbf{A}\,\mathrm{E}\,\mathbf{y} + \mathbf{B}$.

Variances. We can also work with second moments of random vectors. The variance of a vector of random variables is called the *variance-covariance matrix*. It is defined by

$$\mathrm{Var}\,\mathbf{y} = \mathrm{E}((\mathbf{y} - \mathrm{E}\,\mathbf{y})(\mathbf{y} - \mathrm{E}\,\mathbf{y})'). \qquad (2.10)$$

That is, we can express

$$\operatorname{Var}\mathbf{y} = \mathrm{E}\left(\begin{pmatrix} y_1 - \mathrm{E}\,y_1 \\ \vdots \\ y_n - \mathrm{E}\,y_n \end{pmatrix}\begin{pmatrix} y_1 - \mathrm{E}\,y_1 & \cdots & y_n - \mathrm{E}\,y_n \end{pmatrix}\right)$$

$$= \begin{pmatrix} \operatorname{Var} y_1 & \operatorname{Cov}(y_1, y_2) & \cdots & \operatorname{Cov}(y_1, y_n) \\ \operatorname{Cov}(y_2, y_1) & \operatorname{Var} y_2 & \cdots & \operatorname{Cov}(y_2, y_n) \\ \vdots & \vdots & \ddots & \vdots \\ \operatorname{Cov}(y_n, y_1) & \operatorname{Cov}(y_n, y_2) & \cdots & \operatorname{Var} y_n \end{pmatrix},$$

because $\mathrm{E}((y_i - \mathrm{E}\,y_i)(y_j - \mathrm{E}\,y_j)) = \operatorname{Cov}(y_i, y_j)$ for $i \neq j$ and $\operatorname{Cov}(y_i, y_i) = \operatorname{Var} y_i$.

In the case that $y_1, \ldots, y_n$ are mutually uncorrelated, we have that $\operatorname{Cov}(y_i, y_j) = 0$ for $i \neq j$ and thus

$$\operatorname{Var}\mathbf{y} = \begin{pmatrix} \operatorname{Var} y_1 & 0 & \cdots & 0 \\ 0 & \operatorname{Var} y_2 & \cdots & 0 \\ \vdots & \vdots & \ddots & \vdots \\ 0 & 0 & \cdots & \operatorname{Var} y_n \end{pmatrix}.$$

Further, if the variances are identical so that $\operatorname{Var} y_i = \sigma^2$, then we can write $\operatorname{Var}\mathbf{y} = \sigma^2\mathbf{I}$, where $\mathbf{I}$ is the $n \times n$ identity matrix. For example, if $y_1, \ldots, y_n$ are i.i.d., then $\operatorname{Var}\mathbf{y} = \sigma^2\mathbf{I}$.

From equation (2.10), it can be shown that

$$\operatorname{Var}(\mathbf{Ay} + \mathbf{B}) = \operatorname{Var}(\mathbf{Ay}) = \mathbf{A}\,(\operatorname{Var}\mathbf{y})\,\mathbf{A}'. \tag{2.11}$$

For example, if $\mathbf{A} = (a_1, a_2, \ldots, a_n) = \mathbf{a}'$ and $\mathbf{B} = \mathbf{0}$, then equation (2.11) reduces to

$$\operatorname{Var}\left(\sum_{i=1}^{n} a_i y_i\right) = \operatorname{Var}(\mathbf{a}'\mathbf{y}) = \mathbf{a}'\,(\operatorname{Var}\mathbf{y})\,\mathbf{a} = (a_1, a_2, \ldots, a_n)\,(\operatorname{Var}\mathbf{y})\begin{pmatrix} a_1 \\ \vdots \\ a_n \end{pmatrix}$$

$$= \sum_{i=1}^{n} a_i^2 \operatorname{Var} y_i \;+\; 2\sum_{i=2}^{n}\sum_{j=1}^{i-1} a_i a_j \operatorname{Cov}(y_i, y_j).$$

Definition – Multivariate Normal Distribution. A vector of random variables $\mathbf{y} = (y_1, \ldots, y_n)'$ is said to be *multivariate normal* if all linear combinations of the form $\sum_{i=1}^{n} a_i y_i$ are normally distributed. In this case, we write $\mathbf{y} \sim N(\boldsymbol{\mu}, \Sigma)$, where $\boldsymbol{\mu} = \mathrm{E}\,\mathbf{y}$ is the expected value of $\mathbf{y}$ and $\Sigma = \operatorname{Var}\mathbf{y}$ is the variance-covariance matrix of $\mathbf{y}$. From the definition, we have that $\mathbf{y} \sim N(\boldsymbol{\mu}, \Sigma)$ implies that $\mathbf{a}'\mathbf{y} \sim N(\mathbf{a}'\boldsymbol{\mu}, \mathbf{a}'\Sigma\mathbf{a})$. Thus, if y_i are i.i.d., then $\sum_{i=1}^{n} a_i y_i$ is distributed normally with mean $\mu\sum_{i=1}^{n} a_i$ and variance $\sigma^2\sum_{i=1}^{n} a_i^2$.

3

Multiple Linear Regression – I

Chapter Preview. This chapter introduces linear regression in the case of several explanatory variables, known as *multiple linear regression*. Many basic linear regression concepts extend directly, including goodness-of-fit measures such as R^2 and inference using t-statistics. Multiple linear regression models provide a framework for summarizing highly complex, multivariate data. Because this framework requires only linearity in the parameters, we are able to fit models that are nonlinear functions of the explanatory variables, thus providing a wide scope of potential applications.

3.1 Method of Least Squares

Chapter 2 dealt with the problem of a response depending on a single explanatory variable. We now extend the focus of that chapter and study how a response may depend on several explanatory variables.

Ⓡ **Empirical**
Filename is "TermLife"

Example: Term Life Insurance. Like all firms, life insurance companies continually seek new ways to deliver products to the market. Those involved in product development want to know who buys insurance and how much they buy. In economics, this is known as the *demand* side of a market for products. Analysts can readily get information on characteristics of current customers through company databases. Potential customers, those who do not have insurance with the company, are often the main focus for expanding market share.

In this example, we examine the Survey of Consumer Finances (SCF), a nationally representative sample that contains extensive information on assets, liabilities, income, and demographic characteristics of those sampled (potential U.S. customers). We study a random sample of 500 households with positive incomes that were interviewed in the 2004 survey. We initially consider the subset of $n = 275$ families that purchased term life insurance. We address the second portion of the demand question and determine family characteristics that influence the amount of insurance purchased. Chapter 11 will consider the first portion, whether a household purchases insurance, through models where the response is a binary random variable.

For term life insurance, the quantity of insurance is measured by the policy FACE, the amount that the company will pay in the event of the death of the

Table 3.1 Term Life Summary Statistics

Variable	Mean	Median	Standard Deviation	Minimum	Maximum
FACE	747,581	150,000	1,674,362	800	14,000,000
INCOME	208,975	65,000	824,010	260	10,000,000
EDUCATION	14.524	16.000	2.549	2.000	17.000
NUMHH	2.960	3.000	1.493	1.000	9.000
LNFACE	11.990	11.918	1.871	6.685	16.455
LNINCOME	11.149	11.082	1.295	5.561	16.118

named insured. Characteristics that will turn out to be important include annual INCOME; the number of years of EDUCATION of the survey respondent, and the number of household members, NUMHH.

In general, we will consider datasets where there are k explanatory variables and one response variable in a sample of size n. That is, the data consist of

$$\left\{ \begin{array}{l} x_{11}, x_{12}, \ldots, x_{1k}, y_1 \\ x_{21}, x_{22}, \ldots, x_{2k}, y_2 \\ \\ x_{n1}, x_{n2}, \ldots, x_{nk}, y_n \end{array} \right\}.$$

The ith observation corresponds to the ith row, consisting of $(x_{i1}, x_{i2}, \ldots, x_{ik}, y_i)$. For this general case, we take $k+1$ measurements on each entity. For the insurance demand example, $k = 3$ and the data consists of $(x_{11}, x_{12}, x_{13}, y_1), \ldots, (x_{275,1}, x_{275,2}, x_{275,3}, y_{275})$. That is, we use four measurements from each of the $n = 275$ households.

Summarizing the Data

Begin the data analysis by examining each variable in isolation of the others.

We begin the data analysis by examining each variable in isolation of the others. Table 3.1 provides basic summary statistics of the four variables. For FACE and INCOME, we see that the mean is much greater than the median, suggesting that the distribution is skewed to the right. Histograms (not reported here) show that this is the case. It will be useful to also consider their logarithmic transforms, LNFACE and LNINCOME, respectively, which are also reported in Table 3.1.

The next step is to measure the relationship between each x on y, beginning with the scatter plots in Figure 3.1. The left-hand panel is a plot of FACE versus INCOME; with this panel, we see a large clustering in the lower-left-hand corner corresponding to households that have both small incomes and face amounts of insurance. Both variables have skewed distributions and their joint effect is highly nonlinear. The right-hand panel presents the same variables but using logarithmic transforms. In Figure 3.1, we see a relationship that can be more readily approximated with a line.

The term life data are *multivariate* in the sense that several measurements are taken on each household. It is difficult to produce a graph of observations in three

Table 3.2 Term Life Correlations

	NUMHH	EDUCATION	LNINCOME
EDUCATION	−0.064		
LNINCOME	0.179	0.343	
LNFACE	0.288	0.383	0.482

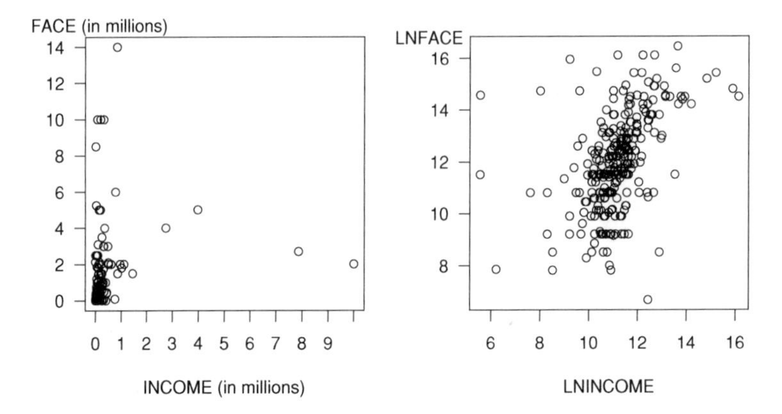

Figure 3.1 Income versus face amount of term life insurance. The left panel is a plot of face versus income, showing a highly nonlinear pattern. In the right-hand panel, face versus income is in natural logarithmic units, suggesting a linear (although variable) pattern.

or more dimensions on a two-dimensional platform, such as a piece of paper, that is not confusing, misleading, or both. To summarize graphically multivariate data in regression applications, consider using a *scatterplot matrix* such as in Figure 3.2. Each square of this figure represents a simple plot of one variable versus another. For each square, the row variable gives the units of the vertical axis and the column variable gives the units of the horizontal axis. The matrix is sometimes called a *half scatterplot matrix* because only the lower-left-hand elements are presented.

The scatterplot matrix can be numerically summarized using a correlation matrix. Each correlation in Table 3.2 corresponds to a square of the scatterplot matrix in Figure 3.2. Analysts often present tables of correlations because they are easy to interpret. However, remember that a correlation coefficient merely measures the extent of linear relationships. Thus, a table of correlations provides a sense of linear relationships but may miss a nonlinear relationship that can be revealed in a scatterplot matrix.

The scatterplot matrix and corresponding correlation matrix are useful devices for summarizing multivariate data. They are easy to produce and to interpret. Still, each device captures only relationships between pairs of variables and cannot quantify relationships among several variables.

Method of Least Squares

Consider this question: can knowledge of education, household size, and income help us understand the demand for insurance? The correlations in Table 3.2 and the graphs in Figures 3.1 and 3.2 suggest that each variable, EDUCATION, NUMHH, and LNINCOME, may be a useful explanatory variable of LNFACE

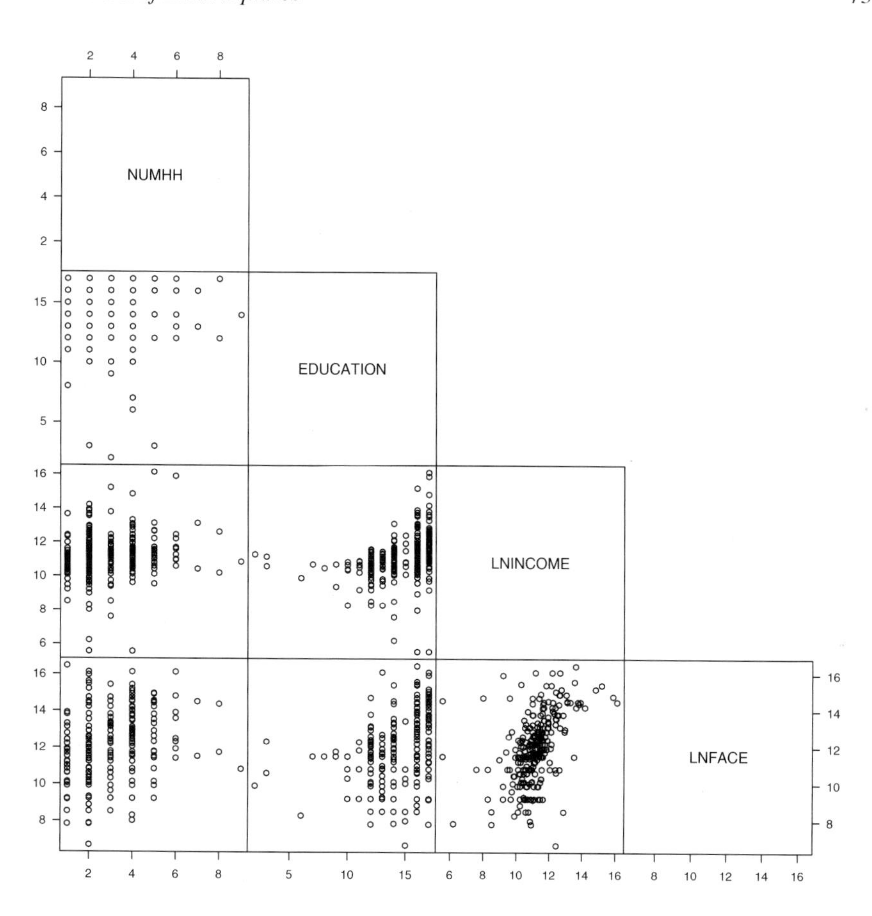

Figure 3.2 Scatterplot matrix of four variables. Each square is a scatter plot.

when taken individually. It seems reasonable to investigate the *joint* effect of these variables on a response.

The geometric concept of a *plane* is used to explore the linear relationship between a response and several explanatory variables. Recall that a plane extends the concept of a line to more than two dimensions. A plane may be defined through an algebraic equation such as

$$y = b_0 + b_1 x_1 + \cdots + b_k x_k.$$

This equation defines a plane in $k + 1$ dimensions. Figure 3.3 shows a plane in three dimensions. For this figure, there is one response variable, LNFACE, and two explanatory variables, EDUCATION and LNINCOME (NUMHH is held fixed). It is difficult to graph more than three dimensions in a meaningful way.

We need a way to determine a plane from the data. The difficulty is that in most regression analysis applications, the number of observations, n, far exceeds the number of observations required to fit a plane, $k + 1$. Thus, it is generally not possible to find a single plane that passes through all n observations. As in Chapter 2, we use the *method of least squares* to determine a plane from the data.

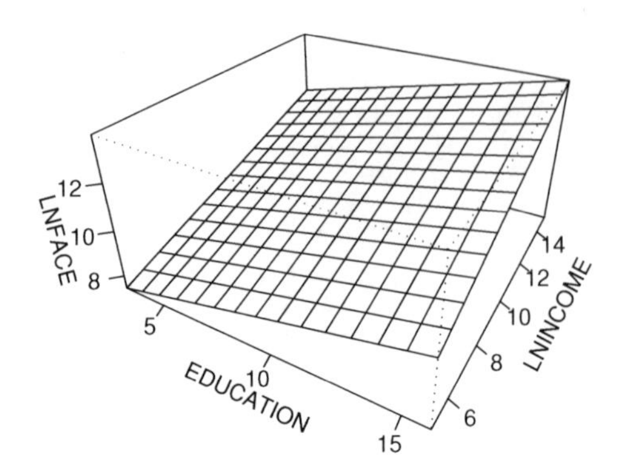

Figure 3.3 An example of a three-dimensional plane.

The method of least squares is based on determining the values of b_0^*, $b_1^*, \ldots, b_k^*$ that minimize the quantity

$$SS(b_0^*, b_1^*, \ldots, b_k^*) = \sum_{i=1}^{n} \left(y_i - \left(b_0^* + b_1^* x_{i1} + \cdots + b_k^* x_{ik}\right)\right)^2. \qquad (3.1)$$

We drop the asterisk notation and use $b_0, b_1, \ldots, b_k$ to denote the best values, known as the *least squares estimates*. With the least squares estimates, define the *least squares*, or *fitted*, *regression plane* as

$$\widehat{y} = b_0 + b_1 x_1 + \cdots + b_k x_k.$$

The least squares estimates are determined by minimizing $SS(b_0^*, b_1^*, \ldots, b_k^*)$. It is difficult to write down the resulting least squares estimators using a simple formula unless one resorts to matrix notation. Because of their importance in applied statistical models, an explicit formula for the estimators is provided here. However, these formulas have been programmed into a wide variety of statistical and spreadsheet software packages. The fact that these packages are readily available allows data analysts to concentrate on the ideas of the estimation procedure instead of focusing on the details of the calculation procedures.

As an example, a regression plane was fit to the term life data where three explanatory variables, x_1 for EDUCATION, x_2 for NUMHH, and x_3 for LNINCOME, were used. The resulting fitted regression plane is

$$\widehat{y} = 2.584 + 0.206 x_1 + 0.306 x_2 + 0.494 x_3. \qquad (3.2)$$

Matrix Notation

Assume that the data are of the form $(x_{i0}, x_{i1}, \ldots, x_{ik}, y_i)$, where $i = 1, \ldots, n$. Here, the variable x_{i0} is associated with the "intercept" term. That is, in most applications, we assume that x_{i0} is identically equal to 1 and thus need not be explicitly represented. However, there are important applications where this is not the case; thus, to express the model in general notation, it is included here.

The data are represented in matrix notation using

$$\mathbf{y} = \begin{pmatrix} y_1 \\ y_2 \\ \vdots \\ y_n \end{pmatrix} \quad \text{and} \quad \mathbf{X} = \begin{pmatrix} x_{10} & x_{11} & \cdots & x_{1k} \\ x_{20} & x_{21} & \cdots & x_{2k} \\ \vdots & \vdots & \ddots & \vdots \\ x_{n0} & x_{n1} & \cdots & x_{nk} \end{pmatrix}.$$

Here, $\mathbf{y}$ is the $n \times 1$ vector of responses and $\mathbf{X}$ is the $n \times (k+1)$ matrix of explanatory variables. We use the matrix algebra convention that lower- and uppercase bold letters represent vectors and matrices, respectively. (If you need to brush up on matrices, review Section 2.11.)

Example: Term Life Insurance, Continued. Recall that y represents the logarithmic face; x_1, years of education; x_2, number of household members; and x_3, logarithmic income. Thus, there are $k = 3$ explanatory variables and $n = 275$ households. The vector of responses and the matrix of explanatory variables are

$$\mathbf{y} = \begin{pmatrix} y_1 \\ y_2 \\ \vdots \\ y_{275} \end{pmatrix} = \begin{pmatrix} 9.904 \\ 11.775 \\ \vdots \\ 9.210 \end{pmatrix} \quad \text{and}$$

$$\mathbf{X} = \begin{pmatrix} 1 & x_{11} & x_{12} & x_{13} \\ 1 & x_{21} & x_{22} & x_{23} \\ \vdots & \vdots & \vdots & \vdots \\ 1 & x_{275,1} & x_{275,2} & x_{275,3} \end{pmatrix} = \begin{pmatrix} 1 & 16 & 3 & 10.669 \\ 1 & 9 & 3 & 9.393 \\ \vdots & \vdots & \vdots & \vdots \\ 1 & 12 & 1 & 10.545 \end{pmatrix}.$$

For example, for the first observation in the dataset, the dependent variable is $y_1 = 9.904$ (corresponding to $\exp(9.904) = \$20{,}000$), for a survey respondent with 16 years of education living in a household with 3 people with logarithmic income of 10.669 ($\exp(10.669) = \$43{,}000$).

Under the least squares estimation principle, our goal is to choose the coefficients $b_0^*, b_1^*, \ldots, b_k^*$ to minimize the sum of squares function $SS(b_0^*, b_1^*, \ldots, b_k^*)$. Using calculus, we return to equation (3.1), take partial derivatives with respect to each coefficient, and set these quantities equal to zero:

$$\begin{aligned} \frac{\partial}{\partial b_j^*} SS\left(b_0^*, b_1^*, \ldots, b_k^*\right) &= \sum_{i=1}^{n} (-2x_{ij}) \left(y_i - \left(b_0^* + b_1^* x_{i1} + \cdots + b_k^* x_{ik}\right)\right) \\ &= 0, \quad \text{for } j = 0, 1, \ldots, k. \end{aligned}$$

This is a system of $k+1$ equations and $k+1$ unknowns that can be readily solved using matrix notation, as follows.

We may express the vector of parameters to be minimized as $\mathbf{b}^* = (b_0^*, b_1^*, \ldots, b_k^*)'$. Using this, the sum of squares can be written as $SS(\mathbf{b}^*) = (\mathbf{y} - \mathbf{X}\mathbf{b}^*)'$

$(\mathbf{y} - \mathbf{X}\mathbf{b}^*)$. Thus, in matrix form, the solution to the minimization problem can be expressed as $(\partial/\partial\mathbf{b}^*)SS(\mathbf{b}^*) = \mathbf{0}$. This solution satisfies the *normal equations*

$$\mathbf{X}'\mathbf{X}\mathbf{b} = \mathbf{X}'\mathbf{y}. \tag{3.3}$$

Here, the asterisk notation (*) has been dropped to denote the fact that $\mathbf{b} = (b_0, b_1, \ldots, b_k)'$ represents the best vector of values in the sense of minimizing $SS(\mathbf{b}^*)$ over all choices of $\mathbf{b}^*$.

The least squares estimator $\mathbf{b}$ need not be unique. However, assuming that the explanatory variables are not linear combinations of one another, we have that $\mathbf{X}'\mathbf{X}$ is invertible. In this case, we can write the unique solution as

$$\mathbf{b} = \left(\mathbf{X}'\mathbf{X}\right)^{-1}\mathbf{X}'\mathbf{y}. \tag{3.4}$$

To illustrate, for the term life example, equation (3.2) yields

$$\mathbf{b} = \begin{pmatrix} b_0 \\ b_1 \\ b_2 \\ b_3 \end{pmatrix} = \begin{pmatrix} 2.584 \\ 0.206 \\ 0.306 \\ 0.494 \end{pmatrix}.$$

3.2 Linear Regression Model and Properties of Estimators

In the previous section, we learned how to use the method of least squares to fit a regression plane with a dataset. This section describes the assumptions underpinning the regression model and some of the resulting properties of the regression coefficient estimators. With the model and the fitted data, we will be able to draw inferences about the sample dataset to a larger population. Moreover, we will later use the regression model assumptions to help us improve the model specification in Chapter 5.

3.2.1 Regression Function

Most of the assumptions of the multiple linear regression model carry over directly from the basic linear regression model assumptions introduced in Section 2.2. The primary difference is that we now summarize the relationship between the response and the explanatory variables through the *regression function*

$$\mathrm{E}\ y = \beta_0 x_0 + \beta_1 x_1 + \cdots + \beta_k x_k, \tag{3.5}$$

which is linear in the parameters $\beta_0, \ldots, \beta_k$. Henceforth, we will use $x_0 = 1$ for the variable associated with the parameter β_0; this is the default in most statistical packages, and most applications of regression include the intercept term β_0. The intercept is the expected value of y when all of the explanatory variables are equal to zero. Although rarely of interest, the term β_0 serves to set the height of the fitted regression plane.

Interpret β_j to be the expected change in y per unit change in x_j assuming all other explanatory variables are held fixed.

In contrast, the other betas are typically important parameters from a regression study. To help interpret them, we initially assume that x_j varies continuously and is not related to the other explanatory variables. Then, we can interpret β_j as the expected change in y per unit change in x_j, assuming all other explanatory variables are held fixed. That is, from calculus, you will recognize that β_j can be interpreted as a partial derivative. Specifically, using equation (3.5), we have

$$\beta_j = \frac{\partial}{\partial x_j} \mathrm{E}\, y.$$

3.2.2 Regression Coefficient Interpretation

Let us examine the regression coefficient estimates from the term life insurance example and focus initially on the *sign* of the coefficients. For example, from equation (3.2), the coefficient associated with NUMHH is $b_2 = 0.306 > 0$. If we consider two households that have the same income and the same level of education, then the larger household (in terms of NUMHH) is expected to demand *more* term life insurance under the regression model. This is a sensible interpretation, as larger households have more dependents for which term life insurance can provide needed financial assets in the event of the untimely death of a breadwinner. The positive coefficient associated with income ($b_3 = 0.494$) is also plausible; households with larger incomes have more disposable dollars to purchase insurance. The positive sign associated with EDUCATION ($b_1 = 0.206$) is also reasonable; more education suggests that respondents are more aware of their insurance needs, other things being equal.

You will also need to interpret the *amount* of the regression coefficient. Consider first the EDUCATION coefficient. Using equation (3.2), fitted values of $\widehat{LNFACE}$ were calculated by allowing EDUCATION to vary and keeping NUMHH and LNINCOME fixed at the sample averages. The results are as follows:

Effects of Small Changes in Education

EDUCATION	14	14.1	14.2	14.3
$\widehat{LNFACE}$	11.883	11.904	11.924	11.945
$\widehat{FACE}$	144,803	147,817	150,893	154,034
$\widehat{FACE}$ % Change		2.081	2.081	2.081

As EDUCATION increases, $\widehat{LNFACE}$ increases. Further, the amount of $\widehat{LNFACE}$ increase is a constant 0.0206. This comes directly from equation (3.2); as EDUCATION increases by 0.1 years, we expect the demand for insurance to increase by 0.0206 logarithmic dollars, holding NUMHH and LNINCOME fixed. This interpretation is correct, but most product development directors are not overly fond of logarithmic dollars. To return to dollars, fitted face values can be calculated through exponentiation as $\widehat{FACE} = \exp(\widehat{LNFACE})$. Moreover, the percentage change can be computed; for example, $100 \times (147{,}817/144{,}803 - 1) \approx 2.08\%$.

This provides another interpretation of the regression coefficient; as EDUCATION increases by 0.1 years, we expect the demand for insurance to increase by 2.08%. This is a simple consequence of calculus, using $\partial \ln y / \partial x = (\partial y / \partial x) / y$; that is, a small change in the logarithmic value of y equals a small change in y as a proportion of y. It is because of this calculus result that we use natural logs instead of common logs in regression analysis. Because this table uses a discrete change in EDUCATION, the 2.08% differs slightly from the continuous result $0.206 \times$ (change in EDUCATION) = 2.06%. However, this proximity is usually regarded as suitable for interpretation purposes.

Continuing this logic, consider small changes in logarithmic income.

Effects of Small Changes in Logarithmic Income

LNINCOME	11	11.1	11.2	11.3
INCOME	59,874	66,171	73,130	80,822
INCOME % Change		10.52	10.52	10.52
$\widehat{LNFACE}$	11.957	12.006	12.055	12.105
$\widehat{FACE}$	155,831	163,722	172,013	180,724
$\widehat{FACE}$ % Change		5.06	5.06	5.06
$\widehat{FACE}$ % Change/INCOME % Change		0.482	0.482	0.482

We can use the same logic to interpret the LNINCOME coefficient in equation (3.2). As logarithmic income increases by 0.1 units, we expect the demand for insurance to increase by 5.06%. This takes care of logarithmic units in the y but not the x. We can use the same logic to say that as logarithmic income increases by 0.1 units, INCOME increases by 10.52%. Thus, a 10.52% change in INCOME corresponds to a 5.06% change in FACE. To summarize, we say that, holding NUMHH and EDUCATION fixed, we expect that a 1% increase in INCOME is associated with a 0.482% increase in $\widehat{FACE}$ (as earlier, this is close to the parameter estimate $b_3 = 0.494$). The coefficient associated with income is known as an *elasticity* in economics. In economics, elasticity is the ratio of the percentage change in one variable to the percentage change in another variable. Mathematically, we summarize this as

$$\frac{\partial \ln y}{\partial \ln x} = \left(\frac{\partial y}{y}\right) \Big/ \left(\frac{\partial x}{x}\right).$$

3.2.3 Model Assumptions

As in Section 2.2 for a single explanatory variable, there are two sets of assumptions that one can use for multiple linear regression. They are equivalents sets, each having comparative advantages as we proceed in our study of regression. The observables representation focuses on variables of interest $(x_{i1}, \ldots, x_{ik}, y_i)$.

The error representation provides a springboard for motivating our goodness-of-fit measures and study of residual analysis. However, the latter set of assumptions focuses on the additive errors case and obscures the sampling basis of the model.

Multiple Linear Regression Model Sampling Assumptions

Observables Representation	Error Representation
F1. $\mathrm{E}\ y_i = \beta_0 + \beta_1 x_{i1} + \cdots + \beta_k x_{ik}$.	E1. $y_i = \beta_0 + \beta_1 x_{i1} + \cdots + \beta_k x_{ik} + \varepsilon_i$.
F2. $\{x_{i1}, \ldots, x_{ik}\}$ are nonstochastic variables.	E2. $\{x_{i1}, \ldots, x_{ik}\}$ are nonstochastic variables.
F3. $\mathrm{Var}\ y_i = \sigma^2$.	E3. $\mathrm{E}\ \varepsilon_i = 0$ and $\mathrm{Var}\ \varepsilon_i = \sigma^2$.
F4. $\{y_i\}$ are independent random variables.	E4. $\{\varepsilon_i\}$ are independent random variables.
F5. $\{y_i\}$ are normally distributed.	E5. $\{\varepsilon_i\}$ are normally distributed.

To further motivate Assumptions F2 and F4, we usually assume that our data have been realized as the result of a stratified sampling scheme, where each unique value of $\{x_{i1}, \ldots, x_{ik}\}$ is treated as a stratum. That is, for each value of $\{x_{i1}, \ldots, x_{ik}\}$, we draw a random sample of responses from a population. Thus, responses in each stratum are independent from one another, as are responses from different strata. Chapter 6 will discuss this sampling basis in further detail.

3.2.4 Properties of Regression Coefficient Estimators

Section 3.1 described the least squares method for estimating regression coefficients. With the regression model assumptions, we can establish some basic properties of these estimators. To do this, from Section 2.11.4, we have that the expectation of a vector is the vector of expectations, so that

$$\mathrm{E}\ \mathbf{y} = \begin{pmatrix} \mathrm{E}\ y_1 \\ \mathrm{E}\ y_2 \\ \vdots \\ \mathrm{E}\ y_n \end{pmatrix}.$$

Further, basic matrix multiplication shows that

$$\mathbf{X}\boldsymbol{\beta} = \begin{pmatrix} 1 & x_{11} & \cdots & x_{1k} \\ 1 & x_{21} & \cdots & x_{2k} \\ \vdots & \vdots & \ddots & \vdots \\ 1 & x_{n,1} & \cdots & x_{n,k} \end{pmatrix} \begin{pmatrix} \beta_0 \\ \beta_1 \\ \vdots \\ \beta_k \end{pmatrix} = \begin{pmatrix} \beta_0 + \beta_1 x_{11} + \cdots + \beta_k x_{1k} \\ \beta_0 + \beta_1 x_{21} + \cdots + \beta_k x_{2k} \\ \vdots \\ \beta_0 + \beta_1 x_{n1} + \cdots + \beta_k x_{nk} \end{pmatrix}.$$

Because the ith row of assumption F1 is $\mathrm{E}\ y_i = \beta_0 + \beta_1 x_{i1} + \cdots + \beta_k x_{ik}$, we can rewrite this assumption in matrix formulation as $\mathrm{E}\ \mathbf{y} = \mathbf{X}\boldsymbol{\beta}$. We are now in a position to state the first important property of least squares regression estimators.

Property 1. Consider a regression model and let Assumptions F1–F4 hold. Then, the estimator $\mathbf{b}$ defined in equation (3.4) is an unbiased estimator of the parameter vector $\boldsymbol{\beta}$.

To establish Property 1, we have that

$$\mathrm{E}\,\mathbf{b} = \mathrm{E}\left((\mathbf{X}'\mathbf{X})^{-1}\mathbf{X}'\mathbf{y}\right) = (\mathbf{X}'\mathbf{X})^{-1}\mathbf{X}'\mathrm{E}\,\mathbf{y} = (\mathbf{X}'\mathbf{X})^{-1}\mathbf{X}'\,(\mathbf{X}\boldsymbol{\beta}) = \boldsymbol{\beta},$$

using matrix multiplication rules. This chapter assumes that $\mathbf{X}'\mathbf{X}$ is invertible. One can also show that the least squares estimator need only be a solution of the normal equations for unbiasedness (not requiring that $\mathbf{X}'\mathbf{X}$ be invertible, see Section 4.7.3). Thus, $\mathbf{b}$ is said to be an *unbiased estimator* of $\boldsymbol{\beta}$. In particular, $\mathrm{E}\,b_j = \beta_j$ for $j = 0, 1, \ldots, k$.

Because independence implies zero covariance, from Assumption F4 we have $\mathrm{Cov}(y_i, y_j) = 0$ for $i \neq j$. From this, Assumption F3 and the definition of the variance of a vector, we have

$$\mathrm{Var}\,\mathbf{y} = \begin{pmatrix} \mathrm{Var}\,y_1 & \mathrm{Cov}(y_1, y_2) & \cdots & \mathrm{Cov}(y_1, y_n) \\ \mathrm{Cov}(y_2, y_1) & \mathrm{Var}\,y_2 & \cdots & \mathrm{Cov}(y_2, y_n) \\ \vdots & \vdots & \ddots & \vdots \\ \mathrm{Cov}(y_n, y_1) & \mathrm{Cov}(y_n, y_2) & \cdots & \mathrm{Var}\,y_n \end{pmatrix}$$

$$= \begin{pmatrix} \sigma^2 & 0 & \cdots & 0 \\ 0 & \sigma^2 & \cdots & 0 \\ \vdots & \vdots & \ddots & \vdots \\ 0 & 0 & \cdots & \sigma^2 \end{pmatrix} = \sigma^2\mathbf{I},$$

where $\mathbf{I}$ is an an $n \times n$ identity matrix. We are now in a position to state the second important property of least squares regression estimators.

> **Property 2.** Consider a regression model and let Assumptions F1-F4 hold. Then, the estimator $\mathbf{b}$ defined in equation (3.4) has variance $\mathrm{Var}\,\mathbf{b} = \sigma^2(\mathbf{X}'\mathbf{X})^{-1}$.

To establish Property 2, we have

$$\mathrm{Var}\,\mathbf{b} = \mathrm{Var}\left((\mathbf{X}'\mathbf{X})^{-1}\mathbf{X}'\mathbf{y}\right) = \left[(\mathbf{X}'\mathbf{X})^{-1}\mathbf{X}'\right]\mathrm{Var}\,(\mathbf{y})\left[\mathbf{X}(\mathbf{X}'\mathbf{X})^{-1}\right]$$
$$= \left[(\mathbf{X}'\mathbf{X})^{-1}\mathbf{X}'\right]\sigma^2\mathbf{I}\left[\mathbf{X}(\mathbf{X}'\mathbf{X})^{-1}\right] = \sigma^2(\mathbf{X}'\mathbf{X})^{-1}\mathbf{X}'\mathbf{X}(\mathbf{X}'\mathbf{X})^{-1} = \sigma^2(\mathbf{X}'\mathbf{X})^{-1},$$

as required. This important property will allow us to measure the precision of the estimator $\mathbf{b}$ when we discuss statistical inference. Specifically, by the definition of the variance of a vector (see Section 2.11.4),

$$\mathrm{Var}\,\mathbf{b} = \begin{pmatrix} \mathrm{Var}\,b_0 & \mathrm{Cov}(b_0, b_1) & \cdots & \mathrm{Cov}(b_0, b_k) \\ \mathrm{Cov}(b_1, b_0) & \mathrm{Var}\,b_1 & \cdots & \mathrm{Cov}(b_1, b_k) \\ \vdots & \vdots & \ddots & \vdots \\ \mathrm{Cov}(b_k, b_0) & \mathrm{Cov}(b_k, b_1) & \cdots & \mathrm{Var}\,b_k \end{pmatrix} = \sigma^2(\mathbf{X}'\mathbf{X})^{-1}. \tag{3.6}$$

Thus, for example, $\mathrm{Var}\,b_j$ is σ^2 times the $(j+1)$st diagonal entry of $(\mathbf{X}'\mathbf{X})^{-1}$. As another example, $\mathrm{Cov}(b_0, b_j)$ is σ^2 times the element in the first row and $(j+1)$st column of $(\mathbf{X}'\mathbf{X})^{-1}$.

Although alternative methods are available that are preferable for specific applications, the least squares estimators have proved effective for many routine data analyses. One desirable characteristic of least squares regression estimators is summarized in the following well-known result.

Gauss-Markov Theorem. Consider the regression model and let Assumptions F1–F4 hold. Then, in the class of estimators that are linear functions of the responses, the least squares estimator **b** defined in equation (3.4) is the minimum variance unbiased estimator of the parameter vector $\boldsymbol{\beta}$.

We have already seen in Property 1 that the least squares estimators are unbiased. The Gauss-Markov theorem states that the least squares estimator is the most precise in that it has the smallest variance. (In a matrix context, *minimum variance* means that if $\mathbf{b}^*$ is any other estimator, then the difference of the variance matrices, Var $\mathbf{b}^*$ − Var $\mathbf{b}$, is nonnegative definite.)

The Gauss-Markov theorem states that the least squares estimator is the most precise in that it has the smallest variance.

An additional important property concerns the distribution of the least squares regression estimators.

Property 3. Consider a regression model and let Assumptions F1–F5 hold. Then, the least squares estimator **b** defined in equation (3.4) is normally distributed.

To establish Property 3, we define the weight vectors, $\mathbf{w}_i = (\mathbf{X}'\mathbf{X})^{-1}(1, x_{i1}, \ldots, x_{ik})'$. With this notation, we note that

$$\mathbf{b} = (\mathbf{X}'\mathbf{X})^{-1}\mathbf{X}'\mathbf{y} = \sum_{i=1}^{n} \mathbf{w}_i y_i,$$

so that **b** is a linear combination of responses. With Assumption F5, the responses are normally distributed. Because linear combinations of normally distributed random variables are normally distributed, we have the conclusion of Property 3. This result underpins much of the statistical inference that will be presented in Sections 3.4 and 4.2.

3.3 Estimation and Goodness of Fit

Residual Standard Deviation

Additional properties of the regression coefficient estimators will be discussed when we focus on statistical inference. We now continue our estimation discussion by providing an estimator of the other parameter in the linear regression model, σ^2.

Our estimator for σ^2 can be developed using the principle of replacing theoretical expectations by sample averages. In examining $\sigma^2 = \mathrm{E}\,(y - \mathrm{E}\,y)^2$, replacing the outer expectation by a sample average suggests using the estimator $n^{-1}\sum_{i=1}^{n}(y_i - \mathrm{E}\,y_i)^2$. Because we do not observe $\mathrm{E}\,y_i = \beta_0 + \beta_1 x_{i1} + \cdots +$

$\beta_k x_{ik}$, we use in its place the corresponding observed quantity $b_0 + b_1 x_{i1} + \cdots + b_k x_{ik} = \widehat{y}_i$. This leads to the following:

Definition. An estimator of σ^2, the *mean square error (MSE)*, is defined as

$$s^2 = \frac{1}{n-(k+1)} \sum_{i=1}^{n} (y_i - \widehat{y}_i)^2 . \tag{3.7}$$

The positive square root, $s = \sqrt{s^2}$, is called the *residual standard deviation.*

This expression generalizes the definition in equation (2.3), which is valid for $k = 1$. It turns out that, by using $n - (k + 1)$ instead of n in the denominator of equation (3.7), s^2 is an unbiased estimator of σ^2. Essentially, by using $\widehat{y}_i$ instead of $\mathrm{E}\, y_i$ in the definition, we have introduced some small dependencies among the deviations from the responses $y_i - \widehat{y}_i$, thus reducing the overall variability. To compensate for this lower variability, we also reduce the denominator in the definition of s^2.

To provide further intuition on the choice of $n - (k + 1)$ in the definition of s^2, we introduced the concept of residuals in the context of multiple linear regression. From Assumption E1, recall that the random errors can be expressed as $\varepsilon_i = y_i - (\beta_0 + \beta_1 x_{i1} + \cdots + \beta_k x_{ik})$. Because the parameters $\beta_0, \ldots, \beta_k$ are not observed, the errors themselves are not observed. Instead, we examine the "estimated errors," or *residuals*, defined by $e_i = y_i - \widehat{y}_i$.

Unlike errors, there exist certain dependencies among the residuals. One dependency is due to the algebraic fact that the average residual is zero. Further, there must be at least $k + 2$ observations for there to be variation in the fit of the plane. If we have only $k + 1$ observations, we can fit a plane to the data perfectly, resulting in no variation in the fit. For example, if $k = 1$, because two observations determine a line, then at least three observations are required to observe any deviation from the line. Because of these dependencies, we have only $n - (k + 1)$ free, or unrestricted, residuals to estimate the variability about the regression plane.

The positive square root of s^2 is our estimator of σ. Using residuals, it can be expressed as

$$s = \sqrt{\frac{1}{n-(k+1)} \sum_{i=1}^{n} e_i^2 .} \tag{3.8}$$

Because it is based on residuals, we refer to s as the *residual standard deviation.* The quantity s is a measure of our "typical error." For this reason, s is also called the *standard error of the estimate.*

The Coefficient of Determination: R^2

To summarize the goodness of fit of the model, as in Chapter 2, we partition the variability into pieces that are "explained" and "unexplained" by the regression fit. Algebraically, the calculations for regression using many variables are similar to the case of using only one variable. Unfortunately, when dealing with many variables, we lose the easy graphical interpretation such as in Figure 2.4.

Begin with the total sum of squared deviations, *Total SS* $= \sum_{i=1}^{n} (y_i - \overline{y})^2$, as our measure of the total variation in the dataset. As in equation (2.1), we may then interpret the equation

$$\underbrace{y_i - \overline{y}}_{\substack{\text{total} \\ \text{deviation}}} = \underbrace{y_i - \widehat{y}_i}_{\substack{\text{unexplained} \\ \text{deviation}}} + \underbrace{\widehat{y}_i - \overline{y}}_{\substack{\text{explained} \\ \text{deviation}}}$$

as the "deviation without knowledge of the explanatory variables equals the deviation not explained by the explanatory variables plus deviation explained by the explanatory variables." Squaring each side and summing over all observations yields

$$\textit{Total SS} = \textit{Error SS} + \textit{Regression SS},$$

where *Error SS* $= \sum_{i=1}^{n} (y_i - \widehat{y}_i)^2$ and *Regression SS* $= \sum_{i=1}^{n} (\widehat{y}_i - \overline{y})^2$. As in Section 2.3 for the one explanatory variable case, the sum of the cross-product terms turns out to be zero.

A statistic that summarizes this relationship is the *coefficient of determination*,

$$R^2 = \frac{\textit{Regression SS}}{\textit{Total SS}}.$$

We interpret R^2 to be the proportion of variability explained by the regression function.

If the model is a desirable one for the data, one would expect a strong relationship between the observed responses and those "expected" under the model, the fitted values. An interesting algebraic fact is the following. If we square the correlation coefficient between the responses and the fitted values, we get the coefficient of determination; that is,

$$R^2 = [r(y, \widehat{y})]^2 .$$

As a result, R, the positive square root of R^2, is called the *multiple correlation coefficient*. It can be interpreted as the correlation between the response and the best linear combination of the explanatory variables, the fitted values. (This relationship is developed using matrix algebra in the technical supplement Section 5.10.1.)

The variability decomposition is also summarized using the *analysis of variance*, or *ANOVA*, table, as follows:

ANOVA Table

Source	Sum of Squares	df	Mean Square
Regression	*Regression SS*	k	*Regression MS*
Error	*Error SS*	$n-(k+1)$	*MSE*
Total	*Total SS*	$n-1$	

Table 3.3 Term Life ANOVA Table

Source	Sum of Squares	df	Mean Square
Regression	328.47	3	109.49
Error	630.43	271	2.326
Total	958.90	274	

The mean square column figures are defined to be the sum of squares figures divided by their respective degrees of freedom. The error degrees of freedom denotes the number of unrestricted residuals. It is this number that we use in our definition of the "average," or mean, square error. That is, we define

$$MSE = Error\ MS = \frac{Error\ SS}{n-(k+1)} = s^2.$$

Similarly, the regression degrees of freedom is the number of explanatory variables. This yields

$$Regression\ MS = \frac{Regression\ SS}{k}.$$

When discussing the coefficient of determination, it can be established that whenever an explanatory variable is added to the model, R^2 never decreases. This is true whether or not the additional variable is useful. We would like a measure of fit that decreases when useless variables are entered into the model as explanatory variables. To circumvent this anomaly, a widely used statistic is the *coefficient of determination adjusted for degrees of freedom*, defined by

$$R_a^2 = 1 - \frac{(Error\ SS)/[n-(k+1)]}{(Total\ SS)/(n-1)} = 1 - \frac{s^2}{s_y^2}. \tag{3.9}$$

To interpret this statistic, note that s_y^2 does not depend on the model or on the model variables. Thus, s^2 and R_a^2 are equivalent measures of model fit. As the model fit improves, then R_a^2 becomes larger and s^2 becomes smaller, and vice versa. Put another way, choosing a model with the smallest s^2 is equivalent to choosing a model with the largest R_a^2.

Example: Term Life Insurance, Continued. To illustrate, Table 3.3 displays the summary statistics for the regression of LNFACE on EDUCATION, NUMHH, and LNINCOME. From the degrees-of-freedom column, we remind ourselves that there are three explanatory variables and 275 observations. As measures of model fit, the coefficient of determination is $R^2 = 34.3\%$ (=328.47/958.90) and the residual standard deviation is $s = 1.525$ ($=\sqrt{2.326}$). If we were to attempt to estimate the logarithmic face amount without knowledge of the explanatory variables EDUCATION, NUMHH, and LNINCOME, then the size of the typical error would be $s_y = 1.871$ ($=\sqrt{958.90/274}$). Thus, by taking advantage of our knowledge of the explanatory variables, we have been able to reduce the size of the typical error. The measure of model fit that compares these two estimates of variability is the adjusted coefficient of determination, $R_a^2 = 1 - 2.326/1.871^2 = 33.6\%$.

Table 3.4 Regression Coefficients from a Model of Female Advantage

Variable	Coefficient	t-Statistic
Intercept	9.904	12.928
Logarithmic number of persons per physician	−0.473	−3.212
Fertility	−0.444	−3.477
Percentage of Hindus and Buddhists	−0.018	−3.196
Soviet Union dummy	4.922	7.235

Source: Lemaire (2002)

Example: Why Do Females Live Longer Than Males? In an article with this title, Lemaire (2002) examined what he called the "female advantage," the difference in life expectancy between women and men. Life expectancies are of interest because they are widely used measures of a nation's health. Lemaire examined data from $n = 169$ countries and found that the average female advantage was 4.51 years worldwide. He sought to explain this difference based on 45 behaviorial measures, variables that capture a nation's degree of economic modernization; social, cultural, and religious mores; geographic position; and quality of health care available.

After a detailed analysis, Lemaire reports coefficients from a regression model that appear in Table 3.4. This regression model explains $R^2 = 61\%$ of the variability. It is a parsimonious model consisting of only $k = 4$ of the original 45 variables.

All variables were strongly statistically significant. The number of persons per physician was also correlated with other variables that capture a country's degree of economic modernization, such as urbanization, number of cars, and percentage working in agriculture. Fertility, the number of births per woman, was highly correlated with education variables in the study, including female illiteracy and female school enrollment. The percentage of Hindus and Buddhists is a social, cultural, and religious variable. The Soviet Union dummy is a geographic variable – it characterizes Eastern European countries that formerly belonged to the Soviet Union. Because of the high degree of collinearity among the 45 candidate variables, other analysts could easily pick an alternative set of variables. Nonetheless, Lemaire's important point was that this simple model explains roughly 61% of the variability from only behaviorial variables, unrelated to biological sex differences.

3.4 Statistical Inference for a Single Coefficient

3.4.1 The t-Test

In many applications, a single variable is of primary interest, and other variables are included in the regression to control for additional sources of variability. To illustrate, a sales agent might be interested in the effect that income has on the quantity of insurance demanded. In a regression analysis, we could also include other explanatory variables such as an individual's sex, occupation, age, size of household, education level, and so on. By including these additional explanatory variables, we hope to gain a better understanding of the relationship between income and insurance demand. To reach sensible conclusions, we will need some rules to decide whether a variable is important.

We respond to the question, Is x_j important? by investigating whether the corresponding slope parameter, β_j, equals zero. The question is whether β_j is zero can be restated in the hypothesis testing framework as "Is $H_0 : \beta_j = 0$ valid?"

We examine the proximity of b_j to zero to determine whether β_j is zero. Because the units of b_j depend on the units of y and x_j, we need to standardize this quantity. In Property 2 and equation (3.6), we saw that Var b_j is σ^2 times the $(j+1)$st diagonal element of $(\mathbf{X}'\mathbf{X})^{-1}$. Replacing σ^2 by the estimator s^2 and taking square roots, we have the following:

Definition. The standard error of b_j can be expressed as

$$se(b_j) = s\sqrt{(j+1)st\ diagonal\ element\ of\ (\mathbf{X}'\mathbf{X})^{-1}}.$$

Recall that a standard error is an estimated standard deviation. To test $H_0 : \beta_j = 0$, we examine the t-ratio, $t(b_j) = b_j / se(b_j)$. We interpret $t(b_j)$ to be the number of standard errors that b_j is away from zero. This is the appropriate quantity because the sampling distribution of $t(b_j)$ can be shown to be the t-distribution with $df = n - (k+1)$ degrees of freedom, under the null hypothesis with the linear regression model assumptions F1–F5. This enables us to construct tests of the null hypothesis such as the following procedure:

Interpret $t(b_j)$ to be the number of standard errors that b_j is away from zero.

Procedure. The *t-test* for a Regression Coefficient (beta).

- The null hypothesis is $H_0 : \beta_j = 0$.
- The alternative hypothesis is $H_a : \beta_j \neq 0$.
- Establish a significance level α (typically but not necessarily 5%).
- Construct the statistic, $t(b_j) = b_j / se(b_j)$.
- Procedure: Reject the null hypothesis in favor of the alternative if $|t(b_j)|$ exceeds a t-value. Here, this t-value is the $(1-\alpha/2)$th percentile from the t-distribution using $df = n - (k+1)$ degrees of freedom, denoted as $t_{n-(k+1),1-\alpha/2}$.

In many applications, the sample size will be large enough so that we may approximate the t-value by the corresponding percentile from the standard normal curve. At the 5% level of significance, this percentile is 1.96. Thus, as a rule of thumb, we can interpret a variable to be important if its t-ratio exceeds two in absolute value.

Rule of thumb: interpret a variable to be important if its t-ratio exceeds two in absolute value.

Table 3.5 Decision-Making Procedures for Testing $H_0 : \beta_j = d$

Alternative Hypothesis (H_a)	Procedure: Reject H_0 in favor of H_a if
$\beta_j > d$	t-ratio $> t_{n-(k+1),1-\alpha}$
$\beta_j < d$	t-ratio $< -t_{n-(k+1),1-\alpha}$
$\beta_j \neq d$	$\lvert t\text{-ratio}\rvert > t_{n-(k+1),1-\alpha/2}$

Notes: The significance level is α. Here, $t_{n-(k+1),1-\alpha}$ is the (1-α)th percentile from the t-distribution using $df = n - (k+1)$ degrees of freedom. The test statistic is t-ratio $= (b_j - d)/se(b_j)$.

Table 3.6 Probability Values for Testing $H_0 : \beta_j = d$

Alternative Hypothesis (H_a)	$\beta_j > d$	$\beta_j < d$	$\beta_j \neq d$
p-Value	$\Pr(t_{n-(k+1)} > t\text{-ratio})$	$\Pr(t_{n-(k+1)} < t\text{-ratio})$	$\Pr(\lvert t_{n-(k+1)}\rvert > \lvert t\text{-ratio}\rvert)$

Notes: Here, $t_{n-(k+1)}$ is a t-distributed random variable with $df = n - (k+1)$ degrees of freedom. The test statistic is t-ratio $= (b_j - d)/se(b_j)$.

Although it is the most common, testing $H_0 : \beta_j = 0$ versus $H_a : \beta_j \neq 0$ is just one of many hypothesis tests that can be performed. Table 3.5 outlines alternative decision-making procedures. These procedures are for testing $H_0 : \beta_j = d$. Here, d is a user-prescribed value that may be equal to zero or any other known value.

Alternatively, one can construct p-values and compare them to given significant levels. The p-value allows the report reader to understand the strength of the deviation from the null hypothesis. Table 3.6 summarizes the procedure for calculating p-values.

Example: Term Life Insurance, Continued. A useful convention when reporting the results of a statistical analysis is to place the standard error of a statistic in parentheses below that statistic. Thus, for example, in our regression of LNFACE on EDUCATION, NUMHH, and LNINCOME, the estimated regression equation is

$$\begin{array}{lccccccc} \widehat{LNFACE} & = 2.584 & + & 0.206\ \text{EDUCATION} & + & 0.306\ \text{NUMHH} & + & 0.494\ \text{LNINCOME}. \\ \text{standard error} & (0.846) & & (0.039) & & (0.063) & & (0.078) \end{array}$$

To illustrate the calculation of the standard errors, first note that, from Table 3.3, the residual standard deviation is $s = 1.525$. Using a statistical package, we have

$$(\mathbf{X}'\mathbf{X})^{-1} = \begin{pmatrix} 0.307975 & -0.004633 & -0.002131 & -0.020697 \\ -0.004633 & 0.000648 & 0.000143 & -0.000467 \\ -0.002131 & 0.000143 & 0.001724 & -0.000453 \\ -0.020697 & -0.000467 & -0.000453 & 0.002585 \end{pmatrix}.$$

To illustrate, we can compute $se(b_3) = s \times \sqrt{0.002585} = 0.078$, as earlier. Calculation of the standard errors, as well as the corresponding t-statistics, is part of the standard output from statistical software and need not be computed by users. Our purpose here is to illustrate the ideas underlying the routine calculations.

With this information, we can immediately compute t-ratios to check to see whether a coefficient associated with an individual variable is significantly different from zero. For example, the t-ratio for the LNINCOME variable is $t(b_3) = 0.494/0.078 = 6.3$. The interpretation is that b_3 is more than four standard errors above zero, and thus LNINCOME is an important variable in the model. More formally, we may be interested in testing the null hypothesis that $H_0 : \beta_3 = 0$ versus $H_0 : \beta_3 \neq 0$. At a 5% level of significance, the t-value is 1.96, because $df = 275 - (1 + 3) = 271$. We thus reject the null in favor of the alternative hypothesis, that logarithmic income (LNINCOME) is important in determining the logarithmic face amount.

3.4.2 Confidence Intervals

Confidence intervals for parameters provide another device for describing the strength of the contribution of the jth explanatory variable. The statistic b_j is a *point estimate* of the parameter β_j. To provide a range of reliability, we use the confidence interval

$$b_j \pm t_{n-(k+1),1-\alpha/2} se(b_j). \tag{3.10}$$

Here, the t-value $t_{n-(k+1),1-\alpha/2}$ is a percentile from the t-distribution with $df = n - (k + 1)$ degrees of freedom. We use the same t-value as in the two-sided hypothesis test. Indeed, there is a duality between the confidence interval and the two-sided hypothesis test. For example, it is not hard to check that if a hypothesized value falls outside the confidence interval, then H_0 will be rejected in favor of H_a. Further, knowledge of the p-value, point estimate, and standard error can be used to determine a confidence interval.

3.4.3 Added Variable Plots

To represent multivariate data graphically, we have seen that a scatterplot matrix is a useful device. However, the major shortcoming of the scatterplot matrix is that it captures relationships only between pairs of variables. When the data can be summarized using a regression model, a graphical device that does not have this shortcoming is an *added variable plot*. The added variable plot is also called a *partial regression plot* because, as we will see, it is constructed in terms of residuals from certain regression fits. We will also see that the added variable plot can be summarized in terms of a partial correlation coefficient, thus providing a link between correlation and regression. To introduce these ideas, we work in the context of the following example:

Table 3.7 Summary Statistics for Each Variable for 37 Refrigerators

Variable	Mean	Median	Standard Deviation	Minimum	Maximum
ECOST	70.51	68.00	9.14	60.00	94.00
RSIZE	13.400	13.200	0.600	12.600	14.700
FSIZE	5.184	5.100	0.938	4.100	7.400
SHELVES	2.514	2.000	1.121	1.000	5.000
FEATURES	3.459	3.000	2.512	1.000	12.000
PRICE	626.4	590.0	139.8	460.0	1200.0

Source: Consumer Reports, July 1992. "Refrigerators: A Comprehensive Guide to the Big White Box."

Table 3.8 Matrix of Correlation Coefficients

	ECOST	RSIZE	FSIZE	SHELVES	FEATURES
RSIZE	0.333				
FSIZE	0.855	0.235			
SHELVES	0.188	0.363	0.251		
FEATURES	0.334	0.096	0.439	0.160	
PRICE	0.522	0.024	0.720	0.400	0.697

Ⓡ **Empirical** *Filename is "Refrigerator"*

Example: Refrigerator Prices. What characteristics of a refrigerator are important in determining its price (PRICE)? We consider here several characteristics of a refrigerator, including the size of the refrigerator in cubic feet (RSIZE), the size of the freezer compartment in cubic feet (FSIZE), the average amount of money spent per year to operate the refrigerator (ECOST, for "energy cost"), the number of shelves in the refrigerator and freezer doors (SHELVES), and the number of features (FEATURES). The features variable includes shelves for cans, see-through crispers, ice makers, egg racks, and so on.

Both consumers and manufacturers are interested in models of refrigerator prices. Other things equal, consumers generally prefer larger refrigerators with lower energy costs that have more features. Because of forces of supply and demand, we would expect consumers to pay more for such refrigerators. A larger refrigerator with lower energy costs that has more features at the similar price is considered a bargain to the consumer. How much extra would the consumer be willing to pay for the additional space? A model of prices for refrigerators on the market provides some insight to this question.

To this end, we analyze data from $n = 37$ refrigerators. Table 3.7 provides the basic summary statistics for the response variable PRICE and the five explanatory variables. From this table, we see that the average refrigerator price is $\overline{y} =$ \$626.40, with standard deviation $s_y =$ \$139.80. Similarly, the average annual amount to operate a refrigerator, or average ECOST, is \$70.51.

To analyze relationships among pairs of variables, Table 3.8 provides a matrix of correlation coefficients. From the table, we see that there are strong linear relationships between PRICE and each of freezer space (FSIZE) and the number

Table 3.9 Fitted Refrigerator Price Model

	Coefficient Estimate	Standard Error	t-Ratio
Intercept	798	271.4	−2.9
ECOST	−6.96	2.275	−3.1
RSIZE	76.5	19.44	3.9
FSIZE	137	23.76	5.8
SHELVES	37.9	9.886	3.8
FEATURES	23.8	4.512	5.3

of FEATURES. Surprisingly, there is also a strong positive correlation between PRICE and ECOST. Recall that ECOST is the energy cost; one might expect that higher-priced refrigerators should enjoy lower energy costs.

A regression model was fit to the data. The fitted regression equation appears in Table 3.9, with $s = 60.65$ and $R^2 = 83.8\%$.

From Table 3.9, the explanatory variables seem to be useful predictors of refrigerator prices. Together, the variables account for 83.8% of the variability. To understand prices, the typical error has dropped from $s_y = \$139.80$ to $s = \$60.65$. The t-ratios for each of the explanatory variables exceeds 2 in absolute value, indicating that each variable is important on an individual basis.

What is surprising about the regression fit is the negative coefficient associated with energy cost. Remember, we can interpret $b_{ECOST} = -6.96$ to mean that, for each dollar increase in ECOST, we expect the PRICE to decrease by \$6.96. This negative relationship conforms to our economic intuition. However, it is surprising that the same dataset has shown us that there is a positive relationship between PRICE and ECOST. This seeming anomaly is because correlation measures relationships only between pairs of variables, though the regression fit can account for several variables simultaneously. To provide more insight into this seeming anomaly, we now introduce the *added variable plot*.

Producing an Added Variable Plot

The added variable plot provides additional links between the regression methodology and more fundamental tools such as scatter plots and correlations. We work in the context of the refrigerator price example to demonstrate the construction of this plot.

Procedure for producing an added variable plot.

(i) Run a regression of PRICE on RSIZE, FSIZE, SHELVES, and FEATURES, omitting ECOST. Compute the residuals from this regression, which we label e_1.

(ii) Run a regression of ECOST on RSIZE, FSIZE, SHELVES, and FEATURES. Compute the residuals from this regression, which we label e_2.

(iii) Plot e_1 versus e_2. This is the added variable plot of PRICE versus ECOST, controlling for the effects of the RSIZE, FSIZE, SHELVES, and FEATURES. This plot appears in Figure 3.4.

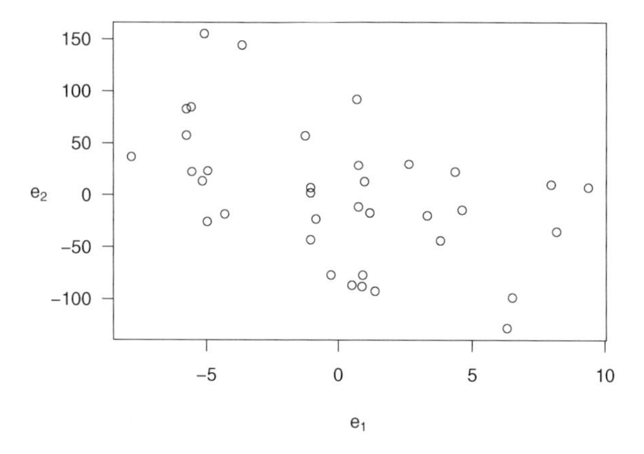

Figure 3.4 An added variable plot. The residuals from the regression of PRICE on the explanatory variables, omitting ECOST, are on the vertical axis. On the horizontal axis are the residuals from the regression fit of ECOST on the other explanatory variables. The correlation coefficient is −0.48.

The error ε can be interpreted as the natural variation in a sample. In many situations, this natural variation is small compared to the patterns evident in the nonrandom regression component. Thus, it is useful to think of the error, $\varepsilon_i = y_i - (\beta_0 + \beta_1 x_{i1} + \cdots + \beta_k x_{ik})$, as the response after controlling for the effects of the explanatory variables. In Section 3.3, we saw that a random error can be approximated by a residual, $e_i = y_i - (b_0 + b_1 x_{i1} + \cdots + b_k x_{ik})$. Thus, in the same way, we may think of a residual as the response after "controlling for" the effects of the explanatory variables.

With this in mind, we can interpret the vertical axis of Figure 3.4 as the refrigerator PRICE controlled for effects of RSIZE, FSIZE, SHELVES, and FEATURES. Similarly, we can interpret the horizontal axis as the ECOST controlled for effects of RSIZE, FSIZE, SHELVES, and FEATURES. The plot then provides a graphical representation of the relation between PRICE and ECOST, after controlling for the other explanatory variables. For comparison, a scatter plot of PRICE and ECOST (not shown here) does not control for other explanatory variables. Thus, it is possible that the positive relationship between PRICE and ECOST is due not to a causal relationship but rather to one or more additional variables that cause both variables to be large.

For example, from Table 3.7, we see that the freezer size (FSIZE) is positively correlated with both ECOST and PRICE. It certainly seems reasonable that increasing the size of a freezer would cause both the energy cost and the price to increase. Rather, the positive correlation may be because large values of FSIZE mean large values of both ECOST and PRICE.

Variables left out of a regression are called *omitted variables*. This omission could cause a serious problem in a regression model fit; regression coefficients could be not only strongly significant when they should not be but also of the incorrect sign. Selecting the proper set of variables to be included in the regression model is an important task; it is the subject of Chapters 5 and 6.

3.4.4 Partial Correlation Coefficients

As we saw in Chapter 2, a correlation statistic is a useful quantity for summarizing plots. The correlation for the added variable plot is called a *partial correlation coefficient*. It is defined to be the correlation between the residuals e_1 and e_2 and is denoted by $r(y, x_j|x_1, \ldots, x_{j-1}, x_{j+1}, \ldots, x_k)$. Because it summarizes an added variable plot, we may interpret $r(y, x_j|x_1, \ldots, x_{j-1}, x_{j+1}, \ldots, x_k))$ to be the correlation between y and x_j, in the presence of the other explanatory variables. To illustrate, the correlation between PRICE and ECOST in the presence of the other explanatory variables is –0.48.

The partial correlation coefficient can also be calculated using

$$r(y, x_j|x_1, \ldots, x_{j-1}, x_{j+1}, \ldots, x_k) = \frac{t(b_j)}{\sqrt{t(b_j)^2 + n - (k+1)}}. \quad (3.11)$$

Here, $t(b_j)$ is the t-ratio for b_j from a regression of y on $x_1, \ldots, x_k$ (including the variable x_j). An important aspect of equation (3.11) is that it allows us to calculate partial correlation coefficients running only one regression. For example, from Table 3.9, the partial correlation between PRICE and ECOST in the presence of the other explanatory variables is $(-3.1)/\sqrt{(-3.1)^2 + 37 - (5+1)} \approx -0.48$.

Calculation of partial correlation coefficients is quicker when using the relationship with the t-ratio, but may fail to detect nonlinear relationships. The information in Table 3.9 allows us to calculate all five partial correlation coefficients in the refrigerator price example after running only one regression. The three-step procedure for producing added variable plots requires ten regressions, two for each of the five explanatory variables. Of course, by producing added variable plots, we can detect nonlinear relationships that are missed by correlation coefficients.

Partial correlation coefficients provide another interpretation for t-ratios. Equation (3.11) shows how to calculate a correlation statistic from a t-ratio, thus providing another link between correlation and regression analysis. Moreover, from equation (3.11), we see that the larger is the t-ratio, the larger is the partial correlation coefficient. That is, a large t-ratio means that there is a large correlation between the response and the explanatory variable, controlling for other explanatory variables. This provides a partial response to the question that is regularly asked by consumers of regression analyses: Which variable is most important?

3.5 Some Special Explanatory Variables

The linear regression function is linear in the parameters but may be a highly nonlinear function of the explanatory variables.

The linear regression model is the basis of a rich family of models. This section provides several examples to illustrate the richness of this family. These examples demonstrate the use of (i) binary variables, (ii) transformation of explanatory variables, and (iii) interaction terms. This section also serves to underscore the meaning of the adjective *linear* in the phrase "linear regression"; the regression function is linear in the parameters but may be a highly nonlinear function of the explanatory variables.

3.5.1 Binary Variables

Categorical variables provide a numerical label for measurements of observations that fall in distinct groups, or *categories*. Because of the grouping, categorical

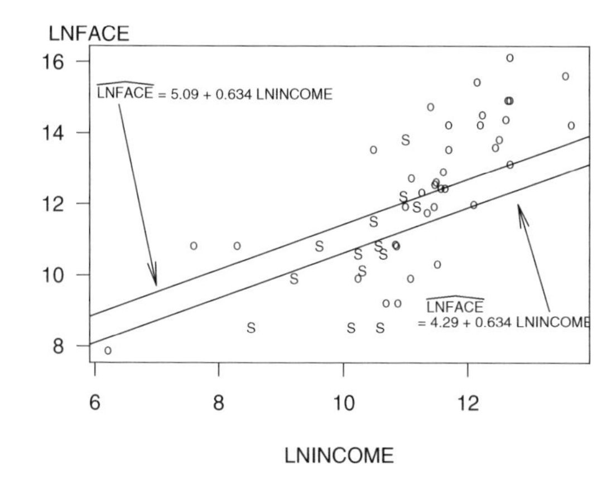

Figure 3.5 Letter plot of LNFACE versus LNINCOME, with the letter code "S" for single and "o" for other. The fitted regression lines have been superimposed. The lower line is for single and the upper line is for other.

variables are discrete and generally take on a finite number of values. We begin our discussion with a categorical variable that can take on one of only two values, a *binary* variable. Further discussion of categorical variables is the topic of Chapter 4.

Example: Term Life Insurance, Continued. We now consider the marital status of the survey respondent. In the Survey of Consumer Finances, respondents can select among several options describing their marital status including "married," "living with a partner," "divorced," and so on. Marital status is not measured continuously but rather takes on values that fall into distinct groups. In this chapter, we group survey respondents according to whether they are single, defined to include those who are separated, divorced, widowed, never married, and are not married nor living with a partner. Chapter 4 will present a more complete analysis of marital status by including additional categories.

Binary explanatory variables are also known as indicator and dummy variables.

The binary variable SINGLE is defined as 1 if the survey respondent is single and 0 otherwise. The variable SINGLE is also known as an *indicator* variable because it indicates whether the respondent is single. Another name for this important type of variable is a *dummy* variable. We could use 0 and 100, or 20 and 36, or any other distinct values. However, 0 and 1 are convenient for the interpretation of the parameter values, discussed subsequently. To streamline the discussion, we now present a model using only LNINCOME and SINGLE as explanatory variables.

For our sample of $n = 275$ households, 57 are single and the other 218 are not. To see the relationships among LNFACE, LNINCOME, and SINGLE, Figure 3.5 introduces a *letter plot* of LNFACE versus LNINCOME, with SINGLE as the code variable. We can see that Figure 3.5 is a scatter plot of LNFACE versus LNINCOME, using 50 randomly selected households from our sample of 275 (for clarity of the graph). However, instead of using the same plotting symbol for each observation, we have coded the symbols so that we can easily understand the behavior of a third variable, SINGLE. In other applications, you may elect to use other plotting symbols such as ♣, ♡, ♠, and so on, or to use different colors, to encode additional information. For this application, the letter codes "S" for single and "o" for other were selected because they remind the reader of the plot of the nature of the coding scheme. Regardless of the coding scheme, the important point is that a letter plot is a useful device for graphically portraying three or more variables in two dimensions. The main restriction is that the additional information must be categorized, such as with binary variables, to make the coding scheme work.

Figure 3.5 suggests that LNFACE is lower for those who are single than for others for a given level of income. Thus, we now consider a regression model, $LNFACE = \beta_0 + \beta_1 LNINCOME + \beta_2 SINGLE + \varepsilon$. The regression function can be written as

$$\mathrm{E}\, y = \begin{cases} \beta_0 + \beta_1 \mathrm{LNINCOME} & \text{for other respondents} \\ \beta_0 + \beta_2 + \beta_1 \mathrm{LNINCOME} & \text{for single respondents} \end{cases}.$$

The interpretation of the model coefficients differs from the continuous variable case. For continuous variables such as LNINCOME, we interpret β_1 as the expected change in y per unit change of logarithmic income, holding other variables fixed. For binary variables such as SINGLE, we interpret β_2 as the expected increase in y when going from the base level of SINGLE (= 0) to the alternative level. Thus, although we have one model for both marital statuses, we can interpret the model using two regression equations, one for each type of marital status. By writing a separate equation for each marital status, we have been able to simplify a complicated multiple regression equation. Sometimes, you will find that it is easier to communicate a series of simple relationships than a single, complex relationship.

Although the interpretation for binary explanatory variables differs from the continuous, the ordinary least squares estimation method remains valid. To illustrate, the fitted version of the preceding model is

$$\begin{array}{llll} \widehat{LNFACE} & = 5.09 \;+ & 0.634\ \mathrm{LNINCOME} & -\ 0.800\ \mathrm{SINGLE}\,. \\ \text{standard error} & \quad (0.89) & \quad (0.078) & \qquad (0.248) \end{array}$$

To interpret $b_2 = -0.800$, we say that we expect the logarithmic face to be smaller by 0.80 for a survey respondent who is single compared to the other category. This assumes that other things, such as income, remain unchanged. For a graphical interpretation, the two fitted regression lines are superimposed in Figure 3.5.

3.5.2 Transforming Explanatory Variables

Regression models have the ability to represent complex, *nonlinear* relationships between the expected response and the explanatory variables. For example, early regression texts, such as Plackett (1960, chapter 6) devote an entire chapter of material to polynomial regression,

$$\mathrm{E}\, y = \beta_0 + \beta_1 x + \beta_2 x^2 + \cdots + \beta_p x^p. \tag{3.12}$$

Here, the idea is that a pth order polynomial in x can be used to approximate general, unknown nonlinear functions of x.

The modern-day treatment of polynomial regression does not require an entire chapter because the model in equation (3.12) can be expressed as a special case of the linear regression model. That is, with the regression function in equation (3.5), $\mathrm{E}\ y = \beta_0 + \beta_1 x_1 + \beta_2 x_2 + \cdots + \beta_k x_k$, we can choose $k = p$ and $x_1 = x, x_2 = x^2, \ldots, x_p = x^p$. Thus, with these choices of explanatory variables, we can model a highly nonlinear function of x.

We are not restricted to powers of x in our choice of transformations. For example, the model $\mathrm{E}\ y = \beta_0 + \beta_1 \ln x$, provides another way to represent a gently sloping curve in x. This model can be written as a special case of the basic linear regression model using $x^* = \ln x$ as the transformed version of x.

Transformations of explanatory variables need not be smooth functions. To illustrate, in some applications, it is useful to categorize a continuous explanatory variable. For example, suppose that x represents the number of years of education, ranging from 0 to 17. If we are relying on information self-reported by our sample of senior citizens, there may be a substantial amount of error in the measurement of x. We could elect to use a less informative but more reliable transform of x such as x^*, a binary variable for finishing 13 years of school (finishing high school). Formally, we would code $x^* = 1$ if $x \geq 13$ and $x^* = 0$ if $x < 13$.

Thus, there are several ways that nonlinear functions of the explanatory variables can be used in the regression model. An example of a nonlinear regression model is $y = \beta_0 + \exp(\beta_1 x) + \varepsilon$. These typically arise in science applications of regressions where there are fundamental scientific principles guiding the complex model development.

3.5.3 Interaction Terms

We have so far discussed how explanatory variables, say, x_1 and x_2, affect the mean response in an additive fashion, that is, $\mathrm{E}\ y = \beta_0 + \beta_1 x_1 + \beta_2 x_2$. Here, we expect y to increase by β_1 per unit increase in x_1, with x_2 held fixed. What if the marginal rate of increase of $\mathrm{E}\ y$ differs for high values of x_2 when compared to low values of x_2? One way to represent this is to create an *interaction variable* $x_3 = x_1 \times x_2$ and consider the model $\mathrm{E}\ y = \beta_0 + \beta_1 x_1 + \beta_2 x_2 + \beta_3 x_3$.

With this model, the change in the expected y per unit change in x_1 now depends on x_2. Formally, we can assess small changes in the regression function as

$$\frac{\partial\, \mathrm{E} y}{\partial x_1} = \frac{\partial}{\partial x_1}(\beta_0 + \beta_1 x_1 + \beta_2 x_2 + \beta_3 x_1 x_2) = \beta_1 + \beta_3 x_2.$$

In this way, we may allow for more complicated functions of x_1 and x_2. Figure 3.6 illustrates this complex structure. From this figure and the preceding calculations, we see that the partial changes of $\mathrm{E}\ y$ due to movement of x_1 depend on the value of x_2. In this way, we say that the partial changes due to each variable are not unrelated but rather "move together."

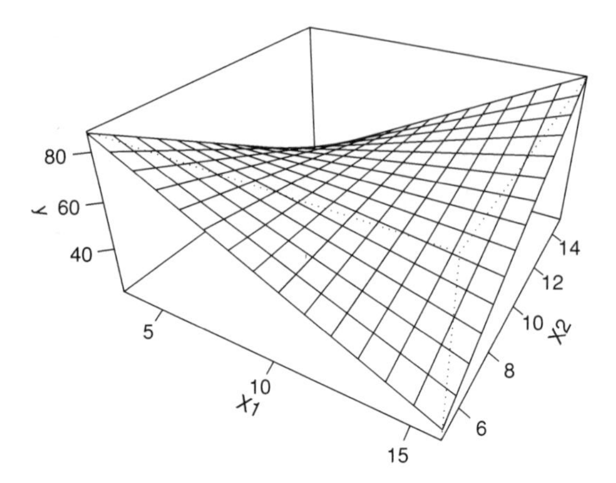

Figure 3.6 Plot of $\mathrm{E}\, y = \beta_0 + \beta_1 x_1 + \beta_2 x_2 + \beta_3 x_1 x_2$ versus x_1 and x_2.

More generally, an interaction term is a variable that is created as a nonlinear function of two or more explanatory variables. These special terms, even though permitting us to explore a rich family of nonlinear functions, can be cast as special cases of the linear regression model. To do this, we simply create the variable of interest and treat this new term as another explanatory variable. Of course, not every variable that we create will be useful. In some instances, the created variable will be so similar to variables already in our model that it will provide no new information. Fortunately, we can use t-tests to check whether the new variable is useful. Further, Chapter 4 will introduce a test to decide whether a group of variables is useful.

The function that we use to create an interaction variable must be more than just a linear combination of other explanatory variables. For example, if we use $x_3 = x_1 + x_2$, we will not be able to estimate all of the parameters. Chapter 5 will introduce some techniques to help avoid situations when one variable is a linear combination of the others.

To give you some exposure to the wide variety of potential applications of special explanatory variables, we now present a series of short examples.

Example: Term Life Insurance, Continued. How do we interpret the interaction of a binary variable with a continuous variable? To illustrate, consider a Term Life regression model, $LNFACE = \beta_0 + \beta_1 \mathrm{LNINCOME} + \beta_2 \mathrm{SINGLE} + \beta_2 \mathrm{LNINCOME*SINGLE} + \varepsilon$. In this model, we have created a third explanatory variable through the interaction of LNINCOME and SINGLE. The regression function can be written as:

$$\mathrm{E}\, y = \begin{cases} \beta_0 + \beta_1 \mathrm{LNINCOME} & \text{for other respondents} \\ \beta_0 + \beta_2 + (\beta_1 + \beta_3)\mathrm{LNINCOME} & \text{for single respondents} \end{cases}.$$

Thus, through this single model with four parameters, we can create two separate regression lines, one for those single and one for others. Figure 3.7 shows the two fitted regression lines for our data.

Table 3.10 Twenty-Three Regression Coefficients from an Expense Cost Model

Variable	Variable: Baseline ($D = 0$)	Variable: Interaction with ($D = 1$)	Variable Squared: Baseline ($D = 0$)	Variable Squared: Interaction with ($D = 1$)
Number of life policies issued (x_1)	−0.454	0.152	0.032	−0.007
Amount of term life insurance sold (x_2)	0.112	−0.206	0.002	0.005
Amount of whole life insurance sold (x_3)	−0.184	0.173	0.008	−0.007
Total annuity considerations (x_4)	0.098	−0.169	−0.003	0.009
Total accident and health premiums (x_5)	−0.171	0.014	0.010	0.002
Intercept	7.726			
Price of labor (PL)	0.553			
Price of capital (PC)	0.102			

Note: x_1 through x_5 are in logarithmic units.
Source: Segal (2002).

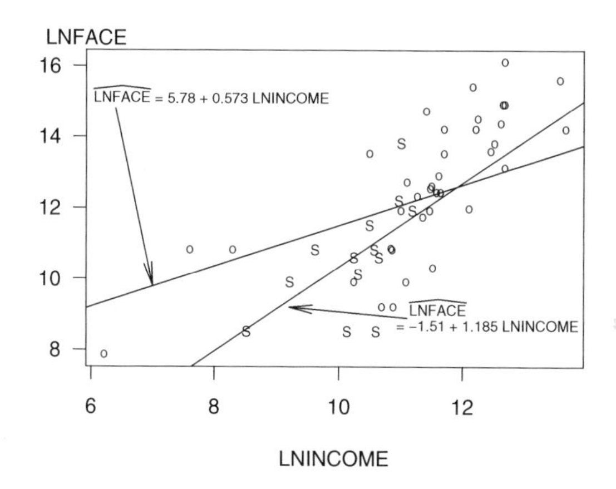

Figure 3.7 Letter plot of LNFACE versus LNINCOME, with the letter code "S" for single and "o" for other. The fitted regression lines have been superimposed. The lower line is for single and the upper line is for other.

Example: Life Insurance Company Expenses. In a well-developed life insurance industry, minimizing expenses is critical for a company's competitive position. Segal (2002) analyzed annual accounting data from more than 100 firms for the period 1995–1998, inclusive, using a data base from the National Association of Insurance Commissioners (NAIC) and other reported information. Segal modeled overall company expenses as a function of firm outputs and the price of inputs. The outputs consist of insurance production, measured by x_1 through x_5, described in Table 3.10. Segal also considered the square of each output, as well as an interaction term with a dummy/binary variable D that indicates whether or not the firm uses a branch company to distribute its products. (In a branch company, field managers are company employees, not independent agents.)

For the price inputs, the price of labor (PL) is defined to be the total cost of employees and agents divided by their number, in logarithmic units. The price of capital (PC) is approximated by the ratio of capital expense to the number of employees and agents, also in logarithmic units. The price of materials consists of expenses other than labor and capital divided by the number of policies sold and terminated during the year. It does not appear directly as an explanatory variable. Rather, Segal took the dependent variable (y) to be total company expenses divided by the price of materials, again in logarithmic units.

With these variable definitions, Segal estimated the following regression function:

$$\mathrm{E}\, y = \beta_0 + \sum_{j=1}^{5} \left(\beta_j x_j + \beta_{j+5} D x_j + \beta_{j+10} x_j^2 + \beta_{j+15} D x_j^2\right) + \beta_{21} PL + \beta_{22} PC.$$

The parameter estimates appear in Table 3.10. For example, the marginal change in $\mathrm{E}\, y$ per unit change in x_1 is

$$\frac{\partial\, \mathrm{E} y}{\partial x_1} = \beta_1 + \beta_6 D + 2\beta_{11} x_1 + 2\beta_{16} D x_1,$$

which is estimated as $-0.454 + 0.152D + (0.064 - 0.014D)x_1$. For these data, the median number of policies issued was $x_1 = 15,944$. At this value of x_1, the estimated marginal change is $-0.454 + 0.152D + (0.064 - 0.014D)$ $\ln(15944) = 0.165 + 0.017D$, or 0.165 for baseline ($D = 0$) and 0.182 for branch ($D = 1$) companies.

These estimates are elasticities, as defined in Section 3.2.2. To interpret these coefficients further, let COST represent total general company expenses and NUMPOL represent the number of life policies issued. Then, for branch ($D = 1$) companies, we have

$$0.182 \approx \frac{\partial y}{\partial x_1} = \frac{\partial \ln COST}{\partial \ln NUMPOL} = \frac{\frac{\partial\, COST}{\partial\, NUMPOL}}{\frac{COST}{NUMPOL}},$$

or $\frac{\partial\, COST}{\partial\, NUMPOL} \approx 0.182 \frac{COST}{NUMPOL}$. The median cost is \$15,992,000, so the marginal cost per policy at these median values is $0.182 \times (15992000/15944) = \$182.55$.

Special Case: Curvilinear Response Functions. We can expand the polynomial functions of an explanatory variable to include several explanatory variables. For example, the expected response, or *response function*, for a second-order model with two explanatory variables is

$$\mathrm{E}y = \beta_0 + \beta_1 x_1 + \beta_2 x_2 + \beta_{11} x_1^2 + \beta_{22} x_2^2 + \beta_{12} x_1 x_2.$$

Figure 3.8 illustrates this response function. Similarly, the response function for a second-order model with three explanatory variables is

$$\mathrm{E}y = \beta_0 + \beta_1 x_1 + \beta_2 x_2 + \beta_3 x_3 + \beta_{11} x_1^2 + \beta_{22} x_2^2 + \beta_{33} x_3^2 + \beta_{12} x_1 x_2 \\ + \beta_{13} x_1 x_3 + \beta_{23} x_2 x_3.$$

When there is more than one explanatory variable, third- and higher-order models are rarely used in applications.

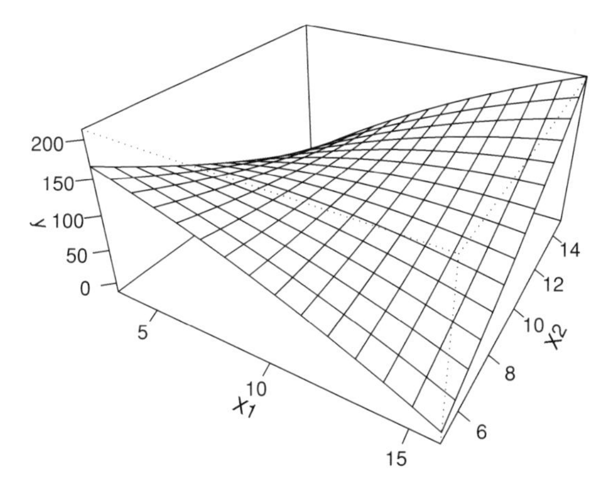

Figure 3.8 Plot of $\mathrm{E}\ y = \beta_0 + \beta_1 x_1 + \beta_2 x_2 + \beta_{11} x_1^2 + \beta_{22} x_2^2 + \beta_{12} x_1 x_2$ versus x_1 and x_2.

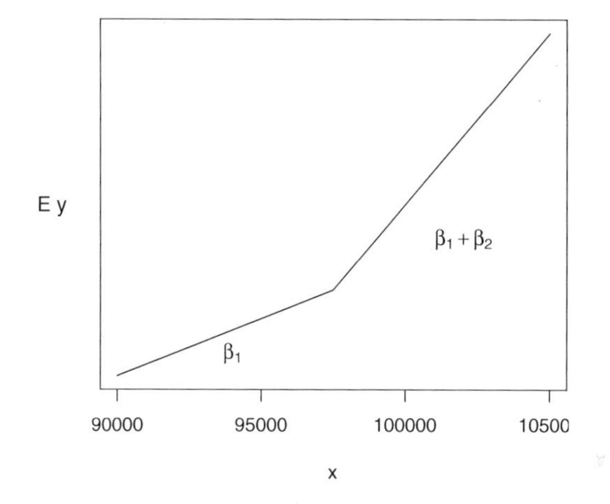

Figure 3.9 The marginal change in E y is lower below \$97,500. The parameter β_2 represents the difference in the slopes.

Special Case: Nonlinear Functions of a Continuous Variable. In some applications, we expect the response to have some abrupt changes in behavior at certain values of an explanatory variable, even if the variable is continuous. For example, suppose that we are trying to model an individual's charitable contributions (y) in terms of his or her wages (x). For 2007 data, a simple model we might entertain is given in Figure 3.9.

A rational for this model is that, in 2007, individuals paid 7.65% of their income for Social Security taxes up to \$97,500. No social security taxes are excised on wages in excess of \$97,500. Thus, one theory is that, for wages in excess of \$97,500, individuals have more disposal income per dollar and thus should be more willing to make charitable contributions.

To model this relationship, define the binary variable z to be zero if $x < 97{,}500$ and to be one if $x \geq 97{,}500$. Define the regression function to be $\mathrm{E}\ y = \beta_0 + \beta_1 x + \beta_2 z(x - 97{,}500)$. This can be written as

$$\mathrm{E}\ y = \begin{cases} \beta_0 + \beta_1 x & x < 97{,}500 \\ \beta_0 - \beta_2(97{,}500) + (\beta_1 + \beta_2)x & x \geq 97{,}500 \end{cases}.$$

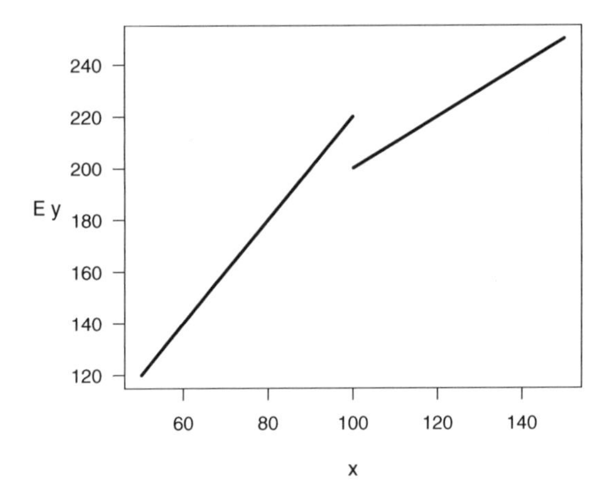

Figure 3.10 Plot of expected commissions (E y) versus number of shares traded (x). The break at $x = 100$ reflects savings in administrative expenses. The lower slope for $x \geq 100$ reflects economies of scales in expenses.

To estimate this model, we would run a regression of y on two explanatory variables, $x_1 = x$ and $x_2 = z \times (x - 97{,}500)$. If $\beta_2 > 0$, then the marginal rate of charitable contributions is higher for incomes exceeding \$97,500.

Figure 3.9 illustrates this relationship, known as *piecewise linear regression* or sometimes a "broken-stick" model. The sharp break in Figure 3.9 at $x = 97{,}500$ is called a *kink*. We have linear relationships above and below the kinks and have used a binary variable to put the two pieces together. We are not restricted to one kink. For example, suppose that we want to do a historical study of federal taxable income for 1992 single filers. Then, there were three tax brackets: the marginal tax rate below \$21,450 was 15%, above \$51,900 was 31%, and in between was 28 percent. For this example, we would use two kinks, at 21,450 and 51,900.

Further, piecewise linear regression is not restricted to continuous response functions. For example, suppose that we are studying the commissions paid to stockbrokers (y) in terms of the number of shares purchased by a client (x). We might expect to see the relationship illustrated in Figure 3.10. Here, the discontinuity at $x = 100$ reflects the administrative expenses of trading in odd lots, as trades of less than 100 shares are called. The lower marginal cost for trades in excess of 100 shares simply reflects the economies of scale for doing business in larger volumes. A regression model of this is $\mathrm{E}\, y = \beta_0 + \beta_1 x + \beta_2 z + \beta_3 z x$, where $z = 0$ if $x < 100$ and $z = 1$ if $x \geq 100$. The regression function depicted in Figure 3.10 is

$$\mathrm{E}\, y = \begin{cases} \beta_0 + \beta_1 x_1 & x < 100 \\ \beta_0 + \beta_2 + (\beta_1 + \beta_3) x_1 & x \geq 100 \end{cases}.$$

3.6 Further Reading and References

For proofs of the Chapter 3 results, see Goldberger (1991). Nonlinear regression models are discussed in, for example, Bates and Watts (1988).

Chapter 3 has introduced the fundamentals of multiple linear regression. Chapter 4 will widen the scope by introducing categorical variables and statistical inference methods for handling several coefficients simultaneously. Chapter 5 will introduce techniques to help you pick appropriate variables in a multiple linear regression model. Chapter 6 is a synthesis chapter, discussing model interpretation, variable selection, and data collection.

Chapter References

Bates, Douglas M., and D. G. Watts (1988). *Nonlinear Regression Analysis and its Applications*. John Wiley and Sons, New York.

Lemaire, Jean (2002). Why do females live longer than males? *North American Actuarial Journal* 6, no. 4, 21–37.

Goldberger, Arthur (1991). *A Course in Econometrics*. Harvard University Press, Cambridge, Massachusetts.

Plackett, R. L. (1960). *Regression Analysis*. Clarendon Press, Oxford, U.K.

Segal, Dan (2002). An economic analysis of life insurance company expenses. *North American Actuarial Journal* 6, no. 4, 81–94.

3.7 Exercises

3.1. Consider a fictitious dataset of $n = 100$ observations with $s_y = 80$. We run a regression with three explanatory variables to get $s = 50$.

a. Calculate the adjusted coefficient of determination, R_a^2.

b. Complete the ANOVA table.

ANOVA Table

Source	Sum of Squares	df	Mean Square
Regression			
Error			
Total			

c. Calculate the (unadjusted) coefficient of determination, R^2.

3.2. Consider a fictitious dataset of $n = 100$ observations with $s_y = 80$. We run a regression with three explanatory variables to get $s = 50$. We also get

$$(\mathbf{X}'\mathbf{X})^{-1} = \begin{pmatrix} 100 & 20 & 20 & 20 \\ 20 & 90 & 30 & 40 \\ 20 & 30 & 80 & 50 \\ 20 & 40 & 50 & 70 \end{pmatrix}.$$

a. Determine the standard error of b_3, $se(b_3)$.

b. Determine the estimated covariance between b_2 and b_3.

c. Determine the estimated correlation between b_2 and b_3.

d. Determine the estimated variance of $4b_2 + 3b_3$.

3.3. Consider the following small fictitious dataset. You will be fitting a regression model to y using two explanatory variables, x_1 and x_2.

i	1	2	3	4
$x_{i,1}$	−1	2	4	6
$x_{i,2}$	0	0	1	1
y_i	0	1	5	8

From the fitted regression model, we have $s = 1.373$ and

$$\mathbf{b} = \begin{pmatrix} 0.15 \\ 0.692 \\ 2.88 \end{pmatrix} \text{ and } (\mathbf{X'X})^{-1} = \begin{pmatrix} 0.53846 & -0.07692 & -0.15385 \\ -0.07692 & 0.15385 & -0.69231 \\ -0.15385 & -0.69231 & 4.11538 \end{pmatrix}.$$

a. Write down the vector of dependent variables, $\mathbf{y}$, and the matrix of explanatory variables, $\mathbf{X}$.

b. Determine the numerical value for $\widehat{y}_3$, the fitted value for the third observation.

c. Determine the numerical value for $se(b_2)$.

d. Determine the numerical value for $t(b_1)$.

Ⓡ **EMPIRICAL**
Filename is "WiscLottery"

3.4. **Wisconsin Lottery.** Section 2.1 described a sample of $n = 50$ geographic areas (Zip codes) containing sales data on the Wisconsin state lottery ($y = SALES$). In that section, sales were analyzed using a basic linear regression model with $x = POP$, the area population, as the explanatory variable. This exercise extends that analysis by introducing additional explanatory variables given in Table 3.11.

Table 3.11 Lottery, Economic, and Demographic Characteristics of 50 Wisconsin Zip Codes

Lottery Characteristics	
SALES	Online Lottery Sales to Individual Consumers
Economic and demographic characteristics	
PERPERHH	Persons per household
MEDSCHYR	Median years of schooling
MEDHVL	Median home value in $1000s for owner-occupied homes
PRCRENT	Percentage of housing that is renter occupied
PRC55P	Percentage of population that is 55 or older
HHMEDAGE	Household median age
MEDINC	Estimated median household income, in $1000s
POP	Population, in thousands

a. Produce a table of summary statistics for all variables. One Zip code (observation 11, Zip = 53211, Shorewood, Wisconsin, a suburb of Milwaukee) appears to have unusually large values of MEDSCHYR and MEDHVL. For this observation, how many standard deviations is the value of MEDSCHYR above the mean? For this observation, how many standard deviations is the value of MEDHVL above the mean?

b. Produce a table of correlations. What three variables are most highly correlated with SALES?
c. Produce a scatterplot matrix of all explanatory variables and SALES. In the plot of MEDSCHYR versus SALES, describe the position of observation 11.
d. Fit a linear model of SALES on all eight explanatory variables. Summarize the fit of this model by citing the residual standard deviation, s; the coefficient of determination, R^2; and its adjusted version, R_a^2.
e. Based on your part (d) model fit, is MEDSCHYR a statistically significant variable? To respond to this question, use a formal test of hypothesis. State your null and alternative hypotheses, decision-making criterion, and decision-making rule.
f. Now fit a more parsimonious model, using SALES as the dependent variable and MEDSCHYR, MEDHVL, and POP as explanatory variables. Summarize the fit of this model by citing the residual standard deviation, s; the coefficient of determination, R^2; and its adjusted version, R_a^2. How do these values compare to the model fit in part (d)?
g. Note that the sign of the regression coefficient associated with MEDSCHYR is now negative. To help interpret this coefficient, compute the corresponding partial correlation coefficient. What is the interpretation of this coefficient?
h. To get further insights into the relation between MEDSCHYR and SALES, produce an added variable plot controlling for the effects of MEDHVL and POP. Check that the correlation associated with this plot agrees with your answer in part (g).
i. Rerun the regression in part (f), after removing observation 11. Cite the basic summary statistics from this regression. For this model fit, is MEDSCHYR a statistically significant variable? To respond to this question, use a formal test of hypothesis. State your null and alternative hypotheses, decision-making criterion, and decision-making rule.
j. Rerun the regression in part (f), after removing observation 9. Cite the basic summary statistics from this regression.

® **EMPIRICAL** *Filename is "NAICExpense"*

3.5. **Insurance Company Expenses.** This exercise considers insurance company data from the NAIC and described in Exercise 1.6.

Table 3.12 describes several variables that can be used to explain expenses. As with Segal's (2002) study of life insurers, firm "outputs" consist of premiums written (for property and casualty, these are subdivided into personal and commercial lines) and losses (subdivided into short and long tail lines). ASSETS and CASH are commonly used measures of the size of a company. GROUP, STOCK, and MUTUAL describe the organizational structure. Firm "inputs" were gathered from the Bureau of Labor Statistics (BLS, from the Occupational Employee Statistics program). WAGESTAFF is calculated as the average wage in the state where the insurance company is headquartered. AGENTWAGE is calculated as the weighted average of annual wages of the brokerage industry, weighted by the percentage of gross premium written in each state.

Table 3.12 Insurer Expense Variables

Variable	Description
NAIC Variables	
EXPENSES	Total expenses incurred, in millions of dollars
LOSSLONG	Losses incurred for long tail lines, in millions of dollars
LOSSSHORT	Losses incurred for short tail lines, in millions of dollars
GPWPERSONAL	Gross premium written for personal lines, in millions of dollars
GPWCOMM	Gross premium written for commercial lines, in millions of dollars
ASSETS	Net admitted assets, in millions of dollars
CASH	Cash and invested assets, in millions of dollars
GROUP	Indicates whether the company is affiliated
STOCK	Indicates whether the company is a stock company
MUTUAL	indicates whether the company is a mutual company
BLS Variables	
STAFFWAGE	Annual average wage of the insurer's administrative staff, in thousands of dollars
AGENTWAGE	Annual average wage of the insurance agent, in thousands of dollars

A preliminary inspection of the data showed that many firms did not report any insurance losses incurred in 2005. For this exercise, we consider the 384 companies with some losses in the file "NAICExpense."

a. Produce summary statistics of the response variable and the (nonbinary) explanatory variables. Note the pattern of skewness for each variable. Note that many variables have negative values.
b. Transform each nonbinary variable through the modified logarithm transform, $\ln(1+x)$. Produce summary statistics of these modified nonbinary explanatory variables. Let LNEXPENSES ($= \ln(1+$ EXPENSES)) denote the modified expense variable.

   For subsequent analysis, use only the modified variables described in part (b).
c. Produce a table of correlations for the nonbinary variables. What three variables are most highly correlated with LNEXPENSES?
d. Provide a boxplot of LNEXPENSES by level of GROUP. Which level of group has higher expenses?
e. Fit a linear model of LNEXPENSES on all eleven explanatory variables. Summarize the fit of this model by citing the residual standard deviation, s; the coefficient of determination, R^2; and its adjusted version, R_a^2.
f. Fit a linear model of LNEXPENSES on a reduced model using eight explanatory variables, dropping CASH, STOCK, and MUTUAL. For the explanatory variables, include assets, GROUP, both versions of losses and gross premiums, as well as the two BLSvariables.

f(i). Summarize the fit of this model by citing s, R^2, and R_a^2.

f(ii). Interpret the coefficient associated with commercial lines gross premiums on the logarithmic scale.

f(iii). Suppose that GPWCOMM increases by $1.00, how much do we expect EXPENSES to increase? Use your answer in part f(ii) and median values of GPWCOMM and EXPENSES for this question.

g. Square each of the two loss and the two gross premium variables. Fit a linear model of LNEXPENSES on a reduced model using twelve explanatory variables, the eight variables in part (f), and the four additional squared terms just created.

g(i). Summarize the fit of this model by citing s, R^2, and R_a^2.

g(ii). Do the quadratic variables appear to be useful explanatory variables?

h. Now omit the two BLS variables, so you are fitting a model of LNEXPENSES on assets, GROUP, both versions of losses and gross premiums, as well as quadratic terms. Summarize the fit of this model by citing s, R^2, and R_a^2. Comment on the number of observations used to fit this model compared to part (f).

i. Drop the quadratic terms in part (g) and add interaction terms with the dummy variable GROUP. Thus, there are now 11 variables, assets, GROUP, both versions of losses and gross premiums, as well as interactions of GROUP with assets and both versions of losses and gross premiums.

i(i). Summarize the fit of this model by citing s, R^2, and R_a^2.

i(ii). Suppose that GPWCOMM increases by $1.00, how much do we expect EXPENSES to increase for GROUP $= 0$ companies? Use the median values of GPWCOMM and EXPENSES of GROUP $= 0$ companies for this question.

i(iii). Suppose that GPWCOMM increases by $1.00, how much do we expect EXPENSES to increase for GROUP $= 1$ companies? Use the median values of GPWCOMM and EXPENSES of GROUP $= 1$ companies for this question.

Ⓡ EMPIRICAL
Filename is "UNLifeExpectancy"

3.6. **National Life Expectancies.** We continue the analysis begun in Exercises 1.7 and 2.22. Now fit a regression model on LIFEEXP using three explanatory variables, FERTILITY, PUBLICEDUCATION, and lnHEALTH (the natural logarithmic transform of PRIVATEHEALTH).

a. Interpret the regression coefficient associated with public education.

b. Interpret the regression coefficient associated with health expenditures without using the logarithmic scale for expenditures.

c. Based on the model fit, is PUBLICEDUCATION a statistically significant variable? To respond to this question, use a formal test of hypothesis. State your null and alternative hypotheses, decision-making criterion, and decision-making rule.

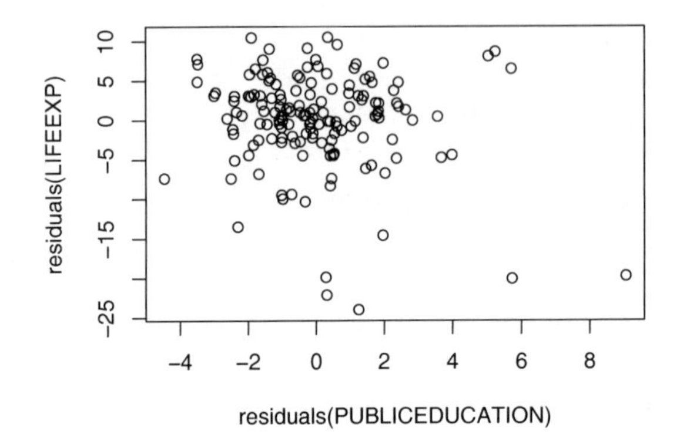

Figure 3.11 Added variable plot of PUBLICEDUCATION versus LIFEEXP, controlling for FERTILITY and lnHEALTH.

d. The negative sign of the PUBLICEDUCATION coefficient is surprising, in that the sign of the correlation between PUBLICEDUCATION and LIFEEXP is positive and intuition suggests a positive relation. To check this result, an added variable plot appears in Figure 3.11.
   - d(i). For an added variable plot, describe its purpose and a method for producing it.
   - d(ii). Calculate the correlation corresponding to the added variable plot that appears in Figure 3.11.

4

Multiple Linear Regression – II

Chapter Preview. This chapter extends the discussion of multiple linear regression by introducing statistical inference for handling several coefficients simultaneously. To motivate this extension, this chapter considers coefficients associated with *categorical variables*. These variables allow us to group observations into distinct categories. This chapter shows how to incorporate categorical variables into regression functions using binary variables, thus widening the scope of potential applications. Statistical inference for several coefficients allows analysts to make decisions about categorical variables and other important applications. Categorical explanatory variables also provide the basis for an ANOVA model, a special type of regression model that permits easier analysis and interpretation.

4.1 The Role of Binary Variables

Categorical variables provide labels for observations to denote membership in distinct groups, or categories. A binary variable is a special case of a categorical variable. To illustrate, a binary variable may tell us whether someone has health insurance. A categorical variable could tell us whether someone has

Categorical variables provide labels for observations to denote membership in distinct groups, or categories.

- Private group insurance (offered by employers and associations),
- Private individual health insurance (through insurance companies),
- Public insurance (e.g., Medicare or Medicaid) or
- No health insurance.

For categorical variables, there may or may not be an ordering of the groups. For health insurance, it is difficult to order these four categories and say which is larger. In contrast, for education, we might group individuals into "low," "intermediate," and "high" years of education. In this case, there is an ordering among groups based on level of educational achievement. As we will see, this ordering may or may not provide information about the dependent variable. *Factor* is another term used for an unordered categorical explanatory variable.

Factor *is another term used for an unordered categorical explanatory variable.*

For ordered categorical variables, analysts typically assign a numerical score to each outcome and treat the variable as if it were continuous. For example, if we had three levels of education, we might employ ranks and use

$$\text{EDUCATION} = \begin{cases} 1 & \text{for low education} \\ 2 & \text{for intermediate education} \\ 3 & \text{for high education.} \end{cases}$$

An alternative would be to use a numerical score that approximates an underlying value of the category. For example, we might use

$$\text{EDUCATION} = \begin{cases} 6 & \text{for low education} \\ 10 & \text{for intermediate education} \\ 14 & \text{for high education.} \end{cases}$$

This gives the approximate number of years of schooling that individuals in each category completed.

The assignment of numerical scores and treating the variable as continuous has important implications for the regression modeling interpretation. Recall that the regression coefficient is the marginal change in the expected response; in this case, the β for education assesses the increase in E y per unit change in EDUCATION. If we record EDUCATION as a rank in a regression model, then the β for education corresponds to the increase in E y moving from EDUCATION $= 1$ to EDUCATION $= 2$ (from low to intermediate); this increase is the same as moving from EDUCATION $= 2$ to EDUCATION $= 3$ (from intermediate to high). Do we want to model this increase as the same? This is an assumption that the analyst makes with this coding of EDUCATION; it may or may not be valid, but it certainly needs to be recognized.

Because of this interpretation of coefficients, analysts rarely use ranks or other numerical scores to summarize *unordered* categorical variables. The most direct way to handle factors in regression is through the use of binary variables. A categorical variable with c levels can be represented using c binary variables, one for each category. For example, suppose that we were uncertain about the direction of the education effect and so decide to treat it as a factor. Then, we could code $c = 3$ binary variables: (1) a variable to indicate low education, (2) one to indicate intermediate education, and (3) one to indicate high education. These binary variables are often known as *dummy variables*. In regression analysis with an intercept term, we use only $c - 1$ of these binary variables; the remaining variable enters implicitly through the intercept term. By identifying a variable as a factor, most statistical software packages will automatically create binary variables for you.

In a linear regression model with an intercept, use c − 1 binary variables to represent a factor with c levels.

Through the use of binary variables, we do not make use of the ordering of categories within a factor. Because no assumption is made regarding the ordering of the categories, for the model fit it does not matter which variable is dropped with regard to the fit of the model. However, it does matter for the interpretation of the regression coefficients. Consider the following example.

Example: Term Life Insurance, Continued. We now return to the marital status of respondents from the Survey of Consumer Finances (SCF). Recall that marital status is not measured continuously but rather takes on values that fall into distinct groups that we treat as unordered. In Chapter 3, we grouped survey respondents according to whether they are "single," where being single includes never married, separated, divorced, widowed, and not married and living

Table 4.1 Summary Statistics of Logarithmic Face by Marital Status

	MARSTAT	Number	Mean	Standard Deviation
Other	0	57	10.958	1.566
Married	1	208	12.329	1.822
Living together	2	10	10.825	2.001
Total		275	11.990	1.871

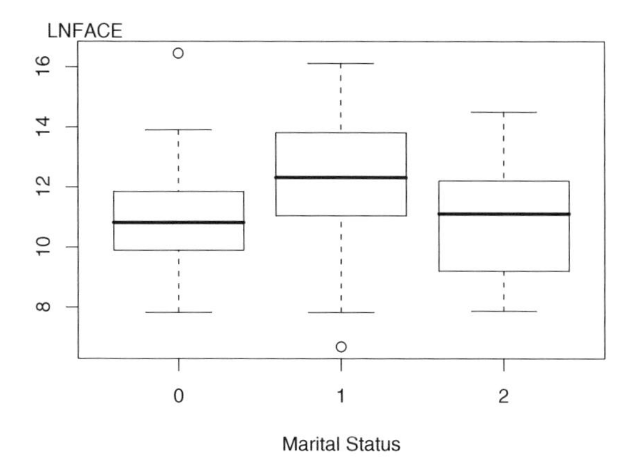

Figure 4.1 Box plots of logarithmic face, by level of marital status.

with a partner. We now supplement this by considering the categorical variable, MARSTAT, which represents the marital status of the survey respondent. This may be:

- 1, for married
- 2, for living with partner
- 0, for other (SCF further breaks down this category into separated, divorced, widowed, never married and inapplicable, persons age 17 or younger, no further persons)

As before, the dependent variable is $y = \text{LNFACE}$, the amount that the company will pay in the event of the death of the named insured (in logarithmic dollars). Table 4.1 summarizes the dependent variable by level of the categorial variable. This table shows that the marital status "married" is the most prevalent in the sample and that those who are married choose to have the most life insurance coverage. Figure 4.1 gives a more complete picture of the distribution of LNFACE for each of the three types of marital status. The table and figure also suggests that those who live together have less life insurance coverage than people in the other two categories.

Are the continuous and categorical variables jointly important determinants of response? To answer this, a regression was run using LNFACE as the response and five explanatory variables, three continuous and two binary (for marital status). Recall that our three continuous explanatory variables are LNINCOME

Table 4.2 Term Life with Marital Status ANOVA Table

Source	Sum of Squares	df	Mean Square
Regression	343.28	5	68.66
Error	615.62	269	2.29
Total	948.90	274	

Note: Residual standard error $s = 1.513$, $R^2 = 35.8\%$, and $R_a^2 = 34.6\%$.

(logarithmic annual income), the number of years of EDUCATION of the survey respondent, and the number of household members (NUMHH).

For the binary variables, first define MAR0 to be the binary variable that is one if MARSTAT $= 0$ and zero otherwise. Similarly, define MAR1 and MAR2 to be binary variables that indicate MARSTAT $= 1$ and MARSTAT $= 2$, respectively. There is a perfect linear dependency among these three binary variables in that MAR0 $+$ MAR1 $+$ MAR2 $= 1$ for any survey respondent. Thus, we need only two of the three. However, there is not a perfect dependency among any two of the three. It turns out that Corr(MAR0,MAR1) $= -0.90$, Corr(MAR0,MAR2) $= -0.10$, and Corr(MAR1,MAR2) $= -0.34$.

A regression model was run using LNINCOME, EDUCATION, NUMHH, MAR0, and MAR2 as explanatory variables. The fitted regression equation turns out to be

$$\begin{aligned}\widehat{y} &= 2.605 + 0.452\text{LNINCOME} + 0.205\text{EDUCATION} + 0.248\text{NUMHH} \\ &\quad - 0.557\text{MAR0} - 0.789\text{MAR2}.\end{aligned}$$

To interpret the regression coefficients associated with marital status, consider a respondent who is married. In this case, then MAR0 $= 0$, MAR1 $= 1$, and MAR2 $= 0$, so that

$$\widehat{y}_m = 2.605 + 0.452\text{LNINCOME} + 0.205\text{EDUCATION} + 0.248\text{NUMHH}.$$

Similarly, if the respondent is coded as "living together," then MAR0 $= 0$, MAR1 $= 0$, and MAR2 $= 1$, and

$$\begin{aligned}\widehat{y}_{lt} &= 2.605 + 0.452\text{LNINCOME} + 0.205\text{EDUCATION} \\ &\quad + 0.248\text{NUMHH} - 0.789.\end{aligned}$$

The difference between $\widehat{y}_m$ and $\widehat{y}_{lt}$ is 0.789. Thus, we may interpret the regression coefficient associated with MAR2, -0.789, to be the difference in fitted values for someone living together compared to a similar person who is married (the omitted category).

Similarly, we can interpret -0.557 to be the difference between the "other" category and the married category, holding other explanatory variables fixed. For the difference in fitted values between the "other" and the "living together" categories, we may use $-0.557 - (-0.789) = 0.232$.

Although the regression was run using MAR0 and MAR2, any two out of the three would produce the same ANOVA Table 4.2. However, the choice of binary

Explanatory Variable	Model 1		Model 2		Model 3	
	Coefficient	t-Ratio	Coefficient	t-Ratio	Coefficient	t-Ratio
LNINCOME	0.452	5.74	0.452	5.74	0.452	5.74
EDUCATION	0.205	5.30	0.205	5.30	0.205	5.30
NUMHH	0.248	3.57	0.248	3.57	0.248	3.57
Intercept	3.395	3.77	2.605	2.74	2.838	3.34
MAR0	−0.557	−2.15	0.232	0.44		
MAR1			0.789	1.59	0.557	2.15
MAR2	−0.789	−1.59			−0.232	−0.44

Table 4.3 Term Life Regression Coefficients with Marital Status

variables does affect the regression coefficients. Table 4.3 shows three models, omitting MAR1, MAR2, and MAR0, respectively. For each fit, the coefficients associated with the continuous variables remain the same. As we have seen, the binary variable interpretations are with respect to the omitted category, known as the *reference level*. Although they change from model to model, the overall interpretation remains the same. That is, if we would like to estimate the difference in coverage between the "other" and the "living together" categories, the estimate would be 0.232, regardless of the model.

Although the three models in Table 4.3 are the same except for different choices of parameters, they do appear different. In particular, the t-ratios differ and give different appearances of statistical significance. For example, both of the t-ratios associated with marital status in Model 2 are less than 2 in absolute value, suggesting that marital status is unimportant. In contrast, both Models 1 and 3 have at least one marital status binary that exceeds 2 in absolute value, suggesting statistical significance. Thus, you can influence the *appearance* of statistical significance by altering the choice of the reference level. To assess the overall importance of marital status (not just each binary variable), Section 4.2 introduces tests of sets of regression coefficients.

The choice of the reference level can influence the appearance of statistical significance.

Example: How Does Cost Sharing in Insurance Plans Affect Expenditures in Health Care? In one of many studies that resulted from the Rand Health Insurance Experiment (HIE) introduced in Section 1.5, Keeler and Rolph (1988) investigated the effects of cost sharing in insurance plans. For this study, 14 health insurance plans were grouped by the coinsurance rate (the percentage paid as out-of-pocket expenditures that varied by 0%, 25%, 50%, and 95%). One of the 95% plans limited annual out-of-pocket outpatient expenditures to \$150 per person (\$450 per family), in effect providing an individual outpatient deductible. This plan was analyzed as a separate group so that there were $c = 5$ categories of insurance plans. In most insurance studies, individuals choose insurance plans, making it difficult to assess cost-sharing effects because of adverse selection. Adverse selection can arise because individuals in poor chronic health are more likely to choose plans with less cost sharing, thus giving the appearance that less coverage leads to greater expenditures. In the Rand HIE, individuals were randomly assigned to plans, thus removing this potential source of bias.

Keeler and Rolph (1988) organized an individual's expenditures into episodes of treatment; each episode contains spending associated with a given bout of illness, a chronic condition, or a procedure. Episodes were classified as hospital, dental, or outpatient; this classification was based primarily on diagnoses, not location of services. Thus, for example, outpatient services preceding or following a hospitalization, as well as related drugs and tests, were included as part of a hospital episode.

For simplicity, here are reported only results for hospital episodes. Although families were randomly assigned to plans, Keeler and Rolph (1988) used regression methods to control for participant attributes and to isolate the effects of plan cost sharing. Table 4.4 summarizes the regression coefficients, based on a sample of $n = 1{,}967$ episode expenditures. In this regression, logarithmic expenditure was the dependent variable.

The cost-sharing categorical variable was decomposed into five binary variables so that no functional form was imposed on the response to insurance. These variables are "Co-ins25," "Co-ins50," and "Co-ins95," for coinsurance rates 25%, 50%, and 95%, respectively, and "Indiv Deductible" for the plan with individual deductibles. The omitted variable is the free insurance plan with 0% coinsurance. The HIE was conducted in six cities; a categorical variable to control for the location was represented with five binary variables, Dayton, Fitchburg, Franklin, Charleston, and Georgetown, with Seattle being the omitted variable. A categorical factor with $c = 6$ levels was used for age and sex; binary variables in the model consisted of "Age 0–2," "Age 3–5," "Age 6–17," "Woman age 18–65," and "Man age 46–65," the omitted category was "Man age 18–45." Other control variables included a health status scale, socioeconomic status, number of medical visits in the year prior to the experiment on a logarithmic scale, and race.

Table 4.4 summarizes the effects of the variables. As noted by Keeler and Rolph, there were large differences by site and age, although the regression only served to summarize $R^2 = 11\%$ of the variability. For the cost-sharing variables, only "Co-ins95" was statistically significant, and this only at the 5% level, not the 1% level.

Keeler and Rolph (1988) examine other types of episode expenditures, as well as the frequency of expenditures. They conclude that cost sharing of health insurance plans has little effect on the amount of expenditures per episode although there are important differences in the frequency of episodes. This is because an episode of treatment comprises two decisions. The amount of treatment is made jointly between the patient and the physician and is largely unaffected by the type of health insurance plan. The decision to seek health-care treatment is made by the patient; this decision-making process is more susceptible to economic incentives in cost-sharing aspects of health insurance plans.

Table 4.4 Coefficients of Episode Expenditures from the Rand HIE

Variable	Regression Coefficient	Variable	Regression Coefficient
Intercept	7.95		
Dayton	0.13*	Co-ins25	0.07
Fitchburg	0.12	Co-ins50	0.02
Franklin	−0.01	Co-ins95	−0.13*
Charleston	0.20*	Indiv Deductible	−0.03
Georgetown	−0.18*		
Health scale	−0.02*	Age 0–2	−0.63**
Socioeconomic status	0.03	Age 3–5	−0.64**
Medical visits	−0.03	Age 6–17	−0.30**
Examination	−0.10*	Woman age 18–65	0.11
Black	0.14*	Man age 46–65	0.26

Note: *Significant at 5%, and **Significant at 1%.
Source: Keeler and Rolph (1988).

4.2 Statistical Inference for Several Coefficients

It can be useful to examine several regression coefficients at the same time. For example, when assessing the effect of a categorical variable with c levels, we need to say something jointly about the $c-1$ binary variables that enter the regression equation. To do this, Section 4.2.1 introduces a method for handling linear combinations of regression coefficients. Section 4.2.2 shows how to test several linear combinations and Section 4.2.3 presents other inference applications.

4.2.1 Sets of Regression Coefficients

Recall that our regression coefficients are specified by $\boldsymbol{\beta} = (\beta_0, \beta_1, \ldots, \beta_k)'$, a $(k+1) \times 1$ vector. It will be convenient to express linear combinations of the regression coefficients using the notation $\mathbf{C}\boldsymbol{\beta}$, where $\mathbf{C}$ is a $p \times (k+1)$ matrix that is user-specified and depends on the application. Some applications involve estimating $\mathbf{C}\boldsymbol{\beta}$. Others involve testing whether $\mathbf{C}\boldsymbol{\beta}$ equals a specific known value (denoted as $\mathbf{d}$). We call $H_0 : \mathbf{C}\boldsymbol{\beta} = \mathbf{d}$ the *general linear hypothesis*. To demonstrate the broad variety of applications in which sets of regression coefficients can be used, we now present a series of special cases.

The general linear hypothesis is denoted as $H_0 : \mathbf{C}\boldsymbol{\beta} = \mathbf{d}$.

Special Case 1: One Regression Coefficient. In Section 3.4, we investigated the importance of a single coefficient, say, β_j. We may express this coefficient as $\mathbf{C}\boldsymbol{\beta}$ by choosing $p = 1$ and $\mathbf{C}$ to be a $1 \times (k+1)$ vector with a one in the $(j+1)$st column and zeros otherwise. These choices result in

$$\mathbf{C}\boldsymbol{\beta} = (0 \ \cdots \ 0 \; 1 \; 0 \ \cdots \ 0) \begin{pmatrix} \beta_0 \\ \vdots \\ \beta_k \end{pmatrix} = \beta_j .$$

Special Case 2: Regression Function. Here, we choose $p = 1$ and $\mathbf{C}$ to be a $1 \times (k+1)$ vector representing the transpose of a set of explanatory variables. These choices result in

$$\mathbf{C}\boldsymbol{\beta} = (x_0, x_1, \ldots, x_k) \begin{pmatrix} \beta_0 \\ \vdots \\ \beta_k \end{pmatrix} = \beta_0 x_0 + \beta_1 x_1 + \cdots + \beta_k x_k = \mathrm{E}\ y,$$

the regression function.

Special Case 3: Linear Combination of Regression Coefficients. When $p = 1$, we use the convention that lowercase, bold letters are vectors and let $\mathbf{C} = \mathbf{c}' = (c_0, \ldots, c_k)'$. In this case, $\mathbf{C}\boldsymbol{\beta}$ is a generic linear combination of regression coefficients

$$\mathbf{C}\boldsymbol{\beta} = \mathbf{c}'\boldsymbol{\beta} = c_0\beta_0 + \cdots + c_k\beta_k.$$

Special Case 4: Testing Equality of Regression Coefficients. Suppose that the interest is in testing $H_0 : \beta_1 = \beta_2$. For this purpose, let $p = 1$, $\mathbf{c}' = (0, 1, -1, 0, \ldots, 0)$, and $\mathbf{d} = 0$. With these choices, we have

$$\mathbf{C}\boldsymbol{\beta} = \mathbf{c}'\boldsymbol{\beta} = (0, 1, -1, 0, \ldots, 0) \begin{pmatrix} \beta_0 \\ \vdots \\ \beta_k \end{pmatrix} = \beta_1 - \beta_2 = 0,$$

so that the general linear hypothesis reduces to $H_0 : \beta_1 = \beta_2$.

Special Case 5: Adequacy of the Model. It is customary in regression analysis to present a test of whether *any* of the explanatory variables are useful for explaining the response. Formally, this is a test of the null hypothesis $H_0 : \beta_1 = \beta_2 = \cdots = \beta_k = 0$. Note that, as a convention, one does not test whether the intercept is zero. To test this using the general linear hypothesis, we choose $p = k$, $\mathbf{d} = (0 \ldots 0)'$ to be a $k \times 1$ vector of zeros and $\mathbf{C}$ to be a $k \times (k+1)$ matrix such that

$$\mathbf{C}\boldsymbol{\beta} = \begin{pmatrix} 0 & 1 & 0 & \cdots & 0 \\ 0 & 0 & 1 & \cdots & 0 \\ \vdots & \vdots & \vdots & \ddots & \vdots \\ 0 & 0 & 0 & \cdots & 1 \end{pmatrix} \begin{pmatrix} \beta_0 \\ \vdots \\ \beta_k \end{pmatrix} = \begin{pmatrix} \beta_1 \\ \vdots \\ \beta_k \end{pmatrix} = \begin{pmatrix} 0 \\ \vdots \\ 0 \end{pmatrix} = \mathbf{d}.$$

Special Case 6: Testing Portions of the Model. Suppose that we are interested in comparing a *full* regression function

$$\mathrm{E}\ y = \beta_0 + \beta_1 x_1 + \cdots + \beta_k x_k + \beta_{k+1} x_{k+1} + \cdots + \beta_{k+p} x_{k+p}$$

to a *reduced* regression function,

$$\mathrm{E}\ y = \beta_0 + \beta_1 x_1 + \cdots + \beta_k x_k.$$

Beginning with the full regression, we see that if the null hypothesis $H_0: \beta_{k+1} = \cdots = \beta_{k+p} = 0$ holds, then we arrive at the reduced regression. To illustrate, the variables $x_{k+1}, \ldots, x_{k+p}$ may refer to several binary variables representing a categorical variable and our interest is in whether the categorical variable is important. To test the importance of the categorical variable, we want to see whether the binary variables $x_{k+1}, \ldots, x_{k+p}$ *jointly* affect the dependent variables.

To test this using the general linear hypothesis, we choose $\mathbf{d}$ and $\mathbf{C}$ such that

$$\mathbf{C}\boldsymbol{\beta} = \begin{pmatrix} 0 & \cdots & 0 & 1 & 0 & \cdots & 0 \\ 0 & \cdots & 0 & 0 & 1 & \cdots & 0 \\ \vdots & \vdots & \vdots & \vdots & \vdots & \ddots & \vdots \\ 0 & \cdots & 0 & 0 & 0 & \cdots & 1 \end{pmatrix} \begin{pmatrix} \beta_0 \\ \vdots \\ \beta_k \\ \beta_{k+1} \\ \vdots \\ \beta_{k+p} \end{pmatrix} = \begin{pmatrix} \beta_{k+1} \\ \vdots \\ \beta_{k+p} \end{pmatrix} = \begin{pmatrix} 0 \\ \vdots \\ 0 \end{pmatrix} = \mathbf{d}.$$

From a list of $k+p$ variables $x_1, \ldots, x_{k+p}$, you may drop any p that you deem appropriate. The additional variables do not need to be the last p in the regression specification. Dropping $x_{k+1}, \ldots, x_{k+p}$ is for notational convenience only.

4.2.2 The General Linear Hypothesis

To recap, the general linear hypothesis can be stated as $H_0: \mathbf{C}\boldsymbol{\beta} = \mathbf{d}$. Here, $\mathbf{C}$ is a $p \times (k+1)$ matrix, $\mathbf{d}$ is a $p \times 1$ vector and both $\mathbf{C}$ and $\mathbf{d}$ are user specified and depend on the application at hand. Although $k+1$ is the number of regression coefficients, p is the number of restrictions under H_0 on these coefficients. (For those readers with knowledge of advanced matrix algebra, p is the rank of $\mathbf{C}$.) This null hypothesis is tested against the alternative $H_a: \mathbf{C}\boldsymbol{\beta} \neq \mathbf{d}$. This may be obvious, but we do require $p \leq k+1$ because we cannot test more constraints than free parameters.

To understand the basis for the testing procedure, we first recall some of the basic properties of the regression coefficient estimators described in Section 3.3. Now, however, our goal is to understand properties of the linear combinations of regression coefficients specified by $\mathbf{C}\boldsymbol{\beta}$. A natural estimator of this quantity is $\mathbf{Cb}$. It is easy to see that $\mathbf{Cb}$ is an unbiased estimator of $\mathbf{C}\boldsymbol{\beta}$, because $\mathrm{E}\,\mathbf{Cb} = \mathbf{C}\mathrm{E}\,\mathbf{b} = \mathbf{C}\boldsymbol{\beta}$. Moreover, the variance is $\mathrm{Var}(\mathbf{Cb}) = \mathbf{C}\mathrm{Var}(\mathbf{b})\mathbf{C}' = \sigma^2\mathbf{C}\left(\mathbf{X}'\mathbf{X}\right)^{-1}\mathbf{C}'$. To assess the difference between $\mathbf{d}$, the hypothesized value of $\mathbf{C}\boldsymbol{\beta}$, and its estimated value, $\mathbf{Cb}$, we use the following statistic:

$$F\text{-ratio} = \frac{(\mathbf{Cb} - \mathbf{d})'\left(\mathbf{C}\left(\mathbf{X}'\mathbf{X}\right)^{-1}\mathbf{C}'\right)^{-1}(\mathbf{Cb} - \mathbf{d})}{p s_{full}^2}. \tag{4.1}$$

Here, s_{full}^2 is the mean square error from the full regression model. Using the theory of linear models, it can be checked that the statistic F-ratio has an F-distribution with numerator degrees of freedom $df_1 = p$ and denominator degrees of freedom

$df_2 = n - (k+1)$. Both the statistic and the theoretical distribution are named for R. A. Fisher, a renowned scientist and statistician who did much to advance statistics as a science in the early half of the twentieth century.

Like the normal and the t-distribution, the F-distribution is a continuous distribution. The F-distribution is the sampling distribution for the F-ratio and is proportional to the ratio of two sum of squares, each of which is positive or zero. Thus, unlike the normal distribution and the t-distribution, the F-distribution takes on only nonnegative values. Recall that the t-distribution is indexed by a single degree-of-freedom parameter. The F-distribution is indexed by two degree of freedom parameters: one for the numerator, df_1, and one for the denominator, df_2. Appendix A3.4 provides additional details.

Appendix A3.4 provides additional details about the F-distribution, including a graph and distribution table.

The test statistic in equation (4.1) is complex in form. Fortunately, there is an alternative that is simpler to implement and to interpret; this alternative is based on the *extra sum of squares principle*.

Procedure for Testing the General Linear Hypothesis

(i) Run the full regression and get the error sum of squares and mean square error, which we label as $(Error\ SS)_{full}$ and s^2_{full}, respectively.

(ii) Consider the model assuming the null hypothesis is true. Run a regression with this model and get the error sum of squares, which we label $(Error\ SS)_{reduced}$.

(iii) Calculate

$$F\text{-ratio} = \frac{(Error\ SS)_{reduced} - (Error\ SS)_{full}}{p s^2_{full}}. \tag{4.2}$$

(iv) Reject the null hypothesis in favor of the alternative if the F-ratio exceeds an F-value. The F-value is a percentile from the F-distribution with $df_1 = p$ and $df_2 = n - (k+1)$ degrees of freedom. The percentile is one minus the significance level of the test. Following our notation with the t-distribution, we denote this percentile as $F_{p,n-(k+1),1-\alpha}$, where α is the significance level.

This procedure is commonly known as an F-test.

Section 4.7.2 provides the mathematical underpinnings. To understand the extra-sum-of-squares principle, recall that the error sum of squares for the full model is determined to be the minimum value of

$$SS(b_0^*, \ldots, b_k^*) = \sum_{i=1}^{n} \left(y_i - \left(b_0^* + \cdots + b_k^* x_{i,k}\right)\right)^2.$$

Here, $SS(b_0^*, \ldots, b_k^*)$ is a function of $b_0^*, \ldots, b_k^*$, and $(Error\ SS)_{full}$ is the minimum over all possible values of $b_0^*, \ldots, b_k^*$. Similarly, $(Error\ SS)_{reduced}$ is the minimum error sum of squares under the constraints in the null hypothesis. Because there are fewer possibilities under the null hypothesis, we have

$$(Error\ SS)_{full} \leq (Error\ SS)_{reduced}. \tag{4.3}$$

To illustrate, consider our first special case, where $H_0 : \beta_j = 0$. In this case, the difference between the full and the reduced models amounts to dropping a variable. A consequence of equation (4.3) is that, when adding variables to a regression model, the error sum of squares never goes up (and, in fact, usually goes down). Thus, adding variables to a regression model increases R^2, the coefficient of determination.

When adding variables to a regression model, the error sum of squares never goes up. The R^2 statistic never goes down.

How large a decrease in the error sum of squares is statistically significant? Intuitively, one can view the F-ratio as the difference in the error sum of squares divided by the number of constraints, $((Error\ SS)_{reduced} - (Error\ SS)_{full})/p$, and then rescaled by the best estimate of the variance term, the s^2, from the full model. Under the null hypothesis, this statistic follows an F-distribution, and we can compare the test statistic to this distribution to see whether it is unusually large.

Using the relationship $Regression\ SS = Total\ SS - Error\ SS$, we can reexpress the difference in the error sum of squares as

$$(Error\ SS)_{reduced} - (Error\ SS)_{full} = (Regression\ SS)_{full} - (Regression\ SS)_{reduced}.$$

This difference is known as a *Type III sum of squares*. When testing the importance of a set of explanatory variables, $x_{k+1}, \ldots, x_{k+p}$, in the presence of $x_1, \ldots, x_k$, you will find that many statistical software packages compute this quantity directly in a single regression run. The advantage of this is that it allows the analyst to perform an F-test using a single regression run instead of two regression runs, as in our four-step procedure described previously.

Example: Term Life Insurance, Continued. Before discussing the logic and the implications of the F-test, let us illustrate the use of it. In the term life insurance example, suppose that we want to understand the impact of marital status. Table 4.3 presented a mixed message in terms of t-ratios; sometimes they were statistically significant and sometimes not. It would be helpful to have a formal test to give a definitive answer, at least in terms of statistical significance. Specifically, we consider a regression model using LNINCOME, EDUCATION, NUMHH, MAR0, and MAR2 as explanatory variables. The model equation is

$$y = \beta_0 + \beta_1 \text{LNINCOME} + \beta_2 \text{EDUCATION} + \beta_3 \text{NUMHH} + \beta_4 \text{MAR0} + \beta_5 \text{MAR2}.$$

Our goal is to test $H_0 : \beta_4 = \beta_5 = 0$.

(i) We begin by running a regression model with all $k + p = 5$ variables. The results were reported in Table 4.2, where we saw that $(Error\ SS)_{full} = 615.62$ and $s^2_{full} = (1.513)^2 = 2.289$.

(ii) The next step is to run the reduced model without MAR0 and MAR2. This was done in Table 3.3 of Chapter 3, where we saw that $(Error\ SS)_{reduced} = 630.43$.

(iii) We then calculate the test statistic

$$F\text{-ratio} = \frac{(Error\ SS)_{reduced} - (Error\ SS)_{full}}{p s^2_{full}} = \frac{630.43 - 615.62}{2 \times 2.289} = 3.235.$$

(iv) The fourth step compares the test statistic to an F-distribution with $df_1 = p = 2$ and $df_2 = n - (k + p + 1) = 269$ degrees of freedom. Using a 5% level of significance, it turns out that the 95th percentile is F-value ≈ 3.029. The corresponding p-value is $\Pr(F > 3.235) = 0.0409$. At the 5% significance level, we reject the null hypothesis $H_0 : \beta_4 = \beta_5 = 0$. This suggests that it is important to use marital status to understand term life insurance coverage, even in the presence of income, education, and number of household members.

Some Special Cases

The general linear hypothesis test is available when you can express one model as a subset of another. For this reason, it useful to think of it as a device for comparing "smaller" to "larger" models. However, the smaller model must be a subset of the larger model. For example, the general linear hypothesis test cannot be used to compare the regression functions $\mathrm{E}\ y = \beta_0 + \beta_7 x_7$ versus $\mathrm{E}\ y = \beta_0 + \beta_1 x_1 + \beta_2 x_2 + \beta_3 x_3 + \beta_4 x_4$. This is because the former, smaller function is not a subset of the latter, larger function.

The general linear hypothesis can be used in many instances, although its use is not always necessary. For example, suppose that we wish to test $H_0 : \beta_k = 0$. We have already seen that this null hypothesis can be examined using the t-ratio test. In this special case, it turns out that $(t\text{-ratio})^2 = F\text{-ratio}$. Thus, these tests are equivalent for testing $H_0 : \beta_k = 0$ versus $H_a : \beta_k \neq 0$. The F-test has the advantage that it works for more than one predictor, whereas the t-test has the advantage that one can consider one-sided alternatives. Thus, both tests are considered useful.

Dividing the numerator and denominator of equation (4.2) by *Total SS*, the test statistic can also be written as

$$F\text{-ratio} = \frac{\left(R^2_{full} - R^2_{reduced}\right)/p}{\left(1 - R^2_{full}\right)/(n - (k+1))}. \tag{4.4}$$

The interpretation of this expression is that the F-ratio measures the drop in the coefficient of determination, R^2.

The expression in equation (4.2) is particularly useful for testing the adequacy of the model, our Special Case 5. In this case, $p = k$, and the regression sum of squares under the reduced model is zero. Thus, we have

$$F\text{-ratio} = \frac{\left((Regression\ SS)_{full}\right)/k}{s^2_{full}} = \frac{(Regression\ MS)_{full}}{(Error\ SS)_{full}}.$$

This test statistic is a regular feature of the ANOVA table for many statistical packages.

For example, in our term life insurance example, testing the adequacy of the model means evaluating $H_0 : \beta_1 = \beta_2 = \beta_3 = \beta_4 = \beta_5 = 0$. From Table 4.2, the F-ratio is 68.66/2.29 = 29.98. With $df_1 = 5$ and $df_2 = 269$, we have that the F-value is approximately 2.248 and the corresponding p-value is $\Pr(F > 29.98) \approx 0$. This leads us to reject strongly the notion that the explanatory variables are not useful in understanding term life insurance coverage, reaffirming what we learned in the graphical and correlation analysis. Any other result would be surprising.

For another expression, dividing by *Total SS*, we may write

$$F\text{-ratio} = \frac{R^2}{1-R^2}\frac{n-(k+1)}{k}.$$

Because both F-ratio and R^2 are measures of model fit, it seems intuitively plausible that they are related in some fashion. A consequence of this relationship is the fact that as R^2 increases, so does the F-ratio and vice versa. The F-ratio is used because its sampling distribution is known under a null hypothesis, so we can make statements about statistical significance. The R^2 measure is used because of the easy interpretations associated with it.

4.2.3 Estimating and Predicting Several Coefficients

Estimating Linear Combinations of Regression Coefficients

In some applications, the main interest is to estimate a linear combination of regression coefficients. To illustrate, recall that, in Section 3.5, we developed a regression function for an individual's charitable contributions (y) in terms of wages (x). In this function, there was an abrupt change in the function at $x = 97{,}500$. To model this, we defined the binary variable z to be zero if $x < 97{,}500$ and to be one if $x \geq 97{,}500$ and the regression function $\mathrm{E}\ y = \beta_0 + \beta_1 x + \beta_2 z(x - 97{,}500)$. Thus, the marginal expected change in contributions per dollar wage change for wages in excess of 97,500 is $\partial\,(\mathrm{E}\ y)\,/\partial x = \beta_1 + \beta_2$.

To estimate $\beta_1 + \beta_2$, a reasonable estimator is $b_1 + b_2$, which is readily available from standard regression software. In addition, we would also like to compute standard errors for $b_1 + b_2$ to be used, for example, in determining a confidence interval for $\beta_1 + \beta_2$. However, b_1 and b_2 are typically correlated so that the calculation of the standard error of $b_1 + b_2$ requires estimation of the covariance between b_1 and b_2.

Estimating $\beta_1 + \beta_2$ is an example of our Special Case 3, which considers linear combinations of regression coefficients of the form $\mathbf{c}'\boldsymbol{\beta} = c_0\beta_0 + c_1\beta_1 + \cdots + c_k\beta_k$. For our charitable contributions example, we would choose $c_1 = c_2 = 1$ and other c's equal to zero.

To estimate $\mathbf{c}'\boldsymbol{\beta}$, we replace the vector of parameters by the vector of estimators and use $\mathbf{c}'\mathbf{b}$. To assess the reliability of this estimator, as in Section 4.2.2, we have that $\mathrm{Var}\left(\mathbf{c}'\mathbf{b}\right) = \sigma^2\mathbf{c}'(\mathbf{X}'\mathbf{X})^{-1}\mathbf{c}$. Thus, we may define the estimated standard deviation, or standard error, of $\mathbf{c}'\mathbf{b}$ to be

$$se\left(\mathbf{c}'\mathbf{b}\right) = s\sqrt{\mathbf{c}'(\mathbf{X}'\mathbf{X})^{-1}\mathbf{c}}.$$

With this quantity, a $100(1-\alpha)\%$ confidence interval for $\mathbf{c}'\boldsymbol{\beta}$ is

$$\mathbf{c}'\mathbf{b} \pm t_{n-(k+1),1-\alpha/2}\, se(\mathbf{c}'\mathbf{b}). \tag{4.5}$$

The confidence interval in equation (4.5) is valid under Assumptions F1–F5. If we choose $\mathbf{c}$ to have a 1 in the $(j+1)$st row and 0 otherwise, then $\mathbf{c}'\boldsymbol{\beta} = \beta_j$, $\mathbf{c}'\mathbf{b} = b_j$ and

$$se(b_j) = s\sqrt{(j+1)st\ diagonal\ element\ of\, (\mathbf{X}'\mathbf{X})^{-1}}.$$

Thus, (4.5) provides a theoretical basis for the individual regression coefficient confidence intervals introduced in Section 3.4's equation (3.10) and generalizes it to arbitrary linear combinations of regression coefficients.

Another important application of equation (4.5) is the choice of $\mathbf{c}$ corresponding to a set of explanatory variables of interest, say, $\mathbf{x}_* = (1, x_{*1}, x_{*2}, \ldots, x_{*k})'$. These may correspond to an observation within the dataset or to a point outside the available data. The parameter of interest, $\mathbf{c}'\boldsymbol{\beta} = \mathbf{x}_*'\boldsymbol{\beta}$, is the expected response or the regression function at that point. Then, $\mathbf{x}_*'\mathbf{b}$ provides a point estimator and equation (4.5) provides the corresponding confidence interval.

Prediction Intervals

Prediction is an inferential goal that is closely related to estimating the regression function at a point. Suppose that, when considering charitable contributions, we know an individual's wages (and thus whether wages are in excess of \$97,500) and want to predict the amount of charitable contributions. In general, we assume that the set of explanatory variables $\mathbf{x}_*$ is known and want to predict the corresponding response, y_*. This new response follows the assumptions as described in Section 3.2. Specifically, the expected response is $\mathrm{E}\ y_* = \mathbf{x}_*'\boldsymbol{\beta}$, $\mathbf{x}_*$ is nonstochastic, $\mathrm{Var}\ y_* = \sigma^2$, and y_* is independent of $\{y_1, \ldots, y_n\}$ and normally distributed. Under these assumptions, a $100(1-\alpha)\%$ prediction interval for y_* is

$$\mathbf{x}_*'\mathbf{b} \pm t_{n-(k+1),1-\alpha/2}\, s\sqrt{1 + \mathbf{x}_*'(\mathbf{X}'\mathbf{X})^{-1}\mathbf{x}_*}. \tag{4.6}$$

Equation (4.6) generalizes the prediction interval introduced in Section 2.4.

4.3 One Factor ANOVA Model

Section 4.1 showed how to incorporate unordered categorical variables, or factors, into a linear regression model through the use of binary variables. Factors are important in social science research; they can be used to classify people by sex, ethnicity, marital status, and so on, or to classify firms by geographic region, organizational structure, and so forth. In studies of insurance, factors are used by insurers to categorize policyholders according to a "risk classification system." Here, the idea is to create groups of policyholders with similar risk characteristics that will have similar claims experience. These groups form the basis of insurance pricing, so that each policyholder is charged an amount that is appropriate to his or her risk category. This process is sometimes known as *segmentation*.

Table 4.5 Automobile Claims Summary Statistics by Risk Class

Class	C1	C11	C1A	C1B	C1C	C2
Number	726	1,151	77	424	38	61
Median (dollars)	948.86	1,013.81	925.48	1,026.73	1,001.73	851.20
Median (in log dollars)	6.855	6.921	6.830	6.934	6.909	6.747
Mean (in log dollars)	6.941	6.952	6.866	6.998	6.786	6.801
Std. dev. (in log dollars)	1.064	1.074	1.072	1.068	1.110	0.948
Class	C6	C7	C71	C72	C7A	C7B
Number	911	913	1,129	85	113	686
Median (dollars)	1,011.24	957.68	960.40	1,231.25	1,139.93	1,113.13
Median (in log dollars)	6.919	6.865	6.867	7.116	7.039	7.015
Mean (in log dollars)	6.926	6.901	6.954	7.183	7.064	7.072
Std. dev. (in log dollars)	1.115	1.058	1.038	0.988	1.021	1.103
Class	C7C	F1	F11	F6	F7	F71
Number	81	29	40	157	59	93
Median (dollars)	1,200.00	1,078.04	774.79	1,105.04	707.40	1,118.73
Median (in log dollars)	7.090	6.983	6.652	7.008	6.562	7.020
Mean (in log dollars)	7.244	7.004	6.804	6.910	6.577	6.935
Std. dev. (in log dollars)	0.944	0.996	1.212	1.193	0.897	0.983

Although factors may be represented as binary variables in a linear regression model, we study one factor models as a separate unit because

- The method of least squares is much simpler, obviating the need to take inverses of high dimensional matrices
- The resulting interpretations of coefficients are more straightforward

The one factor model is still a special case of the linear regression model. Hence, no additional statistical theory is needed to establish its statistical inference capabilities.

To establish notation for the one factor ANOVA model, we now consider the following example.

Ⓡ **Empirical** *Filename is "AutoClaims"*

Example: Automobile Insurance Claims. We examine claims experience from a large U.S. midwestern property and casualty insurer for private passenger automobile insurance. The dependent variable is the amount paid on a closed claim, in dollars (claims that were not closed by the year's end are handled separately). Insurers categorize policyholders according to a risk classification system. This insurer's risk classification system is based on

- Automobile operator characteristics (age, sex, marital status, and whether the primary or occasional driver of a car).
- Vehicle characteristics (city versus farm usage; used to commute to school or work; used for business or pleasure; and if commuting, the approximate distance of the commute).

These factors are summarized by the risk class categorical variable CLASS. Table 4.5 shows 18 risk classes – further classification information is not given here to protect proprietary interests of the insurer.

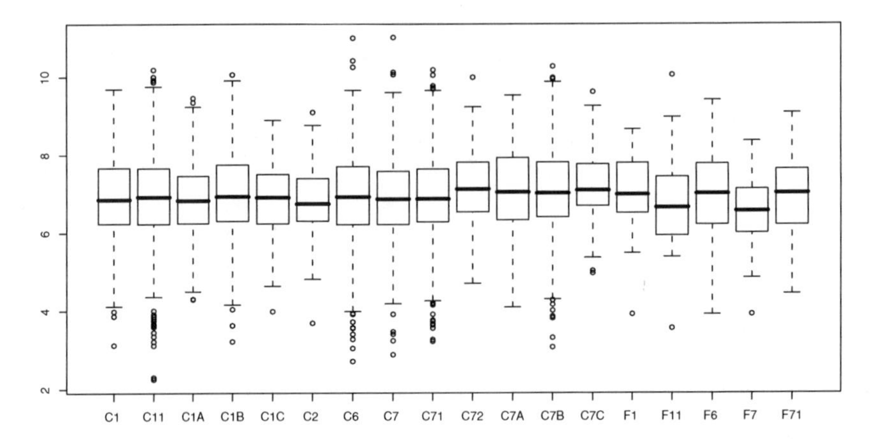

Figure 4.2 Box plots of logarithmic claims by risk class.

Table 4.5 summarizes the results from $n = 6{,}773$ claims for drivers aged 50 and older. We can see the the median claim varies from a low of \$707.40 (CLASS F7) to a high of \$1,231.25 (CLASS C72). The distribution of claims turns out to be skewed, so we consider $y =$ logarithmic claims. The table presents means, medians, and standard deviations. Because the distribution of logarithmic claims is less skewed, means are close to medians. Figure 4.2 shows the distribution of logarithmic claims by risk class.

This section focuses on the risk class (CLASS) as the explanatory variable. We use the notation y_{ij} to mean the ith observation of the jth risk class. For the jth risk class, we assume there are n_j observations. There are $n = n_1 + n_2 + \cdots + n_c$ observations. The data are as follows:

Data for risk class 1	y_{11}	y_{21}	$\cdots$	$y_{n_1,1}$
Data for risk class 2	y_{12}	y_{22}	$\cdots$	$y_{n_2,1}$
.	.	.	$\cdots$	.
Data for risk class c	y_{1c}	y_{2c}	$\cdots$	$y_{n_c,c}$,

where $c = 18$ is the number of levels of the CLASS factor. Because each level of a factor can be arranged in a single row (or column), another term for this type of data is a *one way classification*. Thus, a *one-way model* is another term for a one factor model.

An important summary measure of each level of the factor is the sample average. Let

$$\overline{y}_j = \frac{1}{n_j} \sum_{i=1}^{n_j} y_{ij}$$

denote the average from the jth CLASS.

Model Assumptions and Analysis

The one factor ANOVA model equation is

$$y_{ij} = \mu_j + \varepsilon_{ij} \qquad i = 1, \ldots, n_j, \qquad j = 1, \ldots, c. \tag{4.7}$$

Table 4.6 ANOVA Table for One Factor Model

Source	Sum of Square	df	Mean Square
Factor	*Factor SS*	$c-1$	*Factor MS*
Error	*Error SS*	$n-c$	*Error MS*
Total	*Total SS*	$n-1$	

As with regression models, the random deviations $\{\varepsilon_{ij}\}$ are assumed to be zero mean with constant variance (Assumption E3) and independent of one another (Assumption E4). Because we assume the expected value of each deviation is zero, we have $\mathrm{E}\, y_{ij} = \mu_j$. Thus, we interpret μ_j to be the expected value of the response y_{ij}; that is, the mean μ varies by the jth factor level.

To estimate the parameters $\{\mu_j\}$, as with regression we use the *method of least squares*, introduced in Section 2.1. That is, let μ_j^* be a "candidate" estimate of μ_j. The quantity $SS(\mu_1^*, \ldots, \mu_c^*) = \sum_{j=1}^{c} \sum_{i=1}^{n_j} (y_{ij} - \mu_j^*)^2$ represents the sum of squared deviations of the responses from these candidate estimates. From straightforward algebra, the value of μ_j^* that minimizes this sum of squares is $\bar{y}_j$. Thus, $\bar{y}_j$ is the *least squares estimate* of μ_j.

The least squares estimate of μ_j is $\bar{y}_j$.

To understand the reliability of the estimates, we can partition the variability as in the regression case, presented in Sections 2.3.1 and 3.3. The minimum sum of squared deviations is called the *error sum of squares* and is defined as

$$Error\ SS = SS(\bar{y}_1, \ldots, \bar{y}_c) = \sum_{j=1}^{c} \sum_{i=1}^{n_j} \left(y_{ij} - \bar{y}_j\right)^2 .$$

The total variation in the dataset is summarized by the *total sum of squares*, $Total\ SS = \sum_{j=1}^{c} \sum_{i=1}^{n_j} (y_{ij} - \bar{y})^2$. The difference, called the *factor sum of squares*, can be expressed as

$$\begin{aligned} Factor\ SS &= Total\ SS - Error\ SS \\ &= \sum_{j=1}^{c} \sum_{i=1}^{n_j} (y_{ij} - \bar{y})^2 - \sum_{j=1}^{c} \sum_{i=1}^{n_j} (y_{ij} - \bar{y}_j)^2 = \sum_{j=1}^{c} \sum_{i=1}^{n_j} (\bar{y}_j - \bar{y})^2 \\ &= \sum_{j=1}^{c} n_j (\bar{y}_j - \bar{y})^2 . \end{aligned}$$

The last two equalities follow from algebra manipulation. The *Factor SS* plays the same role as the *Regression SS* in Chapters 2 and 3. The variability decomposition is summarized in Table 4.6.

The conventions for this table are the same as in the regression case. That is, the mean square (MS) column is defined by the sum of squares (SS) column divided by the degrees of freedom (df) column. Thus, *Factor MS* $\equiv$ (*Factor SS*)$/(c-1)$ and *Error MS* $\equiv$ (*Error SS*)$/(n-c)$. We use

$$s^2 = Error\ MS = \frac{1}{n-c} \sum_{j=1}^{c} \sum_{i=1}^{n_j} e_{ij}^2$$

to be our estimate of σ^2, where $e_{ij} = y_{ij} - \bar{y}_j$ is the residual.

Table 4.7 ANOVA Table for Logarithmic Automobile Claims

Source	Sum of Squares	df	Mean Square
CLASS	39.2	17	2.31
Error	7729.0	6755	1.14
Total	7768.2	6772	

With this value for s, it can be shown that the interval estimate for μ_j is

$$\bar{y}_j \pm t_{n-c,1-\alpha/2} \frac{s}{\sqrt{n_j}}. \tag{4.8}$$

Here, the t-value $t_{n-c,1-\alpha/2}$ is a percentile from the t-distribution with $df = n - c$ degrees of freedom.

Example: Automobile Claims, Continued. To illustrate, the ANOVA table summarizing the fit for the automobile claims data appears in Table 4.7. Here, we see that the mean square error is $s^2 = 1.14$.

In automobile ratemaking, one uses the average claims to help set prices for insurance coverages. As an example, for CLASS C72, the average logarithmic claim is 7.183. From equation (4.8), a 95% confidence interval is

$$7.183 \pm (1.96)\frac{\sqrt{1.14}}{\sqrt{85}} = 7.183 \pm 0.227 = (6.952, 7.410).$$

Note that the estimates are in natural logarithmic units. In dollars, our point estimate is $e^{7.183} = \$1{,}316.85$, and our 95% confidence interval is $(e^{6.952}, e^{7.410})$, or ($1,045.24, $1,652.43).

Unlike the usual regression analysis, no matrix calculations are required for the one factor ANOVA decomposition and estimation.

An important feature of the one factor ANOVA decomposition and estimation is the ease of computation. Although the sum of squares appear complex, it is important to note that no matrix calculations are required. Rather, all of the calculations can be done through averages and sums of squares. This been an important consideration historically, before the age of readily available desktop computing. Moreover, insurers can segment their portfolios into hundreds or even thousands of risk classes instead of the 18 used in our automobile claims data. Thus, even today it can be helpful to identify a categorical variable as a factor and let your statistical software use ANOVA estimation techniques. Further, ANOVA estimation also provides for direct interpretation of the results.

Link with Regression

This subsection shows how a one factor ANOVA model can be rewritten as a regression model. To this end, we have seen that both the regression model and one factor ANOVA model use a linear error structure with Assumptions E3 and E4 for identically and independently distributed errors. Similarly, both use the normality assumption E5 for selected inference results (such as confidence intervals). Both employ nonstochastic explanatory variables as in Assumption E2. Both have an additive (mean zero) error term, so the main apparent difference is in the expected response, E y.

For the linear regression model, E y is a linear combination of explanatory variables (Assumption F1). For the one factor ANOVA model, E $y_j = \mu_j$ is a mean that depends on the level of the factor. To equate these two approaches, for the ANOVA factor with c levels, we define c binary variables, $x_1, x_2, \ldots, x_c$. Here, x_j indicates whether an observation falls in the jth level. With these variables, we can rewrite our one factor ANOVA model as

$$y = \mu_1 x_1 + \mu_2 x_2 + \cdots + \mu_c x_c + \varepsilon. \tag{4.9}$$

Thus, we have rewritten the one factor ANOVA expected response as a regression function, although using a no-intercept form (as in equation (3.5)).

The one factor ANOVA is a special case of the regression model, using binary variables from the factor as explanatory variables in the regression function.

The one factor ANOVA is a special case of our usual regression model, using binary variables from the factor as explanatory variables in the regression function. As we have seen, no matrix calculations are needed for least squares estimation. However, one can always use the matrix procedures developed in Chapter 3. Section 4.7.1 shows how our usual matrix expression for regression coefficients ($\mathbf{b} = (\mathbf{X}'\mathbf{X})^{-1}\mathbf{X}'\mathbf{y}$) reduce to the simple estimates $\bar{y}_j$ when using only one categorical variable.

Reparameterization

To include an intercept term, define $\tau_j = \mu_j - \mu$, where μ is an as-yet-unspecified parameter. Because each observation must fall into one of the c categories, we have $x_1 + x_2 + \cdots + x_c = 1$ for each observation. Thus, using $\mu_j = \tau_j + \mu$ in equation (4.9), we have

$$y = \mu + \tau_1 x_1 + \tau_2 x_2 + \cdots + \tau_c x_c + \varepsilon. \tag{4.10}$$

Thus, we have rewritten the model into what appears to be our usual regression format.

We use the τ in lieu of β for historical reasons. ANOVA models were invented by R. A. Fisher in connection with agricultural experiments. Here, the typical setup is to apply several *treatments* to plots of land to quantify crop-yield responses. Thus, the Greek "t", τ, suggests the word *treatment*, another term used to described levels of the factor of interest.

A simpler version of equation (4.10) can be given when we identify the factor level. That is, if we know an observation falls in the jth level, then only x_j is one and the other x's are 0. Thus, a simpler expression for equation (4.10) is

$$y_{ij} = \mu + \tau_j + \varepsilon_{ij}.$$

Comparing equations (4.9) and (4.10), we see that the number of parameters has increased by one. That is, in equation (4.9), there are c parameters, $\mu_1, \ldots, \mu_c$, even though in equation (4.10) there are $c + 1$ parameters, μ and $\tau_1, \ldots, \tau_c$. The model in equation (4.10) is said to be *overparameterized.* It is possible to estimate this model directly, using the general theory of linear models, summarized in Section 4.7.3. In this theory, regression coefficients need not be identifiable. Alternatively, one can make these two expressions equivalent – *restricting* the movement of the parameters in (4.10). We now present two ways of imposing restrictions.

The first type of restriction, usually done in the regression context, is to require that one of the τ's be zero. This amounts to *dropping* one of the explanatory variables. For example, we might use

$$y = \mu + \tau_1 x_1 + \tau_2 x_2 + \cdots + \tau_{c-1} x_{c-1} + \varepsilon, \tag{4.11}$$

dropping x_c. With this formulation, it is easy to fit the model in equation (4.11) using regression statistical software routines because one only needs to run the regression with $c - 1$ explanatory variables. However, one needs to be careful with the interpretation of parameters. To equate the models in (4.9) and (4.10), we need to define $\mu \equiv \mu_c$ and $\tau_j = \mu_j - \mu_c$ for $j = 1, 2, \ldots, c - 1$. That is, the regression intercept term is the mean level of the category dropped, and each regression coefficient is the difference between a mean level and the mean level dropped. It is not necessary to drop the last level c, and indeed, one could drop any level. However, the interpretation of the parameters does depend on the variable dropped. With this restriction, the fitted values are $\hat{\mu} = \hat{\mu}_c = \bar{y}_c$ and $\hat{\tau}_j = \hat{\mu}_j - \hat{\mu}_c = \bar{y}_j - \bar{y}_c$. Recall that the caret (^) stands for an estimated, or fitted, value.

The second type of restriction is to interpret μ as a mean for the entire population. To this end, the usual requirement is $\mu \equiv (1/n)\sum_{j=1}^{c} n_j \mu_j$; that is, μ is a weighted average of means. With this definition, we interpret $\tau_j = \mu_j - \mu$ as treatment differences between a mean level and the population mean. Another way to express this restriction is $\sum_{j=1}^{c} n_j \tau_j = 0$; that is, the (weighted) sum of treatment differences is zero. The disadvantage of this restriction is that it is not readily implementable with a regression routine and a special routine is needed. The advantage is that there is a symmetry in the definitions of the parameters. There is no need to worry about which variable is being dropped from the equation, an important consideration. With this restriction, the fitted values are

$$\hat{\mu} = (1/n)\sum_{j=1}^{c} n_j \hat{\mu}_j = (1/n)\sum_{j=1}^{c} n_j \bar{y}_j = \bar{y} \quad \text{and} \quad \hat{\tau}_j = \hat{\mu}_j - \hat{\mu} = \bar{y}_j - \bar{y}.$$

4.4 Combining Categorical and Continuous Explanatory Variables

There are several ways to combine categorical and continuous explanatory variables. We initially present the case of only one categorical and one continuous variable. We then briefly present the general case, called the *general linear model.* When combining categorical and continuous variable models, we use the

Table 4.8 Several Models That Represent Combinations of One Factor and One Covariate

Model Description	Notation
One factor ANOVA (no covariate model)	$y_{ij} = \mu_j + \varepsilon_{ij}$
Regression with constant intercept and slope (no factor model)	$y_{ij} = \beta_0 + \beta_1 x_{ij} + \varepsilon_{ij}$
Regression with variable intercept and constant slope (analysis-of-covariance model)	$y_{ij} = \beta_{0j} + \beta_1 x_{ij} + \varepsilon_{ij}$
Regression with constant intercept and variable slope	$y_{ij} = \beta_0 + \beta_{1j} x_{ij} + \varepsilon_{ij}$
Regression with variable intercept and slope	$y_{ij} = \beta_{0j} + \beta_{1j} x_{ij} + \varepsilon_{ij}$

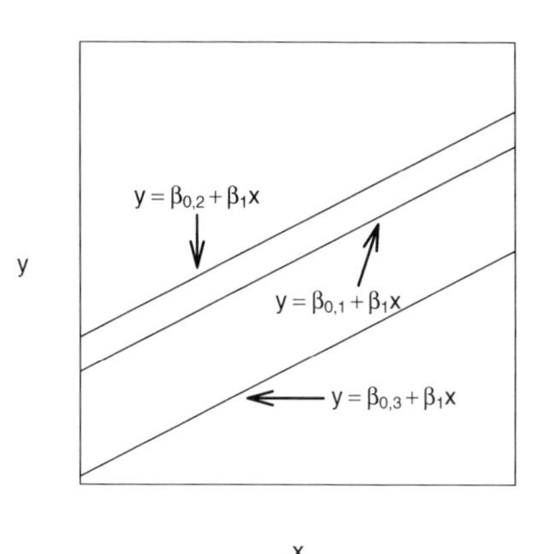

Figure 4.3 Plot of the expected response versus the covariate for the regression model with variable intercept and constant slope.

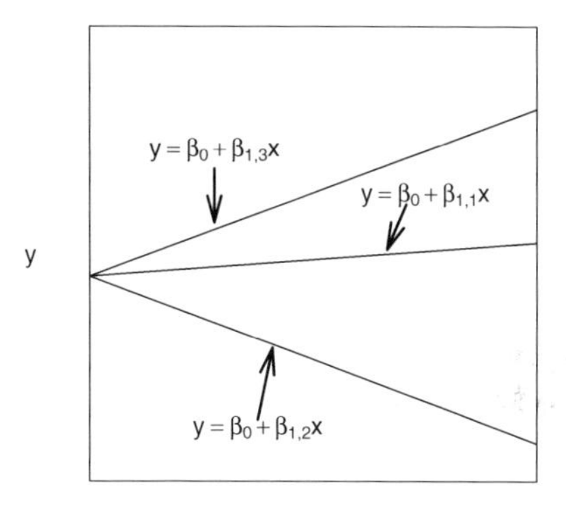

Figure 4.4 Plot of the expected response versus the covariate for the regression model with constant intercept and variable slope.

terminology *factor* for the categorical variable and *covariate* for the continuous variable.

Combining a Factor and Covariate

Let us begin with the simplest models that use a factor and a covariate. In Section 4.3, we introduced the one factor model $y_{ij} = \mu_j + \varepsilon_{ij}$. In Chapter 2, we introduced basic linear regression in terms of one continuous variable, or covariate, using $y_{ij} = \beta_0 + \beta_1 x_{ij} + \varepsilon_{ij}$. Table 4.8 summarizes different approaches that could be used to represent combinations of a factor and covariate.

We can interpret the regression with variable intercept and constant slope to be an additive model, because we are adding the factor effect, β_{0j}, to the covariate effect, $\beta_1 x_{ij}$. Note that we could also use the notation μ_j in lieu of $\beta_{0,j}$ to suggest the presence of a factor effect. This is also know as an *analysis of covariance (ANCOVA) model*. The regression with variable intercept and slope can be thought of as an *interaction model*. Here, both the intercept, β_{0j}, and slope, $\beta_{1,j}$, may vary by level of the factor. In this sense, we interpret the factor and covariate to be "interacting." The model with constant intercept and variable slope is typically not used in practice; it is included here for completeness. With this model, the factor and covariate interact only through the variable slope. Figures 4.3, 4.4, and 4.5 illustrate the expected responses of these models.

For each model presented in Table 4.8, parameter estimates can be calculated using the method of least squares. As usual, this means writing the expected

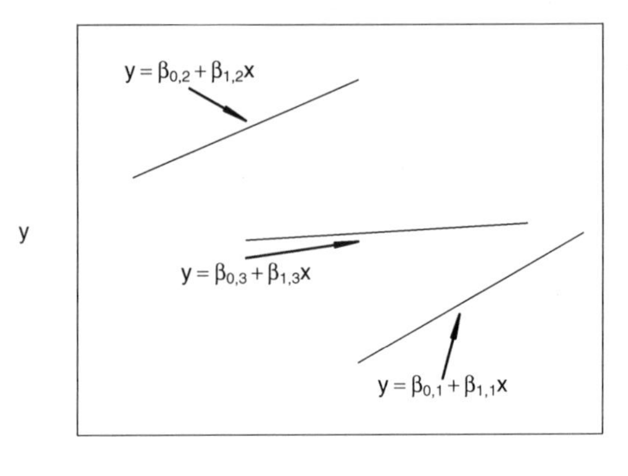

Figure 4.5 Plot of the expected response versus the covariate for the regression model with variable intercept and variable slope.

response, E y_{ij}, as a function of known variables and unknown parameters. For the regression model with variable intercept and constant slope, the least squares estimates can be expressed compactly as

$$b_1 = \frac{\sum_{j=1}^{c} \sum_{i=1}^{n_j} (x_{ij} - \bar{x}_j)(y_{ij} - \bar{y}_j)}{\sum_{j=1}^{c} \sum_{i=1}^{n_j} (x_{ij} - \bar{x}_j)^2}$$

and $b_{0,j} = \bar{y}_j - b_1 \bar{x}_j$. Similarly, the least squares estimates for the regression model with variable intercept and slope can be expressed as

$$b_{1,j} = \frac{\sum_{i=1}^{n_j} (x_{ij} - \bar{x}_j)(y_{ij} - \bar{y}_j)}{\sum_{i=1}^{n_j} (x_{ij} - \bar{x}_j)^2}$$

and $b_{0,j} = \bar{y}_j - b_1 \bar{x}_j$. With these parameter estimates, fitted values may be calculated.

For each model, fitted values are defined as the expected response with the unknown parameters replaced by their least squares estimates. For example, for the regression model with variable intercept and constant slope, the fitted values are $\hat{y}_{ij} = b_{0,j} + b_1 x_{ij}$.

Ⓡ **Empirical**
Filename is "WiscHospCosts"

Example: Wisconsin Hospital Costs. We now study the impact of various predictors on hospital charges in the state of Wisconsin. Identifying predictors of hospital charges can provide direction for hospitals, government, insurers, and consumers in controlling these variables, which in turn leads to better control of hospital costs. The data for the year 1989 were obtained from the Office of Health Care Information, Wisconsin's Department of Health and Human Services. Cross-sectional data are used, which detail the 20 diagnosis related group (DRG) discharge costs for hospitals in Wisconsin, broken down into nine major health service areas and three types of providers (fee for service, health maintenance organization [HMO], and other). Even though there are 540 potential DRG, area, and payer combinations ($20 \times 9 \times 3 = 540$), only 526 combinations were actually realized in the 1989 dataset. Other predictor variables included the

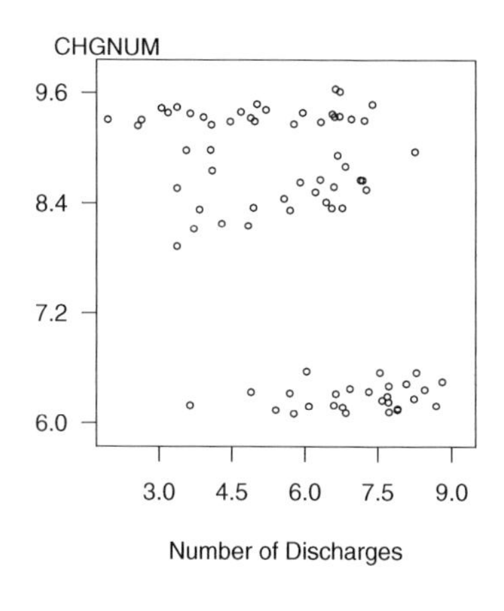

Figure 4.6 Plot of natural logarithm of cost per discharge versus natural logarithm of the number of discharges. This plot suggest a misleading negative relationship.

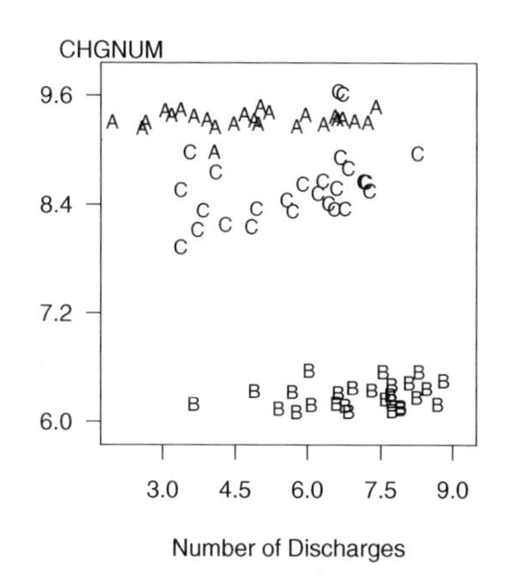

Figure 4.7 Letter plot of natural logarithm of cost per discharge versus natural logarithm of the number of discharges by DRG. Here, A is for DRG #209, B is for DRG #391, and C is for DRG #430.

logarithm of the total number of discharges (NO DSCHG) and total number of hospital beds (NUM BEDS) for each combination. The response variable is the logarithm of total hospital charges per number of discharges (CHGNUM). To streamline the presentation, we now consider only costs associated with three DRGs, DRG #209, DRG #391, and DRG #430.

The covariate, x, is the natural logarithm of the number of discharges. In ideal settings, hospitals with more patients enjoy lower costs because of economies of scale. In nonideal settings, hospitals may not have excess capacity; thus, hospitals with more patients have higher costs. One purpose of this analysis is to investigate the relationship between hospital costs and hospital utilization.

Recall that our measure of hospital charges is the logarithm of costs per discharge (y). The scatter plot in Figure 4.6 gives a preliminary idea of the relationship between y and x. We note that there appears to be a negative relationship between y and x.

The negative relationship between y and x suggested by Figure 4.6 is misleading and is induced by an *omitted variable*, the category of the cost (DRG). To see the joint effect of the categorical variable DRG and the continuous variable x, Figure 4.7 shows a plot of y versus x where the plotting symbols are codes for the level of the categorical variable. From this plot, we see that the level of cost varies by level of the factor DRG. Moreover, for each level of DRG, the slope between y and x is either zero or positive. The slopes are not negative, as suggested by Figure 4.6.

Each of the five models defined in Table 4.8 was fit to this subset of the hospital case study. The summary statistics are in Table 4.9. For this dataset, there are $n = 79$ observations and $c = 3$ levels of the DRG factor. For each model, the model degrees of freedom is the number of model parameters minus one. The error degrees of freedom is the number of observations minus the number of model parameters.

Using binary variables, each of the models in Table 4.8 can be written in a regression format. As we have seen in Section 4.2, when a model can be written as a subset of a larger model, we have formal testing procedures available to decide which model is more appropriate. To illustrate this testing procedure with

Table 4.9 Wisconsin Hospital Cost Models' Goodness of Fit

Model Description	Model Degrees of Freedom	Error Degrees of Freedom	Error Sum of Squares	R^2 (%)	Error Mean Square
One factor ANOVA	2	76	9.396	93.3	0.124
Regression with constant intercept and slope	1	77	115.059	18.2	1.222
Regression with variable intercept and constant slope	3	75	7.482	94.7	0.100
Regression with constant intercept and variable slope	3	75	14.048	90.0	0.187
Regression with variable intercept and slope	5	73	5.458	96.1	0.075

Note: These models represent combinations of one factor and one covariate.

our DRG example, from Table 4.9 and the associated plots, it seems clear that the DRG factor is important. Further, a t-test, not presented here, shows that the covariate x is important. Thus, let's compare the full model $\mathrm{E}\ y_{ij} = \beta_{0,j} + \beta_{1,j}x$ to the reduced model $\mathrm{E}\ y_{ij} = \beta_{0,j} + \beta_1 x$. In other words, is there a different slope for each DRG?

Using the notation from Section 4.2, we call the variable intercept and slope the full model. Under the null hypothesis, $H_0 : \beta_{1,1} = \beta_{1,2} = \beta_{1,3}$, we get the variable intercept, constant slope model. Thus, using the F-ratio in equation (4.2), we have

$$F\text{-ratio} = \frac{(Error\ SS)_{reduced} - (Error\ SS)_{full}}{p s^2_{full}} = \frac{7.482 - 5.458}{2(0.075)} = 13.535.$$

The 95th percentile from the F-distribution with $df_1 = p = 2$ and $df_2 = (df)_{full} = 73$ is approximately 3.13. Thus, this test leads us to reject the null hypothesis and declare the alternative, the regression model with variable intercept and variable slope, to be valid.

Combining Two Factors

We have seen how to combine covariates as well as a covariate and factor, both additively and with interactions. In the same fashion, suppose that we have two factors, say, sex (two levels, male/female) and age (three levels, young/middle/old). Let the corresponding binary variables be x_1 to indicate whether the observation represents a female, x_2 to indicate whether the observation represents a young person and x_3 to indicate whether the observation represents a middle-aged person.

An *additive model* for these two factors may use the regression function

$$\mathrm{E}\ y = \beta_0 + \beta_1 x_1 + \beta_2 x_2 + \beta_3 x_3.$$

Table 4.10 Regression Function for a Two Factor Model with Interactions

Sex	Age	x_1	x_2	x_3	x_4	x_5	Regression Function (4.12)
Male	Young	0	1	0	0	0	$\beta_0 + \beta_2$
Male	Middle	0	0	1	0	0	$\beta_0 + \beta_3$
Male	Old	0	0	0	0	0	β_0
Female	Young	1	1	0	1	0	$\beta_0 + \beta_1 + \beta_2 + \beta_4$
Female	Middle	1	0	1	0	1	$\beta_0 + \beta_1 + \beta_3 + \beta_5$
Female	Old	1	0	0	0	0	$\beta_0 + \beta_1$

As we have seen, this model is simple to interpret. For example, we can interpret β_1 as the sex effect, holding age constant.

We can also incorporate two interaction terms, x_1x_2 and x_1x_3. Using all five explanatory variables yields the regression function

$$\mathrm{E}\ y = \beta_0 + \beta_1 x_1 + \beta_2 x_2 + \beta_3 x_3 + \beta_4 x_1 x_2 + \beta_5 x_1 x_3. \tag{4.12}$$

Here, the variables x_1, x_2 and x_3 are known as the *main effects*. Table 4.10 helps interpret this equation. Specifically, there are six types of people that we can encounter, men and women who are young, middle aged, or old. We have six parameters in equation (4.12). Table 4.10 provides the link between the parameters and the types of people. By using the interaction terms, we do not impose any prior specifications on the additive effects of each factor. In Table 4.10, we see that the interpretation of the regression coefficients in equation (4.12) is not straightforward. However, using the additive model with interaction terms is equivalent to creating a new categorial variable with six levels, one for each type of person. If the interaction terms are critical in your study, you may wish to create a new factor that incorporates the interaction terms simply for ease of interpretation.

Extensions to more than two factors follow in a similar fashion. For example, suppose that you are examining the behavior of firms with headquarters in 10 geographic regions, two organizational structures (profit versus nonprofit), and four years of data. If you decide to treat each variable as a factor and want to model all interaction terms, then this is equivalent to a factor with $10 \times 2 \times 4 = 80$ levels. Models with interaction terms can have a substantial number of parameters and the analyst must be prudent when specifying interactions to be considered.

General Linear Model

The general linear model extends the linear regression model in two ways. First, explanatory variables may be continuous, categorical, or a combination. The only restriction is that they enter linearly such that the resulting regression function

$$\mathrm{E}\ y = \beta_0 + \beta_1 x_1 + \cdots + \beta_k x_k \tag{4.13}$$

is a linear combination of coefficients. As we have seen, we can square continuous variables or take other nonlinear transforms (e.g., logarithms) and use binary variables to represent categorical variables, so this restriction, as the name suggests, allows for a broad class of general functions to represent data.

The second extension is that the explanatory variables may be linear combinations of one another in the general linear model. Because of this, in the general linear model case, the parameter estimates need not be unique. However, an important feature of the general linear model is that the resulting fitted values turn out to be unique, using the method of least squares.

For example, in Section 4.3, we saw that the one factor ANOVA model could be expressed as a regression model with c indicator variables. However, if we had attempted to estimate the model in equation (4.10), the method of least squares would not have arrived at a unique set of regression coefficient estimates. The reason is that, in equation (4.10), each explanatory variable can be expressed as a linear combination of the others. For example, observe that $x_c = 1 - (x_1 + x_2 + \cdots + x_{c-1})$.

The fact that parameter estimates are not unique is a drawback but not an overwhelming one. The assumption that the explanatory variables are not linear combinations of one another means that we can compute unique estimates of the regression coefficients using the method of least squares. In terms of matrices, because the explanatory variables are not linear combinations of one another, the matrix $\mathbf{X}'\mathbf{X}$ is not invertible.

Specifically, suppose that we are considering the regression function in equation (4.13) and, using the method of least squares, our regression coefficient estimates are $b_0^o, b_1^o, \ldots, b_k^o$. This set of regression coefficients estimates minimizes our error sum of squares, but there may be other sets of coefficients that also minimize the error sum of squares. The fitted values are computed as $\hat{y}_i = b_0^o + b_1^o x_{i1} + \cdots + b_k^o x_{ik}$. It can be shown that the resulting fitted values are unique, in the sense that any set of coefficients that minimize the error sum of squares produce the same fitted values (see Section 4.7.3).

Thus, for a set of data and a specified general linear model, fitted values are unique. Because residuals are computed as observed responses minus fitted values, we have that the residuals are unique. Because residuals are unique, the error sums of squares are unique. Thus, it seems reasonable, and is true, that we can use the general test of hypotheses described in Section 4.2 to decide whether collections of explanatory variables are important.

To summarize, for general linear models, parameter estimates may not be unique and thus not meaningful. An important part of regression models is the interpretation of regression coefficients. This interpretation is not necessarily available in the general linear model context. However, for general linear models, we may still discuss the important of an individual variable or collection of variables through partial F-tests. Further, fitted values, and the corresponding exercise of prediction, works in the general linear model context. The advantage of the general linear model context is that we need not worry about the type of restrictions to impose on the parameters. Although not the subject of this text, this advantage is particularly important in complicated experimental designs used in the life sciences. You will find that general linear model estimation routines are widely available in statistical software packages available on the market today.

4.5 Further Reading and References

There are several good linear model books that focus on categorical variables and analysis of variance techniques. Hocking (2003) and Searle (1987) are good examples.

Chapter References

Hocking, Ronald R. (2003). *Methods and Applications of Linear Models: Regression and the Analysis of Variance*. John Wiley and Sons, New York.

Keeler, Emmett B., and John E. Rolph (1988). The demand for episodes of treatment in the Health Insurance Experiment. *Journal of Health Economics* 7: 337–67.

Searle, Shayle R. (1987). *Linear Models for Unbalanced Data*. John Wiley and Sons, New York.

4.6 Exercises

4.1. In this exercise, we consider relating two statistics that summarize how well a regression model fits, the F-ratio, and R^2, the coefficient of determination. (Here, the F-ratio is the statistic used to test model adequacy, not a partial F statistic.)

a. Write down R^2 in terms of both *Error SS* and *Regression SS*.

b. Write down F-ratio in terms of *Error SS*, *Regression SS*, k, and n.

c. Establish the algebraic relationship

$$F\text{-ratio} = \frac{R^2}{1 - R^2} \frac{n - (k + 1)}{k}.$$

d. Suppose that $n = 40$, $k = 5$, and $R^2 = 0.20$. Calculate the F-ratio. Perform the usual test of model adequacy to determine whether the five explanatory variables jointly and significantly affect the response variable.

e. Suppose that $n = 400$ (not 40), $k = 5$, and $R^2 = 0.20$. Calculate the F-ratio. Perform the usual test of model adequacy to determine whether the five explanatory variables jointly and significantly affect the response variable.

Ⓡ **Empirical**
Filename is "HospitalCosts"

4.2. **Hospital Costs.** This exercise considers hospital expenditures data provided by the U.S. Agency for Healthcare Research and Quality (AHRQ) and described in Exercise 1.4.

a. Produce a scatterplot, correlation, and linear regression of LNTOTCHG on AGE. Is AGE a significant predictor of LNTOTCHG?

b. You are concerned that newborns follow a different pattern than other ages do. Create a binary variable that indicates whether AGE equals zero. Run a regression using this binary variable and AGE as explanatory variables. Is the binary variable statistically significant?

c. Now examine the sex effect, using the binary variable FEMALE, which is one if the patient is female and zero otherwise. Run a regression using AGE and FEMALE as explanatory variables. Run a second regression running the two variables with an interaction term. Comment on whether the gender effect is important in either model.

d. Now consider the type of admission, APRDRG, an acronym for "all patient refined diagnostic related group." This is a categorical explanatory variable that provides information on the type of hospital admission. There are several hundred levels of this category. For example, level 640 represents admission for a normal newborn, with neonatal weight greater than or equal to 2.5 kilograms. As another example, level 225 represents admission resulting in an appendectomy.

d(i). Run a one factor ANOVA model, using APRDRG to predict LNTOTCHG. Examine the R^2 from this model and compare it to the coefficient of determination of the linear regression model of LNTOTCHG on AGE. On the basis of this comparison, which model do you think is preferred?

d(ii). For the one factor model in part d(i), provide a 95% confidence interval for LNTOTCHG for level 225 corresponding to an appendectomy. Convert your final answer from logarithmic dollars to dollars via exponentiation.

d(iii). Run a regression model of APRDRG, FEMALE, and AGE on LNTOTCHG. State whether AGE is a statistically significant predictor of LNTOTCHG. State whether FEMALE is a statistically significant predictor of LNTOTCHG.

Ⓡ **EMPIRICAL**
Filename is "WiscNursingHome"

4.3. **Nursing Home Utilization.** This exercise considers nursing home data provided by the Wisconsin Department of Health and Family Services (DHFS) and described in Exercises 1.2, 2.10, and 2.20.

In addition to the size variables, we also have information on several binary variables. The variable URBAN is used to indicate the facility's location. It is one if the facility is located in an urban environment and zero otherwise. The variable MCERT indicates whether the facility is Medicare certified. Most, but not all, nursing homes are certified to provide Medicare-funded care. There are three organizational structures for nursing homes. They are government (state, counties, municipalities), for-profit businesses, and tax-exempt organizations. Periodically, facilities may change ownership and, less frequently, ownership type. We create two binary variables PRO and TAXEXEMPT to denote for-profit business and tax-exempt organizations, respectively. Some nursing homes opt not to purchase private insurance coverage for their employees. Instead, such facilities directly provide insurance and pension benefits to their employees; this is referred to as "self-funding of insurance." We use binary variable SELFFUNDINS to denote it.

You decide to examine the relationship between LOGTPY(y) and the explanatory variables. Use cost-report year 2001 data, and do the following analysis:

a. There are three levels of organizational structures, but we only use two binary variables (PRO and TAXEXEMPT). Explain why.
b. Run a one-way analysis of variance using TAXEXEMPT as the factor. Decide whether tax-exempt is an important factor in determining LOGTPY. State your null hypothesis, alternative hypothesis, and all components of the decision-making rule. Use a 5% level of significance.
c. Run a one-way analysis of variance using MCERT as the factor. Decide whether location is an important factor in determining LOGTPY.
    c(i). Provide a point estimate of LOGTPY for a nursing facility that is not Medicare certified.
    c(ii). Provide a 95% confidence interval for your point estimate in part c(i).
d. Run a regression model using the binary variables, URBAN, PRO, TAXEXEMPT, SELFFUNDINS, and MCERT. Find R^2. Which variables are statistically significant?
e. Run a regression model using all explanatory variables, LOGNUMBED, LOGSQRFOOT, URBAN, PRO, TAXEXEMPT, SELFFUNDINS, and MCERT. Find R^2. Which variables are statistically significant?
    e(i). Calculate the partial correlation between LOGTPY and LOGSQRFOOT. Compare this to the correlation between LOGTPY and LOGSQRFOOT. Explain why the partial correlation is small.
    e(ii). Compare the low level of the t-ratios (for testing the importance of individual regression coefficients) and the high level of the F-ratio (for testing model adequacy). Describe the seeming inconsistency, and provide an explanation for this inconsistency.

4.4. **Automobile Insurance Claims.** Refer to Exercise 1.3.

Ⓡ **Empirical**
Filename is "AutoClaims"

a. Run a regression of LNPAID on AGE. Is AGE a statistically significant variable? To respond to this question, use a formal test of hypothesis. State your null and alternative hypotheses, decision-making criterion, and decision-making rule. Also comment on the goodness of fit of this variable.
b. Consider using class as a single explanatory variable. Use the one factor to estimate the model and respond to the following questions.
    b(i). What is the point estimate of claims in class C7, drivers aged 50–69, driving to work or school, fewer than 30 miles per week with annual mileage of less than 7500, in natural logarithmic units?
    b(ii). Determine the corresponding 95% confidence interval of expected claims, in natural logarithmic units.
    b(iii). Convert the 95% confidence interval of expected claims that you determined in part b(ii) to dollars.
c. Run a regression of LNPAID on AGE, GENDER, and the categorical variables STATE CODE and CLASS.

c(i). Is GENDER a statistically significant variable? To respond to this question, use a formal test of hypothesis. State your null and alternative hypotheses, decision-making criterion, and decision-making rule.

c(ii). Is CLASS a statistically significant variable? To respond to this question, use a formal test of hypothesis. State your null and alternative hypotheses, decision-making criterion, and decision-making rule.

c(iii). Use the model to provide a point estimate of claims in dollars (not log dollars) for a male age 60 in state 2 in class C7.

c(iv). Write down the coefficient associated with class C7 and interpret this coefficient.

Ⓡ **EMPIRICAL**
Filename is "WiscLottery"

4.5. **Wisconsin Lottery Sales.** This exercise considers state of Wisconsin lottery sales data (described in Section 2.1 and examined in Exercise 3.4).

Part 1: You decide to examine the relationship between SALES (y) and all eight explanatory variables (PERPERHH, MEDSCHYR, MEDHVL, PRCRENT, PRC55P, HHMEDAGE, MEDINC, and POP).

a. Fit a regression model of SALES on all eight explanatory variables.

b. Find R^2.

b(i). Use it to calculate the correlation coefficient between the observed and fitted values.

b(ii). You want to use R^2 to test the adequacy of the model in part (a). Use a formal test of hypothesis. State your null and alternative hypothesis, decision-making criterion, and decision-making rules.

c. Test whether POP, MEDSCHYR, and MEDHVL are jointly important explanatory variables for understanding SALES.

Part 2: After the preliminary analysis in part 1, you decide to examine the relationship between SALES(y) and POP, MEDSCHYR, and MEDHVL.

a. Fit a regression model of SALES on these three explanatory variables.

b. Has the coefficient of determination decreased from the eight-variable regression model to the three-variable model? Does this mean that the model is not improved or does it provide little information? Explain your response.

c. To state formally whether one should use the three- or eight-variable model, use a partial F-test. State your null and alternative hypotheses, decision-making criterion, and decision-making rules.

Ⓡ **EMPIRICAL**
Filename is "NAICExpense"

4.6. **Insurance Company Expenses.** This exercise considers insurance company data from the NAIC and described in Exercises 1.6 and 3.5.

a. Are the quadratic terms important?

Consider a linear model of LNEXPENSES on 12 explanatory variables. For the explanatory variables, include assets, GROUP, both versions of losses and gross premiums, as well as the two BLS variables. Also include the square each of the two loss and the two gross premium variables.

Test whether the four squared terms are jointly statistically significant, using a partial F-test. State your null and alternative hypotheses, decision-making criterion, and decision-making rules.

b. Are the interaction terms with GROUP important?

Omit the two BLS variables, so that now there are 11 variables, assets, GROUP, both versions of losses and gross premiums, as well as interactions of GROUP with assets and both versions of losses and gross premiums.

Test whether the five interaction terms are jointly statistically significant, using a partial F-test. State your null and alternative hypotheses, decision-making criterion, and decision-making rules.

c. You are examining a company that is not in the sample with values LOSSLONG = 0.025, LOSSSHORT = 0.040, GPWPERSONAL = 0.050, GPWCOMM = 0.120, ASSETS = 0.400,CASH = 0.350, and GROUP = 1.

Use the 11-variable-interaction model in part (b) to produce a 95% prediction interval for this company.

4.7. **National Life Expectancies.** We continue the analysis begun in Exercises 1.7, 2.22, and 3.6.

Ⓡ **EMPIRICAL** *Filename is "UNLifeExpectancy"*

a. Consider the regression using three explanatory variables, FERTILITY, PUBLICEDUCATION, and LNHEALTH that you did in Exercise 3.3.6. Test whether PUBLICEDUCATION and LNHEALTH are jointly statistically significant, using a partial F-test. State your null and alternative hypotheses, decision-making criterion, and decision-making rules. (Hint: use the coefficient of determination form for calculating the test statistic.) Provide an approximate p-value for the test.

b. We now introduce the REGION variable, summarized in Table 4.11. A box plot of life expectancies versus REGION is given in Figure 4.8. Describe what we learn from the table and box plot about the effect of REGION on LIFEEXP.

Table 4.11 Average Life Expectancy by Region

REGION	Region Description	Number	Mean
1	Arab states	13	71.9
2	East Asia and the Pacific	17	69.1
3	Latin American and the Carribean	25	72.8
4	South Asia	7	65.1
5	Southern Europe	3	67.4
6	Sub-Saharan Africa	38	52.2
7	Central and Eastern Europe	24	71.6
8	High-income OECD	23	79.6
	All	150	67.4

Note: OECD stands for Organization for Economic Co-operation and Development.

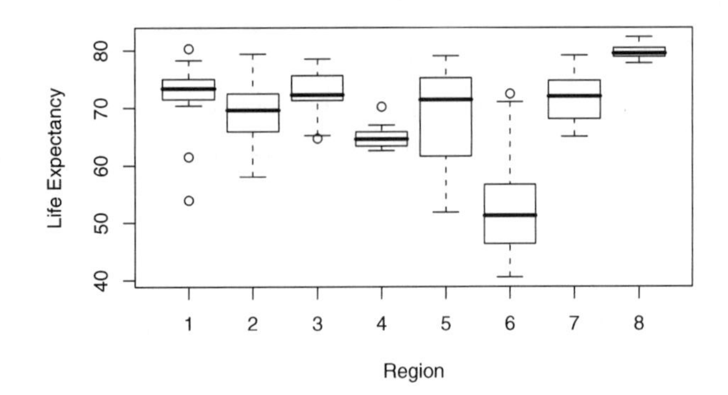

Figure 4.8 Box plots of LIFEEXP by REGION.

c. Fit a regression model using only the factor REGION. Is REGION a statistically significant determinant of LIFEEXP? State your null and alternative hypotheses, decision-making criterion, and decision-making rules.

d. Fit a regression model using three explanatory variables, FERTILITY, PUBLICEDUCATION, and LNHEALTH, as well as the categorical variable REGION.

d(i). You are examining a country that is not in the sample with values FERTILITY $= 2.0$, PUBLICEDUCATION $= 5.0$, and LNHEALTH $= 1.0$. Produce two predicted life expectancy values by assuming that the country is from (1) an Arab state and (2) sub-Saharan Africa.

d(ii). Provide a 95% confidence interval for the difference in life expectancies between an Arab state and a country from sub-Saharan Africa.

d(iii). Provide the (usual ordinary least squares) point estimate for the difference in life expectancies between a country from sub-Saharan Africa and a high-income OECD country.

4.7 Technical Supplement – Matrix Expressions

4.7.1 Expressing Models with Categorical Variables in Matrix Form

Chapter 3 showed how to write the regression model equation in the form $\mathbf{y} = \mathbf{X}\boldsymbol{\beta} + \boldsymbol{\varepsilon}$, where $\mathbf{X}$ is a matrix of explanatory variables. This form permits straightforward calculation of regression coefficients, $\mathbf{b} = \left(\mathbf{X}'\mathbf{X}\right)^{-1}\mathbf{X}'\mathbf{y}$. This section shows how the model and calculations reduce to simpler expressions when the explanatory variables are categorical.

One Categorical Variable Model

Consider the model with one categorical variable introduced in Section 4.3 with c levels of the categorical variable. From equation (4.9), this model can be written as

$$\mathbf{y} = \begin{bmatrix} y_{1,1} \\ \cdot \\ \cdot \\ \cdot \\ y_{n_1,1} \\ \cdot \\ \cdot \\ \cdot \\ y_{1,c} \\ \cdot \\ \cdot \\ \cdot \\ y_{n_c,c} \end{bmatrix} = \begin{bmatrix} 1 & 0 & \cdots & 0 \\ \cdot & \cdot & \cdots & \cdot \\ \cdot & \cdot & \cdots & \cdot \\ \cdot & \cdot & \cdots & \cdot \\ 1 & 0 & \cdots & \cdot \\ \cdot & \cdot & \cdots & \cdot \\ \cdot & \cdot & \cdots & \cdot \\ \cdot & \cdot & \cdots & \cdot \\ 0 & 0 & \cdots & 1 \\ \cdot & \cdot & \cdots & \cdot \\ \cdot & \cdot & \cdots & \cdot \\ \cdot & \cdot & \cdots & \cdot \\ 0 & 0 & \cdots & 1 \end{bmatrix} \begin{bmatrix} \mu_1 \\ \cdot \\ \cdot \\ \cdot \\ \mu_c \end{bmatrix} + \begin{bmatrix} \varepsilon_{1,1} \\ \cdot \\ \cdot \\ \cdot \\ \varepsilon_{n_1,1} \\ \cdot \\ \cdot \\ \cdot \\ \varepsilon_{1,c} \\ \cdot \\ \cdot \\ \cdot \\ \varepsilon_{n_c,c} \end{bmatrix} = \mathbf{X}\boldsymbol{\beta} + \boldsymbol{\varepsilon}. \tag{4.13}$$

To make the notation more compact, we write $\mathbf{0}$ and $\mathbf{1}$ for a column of zeros and ones, respectively. With this convention, another way to express equation (4.13) is

$$\mathbf{y} = \begin{bmatrix} \mathbf{1}_1 & \mathbf{0}_1 & \cdots & \mathbf{0}_1 \\ \mathbf{0}_2 & \mathbf{1}_2 & \cdots & \mathbf{0}_2 \\ \cdot & \cdot & \cdots & \cdot \\ \cdot & \cdot & \cdots & \cdot \\ \cdot & \cdot & \cdots & \cdot \\ \mathbf{0}_c & \mathbf{0}_c & \cdots & \mathbf{1}_c \end{bmatrix} \begin{bmatrix} \mu_1 \\ \mu_2 \\ \cdot \\ \cdot \\ \cdot \\ \mu_c \end{bmatrix} + \boldsymbol{\varepsilon} = \mathbf{X}\boldsymbol{\beta} + \boldsymbol{\varepsilon}. \tag{4.14}$$

Here, $\mathbf{0}_1$ and $\mathbf{1}_1$ stand for vector columns of length n_1 of zeros and ones, respectively, and similarly for $\mathbf{0}_2, \mathbf{1}_2, \ldots, \mathbf{0}_c, \mathbf{1}_c$.

Equation (4.14) allows us to apply the machinery developed for the regression model to the model with one categorical variable. As an intermediate calculation, we have

$$\begin{aligned} (\mathbf{X}'\mathbf{X})^{-1} &= \left(\begin{bmatrix} \mathbf{1}_1 & \mathbf{0}_2 & \cdots & \mathbf{0}_c \\ \mathbf{0}_1 & \mathbf{1}_2 & \cdots & \mathbf{0}_c \\ \cdot & \cdot & \cdots & \cdot \\ \cdot & \cdot & \cdots & \cdot \\ \cdot & \cdot & \cdots & \cdot \\ \mathbf{0}_1 & \mathbf{0}_2 & \cdots & \mathbf{1}_c \end{bmatrix}' \begin{bmatrix} \mathbf{1}_1 & \mathbf{0}_1 & \cdots & \mathbf{0}_1 \\ \mathbf{0}_2 & \mathbf{1}_2 & \cdots & \mathbf{0}_2 \\ \cdot & \cdot & \cdots & \cdot \\ \cdot & \cdot & \cdots & \cdot \\ \cdot & \cdot & \cdots & \cdot \\ \mathbf{0}_c & \mathbf{0}_c & \cdots & \mathbf{1}_c \end{bmatrix} \right)^{-1} \\ &= \begin{bmatrix} n_1 & 0 & \cdots & 0 \\ 0 & n_2 & \cdots & 0 \\ \cdot & \cdot & \cdots & \cdot \\ \cdot & \cdot & \cdots & \cdot \\ \cdot & \cdot & \cdots & \cdot \\ 0 & 0 & \cdots & n_c \end{bmatrix}^{-1} = \begin{bmatrix} \frac{1}{n_1} & 0 & \cdots & 0 \\ 0 & \frac{1}{n_2} & \cdots & 0 \\ \cdot & \cdot & \cdots & \cdot \\ \cdot & \cdot & \cdots & \cdot \\ \cdot & \cdot & \cdots & \cdot \\ 0 & 0 & \cdots & \frac{1}{n_c} \end{bmatrix}. \end{aligned} \tag{4.15}$$

Thus, the parameter estimates are

$$\mathbf{b} = \begin{bmatrix} \hat{\mu}_1 \\ \cdot \\ \cdot \\ \cdot \\ \hat{\mu}_c \end{bmatrix} = (\mathbf{X}'\mathbf{X})^{-1}\mathbf{X}'\mathbf{y} = \begin{bmatrix} \frac{1}{n_1} & 0 & \cdots & 0 \\ 0 & \frac{1}{n_2} & \cdots & 0 \\ \cdot & \cdot & \cdots & \cdot \\ \cdot & \cdot & \cdots & \cdot \\ \cdot & \cdot & \cdots & \cdot \\ 0 & 0 & \cdots & \frac{1}{n_c} \end{bmatrix} \begin{bmatrix} \mathbf{1}_1 & \mathbf{0}_2 & \cdots & \mathbf{0}_c \\ \mathbf{0}_1 & \mathbf{1}_2 & \cdots & \mathbf{0}_c \\ \cdot & \cdot & \cdots & \cdot \\ \cdot & \cdot & \cdots & \cdot \\ \cdot & \cdot & \cdots & \cdot \\ \mathbf{0}_1 & \mathbf{0}_2 & \cdots & \mathbf{1}_c \end{bmatrix}' \begin{bmatrix} y_{1,1} \\ \cdot \\ \cdot \\ \cdot \\ y_{n_1,1} \\ \cdot \\ \cdot \\ \cdot \\ y_{1,c} \\ \cdot \\ \cdot \\ \cdot \\ y_{n_c,c} \end{bmatrix}$$

$$= \begin{bmatrix} \frac{1}{n_1} & 0 & \cdots & 0 \\ 0 & \frac{1}{n_2} & \cdots & 0 \\ \cdot & \cdot & \cdots & \cdot \\ \cdot & \cdot & \cdots & \cdot \\ \cdot & \cdot & \cdots & \cdot \\ 0 & 0 & \cdots & \frac{1}{n_c} \end{bmatrix} \begin{bmatrix} \sum_{i=1}^{n_1} y_{i1} \\ \cdot \\ \cdot \\ \cdot \\ \sum_{i=1}^{n_c} y_{ic} \end{bmatrix} = \begin{bmatrix} \bar{y}_1 \\ \cdot \\ \cdot \\ \cdot \\ \bar{y}_c \end{bmatrix}. \tag{4.16}$$

We have seen that the least squares estimate of μ_j, $\bar{y}_j$, can been obtained directly from equation (4.9). By rewriting the model in matrix regression notation, we can appeal to linear regression model results and need not prove properties of models with categorical variables from first principles. That is, because this model is in regression format, we immediately have all the properties of the regression model.

Telling your software package that a variable is categorical can mean more efficient calculations, as in the calculations of the least squares regression estimates in equation (4.16). Further, calculation of other quantities can also be done more directly. As another example, we have that the standard error of $\hat{\mu}_j$ is

$$se(\hat{\mu}_j) = s\sqrt{j\text{th } \textit{diagonal element of } (\mathbf{X}'\mathbf{X})^{-1}} = s/\sqrt{n_j}.$$

One Categorical and One Continuous Variable Model

As another illustration, we consider the variable intercept and constant slope model in Table 4.8. This can be expressed as $\mathbf{y} = \mathbf{X}\boldsymbol{\beta} + \boldsymbol{\varepsilon}$, where

$$\mathbf{X} = \begin{bmatrix} \mathbf{1}_1 & \mathbf{0}_1 & \cdots & \mathbf{0}_1 & \mathbf{x}_1 \\ \mathbf{0}_2 & \mathbf{1}_2 & \cdots & \mathbf{0}_2 & \mathbf{x}_2 \\ \cdot & \cdot & \cdots & \cdot & \cdot \\ \cdot & \cdot & \cdots & \cdot & \cdot \\ \cdot & \cdot & \cdots & \cdot & \cdot \\ \mathbf{0}_c & \mathbf{0}_c & \cdots & \mathbf{1}_c & \mathbf{x}_c \end{bmatrix} \quad \text{and} \quad \boldsymbol{\beta} = \begin{bmatrix} \beta_{01} \\ \beta_{02} \\ \cdot \\ \cdot \\ \cdot \\ \beta_{0c} \\ \beta_1 \end{bmatrix}. \tag{4.17}$$

As before, $\mathbf{0}_j$ and $\mathbf{1}_j$ stand for vector columns of length n_j of zeros and ones, respectively. Further, $\mathbf{x}_j = (x_{1j}, x_{2j}, \ldots, x_{n_j,j})'$ is the column of the continuous variable at the jth level. Now, straightforward matrix algebra techniques provide the least squares estimates.

4.7.2 Calculating Least Squares Recursively

When computing regression coefficients using least squares, $\mathbf{b} = \left(\mathbf{X}'\mathbf{X}\right)^{-1}\mathbf{X}'\mathbf{y}$, for some applications, the dimension of $\mathbf{X}'\mathbf{X}$ can be large, causing computational difficulties. Fortunately, for some problems, the computations can be partitioned into smaller problems that can be solved recursively.

Recursive Least Squares Calculation. Suppose that the regression function can be written as

$$\mathrm{E}\,\mathbf{y} = \mathbf{X}\boldsymbol{\beta} = (\mathbf{X}_1 : \mathbf{X}_2)\begin{pmatrix}\boldsymbol{\beta}_1\\ \boldsymbol{\beta}_2\end{pmatrix}, \tag{4.18}$$

where $\mathbf{X}_1$ has dimensions $n \times k_1$, $\mathbf{X}_2$ has dimensions $n \times k_2$, $k_1 + k_2 = k$, $\boldsymbol{\beta}_1$ has dimensions $k_1 \times 1$, and $\boldsymbol{\beta}_2$ has dimensions $k_2 \times 1$. Define $\mathbf{Q}_1 = \mathbf{I} - \mathbf{X}_1\left(\mathbf{X}_1'\mathbf{X}_1\right)^{-1}\mathbf{X}_1'$. Then, the least squares estimator can be computed as

$$\mathbf{b} = \begin{pmatrix}\mathbf{b}_1\\ \mathbf{b}_2\end{pmatrix} = \begin{pmatrix}(\mathbf{X}_1'\mathbf{X}_1)^{-1}\mathbf{X}_1'(\mathbf{y} - \mathbf{X}_2\mathbf{b}_2)\\ \left(\mathbf{X}_2'\mathbf{Q}_1\mathbf{X}_2\right)^{-1}\mathbf{X}_2'\mathbf{Q}_1\mathbf{y}\end{pmatrix}. \tag{4.19}$$

Equation (4.19) provides the first step in the recursion. It can easily be iterated to allow for a more detailed decomposition of $\mathbf{X}$.

Special Case: One Categorical and One Continuous Variable Model. To illustrate the relevance of equation (4.19), let us return to the model summarized in equation (4.17). Here, we saw that the dimension of $\mathbf{X}$ is $n \times (c+1)$, and so the dimension of $\mathbf{X}'\mathbf{X}$ is $(c+1) \times (c+1)$. Taking the inverse of this matrix could be difficult if c is large, say, in the thousands. To apply equation (4.19), we define

$$\mathbf{X}_1 = \begin{bmatrix}\mathbf{1}_1 & \mathbf{0}_1 & \cdots & \mathbf{0}_1\\ \mathbf{0}_2 & \mathbf{1}_2 & \cdots & \mathbf{0}_2\\ \cdot & \cdot & \cdots & \cdot\\ \cdot & \cdot & \cdots & \cdot\\ \cdot & \cdot & \cdots & \cdot\\ \mathbf{0}_c & \mathbf{0}_c & \cdots & \mathbf{1}_c\end{bmatrix} \quad \text{and} \quad \mathbf{X}_2 = \begin{bmatrix}\mathbf{x}_1\\ \mathbf{x}_2\\ \cdot\\ \cdot\\ \cdot\\ \mathbf{x}_c\end{bmatrix}.$$

In this case, we have seen in equation (4.15) how it is straightforward to compute $\left(\mathbf{X}_1'\mathbf{X}_1\right)^{-1}$ without requiring matrix inversion. This means that calculating $\mathbf{Q}_1$ is also straightforward. With this, we can compute $\mathbf{X}_2'\mathbf{Q}_1\mathbf{X}_2$ and, because it is a scalar, immediately get its inverse. This gives us $\mathbf{b}_2$ and then we use this result to calculate $\mathbf{b}_1$. Although this procedure is not as direct as the our usual expression $\mathbf{b} = \left(\mathbf{X}'\mathbf{X}\right)^{-1}\mathbf{X}'\mathbf{y}$, it can be much more computationally efficient.

To establish equation (4.19), we use two results that are standard matrix algebra results on inverses of partitioned matrices.

Partitioned Matrix Results
Suppose that we can partition the $(p+q)\times(p+q)$ matrix $\mathbf{B}$ as

$$\mathbf{B} = \begin{bmatrix} \mathbf{B}_{11} & \mathbf{B}_{12} \\ \mathbf{B}'_{12} & \mathbf{B}_{22} \end{bmatrix},$$

where $\mathbf{B}_{11}$ is a $p \times p$ invertible matrix, $\mathbf{B}_{22}$ is a $q \times q$ invertible matrix, and $\mathbf{B}_{12}$ is a $p \times q$ matrix. Then

$$\mathbf{B}^{-1} = \begin{bmatrix} \mathbf{C}_{11}^{-1} & -\mathbf{B}_{11}^{-1}\mathbf{B}_{12}\mathbf{C}_{22}^{-1} \\ -\mathbf{C}_{22}^{-1}\mathbf{B}'_{12}\mathbf{B}_{11}^{-1} & \mathbf{C}_{22}^{-1} \end{bmatrix}, \tag{4.20}$$

where $\mathbf{C}_{11} = \mathbf{B}_{11} - \mathbf{B}_{12}\mathbf{B}_{22}^{-1}\mathbf{B}'_{12}$ and $\mathbf{C}_{22} = \mathbf{B}_{22} - \mathbf{B}'_{12}\mathbf{B}_{11}^{-1}\mathbf{B}_{12}$. To check equation (4.20), simply multiply $\mathbf{B}^{-1}$ by $\mathbf{B}$ to get $\mathbf{I}$, the identity matrix. Further,

$$\mathbf{C}_{11}^{-1} = \mathbf{B}_{11}^{-1} + \mathbf{B}_{11}^{-1}\mathbf{B}_{12}\mathbf{C}_{22}^{-1}\mathbf{B}'_{12}\mathbf{B}_{11}^{-1}. \tag{4.21}$$

Now, we first write the least squares estimator as

$$\mathbf{b} = (\mathbf{X}'\mathbf{X})^{-1}\mathbf{X}'\mathbf{y} = \left(\begin{pmatrix} \mathbf{X}'_1 \\ \mathbf{X}'_2 \end{pmatrix}(\mathbf{X}_1 : \mathbf{X}_2)\right)^{-1}\begin{pmatrix} \mathbf{X}'_1 \\ \mathbf{X}'_2 \end{pmatrix}\mathbf{y}$$

$$= \begin{pmatrix} \mathbf{X}'_1\mathbf{X}_1 & \mathbf{X}'_1\mathbf{X}_2 \\ \mathbf{X}'_2\mathbf{X}_1 & \mathbf{X}'_2\mathbf{X}_2 \end{pmatrix}^{-1}\begin{pmatrix} \mathbf{X}'_1\mathbf{y} \\ \mathbf{X}'_2\mathbf{y} \end{pmatrix} = \begin{pmatrix} \mathbf{b}_1 \\ \mathbf{b}_2 \end{pmatrix}.$$

To apply the partitioned matrix results, we define

$$\mathbf{Q}_j = \mathbf{I} - \mathbf{X}_j\left(\mathbf{X}'_j\mathbf{X}_j\right)^{-1}\mathbf{X}'_j,$$

$j = 1, 2$, and $\mathbf{B}_{j,k} = \mathbf{X}'_j\mathbf{X}_k$ for $j, k = 1, 2$. This means that $\mathbf{C}_{11} = \mathbf{X}'_1\mathbf{X}_1 - \mathbf{X}'_1\mathbf{X}_2(\mathbf{X}'_2\mathbf{X}_2)^{-1}\mathbf{X}'_2\mathbf{X}'_1 = \mathbf{X}'_1\mathbf{Q}_2\mathbf{X}_1$, and similarly $\mathbf{C}_{22} = \mathbf{X}'_2\mathbf{Q}_1\mathbf{X}_2$. From the second row, we have

$$\begin{aligned} \mathbf{b}_2 &= \mathbf{C}_{22}^{-1}\left(-\mathbf{B}'_{12}\mathbf{B}_{11}^{-1}\mathbf{X}'_1\mathbf{y} + \mathbf{X}'_2\mathbf{y}\right) \\ &= \left(\mathbf{X}'_2\mathbf{Q}_1\mathbf{X}_2\right)^{-1}\left(-\mathbf{X}'_2\mathbf{X}_1(\mathbf{X}'_1\mathbf{X}_1)^{-1}\mathbf{X}'_1\mathbf{y} + \mathbf{X}'_2\mathbf{y}\right) \\ &= \left(\mathbf{X}'_2\mathbf{Q}_1\mathbf{X}_2\right)^{-1}\mathbf{X}'_2\mathbf{Q}_1\mathbf{y}. \end{aligned}$$

From the first row,

$$\begin{aligned}
\mathbf{b}_1 &= \mathbf{C}_{11}^{-1}\mathbf{X}_1'\mathbf{y} - \mathbf{B}_{11}^{-1}\mathbf{B}_{12}\mathbf{C}_{22}^{-1}\mathbf{X}_2'\mathbf{y} \\
&= \left(\mathbf{B}_{11}^{-1} + \mathbf{B}_{11}^{-1}\mathbf{B}_{12}\mathbf{C}_{22}^{-1}\mathbf{B}_{21}\mathbf{B}_{11}^{-1}\right)\mathbf{X}_1'\mathbf{y} - \mathbf{B}_{11}^{-1}\mathbf{B}_{12}\mathbf{C}_{22}^{-1}\mathbf{X}_2'\mathbf{y} \\
&= \mathbf{B}_{11}^{-1}\mathbf{X}_1'\mathbf{y} - \mathbf{B}_{11}^{-1}\mathbf{B}_{12}\mathbf{C}_{22}^{-1}\left(-\mathbf{B}_{21}\mathbf{B}_{11}^{-1}\mathbf{X}_1'\mathbf{y} + \mathbf{X}_2'\mathbf{y}\right) \\
&= \mathbf{B}_{11}^{-1}\mathbf{X}_1'\mathbf{y} - \mathbf{B}_{11}^{-1}\mathbf{B}_{12}\mathbf{b}_2 \\
&= (\mathbf{X}_1'\mathbf{X}_1)^{-1}\mathbf{X}_1'\mathbf{y} - (\mathbf{X}_1'\mathbf{X}_1)^{-1}\mathbf{X}_1'\mathbf{X}_2\mathbf{b}_2 \\
&= (\mathbf{X}_1'\mathbf{X}_1)^{-1}\mathbf{X}_1'(\mathbf{y} - \mathbf{X}_2\mathbf{b}_2).
\end{aligned}$$

This establishes equation (4.19).

Reparameterized Model

For the partitioned regression function in equation 4.18, define $\mathbf{A} = \left(\mathbf{X}_1'\mathbf{X}_1\right)^{-1}\mathbf{X}_1'\mathbf{X}_2$ and $\mathbf{E}_2 = \mathbf{X}_2 - \mathbf{X}_1\mathbf{A}$. If one were to run a "multivariate" regression using $\mathbf{X}_2$ as the response and $\mathbf{X}_1$ as explanatory variables, then the parameter estimates would be $\mathbf{A}$ and the residuals $\mathbf{E}_2$.

With these definitions, use equation (4.18) to define the reparameterized regression model

$$\begin{aligned}
\mathbf{y} &= \mathbf{X}_1\boldsymbol{\beta}_1 + \mathbf{X}_2\boldsymbol{\beta}_2 + \boldsymbol{\varepsilon} = \mathbf{X}_1\boldsymbol{\beta}_1 + (\mathbf{E}_2 + \mathbf{X}_1\mathbf{A})\boldsymbol{\beta}_2 + \boldsymbol{\varepsilon} \\
&= \mathbf{X}_1\boldsymbol{\alpha}_1 + \mathbf{E}_2\boldsymbol{\beta}_2 + \boldsymbol{\varepsilon}, \qquad (4.22)
\end{aligned}$$

where $\boldsymbol{\alpha}_1 = \boldsymbol{\beta}_1 + \mathbf{A}\boldsymbol{\beta}_2$ is a new vector of parameters. The reason for introducing this new parameterization is that now the vector of explanatory variables is *orthogonal* to the other explanatory variables; that is, straightforward algebra shows that $\mathbf{X}_1'\mathbf{E}_2 = \mathbf{0}$.

By equation (4.19), the vector of least squares estimates is

$$\mathbf{a} = \begin{bmatrix} \mathbf{a}_1 \\ \mathbf{b}_2 \end{bmatrix} = \left(\begin{bmatrix} \mathbf{X}_1' \\ \mathbf{E}_2' \end{bmatrix} \begin{bmatrix} \mathbf{X}_1 & \mathbf{E}_2 \end{bmatrix}\right)^{-1} \begin{bmatrix} \mathbf{X}_1' \\ \mathbf{E}_2' \end{bmatrix} \mathbf{y} = \begin{bmatrix} \left(\mathbf{X}_1'\mathbf{X}_1\right)^{-1}\mathbf{X}_1'\mathbf{y} \\ \left(\mathbf{E}_2'\mathbf{E}_2\right)^{-1}\mathbf{E}_2'\mathbf{y} \end{bmatrix}. \qquad (4.23)$$

Extra Sum of Squares

Suppose that we wish to consider the increase in the error sum of squares going from a *reduced* model

$$\mathbf{y} = \mathbf{X}_1\boldsymbol{\beta}_1 + \boldsymbol{\varepsilon}$$

to a *full* model

$$\mathbf{y} = \mathbf{X}_1\boldsymbol{\beta}_1 + \mathbf{X}_2\boldsymbol{\beta}_2 + \boldsymbol{\varepsilon}.$$

For the reduced model, the error sum of squares is

$$(Error\ SS)_{reduced} = \mathbf{y}'\mathbf{y} - \mathbf{y}'\mathbf{X}_1(\mathbf{X}_1'\mathbf{X}_1)^{-1}\mathbf{X}_1'\mathbf{y}. \qquad (4.24)$$

Using the reparameterized version of the full model, the error sum of squares is

$$(Error\ SS)_{full} = \mathbf{y}'\mathbf{y} - \mathbf{a}' \begin{bmatrix} \mathbf{X}_1' \\ \mathbf{E}_2' \end{bmatrix} \mathbf{y} = \mathbf{y}'\mathbf{y} - \begin{bmatrix} (\mathbf{X}_1'\mathbf{X}_1)^{-1}\mathbf{X}_1'\mathbf{y} \\ (\mathbf{E}_2'\mathbf{E}_2)^{-1}\mathbf{E}_2'\mathbf{y} \end{bmatrix}' \begin{bmatrix} \mathbf{X}_1'\mathbf{y} \\ \mathbf{E}_2'\mathbf{y} \end{bmatrix}$$

$$= \mathbf{y}'\mathbf{y} - \mathbf{y}'\mathbf{X}_1(\mathbf{X}_1'\mathbf{X}_1)^{-1}\mathbf{X}_1'\mathbf{y} - \mathbf{y}'\mathbf{E}_2(\mathbf{E}_2'\mathbf{E}_2)^{-1}\mathbf{E}_2'\mathbf{y}. \qquad (4.25)$$

Thus, the reduction in the error sum of squares by adding $\mathbf{X}_2$ to the model is

$$(Error\ SS)_{reduced} - (Error\ SS)_{full} = \mathbf{y}'\mathbf{E}_2(\mathbf{E}_2'\mathbf{E}_2)^{-1}\mathbf{E}_2'\mathbf{y}. \qquad (4.26)$$

As noted in Section 4.3, the quantity $(Error\ SS)_{reduced} - (Error\ SS)_{full}$ is called the *extra sum of squares*, or Type III sum of squares. It is produced automatically by some statistical software packages, thus obviating the need to run separate regressions.

4.7.3 General Linear Model

Recall the general linear model from Section 4.4. That is, we use

$$y_i = \beta_0 x_{i0} + \beta_1 x_{i1} + \cdots + \beta_k x_{ik} + \varepsilon_i,$$

or, in matrix notation, $\mathbf{y} = \mathbf{X}\boldsymbol{\beta} + \boldsymbol{\varepsilon}$. As before, we use Assumptions F1–F4 (or E1–E4) so that the disturbance terms are i.i.d mean zero with common variance σ^2 and the explanatory variables $\{x_{i0}, x_{i1}, x_{i2}, \ldots, x_{ik}\}$ are nonstochastic.

In the general linear model, we do not require that $\mathbf{X}'\mathbf{X}$ be invertible. As we have seen in Chapter 4, an important reason for this generalization relates to handling categorical variables. That is, to use categorical variables, they are generally recoded using binary variables. For this recoding, generally some type of restrictions need to be made on the set of parameters associated with the indicator variables. However, it is not always clear what type of restrictions are the most intuitive. By expressing the model without requiring that $\mathbf{X}'\mathbf{X}$ be invertible, the restrictions can be imposed after the estimation is done, not before.

Normal Equations

Even when $\mathbf{X}'\mathbf{X}$ is not invertible, solutions to the normal equations still provide least squares estimates of $\boldsymbol{\beta}$. That is, the sum of squares is

$$SS(\mathbf{b}^*) = (\mathbf{y} - \mathbf{X}\mathbf{b}^*)'(\mathbf{y} - \mathbf{X}\mathbf{b}^*),$$

where $\mathbf{b}^* = (b_0^*, b_1^*, \ldots, b_k^*)'$ is a vector of candidate estimates. Solutions of the normal equations are those vectors $\mathbf{b}^\circ$ that satisfy the normal equations

$$\mathbf{X}'\mathbf{X}\mathbf{b}^\circ = \mathbf{X}'\mathbf{y}. \qquad (4.27)$$

We use the notation $^\circ$ to remind ourselves that $\mathbf{b}^\circ$ need not be unique. However, it is a minimizer of the sum of squares. To see this, consider another candidate vector $\mathbf{b}^*$ and note that $SS(\mathbf{b}^*) = \mathbf{y}'\mathbf{y} - 2\mathbf{b}^{*\prime}\mathbf{X}'\mathbf{y} + \mathbf{b}^{*\prime}\mathbf{X}'\mathbf{X}\mathbf{b}^*$. Then, using equation

(4.27), we have

$$\begin{aligned} SS(\mathbf{b}^*) - SS(\mathbf{b}^\circ) &= -2\mathbf{b}^{*\prime}\mathbf{X}'\mathbf{y} + \mathbf{b}^{*\prime}\mathbf{X}'\mathbf{X}\mathbf{b}^* - (-2\mathbf{b}^{\circ\prime}\mathbf{X}\mathbf{y} + \mathbf{b}^{\circ\prime}\mathbf{X}'\mathbf{X}\mathbf{b}^\circ) \\ &= -2\mathbf{b}^{*\prime}\mathbf{X}\mathbf{b}^\circ + \mathbf{b}^{*\prime}\mathbf{X}'\mathbf{X}\mathbf{b}^* + \mathbf{b}^{\circ\prime}\mathbf{X}'\mathbf{X}\mathbf{b}^\circ \\ &= (\mathbf{b}^* - \mathbf{b}^\circ)'\mathbf{X}'\mathbf{X}(\mathbf{b}^* - \mathbf{b}^\circ) = \mathbf{z}'\mathbf{z} \geq 0, \end{aligned}$$

where $\mathbf{z} = \mathbf{X}(\mathbf{b}^* - \mathbf{b}^\circ)$. Thus, any other candidate $\mathbf{b}^*$ yields a sum of squares at least as large as $SS(\mathbf{b}^\circ)$.

Unique Fitted Values

Despite the fact that there may be (infinitely) many solutions to the normal equations, the resulting fitted values, $\hat{\mathbf{y}} = \mathbf{X}\mathbf{b}^\circ$, are unique. To see this, suppose that $\mathbf{b}_1^\circ$ and $\mathbf{b}_2^\circ$ are two different solutions of equation (4.27). Let $\hat{\mathbf{y}}_1 = \mathbf{X}\mathbf{b}_1^\circ$ and $\hat{\mathbf{y}}_2 = \mathbf{X}\mathbf{b}_2^\circ$ denote the vectors of fitted values generated by these estimates. Then

$$(\hat{\mathbf{y}}_1 - \hat{\mathbf{y}}_2)'(\hat{\mathbf{y}}_1 - \hat{\mathbf{y}}_2) = (\mathbf{b}_1^\circ - \mathbf{b}_2^\circ)'\mathbf{X}'\mathbf{X}(\mathbf{b}_1^\circ - \mathbf{b}_2^\circ) = 0,$$

because $\mathbf{X}'\mathbf{X}(\mathbf{b}_1^\circ - \mathbf{b}_2^\circ) = \mathbf{X}'\mathbf{y} - \mathbf{X}'\mathbf{y} = \mathbf{0}$, from equation (4.27). Hence, $\hat{\mathbf{y}}_1 = \hat{\mathbf{y}}_2$ for any choice of $\mathbf{b}_1^\circ$ and $\mathbf{b}_2^\circ$, thus establishing the uniqueness of the fitted values.

Because the fitted values are unique, the residuals are also unique. Thus, the error sum of squares and estimates of variability (e.g., s^2) are also unique.

Generalized Inverses

A *generalized inverse* of a matrix $\mathbf{A}$ is a matrix $\mathbf{B}$ such that $\mathbf{ABA} = \mathbf{A}$. We use the notation $\mathbf{A}^-$ to denote the generalized inverse of $\mathbf{A}$. In the case that $\mathbf{A}$ is invertible, then $\mathbf{A}^-$ is unique and equals $\mathbf{A}^{-1}$. Although there are several definitions of generalized inverses, the foregoing definition suffices for our purposes. See Searle (1987) for further discussion of alternative definitions of generalized inverses.

With this definition, it can be shown that a solution to the equation $\mathbf{Ab} = \mathbf{c}$ can be expressed as $\mathbf{b} = \mathbf{A}^-\mathbf{c}$. Thus, we can express a least squares estimate of $\boldsymbol{\beta}$ as $\mathbf{b}^\circ = (\mathbf{X}'\mathbf{X})^-\mathbf{X}'\mathbf{y}$. Statistical software packages can calculate versions of $(\mathbf{X}'\mathbf{X})^-$ and thus generate $\mathbf{b}^\circ$.

Estimable Functions

We have seen that each fitted value $\hat{y}_i$ is unique. Because fitted values are simply linear combinations of parameters estimates, it seems reasonable to ask what other linear combinations of parameter estimates are unique. To this end, we say that $\mathbf{C}\boldsymbol{\beta}$ is an *estimable function* of parameters if $\mathbf{C}\mathbf{b}^\circ$ does not depend (is *invariant*) to the choice of $\mathbf{b}^\circ$. Because fitted values are invariant to the choice of $\mathbf{b}^\circ$, we have that $\mathbf{X} = \mathbf{C}$ produces one type of estimable function. Interestingly, it turns out that all estimable functions are of the form $\mathbf{LXb}^\circ$; that is, $\mathbf{C} = \mathbf{LX}$. See Searle (1987, p. 284) for a demonstration of this. Thus, all estimable function are linear combinations of fitted values; that is, $\mathbf{LXb}^\circ = \mathbf{L}\hat{\mathbf{y}}$.

Estimable functions are unbiased and a variance that does not depend on the choice of the generalized inverse. That is, it can be shown that E $\mathbf{Cb}^\circ = \mathbf{C}\boldsymbol{\beta}$, and Var $\mathbf{Cb}^\circ = \sigma^2\mathbf{C}(\mathbf{X}'\mathbf{X})^-\mathbf{C}'$ does not depend on the choice of $(\mathbf{X}'\mathbf{X})^-$.

Testable Hypotheses

As described in Section 4.2, if is often of interest to test H$_0$: $\mathbf{C}\boldsymbol{\beta} = \mathbf{d}$, where $\mathbf{d}$ is a specified vector. This hypothesis is said to be *testable* if $\mathbf{C}\boldsymbol{\beta}$ is an estimable function, $\mathbf{C}$ is of full row rank, and the rank of $\mathbf{C}$ is less than the rank of $\mathbf{X}$. For consistency with the notation of Section 4.2, let p be the rank of $\mathbf{C}$ and $k+1$ be the rank of $\mathbf{X}$. Recall that the rank of a matrix is the smaller of the number of linearly independent rows and linearly independent columns. When we say that $\mathbf{C}$ has full row rank, we mean that there are p rows in $\mathbf{C}$, so that the number of rows equals the rank.

General Linear Hypothesis

As in Section 4.2, the test statistic for examining H_0: $\mathbf{C}\boldsymbol{\beta} = \mathbf{d}$ is

$$F\text{-ratio} = \frac{(\mathbf{Cb}^\circ - \mathbf{d})'(\mathbf{C}(\mathbf{X}'\mathbf{X})^-\mathbf{C}')^{-1}(\mathbf{Cb}^\circ - \mathbf{d})}{p s_{full}^2}.$$

Note that the statistic F-ratio does not depend on the choice of $\mathbf{b}^\circ$ because $\mathbf{Cb}^\circ$ is invariant to $\mathbf{b}^\circ$. If H$_0$: $\mathbf{C}\boldsymbol{\beta} = \mathbf{d}$ is a testable hypothesis and the errors ε_i are i.i.d. N$(0, \sigma^2)$, then the F-ratio has an F-distribution with $df_1 = p$ and $df_2 = n - (k+1)$.

One Categorical Variable Model

We now illustrate the general linear model by considering an overparameterized version of the one-factor model that appears in equation (4.10) using

$$y_{ij} = \mu + \tau_j + e_{ij} = \mu + \tau_1 x_{i1} + \tau_2 x_{i2} + \cdots + \tau_c x_{ic} + \varepsilon_{ij}.$$

At this point, we do not impose additional restrictions in the parameters. As with equation (4.13), this can be written in matrix form as

$$\mathbf{y} = \begin{bmatrix} \mathbf{1}_1 & \mathbf{1}_1 & \mathbf{0}_1 & \cdots & \mathbf{0}_1 \\ \mathbf{1}_2 & \mathbf{0}_2 & \mathbf{1}_2 & \cdots & \mathbf{0}_2 \\ \cdot & \cdot & \cdot & \cdots & \cdot \\ \mathbf{1}_c & \mathbf{0}_c & \mathbf{0}_c & \cdots & \mathbf{1}_c \end{bmatrix} \begin{bmatrix} \mu \\ \tau_1 \\ \cdot \\ \cdot \\ \cdot \\ \tau_c \end{bmatrix} + \boldsymbol{\varepsilon} = \mathbf{X}\boldsymbol{\beta} + \boldsymbol{\varepsilon}.$$

Thus, the $\mathbf{X}'\mathbf{X}$ matrix is

$$\mathbf{X}'\mathbf{X} = \begin{bmatrix} n & n_1 & n_2 & \cdots & n_c \\ n_1 & n_1 & 0 & \cdots & 0 \\ n_2 & 0 & n_2 & \cdots & 0 \\ \cdot & \cdot & \cdot & \cdots & \cdot \\ \cdot & \cdot & \cdot & \cdots & \cdot \\ \cdot & \cdot & \cdot & \cdots & \cdot \\ n_c & 0 & 0 & \cdots & n_c \end{bmatrix},$$

where $n = n_1 + n_2 + \cdots + n_c$. This matrix is not invertible. To see this, note that adding the last c rows together yields the first row. Thus, the last c rows are an exact linear combination of the first row, meaning that the matrix is not full rank.

The (nonunique) least squares estimates can be expressed as

$$\mathbf{b}^{\circ} = \begin{bmatrix} \mu^{\circ} \\ \tau_1^{\circ} \\ \cdot \\ \cdot \\ \cdot \\ \tau_c^{\circ} \end{bmatrix} = (\mathbf{X}'\mathbf{X})^{-}\mathbf{X}'\mathbf{y}.$$

Estimable functions are linear combinations of fitted values. Because fitted values are $\hat{y}_{ij} = \bar{y}_j$, estimable functions can be expressed as $L = \sum_{j=1}^{c} a_i \bar{y}_j$, where $a_1, \ldots, a_c$ are constants. This linear combination of fitted values is an unbiased estimator of $\mathrm{E}\, L = \sum_{i=1}^{c} a_i(\mu + \tau_i)$.

Thus, for example, by choosing $a_1 = 1$, and the other $a_i = 0$, we see that $\mu + \tau_1$ is estimable. As another example, by choosing $a_1 = 1$, $a_2 = -1$, and the other $a_i = 0$, we see that $\tau_1 - \tau_2$ is estimable. It can be shown that μ is not an estimable parameter without further restrictions on $\tau_1, \ldots, \tau_c$.

5

Variable Selection

Chapter Preview. This chapter describes tools and techniques to help you select variables to enter into a linear regression model, beginning with an iterative model selection process. In applications with many potential explanatory variables, automatic variable selection procedures will help you quickly evaluate many models. Nonetheless, automatic procedures have serious limitations, including the inability to account properly for nonlinearities such as the impact of unusual points; this chapter expands on the discussion in Chapter 2 of unusual points. It also describes collinearity, a common feature of regression data where explanatory variables are linearly related to one another. Other topics that affect variable selection, including heteroscedasticity and out-of-sample validation, are also introduced.

5.1 An Iterative Approach to Data Analysis and Modeling

In our introduction of basic linear regression in Chapter 2, we examined the data graphically, hypothesized a model structure, and compared the data to a candidate model to formulate an improved model. Box (1980) describes this as an *iterative process*, which is shown in Figure 5.1.

Diagnostic checking reveals symptoms of mistakes made in previous specification steps and provides ways to correct the mistakes.

This iterative process provides a useful recipe for structuring the task of specifying a model to represent a set of data. The first step, the model formulation stage, is accomplished by examining the data graphically and using prior knowledge of relationships, such as from economic theory or standard industry practice. The second step in the iteration is based on the assumptions of the specified model. These assumptions must be consistent with the data to make valid use of the model. The third step, *diagnostic checking*, is also known as *data and model*

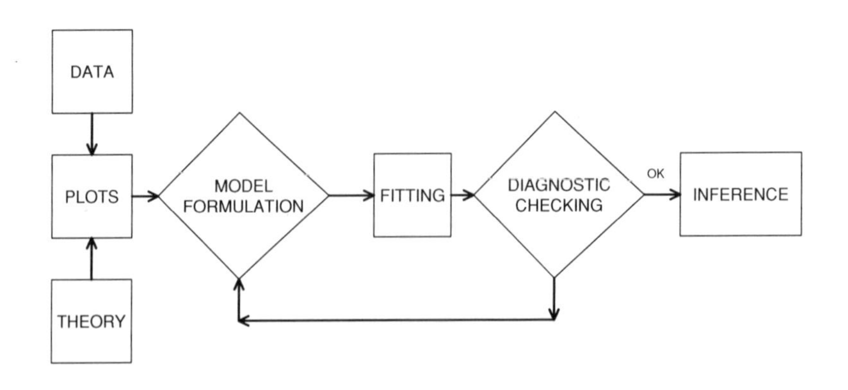

Figure 5.1 The iterative model specification process.

Table 5.1 Sixteen Possible Models

Model			Count
$\mathrm{E}\, y = \beta_0$			1 model with no independent variables
$\mathrm{E}\, y = \beta_0 + \beta_1 x_i$,	$i =$	1, 2, 3, 4	4 models with one independent variable
$\mathrm{E}\, y = \beta_0 + \beta_1 x_i + \beta_2 x_j$,	$(i, j) =$	(1, 2), (1, 3), (1, 4), (2, 3), (2, 4), (3, 4)	6 models with two independent variables
$\mathrm{E}\, y = \beta_0 + \beta_1 x_1 + \beta_2 x_j + \beta_3 x_k$,	$(i, j, k) =$	(1, 2, 3), (1, 2, 4), (1, 3, 4), (2, 3, 4)	4 models with three independent variables
$\mathrm{E}\, y = \beta_0 + \beta_1 x_1 + \beta_2 x_2 + \beta_3 x_3 + \beta_4 x_4$			1 model with all independent variables

criticism; the data and model must be consistent with one another before additional inferences can be made. Diagnostic checking is an important part of the model formulation; it can reveal mistakes made in previous steps and provide ways to correct them.

The iterative process also emphasizes the skills you need to make regression analysis work. First, you need a willingness to summarize information numerically and portray this information graphically. Second, it is important to develop an understanding of model properties. You should understand how a theoretical model behaves to match a set of data to it. Third, understanding theoretical properties of the model are also important for inferring general relationships based on the behavior of the data.

5.2 Automatic Variable Selection Procedures

Business and economics relationships are complicated; there are typically many variables that can serve as useful predictors of the dependent variable. In searching for a suitable relationship, there is a large number of potential models that are based on linear combinations of explanatory variables and an infinite number that can be formed from nonlinear combinations. To search among models based on linear combinations, several automatic procedures are available to select variables to be included in the model. These automatic procedures are easy to use and will suggest one or more models that you can explore in further detail.

To illustrate how large the potential number of linear models is, suppose that there are only four variables, x_1, x_2, x_3, and x_4, under consideration for fitting a model to y. Without any consideration of multiplication or other nonlinear combinations of explanatory variables, how many possible models are there? Table 5.1 shows that the answer is 16.

If there were only three explanatory variables, then you can use the same logic to verify that there are eight possible models. Extrapolating from these two examples, how many linear models will there be if there are ten explanatory variables? The answer is 1024, which is quite a few. In general, the answer is 2^k, where k is the number of explanatory variables. For example, 2^3 is 8, 2^4 is 16, and so on.

In any case, for a moderately large number of explanatory variables, there are many potential models that are based on linear combinations of explanatory variables. We would like a procedure to search quickly through these potential models to give us more time to think about other interesting aspects of model selection. *Stepwise regression* are procedures that employ t-tests to check the "significance" of explanatory variables entered into, or deleted from, the model.

To begin, in the *forward selection* version of stepwise regression, variables are added one at a time. In the first stage, out of all the candidate variables, the one that is most statistically significant is added to the model. At the next stage, with the first stage variable already included, the next most statistically significant variable is added. This procedure is repeated until all statistically significant variables have been added. Here, statistical significance is typically assessed using a variable's t-ratio – the cutoff for statistical significance is typically a predetermined t-value (e.g., two, corresponding to an approximate 95% significance level).

The *backward selection* version works in a similar manner, except that all variables are included in the initial stage and then dropped one at a time (instead of added).

More generally, an algorithm that adds and deletes variables at each stage is sometimes known as the *stepwise regression algorithm*.

Stepwise Regression Algorithm. Suppose that the analyst has identified one variable as the response, y, and k potential explanatory variables, $x_1, x_2, \ldots, x_k$.

(i) Consider all possible regressions using one explanatory variable. For each of the k regressions, compute $t(b_1)$, the t-ratio for the slope. Choose that variable with the largest t-ratio. If the t-ratio does not exceed a prespecified t-value (e.g., two), then do not choose any variables and halt the procedure.
(ii) Add a variable to the model from the previous step. The variable to enter is the one that makes the largest significant contribution. To determine the size of contribution, use the absolute value of the variable's t-ratio. To enter, the t-ratio must exceed a specified t-value in absolute value.
(iii) Delete a variable to the model from the previous step. The variable to be removed is the one that makes the smallest contribution. To determine the size of contribution, use the absolute value of the variable's t-ratio. To be removed, the t-ratio must be less than a specified t-value in absolute value.
(iv) Repeat steps (ii) and (iii) until all possible additions and deletions are performed.

When implementing this routine, some statistical software packages use an F-test in lieu of t-tests. Recall that when only one variable is being considered, $(t\text{-ratio})^2 = F\text{-ratio}$ and thus the procedures are equivalent.

This algorithm is useful in that it quickly searches through a number of candidate models. However, there are several drawbacks:

1. The procedure "snoops" through a large number of models and may fit the data "too well."
2. There is no guarantee that the selected model is the best. The algorithm does not consider models that are based on nonlinear combinations of explanatory variables. It also ignores the presence of outliers and high leverage points.
3. In addition, the algorithm does not even search all 2^k possible linear regressions.
4. The algorithm uses one criterion, a t-ratio, and does not consider other criteria such as s, R^2, R_a^2, and so on.
5. There is a sequence of significance tests involved. Thus, the significance level that determines the t-value is not meaningful.
6. By considering each variable separately, the algorithm does not take into account the joint effect of explanatory variables.
7. Purely automatic procedures may not take into account an investigator's special knowledge.

Many of the criticisms of the basic stepwise regression algorithm can be addressed with modern computing software that is now widely available. We now consider each drawback, in reverse order. To respond to drawback number (7), many statistical software routines have options for forcing variables into a model equation. In this way, if other evidence indicates that one or more variables should be included in the model, then the investigator can force the inclusion of these variables.

For drawback number (6), in Section 5.5.4 on *suppressor variables*, we will provide examples of variables that do not have important individual effects but are important when considered jointly. These combinations of variables may not be detected with the basic algorithm but will be detected with the backward selection algorithm. Because the backward procedure starts with all variables, it will detect, and retain, variables that are jointly important.

Drawback number (5) is really a suggestion about the way to use stepwise regression. Bendel and Afifi (1977) suggested using a cutoff smaller than you ordinarily might. For example, in lieu of using t-value $= 2$, corresponding approximately to a 5% significance level, consider using t-value $= 1.645$, corresponding approximately to a 10% significance level. In this way, there is less chance of screening out variables that may be important. A lower bound, but still a good choice for exploratory work, is a cutoff as small as t-value $= 1$. This choice is motivated by an algebraic result: when a variable enters a model, s will decrease if the t-ratio exceeds one in absolute value.

When a variable enters a model, s will decrease if the t-ratio exceeds one in absolute value.

To address drawbacks (3) and (4), we now introduce the *best regressions* routine. Best regressions is a useful algorithm that is now widely available in statistical software packages. The best regression algorithm searches over all possible combinations of explanatory variables, unlike stepwise regression, which adds and deletes one variable at a time. For example, suppose that there are four possible explanatory variables, x_1, x_2, x_3, and x_4, and the user would like to know what is the best two variable model. The best regression algorithm searches over all six models of the form $\mathrm{E}\ y = \beta_0 + \beta_1 x_i + \beta_2 x_j$. Typically, a best regression routine recommends one or two models for each p coefficient model, where p is a number that is user specified. Because it has specified the number of coefficients to enter the model, it does not matter which of the criteria we use: R^2, R_a^2, or s.

The best regression algorithm performs its search by a clever use of the algebraic fact that, when a variable is added to the model, the error sum of squares does not increase. Because of this fact, certain combinations of variables included in the model need not be computed. An important drawback of this algorithm is that it can take a considerable amount of time when the number of variables considered is large.

Users of regression do not always appreciate the depth of drawback (1), *data snooping*. Data snooping occurs when the analyst fits a great number of models to a dataset. We will address the problem of data snooping in Section 5.6.2 on model validation. Here, we illustrate the effect of data snooping in stepwise regression.

Example: Data Snooping in Stepwise Regression. The idea of this illustration is due to Rencher and Pun (1980). Consider $n = 100$ observations of y and 50 explanatory variables, $x_1, x_2, \ldots, x_{50}$. The data we consider here were simulated using independent standard normal random variates. Because the variables were simulated independently, we are working under the null hypothesis of no relation between the response and the explanatory variables; that is, H_0: $\beta_1 = \beta_2 = \cdots = \beta_{50} = 0$. Indeed, when the model with all 50 explanatory variables was fit, it turns out that $s = 1.142$, $R^2 = 46.2\%$, and F-ratio = (*Regression MS*)/(*Error MS*) = 0.84. Using an F-distribution with $df_1 = 50$ and $df_2 = 49$, the 95th percentile is 1.604. In fact, 0.84 is the 27th percentile of this distribution, indicating that the p-value is 0.73. Thus, as expected, the data are in congruence with H_0.

Next, a stepwise regression with t-value $= 2$ was performed. Two variables were retained by this procedure, yielding a model with $s = 1.05$, $R^2 = 9.5\%$, and F-ratio $= 5.09$. For an F-distribution with $df_1 = 2$ and $df_2 = 97$, the 95th percentile is F-value $= 3.09$. This indicates that the two variables are statistically significant predictors of y. At first glance, this result is surprising. The data were generated so that y is unrelated to the explanatory variables. However, because F-ratio $>$ F-value, the F-test indicates that two explanatory variables are significantly related to y. The reason is that stepwise regression has performed many hypothesis tests on the data. For example, in Step 1, 50 tests were performed to find significant variables. Recall that a 5% level means that we expect to make roughly 1 mistake in 20. Thus, with 50 tests, we expect to find $50(0.05) = 2.5$ "significant" variables, even under the null hypothesis of no relationship between y and the explanatory variables.

To continue, a stepwise regression with t-value = 1.645 was performed. Six variables were retained by this procedure, yielding a model with $s = 0.99$, $R^2 =$ 22.9%, and F-ratio = 4.61. As before, an F-test indicates a significant relationship between the response and these six explanatory variables.

To summarize, using simulation, we constructed a dataset so that the explanatory variables have no relationship with the response. However, when using stepwise regression to examine the data, we "found" seemingly significant relationships between the response and certain subsets of the explanatory variables. This example illustrates a general caveat in model selection: when explanatory variables are selected using the data, t-ratios and F-ratios will be too large, thus overstating the importance of variables in the model.

When explanatory variables are selected using the data, t-ratios and F-ratios will be too large, thus overstating the importance of variables in the model.

Stepwise regression and best regressions are examples of *automatic variable selection procedures*. In your modeling work, you will find these procedures to be useful because they can quickly search through several candidate models. However, these procedures ignore nonlinear alternatives and the effect of outliers and high leverage points. The main point of the procedures is to mechanize certain routine tasks. This automatic selection approach can be extended, and indeed, there are a number of so-called "expert systems" available in the market. For example, algorithms are available that "automatically" handle unusual points such as outliers and high leverage points. A model suggested by automatic variable selection procedures should be subject to the same careful diagnostic checking procedures as a model arrived at by any other means.

A model suggested by automatic variable selection procedures should be subject to the same careful diagnostic checking procedures as a model arrived at by any other means.

5.3 Residual Analysis

Recall the role of a residual in the linear regression model introduced in Section 2.6. A residual is a response minus the corresponding fitted value under the model. Because the model summarizes the linear effect of several explanatory variables, we can think of a residual as a response controlled for values of the explanatory variables. If the model is an adequate representation of the data, then residuals should closely approximate random errors. Random errors are used to represent the natural variation in the model; they represent the result of an unpredictable mechanism. Thus, to the extent that residuals resemble random errors, there should be no discernible patterns in the residuals. Patterns in the residuals indicate the presence of additional information that we hope to incorporate into the model. A lack of patterns in the residuals indicates that the model seems to account for the primary relationships in the data.

Patterns in the residuals indicate the presence of additional information that we hope to incorporate into the model. A lack of patterns in the residuals indicates that the model seems to account for the primary relationships in the data.

5.3.1 Residuals

There are at least four types of patterns that can be uncovered through the residual analysis. In this section, we discuss the first two: residuals that are unusual and those that are related to other explanatory variables. We then introduce the third type, residuals that display a heteroscedastic pattern, in Section 5.7. In our study of time series data that begins in Chapter 7, we will introduce the fourth type, residuals that display patterns through time.

When examining residuals, it is usually easier to work with a *standardized residual*, a residual that has been rescaled to be dimensionless. We generally work with standardized residuals because we achieve some carryover of experience from one dataset to another and may thus focus on relationships of interest. By using standardized residuals, we can train ourselves to look at a variety of residual plots and immediately recognize an unusual point when working in standard units.

There are a number of ways to define a standardized residual. Using $e_i = y_i - \hat{y}_i$ as the ith residual, here are three commonly used definitions:

$$\text{(a)}\ \frac{e_i}{s}, \quad \text{(b)}\ \frac{e_i}{s\sqrt{1-h_{ii}}}, \quad \text{and} \quad \text{(c)}\frac{e_i}{s_{(i)}\sqrt{1-h_{ii}}}. \tag{5.1}$$

Here, h_{ii} is the ith leverage. It is calculated according to values of the explanatory variables and will be defined in Section 5.4.1. Recall that s is the residual standard deviation (defined in equation 3.8). Similarly, define $s_{(i)}$ to be the residual standard deviation when running a regression after having deleted the ith observation.

Now, the first definition in (a) is simple and easy to explain. An easy calculation shows that the sample standard deviation of the residuals is approximately s (one reason that s is often referred to as the residual standard deviation). Thus, it seems reasonable to standardize residuals by dividing by s.

The second choice presented in (b), although more complex, is more precise. The variance of the ith residual is

$$\text{Var}(e_i) = \sigma^2(1 - h_{ii}).$$

This result will be established in equation (5.15) of Section 5.10. Note that this variance is smaller than the variance of the error term, Var $(\varepsilon_i) = \sigma^2$. Now, we can replace σ by its estimate, s. Then, this result leads to using the quantity $s(1 - h_{ii})^{1/2}$ as an estimated standard deviation, or standard error, for e_i. Thus, we define the standard error of e_i to be

$$se(e_i) = s\sqrt{1 - h_{ii}}.$$

Following the conventions introduced in Section 2.6, in this text, we use $e_i/se(e_i)$ to be our *standardized residual.*

The third choice presented in (c) is a modification of (b) and is known as a *studentized residual.* As emphasized in Section 5.3.2, an important use of residuals is to identify unusually large responses. Now, suppose that the ith response is unusually large and that this is measured through its residual. This unusually large residual will also cause the value of s to be large. Because the large effect appears in both the numerator and the denominator, the standardized residual may not detect this unusual response. However, this large response will not inflate $s_{(i)}$ because it is constructed after having deleted the ith observation. Thus, when using studentized residuals we get a better measure of observations that have unusually large residuals. By omitting this observation from the estimate of σ, the size of the observation affects only the numerator e_i and not the denominator $s_{(i)}$.

As another advantage, studentized residuals follow a t-distribution with $n - (k + 1)$ degrees of freedom, assuming the errors are normally distributed (assumption E5). This knowledge of the precise distribution helps us assess the degree of model fit and is particularly useful in small samples. It is this relationship with the "student's" t-distribution that suggests the name *studentized* residuals.

5.3.2 Using Residuals to Identify Outliers

A commonly used rule of thumb is to mark an observation as an outlier if its standardized residual exceeds two in absolute value.

An important role of residual analysis is to identify outliers. An outlier is an observation that is not well fit by the model; these are observations where the residual is unusually large. A rule of thumb used by many statistical packages is that an observation is marked as an outlier if the standardized residual exceeds two in absolute value. To the extent that the distribution of standardized residuals mimics the standard normal curve, we expect about only one in twenty observations, or 95%, to exceed two in absolute value and very few observations to exceed three.

Outliers provide a signal that an observation should be investigated to understand special causes associated with the point. An outlier is an observation that seems unusual with respect to the rest of the dataset. It is often the case that the reason for this atypical behavior is uncovered after additional investigation. Indeed, this may be the primary purpose of the regression analysis of a data set.

Consider a simple example of so-called *performance analysis*. Suppose we have available a sample of n salespeople and are trying to understand each person's second-year sales based on their first-year sales. To a certain extent, we expect that higher first-year sales are associated with higher second-year sales. High sales may be due to a salesperson's natural ability, ambition, good territory, and so on. First-year sales may be thought of as a proxy variable that summarizes these factors. We expect variation in sales performance both cross-sectionally and across years. It is interesting when one salesperson performs unusually well (or poorly) in the second year compared to their first-year performance. Residuals provide a formal mechanism for evaluating second-year sales after controlling for the effects of first-year sales.

There are a number of options available for handling outliers.

Options for Handling Outliers

- Include the observation in the usual summary statistics but comment on its effects. An outlier may be large but not so large as to skew the results of the entire analysis. If no special causes for this unusual observation can be determined, then the observation may simply reflect the variability in the data.
- Delete the observation from the dataset. The observation may be determined to be unrepresentative of the population from which the sample is drawn. If this is the case, then there may be little information contained in the observation that can be used to make general statements about the population. This option means that we would omit the observation from the regression summary statistics and discuss it in our report as a separate case.
- Create a binary variable to indicate the presence of an outlier. If one or several special causes have been identified to explain an outlier, then these causes could be introduced into the modeling procedure formally by introducing a variable to indicate the presence (or absence) of these causes. This approach is similar to point deletion but allows the outlier to be formally included in the model formulation so that, if additional observations arise that are affected by the same causes, then they can be handled on an automatic basis.

5.3.3 Using Residuals to Select Explanatory Variables

Another important role of residual analysis is to help identify additional explanatory variables that may be used to improve the formulation of the model. If we have specified the model correctly, then residuals should resemble random errors and contain no discernible patterns. Thus, when comparing residuals to explanatory variables, we do not expect any relationships. If we do detect a relationship, then this suggests the need to control for this additional variable. This can be accomplished by introducing the additional variable into the regression model.

Relationships between residuals and explanatory variables can be quickly established using correlation statistics. However, if an explanatory variable is already included in the regression model, then the correlation between the residuals and an explanatory variable will be zero (Section 5.10.1 provides the algebraic demonstration). It is a good idea to reinforce this correlation with a scatter plot. A plot of residuals versus explanatory variables not only will reinforce graphically the correlation statistic but also will serve to detect potential nonlinear relationships. For example, a quadratic relationship can be detected using a scatter plot, not a correlation statistic.

If you detect a relationship between the residuals from a preliminary model fit and an additional explanatory variable, then introducing this additional variable will not always improve your model specification. The reason is that the additional variable may be linearly related to the variables that are already in the model.

If you want a guarantee that adding a variable will improve your model, then construct an added variable plot (from Section 3.4.3).

To summarize, after a preliminary model fit, you should do the following:

- Calculate summary statistics and display the distribution of (standardized) residuals to identify outliers.
- Calculate the correlation between the (standardized) residuals and additional explanatory variables to search for linear relationships.
- Create scatter plots between the (standardized) residuals and additional explanatory variables to search for nonlinear relationships.

Ⓡ **EMPIRICAL**
Filename is "Liquidity"

Example: Stock Market Liquidity. An investor's decision to purchase a stock is generally made with a number of criteria in mind. First, investors usually look for a high expected return. A second criterion is the riskiness of a stock, which can be measured through the variability of the returns. Third, many investors are concerned with the length of time that they are committing their capital with the purchase of a security. Many income stocks, such as utilities, regularly return portions of capital investments in the form of dividends. Other stocks, particularly growth stocks, return nothing until the sale of the security. Thus, the average length of investment in a security is another criterion. Fourth, investors are concerned with the ability to sell the stock at any time convenient to the investor. We refer to this fourth criterion as the *liquidity* of the stock. The more liquid is the stock, the easier it is to sell. To measure the liquidity, in this study, we use the number of shares traded on an exchange over a specified period of time (called the VOLUME). We are interested in studying the relationship between the volume and other financial characteristics of a stock.

We begin this study with 126 companies whose options were traded on December 3, 1984. The stock data were obtained from Francis Emory Fitch Inc. for the period from December 3, 1984, to February 28, 1985. For the trading activity variables, we examine

- The three-month total trading volume (VOLUME, in millions of shares)
- The three-month total number of transactions (NTRAN)
- The average time between transactions (AVGT, measured in minutes)

For the firm size variables, we use:

- Opening stock price on January 2, 1985 (PRICE),
- The number of outstanding shares on December 31, 1984 (SHARE, in millions of shares)
- The market equity value (VALUE, in billions of dollars) obtained by taking the product of PRICE and SHARE.

Finally, for the financial leverage, we examine the debt-to-equity ratio (DEB_EQ) obtained from the Compustat Industrial Tape and the Moody's manual. The data

Table 5.2 Summary Statistics of the Stock Liquidity Variables

	Mean	Median	Standard Deviation	Minimum	Maximum
VOLUME	13.423	11.556	10.632	0.658	64.572
AVGT	5.441	4.284	3.853	0.590	20.772
NTRAN	6436	5071	5310	999	36420
PRICE	38.80	34.37	21.37	9.12	122.37
SHARE	94.7	53.8	115.1	6.7	783.1
VALUE	4.116	2.065	8.157	0.115	75.437
DEB_EQ	2.697	1.105	6.509	0.185	53.628

Source: Francis Emory Fitch Inc., Standard & Poor's Compustat, and University of Chicago's Center for Research on Security Prices.

Table 5.3 Correlation Matrix of the Stock Liquidity

	AVGT	NTRAN	PRICE	SHARE	VALUE	DEB_EQ
NTRAN	−0.668					
PRICE	−0.128	0.190				
SHARE	−0.429	0.817	0.177			
VALUE	−0.318	0.760	0.457	0.829		
DEB_EQ	0.094	−0.092	−0.038	−0.077	−0.077	
VOLUME	−0.674	0.913	0.168	0.773	0.702	−0.052

in SHARE are obtained from the Center for Research in Security Prices' (CRSP) monthly tape.

After examining some preliminary summary statistics of the data, three companies were deleted because they had either an unusually large volume or high price. They are Teledyne and Capital Cities Communication, whose prices were more than four times the average price of the remaining companies, and AT&T, whose total volume was more than seven times than the average total volume of the remaining companies. After additional investigation, the details of which are not presented here, these companies were deleted because they seemed to represent special circumstances that we did not want to model. Table 5.2 summarizes the descriptive statistics based on the remaining $n = 123$ companies. For example, from Table 5.2, we see that the average time between transactions is about five minutes and this time ranges from a minimum of less than 1 minute to a maximum of about 20 minutes.

Table 5.3 reports the correlation coefficients and Figure 5.2 provides the corresponding scatterplot matrix. If you have a background in finance, you will find it interesting to note that the financial leverage, measured by DEB_EQ, does not seem to be related to the other variables. From the scatterplot and correlation matrix, we see a strong relationship between VOLUME and the size of the firm as measured by SHARE and VALUE. Further, the three trading activity variables, VOLUME, AVGT, and NTRAN, are all highly related to one another.

Figure 5.2 shows that the variable AVGT is inversely related to VOLUME and NTRAN is inversely related to AVGT. In fact, it turned out the correlation

Table 5.4 First Table of Correlations

	AVGT	PRICE	SHARE	VALUE	DEB_EQ
RESID	−0.155	−0.017	0.055	0.007	0.078

Note: The residuals were created from a regression of VOLUME on NTRAN.

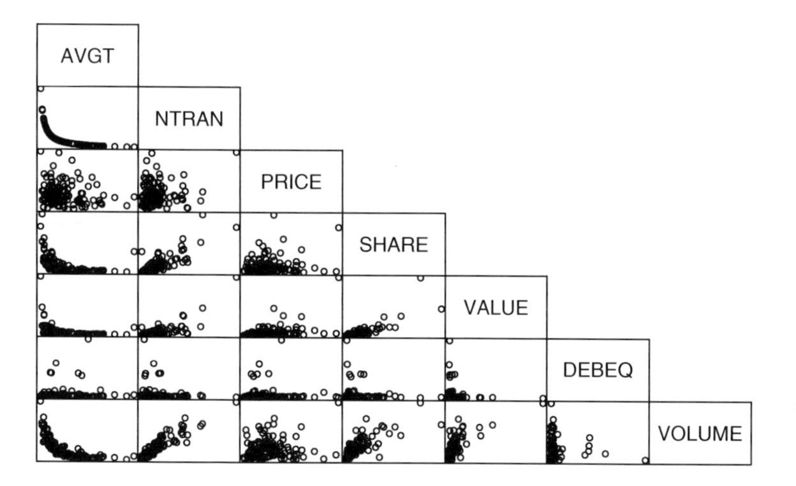

Figure 5.2 Scatterplot matrix for stock liquidity variables. The number of transactions variable (NTRAN) appears to be strongly related to the VOLUME of shares traded and inversely related to AVGT.

between the average time between transactions and the reciprocal of the number of transactions was 99.98%! This is not so surprising when one thinks about how AVGT might be calculated. For example, on the New York Stock Exchange, the market is open from 10:00 a.m. to 4:00 p.m. For each stock on a particular day, the average time between transactions times the number of transactions is nearly equal to 360 minutes (or 6 hours). Thus, except for rounding errors because transactions are recorded only to the nearest minute, there is a perfect linear relationship between AVGT and the reciprocal of NTRAN.

To begin to understand the liquidity measure VOLUME, we first fit a regression model using NTRAN as an explanatory variable. The fitted regression model is:

$$\begin{array}{lccc} \text{VOLUME} & = & 1.65 & +0.00183\text{NTRAN} \\ \text{standard error} & & (0.0018) & (0.000074) \end{array}$$

with $R^2 = 83.4\%$ and $s = 4.35$. Note that the t-ratio for the slope associated with NTRAN is $t(b_1) = b_1/se(b_1) = 0.00183/0.000074 = 24.7$, indicating strong statistical significance. Residuals were computed using this estimated model. To determine whether the residuals are related to the other explanatory variables, Table 5.4 shows correlations.

The correlation between the residual and AVGT and the scatter plot (not given here) indicates that there may be some information in the variable AVGT in the residual. Thus, it seems sensible to use AVGT directly in the regression model. Remember that we are interpreting the residual as the value of VOLUME having controlled for the effect of NTRAN.

Table 5.5 Second Table of Correlations

	PRICE	SHARE	VALUE	DEB_EQ
RESID	−0.015	0.096	0.071	0.089

Note: The residuals were created from a regression of VOLUME on NTRAN and AVGT.

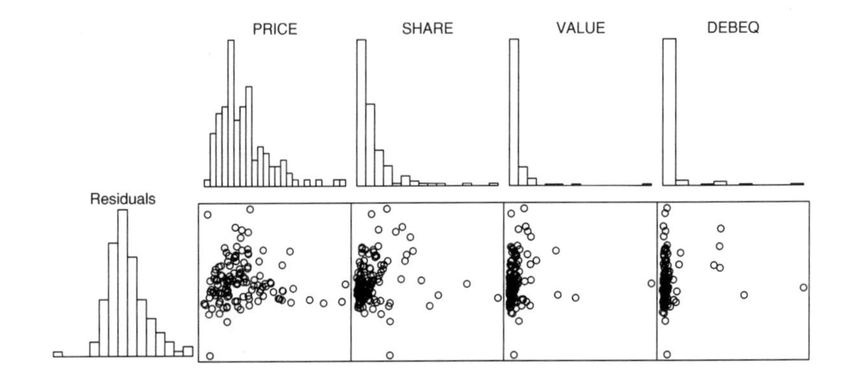

Figure 5.3 Scatterplot matrix of the residuals from the regression of VOLUME on NTRAN and AVGT on the vertical axis and the remaining predictor variables on the horizontal axis.

We next fit a regression model using NTRAN and AVGT as an explanatory variables. The fitted regression model is:

$$\begin{array}{lcccc} \text{VOLUME} & = & 4.41 & -0.322\,\text{AVGT} & +0.00167\text{NTRAN} \\ \text{standard error} & & (1.30) & (0.135) & (0.000098) \end{array}$$

with $R^2 = 84.2\%$ and $s = 4.26$. Given the t-ratio for AVGT, $t(b_{AVGT}) = (-0.322)/0.135 = -2.39$, it seems that AVGT is a useful explanatory variable in the model. Note also that s has decreased, indicating that R_a^2 has increased.

Table 5.5 provides correlations between the model residuals and other potential explanatory variables and indicates that there does not seem to be much additional information in the explanatory variables. This is reaffirmed by the corresponding table of scatter plots in Figure 5.3. The histograms in Figure 5.3 suggest that, although the distribution of the residuals is fairly symmetric, the distribution of each explanatory variable is skewed. Because of this, transformations of the explanatory variables were explored. This line of thought provided no real improvements and thus the details are not provided here.

5.4 Influential Points

Not all points are created equal – in this section, we will see that specific observations potentially have a disproportionate effect on the overall regression fit. We will call such points *influential*. This is not too surprising; we have already seen that regression coefficients estimates are *weighted* sums of responses (see

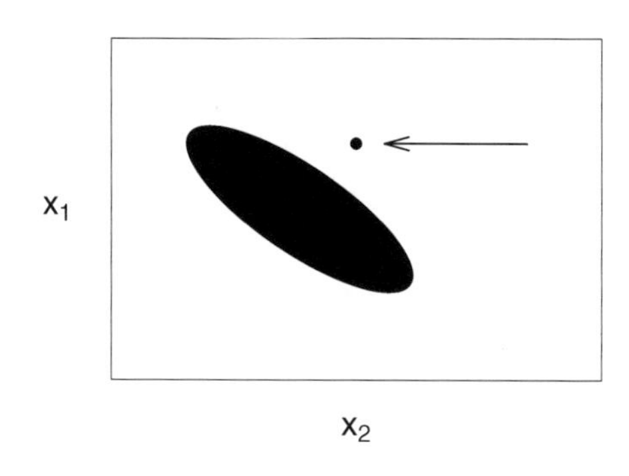

Figure 5.4 The ellipsoid represents most of the data. The arrow marks an unusual point.

Section 3.2.4). Some observations have heavier weights than others and thus have a greater influence on the regression coefficient estimates. Of course, simply because an observation is influential does not mean that it is incorrect or that its impact on the model is misleading. As analysts, we would simply like to know whether our fitted model is sensitive to mild changes, such as the removal of a single point, so that we are comfortable generalizing our results from the sample to a larger population.

To assess influence, we think of observations as unusual responses, given a set of explanatory variables, or as having an unusual set of explanatory variables. We have already seen in Section 5.3 how to assess unusual responses using residuals. This section focuses on unusual sets of explanatory variables.

5.4.1 Leverage

We introduced this topic in Section 2.6, where we called an observation with an unusual explanatory variable a *high leverage point*. With more than one explanatory variable, determining whether an observation is a high leverage point is not as straightforward. For example, it is possible for an observation to be "not unusual" for any single variable and yet still be unusual in the space of explanatory variables. Consider the fictitious data set represented in Figure 5.4. Visually, it seems clear that the point marked in the upper-right-hand corner is unusual. However, it is not unusual when examining the histogram of either x_1 or x_2. It is only unusual when the explanatory variables are considered jointly.

For two explanatory variables, this is apparent when examining the data graphically. Because it is difficult to examine graphically data having more than two explanatory variables, we need a numerical procedure to assess leverage.

To define the concept of leverage in multiple linear regression, we use some concepts from matrix algebra. Specifically, in Section 3.1, we showed that the vector of least squares regression coefficients could be calculated using $\mathbf{b} = (\mathbf{X}'\mathbf{X})^{-1}\mathbf{X}'\mathbf{y}$. Thus, we can express the vector of fitted values $\hat{\mathbf{y}} = (\hat{y}_1, \ldots, \hat{y}_n)'$ as

$$\hat{\mathbf{y}} = \mathbf{X}\mathbf{b} \tag{5.2}$$

Similarly, the vector of residuals is the vector of response minus the vector of fitted values, that is, $\mathbf{e} = \mathbf{y} - \hat{\mathbf{y}}$.

From expression for the regression coefficients $\mathbf{b}$ in equation (3.4), we have $\hat{\mathbf{y}} = \mathbf{X}(\mathbf{X}'\mathbf{X})^{-1}\mathbf{X}'\mathbf{y}$. This equation suggests defining $\mathbf{H} = \mathbf{X}(\mathbf{X}'\mathbf{X})^{-1}\mathbf{X}'$, so that $\hat{\mathbf{y}} = \mathbf{H}\mathbf{y}$. From this, the matrix $\mathbf{H}$ is said to *project* the vector of responses $\mathbf{y}$ onto the vector of fitted values $\hat{\mathbf{y}}$. Alternatively, you may think of $\mathbf{H}$ as the matrix that puts the "hat," or caret, on $\mathbf{y}$. From the ith row of the vector equation $\hat{\mathbf{y}} = \mathbf{H}\mathbf{y}$, we have

$$\hat{y}_i = h_{i1}y_1 + h_{i2}y_2 + \cdots + h_{ii}y_i + \cdots + h_{in}y_n.$$

Here, h_{ij} is the number in the ith row and jth column of $\mathbf{H}$. From this expression, we see that the greater is h_{ii}, the greater is the effect that the ith response (y_i) has on the corresponding fitted value ($\hat{y}_i$). Thus, we call h_{ii} to be the *leverage* for the ith observation. Because h_{ii} is the ith diagonal element of $\mathbf{H}$, a direct expression for h_{ii} is

$$h_{ii} = \mathbf{x}_i'(\mathbf{X}'\mathbf{X})^{-1}\mathbf{x}_i \tag{5.3}$$

where $\mathbf{x}_i = (x_{i0}, x_{i1}, \ldots, x_{ik})'$. Because the values h_{ii} are calculated based on the explanatory variables, the values of the response variable do not affect the calculation of leverages.

Large leverage values indicate that an observation may exhibit a disproportionate effect on the fit, essentially because it is distant from the other observations (when looking at the space of explanatory variables). How large is large? Some guidelines are available from matrix algebra, where we have that

$$\frac{1}{n} \leq h_{ii} \leq 1$$

and

$$\bar{h} = \frac{1}{n}\sum_{i=1}^{n} h_{ii} = \frac{k+1}{n}.$$

Thus, each leverage is bounded by n^{-1} and 1 and the average leverage equals the number of regression coefficients divided by the number of observations. From these and related arguments, we use a widely adopted convention and declare an observation to be a *high leverage point* if the leverage exceeds three times the average, that is, if $h_{ii} > 3(k+1)/n$.

Having identified high leverage points, as with outliers, it is important for the analyst to search for special causes that may have produced these unusual points. To illustrate, in Section 2.7, we identified the 1987 market crash as the reason behind the high leverage point. Further, high leverage points are often due to clerical errors in coding the data, which may or may not be easy to rectify. In general, the options for dealing with high leverage points are similar to those available for dealing with outliers.

Options for Handling High Leverage Points

(i) Include the observation in the summary statistics but comment on its effect. For example, an observation may barely exceed a cutoff and its effect may not be important in the overall analysis.
(ii) Delete the observation from the dataset. Again, the basic rationale for doing so is that the observation is deemed not representative of some larger population. An intermediate course of action between (i) and (ii) is to present the analysis both with and without the high leverage point. In this way, the impact of the point is fully demonstrated and the reader of your analysis may decide which option is more appropriate.
(iii) Choose another variable to represent the information. In some instances, another explanatory variables will be available to serve as a replacement. For example, in an apartment rents example, we could use the number of bedrooms to replace a square footage variable as a measure of apartment size. Although an apartment's square footage may be unusually large, thus causing it to be a high leverage point, it may have one, two, or three bedrooms, depending on the sample examined.
(iv) Use a nonlinear transformation of an explanatory variable. To illustrate, with our stock liquidity example in Section 5.3.3, we can transform the debt-to-equity DEB_EQ continuous variable into a variable that indicates the presence of "high" debt-to-equity. For example, we might code $\text{DE\_IND} = 1$ if $\text{DEB\_EQ} > 5$ and $\text{DE\_IND} = 0$ if $\text{DEB\_EQ} \leq 5$. With this recoding, we still retain information on the financial leverage of a company without allowing the large values of DEB_EQ to drive the regression fit.

Some analysts use robust estimation methodologies as an alternative to least squares estimation. The basic idea of robust techniques is to reduce the effect of any particular observation. These techniques are useful in reducing the effect of both outliers and high leverage points. This tactic may be viewed as intermediate between one extreme procedure that ignores the effect of unusual points and another extreme that gives unusual points full credibility by deleting them from the dataset. The word *robust* is meant to suggest that these estimation methodologies are "healthy" even when attacked by an occasional bad observation (a germ). We have seen that this is not true for least squares estimation.

5.4.2 Cook's Distance

To quantify the influence of a point, a measure that considers both the response and the explanatory variables is *Cook's distance*. This distance, D_i, is

Table 5.6 Measures of Three Types of Unusual Points

Observation	Standardized residual $e/se(e)$	Leverage h	Cook's distance D
A	4.00	.067	.577
B	.77	.550	.363
C	−4.01	.550	9.832

defined as

$$D_i = \frac{\sum_{j=1}^{n}(\hat{y}_j - \hat{y}_{j(i)})^2}{(k+1)s^2} \tag{5.4}$$

$$= \left(\frac{e_i}{se(e_i)}\right)^2 \frac{h_{ii}}{(k+1)(1-h_{ii})}.$$

The first expression provides a definition. Here, $\hat{y}_{j(i)}$ is the prediction of the jth observation, computed leaving the ith observation out of the regression fit. To measure the impact of the ith observation, we compare the fitted values with and without the ith observation. Each difference is then squared and summed over all observations to summarize the impact.

The second equation provides another interpretation of the distance D_i. The first part, $(e_i/se(e_i))^2$, is the square of the ith standardized residual. The second part, $h_{ii}/((k+1)(1-h_{ii}))$, is attributable solely to the leverage. Thus, the distance D_i is composed of a measure for outliers times a measure for leverage. In this way, Cook's distance accounts for both the response and the explanatory variables. Section 5.10.3 establishes the validity of equation (5.4).

To get an idea of the expected size of D_i for a point that is not unusual, recall that we expect the standardized residuals to be about 1 and the leverage h_{ii} to be about $(k+1)/n$. Thus, we anticipate that D_i should be about $1/n$. Another rule of thumb is to compare D_i to an F-distribution with $df_1 = k+1$ and $df_2 = n-(k+1)$ degrees of freedom. Values of D_i that are greater than this distribution merit attention.

Example: Outliers and High Leverage Points – Continued. To illustrate, we return to our example in Section 2.6. In this example, we considered 19 "good," or base, points plus each of the three types of unusual points, labeled A, B, and C. Table 5.6 summarizes the calculations.

As noted in Section 2.6, from the standardized residual column, we see that both point A and point C are outliers. To judge the size of the leverages, because there are $n = 20$ points, the leverages are bounded by 0.05 and 1.00, with the average leverage being $\bar{h} = 2/20 = 0.10$. Using 0.3 ($= 3 \times \bar{h}$) as a cutoff, points B and C are high leverage points. Note that their values are the same. This is because, from Figure 2.7, the values of the explanatory variables are the same and only the response variable has been changed. The column for Cook's distance captures both types of unusual behavior. Because the typical value of D_i is $1/n$ or 0.05, Cook's distance provides one statistic to alert us to the fact that each point is unusual in one respect or another. In particular, point C has a very large D_i, reflecting the fact that it is both an outlier and a high leverage point. The 95th percentile of an F-distribution with $df_1 = 2$ and $df_2 = 18$ is 3.555. The fact that point C has a value of D_i that well exceeds this cutoff indicates the substantial influence of this point.

5.5 Collinearity

5.5.1 What Is Collinearity?

Collinearity, or *multicollinearity*, occurs when one explanatory variable is, or nearly is, a linear combination of the other explanatory variables. Intuitively, with collinear data it is useful to think of explanatory variables as highly correlated with one another. If an explanatory variable is collinear, then the question arises as to whether it is redundant, that is, whether the variable provides little additional information beyond the information in the other explanatory variables. The issues are: Is collinearity important? If so, how does it affect our model fit and how do we detect it? To address the first question, consider a somewhat pathological example.

Example: Perfectly Correlated Explanatory Variables. Joe Finance was asked to fit the model E $y = \beta_0 + \beta_1 x_1 + \beta_2 x_2$ to a dataset. His resulting fitted model was $\hat{y} = -87 + x_1 + 18x_2$. The dataset under consideration is:

i	1	2	3	4
y_i	23	83	63	103
x_{i1}	2	8	6	10
x_{i2}	6	9	8	10

Joe checked the fit for each observation. Joe was very happy because he fit the data perfectly! For example, for the third observation the fitted value is $\hat{y}_3 = -87 + 6 + 18(8) = 63$, which is equal to the third response, y_3. Because the response equals the fitted value, the residual is zero. You may check that this is true of each observation and thus the R^2 turned out to be 100%.

However, Jane Actuary came along and fit the model $\hat{y} = -7 + 9x_1 + 2x_2$. Jane performed the same careful checks that Joe did and also got a perfect fit ($R^2 = 1$). Who is right?

The answer is both and neither. There are, in fact, an infinite number of fits. This is because of the perfect relationship $x_2 = 5 + x_1/2$ between the two explanatory variables.

This example illustrates some important facts about collinearity.

Collinearity Facts

- Collinearity precludes us neither from getting good fits nor from making predictions of new observations. Note that in the prior example, we got perfect fits.
- Estimates of error variances and, therefore, tests of model adequacy, are still reliable.
- In cases of serious collinearity, standard errors of individual regression coefficients are greater than in cases where, other things equal, serious collinearity does not exist. With large standard errors, individual regression coefficients may not be meaningful. Further, because a large standard error means that the corresponding t-ratio is small, it is difficult to detect the importance of a variable.

To detect collinearity, begin with a matrix of correlation coefficients of the explanatory variables. This matrix is simple to create, easy to interpret, and quickly captures linear relationships between pairs of variables. A scatterplot matrix provides a visual reinforcement of the summary statistics in the correlation matrix.

5.5.2 Variance Inflation Factors

Correlation and scatterplot matrices capture only relationships between pairs of variables. To capture more complex relationships among several variables, we introduce the *variance inflation factor (VIF)*. To define a VIF, suppose that the set of explanatory variables is labeled $x_1, x_2, \ldots, x_k$. Now, run the regression using x_j as the response and the other x's $(x_1, x_2, \ldots, x_{j-1}, x_{j+1}, \ldots, x_k)$ as the explanatory variables. Denote the coefficient of determination from this regression by R_j^2. We interpret $R_j = \sqrt{R_j^2}$ as the multiple correlation coefficient between x_j and linear combinations of the other x's. From this coefficient of determination, we define the variance inflation factor

$$VIF_j = \frac{1}{1 - R_j^2}, \quad \text{for } j = 1, 2, \ldots, k.$$

A larger R_j^2 results in a larger VIF_j; this means greater collinearity between x_j and the other x's. Now, R_j^2 alone is enough to capture the linear relationship of interest. However, we use VIF_j in lieu of R_j^2 as our measure for collinearity because of the algebraic relationship

$$se(b_j) = s \frac{\sqrt{VIF_j}}{s_{x_j}\sqrt{n-1}}. \tag{5.5}$$

Here, $se(b_j)$ and s are standard errors and residual standard deviation from a full regression fit of y on $x_1, \ldots, x_k$. Further, $s_{x_j} = \sqrt{(n-1)^{-1}\sum_{i=1}^{n}(x_{ij} - \bar{x}_j)^2}$ is

x_j	s_{x_j}	b_j	$se(b_j)$	$t(b_j)$	VIF_j
PRICE	21.37	−0.022	0.035	−0.63	1.5
SHARE	115.1	0.054	0.010	5.19	3.8
VALUE	8.157	0.313	0.162	1.94	4.7

Table 5.7 Statistics from a Regression of VOLUME on PRICE, SHARE and VALUE

the sample standard deviation of the jth variable x_j. Section 5.10.3 provides a verification of equation (5.5).

Thus, a larger VIF_j results in a larger standard error associated with the jth slope, b_j. Recall that $se(b_j)$ is s times the square root of the $(j+1)$st diagonal element of $(\mathbf{X}'\mathbf{X})^{-1}$. The idea is that when collinearity occurs, the matrix $\mathbf{X}'\mathbf{X}$ has properties similar to the number zero. When we attempt to calculate the inverse of $\mathbf{X}'\mathbf{X}$, this is analogous to dividing by zero for scalar numbers. As a rule of thumb, when VIF_j exceeds 10 (which is equivalent to $R_j^2 > 90\%$), we say that severe collinearity exists. This may signal a need for action. *Tolerance*, defined as the reciprocal of the variance inflation factor, is another measure of collinearity used by some analysts.

A commonly used rule of thumb is that $VIF_j > 10$ is a signal that severe collinearity exists.

For example, with $k = 2$ explanatory variables in the model, then R_1^2 is the squared correlation between the two explanatory variables, say, r_{12}^2. Then, from equation (5.5), we have that $se(b_j) = s\left(s_{x_j}\sqrt{n-1}\right)^{-1}\left(1 - r_{12}^2\right)^{-1/2}$, for $j = 1, 2$. As the correlation approaches one in absolute value, $|r_{12}| \to 1$, then the standard error becomes large meaning that the corresponding t-statistic becomes small. In summary, a high *VIF* may mean small t-statistics even though variables are important. Further, one can check that the correlation between b_1 and b_2 is $-r_{12}$, indicating that the coefficient estimates are highly correlated.

Example: Stock Market Liquidity, Continued. As an example, consider a regression of VOLUME on PRICE, SHARE, and VALUE. Unlike the explanatory variables considered in Section 5.3.3, these three explanatory variables are not measures of trading activity. From a regression fit, we have $R^2 = 61\%$ and $s = 6.72$. The statistics associated with the regression coefficients are in Table 5.7.

You may check that the relationship in equation (5.5) is valid for each of the explanatory variables in Table 5.7. Because each *VIF* statistic is less than 10, there is little reason to suspect severe collinearity. This is interesting because you may recall that there is a perfect relationship among PRICE, SHARE, and VALUE in that we defined the market value to be VALUE = PRICE × SHARE. However, the relationship is multiplicative, and hence is nonlinear. Because the variables are not linearly related, it is valid to enter all three into the regression model. From a financial perspective, the variable VALUE is important because it measures the worth of a firm. From a statistical perspective, the variable VALUE quantifies the interaction between PRICE and SHARE (interaction variables were introduced in Section 3.5.3).

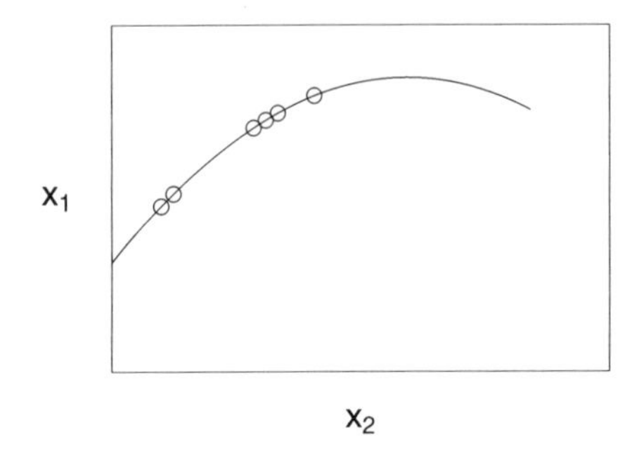

Figure 5.5 The relationship between x_1 and x_2 is nonlinear. However, over the region sampled, the variables have close to a linear relationship.

For collinearity, we are interested only in detecting linear trends, so nonlinear relationships between variables are not an issue here. For example, we have seen that it is sometimes useful to retain both an explanatory variable (x) and its square (x^2), despite the fact that there is a perfect (nonlinear) relationship between the two. Still, we must check that nonlinear relationships are not approximately linear over the sampling region. Even though the relationship is theoretically nonlinear, if it is close to linear for our available sample, then problems of collinearity might arise. Figure 5.5 illustrates this situation.

What can we do in the presence of collinearity? One option is to center each variable by subtracting its average and dividing by its standard deviation. For example, create a new variable $x^*_{ij} = (x_{ij} - \bar{x}_j)/s_{x_j}$. Occasionally, one variable appears as millions of units and another variable appears as fractions of units. Compared to the first mentioned variable, the second variable is close to a constant column of zeros (in that computers typically retain a finite number of digits). If this is true, then the second variable looks much like a linear shift of the constant column of ones corresponding to the intercept. This is a problem because, with the least squares operations, we are implicitly squaring numbers that can make these columns appear even more similar.

This problem is simply a computational one and is easy to rectify. Simply recode the variables so that the units are of similar order of magnitude. Some data analysts automatically center all variables to avoid these problems. This is a legitimate approach because regression techniques search for linear relationships; location and scale shifts do not affect linear relationships.

Another option is to simply not explicitly account for collinearity in the analysis but to discuss some of its implications when interpreting the results of the regression analysis. This approach is probably the most commonly adopted one. It is a fact of life that, when dealing with business and economic data, collinearity tends to exist among variables. Because the data tends to be observational in lieu of experimental in nature, there is little that the analyst can do to avoid this situation.

When severe collinearity exists, often the only option is to remove one or more variables from the regression equation.

In the best-case situation, an auxiliary variable that provides similar information and eases the collinearity problem is available to replace a variable. Similar to our discussion of high leverage points, a transformed version of the explanatory variable may also be a useful substitute. In some situations, such an ideal replacement is not available and we are forced to remove one or more variables.

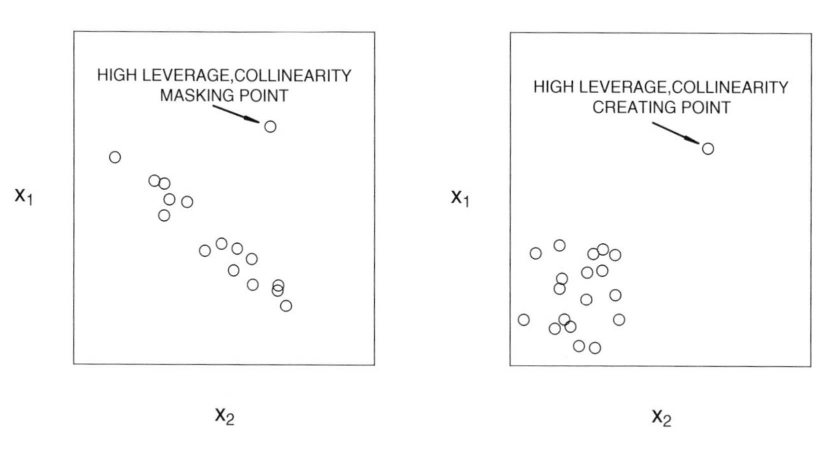

Figure 5.6 With the exception of the marked point, x_1 and x_2 are highly linearly related.

Figure 5.7 The highly linear relationship between x_1 and x_2 is primarily due to the marked point.

Deciding which variables to remove is a difficult choice. When deciding among variables, often the choice will be dictated by the investigator's judgement as to which is the most relevant set of variables.

5.5.3 Collinearity and Leverage

Measures of collinearity and leverage share common characteristics and yet are designed to capture different aspects of a dataset. Both are useful for data and model criticism; they are applied after a preliminary model fit with the objective of improving model specification. Further, both are calculated using only the explanatory variables; values of the responses do not enter into either calculation.

Our measure of collinearity, the variance inflation factor, is designed to help with model criticism. It is a measure calculated for each explanatory variable, designed to explain the relationship with other explanatory variables.

The leverage statistic is designed to help us with data criticism. It is a measure calculated for each observation to help us explain how unusual an observation is with respect to other observations.

Collinearity may be masked, or induced, by high leverage points, as pointed out by Mason and Gunst (1985) and Hadi (1988). Figures 5.6 and 5.7 provide illustrations of each case. These simple examples underscore an important point; data criticism and model criticism are not separate exercises.

The examples in Figures 5.6 and 5.7 also help us to see one way in which high leverage points may affect standard errors of regression coefficients. Recall that, in Section 5.4.1, we saw that high leverage points may affect the model fitted values. In Figures 5.6 and 5.7, we see that high leverage points affect collinearity. Thus, from equation (5.5), we have that high leverage points can also affect our standard errors of regression coefficients.

5.5.4 Suppressor Variables

As we have seen, severe collinearity can seriously inflate standard errors of regression coefficients. Because we rely on these standard errors for judging

Table 5.8 Correlation Matrix for the Suppressor Example

	x_1	x_2
x_2	0.972	
y	0.188	−0.022

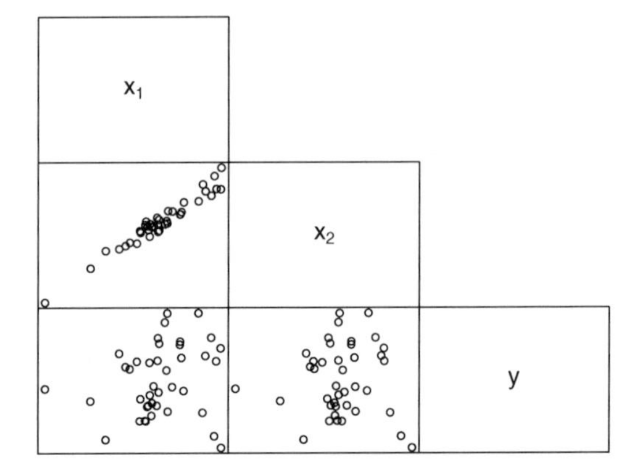

Figure 5.8 Scatterplot matrix of a response and two explanatory variable for the suppressor variable example.

the usefulness of explanatory variables, our model selection procedures and inferences may be deficient in the presence of severe collinearity. Despite these drawbacks, mild collinearity in a dataset should not be viewed as a deficiency of the dataset; it is simply an attribute of the available explanatory variables.

Even if one explanatory variable is nearly a linear combination of the others, that does not necessarily mean that the information that it provides is redundant. To illustrate, we now consider a *suppressor variable*, an explanatory variable that increases the importance of other explanatory variables when included in the model.

Example: Suppressor Variable. Figure 5.8 shows a scatterplot matrix of a hypothetical dataset of 50 observations. This dataset contains a response and two explanatory variables. Table 5.8 provides the corresponding matrix of correlation coefficients. There, we see that the two explanatory variables are highly correlated. Now recall, for regression with one explanatory variable, that the correlation coefficient squared is the coefficient of determination. Thus, using Table 5.8, for a regression of y on x_1, the coefficient of determination is $(0.188)^2 = 3.5\%$. Similarly, for a regression of y on x_2, the coefficient of determination is $(-0.022)^2 = 0.04\%$. However, for a regression of y on x_1 and x_2, the coefficient of determination turns out to be a surprisingly high 80.7%. The interpretation is that individually, both x_1 and x_2 have little impact on y. However, when taken jointly, the two explanatory variables have a significant effect on y. Although Table 5.8 shows that x_1 and x_2 are strongly linearly related, this relationship does not mean that x_1 and x_2 provide the same information. In fact, in this example the two variables complement one another.

5.5.5 Orthogonal Variables

Another way to understand the impact of collinearity is to study the case when there are *no* relationships among sets of explanatory variables. Mathematically, two matrices $\mathbf{X}_1$ and $\mathbf{X}_2$ are said to be *orthogonal* if $\mathbf{X}_1'\mathbf{X}_2 = \mathbf{0}$. Intuitively, because we generally work with centered variables (with zero averages), this means that each column of $\mathbf{X}_1$ is uncorrelated with each column of $\mathbf{X}_2$. Although they are unlikely to occur with observational data in the social sciences, when designing experimental treatments or constructing high degree polynomials, applications of orthogonal variables are regularly used (see, e.g., Hocking, 2003). For our purposes, we will work with orthogonal variables simply to understand the logical consequences of a total lack of collinearity.

Suppose that $\mathbf{x}_2$ is a vector of explanatory variables that is orthogonal to $\mathbf{X}_1$, where $\mathbf{X}_1$ is a matrix of explanatory variables that includes the intercept. Then, it is straightforward to check that the addition of $\mathbf{x}_2$ to the regression equation does not change the fit for coefficients corresponding to $\mathbf{X}_1$. That is, without $\mathbf{x}_2$, the coefficients corresponding to $\mathbf{X}_1$ would be calculated as $\mathbf{b}_1 = \left(\mathbf{X}_1'\mathbf{X}_1\right)^{-1}\mathbf{X}_1'\mathbf{y}$. Using the orthogonal $\mathbf{x}_2$ as part of the least squares calculation would not change the result for $\mathbf{b}_1$ (see the recursive least squares calculation in Section 4.7.2).

Further, the variance inflation factor for $\mathbf{x}_2$ is 1, indicating that the standard error is unaffected by the other explanatory variables. In the same vein, the reduction in the error sum of squares by adding the orthogonal variable $\mathbf{x}_2$ is due to only that variable and not its interaction with other variables in $\mathbf{X}_1$.

Orthogonal variables can be created for observational social science data (as well as other collinear data) using the method of *principal components*. With this method, one uses a linear transformation of the matrix of explanatory variables of the form, $\mathbf{X}^* = \mathbf{XP}$, so that the resulting matrix $\mathbf{X}^*$ is composed of orthogonal columns. The transformed regression function is $\mathrm{E}\,\mathbf{y} = \mathbf{X}\boldsymbol{\beta} = \mathbf{XPP}^{-1}\boldsymbol{\beta} = \mathbf{X}^*\boldsymbol{\beta}^*$, where $\boldsymbol{\beta}^* = \mathbf{P}^{-1}\boldsymbol{\beta}$ is the set of new regression coefficients. Estimation proceeds as previously, with the orthogonal set of explanatory variables. By choosing the matrix $\mathbf{P}$ appropriately, each column of $\mathbf{X}^*$ has an identifiable contribution. Thus, we can readily use variable selection techniques to identify the principal components portions of $\mathbf{X}^*$ to use in the regression equation. Principal components regression is a widely used method in some application areas, such as psychology. It can easily address highly collinear data in a disciplined manner. The main drawback of this technique is that the resulting parameter estimates are difficult to interpret.

5.6 Selection Criteria

5.6.1 Goodness of Fit

How well does the model fit the data? Criteria that measure the proximity of the fitted model and realized data are known as *goodness-of-fit* statistics. Specifically, we interpret the fitted value $\hat{y}_i$ to be the best model approximation of the ith observation and compare it to the actual value y_i. In linear regression, we examine the difference through the residual $e_i = y_i - \hat{y}_i$; small residuals imply a good model fit. We have quantified this through the size of the typical error (s), include the coefficient of determination (R^2) and an adjusted version (R_a^2).

For nonlinear models, we will need additional measures, and it is helpful to introduce these measures in this simpler linear case. One such measure is *Akaike's information criterion* (AIC), which will be defined in terms of likelihood fits in Section 11.9.4. For linear regression, it reduces to

$$AIC = n \ln(s^2) + n \ln(2\pi) + n + 3 + k. \tag{5.6}$$

For model comparison, the smaller is the *AIC*, the better is the fit. Comparing models with the same number of variables (k) means that selecting a model with small values of *AIC* leads to the same choice as selecting a model with small values of the residual standard deviation s. Further, a small number of parameters means a small value of *AIC*, other things being equal. The idea is that this measure balances the fit ($n \ln(s^2)$) with a penalty for complexity (the number of parameters, $k + 2$). Statistical packages often omit constants such as $n \ln(2\pi)$ and $n + 3$ when reporting *AIC* because they do not matter when comparing models.

Section 11.9.4 will introduce another measure, the *Bayes information criterion* (*BIC*), which gives a smaller weight to the penalty for complexity. A third goodness-of-fit measure that is used in linear regression model is the C_p statistic. To define this statistic, assume that we have available k explanatory variables $x_1, \ldots, x_k$ and run a regression to get s_{full}^2 as the mean square error. Now, suppose that we are considering using only $p - 1$ explanatory variables so that there are p regression coefficients. With these $p - 1$ explanatory variables, we run a regression to get the error sum of squares $(Error\ SS)_p$. Thus, we are in the position to define

$$C_p = \frac{(Error\ SS)_p}{s_{full}^2} - n + 2p.$$

As a selection criterion, we choose the model with a "small" C_p coefficient, where small is taken to be relative to p. In general, models with smaller values of C_p are more desirable.

Like the *AIC* and *BIC* statistics, the C_p statistic strikes a balance between the model fit and complexity. That is, each statistic summarizes the trade-off between model fit and complexity, although with different weights. For most datasets, they recommend the same model, so an analyst can report any or all three statistics. However, for some applications, they lead to different recommended models. In this case, the analyst needs to rely more heavily on non-data driven criteria for model selection (which are always important in any regression application).

5.6.2 Model Validation

Model validation is the process of confirming that our proposed model is appropriate, especially in light of the purposes of the investigation. Recall the iterative model formulation selection process described in Section 5.1. An important criticism of this iterative process is that it is guilty of *data snooping*, that is, of fitting a great number of models to a single set of data. As we saw in Section 5.2 on data snooping in stepwise regression, by looking at a large number of models we may overfit the data and understate the natural variation in our representation.

We can respond to this criticism by using a technique called *out-of-sample validation*. The ideal situation is to have available two sets of data, one for model development and one for model validation. We initially develop one, or several, models on a first dataset. The models developed from the first set of data are called our *candidate* models. Then, the relative performance of the candidate models could be measured on a second set of data. In this way, the data used to validate the model is unaffected by the procedures used to formulate the model.

Unfortunately, rarely will two sets of data be available to the investigator. However, we can implement the validation process by splitting the dataset into two subsamples. We call these the *model development* and *validation subsamples*, respectively. They are also known as *training* and *testing* samples, respectively. To see how the process works in the linear regression context, consider the following procedure.

Out-of-Sample Validation Procedure

(i) Begin with a sample size of n and divide it into two subsamples, the model development and validation subsamples. Let n_1 and n_2 denote the size of each subsample. In cross-sectional regression, do this split using a random sampling mechanism. Use the notation $i = 1, \ldots, n_1$ to represent observations from the model development subsample and $i = n_1 + 1, \ldots, n_1 + n_2 = n$ for the observations from the validation subsample. Figure 5.9 illustrates this procedure.

(ii) Using the model development subsample, fit a candidate model to the data set $i = 1, \ldots, n_1$.

(iii) Using the model created in step (ii) and the explanatory variables from the validation subsample, predict the dependent variables in the validation subsample, $\hat{y}_i$, where $i = n_1 + 1, \ldots, n_1 + n_2$. (To get these predictions, you may need to transform the dependent variables back to the original scale.)

(iv) Assess the proximity of the predictions to the held-out data. One measure is the *sum of squared prediction errors*:

$$SSPE = \sum_{i=n_1+1}^{n_1+n_2} (y_i - \hat{y}_i)^2 \tag{5.7}$$

Repeat steps (ii) through (iv) for each candidate model. Choose the model with the smallest *SSPE*.

There are a number of criticisms of the *SSPE*. First, it is clear that it takes a considerable amount of time and effort to calculate this statistic for each of several candidate models. However, as with many statistical techniques, this is

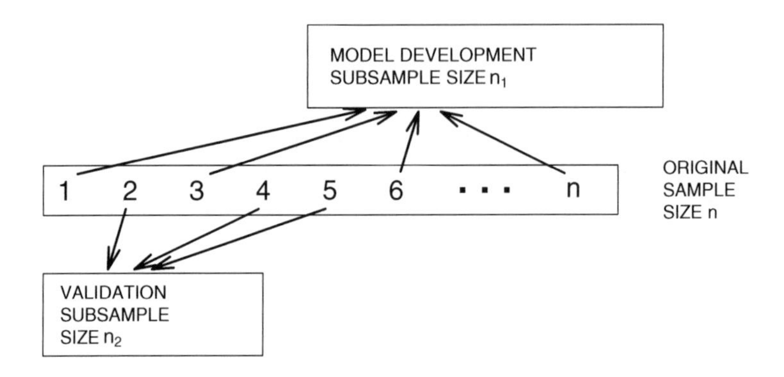

Figure 5.9 For model validation, a dataset of size n is randomly split into two subsamples.

merely a matter of having specialized statistical software available to perform the steps described earlier. Second, because the statistic itself is based on a random subset of the sample, its value will vary from analyst to analyst. This objection could be overcome by using the first n_1 observations from the sample. In most applications this is not done in case there is a lurking relationship in the order of the observations. Third, and perhaps most important, is that the choice of the relative subset sizes, n_1 and n_2, is not clear. Various researchers recommend different proportions for the allocation. Snee (1977) suggests that data splitting not be done unless the sample size is moderately large, specifically, $n \geq 2(k+1)+20$. The guidelines of Picard and Berk (1990) show that the greater the number of parameters to be estimated, the greater the proportion of observations needed for the model development subsample. As a rule of thumb, for datasets with 100 or fewer observations, use about 25–35% of the sample for out-of-sample validation. For datasets with 500 or more observations, use 50% of the sample for out-of-sample validation. Hastie, Tibshirani, and Friedman (2001) remark that a typical split is 50% for development and/or training, 25% for validation, and the remaining 25% for a third stage for further validation that they call *testing*.

Because of these criticisms, several variants of the basic out-of-sample validation process are used by analysts. Although there is no theoretically best procedure, it is widely agreed that model validation is an important part of confirming the usefulness of a model.

5.6.3 Cross-Validation

Cross-validation is the technique of model validation that splits the data into two disjoint sets. Section 5.6.2 discussed out-of-sample validation where the data was split randomly into two subsets both containing a sizable percentage of data. Another popular method is *leave-one-out* cross-validation, where the validation sample consists of a single observation and the development sample is based on the remainder of the dataset.

Especially for small sample sizes, an attractive leave-one-out cross-validation statistic is *PRESS*, the *predicted residual sum of squares*. To define the statistic, consider the following procedure in which we suppose that a candidate model is available.

PRESS Validation Procedure

(i) From the full sample, omit the ith point and use the remaining $n - 1$ observations to compute regression coefficients.
(ii) Use the regression coefficients computed in step one and the explanatory variables for the ith observation to compute the predicted response, $\hat{y}_{(i)}$. This part of the procedure is similar to the calculation of the *SSPE* statistic with $n_1 = n - 1$ and $n_2 = 1$.
(iii) Now, repeat (i) and (ii) for $i = 1, \ldots, n$. Summarizing, define

$$PRESS = \sum_{i=1}^{n} (y_i - \hat{y}_{(i)})^2. \tag{5.8}$$

As with *SSPE*, this statistic is calculated for each of several competing models. Under this criterion, we choose the model with the smallest *PRESS*.

Given this definition, the statistic seems very computationally intensive in that it requires n regression fits to evaluate it. To address this, interested readers will find that Section 5.10.2 establishes

$$y_i - \hat{y}_{(i)} = \frac{e_i}{1 - h_{ii}}. \tag{5.9}$$

Here, e_i and h_{ii} represent the ith residual and leverage from the regression fit using the complete dataset. This yields

$$PRESS = \sum_{i=1}^{n} \left(\frac{e_i}{1 - h_{ii}} \right)^2, \tag{5.10}$$

which is a much easier computational formula. Thus, the *PRESS* statistic is less computationally intensive than *SSPE*.

Another important advantage of this statistic, when compared to *SSPE*, is that we do not need to make an arbitrary choice as to our relative subset sizes split. Indeed, because we are performing an "out-of-sample" validation for each observation, it can be argued that this procedure is more efficient, an especially important consideration when the sample size is small (say, fewer than 50 observations). A disadvantage is that because the model is refit for each point deleted, *PRESS* does not enjoy the appearance of independence between the estimation and prediction aspects, unlike *SSPE*.

5.7 Heteroscedasticity

In most regression applications, the goal is to understand determinants of the regression function $\mathrm{E}\, y_i = \mathbf{x}_i' \boldsymbol{\beta} = \mu_i$. Our ability to understand the mean is strongly influenced by the amount of spread from the mean that we quantify using the variance $\mathrm{E}(y_i - \mu_i)^2$. In some applications, such as when I weigh myself on a scale, there is relatively little variability; repeated measurements yield almost the same result. In other applications, such as the time it takes me

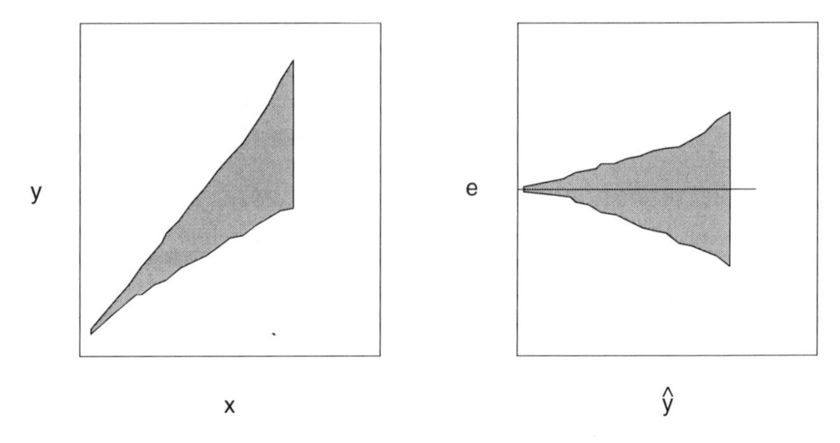

Figure 5.10 The shaded area represents the data. The line is the true regression line.

Figure 5.11 Residuals plotted versus the fitted values for the data in Figure 5.10.

to fly to New York, repeated measurements yield substantial variability and are fraught with inherent uncertainty.

The amount of uncertainty can also vary on a case-by-case basis. We denote the case of "varying variability" with the notation $\sigma_i^2 = \mathrm{E}\,(y_i - \mu_i)^2$. When the variability varies by observation, this is known as *heteroscedasticity* for "different scatter." In contrast, the usual assumption of common variability (assumption E3/F3 in Section 3.2) is called *homoscedasticity*, meaning "same scatter."

Our estimation strategies depend on the extent of heteroscedasticity. For datasets with only a mild amount of heteroscedasticity, one can use least squares to estimate the regression coefficients, perhaps combined with an adjustment for the standard errors (described in Section 5.7.2). This is because least squares estimators are unbiased even in the presence of heteroscedasticity (see property 1 in Section 3.2).

However, with heteroscedastic dependent variables, the Gauss-Markov theorem no longer applies and so the least squares estimators are not guaranteed to be optimal. In cases of severe heteroscedasticity, alternative estimators are used, the most common being those based on transformations of the dependent variable, as will be described in Section 5.7.4.

5.7.1 Detecting Heteroscedasticity

To decide a strategy for handling potential heteroscedasticity, we must first assess, or detect, its presence.

To detect heteroscedasticity, plot the residuals versus the fitted values.

To detect heteroscedasticity graphically, a good idea is to perform a preliminary regression fit of the data and plot the residuals versus the fitted values. To illustrate, Figure 5.10 is a plot of a fictitious dataset with one explanatory variable where the scatter increases as the explanatory variable increases. A least squares regression was performed – residuals and fitted values were computed. Figure 5.11 is an example of a plot of residuals versus fitted values. The preliminary regression fit removes many of the major patterns in the data and leaves the eye free to concentrate on other patterns that may influence the fit. We plot residuals versus fitted values because the fitted values are an approximation of the expected value of the response and, in many situations, the variability grows with the expected response.

More formal tests of heteroscedasticity are also available in the regression literature. To illustrate, let us consider a test due to Breusch and Pagan (1980). Specifically, this test examines the alternative hypothesis H_a: Var $y_i = \sigma^2 + \mathbf{z}_i'\boldsymbol{\gamma}$, where $\mathbf{z}_i$ is a known vector of variables and $\boldsymbol{\gamma}$ is a p-dimensional vector of parameters. Thus, the null hypothesis is $H_0: \ \boldsymbol{\gamma} = \mathbf{0}$, which is equivalent to homoscedasticity, Var $y_i = \sigma^2$.

Procedure to Test for Heteroscedasticity

(i) Fit a regression model and calculate the model residuals, e_i.
(ii) Calculate squared standardized residuals, $e_i^{*2} = e_i^2/s^2$.
(iii) Fit a regression model of e_i^{*2} on $\mathbf{z}_i$.
(iv) The test statistic is $LM = (Regress\ SS_z)/2$, where $Regress\ SS_z$ is the regression sum of squares from the model fit in step (iii).
(v) Reject the null hypothesis if LM exceeds a percentile from a chi-square distribution with p degrees of freedom. The percentile is one minus the significance level of the test.

Here, we use LM to denote the test statistic because Breusch and Pagan derived it as a Lagrange multiplier statistic; see Breusch and Pagan (1980) for more details.

5.7.2 Heteroscedasticity-Consistent Standard Errors

For data sets with only mild heteroscedasticity, a sensible strategy is to employ least squares estimators of the regression coefficients and to adjust the calculation of standard errors to account for the heteroscedasticity.

From the Section 3.2 on properties, we saw that least squares regression coefficients could be written as $\mathbf{b} = \sum_{i=1}^n \mathbf{w}_i y_i$, where $\mathbf{w}_i = \left(\mathbf{X}'\mathbf{X}\right)^{-1} \mathbf{x}_i$. Thus, with $\sigma_i^2 = $ Var y_i, we have

$$\text{Var}\ \mathbf{b} = \sum_{i=1}^n \mathbf{w}_i \mathbf{w}_i' \sigma_i^2 = \left(\mathbf{X}'\mathbf{X}\right)^{-1} \left(\sum_{i=1}^n \sigma_i^2 \mathbf{x}_i \mathbf{x}_i' \right) \left(\mathbf{X}'\mathbf{X}\right)^{-1}. \tag{5.11}$$

This quantity is known except for σ_i^2. We can compute residuals using the least squares regression coefficients as $e_i = y_i - \mathbf{x}_i'\mathbf{b}$. With these, we may define the *empirical*, or *robust*, estimate of the variance covariance matrix as

$$\widehat{\text{Var}\ \mathbf{b}} = \left(\mathbf{X}'\mathbf{X}\right)^{-1} \left(\sum_{i=1}^n e_i^2 \mathbf{x}_i \mathbf{x}_i' \right) \left(\mathbf{X}'\mathbf{X}\right)^{-1}.$$

The corresponding "heteroscedasticity-consistent" standard errors are

$$se_r(b_j) = \sqrt{(j+1)^{st}\ diagonal\ element\ of\ \widehat{\text{Var}\ \mathbf{b}}}. \tag{5.12}$$

The logic behind this estimator is that each squared residual, e_i^2 may be a poor estimate of σ_i^2. However, our interest is estimating a (weighted) sum of variances in equation (5.11); estimating the sum is a much easier task than estimating any individual variance estimate.

Robust, or heteroscedasticity-consistent, standard errors are widely available in statistical software packages. Here, you will also see alternative definitions of residuals employed, as in Section 5.3.1. If your statistical package offers options, the robust estimator using studentized residuals is generally preferred.

5.7.3 Weighted Least Squares

The least squares estimators are less useful for data sets with severe heteroscedasticity. One strategy is to use a variation of least squares estimation by *weighting* observations. The idea is that, when minimizing the sum of squared errors using heteroscedastic data, the expected variability of some observations is smaller than others. Intuitively, it seems reasonable that the smaller the variability of the response, the more reliable that response and the greater weight that it should receive in the minimization procedure. *Weighted least squares* is a technique that accounts for this "varying variability."

Specifically, we use Section 3.2.3 assumptions E1, E2 and E4, with E3 replaced by $\mathrm{E}\, \varepsilon_i = 0$ and $\mathrm{Var}\, \varepsilon_i = \sigma^2/w_i$, so that the variability is proportional to a known weight w_i. For example, if unit of analysis i represents a geographical entity such as a state, you might use the number of people in the state as a weight. Or if i represents a firm, you might use firm assets for the weighting variable. Larger values of w_i indicate a more precise response variable through the smaller variability. In actuarial applications, weights are used to account for an exposure such as the amount of insurance premium, number of employees, size of the payroll, number of insured vehicles and so forth (further discussion is in Chapter 18).

This model can be readily converted to the "ordinary" least squares problem by multiplying all regression variables by $\sqrt{w_i}$. That is, if we define $y_i^* = y_i \times \sqrt{w_i}$ and $x_{ij}^* = x_{ij} \times \sqrt{w_i}$, then from assumption E1 we have

$$\begin{aligned} y_i^* = y_i \times \sqrt{w_i} &= (\beta_0 x_{i0} + \beta_1 x_{i1} + \cdots + \beta_k x_{ik} + \varepsilon_i)\sqrt{w_i} \\ &= \beta_0 x_{i0}^* + \beta_1 x_{i1}^* + \cdots + \beta_k x_{ik}^* + \varepsilon_i^*, \end{aligned}$$

where $\varepsilon_i^* = \varepsilon_i \times \sqrt{w_i}$ has homoscedastic variance σ^2. Thus, with the rescaled variables, all inference can proceed as earlier.

This work has been automated in statistical packages where the user merely specifies the weights w_i and the package does the rest. In terms of matrix algebra, this procedure can be accomplished by defining an $n \times n$ weight matrix $\mathbf{W} = diag(w_i)$ so that the ith diagonal element of $\mathbf{W}$ is w_i. Extending equation (3.14), for example, the weighted least squares estimates can be expressed as

$$\mathbf{b}_{WLS} = \left(\mathbf{X}'\mathbf{W}\mathbf{X}\right)^{-1} \mathbf{X}'\mathbf{W}\mathbf{y}. \tag{5.13}$$

Additional discussions of weighted least squares estimation will be presented in Section 15.1.1.

5.7.4 Transformations

Another approach that handles severe heteroscedasticity, introduced in Section 1.3, is to transform the dependent variable, typically with a logarithmic transformation of the form $y^* = \ln y$. As we saw in Section 1.3, transformations can serve to "shrink" spread-out data and symmetrize a distribution. Through a change of scale, a transformation also changes the variability, potentially altering a heteroscedastic dataset into a homoscedastic one. This is both a strength and limitation of the transformation approach – a transformation simultaneously affects both the distribution and the heteroscedasticity.

The transformation of the dependent variable affects both the skewness of the distribution and the heteroscedasticity.

Power transformations, such as the logarithmic transform, are most useful when the variability of the data grows with the mean. In this case, the transform will serve to "shrink" the data to a scale that appears to be homoscedastic. Conversely, because transformations are monotonic functions, they will not help with patterns of variability that are nonmonotonic. Further, if your data is reasonably symmetric but heteroscedastic, a transformation will not be useful because any choice that mitigates the heteroscedasticity will skew the distribution.

When data are nonpositive, it is common to add a constant to each observation so that all observations are positive prior to transformation. For example, the transform $\ln(1 + y)$ accommodates the presence of zeros. One can also multiply by a constant so that the approximate original units are retained. For example, the transform $100 \ln(1 + y/100)$ may be applied to percentage data, where negative percentages sometimes appear.

Our discussions of transformations have focussed on transforming dependent variables. As noted in Section 3.5, transformations of explanatory variables are also possible. This is because the regression assumptions condition on explanatory variables (Section 3.2.3). Some analysts prefer to transform variables to approximate normality, thinking of multivariate normal distributions as a foundation for regression analysis. Others are reluctant to transform explanatory variables because of the difficulties in interpreting resulting models. The approach taken here is to use transforms that are readily interpretable, such as those introduced in Section 3.5. Other transforms are certainly candidates to include in a selected model but they should provide substantial dividends in terms of fit or predictive power if they are difficult to communicate.

5.8 Further Reading and References

Long and Ervin (2000) gather compelling evidence for the use of alternative heteroscedasticity-consistent estimators of standard errors that have better finite sample performance than the classic versions. The large sample properties of empirical estimators have been established by Eicker (1967), Huber (1967), and White (1980) in the linear regression case. For the linear regression case, MacKinnon and White (1985) suggest alternatives that provide superior small-sample properties. For small samples, the evidence is based on (1) the biasedness of the estimators, (2) their motivation as jackknife estimators and (3) their performance in simulation studies.

Other measures of collinearity based on matrix algebra concepts involving eigenvalues, such as condition numbers and condition indices, are used by some analysts. See Belseley, Kuh, and Welsch (1980) for a solid treatment of collinearity and regression diagnostics. Hocking (2003) provides additional background reading on collinearity and principal components. See Carroll and Ruppert (1988) for further discussions of transformations in regression.

Hastie, Tibshirani, and Friedman (2001) give an advanced discussion of model selection issues, focusing on predictive aspects of models in the language of machine learning.

Chapter References

Belseley, David A., Edwin Kuh, and Roy E. Welsch (1980). *Regression Diagnostics: Identifying Influential Data and Sources of Collinearity*. Wiley, New York.

Bendel, R. B., and A. A. Afifi (1977). Comparison of stopping rules in forward "stepwise" regression. *Journal of the American Statistical Association* 72, 46–53.

Box, George E. P. (1980). Sampling and Bayes inference in scientific modeling and robustness (with discussion). *Journal of the Royal Statistical Society*, Ser. A, 143, 383–430.

Breusch, T. S., and A. R. Pagan (1980). The Lagrange multiplier test and its applications to model specification in econometrics. *Review of Economic Studies*, 47, 239–53.

Carroll, Raymond J., and David Ruppert (1988). *Transformation and Weighting in Regression*, Chapman-Hall, New York.

Eicker, F. (1967), Limit theorems for regressions with unequal and dependent errors. *Proceedings of the Fifth Berkeley Symposium on Mathematical Statistics and Probability*, L. M. LeCam and J. Neyman, eds. University of California Press Berkeley, CA, 1:59–82.

Hadi, A. S. (1988). Diagnosing collinearity-influential observations. *Computational Statistics and Data Analysis* 7, 143–59.

Hastie, Trevor, Robert Tibshirani, and Jerome Friedman (2001). *The Elements of Statistical Learning: Data Mining, Inference and Prediction.* Springer-Verlag, New York.

Hocking, Ronald R. (2003). *Methods and Applications of Linear Models: Regression and the Analysis of Variance*. Wiley, New York.

Huber, P. J. (1967). The behaviour of maximum likelihood estimators under non-standard conditions. *Proceedings of the Fifth Berkeley Symposium on Mathematical Statistics and Probability*, L. M. LeCam and J. Neyman, eds. University of California Press Berkeley, CA, 1:221–33.

Long, J. S., and L. H. Ervin (2000). Using heteroscedasticity consistent standard errors in the linear regression model. *American Statistician* 54, 217–24.

MacKinnon, J. G., and H. White (1985). Some heteroskedasticity consistent covariance matrix estimators with improved finite sample properties. *Journal of Econometrics* 29, 53–7.

Mason, R. L., and R. F. Gunst (1985). Outlier-induced collinearities. *Technometrics* 27, 401–7.

Picard, R. R., and K. N. Berk (1990). Data splitting. *American Statistician* 44, 140–47.

Rencher, A. C., and F. C. Pun (1980). Inflation of R^2 in best subset regression. *Technometrics* 22, 49–53.

Snee, R. D. (1977). Validation of regression models. Methods and examples. *Technometrics* 19, 415–28.

5.9 Exercises

5.1. You are doing regression with one explanatory variable and so consider the basic linear regression model $y_i = \beta_0 + \beta_1 x_i + \varepsilon_i$.

a. Show that the ith leverage can be simplified to

$$h_{ii} = \frac{1}{n} + \frac{(x_i - \overline{x})^2}{(n-1)s_x^2}.$$

b. Show that $\overline{h} = 2/n$.
c. Suppose that $h_{ii} = 6/n$. How many standard deviations is x_i away (either above or below) from the mean?

5.2. Consider the output of a regression using one explanatory variable on $n = 3$ observations. The residuals and leverages are:

i	1	2	3
Residuals e_i	3.181	−6.362	3.181
Leverages h_{ii}	0.8333	0.3333	0.8333

Compute the *PRESS* statistic.

Ⓡ **Empirical**
Filename is "UNLifeExpectancy"

5.3. **National Life Expectancies.** We continue the analysis begun in Exercises 1.7, 2.22, 3.6, and 4.7. The focus of this exercise is variable selection.
a. Begin with the data from $n = 185$ countries throughout the world that have valid (nonmissing) life expectancies. Plot the life expectancy versus the gross domestic product and private expenditures on health. From these plots, describe why it is desirable to use logarithmic transforms, lnGDP and lnHEALTH, respectively. Also plot life expectancy versus lnGDP and lnHEALTH to confirm your intuition.
b. Use a stepwise regression algorithm to help you select a model. Do not consider the variables RESEARCHERS, SMOKING, and FEMALEBOSS, as these have many missing values. For the remaining variables, use only the observations without any missing values. Do this twice, with and without the categorical variable REGION.
c. Return to the full dataset of $n = 185$ countries and run a regression model using FERTILITY, PUBLICEDUCATION, and lnHEALTH as explanatory variables.
c(i). Provide histograms of standardized residuals and leverages.
c(ii). Identify the standardized residual and leverage associated with Lesotho, formerly Basutoland, a kingdom surrounded by South Africa. Is this observation an outlier, a high leverage point, or both?
c(iii). Rerun the regression without Lesotho. Cite any differences in the statistical coefficients between this model and the one in part c(i).

Ⓡ **Empirical**
Filename is "TermLife"

5.4. **Term Life Insurance.** We continue our study of term life insurance demand from Chapters 3 and 4. Specifically, we examine the 2004 Survey of Consumer Finances (SCF), a nationally representative sample that contains extensive information on assets, liabilities, income, and demographic characteristics of those sampled (potential U.S. customers). We study a random sample of 500 families with positive incomes. From the sample of 500, we initially consider a subsample of $n = 275$ families that purchased term life insurance.

Consider a linear regression of LNINCOME, EDUCATION, NUMHH, MARSTAT, AGE, and GENDER on LNFACE.

a. Collinearity. Not all of the variables turned out to be statistically significant. To investigate one possible explanation, calculate variance inflation factors.
   - a(i). Briefly explain the idea of collinearity and a variance inflation factor.
   - a(ii). What constitutes a large variance inflation factor?
   - a(iii). If a large variance inflation factor is detected, what possible courses of action do we have to address this aspect of the data?
   - a(iv). Supplement the variance inflation factor statistics with a table of correlations of explanatory variables. Given these statistics, is collinearity an issue with this fitted model? Why or why not?

b. Unusual Points. Sometimes a poor model fit can be due to unusual points.
   - b(i). Define the idea of leverage for an observation.
   - b(ii). For this fitted model, give standard rules of thumbs for identifying points with unusual leverage. Identify any unusual points from the attached summary statistics.
   - b(iii). An analyst is concerned with leverage values for this fitted model and suggests using FACE as the dependent variable instead of LNFACE. Describe how leverage values would change using this alternative dependent variable.

c. Residual Analysis. We can learn how to improve model fits from analyses of residuals.
   - c(i). Provide a plot of residuals versus fitted values. What do we hope to learn from this type of plot? Does this plot display any model inadequacies?
   - c(ii). Provide a *qq* plot of residuals. What do we hope to learn from this type of plot? Does this plot display any model inadequacies?
   - c(iii). Provide a plot of residuals versus leverages. What do we hope to learn from this type of plot? Does this plot display any model inadequacies?

d. Stepwise Regression. Run a stepwise regression algorithm. Suppose that this algorithm suggests a model using LNINCOME, EDUCATION, NUMHH, and GENDER as explanatory variables to predict the dependent variable LNFACE.
   - d(i). What is the purpose of stepwise regression?
   - d(ii). Describe two important drawbacks of stepwise regression algorithms.

5.10 Technical Supplements for Chapter 5

5.10.1 Projection Matrix

Hat Matrix. We define the hat matrix to be $\mathbf{H} = \mathbf{X}(\mathbf{X}'\mathbf{X})^{-1}\mathbf{X}'$, so that $\hat{\mathbf{y}} = \mathbf{Xb} = \mathbf{Hy}$. From this, the matrix $\mathbf{H}$ is said to *project* the vector of responses $\mathbf{y}$ onto the vector of fitted values $\hat{\mathbf{y}}$.

Because $\mathbf{H}' = \mathbf{H}$, the hat matrix is symmetric. Further, it is also an *idempotent* matrix due to the property $\mathbf{HH} = \mathbf{H}$. To see this, we have $\mathbf{HH} = (\mathbf{X}(\mathbf{X}'\mathbf{X})^{-1}\mathbf{X}')(\mathbf{X}(\mathbf{X}'\mathbf{X})^{-1}\mathbf{X}') = \mathbf{X}(\mathbf{X}'\mathbf{X})^{-1}(\mathbf{X}'\mathbf{X})(\mathbf{X}'\mathbf{X})^{-1}\mathbf{X}' = \mathbf{X}(\mathbf{X}'\mathbf{X})^{-1}\mathbf{X}' = \mathbf{H}$. Similarly, it is easy to check that $\mathbf{I} - \mathbf{H}$ is idempotent. Because $\mathbf{H}$ is idempotent, from some results in matrix algebra, it is straightforward to show that $\sum_{i=1}^{n} h_{ii} = k + 1$. As discussed in Section 5.4.1, we use our bounds and the average leverage, $\bar{h} = (k+1)/n$, to help identify observations with unusually high leverage.

Variance of Residuals. Using the model equation $\mathbf{y} = \mathbf{X}\boldsymbol{\beta} + \boldsymbol{\varepsilon}$, we can express the vector of residuals as

$$\mathbf{e} = \mathbf{y} - \hat{\mathbf{y}} = \mathbf{y} - \mathbf{H}\mathbf{y} = (\mathbf{I} - \mathbf{H})(\mathbf{X}\boldsymbol{\beta} + \boldsymbol{\varepsilon}) = (\mathbf{I} - \mathbf{H})\boldsymbol{\varepsilon}. \tag{5.14}$$

The last equality is because $(\mathbf{I} - \mathbf{H})\mathbf{X} = \mathbf{X} - \mathbf{HX} = \mathbf{X} - \mathbf{X} = \mathbf{0}$. Using $\mathrm{Var}\ \boldsymbol{\varepsilon} = \sigma^2\mathbf{I}$, we have

$$\begin{aligned}\mathrm{Var}\ \mathbf{e} &= \mathrm{Var}\ [(\mathbf{I} - \mathbf{H})\boldsymbol{\varepsilon}] = (\mathbf{I} - \mathbf{H})\mathrm{Var}\ \boldsymbol{\varepsilon}(\mathbf{I} - \mathbf{H}) \\ &= \sigma^2(\mathbf{I} - \mathbf{H})\mathbf{I}(\mathbf{I} - \mathbf{H}) = \sigma^2(\mathbf{I} - \mathbf{H}).\end{aligned}$$

The last equality comes from the fact that $\mathbf{I} - \mathbf{H}$ is idempotent. Thus, we have that

$$\mathrm{Var}\ e_i = \sigma^2(1 - h_{ii}) \quad \text{and} \quad \mathrm{Cov}(e_i, e_j) = -\sigma^2 h_{ij}. \tag{5.15}$$

Thus, although the true errors $\boldsymbol{\varepsilon}$ are uncorrelated, there is a small negative correlation among residuals $\mathbf{e}$.

Dominance of the Error in the Residual. Examining the ith row of equation (5.14), we have that the ith residual

$$e_i = \varepsilon_i - \sum_{j=1}^{n} h_{ij}\varepsilon_j \tag{5.16}$$

can be expressed as a linear combination of independent errors. The relation $\mathbf{H} = \mathbf{HH}$ yields

$$h_{ii} = \sum_{j=1}^{n} h_{ij}^2. \tag{5.17}$$

Because h_{ii} is, on average, $(k+1)/n$, this indicates that each h_{ij} is small relative to 1. Thus, when interpreting equation (5.16), we say that most of the information in e_i is due to ε_i.

Correlations with Residuals. First define $\mathbf{x}^j = (x_{1j}, x_{2j}, \ldots, x_{nj})'$ to be the column representing the jth variable. With this notation, we can partition the matrix of explanatory variables as $\mathbf{X} = \left(\mathbf{x}^0, \mathbf{x}^1, \ldots, \mathbf{x}^k\right)$. Now, examining the jth column of the relation $(\mathbf{I} - \mathbf{H})\mathbf{X} = \mathbf{0}$, we have $(\mathbf{I} - \mathbf{H})\mathbf{x}^j = \mathbf{0}$. With $\mathbf{e} = (\mathbf{I} - \mathbf{H})\boldsymbol{\varepsilon}$, this yields $\mathbf{e}'\mathbf{x}^j = \boldsymbol{\varepsilon}'(\mathbf{I} - \mathbf{H})\mathbf{x}^j = 0$, for $j = 0, 1, \ldots, k$. This result has several implications. If the intercept is in the model, then $\mathbf{x}^0 = (1, 1, \ldots, 1)'$ is a vector of ones. Here, $\mathbf{e}'\mathbf{x}^0 = 0$ means that $\sum_{i=1}^{n} e_i = 0$ or, the average residual is zero. Further, because $\mathbf{e}'\mathbf{x}^j = 0$, it is easy to check that the sample correlation between $\mathbf{e}$ and $\mathbf{x}^j$ is zero. Along the same line, we also have that $\mathbf{e}'\hat{\mathbf{y}} = \mathbf{e}'(\mathbf{I} - \mathbf{H})\mathbf{Xb} = \mathbf{0}$.

When a vector of ones is present, then the average residual is zero.

Thus, using the same argument as above, the sample correlation between $\mathbf{e}$ and $\hat{\mathbf{y}}$ is zero.

When a vector of ones is present, then the average fitted value is $\bar{y}$.

Multiple Correlation Coefficient. For an example of a nonzero correlation, consider $r(\mathbf{y}, \hat{\mathbf{y}})$, the sample correlation between $\mathbf{y}$ and $\hat{\mathbf{y}}$. Because $(\mathbf{I} - \mathbf{H})\mathbf{x}^0 = \mathbf{0}$, we have $\mathbf{x}^0 = \mathbf{H}\mathbf{x}^0$ and thus, $\hat{\mathbf{y}}'\mathbf{x}^0 = \mathbf{y}'\mathbf{H}\mathbf{x}^0 = \mathbf{y}'\mathbf{x}^0$. Assuming $\mathbf{x}^0 = (1, 1, \ldots, 1)'$, this means that $\sum_{i=1}^n \hat{y}_i = \sum_{i=1}^n y_i$, so that the average fitted value is $\bar{y}$. Now,

$$r(\mathbf{y}, \hat{\mathbf{y}}) = \frac{\sum_{i=1}^n (y_i - \bar{y})(\hat{y}_i - \bar{y})}{(n-1)s_y s_{\hat{y}}}.$$

Recall that $(n-1)s_y^2 = \sum_{i=1}^n (y_i - \bar{y})^2 =$ Total SS and $(n-1)s_{\hat{y}}^2 = \sum_{i=1}^n (\hat{y}_i - \bar{y})^2 =$ Regress SS. Further, with $\mathbf{x}^0 = (1, 1, \ldots, 1)'$,

$$\sum_{i=1}^n (y_i - \bar{y})(\hat{y}_i - \bar{y}) = (\mathbf{y} - \bar{y}\mathbf{x}^0)'(\hat{\mathbf{y}} - \bar{y}\mathbf{x}^0) = \mathbf{y}'\hat{\mathbf{y}} - \bar{y}^2\mathbf{x}^{0\prime}\mathbf{x}^0$$

$$= \mathbf{y}'\mathbf{X}\mathbf{b} - n\bar{y}^2 = \text{Regress SS}.$$

This yields

$$r(\mathbf{y}, \hat{\mathbf{y}}) = \frac{\text{Regress SS}}{\sqrt{(\text{Total SS})\,(\text{Regress SS})}} = \sqrt{\frac{\text{Regress SS}}{\text{Total SS}}} = \sqrt{R^2}. \qquad (5.18)$$

That is, the coefficient of determination can be interpreted as the square root of the correlation between the observed and fitted responses.

5.10.2 Leave-One-Out Statistics

Notation. To test the sensitivity of regression quantities, there are a number of statistics of interest that are based on the notion of "leaving out," or omitting, an observation. To this end, the subscript notation (i) means to *leave out* the ith observation. For example, omitting the row of explanatory variables $\mathbf{x}_i' = (x_{i0}, x_{i1}, \ldots, x_{ik})$ from $\mathbf{X}$ yields $\mathbf{X}_{(i)}$, a $(n-1) \times (k+1)$ matrix of explanatory variables. Similarly, $\mathbf{y}_{(i)}$ is a $(n-1) \times 1$ vector, based on removing the ith row from $\mathbf{y}$.

Basic Matrix Result. Suppose that $\mathbf{A}$ is an invertible, $p \times p$ matrix and $\mathbf{z}$ is a $p \times 1$ vector. The following result from matrix algebra provides an important tool for understanding leave one out statistics in linear regression analysis.

$$\left(\mathbf{A} - \mathbf{z}\mathbf{z}'\right)^{-1} = \mathbf{A}^{-1} + \frac{\mathbf{A}^{-1}\mathbf{z}\mathbf{z}'\mathbf{A}^{-1}}{1 - \mathbf{z}'\mathbf{A}^{-1}\mathbf{z}}. \qquad (5.19)$$

To check this result, simply multiply $\mathbf{A} - \mathbf{z}\mathbf{z}'$ by the right-hand side of equation (5.19) to get $\mathbf{I}$, the identity matrix.

Vector of Regression Coefficients. Omitting the ith observation, our new vector of regression coefficients is $\mathbf{b}_{(i)} = \left(\mathbf{X}_{(i)}'\mathbf{X}_{(i)}\right)^{-1}\mathbf{X}_{(i)}'\mathbf{y}_{(i)}$. An alternative expression for $\mathbf{b}_{(i)}$ that is simpler to compute turns out to be

$$\mathbf{b}_{(i)} = \mathbf{b} - \frac{\left(\mathbf{X}'\mathbf{X}\right)^{-1}\mathbf{x}_i e_i}{1 - h_{ii}} \tag{5.20}$$

To verify equation (5.20), first use equation (5.19) with $\mathbf{A} = \mathbf{X}'\mathbf{X}$ and $\mathbf{z} = \mathbf{x}_i$ to get

$$\left(\mathbf{X}'_{(i)}\mathbf{X}_{(i)}\right)^{-1} = (\mathbf{X}'\mathbf{X} - \mathbf{x}_i\mathbf{x}'_i)^{-1} = (\mathbf{X}'\mathbf{X})^{-1} + \frac{\left(\mathbf{X}'\mathbf{X}\right)^{-1}\mathbf{x}_i\mathbf{x}'_i\left(\mathbf{X}'\mathbf{X}\right)^{-1}}{1 - h_{ii}},$$

where, from equation (5.3), we have $h_{ii} = \mathbf{x}'_i(\mathbf{X}'\mathbf{X})^{-1}\mathbf{x}_i$. Multiplying each side by $\mathbf{X}'_{(i)}\mathbf{y}_{(i)} = \mathbf{X}'\mathbf{y} - \mathbf{x}_i y_i$ yields

$$\begin{aligned}
\mathbf{b}_{(i)} &= \left(\mathbf{X}'_{(i)}\mathbf{X}_{(i)}\right)^{-1}\mathbf{X}'_{(i)}\mathbf{y}_{(i)} = \left((\mathbf{X}'\mathbf{X})^{-1} + \frac{\left(\mathbf{X}'\mathbf{X}\right)^{-1}\mathbf{x}_i\mathbf{x}'_i\left(\mathbf{X}'\mathbf{X}\right)^{-1}}{1 - h_{ii}}\right)\left(\mathbf{X}'\mathbf{y} - \mathbf{x}_i y_i\right) \\
&= \mathbf{b} - \left(\mathbf{X}'\mathbf{X}\right)^{-1}\mathbf{x}_i y_i + \frac{\left(\mathbf{X}'\mathbf{X}\right)^{-1}\mathbf{x}_i\mathbf{x}'_i\mathbf{b} - \left(\mathbf{X}'\mathbf{X}\right)^{-1}\mathbf{x}_i\mathbf{x}'_i\left(\mathbf{X}'\mathbf{X}\right)^{-1}\mathbf{x}_i y_i}{1 - h_{ii}} \\
&= \mathbf{b} - \frac{(1 - h_{ii})\left(\mathbf{X}'\mathbf{X}\right)^{-1}\mathbf{x}_i y_i - \left(\mathbf{X}'\mathbf{X}\right)^{-1}\mathbf{x}_i\mathbf{x}'_i\mathbf{b} - \left(\mathbf{X}'\mathbf{X}\right)^{-1}\mathbf{x}_i h_{ii} y_i}{1 - h_{ii}} \\
&= \mathbf{b} - \frac{\left(\mathbf{X}'\mathbf{X}\right)^{-1}\mathbf{x}_i y_i - \left(\mathbf{X}'\mathbf{X}\right)^{-1}\mathbf{x}_i\mathbf{x}'_i\mathbf{b}}{1 - h_{ii}} = \mathbf{b} - \frac{\left(\mathbf{X}'\mathbf{X}\right)^{-1}\mathbf{x}_i e_i}{1 - h_{ii}}.
\end{aligned}$$

This establishes equation (5.20).

Cook's Distance. To measure the effect, or *influence*, of omitting the ith observation, Cook examined the difference between fitted values with and without the observation. We define Cook's distance to be

$$D_i = \frac{\left(\hat{\mathbf{y}} - \hat{\mathbf{y}}_{(i)}\right)'\left(\hat{\mathbf{y}} - \hat{\mathbf{y}}_{(i)}\right)}{(k+1)s^2},$$

where $\hat{\mathbf{y}}_{(i)} = \mathbf{X}\mathbf{b}_{(i)}$ is the vector of fitted values calculated omitting the ith point. Using equation (5.20) and $\hat{\mathbf{y}} = \mathbf{X}\mathbf{b}$, an alternative expression for Cook's distance is

$$\begin{aligned}
D_i &= \frac{\left(\mathbf{b} - \mathbf{b}_{(i)}\right)'\left(\mathbf{X}'\mathbf{X}\right)\left(\mathbf{b} - \mathbf{b}_{(i)}\right)}{(k+1)s^2} \\
&= \frac{e_i^2}{(1 - h_{ii})^2}\frac{\mathbf{x}'_i\left(\mathbf{X}'\mathbf{X}\right)^{-1}\left(\mathbf{X}'\mathbf{X}\right)\left(\mathbf{X}'\mathbf{X}\right)^{-1}\mathbf{x}_i}{(k+1)s^2} \\
&= \frac{e_i^2}{(1 - h_{ii})^2}\frac{h_{ii}}{(k+1)s^2} = \left(\frac{e_i^2}{s\sqrt{1 - h_{ii}}}\right)^2\frac{h_{ii}}{(k+1)(1 - h_{ii})}.
\end{aligned}$$

This result is not only useful computationally, it also serves to decompose the statistic into the part due to the standardized residual, $\left(e_i / \left(s\,(1 - h_{ii})^{1/2}\right)\right)^2$, and due to the leverage, $h_{ii} / \left((k+1)\,(1 - h_{ii})\right)$.

Leave-One-Out Residual. The leave-one-out residual is defined by $e_{(i)} = y_i - \mathbf{x}_i'\mathbf{b}_{(i)}$. It is used in computing the *PRESS* statistic, described in Section 5.6.3. A simple computational expression is $e_{(i)} = e_i/(1-h_{ii})$. To verify this, use equation (5.20) to get

$$e_{(i)} = y_i - \mathbf{x}_i'\mathbf{b}_{(i)} = y_i - \mathbf{x}_i'\left(\mathbf{b} - \frac{\left(\mathbf{X}'\mathbf{X}\right)^{-1}\mathbf{x}_i e_i}{1-h_{ii}}\right)$$

$$= e_i + \frac{\mathbf{x}_i\left(\mathbf{X}'\mathbf{X}\right)^{-1}\mathbf{x}_i e_i}{1-h_{ii}} = e_i + \frac{h_{ii}e_i}{1-h_{ii}} = \frac{e_i}{1-h_{ii}}.$$

Leave-One-Out Variance Estimate. The leave-one-out estimate of the variance is defined by $s_{(i)}^2 = ((n-1)-(k+1))^{-1}\sum_{j\neq i}\left(y_j - \mathbf{x}_j'\mathbf{b}_{(i)}\right)^2$. It is used in the definition of the *studentized residual*, defined in Section 5.3.1. A simple computational expression is given by

$$s_{(i)}^2 = \frac{(n-(k+1))s^2 - \frac{e_i^2}{1-h_{ii}}}{(n-1)-(k+1)}. \tag{5.21}$$

To see this, first note that from equation (5.14), we have $\mathbf{He} = \mathbf{H}(\mathbf{I}-\mathbf{H})\boldsymbol{\varepsilon} = \mathbf{0}$, because $\mathbf{H} = \mathbf{HH}$. In particular, from the ith row of $\mathbf{He} = \mathbf{0}$, we have $\sum_{j=1}^n h_{ij}e_j = 0$. Now, using equations (5.17) and (5.20), we have

$$\sum_{j\neq i}\left(y_j - \mathbf{x}_j'\mathbf{b}_{(i)}\right)^2 = \sum_{j=1}^n \left(y_j - \mathbf{x}_j'\mathbf{b}_{(i)}\right)^2 - \left(y_i - \mathbf{x}_i'\mathbf{b}_{(i)}\right)^2$$

$$= \sum_{j=1}^n \left(y_j - \mathbf{x}_j'\mathbf{b} + \frac{\mathbf{x}_j'(\mathbf{X}'\mathbf{X})^{-1}\mathbf{x}_i e_i}{1-h_{ii}}\right) - e_{(i)}^2$$

$$= \sum_{j=1}^n (e_j + \frac{h_{ij}e_i}{1-h_{ii}})^2 - \frac{e_i^2}{(1-h_{ii})^2}$$

$$= \sum_{j=1}^n e_j^2 + 0 + \frac{e_i^2}{(1-h_{ii})^2}h_{ii} - \frac{e_i^2}{(1-h_{ii})^2}$$

$$= \sum_{j=1}^n e_j^2 - \frac{e_i^2}{1-h_{ii}} = (n-(k+1))s^2 - \frac{e_i^2}{1-h_{ii}}.$$

This establishes equation (5.21).

5.10.3 Omitting Variables

Notation. To measure the effect on regression quantities, there are a number of statistics of interest that are based on the notion of omitting an explanatory variable. To this end, the superscript notation (j) means to omit the jth variable, where $j = 0, 1, \ldots, k$. First, recall that $\mathbf{x}^j = (x_{1j}, x_{2j}, \ldots, x_{nj})'$ is the column representing the j th variable. Further, define $\mathbf{X}^{(j)}$ to be the $n \times k$ matrix of explanatory variables defined by removing $\mathbf{x}^j$ from $\mathbf{X}$. For example, taking $j = k$, we often partition $\mathbf{X}$ as $\mathbf{X} = \left(\mathbf{X}^{(k)} : \mathbf{x}^k\right)$. Employing the results of Section 4.7.2, we will use $\mathbf{X}^{(k)} = \mathbf{X}_1$ and $\mathbf{x}^k = \mathbf{X}_2$.

Variance Inflation Factor. We first would like to establish the relationship between the definition of the standard error of b_j given by

$$se(b_j) = s\sqrt{(j+1)\text{th } \textit{diagonal element} \text{ of } (\mathbf{X}'\mathbf{X})^{-1}}$$

and the relationship involving the variance inflation factor,

$$se(b_j) = s\frac{\sqrt{VIF_j}}{s_{x_j}\sqrt{n-1}}.$$

By symmetry of the independent variables, we need consider only the case where $j = k$. Thus, we would like to establish

$$(k+1)\text{st diagonal element of } (\mathbf{X}'\mathbf{X})^{-1} = VIF_k/((n-1)s_{x_k}^2). \tag{5.22}$$

First consider the reparameterized model in equation (4.22). From equation (4.23), we can express the regression coefficient estimate $b_k = (\mathbf{e}_1'\mathbf{y})/(\mathbf{e}_1'\mathbf{e}_1)$. From equation (4.23), we have that Var $b_k = \sigma^2(\mathbf{E}_2'\mathbf{E}_2)^{-1}$ and thus

$$se(b_k) = s(\mathbf{E}_2'\mathbf{E}_2)^{-1/2}. \tag{5.23}$$

Thus, the $(\mathbf{E}_2'\mathbf{E}_2)^{-1}$ is $(k+1)$st diagonal element of

$$\left(\begin{bmatrix}\mathbf{X}_1'\\ \mathbf{E}_2'\end{bmatrix}\begin{bmatrix}\mathbf{X}_1 & \mathbf{E}_2\end{bmatrix}\right)^{-1}$$

and is also the $(k+1)$st diagonal element of $(\mathbf{X}'\mathbf{X})^{-1}$. Alternatively, this can be verified directly using the partitioned matrix inverse in equation (4.19).

Now, suppose that we run a regression using $\mathbf{x}^k = \mathbf{X}_2$ as the response vector and $\mathbf{X}^{(k)} = \mathbf{X}_1$ as the matrix of explanatory variables. As noted in equation (4.22), $\mathbf{E}_2$ represents the "residuals" from this regression and thus $\mathbf{E}_2'\mathbf{E}_2$ represents the error sum of squares. For this regression, the total sum of squares is $\sum_{i=1}^n (x_{ik} - \bar{x}_k)^2 = (n-1)s_{x_k}^2$ and the coefficient of determination is R_k^2. Thus,

$$\mathbf{E}_2'\mathbf{E}_2 = \text{Error SS} = \text{Total SS}(1 - R_k^2) = (n-1)s_{x_k}^2/VIF_k.$$

This establishes equation (5.22).

Establishing $\mathbf{t^2 = F}$. For testing the null hypothesis H$_0$: $\beta_k = 0$, the material in Section 3.4.1 provides a description of a test based on the t-statistic, $t(b_k) = b_k/se(b_k)$. An alternative test procedure, described in Sections 4.2.2, uses the test statistic

$$F\text{-ratio} = \frac{(Error\ SS)_{reduced} - (Error\ SS)_{full}}{p \times (Error\ MS)_{full}} = \frac{\left(\mathbf{E}_2'\mathbf{y}\right)^2}{s^2\mathbf{E}_2'\mathbf{E}_2}$$

from equation (4.26). Alternatively, from equations (4.23) and (5.23), we have

$$t(b_k) = \frac{b_k}{se(b_k)} = \frac{(\mathbf{E}_2'\mathbf{y}) \,/\, (\mathbf{E}_2'\mathbf{E}_2)}{s/\sqrt{\mathbf{E}_2'\mathbf{E}_2}} = \frac{(\mathbf{E}_2'\mathbf{y})}{s\sqrt{\mathbf{E}_2'\mathbf{E}_2}}. \tag{5.24}$$

Thus, $t(b_k)^2 = F$-ratio.

Partial Correlation Coefficients. From the full regression model $\mathbf{y} = \mathbf{X}^{(k)}\boldsymbol{\beta}^{(k)} + \mathbf{x}_k\beta_k + \boldsymbol{\varepsilon}$, consider two separate regressions. A regression using $\mathbf{x}^k$ as the response vector and $\mathbf{X}^{(k)}$ as the matrix of explanatory variables yields the residuals $\mathbf{E}_2$. Similarly, a regression $\mathbf{y}$ as the response vector and $\mathbf{X}^{(k)}$ as the matrix of explanatory variables yields the residuals

$$\mathbf{E}_1 = \mathbf{y} - \mathbf{X}^{(k)}\left(\mathbf{X}^{(k)\prime}\mathbf{X}^{(k)}\right)^{-1}\mathbf{X}^{(k)}\mathbf{y}.$$

If $x^0 = (1, 1, \ldots, 1)'$, then the average of $\mathbf{E}_1$ and $\mathbf{E}_2$ is zero. In this case, the sample correlation between $\mathbf{E}_1$ and $\mathbf{E}_2$ is

$$r(\mathbf{E}_1, \mathbf{E}_2) = \frac{\sum_{i=1}^n E_{1i}E_{2i}}{\sqrt{\left(\sum_{i=1}^n E_{i1}^2\right)\left(\sum_{i=1}^n E_{i2}^2\right)}} = \frac{\mathbf{E}_1'\mathbf{E}_2}{\sqrt{(\mathbf{E}_1'\mathbf{E}_1)(\mathbf{E}_2'\mathbf{E}_2)}}.$$

Because $\mathbf{E}_2$ is a vector of residuals using $\mathbf{X}^{(k)}$ as the matrix of explanatory variables, we have that $\mathbf{E}_2'\mathbf{X}^{(k)} = 0$. Thus, for the numerator, we have $\mathbf{E}_2'\mathbf{E}_1 = \mathbf{E}_2'(\mathbf{y} - \mathbf{X}^{(k)}(\mathbf{X}^{(k)\prime}\mathbf{X}^{(k)})^{-1}\mathbf{X}^{(k)}\mathbf{y}) = \mathbf{E}_2'\mathbf{y}$. From equations (4.24) and (4.25), we have that

$$\begin{aligned}(n-(k+1))s^2 = (Error\ SS)_{full} &= \mathbf{E}_1'\mathbf{E}_1 - (\mathbf{E}_1'\mathbf{y})^2 \,/\, (\mathbf{E}_2'\mathbf{E}_2)\\ &= \mathbf{E}_1'\mathbf{E}_1 - (\mathbf{E}_1'\mathbf{E}_2)^2 \,/\, (\mathbf{E}_2'\mathbf{E}_2).\end{aligned}$$

Thus, from equation (5.24)

$$\begin{aligned}\frac{t(b_k)}{\sqrt{t(b_k)^2 + n - (k+1)}} &= \frac{\mathbf{E}_2'\mathbf{y} / \left(s\sqrt{\mathbf{E}_2'\mathbf{E}_2}\right)}{\sqrt{\frac{(\mathbf{E}_2'\mathbf{y})^2}{s^2\mathbf{E}_2'\mathbf{E}_2} + n - (k+1)}}\\ &= \frac{\mathbf{E}_2'\mathbf{y}}{\sqrt{(\mathbf{E}_2'\mathbf{y})^2 + \mathbf{E}_2'\mathbf{E}_2 s^2(n-(k+1))}}\\ &= \frac{\mathbf{E}_2'\mathbf{E}_1}{\sqrt{(\mathbf{E}_2'\mathbf{E}_1)^2 + \mathbf{E}_2'\mathbf{E}_2\left(\mathbf{E}_1'\mathbf{E}_1 - \frac{(\mathbf{E}_2'\mathbf{E}_1)^2}{\mathbf{E}_2'\mathbf{E}_2}\right)}}\\ &= \frac{\mathbf{E}_1'\mathbf{E}_2}{\sqrt{(\mathbf{E}_1'\mathbf{E}_1)(\mathbf{E}_2'\mathbf{E}_2)}} = r(\mathbf{E}_1, \mathbf{E}_2).\end{aligned}$$

This establishes the relationship between the partial correlation coefficient and the t-ratio statistic.

6

Interpreting Regression Results

Chapter Preview. A regression analyst collects data, selects a model, and then reports on the findings of the study, in that order. This chapter considers these three topics in *reverse* order, emphasizing how each stage of the study is influenced by preceding steps. An application, determining a firm's characteristics that influence its effectiveness in managing risk, illustrates the regression modeling process from start to finish.

Studying a problem using a regression modeling process involves a substantial commitment of time and energy. One must first embrace the concept of *statistical thinking*, a willingness to use data actively as part of a decision-making process. Second, one must appreciate the usefulness of a model that is used to approximate a real situation. Having made this substantial commitment, there is a natural tendency to "oversell" the results of statistical methods such as regression analysis. By overselling any set of ideas, consumers eventually become disappointed when the results do not live up to their expectations. This chapter begins in Section 6.1 by summarizing what we can reasonably expect to learn from regression modeling.

"All models are wrong, but some are useful" (Box, 1979).

Models are designed to be much simpler than relationships among entities that exist in the real world. A model is merely an approximation of reality. As stated by George Box (1979), "All models are wrong, but some are useful." Developing the model, the subject of Chapter 5, is part of the art of statistics. Although the principles of variable selection are widely accepted, the application of these principles can vary considerably among analysts. The resulting product has certain aesthetic values and is by no means predetermined. Statistics can be thought of as the art of reasoning with data. Section 6.2 will underscore the importance of variable selection.

Model formulation and data collection form the first stage of the modeling process. Students of statistics are usually surprised at the difficulty of relating ideas about relationships to available data. These difficulties include a lack of readily available data and the need to use certain data as proxies for ideal information that is not available numerically. Section 6.3 will describe several types of difficulties that can arise when collecting data. Section 6.4 will describe some models to alleviate these difficulties.

6.1 What the Modeling Process Tells Us

Model inference is the final stage of the modeling process. By studying the behavior of models, we hope to learn something about the real world. Models serve to impose an order on reality and provide a basis for understanding reality through the nature of the imposed order. Further, statistical models are based on reasoning with the available data from a sample. Thus, models serve as an important guide for predicting the behavior of observations outside the available sample.

6.1.1 Interpreting Individual Effects

When interpreting results from multiple regression, the main goal is often to convey the importance of individual variables, or effects, on an outcome of interest. The interpretation depends on whether the effects are substantively significant, statistically significant, and causal.

Substantive Significance

Readers of a regression study first want to understand the direction and magnitude of individual effects. Do females have more or fewer claims than males in a study of insurance claims? If fewer, by how many? You can give answers to these questions through a table of regression coefficients. Moreover, to give a sense of the reliability of the estimates, you may also wish to include the standard error or a confidence interval, as introduced in Section 3.4.2.

Recall that regression coefficients are estimates of partial derivatives of the regression function

$$\mathrm{E}\, y = \beta_0 + \beta_1 x_1 + \cdots + \beta_k x_k.$$

When interpreting coefficients for continuous explanatory variables, it is helpful to do so in terms of meaningful changes of each x. For example, if population is an explanatory variable, we may talk about the expected change in y per 1000 or 1 million change in population. Moreover, when interpreting regression coefficients, comment on their substantive significance. For example, suppose that we find a difference in claims between males and females but the estimated difference is only 1% of expected claims. This difference may well be statistically significant but not economically meaningful. Substantive significance refers to importance in the field of inquiry; in actuarial science, this is typically financial or economic significance but could also be nonmonetary, such as effects on future life expectancy.

Substantive significance refers to importance in the field of inquiry.

Statistical Significance

Are the effects due to chance? The hypothesis testing machinery introduced in Section 3.4.1 provides a formal mechanism for answering this question. Tests of hypotheses are useful in that they provide a formal, agreed-on standard, for deciding whether a variable provides an important contribution to an expected response. When interpreting results, typically researchers cite a t-ratio or a p-value to demonstrate statistical significance.

In some situations, it is of interest to comment on variables that are *not* statistically significant. Effects that are not statistically significant have standard errors that are large relative to the regression coefficients. In Section 5.5.2, we expressed this standard error as

$$se(b_j) = s \frac{\sqrt{VIF_j}}{s_{x_j}\sqrt{n-1}}. \tag{6.1}$$

One possible explanation for a lack of statistical significance is a large variation in the disturbance term. By expressing the standard error in this form, we see that the larger the natural variation, as measured by s, the more difficult it is to reject the null hypothesis of no effect (H_0), other things being equal.

A second possible explanation for the lack of statistical significant is the high collinearity, as measured by VIF_j. A variable may be be confounded with other variables such that, from the data being analyzed, it is impossible to distinguish the effects of one variable from another.

A third possible explanation is the sample size. Suppose that a mechanism similar to draws from a stable population is used to observe the explanatory variables. Then, the standard deviation of x_j, s_{x_j}, should be stable as the number of draws increases. Similarly, so should R_j^2 and s^2. Then, the standard error $se(b_j)$ should decrease as the sample size, n, increases. Conversely, a smaller sample size means a larger standard error, other things being equal. This means that we may not be able to detect the importance of variables in small or moderate size samples.

Thus, in an ideal world, if you do not detect statistical significance where it was hypothesized (and fully expected), you could (1) get a more precise measure of y, thus reducing its natural variability; (2) redesign the sample collection scheme so that the relevant explanatory variables are less redundant; and (3) collect more data. Typically, these options are not available with observational data but it can nonetheless be helpful to point out the next steps in a research program.

Large samples provide an opportunity to detect the importance of variables that might go unnoticed in small samples.

Analysts occasionally observe statistically significant relationships that were not anticipated – these could be due to a large sample size. Previously, we noted that a small sample may not provide enough information to detect meaningful relationships. The flip side of this argument is that, for large samples, we have an opportunity to detect the importance of variables that might go unnoticed in small or even moderate-sized samples. Unfortunately, it also means that variables with small parameter coefficients, that contribute little to understanding the variation in the response, can be judged to be significant using our decision-making procedures. This serves to highlight the difference between substantive and statistical significance – particularly for large samples, investigators encounter variables that are *statistically significant but practically unimportant.* In these cases, it can

Variables can be statistically significant but practically unimportant.

be prudent for the investigator to omit variables from the model specification when their presence is not in accord with accepted theory, even if they are judged statistically significant.

Causal Effects

If we change x, would y change? As students of basic sciences, we learned principles involving actions and reactions. Adding mass to a ball in motion increases the force of its impact into a wall. However, in the social sciences, relationships are probabilistic, not deterministic, and hence more subtle. For example, as age (x) increases, the one-year probability of death (y) increases for most human mortality curves. Understanding causality, even probabilistic, is the root of all science and provides the basis for informed decision making.

It is important to acknowledge that causal processes generally cannot be demonstrated exclusively from the data; the data can only present relevant empirical evidence serving as a link in a chain of reasoning about causal mechanisms. For causality, there are three necessary conditions: (1) statistical association between variables, (2) appropriate time order, and (3) the elimination of alternative hypotheses or establishment of a formal causal mechanism.

As an example, recall the Section 1.1 Galton study relating adult children's height (y) to an index of parents' height (x). For this study, it was clear that there was a strong statistical association between x and y. The demographics also make it clear that the parents measurements (x) precedes the children measurements (y). What is uncertain is the causal mechanism. For example, in Section 1.5, we cited the possibility that an omitted variable, family diet, could be influencing both x and y. Evidence and theories from human biology and genetics are needed to establish a formal causal mechanism.

Example: Race, Redlining, and Automobile Insurance Prices. In an article with this title, Harrington and Niehaus (1998) investigated whether insurance companies engaged in (racial) discriminatory behavior, often known as *redlining*. Racial discrimination is illegal and insurance companies may not use race in determining prices. The term *redlining* refers to the practice of drawing red lines on a map to indicate areas that insurers will not serve, areas typically containing a high proportion of minorities.

To investigate whether there exists racial discrimination in insurance pricing, Harrington and Niehaus gathered private passenger premiums and claims data from the Missouri Department of Insurance for the period 1988–92. Although insurance companies do not keep race or ethnicity information in their premiums and claims data, such information is available at the Zip code level from the U.S. Census Bureau. By aggregating premiums and claims up to the Zip code level, Harrington and Niehaus were able to assess whether areas with a higher percentage of African Americans paid more for insurance (PCTBLACK).

Table 6.1 Loss Ratio Regression Results

Variable	Description	Regression Coefficient	t-Statistic
Intercept		1.98	2.73
PCTBLACK	Proportion of population black	0.11	0.63
ln TOTPOP	Logarithmic total population	–0.10	–4.43
PCT1824	Percentage of population between 18 and 24	–0.23	–0.50
PCT55UP	Percentage of population 55 or older	–0.47	–1.76
MARRIED	Percentage of population married	–0.32	–0.90
PCTUNEMP	Percentage of population unemployed	0.11	0.10
ln AVCARV	Logarithmic average car value insured	–0.87	–3.26
R_a^2		0.11	

Source: Harrington and Niehaus (1998).

A widely used pricing measure is the loss ratio, defined to be the ratio of claims to premiums. This measures insurers' profitably; if racial discrimination exists in pricing, one would expect to see a low loss ratio in areas with a high proportion of minorities. Harrington and Niehaus (1998) used this as the dependent variable, after taking logarithms to address the skewness in the loss ratio distribution.

Harrington and Niehaus (1998) studied 270 Zip codes surrounding six major cities in Missouri where there were large concentrations of minorities. Table 6.1 reports findings from comprehensive coverage, although the authors also investigated collision and liability coverage. In addition to the primary variable of interest, PCTBLACK, a few control variables relating to age distribution (PCT1824 and PCT55P), marital status (MARRIED), population (ln TOTPOP) and income (PCTUNEMP) were introduced. Policy size was measured indirectly through an average car value (ln AVCARV).

Table 6.1 reports that only policy size and population are statistically significant determinants of loss ratios. In fact, the coefficient associated with PCT-BLACK has a positive sign, indicating that premiums are lower in areas with high concentrations of minorities (although, not significant). In an efficient insurance market, we would expect prices to be closely aligned with claims and that few broad patterns exist.

Certainly, the findings of Harrington and Niehaus (1998) are inconsistent with the hypothesis of racial discrimination in pricing. Establishing a lack of statistical significance is typically more difficult than establishing significance. In the paper by Harrington and Niehaus (1998), there are many alternative model specifications that assess the robustness of their findings to different variable selection procedures and different data subsets. Table 6.1 reports coefficient estimators and standard errors calculated using weighted least squares, with population size as weights. The authors also ran (ordinary) least squares, with robust standard errors, achieving similar results.

6.1.2 Other Interpretations

When taken collectively, linear combinations of the regression coefficients can be interpreted as the regression function

$$\mathrm{E}\, y = \beta_0 + \beta_1 x_1 + \cdots + \beta_k x_k .$$

When reporting regression results, readers want to know how well the model fits the data. Section 5.6.1 summarized several goodness-of-fit statistics that are routinely reported in regression investigations.

Regression Function and Pricing
When evaluating insurance claims data, the regression function represents expected claims and hence forms the basis of the pricing function. (See the example in Chapter 4.) In this case, the shape of the regression function and levels for key combinations of explanatory variables are of interest.

Benchmarking Studies
In some investigations, the main purpose may be to determine whether a specific observation is "in line" with the others available. For example, in Chapter 20, we will examine CEO salaries. The main purpose of such an analysis could have been to see whether a person's salary is high or low compared to others in the sample, *controlling for* characteristics such as industry and years of experience. The residual summarizes the deviation of the response from that expected under the model. If the residual is unusually large or small, then we interpret this to mean that there are unusual circumstances associated with this observation. This analysis does not suggest the nature nor the causes of these circumstances. It merely states that the observation is unusual with respect to others in the sample. For some investigations, such as for litigation concerning compensation packages, this is a powerful statement.

Prediction
Many actuarial applications concern prediction, where the interest is on describing the distribution of a random variable that is not yet realized. When setting reserves, insurance company actuaries are establishing liabilities for future claims that they predict will be realized, and thus becoming eventual expenses of the company. Prediction, or *forecasting*, is the main motivation of most analyses of time series data, the subject of Chapters 7–10.

Prediction of a single random variable in the multiple linear regression context was introduced in Section 4.2.3. Here, we assumed that we have available a given set of characteristics, $\mathbf{x}_* = (1, x_{*1}, \ldots, x_{*k})'$. According to our model, the new response is

$$y_* = \beta_0 + \beta_1 x_{*1} + \cdots + \beta_k x_{*k} + \varepsilon_* .$$

We use as our point predictor

$$\hat{y}_* = b_0 + b_1 x_{*1} + \cdots + b_k x_{*k} .$$

As in Section 2.5.3, we can decompose the prediction error into the estimation error plus the random error, as follows:

$$\underbrace{y^* - \widehat{y}^*}_{\text{prediction error}} = \underbrace{\beta_0 - b_0 + (\beta_1 - b_1)x_{*1} + \cdots + (\beta_k - b_k)x_{*k}}_{\text{error in estimating the regression function at } x_{*1}, \ldots, x_{*k}} + \underbrace{\varepsilon^*}_{\text{additional deviation}}$$

This decomposition allows us to provide a distribution for the prediction error. It is customary to assume approximate normality. With this additional assumption, we summarize this distribution using a prediction interval

$$\hat{y}_* \pm t_{n-(k+1),1-\alpha/2}\, se(pred), \tag{6.2}$$

where

$$se(pred) = s\sqrt{1 + \mathbf{x}_*'(\mathbf{X}'\mathbf{X})^{-1}\mathbf{x}_*}.$$

Here, the t-value $t_{n-(k+1),1-\alpha/2}$ is a percentile from the t-distribution with $df = n - (k+1)$ degrees of freedom. This extends equation (2.7).

Communicating the range of likely outcomes is an important goal. When analyzing data, there may be several alternative prediction techniques available. Even within the class of regression models, each of several candidate models will produce a different prediction. It is important to provide a distribution, or range, of potential errors. Naive consumers can easily become disappointed with the results of predictions from regression models. These consumers are told (correctly) that the regression model is optimal, based on certain well-defined criteria, and are then provided with a point prediction, such as $\hat{y}_*$. Without knowledge of an interval, the consumer has expectations for the performance of the prediction, usually greater than are warranted by information available in the sample. A prediction interval provides not only a single optimal point prediction but also a range of reliability.

When making the predictions, there is an important assumption that the new observation follows the same model as that used in the sample. Thus, the basic conditions about the distribution of the errors should remain unchanged for new observation. It is also important that the level of the predictor variables, $x_{*1}, \ldots, x_{*k}$, is similar to those in the available sample. If one or several of the predictor variables differs dramatically from those in the available sample, then the resulting prediction can perform poorly. For example, it would be imprudent to use the model developed in Sections 2.1 through 2.3 to predict a region's lottery with a population of $x_* = 400{,}000$, more than ten times the largest population in our sample. Even though it would be easy to plug $x_* = 400{,}000$ into our formulas, the result would have little intuitive appeal. Extrapolating relationships beyond the observed data requires expertise with the nature of the data as well as the statistical methodology. In Section 6.3, we will identify this problem as a potential bias due to the sampling region.

6.2 The Importance of Variable Selection

On the one hand, choosing a theoretical model to represent precisely real-world events is probably an impossible task. On the other hand, choosing a model to represent approximately the real world is an important practical matter. The closer our model is to the real world, the more accurate are the statements that we make, which is suggested by the model. Although we cannot get the right model, we may be able to select a useful, or at least adequate, model.

Users of statistics, from the raw beginner to the seasoned expert, will always select an inadequate model from time to time. The key question is, How important is it to select an adequate model? Although not every kind of mistake can be accounted for in advance, there are some guiding principles that are useful to keep in mind when selecting a model.

6.2.1 Overfitting the Model

This type of mistake occurs when superfluous, or extraneous, variables are added to the specified model. If only a small number of extraneous variables, such as one or two, are added, then this type of error will probably not dramatically skew most of the types of conclusions that might be reached from the fitted model. For example, we know that when we add a variable to the model, the error sum of squares does not increase. If the variable is extraneous, then the error sum of squares will not get appreciably smaller either. In fact, adding an extraneous variable can increase s^2 because the denominator is smaller by one degree of freedom. However, for data sets of moderate sample size, the effect is minimal. Adding several extraneous variables can inflate s^2 appreciably, however. Further, there is the possibility that adding extraneous explanatory variables will induce, or worsen, the presence of collinearity.

A more important point is that, by adding extraneous variables, our regression coefficient estimates remain *unbiased*. Consider the following example.

Example: Regression Using One Explanatory Variable. Assume that the true model of the responses is

$$y_i = \beta_0 + \varepsilon_i, \quad i = 1, \ldots, n.$$

Under this model, the level of a generic explanatory variable x does not affect the value of the response y. If we were to predict the response at any level of x, the prediction would have expected value β_0. However, suppose that we mistakenly fit the model

$$y_i = \beta_0^* + \beta_1^* x_i + \varepsilon_i^*.$$

With this model, the prediction at a generic level x is $b_0^* + b_1^* x$ where b_0^* and b_1^* are the usual least squares estimates of β_0^* and β_1^*, respectively. It is not to hard to confirm that

$$Bias = \mathrm{E}\,(b_0^* + b_1^* x) - \mathrm{E}\,y = 0,$$

where the expectations are calculated using the true model. Thus, by using a slightly larger model than we should have, we did not pay for it in terms of making a persistent, long-term error such as represented by the bias. The price of making this mistake is that our standard error is slightly higher than it would be if we had chosen the correct model.

6.2.2 Underfitting the Model

This type of error occurs when important variables are omitted from the model specification; it is more serious than overfitting. Omitting important variables can cause appreciable amounts of bias in our resulting estimates. Further, because of the bias, the resulting estimates of s^2 are larger than need be. A larger s inflates our prediction intervals and produces inaccurate tests of hypotheses concerning the importance of explanatory variables. To see the effects of underfitting a model, we return to the previous example.

Example: Regression Using One Explanatory Variable, Continued. We now reverse the roles of the models described earlier. Assume that the true model is

$$y_i = \beta_0 + \beta_1 x_i + \varepsilon_i,$$

and that we mistakenly fit the model,

$$y_i = \beta_0^* + \varepsilon_i^*.$$

Thus, we have inadvertently omitted the effects of the explanatory variable x. With the fitted model, we would use $\bar{y}$ for our prediction at a generic level of x. From the true model, we have $\bar{y} = \beta_0 + \beta_1 \bar{x} + \bar{\varepsilon}$. The bias of the prediction at x is

$$\begin{aligned} Bias &= \mathrm{E}\,\bar{y} - \mathrm{E}\,(\beta_0 + \beta_1 x + \varepsilon) \\ &= \mathrm{E}\,(\beta_0 + \beta_1 \bar{x} + \bar{\varepsilon}) - (\beta_0 + \beta_1 x) = \beta_1(\bar{x} - x). \end{aligned}$$

If β_1 is positive, then we underpredict for large values of x, resulting in a negative bias, and overpredict for small values of x (relative to $\overline{x}$). Thus, there is a persistent, long-term error in omitting the explanatory variable x. Similarly, one can check that this type of error produces biased regression parameter estimates and an inflated value of s^2.

Of course, no one wants to overfit or underfit the model. However, data from the social sciences are often messy, and it can be hard to know whether to include a variable in the model. When selecting variables, analysts are often guided by the principle of parsimony, also known as Occam's razor, which states that when there are several possible explanations for a phenomenon, use the simplest one. There are several arguments for preferring simpler models:

Occam's razor: When there are several possible explanations for a phenomenon, use the simplest one.

- A simpler explanation is easier to interpret.
- Simple models, also known as *parsimonious* models, often do well on fitting out-of-sample data.
- Extraneous variables can cause problems of collinearity, leading to difficulty in interpreting individual coefficients.

The contrasting viewpoint can be summarized in a quote often attributed to Albert Einstein, that states that we should use "the simplest model possible, but no simpler." This section demonstrates that underfitting a model, by omitting important variables, is typically a more serious error than including extraneous variables that add little to our ability to explain the data. Including extraneous variables decreases the degrees of freedom and increases the estimate of variability, typically of less concern in actuarial applications.

When in doubt, leave the variable in.

6.3 The Importance of Data Collection

The regression modeling process starts with collecting data. Having studied the results, and the variable selection process, we can now discuss the inputs to the process. Not surprisingly, there is a long list of potential pitfalls that are frequently encountered when collecting regression data. In this section, we identify the major pitfalls and provide some avenues for avoiding these pitfalls.

6.3.1 Sampling Frame Error and Adverse Selection

Sampling frame error occurs when the sampling frame, the list from which the sample is drawn, is not an adequate approximation of the population of interest. In the end, a sample must be a representative subset of a larger population, or universe, of interest. If the sample is not representative, taking a larger sample does not eliminate bias; you simply repeat the same mistake over again and again.

Example: Literary Digest Poll. Perhaps the most widely known example of sampling frame error is from the 1936 *Literary Digest* poll. This poll was conducted to predict the winner of the 1936 U.S. presidential election. The two leading candidates were Franklin D. Roosevelt, the Democrat, and Alfred Landon, the Republican. *Literary Digest*, a prominent magazine at the time, conducted a survey of 10 million voters. Of those polled, 2.4 million responded, predicting a "landslide" Landon victory by a 57% to 43% margin. However, the actual election resulted in an overwhelming Roosevelt victory, by a 62% to 38% margin. What went wrong?

There were a number of problems with the *Literary Digest* survey. Perhaps the most important was the sampling frame error. To develop their sampling frame, *Literary Digest* used addresses from telephone books and membership lists of clubs. In 1936, the United States was in the depths of the Great Depression; telephones and club memberships were a luxury that only upper-income individuals could afford. Thus, *Literary Digests*'s list included an unrepresentative number of upper-income individuals. In previous presidential elections conducted by *Literary Digest*, the rich and poor tended to vote along similar lines and this was not a problem. However, economic problems were top political issues in the 1936 presidential election. As it turned out, the poor tended to vote for Roosevelt and the rich tended to vote for Landon. As a result, the *Literary Digest* poll results were grossly mistaken. Taking a large sample, even of size 2.4 million, did not help; the basic mistake was repeated over and over again.

Sampling frame bias occurs when the sample is not a representative subset of the population of interest. When analyzing insurance company data, this bias can arise due to *adverse selection*. In many insurance markets, companies design and price contracts and policyholders decide whether to enter a contractual agreement (actually, policyholders "apply" for insurance, so insurers also have a right not to enter into the agreement). Thus, someone is more likely to enter into an agreement if they believe that the insurer is underpricing their risk, especially in light of policyholder characteristics that are not observed by the insurer. For example, it is well known that mortality experience of a sample of purchasers of life annuities is not representative of the overall population; people who purchase annuities tend to be healthy relative to the overall population. You would not purchase a life annuity that pays a periodic benefit while living if you were in poor health and thought that your probability of a long life to be low. Adverse selection arises because "bad risks," those with higher than expected claims, are more likely to enter into contracts than corresponding "good risks." Here, the expectation is developed on the basis of characteristics (explanatory variables) that can be observed by the insurer.

Of course, there is a large market for annuities and other forms of insurance in which adverse selection exists. Insurance companies can price these markets appropriately by redefining their "population of interest" to be not the general population but the population of potential policyholders. Thus, for example, in pricing annuities, insurers use annuitant mortality data, not data for the overall population. In this way, they can avoid potential mismatches between the population and sample. More generally, the experience of almost any company differs from the overall population due to underwriting standards and sales philosophies. Some companies seek "preferred risks" by offering educational discounts, good driving bonuses, and so forth, whereas other seek high-risk insureds. The company's

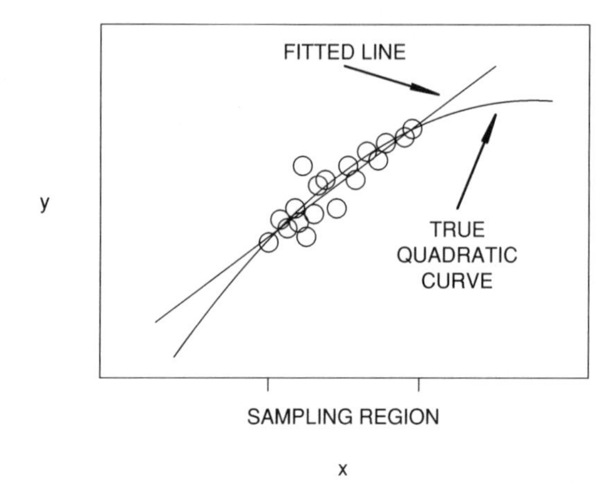

Figure 6.1 Extrapolation outside of the sampling region may be biased.

sample of insureds will differ from the overall population and the extent of the difference can be an interesting aspect to quantify in an analysis.

Sampling frame bias can be particularly important when a company seeks to market a new product for which it has no experience data. Identifying a target market and its relation to the overall population is an important aspect of a market development plan.

6.3.2 Limited Sampling Regions

A limited sampling region can give rise to potential bias when we try to extrapolate outside of the sampling region. To illustrate, consider Figure 6.1. Here, based on the data in the sampling region, a line may seem to be an appropriate representation. However, if a quadratic curve is the true expected response, any forecast that is far from the sampling region will be seriously biased.

Another pitfall due to a limited sampling region, although not a bias, that can arise is the difficulty in estimating a regression coefficient. In Chapter 5, we saw that a smaller spread of a variable, other things equal, means a less reliable estimate of the slope coefficient associated with that variable. That is, from Section 5.5.2 or equation 6.1, we see that the smaller is the spread of x_j, as measured by s_{x_j}, the larger is the standard error of b_j, $se(b_j)$. Taken to the extreme, where $s_{x_j} = 0$, we might have a situation such as illustrated in Figure 6.2. For the extreme situation illustrated in Figure 6.2, there is not enough variation in x to estimate the corresponding slope parameter.

6.3.3 Limited Dependent Variables, Censoring, and Truncation

In some applications, the dependent variable is constrained to fall within certain regions. To see why this is a problem, first recall that under the linear regression model, the dependent variable equals the regression function plus a random error. Typically, the random error is assumed to be approximately normally distributed, so that the response varies continuously. However, if the outcomes of the dependent variable are restricted, or limited, then the outcomes are not purely

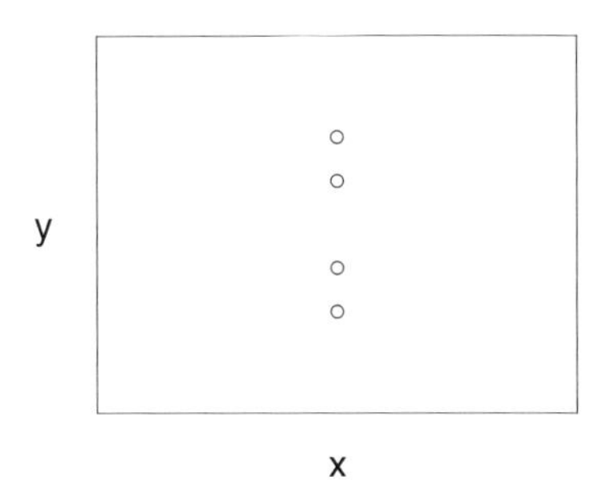

Figure 6.2 The lack of variation in x means that we cannot fit a unique line relating x and y.

continuous. This means that our assumption of normal errors is not strictly correct and may not even be a good approximation.

To illustrate, Figure 16.1 shows a plot of individual's income (x) versus amount of insurance purchased (y). The sample in this plot represents two subsamples, those who purchased insurance, corresponding to $y > 0$, and those who did not, corresponding to "price" $y = 0$. Fitting a single line to these data would misinform users about the effects of x on y.

If we dealt with only those who purchased insurance, then we still would have an implicit lower bound of zero (if an insurance price must exceed zero). However, prices need not be close to this bound for a given sampling region and thus not represent an important practical problem. By including several individuals who did not purchase insurance (and thus spent $0 on insurance), our sampling region now clearly includes this lower bound.

There are several ways in which dependent variables can be restricted, or *censored*. Figure 16.1 illustrates the case in which the value of y may be no lower than zero. As another example, insurance claims are often restricted to be less than or equal to an upper limit specified in the insurance policy. If censoring is severe, ordinary least squares produces biased results. Specialized approaches, known as *censored regression* models, are described in Chapter 15 to handle this problem.

Figure 6.4 illustrates another commonly encountered limitation on the value of the dependent variable. For this illustration, suppose that y represents an insured loss and that d represents the deductible on an insurance policy. In this scenario, it is common practice for insurers to not record losses below d (they are typically not reported by policyholders). In this case, the data are said to be *truncated*. Not surprisingly, *truncated regression models* are available to handle this situation. As a rule of thumb, truncated data represent a more serious source of bias than censored data. When data are truncated, we do not have values of dependent variables and thus have less information than when the data are censored. See Chapter 15 for further discussion.

6.3.4 Omitted and Endogenous Variables

Of course, analysts prefer to include all important variables. However, a common problem is that we may not have the resources nor the foresight to gather and analyze all the relevant data. Further, sometimes we are prohibited from including variables. For example, in insurance rating we are typically precluded from

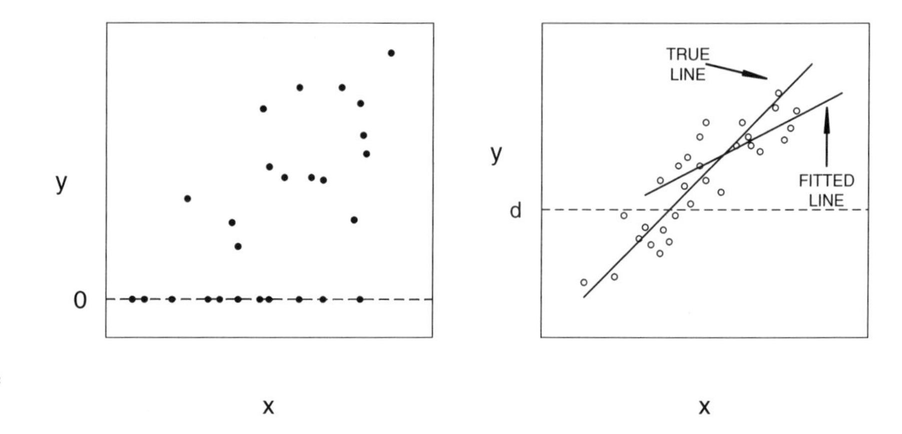

Figure 6.3 When individuals do not purchase anything, they are recorded as $y = 0$ sales.

Figure 6.4 If the responses below the horizontal line at $y = d$ are omitted, then the fitted regression line can be very different from the true regression line.

using ethnicity as a rating variable. Further, there are many mortality and other decrement tables that are "unisex," that is, blind to gender.

Omitting important variables can affect our ability to fit the regression function; this can affect in-sample (explanation) as well as out-of-sample (prediction) performance. If the omitted variable is uncorrelated with other explanatory variables, then the omission will not affect estimation of regression coefficients. Typically this is not the case. The Section 3.4.3 refrigerator example illustrates a serious case where the direction of a statistically significant result was reversed based on the presence of an explanatory variable. In this example, we found that a cross-section of refrigerators displayed a significantly positive correlation between price and the annual energy cost of operating the refrigerator. This positive correlation was counterintuitive because we would hope that higher prices would mean lower annual expenditures in operating a refrigerator. However, when we included several additional variables, in particular, measures of the size of a refrigerator, we found a significantly negative relationship between price and energy costs. Again, by omitting these additional variables, there was an important bias when using regression to understand the relationship between price and energy costs.

Omitted variables can lead to the presence of endogenous explanatory variables. An exogenous variable is one that can be taken as "given" for the purposes at hand. An endogenous variable is one that fails the exogeneity requirement. An omitted variable can affect both the y and the x and in this sense induce a relationship between the two variables. If the relationship between x and y is due to an omitted variable, it is difficult to condition on the x when estimating a model for y.

Up to now, the explanatory variables have been treated as non-stochastic. For many social science applications, it is more intuitive to consider the x's to be stochastic, and perform inference conditional on their realizations. For example, under common sampling schemes, we can estimate the conditional regression function

$$\mathrm{E}\ (y|x_1, \ldots, x_k) = \beta_0 + \beta_1 x_1 + \cdots + \beta_k x_k .$$

This is known as a *sampling-based* model.

In the economics literature, Goldberger (1972) defines a *structural model* as a stochastic model representing a causal relationship, not a relationship that simply captures statistical associations. Structural models can readily contain endogenous explanatory variables. To illustrate, we consider an example relating claims and premiums. For many lines of business, premium classes are simply nonlinear functions of exogenous factors such as age, gender, and so forth. For other lines of business, premiums charged are a function of prior claims history. Consider model equations that relate one's claims ($y_{it}, t = 1, 2$) to premiums ($x_{it}, t = 1, 2$):

$$y_{i2} = \beta_{0,C} + \beta_{1,C} y_{i1} + \beta_{2,C} x_{i2} + \varepsilon_{i1}$$

$$x_{i2} = \beta_{0,P} + \beta_{1,P} y_{i1} + \beta_{2,P} x_{i1} + \varepsilon_{i2}.$$

In this model, current period ($t = 2$) claims and premiums are affected by the prior period's claims and premiums. This is an example of a *structural equations* model that requires special estimation techniques. Our usual estimation procedures are biased!

Example: Race, Redlining, and Automobile Insurance Prices, Continued. Although Harrington and Niehaus (1998) did not find racial discrimination in insurance pricing, their results on access to insurance were inconclusive. Insurers offer standard and preferred risk contracts to applicants that meet restrictive underwriting standards, as compared to substandard risk contracts where underwriting standards are more relaxed. Expected claims are lower for standard and preferred risk contracts, and so premiums are lower, than for substandard contracts. Harrington and Niehaus examined the proportion of applicants offered substandard contracts, NSSHARE, and found it significantly, positively related to PCTBLACK, the proportion of population that is African American. This suggests evidence of racial discrimination; they state this to be an inappropriate interpretation due to omitted variable bias.

Harrington and Niehaus argue that the proportion of applicants offered substandard contracts should be positively related to expected claim costs. Further, expected claim costs are strongly related to PCTBLACK, because minorities in the sample tended to be lower income. Thus, unobserved variables such as income tend to drive the positive relationship between NSSHARE and PCTBLACK. Because the data are analyzed at the Zip code level and not at the individual level, the potential omitted variable bias rendered the analysis inconclusive.

6.3.5 Missing Data

In the data examples, illustrations, case studies, and exercises of this text, there are many instances where certain data are unavailable for analysis, or *missing*. In every instance, the data were not carelessly lost but were unavailable because of substantive reasons associated with the data collection. For example, when we examined stock returns from a cross-section of companies, we saw that some companies did not have an average five-year earnings-per-share figure. The reason was simply that they had not been in existence for five years. As another example, when examining life expectancies, some countries did not report the total fertility rate because they lacked administrative resources to capture this data. Missing data are an inescapable aspect of analyzing data in the social sciences.

When the reason for the lack of availability of data is unrelated to actual data values, the data are said to be *missing at random*. There are a variety of techniques for handling missing at random data, none of which is clearly superior to the others. One "technique" is to simply ignore the problem. Hence, missing at random is sometimes called the *ignorable case* of missing data.

If there are only a few missing data, compared to the total number available, a widely employed strategy is to delete the observations corresponding to the missing data. Assuming that the data are missing at random, little information is lost by deleting a small portion of the data. Further, with this strategy, we need not make additional assumptions about the relationships among the data.

If the missing data are primarily from one variable, we can consider omitting this variable. Here, the motivation is that we lose less information when omitting this variable as compared to retaining the variable but losing the observations associated with the missing data.

Another strategy is to fill in, or *impute*, missing data. There are many variations of the imputation strategy. All assume some type of relationships among the variables in addition to the regression model assumptions. Although these methods yield reasonable results, note that any type of filled-in values do not yield the same inherent variability as the real data. Thus, results of analyses based on imputed values often reflect less variability than those with real data.

Example: Insurance Company Expenses, Continued. When examining company financial information, analysts commonly are forced to omit substantial amounts of information when using regression models to search for relationships. To illustrate, Segal (2002) examined life insurance financial statements from data provided by the National Association of Insurance Commissioners (NAIC). He initially considered 733 firm-year observations over the period 1995–8. However, 154 observations were excluded because of inconsistent or negative premiums, benefits and other important explanatory variables. Small companies representing 131 observations were also excluded. Small companies consist of fewer than 10 employees and agents, operating costs less than \$1 million or fewer than 1000 life policies sold. The resulting sample was $n = 448$ observations. The sample restrictions were based on explanatory variables – this procedure does not necessarily bias results. Segal (2002) argued that his final sample remained representative of the population of interest. There were about 110 firms in each of 1995–8. In 1998, aggregate assets of the firms in the sample represent approximately \$650 billion, a third of the life insurance industry.

6.4 Missing Data Models

To understand the mechanisms that lead to unplanned nonresponse, we model it stochastically. Let r_i be a binary variable for the ith observation, with a one indicating that this response is observed and a zero indicating that the response is missing. Let $\mathbf{r} = (r_1, \ldots, r_n)'$ summarize the data availability for all subjects. The interest is in whether the responses influence the missing data mechanism. For notation, we use $\mathbf{Y} = (y_1, \ldots, y_n)'$ to be the collection of all potentially observed responses.

6.4.1 Missing at Random

In the case where $\mathbf{Y}$ does not affect the distribution of $\mathbf{r}$, we follow Rubin (1976) and call this case *missing completely at random (MCAR)*. Specifically, the missing data are MCAR if $\mathrm{f}(\mathbf{r}|\mathbf{Y}) = \mathrm{f}(\mathbf{r})$, where f(.) is a generic probability mass function. An extension of this idea is in Little (1995), where the adjective *covariate dependent* is added when $\mathbf{Y}$ does not affect the distribution of $\mathbf{r}$, conditional on the covariates. If the covariates are summarized as $\mathbf{X}$, then the condition corresponds to the relation $\mathrm{f}(\mathbf{r}|\mathbf{Y}, \mathbf{X}) = \mathrm{f}(\mathbf{r}|\mathbf{X})$. To illustrate this point, consider an example of Little and Rubin (1987), where $\mathbf{X}$ corresponds to age and $\mathbf{Y}$ corresponds to income of all potential observations. If the probability of being missing does not depend on income, then the missing data are MCAR. If the probability of being missing varies by age but not by income over observations within an age group, then the missing data are covariate-dependent MCAR. Under the latter specification, it is possible for the missing data to vary by income. For example, younger people may be less likely to respond to a survey. This shows that the missing-at-random feature depends on the purpose of the analysis. Specifically, it is possible that an analysis of the joint effects of age and income may encounter serious patterns of missing data, whereas an analysis of income controlled for age suffers from no serious bias patterns.

Little and Rubin (1987) advocate modeling the missing data mechanisms. To illustrate, consider a likelihood approach using a selection model for the missing data mechanism. Now, partition $\mathbf{Y}$ into observed and missing components using the notation $\mathbf{Y} = \{\mathbf{Y}_{obs}, \mathbf{Y}_{miss}\}$. With the likelihood approach, we base inference on the observed random variables. Thus, we use a likelihood proportional to the joint function $\mathrm{f}(\mathbf{r}, \mathbf{Y}_{obs})$. We also specify a *selection model* by specifying the conditional mass function $\mathrm{f}(\mathbf{r}|\mathbf{Y})$.

Suppose that the observed responses and the selection model distributions are characterized by a vectors of parameters $\boldsymbol{\theta}$ and $\boldsymbol{\psi}$, respectively. Then, with the relation $\mathrm{f}(\mathbf{r}, \mathbf{Y}_{obs}, \boldsymbol{\theta}, \boldsymbol{\psi}) = \mathrm{f}(\mathbf{Y}_{obs}, \boldsymbol{\theta}) \times \mathrm{f}(\mathbf{r}|\mathbf{Y}_{obs}, \boldsymbol{\psi})$, we can express the log likelihood of the observed random variables as

$$L(\boldsymbol{\theta}, \boldsymbol{\psi}) = \ln \mathrm{f}(\mathbf{r}, \mathbf{Y}_{obs}, \boldsymbol{\theta}, \boldsymbol{\psi}) = \ln \mathrm{f}(\mathbf{Y}_{obs}, \boldsymbol{\theta}) + \ln \mathrm{f}(\mathbf{r}|\mathbf{Y}_{obs}, \boldsymbol{\psi}).$$

(See Section 11.9 for a refresher on likelihood inference.) In the case that the data are MCAR, then $f(\mathbf{r}|\mathbf{Y}_{obs}, \boldsymbol{\psi}) = f(\mathbf{r}|\boldsymbol{\psi})$ does not depend on $\mathbf{Y}_{obs}$. Little and Rubin (1987) also consider the case where the selection mechanism model distribution does not depend on $\mathbf{Y}_{miss}$ but may depend on $\mathbf{Y}_{obs}$. In this case, they call the data *missing at random* (MAR).

In both the MAR and the MCAR cases, we see that the likelihood may be maximized over the parameters, separately for each case. In particular, if one is interested only in the maximum likelihood estimator of $\boldsymbol{\theta}$, then the selection model mechanism may be "ignored." Hence, both situations are often referred to as the *ignorable case*.

Example: Dental Expenditures. Let y represent a household's annual dental expenditure and x represent income. Consider the following five selection mechanisms.

1. The household is not selected (missing) with probability without regard to the level of dental expenditure. In this case, the selection mechanism is MCAR.
2. The household is not selected if the dental expenditure is less than \$100. In this case, the selection mechanism depends on the observed and missing response. The selection mechanism cannot be ignored.
3. The household is not selected if income is less than \$20,000. In this case, the selection mechanism is MCAR, covariate dependent. That is, assuming that the purpose of the analysis is to understand dental expenditures conditional on knowledge of income, stratifying on the basis of income does not seriously bias the analysis.
4. The probability of a household being selected increases with dental expenditure. For example, suppose the probability of being selected is a linear function of $\exp(\psi y_i)/(1+\exp(\psi y_i))$. In this case, the selection mechanism depends on the observed and missing response. The selection mechanism cannot be ignored.
5. The household is followed over $T = 2$ periods. In the second period, a household is not selected if the first period expenditure is less than \$100. In this case, the selection mechanism is MAR. That is, the selection mechanism is based on an observed response.

The second and fourth selection mechanisms represent situations in which the selection mechanism must be explicitly modeled; these are nonignorable cases. In these situations without explicit adjustments, procedures that ignore the selection effect may produce seriously biased results. To illustrate a correction for selection bias in a simple case, we outline an example from Little and Rubin (1987). Section 6.4.2 describes additional mechanisms.

Example: Historical Heights. Little and Rubin (1987) discuss data from Wachter and Trusell (1982) on y, the height of men recruited to serve in the military. The sample is subject to censoring, in that minimum height standards were imposed for admission to the military. Thus, the selection mechanism is

$$r_i = \begin{cases} 1 & y_i > c_i \\ 0 & \text{otherwise} \end{cases},$$

where c_i is the known minimum height standard imposed at the time of recruitment. The selection mechanism is nonignorable because it depends on the individual's height, y.

For this example, additional information is available to provide reliable model inference. Specifically, given other studies of male heights, we can assume that the population of heights is normally distributed. Thus, the likelihood of the observables can be written down and inference can proceed directly. To illustrate, suppose that $c_i = c$ is constant. Let μ and σ denote the mean and standard deviation of y. Further suppose that we have a random sample of $n + m$ men in which m men fall below the minimum standard height c, and we observe $\mathbf{Y}_{obs} = (y_1, \ldots, y_n)'$. The joint distribution for observables is

$$\begin{aligned} \mathrm{f}(\mathbf{r}, \mathbf{Y}_{obs}, \mu, \sigma) &= \mathrm{f}(\mathbf{Y}_{obs}, \mu, \sigma) \times \mathrm{f}(\mathbf{r} | \mathbf{Y}_{obs}) \\ &= \left\{ \prod_{i=1}^{n} \mathrm{f}(y_i | y_i > c) \times \Pr(y_i > c) \right\} \times \{\Pr(y_i \leq c)\}^m . \end{aligned}$$

Now, let ϕ and Φ represent the density and distribution function for the standard normal distribution. Thus, the log likelihood is

$$\begin{aligned} L(\mu, \sigma) &= \ln \mathrm{f}(\mathbf{r}, \mathbf{Y}_{obs}, \mu, \sigma) \\ &= \sum_{i=1}^{n} \ln \left\{ \frac{1}{\sigma} \phi \left(\frac{y_i - \mu}{\sigma} \right) \right\} + m \ln \left\{ \Phi \left(\frac{c - \mu}{\sigma} \right) \right\}. \end{aligned}$$

This is easy to maximize in μ and σ. If one ignored the censoring mechanisms, then one would derive estimates of the observed data from the "log likelihood,"

$$\sum_{i=1}^{n} \ln \left\{ \frac{1}{\sigma} \phi \left(\frac{y_i - \mu}{\sigma} \right) \right\},$$

which yields different, and biased, results.

6.4.2 Nonignorable Missing Data

For nonignorable missing data, Little (1995) recommends the following:

- Avoid missing responses whenever possible by using appropriate follow-up procedures.

- Collect covariates that are useful for predicting missing values.
- Collect as much information as possible regarding the nature of the missing data mechanism.

For the last point, if little is known about the missing data mechanism, then it is difficult to employ a robust statistical procedure to correct for the selection bias.

There are many models of missing data mechanisms. A general overview appears in Little and Rubin (1987). Little (1995) surveys the problem of attrition. Rather than survey this developing literature, we give a widely used model of nonignorable missing data.

Heckman Two-Stage Procedure

Heckman (1976) assumes that the sampling response mechanism is governed by the latent (unobserved) variable r_i^*, where

$$r_i^* = \mathbf{z}_i' \boldsymbol{\gamma} + \eta_i .$$

The variables in $\mathbf{z}_i$ may or may not include the variables in $\mathbf{x}_i$. We observe y_i if $r_i^* > 0$, that is, if r_i^* crosses the threshold 0. Thus, we observe

$$r_i = \begin{cases} 1 & r_i^* > 0 \\ 0 & \text{otherwise} \end{cases} .$$

To complete the specification, we assume that (ε_i, η_i) are identically and independently distributed, and that the joint distribution of (ε_i, η_i) is bivariate normal with means zero, variances σ^2 and σ_η^2, and correlation ρ. Note that if the correlation parameter ρ equals zero, then the response and selection models are independent. In this case, the data are MCAR and the usual estimation procedures are unbiased and asymptotically efficient.

Under these assumptions, basic multivariate normal calculations show that

$$\mathrm{E}\,(y_i | r_i^* > 0) = \mathbf{x}_i' \boldsymbol{\beta} + \beta_\lambda \lambda(\mathbf{z}_i' \boldsymbol{\gamma}),$$

where $\beta_\lambda = \rho\sigma$ and $\lambda(a) = \phi(a)/\Phi(a)$. Here, $\lambda(\cdot)$ is the inverse of the so-called *Mills ratio*. This calculation suggests the following two-step procedure for estimating the parameters of interest.

Heckman's Two-Stage Procedure

(i) Use the data $(r_i, \mathbf{z}_i)$ and a probit regression model to estimate $\boldsymbol{\gamma}$. Call this estimator $\mathbf{g}_H$.

(ii) Use the estimator from stage (i) to create a new explanatory variable, $x_{i,K+1} = \lambda(\mathbf{z}_i' \mathbf{g}_H)$. Run a regression model using the K explanatory variables $\mathbf{x}_i$ as well as the additional explanatory variable $x_{i,K+1}$. Use $\mathbf{b}_H$ and $b_{\lambda,H}$ to denote the estimators of $\boldsymbol{\beta}$ and β_λ, respectively.

Chapter 11 will introduce probit regressions. Note that the two-step method does not work in the absence of covariates to predict the response and, for practical purposes, requires variables in $\mathbf{z}$ that are not in $\mathbf{x}$ (see Little and Rubin, 1987).

To test for selection bias, we can test the null hypothesis $H_0 : \beta_\lambda = 0$ in the second stage because of the relation $\beta_\lambda = \rho\sigma$. When conducting this test, one should use heteroscedasticity-corrected standard errors because the conditional variance $\mathrm{Var}(y_i | r_i^* > 0)$ depends on the observation i. Specifically, $\mathrm{Var}(y_i | r_i^* > 0) = \sigma^2(1 - \rho^2 \delta_i)$, where $\delta_i = \lambda_i(\lambda_i + \mathbf{z}_i'\boldsymbol{\gamma})$ and $\lambda_i = \phi(\mathbf{z}_i'\boldsymbol{\gamma})/\Phi(\mathbf{z}_i'\boldsymbol{\gamma})$.

This procedure assumes normality for the selection latent variables to form the augmented variables. Other distribution forms are available in the literature, including the logistic and uniform distributions. A deeper criticism, raised by Little (1985), is that the procedure relies heavily on assumptions that cannot be tested using the data available. This criticism is analogous to the historical heights example in which we relied heavily on the normal curve to infer the distribution of heights below the censoring point. Despite these criticisms, Heckman's procedure is widely used in the social sciences.

EM Algorithm

Section 6.4.2 has focused on introducing specific models of nonignorable nonresponse. General robust models of nonresponse are not available. Rather, a more appropriate strategy is to focus on a specific situation, to collect as much information as possible regarding the nature of the selection problem, and then to develop a model for this specific selection problem.

The EM algorithm is a computational device for computing model parameters. Although specific to each model, it has found applications in a wide variety of models involving missing data. Computationally, the algorithm iterates between the "E," for conditional expectation, and "M," for maximization, steps. The E step finds the conditional expectation of the missing data given the observed data and current values of the estimated parameters. This is analogous to the time-honored tradition of imputing missing data. A key innovation of the EM algorithm is that one imputes sufficient statistics for missing values, not individual data points. For the M step, one updates parameter estimates by maximizing an observed log likelihood. Both the sufficient statistics and the log likelihood depend on the model specification.

Many introductions of the EM algorithm are available in the literature. Little and Rubin (1987) provide a detailed treatment.

6.5 Application: Risk Managers' Cost-Effectiveness

This section examines data from a survey on the cost-effectiveness of risk management practices. Risk management practices are activities undertaken by a firm to minimize the potential cost of future losses, such as the event of a fire in a warehouse or an accident that injures employees. This section develops a model that can be used to make statements about cost of managing risks.

An outline of the regression modeling process is as follows. We begin by providing an introduction to the problem and giving some brief background on the data. Certain prior theories will lead us to present a preliminary model fit. Using diagnostic techniques, it will be evident that several assumptions underpinning this model are not in accord with the data. This will lead us to go back to the beginning and start the analysis from scratch. What we learn from a detailed examination of the data will lead us to postulate some revised models. Finally, to communicate certain aspects of the new model, we will explore graphical presentations of the recommended model.

Introduction

Ⓡ **Empirical**
Filename is "RiskSurvey"

The data for this study were provided by Professor Joan Schmit and are discussed in more detail in the paper "Cost Effectiveness of Risk Management Practices" (Schmit and Roth, 1990). The data are from a questionnaire that was sent to 374 risk managers of large U.S.-based organizations. The purpose of the study was to relate cost-effectiveness to management's philosophy of controlling the company's exposure to various property and casualty losses, after adjusting for company effects such as size and industry type.

First, some caveats. Survey data are often based on samples of convenience, not probability samples. As with all observational data sets, regression methodology is a useful tool for summarizing data. However, we must be careful when making inferences based on this type of data set. For this particular survey, 162 managers returned completed surveys, resulting in a good response rate of 43%. However, for the variables included in the analysis (defined subsequently), only 73 forms were completed, resulting in a complete response rate of 20%. Why such a dramatic difference? Managers, like most people, typically do not mind responding to queries about their attitudes or opinions about various issues. When questioned about hard facts, in this case, company asset size or insurance premiums, either they considered the information proprietary and were reluctant to respond even when guaranteed anonymity or they simply were not willing to take the time to look up the information. From a surveyor's standpoint, this is unfortunate because typically "attitudinal" data are fuzzy (high variance compared to the mean) as compared to hard financial data. The trade-off is that the latter data are often hard to obtain. In fact, for this survey, several pre-questionnaires were sent to ascertain managers' willingness to answer specific questions. From the pre-questionnaires, the researchers severely reduced the number of financial questions that they intended to ask.

A measure of risk management cost-effectiveness, FIRMCOST, is the dependent variable. This variable is defined as total property and casualty premiums and uninsured losses as a percentage of total assets. It is a proxy for annual expenditures associated with insurable events, standardized by company size. Here, for the financial variables, ASSUME is the per occurrence retention amount as a percentage of total assets, CAP indicates whether the company owns a captive insurance company, SIZELOG is the logarithm of total assets, and INDCOST is a measure of the firm's industry risk. Attitudinal variables include CENTRAL, a measure of the importance of the local managers in choosing the amount of risk to be retained, and SOPH, a measure of the degree of importance in using analytical tools, such as regression, in making risk management decisions.

In their paper, the researchers described several weaknesses of the definitions used but argue that the definitions provide useful information, given the willingness of risk managers to obtain reliable information. The researchers also described several theories concerning relationships that can be confirmed by the data. Specifically, they hypothesized the following:

- There exists an inverse relationship between risk retention (ASSUME) and cost (FIRMCOST). The idea behind this theory is that larger retention amounts should mean lower expenses to a firm, resulting in lower costs.
- The use of a captive insurance company (CAP) results in lower costs. Presumably, a captive is used only when cost-effective and consequently, this variable should indicate lower costs if used effectively.
- There exists an inverse relationship between the measure of centralization (CENTRAL) and cost (FIRMCOST). Presumably, local managers are able to make more cost-effective decisions because they are more familiar with local circumstances regarding risk management than are centrally located managers.
- There exists an inverse relationship between the measure of sophistication (SOPH) and cost (FIRMCOST). Presumably, more sophisticated analytical tools help firms to manage risk better, resulting in lower costs.

Preliminary Analysis

To test the theories described previously, the regression analysis framework can be used. To do this, posit the model

$$\begin{aligned}\text{FIRMCOST} &= \beta_0 + \beta_1\text{ASSUME} + \beta_2\text{CAP} + \beta_3\text{SIZELOG}\\ &\quad + \beta_4\text{INDCOST} + \beta_5\text{CENTRAL} + \beta_6\text{SOPH} + \varepsilon.\end{aligned}$$

With this model, each theory can be interpreted in terms of regression coefficients. For example, β_1 can be interpreted as the expected change in cost per unit change in retention level (ASSUME). Thus, if the first hypothesis is true, we expect β_1 to be negative. To test this, we can estimate b_1 and use our tests of hypotheses machinery to decide whether b_1 is significantly less than zero. The variables SIZELOG and INDCOST are included in the model to control for the effects of these variables. These variables are not directly under a risk manager's control and thus are not of primary interest. However, inclusion of these variables can account for an important part of the variability.

Data from 73 managers was fit using this regression model. Table 6.2 summarizes the fitted model.

The adjusted coefficient of determination is $R_a^2 = 18.8\%$, the F-ratio is 3.78, and the residual standard deviation is $s = 14.56$.

On the basis of the summary statistics from the regression model, we can conclude that the measures of centralization and sophistication do not have an impact on our measure of cost-effectiveness. For both of these variables, the t-ratio is low, less than 1.0 in absolute value. The effect of risk retention seems only somewhat important. The coefficient has the appropriate sign, although it is only 1.35 standard errors below zero. This would not be considered statistically significant at the 5% level, although it would be at the 10% level (the p-value is

Table 6.2 Regression Results from a Preliminary Model Fit

Variable	Coefficient	Standard Error	t-Statistic
INTERCEPT	59.76	19.1	3.13
ASSUME	–0.300	0.222	–1.35
CAP	5.50	3.85	1.43
SIZELOG	–6.84	1.92	–3.56
INDCOST	23.08	8.30	2.78
CENTRAL	0.133	1.44	0.89
SOPH	–0.137	0.347	–0.39

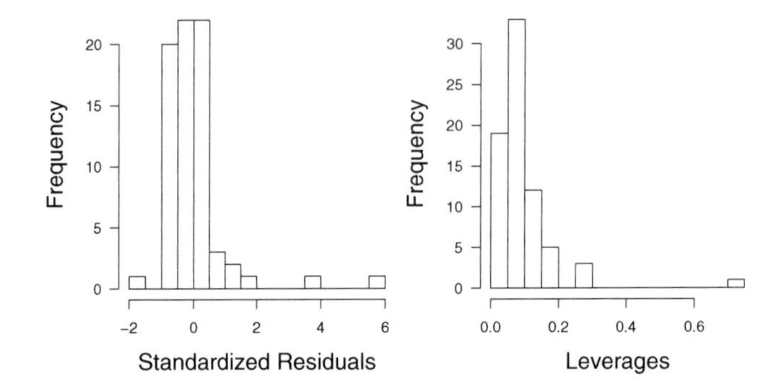

Figure 6.5 Histograms of standardized residuals and leverages from a preliminary regression model fit.

9%). Perhaps most perplexing is the coefficient associated with the CAP variable. We theorized that this coefficient would be negative. However, in our analysis of the data, the coefficient turns out to be positive and is 1.43 standard errors above zero. This leads us not only to disaffirm our theory but also to search for new ideas that are in accord with the information learned from the data. Schmit and Roth (1990) suggest reasons that may help us interpret the results of our hypothesis-testing procedures. For example, they suggest that managers in the sample may not have the most sophisticated tools available to them when managing risks, resulting in an insignificant coefficient associated with SOPH. They also discussed alternative suggestions and interpretations for the other results of the tests of hypotheses.

How robust is this model? Section 6.2 emphasized some of the dangers of working with an inadequate model. Some readers may be uncomfortable with the model selected because two out of the six variables have t-ratios less of than 1 in absolute value and four out of six have t-ratios of less than 1.5 in absolute value. Perhaps even more important, histograms of the standardized residuals and leverages, in Figure 6.5, show several observations to be outliers and high leverage points. To illustrate, the largest residual turns out to be $e_{15} = 83.73$. The error sum of squares is $Error\ SS = (n-(k+1))s^2 = (73–7)(14.56)^2 =$ 13,987. Thus, the 15th observation represents 50.1% of the error sum of squares ($= 83.73^2/13{,}987$), suggesting that this 1 observation of 73 has a dominant impact on the model fit. Further, plots of standardized residuals versus fitted

	Mean	Median	Standard Deviation	Minimum	Maximum
FIRMCOST	10.97	6.08	16.16	0.20	97.55
ASSUME	2.574	0.510	8.445	0.000	61.820
CAP	0.342	0.000	0.478	0.000	1.000
SIZELOG	8.332	8.270	0.963	5.270	10.600
INDCOST	0.418	0.340	0.216	0.090	1.220
CENTRAL	2.247	2.200	1.256	1.000	5.000
SOPH	21.192	23.00	5.304	5.000	31.000

Source: Schmit and Roth, 1990.

Table 6.3 Summary Statistics of $n = 73$ Risk Management Surveys

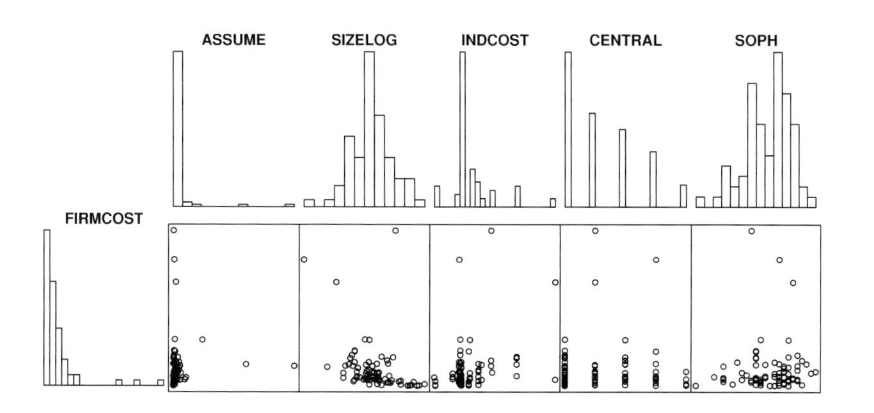

Figure 6.6 Histograms and scatterplots of FIRMCOST and several explanatory variables. The distributions of FIRMCOST and ASSUME are heavily skewed to the right. There is a negative, though nonlinear relationship between FIRMCOST and SIZELOG.

values, not presented here, displayed evidence of heteroscedastic residuals. On the basis of these observations, it seems reasonable to assess the robustness of the model.

Back to the Basics

To get a better understanding of the data, we begin by examining the basic summary statistics in Table 6.3 and corresponding histograms in Figure 6.6. From Table 6.3, the largest value of FIRMCOST is 97.55, which is more than five standard deviations above the mean [$10.97 + 5(16.16) = 91.77$]. An examination of the data shows that this point is observation 15, the same observation that was an outlier in the preliminary regression fit. However, the histogram of FIRMCOST in Figure 6.6 reveals that this is not the only unusual point. Two other observations have unusually large values of FIRMCOST, resulting in a distribution that is skewed to the right. The histogram, in Figure 6.6, of the ASSUME variable shows that this distribution is also skewed to the right, possibly solely because of two large observations. From the basic summary statistics in Table 6.3, we see that the largest value of ASSUME is more than seven standard deviations above the mean. This observation may well turn out to be influential in subsequent

Table 6.4 Table of Means by Level of CAP

	n	FIRMCOST	ASSUME	SIZELOG	INDCOST	CENTRAL	SOPH	COSTLOG
CAP = 0	48	9.954	1.175	8.197	0.399	2.250	21.521	1.820
CAP = 1	25	12.931	5.258	8.592	0.455	2.240	20.560	1.595
TOTAL	73	10.973	2.574	8.332	0.418	2.247	21.192	1.743

Table 6.5 Correlation Matrix

	COSTLOG	FIRMCOST	ASSUME	CAP	SIZELOG	INDCOST	CENTRAL
FIRMCOST	0.713						
ASSUME	0.165	0.039					
CAP	−0.088	0.088	0.231				
SIZELOG	−0.637	−0.366	−0.209	0.196			
INDCOST	0.395	0.326	0.249	0.122	−0.102		
CENTRAL	−0.054	0.014	−0.068	−0.004	−0.080	−0.085	
SOPH	0.144	0.048	0.062	−0.087	−0.209	0.093	0.283

regression model fitting. The scatter plot of FIRMCOST versus ASSUME in Figure 6.6 tells us that the observation with the largest value of FIRMCOST is not the same as the observation with the largest value of ASSUME.

From the histograms of SIZELOG, INDCOST, CENTRAL, and SOPH, we see that the distributions are not heavily skewed. Taking logarithms of the size of total company assets has served to make the distribution more symmetric than in the original units. From the histogram and summary statistics, we see that CENTRAL is a discrete variable, taking on values one through five. The other discrete variable is CAP, a binary variable taking values only zero and one. The histogram and scatter plot corresponding to CAP are not presented here. It is more informative to provide a *table of means* of each variable by levels of CAP, as in Table 6.4. From this table, we see that 25 of the 73 companies surveyed own captive insurers. Further, on the one hand, the average FIRMCOST for those companies with captive insurers (CAP = 1) is larger than those without (CAP = 0). On the other hand, when moving to the logarithmic scale, the opposite is true; that is, average COSTLOG for those companies with captive insurers (CAP = 1) is larger than for those without (CAP = 0).

When examining relationships between pairs of variables, in Figure 6.6, we see some of the relationships that were evident from preliminary regression fit. There is an inverse relationship between FIRMCOST and SIZELOG, and the scatter plot suggests that the relationship may be nonlinear. There is also a mild positive relationship between FIRMCOST and INDCOST and no apparent relationships between FIRMCOST and any of the other explanatory variables. These observations are reinforced by the table of correlations given in Table 6.5. Note that the table masks a feature that is evident in the scatter plots: the effect of the unusually large observations.

Because of the skewness of the distribution and the effect of the unusually large observations, a transformation of the response variable might lead to fruitful results. Figure 6.7 is the histogram of COSTLOG, defined to be the logarithm

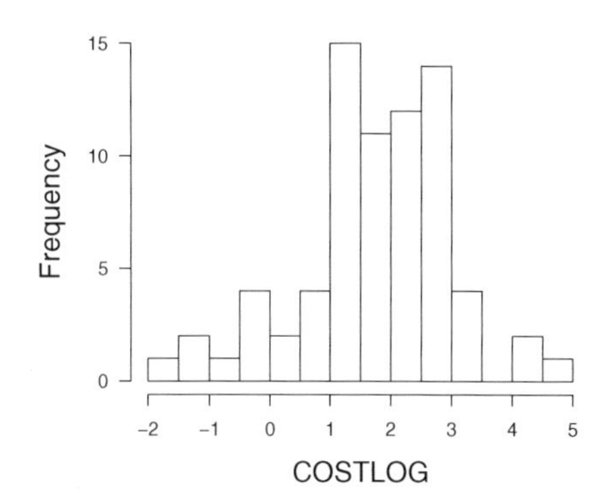

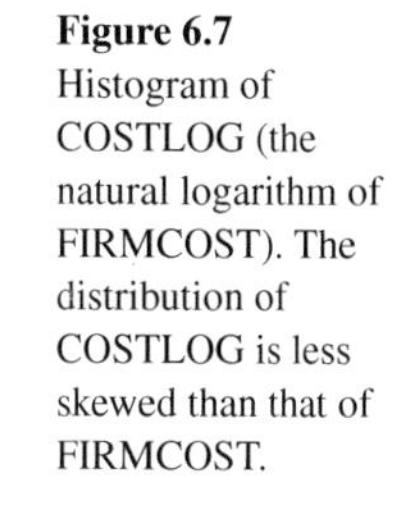

Figure 6.7 Histogram of COSTLOG (the natural logarithm of FIRMCOST). The distribution of COSTLOG is less skewed than that of FIRMCOST.

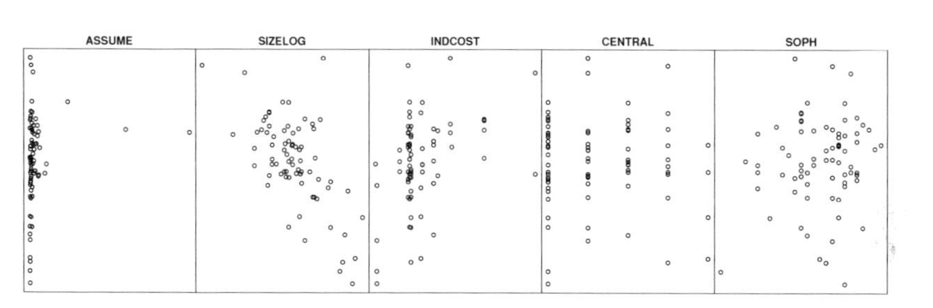

Figure 6.8 Scatterplots of COSTLOG versus several explanatory variables. There is a negative relationship between COSTLOG and SIZELOG and a mild positive relationship between COSTLOG and INDCOST.

of FIRMCOST. The distribution is much less skewed than the distribution of FIRMCOST. The variable COSTLOG was also included in the correlation matrix in Table 6.5. From that table, the relationship between SIZELOG appears to be stronger with COSTLOG than with FIRMCOST. Figure 6.8 shows several scatter plots illustrating the relationship between COSTLOG and the explanatory variables. The relationship between COSTLOG and SIZELOG appears to be linear. It is easier to interpret these scatter plots than those in Figure 6.6 because of the absence of the large unusual values of the dependent variable.

Some New Models

Now we explore the use of COSTLOG as the dependent variable. This line of thought is based on the work in the previous subsection and the plots of residuals from the preliminary regression fit. As a first step, we fit a model with all explanatory variables. Thus, this model is the same as the preliminary regression fit except it uses COSTLOG in lieu of FIRMCOST as the dependent variable. This model serves as a useful benchmark for our subsequent work. Table 6.6 summarizes the fit.

Here, $R_a^2 = 48\%$, F-ratio $= 12.1$, and $s = 0.882$. Figure 6.9 shows that the distribution of standardized residuals is less skewed than the corresponding distribution in Figure 6.5. The distribution of leverages shows that there are still highly influential observations. (As a matter of fact, the distribution of leverages appears to be the same as in Figure 6.5. Why?) Four of the six variables have t-ratios of less than one in absolute value, suggesting that we continue our search for a better model.

Table 6.6 Regression Results: COSTLOG as Dependent Variable

Variable	Coefficient	Standard Error	t-Statistic
INTERCEPT	7.64	1.16	6.62
ASSUME	−0.008	0.013	−0.61
CAP	0.015	0.233	0.06
SIZELOG	−0.787	0.117	−6.75
INDCOST	1.90	0.503	3.79
CENTRAL	−0.080	0.087	−0.92
SOPH	0.002	0.021	0.12

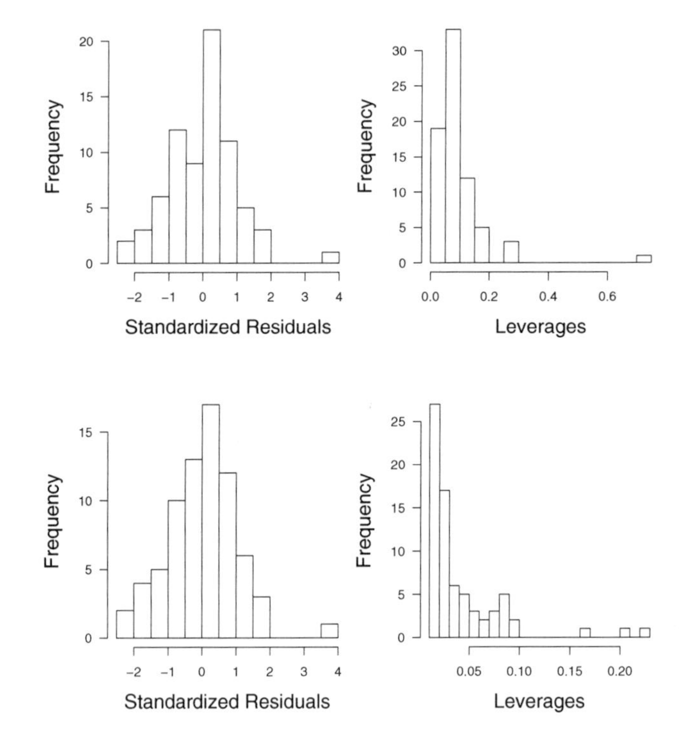

Figure 6.9 Histograms of standardized residuals and leverages using COSTLOG as the dependent variable.

Figure 6.10 Histograms of standardized residuals and leverages using SIZELOG and INDCOST as explanatory variables.

To continue the search, we can run a stepwise regression (although the output is not reproduced here). The output from this search technique, as well as the foregoing fitted regression model, suggests using the variables SIZELOG and INDCOST to explain the dependent variable COSTLOG.

We can run regression using SIZELOG and INDCOST as explanatory variables. From Figure 6.10, we see that the size and shape of the distribution of standardized residuals are similar to those in Figure 6.9. The leverages are much smaller, reflecting the elimination of several explanatory variables from the model. Remember that the average leverage is $\bar{h} = (k+1)/n = 3/73 \approx 0.04$. Thus, we still have three points that exceed three times the average and thus are considered high leverage points.

Plots of residuals versus the explanatory variables reveal some mild patterns. The scatter plot of residuals versus INDCOST, in Figure 6.11, displays a mild

Table 6.7 Regression Results with a Quadratic term in INDCOST

Variable	Coefficient	Standard Error	t-Statistic
INTERCEPT	6.35	0.953	6.67
SIZELOG	−0.773	0.101	−7.63
INDCOST	6.26	1.61	3.89
INDCOST^2	−3.58	1.27	−2.83

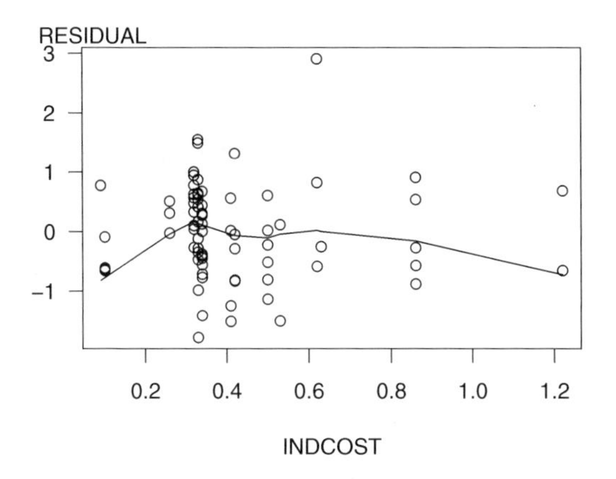

Figure 6.11 Scatter plot of residuals versus INDCOST. The smooth fitted curve (using locally weighted scatterplot smoothing) suggests a quadratic term in INDCOST.

quadratic trend in INDCOST. To determine whether this trend was important, the variable INDCOST was squared and used as an explanatory variable in a regression model. The results of this fit are in Table 6.7.

From the t-ratio associated with (INDCOST)^2, we see that the variable seems to be important. The sign is reasonable, indicating that the rate of increase of COSTLOG decreases as INDCOST increases. That is, the expected change in COSTLOG per unit change of INDCOST is positive and decreases as INDCOST increases.

Further diagnostic checks of the model revealed no additional patterns. Thus, from the data available, we cannot affirm any of the four hypotheses that were described previously in the Introduction subsection. This is not to say that these variables are not important. We are simply stating that the natural variability of the data was great enough to obscure any relationships that might exist. We have established, however, the importance of the size of the firm and the firm's industry risk.

Figure 6.12 graphically summarizes the estimated relationships among these variables. In particular, in the lower-right-hand panel, we see that, for most of the firms in the sample, FIRMCOST was relatively stable. However, for small firms, as measured by SIZELOG, the industry risk, as measured by INDCOST, was particularly important. For small firms, we see that the fitted FIRMCOST increases as the variable INDCOST increases, with the rate of increase leveling off. Although the model theoretically predicts FIRMCOST to decrease with a large INDCOST (>1.2), no small firms were actually in this area of the data region.

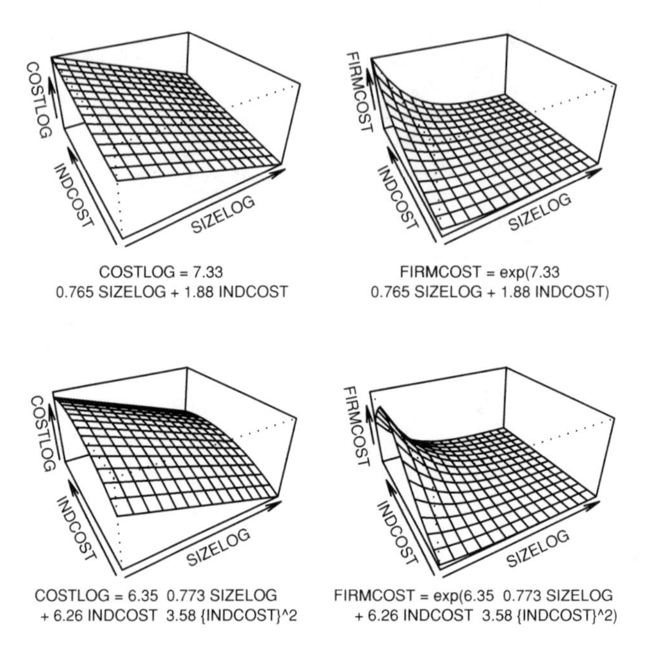

Figure 6.12 Graph of four fitted models versus INDCOST and SIZELOG.

6.6 Further Reading and References

This chapter concludes our Part I, an introduction to linear regression. To learn more about linear regression, see Section 1.7 for references to alternative statistics books that introduce the topic. You may also be interested in a more technical presentation, such as the classic work by Seber (1977) or a more recent work by Abraham and Ledolter (2006). For other approaches, texts such as that of Wooldridge (2009) provide an econometrics perspective where the emphasis is on introducing regression in the context of economic theory. Alternatively, books such as that of Agresti and Finlay (2008) give an introduction from a broader social science perspective.

There are many explanations of regression for readers with different perspectives and levels of quantitative training, this provides further evidence that this is an important topic for actuaries and other financial risk managers to understand. The other way of getting further insights into linear regression is to see it applied in a time series context in Part II of this book or in extensions to nonlinear modeling in Part III.

See Bollen (1989) for a classic introduction to structural equations modeling.

Chapter References

Abraham, Bova, and Johannes Ledolter (2006). *Introduction to Regression Modeling*. Thomson Higher Education, Belmont, California.

Agresti, Alan, and Barbara Finlay (2008). *Statistical Methods for the Social Sciences, 4th ed.* Prentice Hall, Upper Saddle, New Jersey.

Bollen, Kenneth A. (1989). *Structural Equations with Latent Variables*. Wiley, New York.

Box, George E. P. (1979). Robustness in the strategy of scientific model building. In *Robustness in Statistics*, ed. R. Launer and G. Wilderson (eds.), 201–36, Academic Press, New York.

Faraway, Julian J. (2005). *Linear Models with R*. Chapman & Hall/CRC, Boca Raton, Florida.
Fienberg, S. E. (1985). Insurance availability in Chicago. In *Data: A Collection of Problems from Many Fields for the Student and Research Worker*, ed. D. F. Andrews and A. M. Herzberg, 487–412, Springer-Verlag, New York.
Goldberger, Arthur S. (1972). Structural equation methods in the social sciences. *Econometrica* 40, 979–1001.
Harrington, Scott E., and Greg Niehaus (1998). Race, redlining and automobile insurance prices. *Journal of Business* 71, no. 3, 439–69.
Heckman, J. J. (1976). The common structure of statistical models of truncation, sample selection and limited dependent variables, and a simple estimator for such models. *Annals of Economic and Social Measurement* 5, 475–92.
Little, R. J. (1995). Modelling the drop-out mechanism in repeated-measures studies. *Journal of the American Statistical Association* 90, 1112–21.
Little, R. J., and Rubin, D. B. (1987). *Statistical Analysis with Missing Data*. John Wiley, New York.
Roberts, Harry V. (1990). Business and economic statistics (with discussion). *Statistical Science* 4, 372–402.
Rubin, D. R. (1976). Inference and missing data. *Biometrika* 63, 581–92.
Schmit, Joan T., and K. Roth (1990). Cost effectiveness of risk management practices. *Journal of Risk and Insurance* 57, no. 3, 455–70.
Seber, G. A. F. (1977). *Linear Regression Analysis*. John Wiley, New York.
Wachter, K. W., and J. Trusell (1982). Estimating historical heights. *Journal of the American Statistical Association* 77, 279–301.
Wooldridge, Jeffrey (2009). *Introductory Econometrics: A Modern Approach, 4th ed.* South-Western Publishing, Mason, Ohio.

6.7 Exercises

Ⓡ **Empirical**
Filename is "Chicago"

6.1. **Insurance Redlining.** Do insurance companies use race as a determining factor when making insurance available? Fienberg (1985) gathered data from a report issued by the U.S. Commission on Civil Rights about the number of homeowners and residential fire insurance policies issued in Chicago over the months of December 1977 through February 1978. Policies issued were categorized as part of either the standard, voluntary market or the substandard, involuntary market. The involuntary market consists of fair access to insurance requirements (FAIR) plans; these are state insurance programs sometimes subsidized by private companies. These plans provide insurance to people who would otherwise be denied insurance on their property because of high-risk problems. The main purpose is to understand the relationship between insurance activity and the variable race, the percentage minority. Data are available for $n = 47$ Zip codes in the Chicago area. These data have also been analyzed by Faraway (2005).

To help control for the size of the expected loss, Fienberg also gathered theft and fire data from Chicago's police and fire departments. Another variable that gives some information about loss size is the age of the house. The median income, from the Census Bureau, gives indirect information on the size of the expected loss and on whether the applicant can afford insurance. Table 6.8 provides more details on these variables.

Table 6.8 Insurance Availability in Chicago

Variable	Description	Mean
row.names	Zip (postal) code	
race	Racial composition in percent minority	35.0
fire	Fires per 1,000 housing units	12.3
theft	Thefts per 1,000 population	32.4
age	Percentage of housing units built in or before 1939	60.3
volact	New homeowner policies plus renewals, minus cancelations and nonrenewals per 100 housing units	6.53
involact	New FAIR plan policies and renewals per 100 housing units	0.615
income	Median family income	10,696

Source: Fienberg (1985).

a. Produce summary statistics of all variables, noting patterns of skewness for each variable.
b. Create a scatterplot matrix of volact, involact, and race. Comment on the three pairwise relationships. Are the patterns consistent with a hypothesis of racial discrimination?
c. To understand relationships among the variables, produce a table of correlations.
d. Fit a linear model using volact as the dependent variable and race, fire theft, age and income as explanatory variables.
   d(i). Comment on the sign and statistical significance of the coefficient associated with race.
   d(ii). Two Zip codes turn out to have high leverage. Repeat your analysis after deleting these two observations. Has the significance of the race variable changed? What about the other explanatory variables?
e. Repeat the analysis in part (d) using involact as the dependent variable.
f. Define proportion to be involact/(volact + involact). Repeat the analysis in part (d) using proportion as the dependent variable.
g. The same two Zip codes have high leverage in parts (d), (e), and (f). Why?
h. This analysis is done at the Zip-code level, not the individual level. As emphasized by Harrington and Niehaus (1998), this introduces substantial potential omitted variable bias. What variables have been omitted from the analysis that you think might affect homeowners' insurance availability and race?
i. Fienberg notes that proximity of one Zip code to another may affect the dependence of observations. Describe how you might incorporate spatial relations into a regression analysis.

6.2. **Gender Equity in Faculty Pay.** The University of Wisconsin at Madison completed a study titled "Gender Equity Study of Faculty Pay," dated June 5, 1992. The main purpose of the study was to determine whether women are treated unfairly in salary determinations at a major research university in the United States. To this end, the committee that issued the report studied salaries in the year 1990 of 1,898 faculty members in the university. It is well known that men are paid more than women. In fact, the mean 1990 salary for the 1,528 male faculty members is $54,478, which is 28% higher than the mean 1990 salary for female faculty members, which is $43,315. However, it is argued that male faculty members are, in general, more senior (average years of experience is 18.8) than female faculty members (average years of experience is 11.9) and thus deserved higher pay. When comparing salaries of full professors (thus controlling for years of experience), male faculty members earned about 13% more than their female counterparts. Even so, it is generally agreed that fields in demand must offer higher salaries to maintain a world-class faculty. For example, salaries in engineering are higher than salaries in humanities simply because faculty in engineering have many more employment opportunities outside of academia than faculty in humanities. Thus, when considering salaries, one must also control for department.

To control for these variables, a faculty study reports a regression analysis using the logarithm of salary as the dependent variable. The explanatory variables included information on race, sex, rank (assistant professor or instructor, associate professor, or full professor), several measures of years of experience, 98 different categories of departments, and a measure of salary differential by department. There were 109 explanatory variables in all (including 97 departmental binary variables), of which 12 were nondepartmental variables. Table 6.9 reports variable definitions, parameter estimates, and t-ratios for the 12 nondepartmental variables. The ANOVA in Table 6.10 summarizes the regression fit.

Table 6.9 Nondepartmental Variables and Parameter Estimates

Explanatory Variable	Variable Description	Parameter Estimate	t-Ratio
INTERCEPT		10.746	261.1
GENDER	= 1 if male, 0 otherwise	0.016	1.86
RACE	= 1 if white, 0 otherwise	–0.029	–2.44
FULL	= 1 if a full professor, 0 otherwise	0.186	16.42
ASSISTANT	= 1 if an assistant professor, 0 otherwise	–0.205	–15.93
ANYDOC	= 1 if has a terminal degree such as a Ph.D.	0.022	1.11
COHORT1	= 1 if hired before 1969, 0 otherwise	–0.102	–4.84
COHORT2	= 1 if hired 1969–85, 0 otherwise	–0.046	–3.48
FULLYEARS	Number of years as a full professor at UW	0.012	12.84
ASSOCYEARS	Number of years as an associate professor at UW	–0.012	–8.65
ASSISYEARS	Number of years as an assistant professor or an instructor at UW	0.002	0.91
DIFYRS	Number of years since receiving a terminal degree before arriving at UW	0.004	4.46
MRKTRATIO	Natural logarithm of a market ratio, defined as the ratio of the average salary at peer institutions for a given discipline and rank	0.665	7.64

Source: "Gender Equity Study of Faculty Pay," June 5, 1992, University of Wisconsin at Madison.

Table 6.10 Faculty Pay ANOVA Table

Source	Sum of Squares	df	Mean Square	F-Ratio
Regression	114.048	109	1.0463	62.943
Error	29.739	1789	0.0166	
Total	143.788	1898		

a. Suppose that a female faculty member in the chemistry department believes that her salary is lower than it should be. Briefly describe how this study can be used as a basis for performance evaluation.
b. From this study, do you think that salaries of women are significantly lower than those of men?
   b(i). Cite statistical arguments supporting the fact that men are not paid significantly more than women.
   b(ii). Cite statistical arguments supporting the fact that men are paid significantly more than women.
   b(iii). Suppose that you decide that women are paid less than men. From this study, how much would you raise female faculty members salaries to be on par with those of their male counterparts?

6.8 Technical Supplements for Chapter 6

6.8.1 Effects of Model Misspecification

Notation. Partition the matrix of explanatory variables $\mathbf{X}$ into two submatrices, each with n rows, so that $\mathbf{X} = (\mathbf{X}_1 : \mathbf{X}_2)$. For convenience, assume that $\mathbf{X}_1$ is an $n \times p$ matrix. Similarly, partition the vector of parameters $\boldsymbol{\beta} = \left(\boldsymbol{\beta}_1', \boldsymbol{\beta}_2'\right)'$ such that $\mathbf{X}\boldsymbol{\beta} = \mathbf{X}_1\boldsymbol{\beta}_1 + \mathbf{X}_2\boldsymbol{\beta}_2$. We compare the full, or "long," model

$$\mathbf{y} = \mathbf{X}\boldsymbol{\beta} + \boldsymbol{\varepsilon} = \mathbf{X}_1\boldsymbol{\beta}_1 + \mathbf{X}_2\boldsymbol{\beta}_2 + \boldsymbol{\varepsilon}$$

to the reduced, or "short," model

$$\mathbf{y} = \mathbf{X}_1\boldsymbol{\beta}_1 + \boldsymbol{\varepsilon}.$$

This simply generalizes the setup earlier to allow for omitting several variables.

Effect of Underfitting. Suppose that the true representation is the long model but we mistakenly run the short model. Our parameter estimates when running the short model are given by $\mathbf{b}_1 = (\mathbf{X}_1'\mathbf{X}_1)^{-1}\mathbf{X}_1'\mathbf{y}$. These estimates are biased because

$$\begin{aligned}\text{Bias} &= \mathrm{E}\,\mathbf{b}_1 - \boldsymbol{\beta}_1 = \mathrm{E}(\mathbf{X}_1'\mathbf{X}_1)^{-1}\mathbf{X}_1'\mathbf{y} - \boldsymbol{\beta}_1 = (\mathbf{X}_1'\mathbf{X}_1)^{-1}\mathbf{X}_1'\mathrm{E}\,\mathbf{y} - \boldsymbol{\beta}_1 \\ &= (\mathbf{X}_1'\mathbf{X}_1)^{-1}\mathbf{X}_1'\left(\mathbf{X}_1\boldsymbol{\beta}_1 + \mathbf{X}_2\boldsymbol{\beta}_2\right) - \boldsymbol{\beta}_1 = (\mathbf{X}_1'\mathbf{X}_1)^{-1}\mathbf{X}_1'\mathbf{X}_2\boldsymbol{\beta}_2 = \mathbf{A}\boldsymbol{\beta}_2.\end{aligned}$$

Here, $\mathbf{A} = (\mathbf{X}_1'\mathbf{X}_1)^{-1}\mathbf{X}_1'\mathbf{X}_2$ is the *alias*, or bias, matrix. When running the short model, the estimated variance is $s_1^2 = (\mathbf{y}'\mathbf{y} - \mathbf{b}_1'\mathbf{X}_1'\mathbf{y})/(n-p)$. It can be shown that

$$\mathrm{E}\, s_1^2 = \sigma^2 + (n-p)^{-1} \boldsymbol{\beta}_2' \left(\mathbf{X}_2'\mathbf{X}_2 - \mathbf{X}_2'\mathbf{X}_1(\mathbf{X}_1'\mathbf{X}_1)^{-1}\mathbf{X}_1'\mathbf{X}_2 \right) \boldsymbol{\beta}_2. \tag{6.1}$$

Thus, s_1^2 is an overbiased estimate of σ^2.

Let $\mathbf{x}_{1i}'$ and $\mathbf{x}_{2i}'$ be the ith rows of $\mathbf{X}_1$ and $\mathbf{X}_2$, respectively. Using the fitted short model, the ith fitted value is $\hat{y}_{1i} = \mathbf{x}_{1i}'\mathbf{b}_1$. The true ith expected response is $\mathrm{E}\ \hat{y}_{1i} = \mathbf{x}_{1i}'\boldsymbol{\beta}_1 + \mathbf{x}_{2i}'\boldsymbol{\beta}_2$. Thus, the bias of the ith fitted value is

$$\begin{aligned}
\mathrm{Bias}(\hat{y}_{1i}) &= \mathrm{E}\ \hat{y}_{1i} - \mathrm{E}\ y_i = \mathbf{x}_{1i}'\mathrm{E}\ \mathbf{b}_1 - \left(\mathbf{x}_{1i}'\boldsymbol{\beta}_1 + \mathbf{x}_{2i}'\boldsymbol{\beta}_2\right) \\
&= \mathbf{x}_{1i}'(\boldsymbol{\beta}_1 + \mathbf{A}\boldsymbol{\beta}_2) - \left(\mathbf{x}_{1i}'\boldsymbol{\beta}_1 + \mathbf{x}_{2i}'\boldsymbol{\beta}_2\right) = (\mathbf{x}_{1i}'\mathbf{A} - \mathbf{x}_{2i}')\boldsymbol{\beta}_2.
\end{aligned}$$

Using this and equation (6.1), straightforward algebra shows that

$$\mathrm{E}\, s_1^2 = \sigma^2 + (n-p)^{-1} \sum_{i=1}^{n} (\mathrm{Bias}(\hat{y}_{1i}))^2. \tag{6.2}$$

Effect of Overfitting. Now suppose that the true representation is the short model but we mistakenly use the large model. With the alias matrix $\mathbf{A} = (\mathbf{X}_1'\mathbf{X}_1)^{-1}\mathbf{X}_1'\mathbf{X}_2$, we can *reparameterize* the long model

$$\mathbf{y} = \mathbf{X}_1\boldsymbol{\beta}_1 + \mathbf{X}_2\boldsymbol{\beta}_2 + \boldsymbol{\varepsilon} = \mathbf{X}_1\left(\boldsymbol{\beta}_1 + \mathbf{A}\boldsymbol{\beta}_2\right) + \mathbf{E}_1\boldsymbol{\beta}_2 + \boldsymbol{\varepsilon} = \mathbf{X}_1\boldsymbol{\alpha}_1 + \mathbf{E}_1\boldsymbol{\beta}_2 + \boldsymbol{\varepsilon},$$

where $\mathbf{E}_1 = \mathbf{X}_2 - \mathbf{X}_1\mathbf{A}$ and $\boldsymbol{\alpha}_1 = \boldsymbol{\beta}_1 + \mathbf{A}\boldsymbol{\beta}_2$. The advantage of this new parameterization is that $\mathbf{X}_1$ is orthogonal to $\mathbf{E}_1$ because $\mathbf{X}_1'\mathbf{E}_1 = \mathbf{X}_1'(\mathbf{X}_2 - \mathbf{X}_1\mathbf{A}) = \mathbf{0}$. With $\mathbf{X}^* = (\mathbf{X}_1 : \mathbf{E}_1)$ and $\boldsymbol{\alpha} = (\boldsymbol{\alpha}_1'\boldsymbol{\beta}_1')'$, the vector of least squares estimates is

$$\begin{aligned}
\mathbf{a} = \begin{bmatrix} \mathbf{a}_1 \\ \mathbf{b}_1 \end{bmatrix} &= \left(\mathbf{X}^{*\prime}\mathbf{X}^*\right)^{-1} \mathbf{X}^{*\prime}\mathbf{y} \\
&= \begin{bmatrix} (\mathbf{X}_1'\mathbf{X}_1)^{-1} & 0 \\ 0 & (\mathbf{E}_1'\mathbf{E}_1)^{-1} \end{bmatrix} \begin{bmatrix} \mathbf{X}_1'\mathbf{y} \\ \mathbf{E}_1'\mathbf{y} \end{bmatrix} = \begin{bmatrix} (\mathbf{X}_1'\mathbf{X}_1)^{-1}\mathbf{X}_1'\mathbf{y} \\ (\mathbf{E}_1'\mathbf{E}_1)^{-1}\mathbf{E}_1'\mathbf{y} \end{bmatrix}.
\end{aligned}$$

From the true (short) model, $\mathrm{E}\ \mathbf{y} = \mathbf{X}_1\boldsymbol{\beta}_1$, we have that $\mathrm{E}\ \mathbf{b}_2 = (\mathbf{E}_1'\mathbf{E}_1)^{-1}\mathbf{E}_1'\mathrm{E}\ \mathbf{y} = (\mathbf{E}_1'\mathbf{E}_1)^{-1}\mathbf{E}_1'\mathrm{E}\ (\mathbf{X}_1\beta_1) = \mathbf{0}$, because $\mathbf{X}_1'\mathbf{E}_1 = \mathbf{0}$. The least squares estimate of $\boldsymbol{\beta}_1$ is $\mathbf{b}_1 = \mathbf{a}_1 - \mathbf{A}\mathbf{b}_2$. Because $\mathrm{E}\ \mathbf{a}_1 = (\mathbf{X}_1'\mathbf{X}_1)^{-1}\mathbf{X}_1'\mathrm{E}\ \mathbf{y} = \boldsymbol{\beta}_1$ in the short model, we have $\mathrm{E}\ \mathbf{b}_1 = \mathrm{E}\ \mathbf{a}_1 - \mathbf{A}\mathrm{E}\ \mathbf{b}_2 = \boldsymbol{\beta}_1 - \mathbf{0} = \boldsymbol{\beta}_1$. Thus, even though we mistakenly run the long model, $\mathbf{b}_1$ is still an unbiased estimator of $\boldsymbol{\beta}_1$ and $\mathbf{b}_2$ is an unbiased estimator of $\mathbf{0}$. Thus, there is no bias in the ith fitted value because $\mathrm{E}\ \hat{y}_i = \mathrm{E}\ (\mathbf{x}_{1i}'\mathbf{b}_1 + \mathbf{x}_{2i}'\mathbf{b}_2) = \mathbf{x}_{1i}'\boldsymbol{\beta}_1 = \mathrm{E}\ y_i$.

C_p Statistic. Suppose initially that the true representation is the long model but we mistakenly use the short model. The ith fitted value is $\hat{y}_{1i} = \mathbf{x}_{1i}'\mathbf{b}_1$, which has mean square error

$$\mathrm{MSE}\ \hat{y}_{1i} = \mathrm{E}(\hat{y}_{1i} - \mathrm{E}\ \hat{y}_{1i})^2 = \mathrm{Var}\ \hat{y}_{1i} + (\mathrm{Bias}\ \hat{y}_{1i})^2.$$

For the first part, we have that $\mathrm{Var}\ \hat{y}_{1i} = \mathrm{Var}\left(\mathbf{x}_{1i}'\mathbf{b}_1\right) = \mathrm{Var}\left(\mathbf{x}_{1i}'(\mathbf{X}_1'\mathbf{X}_1)^{-1}\mathbf{X}_1'\mathbf{y}\right) = \sigma^2\mathbf{x}_{1i}(\mathbf{X}_1'\mathbf{X}_1)^{-1}\mathbf{x}_{1i}'$. We can think of $\mathbf{x}_{1i}(\mathbf{X}_1'\mathbf{X}_1)^{-1}\mathbf{x}_{1i}'$ as the ith leverage, as in equation (5.3). Thus, $\sum_{i=1}^{n} \mathbf{x}_{1i}(\mathbf{X}_1'\mathbf{X}_1)^{-1}\mathbf{x}_{1i}' = p$, the number of columns of $\mathbf{X}_1$.

With this, we can define the *standardized total error*

$$\frac{\sum_{i=1}^{n} \text{MSE } \hat{y}_{1i}}{\sigma^2} = \frac{\sum_{i=1}^{n} \left(\text{Var } \hat{y}_{1i} + (\text{Bias } \hat{y}_{1i})^2\right)}{\sigma^2}$$

$$= \frac{\sigma^2 \sum_{i=1}^{n} \left(\mathbf{x}_{1i}(\mathbf{X}_1'\mathbf{X}_1)^{-1}\mathbf{x}_{1i}' + (\text{Bias } \hat{y}_{1i})^2\right)}{\sigma^2}$$

$$= p + \sigma^{-2} \sum_{i=1}^{n} (\text{Bias } \hat{y}_{1i})^2 .$$

Now, if σ^2 is known, from equation (6.2), an unbiased estimate of the standardized total error is $p + (n - p)(s_1^2 - \sigma^2)/\sigma^2$. Because σ^2 is unknown, it must be estimated. If we are not sure whether the long or short model is the appropriate representation, a conservative choice is to use s^2 from the long, or full, model. Even if the short model is the true model, s^2 from the long model is still an unbiased estimate of σ^2. Thus, we define

$$C_p = p + (n - p)(s_1^2 - s^2)/s^2.$$

If the short model is correct, then $\text{E } s_1^2 = \text{E } s^2 = \sigma^2$ and $\text{E } C_p \approx p$. If the long model is true, then $\text{E } s_1^2 > \sigma^2$ and $\text{E } C_p > p$.

Part II

Topics in Time Series

7

Modeling Trends

Chapter Preview. This chapter begins our study of time series data by introducing techniques to account for major patterns, or trends, in data that evolve over time. The focus is on how regression techniques developed in earlier chapters can be used to model trends. Further, new techniques, such differencing data, allow us to naturally introduce a random walk, an important model of efficient financial markets.

7.1 Introduction

Time Series and Stochastic Processes

Business firms are not defined by physical structures such as the solid stone bank building that symbolizes financial security. Nor are businesses defined by space-alien invader toys that they manufacture for children. Businesses comprise several complex, interrelated processes. A *process* is a series of actions or operations that lead to a particular end.

Processes not only are the building blocks of businesses but also provide the foundations for our everyday lives. We may go to work or school every day, practice martial arts, or study statistics. These are regular sequences of activities that define us. In this text, our interest is in modeling *stochastic processes*, defined as ordered collections of random variables that quantify a process of interest.

A time series is a single measurement of a process that yields a variable over time, denoted by $y_1, \ldots, y_T$.

Some processes evolve over time, such as daily trips to work or school, or the quarterly earnings of a firm. We use the term *longitudinal data* for measurements of a process that evolves over time. A single measurement of a process yields a variable over time, denoted by $y_1, \ldots, y_T$ and referred to as a *time series*. In this portion of the text, we follow common practice and use T to denote the number of observations available (instead of n). Chapter 10 will describe another type of longitudinal data for which we examine a cross-section of entities, such as firms, and examine their evolution over time. This type of data is also known as *panel data*.

Collections of random variables may be based on orderings other than time. For example, hurricane claim damages are recorded at the place where the damage has occurred and thus are ordered spatially. The evaluation of an oil-drilling project requires taking samples of the earth at various longitudes, latitudes, and depths. This yields observations ordered by the three dimensions of space but not time. For another example, the study of holes in the ozone layer requires

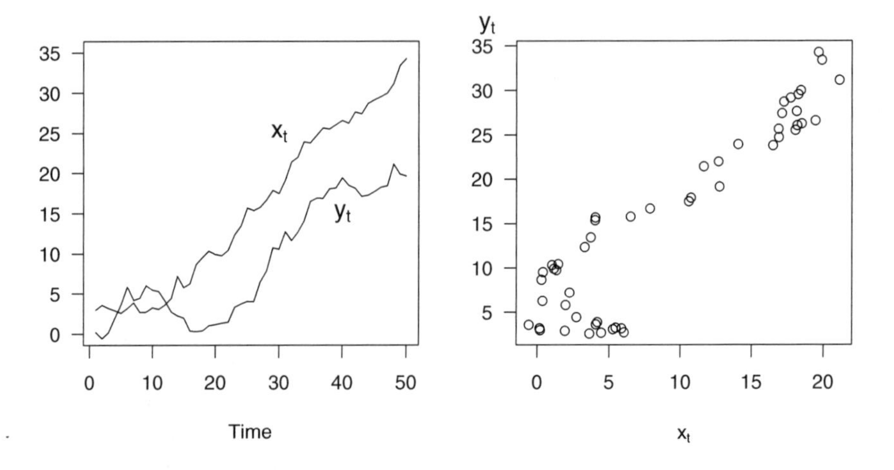

Figure 7.1 Spurious regressions. The left-hand panel shows two time series that are increasing over time. The right-hand panel shows a scatter plot of the two series, suggesting a positive relationship between the two. The relationship is spurious in the sense that both series are driven by growth over time, not by their positive dependence on each other.

taking atmospheric measurements. Because the interest is in the trend of ozone depletion, the measurements are taken at various longitudes, latitudes, heights, and time. Although we consider only processes ordered by time, in other studies of longitudinal data, you may see alternatives orderings. Data that are not ordered are called *cross-sectional.*

Cross-sectional data are not ordered by time.

Time Series versus Causal Models

Regression methods can be used to summarize many time series data sets. However, simply using regression techniques without establishing an appropriate context can be disastrous. This concept is reinforced by an example based on Granger and Newbold's (1974) work.

Example: Spurious Regression. Let $\{\varepsilon_{x,t}\}$ and $\{\varepsilon_{y,t}\}$ be two independent sequences, each of which also has a standard normal distribution. From these, recursively construct the variables $x_t = 0.5 + x_{t-1} + \varepsilon_{x,t}$ and $y_t = 0.5 + y_{t-1} + \varepsilon_{y,t}$, using the initial conditions $x_0 = y_0 = 0$. (In Section 7.3, we will identify x_t and y_t as random walk models.) Figure 7.1 shows a realization of $\{x_t\}$ and $\{y_t\}$, generated for $T = 50$ observations using simulation. The left-hand panel shows the growth of each series over time – the increasing nature is due to the addition of 0.5 at each time point. The right-hand panel shows a strong relationship between $\{x_t\}$ and $\{y_t\}$ – the correlation between the two series turns out to be 0.92. This is despite the fact that the two series were generated *independently.* Their apparent relationship, said to be *spurious*, is because both are related to the growth over time.

In a longitudinal context, regression models of the form

$$y_t = \beta_0 + \beta_1 x_t + \varepsilon_t$$

are known as *causal models*. Causal models are regularly employed in econometrics, where it is assumed that economic theory provides the information needed to specify the causal relationship (x "causes" y). In contrast, statistical models can only validate empirical relationships ("correlation, not causation"). In the spurious regression example, both variables evolve over time, and so a model of how one variable influences another needs to account for time patterns of both the left- and the right-hand-side variables. Specifying causal models for actuarial applications can be difficult for this reason – time series patterns in the explanatory variables may mask or induce a significant relationship with the dependent variable. In contrast, regression modeling can be readily applied when explanatory variables are simply functions of time, the topic of the next section. This is because functions of time are deterministic and so do not exhibit time series patterns.

Regression models can be readily applied when explanatory variables are functions of time.

Causal models also suffer from the drawback that their applications are limited for forecasting purposes. This is because to make a forecast of a future realization of the series, for example y_{T+2}, one needs to have knowledge (or a good forecast) of x_{T+2}, the value of the explanatory variable at time $T+2$. If x is a known function of time (as in the next section), then this is not an issue. Another possibility is to use a lagged value of x such as $y_t = \beta_0 + \beta_1 x_{t-1} + \varepsilon_t$, so that one-step predictors are possible (we can use the equation to predict y_{T+1} because x_T is known at time T).

7.2 Fitting Trends in Time

Understanding Patterns over Time

A forecast is a prediction of a future value of a time series.

Forecasting is about predicting future realizations of a time series. Over the years, analysts have found it convenient to decompose a series into three types of patterns: trends in time (T_t); seasonal (S_t); and random, or irregular, patterns (ε_t). A series can then be forecast by extrapolating each of the three patterns. The trend is that part of a series that corresponds to a long-term, slow evolution of the series. This is the most important part for long-term forecasts. The seasonal part of the series corresponds to aspects that repeat itself periodically, say, over a year. The irregular patterns of a series are short-term movements that are typically harder to anticipate.

Analysts typically combine these patterns in two ways: in an additive fashion,

$$y_t = T_t + S_t + \varepsilon_t, \tag{7.1}$$

or in a multiplicative fashion,

$$y_t = T_t \times S_t + \varepsilon_t. \tag{7.2}$$

Models without seasonal components can be readily handled by using $S_t = 0$ for the additive model in equation (7.1) and $S_t = 1$ for the multiplicative model in equation (7.2). If the model is purely multiplicative such that $y_t = T_t \times S_t \times \varepsilon_t$, then it can be converted to an additive model by taking logarithms of both sides.

A plot of y_t versus t is called a time series plot.

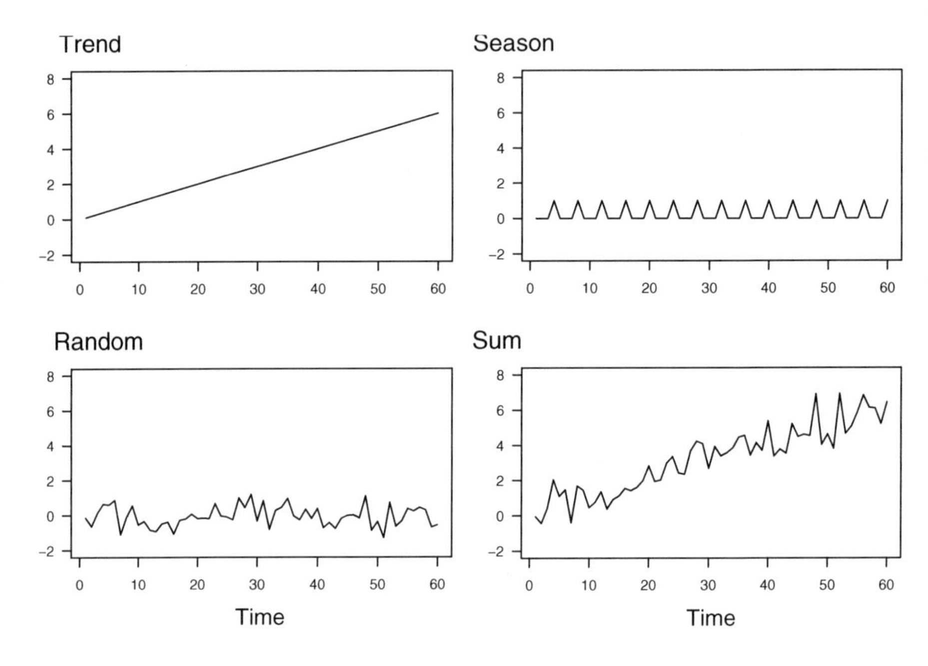

Figure 7.2 Time Series Plots of Response Components. The linear trend component appears in the upper-left-hand panel, the seasonal trend in the upper-right panel, and the random variation in the lower-left-hand panel. The sum of the three components appears in the lower-right-hand panel.

It is instructive to see how these three components can be combined to form a series of interest. Consider the three components in Figure 7.2. Under the additive model, the trend, seasonal, and random variation components are combined to form the series that appears in the lower-right-hand panel. A plot of y_t versus t is called a *time series plot*. In time series plots, the convention is to connect adjacent points using a line to help us detect patterns over time.

When analyzing data, the graph in the lower-right-hand panel is the first type of plot that we will examine. The goal of the analysis is to go backward – that is, we want to decompose the series into the three components. Each component can then be forecast, which will provide us with forecasts that are reasonable and easy to interpret.

Fitting Trends in Time

The simplest type of time trend is a complete lack of trend. Assuming that the observations are identically and independently distributed (i.i.d.), then we could use the model

$$y_t = \beta_0 + \varepsilon_t.$$

For example, if we are observing a game of chance, such as bets placed on the roll of two dice, then we typically model this as an i.i.d. series.

The linear trend in time model is a regression model with a straight line in time as the regression function.

Fitting polynomial functions of time is another type of trend that is easy to interpret and fit to the data. We begin with a straight line for our polynomial function of time, yielding the *linear trend in time model*,

$$y_t = \beta_0 + \beta_1 t + \varepsilon_t. \tag{7.3}$$

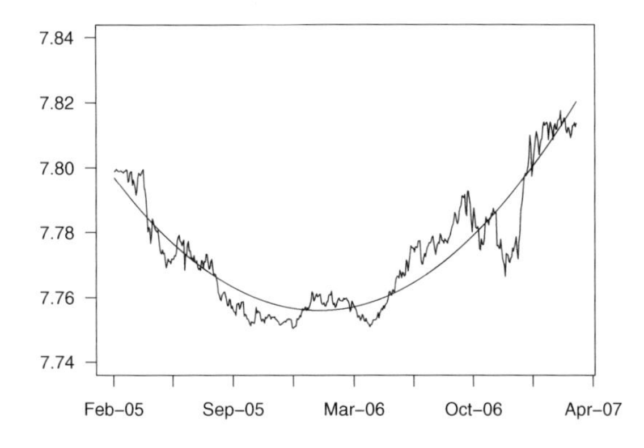

Figure 7.3 Time series plot of Hong Kong exchange rates with fitted values superimposed. The fitted values are from a regression using a quadratic time trend. *Source:* Foreign Exchange Rates (Federal Reserve, H10 report).

Similarly, regression techniques can be used to fit other functions that represent trends in time. Equation (7.3) is easily extended to handle a *quadratic trend in time*,

$$y_t = \beta_0 + \beta_1 t + \beta_2 t^2 + \varepsilon_t,$$

or a higher-order polynomial.

Ⓡ **Empirical** *Filename is "HKExchange"*

Example: Hong Kong Exchange Rates. For travelers and firms, exchange rates are an important part of the monetary economy. The exchange rate that we consider here is the number of Hong Kong dollars that one can purchase for one U.S. dollar. We have $T = 502$ daily observations for the period April 1, 2005, through May 31, 2007, which were obtained from the Federal Reserve (H10 report). Figure 7.3 provides a time series plot of the Hong Kong exchange rate.

Figure 7.3 shows a clear quadratic trend in the data. To handle this trend, we use $t = 1, \ldots, 502$, as an explanatory variable to indicate the time period. The fitted regression equation turns out to be:

$$\begin{array}{lccc} \widehat{INDEX}_t = & 7.797 & -3.68 \times 10^{-4} t & +8.269 \times 10^{-7} t^2 \\ t\text{-statistics} & (8{,}531.9) & (-44.0) & (51.2) \end{array}.$$

The coefficient of determination is a healthy $R^2 = 92.9\%$, and the standard deviation estimate has dropped from $s_y = 0.0183$ to $s = 0.0068$ (our residual standard deviation). Figure 7.3 shows the relationship between the data and fitted values through the time series plot of the exchange rate with the fitted values superimposed. To apply these regression results to the forecasting problem, suppose that we wanted to predict the exchange rate for April 1, 2007, or $t = 503$. Our prediction is

$$\widehat{INDEX}_{503} = 7.797 - 3.68 \times 10^{-4}(503) + 8.269 \times 10^{-7}(503)^2 = 7.8208.$$

The overall conclusion is that the regression model using a quadratic term in time t as an explanatory variable fits the data well. A close inspection of Figure 7.3, however, reveals that there are patterns in the residuals where the

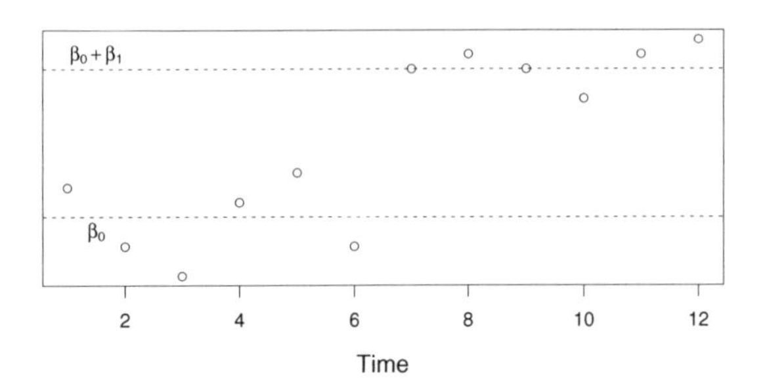

Figure 7.4 Time series plot of interest rates. There is a clear shift in the rates due to a change in the economy. This shift can be measured using a regression model with an explanatory variable to indicate the change.

responses are in some places consistently higher and in other places consistently lower than the fitted values. These patterns suggest that we can improve on the model specification. One way would be to introduce a higher-order polynomial model in time. In Section 7.3, we will argue that the random walk is an even better model for this data.

Other nonlinear functions of time may also be useful. To illustrate, we might study some measure of interest rates over time (y_t) and be interested in the effect of a change in the economy (e.g., the advent of a war). Define z_t to be a binary variable that is zero before the change occurs and one during and after the change. Consider the model

$$y_t = \beta_0 + \beta_1 z_t + \varepsilon_t. \tag{7.4}$$

Thus, using

$$\mathrm{E}\, y_t = \begin{cases} \beta_0 + \beta_1 & \textit{if } z_t = 1 \\ \beta_0 & \textit{if } z_t = 0 \end{cases},$$

the parameter β_1 captures the expected change in interest rates due to the change in the economy (see Figure 7.4).

Example: Regime-Switching Models of Long-Term Stock Returns. With the assumption of normality, we can write the model in equation (7.4) as

$$y_t \sim \begin{cases} N(\mu_1, \sigma^2) & t < t_0 \\ N(\mu_2, \sigma^2) & t \geq t_0 \end{cases},$$

where $\mu_1 = \beta_0$, $\mu_2 = \beta_0 + \beta_1$, and t_0 is the change point. A *regime-switching model* generalizes this concept, primarily by assuming that the change point is not known. Instead, one assumes that there exists a transition mechanism that allows us to shift from one regime to another with a probability that is typically estimated from the data. In this model, there is a finite number of states, or regimes. Within each regime, a probabilistic model is specified, such as the (conditionally) independent normal distribution ($N(\mu_2, \sigma^2)$). One could also specify an autoregressive or conditionally autoregressive model, which we will define in Chapter 8. Further, there is a conditional probability of transiting from one state to another (so-called Markov transition probabilities).

Hardy (2001) introduced regime-switching models to the actuarial literature where the dependent variable of interest was the long-term stock market return as measured by monthly returns on the Standard and Poor's 500 and the Toronto Stock Exchange 300. Hardy considered two and three regime models for data over 1956 to 1999, inclusive. Hardy showed how to use the parameter estimates from the regime-switching model to compute option prices and risk measures for equity-linked insurance contracts.

Fitting Seasonal Trends

Regular periodic behavior is often found in business and economic data. Because such periodicity is often tied to the climate, such trends are called *seasonal components*. Seasonal trends can be modeled using the same techniques as for regular, or aperiodic, trends. The following example shows how to capture periodic behavior using seasonal binary variables.

Example: Trends in Voting. On any given Election Day, the number of voters who actually turn out to voting booths depends on a number of factors: the publicity that an election race has received; the issues that are debated as part of the race; other issues facing voters on Election Day; and nonpolitical factors, such as the weather. Potential political candidates base their projections of campaign financing and chances of winning an election on forecasts of the number of voters who will actually participate in an election. Decisions as to whether to participate as a candidate must be made well in advance; generally, so far in advance that well-known factors such as the weather on Election Day cannot be used to generate forecasts.

We consider here the number of Wisconsin voters who participated in statewide elections over the period 1920 through 1990. Although the interest is in forecasting the actual number of voters, we consider voters as a percentage of the qualified voting public. Dividing by the qualified voting public controls for the size of the population of voters; this enhances comparability between the early and latter parts of the series. Because mortality trends are relatively stable, reliable projections of the qualified voting public can be readily attained. Forecasts of the percentage can then be multiplied by projections of the voting public to obtain forecasts of the actual voter turnout.

To specify a model, we examine Figure 7.5, a time series plot of the voter turnout as a percentage of the qualified voting public. This figure displays the low voter turnout in the early part of the series, followed by higher turnout in the 1950s and 1960s, followed by a lower turnout in the 1980s. This pattern can be modeled using, for example, a quadratic trend in time. The figure also displays a much higher turnout in presidential election years. This periodic, or seasonal,

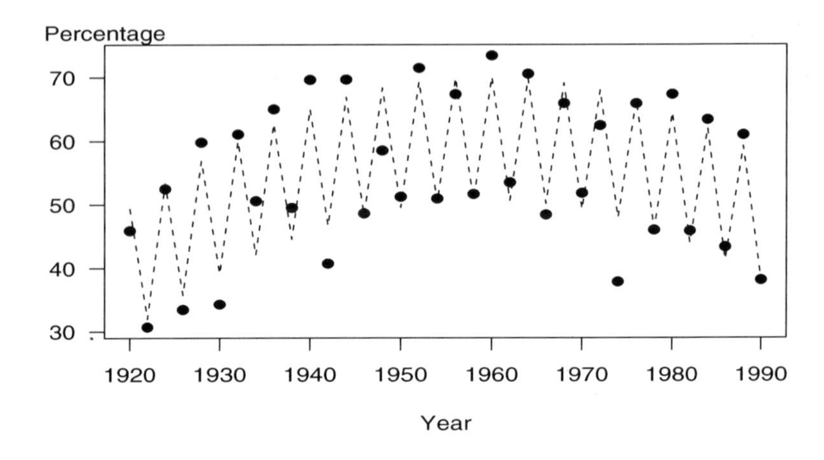

Figure 7.5 Wisconsin Voters as a Percentage of the Qualified Voting Public, by Year. The opaque circles represent the actual voting percentages. The dashed lines represent the fitted trend, using a quadratic trend in time plus a binary variable to indicate a presidential election year.

component can be modeled using a binary variable. A candidate model is

$$y_t = \beta_0 + \beta_1 t + \beta_2 t^2 + \beta_3 z_t + \varepsilon_t,$$

where

$$z_t = \begin{cases} 1 & \text{if presidential election year} \\ 0 & \text{otherwise} \end{cases}.$$

Here, $\beta_3 z_t$ captures the seasonal component in this model.

Regression was used to fit the model. The fitted model provided a good fit of the data – the coefficient of determination from the fit was $R^2 = 89.6\%$. Figure 7.5 shows a strong relationship between the fitted and actual values.

The voting trend example demonstrates the use of binary variables to capture seasonal components. Similarly, seasonal effects can be represented using categorical variables, such as

$$z_t = \begin{cases} 1 & \text{if spring} \\ 2 & \text{if summer} \\ 3 & \text{if fall} \\ 4 & \text{if winter} \end{cases}.$$

Another way to capture seasonal effects is through the use of trigonometric functions. Further discussion of the use of trigonometric functions to handle seasonal components is in Section 9.3.

Removal of seasonal patterns is known as seasonal adjustment.

Removal of seasonal patterns is known as *seasonal adjustment*. This strategy is appropriate in public policy situations where interest centers on interpreting the resulting seasonally adjusted series. For example, government agencies typically report industrial manufacturing revenues in terms of seasonally adjusted numbers, with the understanding that known holiday- and weather-related patterns are accounted for when reporting growth. However, for most actuarial and risk management applications, the interest is typically in forecasting the variation of the entire series, not just the seasonally adjusted portion.

Reliability of Time Series Forecasts

Time series forecasts are sometimes called *naive* forecasts. The adjective "naive" is somewhat ironic because many time series forecasting techniques are technical in nature and complex to compute. However, these forecasts are based on extrapolating a single series of observations. Thus, they are naive in the sense that the forecasts ignore other sources of information that may be available to the forecaster and users of the forecasts. Despite ignoring this possibly important information, time series forecasts are useful in that they provide an objective benchmark that other forecasts and expert opinions can be compared against.

Projections should provide a user with a sense of the reliability of the forecast.

Projections should provide a user with a sense of the reliability of the forecast. One way of quantifying this is to provide forecasts under "low-intermediate-high" sets of assumptions. For example, if we are forecasting the national debt, we might do so under three scenarios of the future performance of the economy. Alternatively, we can calculate prediction intervals using many of the models for forecasting that are discussed in this text. Prediction intervals provide a measure of reliability that can be interpreted in a familiar probabilistic sense. Further, by varying the desired level of confidence, the prediction intervals vary, thus allowing us to respond to "what-if" types of questions.

For example, in Figure 21.10, you will find a comparison of low-intermediate-high projections to prediction intervals for forecasts of the inflation rate (consumer price index, CPI) used in projecting Social Security funds. The low-intermediate-high projections are based on a range of expert opinions and thus reflect variability of the forecasters. The prediction intervals reflect innovation uncertainty in the model (assuming that the model is correct). Both ranges give the user a sense of reliability of the forecasts, though in different ways.

Prediction intervals have the additional advantage of quantifying the fact that forecasts become less reliable the farther that we forecast into the future. Even with cross-sectional data, we saw that the farther we were from the main part of the data, the less confident we were in our predictions. This is also true in forecasting for longitudinal data. It is important to communicate this to consumers of forecasts, and prediction intervals are a convenient way of doing so.

In forecasting, the primary concern is for the most recent part of the series.

In summary, regression analysis using various functions of time as explanatory variables is a simple yet powerful tool for forecasting longitudinal data. It does, however, have drawbacks. Because we are fitting a curve to the entire dataset, there is no guarantee that the fit for the most recent part of the data will be adequate. That is, in forecasting, the primary concern is for the most recent part of the series. We know that regression analysis estimates give the most weight to observations with unusually large explanatory variables. To illustrate, using a linear trend in time model, this means giving the most weight to observations at the end and at the beginning of the series. Using a model that gives large weight to observations at the beginning of the series is viewed with suspicion by forecasters. This drawback of regression analysis motivates us to introduce additional forecasting tools. (Section 9.1 develops this point further.)

7.3 Stationarity and Random Walk Models

A basic concern with processes that evolve over time is the *stability* of the process. For example: "Is it taking me longer to get to work since they put in the new stop light?" "Have quarterly earnings improved since the new CEO took over?" We measure processes to improve or manage their performance and to forecast their future. Because stability is a fundamental concern, we will work with a special kind of stability called *stationarity.*

Definition. Stationarity is the formal mathematical concept corresponding to the stability of a time series of data. A series is said to be (weakly) stationary if

- The mean $\mathrm{E}\ y_t$ does not depend on t
- The covariance between y_s and y_t depends only on the difference between time units, $|t-s|$.

Thus, for example, under weak stationarity $\mathrm{E}\ y_4 = \mathrm{E}\ y_8$ because the means do not depend on time and thus are equal. Further, $\mathrm{Cov}(y_4, y_6) = \mathrm{Cov}(y_6, y_8)$, because y_4 and y_6 are two time units apart, as are y_6 and y_8. As another implication of the second condition, note that $\sigma^2 = \mathrm{Cov}(y_t, y_t) = \mathrm{Cov}(y_s, y_s) = \sigma^2$. Thus, a weakly stationary series has a constant mean as well as a constant variance (homoscedastic). Another type of stationarity known as *strict*, or *strong, stationarity* requires that the entire distribution of y_t be constant over time, not just the mean and the variance.

A weakly stationary series has a constant mean and a constant variance (homoscedastic).

White Noise

The link between longitudinal and cross-sectional models can be established through the notion of a *white noise* process. A white noise process is a stationary process that displays no apparent patterns through time. More formally, a white noise process is simply a series that is i.i.d., identically and independently distributed. A white noise process is only one type of stationary process – Chapter 8 will introduce another type, an autoregressive model.

A white noise process is a stationary process that displays no apparent patterns through time – it is i.i.d.

A special feature of the white noise process is that forecasts do not depend on how far into the future we want to forecast. Suppose that a series of observations, $y_1, \ldots, y_T$, has been identified as a white noise process. Let $\overline{y}$ and s_y denote the sample average and standard deviation, respectively. A forecast of an observation in the future, say y_{T+l}, for l lead-time units in the future, is $\overline{y}$. Further, a forecast interval is

$$\overline{y} \pm t_{T-1,1-\alpha/2}\, s_y \sqrt{1+\frac{1}{T}}. \qquad (7.5)$$

t	1	2	3	4	5
c_t^*	10	9	7	5	7
c_t	3	2	0	−2	0
y_t	103	105	105	103	103

Table 7.1 Winnings for 5 of the 50 Rolls

In time series applications, because the sample size T is typically relatively large, we use the approximate 95% prediction interval $\overline{y} \pm 2s_y$. This approximate forecast interval ignores the parameter uncertainty in using $\overline{y}$ and s_y to estimate the mean E y and standard deviation σ of the series. Instead, it emphasizes the uncertainty in future realizations of the series (known as *innovation uncertainty*). Note that this interval does *not* depend on the choice of l, the number of lead units that we forecast into the future.

The white noise model is both the least important and the most important of time series models.

The white noise model is both the least important and the most important of time series models. It is the least important in the sense that the model assumes that the observations are unrelated to one another, an unlikely event for most series of interest. It is the most important because our modeling efforts are directed toward reducing a series to a white noise process. In time series analysis, the procedure for reducing a series to a white noise process is called a *filter*. After all patterns have been filtered from the data, the uncertainty is said to be *irreducible*.

Random Walk

We now introduce the *random walk model*. For this time series model, we will show how to filter the data simply by taking differences.

To illustrate, suppose that you play a simple game based on the roll of two dice. To play the game, you must pay $7 each time you roll the dice. You receive the number of dollars corresponding to the sum of the two dice, c_t^*. Let c_t denote your winnings on each roll, so that $c_t = c_t^* - 7$. Assuming that the rolls are independent and come from the same distribution, the series $\{c_t\}$ is a white noise process.

Assume that you start with initial capital of $y_0 = \$100$. Let y_t denote the sum of capital after the tth roll. Note that y_t is determined recursively by $y_t = y_{t-1} + c_t$. For example, because you won $3 on the first roll, $t = 1$, you now have capital $y_1 = y_0 + c_1$, or $103 = 100 + 3$. Table 7.1 shows the results for the first five throws. Figure 7.6 is a time series plot of the sums, y_t, for 50 throws.

The partial sums of a white noise process define a random walk model. For example, the series $\{y_1, \ldots, y_{50}\}$ in Figure 7.6 is a realization of the random walk model. The phrase *partial sum* is used because each observation, y_t, was created by summing the winnings up to time t. For this example, winnings, c_t, are a white noise process because the amount returned, c_t^*, is i.i.d. In our example, your winnings from each roll of the dice are represented using a white noise process. Whether you win on one roll of the dice has no influence on the outcome of the next, or previous, roll of the dice. In contrast, your amount of capital at

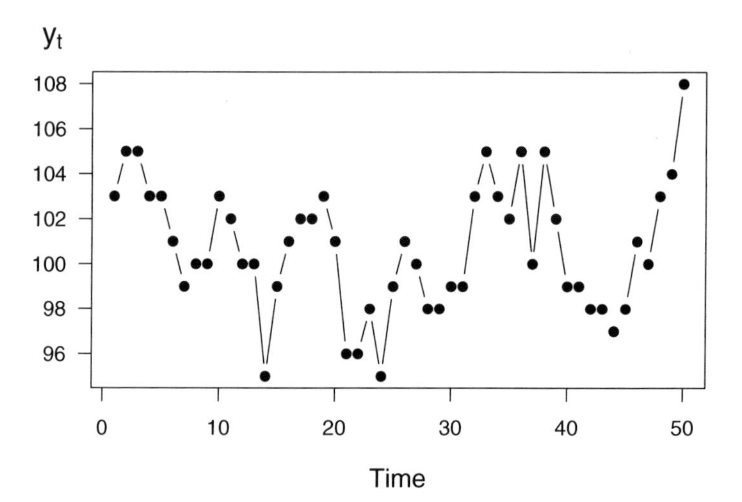

Figure 7.6 Time series plot of the sum of capital.

any roll of the dice is highly related to the amount of capital after the next roll or before the previous roll. Your amount of capital after each roll of the dice is represented by a random walk model.

7.4 Inference Using Random Walk Models

The random walk is a commonly used time series model. To see how it can be applied, we first discuss a few model properties. These properties are then used to forecast and identify a series as a random walk. Finally, this section compares the random walk to a competitor, the linear trend in time model.

Model Properties

To state the properties of the random walk, we first recap some definitions. Let $c_1, \ldots, c_T$ be T observations from a white noise process. A random walk can be expressed recursively as

$$y_t = y_{t-1} + c_t. \tag{7.6}$$

By repeated substitution, we have

$$y_t = c_t + y_{t-1} = c_t + (c_{t-1} + y_{t-2}) = \cdots$$

If we use y_0 as the initial level, then we can express the random walk as

$$y_t = y_0 + c_1 + \cdots + c_t. \tag{7.7}$$

A random walk is the partial sum of a white noise process.

Equation (7.7) shows that a random walk is the partial sum of a white noise process.

The random walk is *not* a stationary process because the variability, and possibly the mean, depends on the time point at which the series is observed.

Taking the expectation and variance of equation (7.7) yields the mean level and variability of the random walk process:

$$\mathrm{E}\ y_t = y_0 + t\mu_c \quad \text{and} \quad \mathrm{Var}\ y_t = t\sigma_c^2,$$

where $\mathrm{E}\ c_t = \mu_c$ and $\mathrm{Var}\ c_t = \sigma_c^2$. Hence, as long as there is some variability in the white noise process ($\sigma_c^2 > 0$), the random walk is nonstationary in the variance. Further, if $\mu_c \neq 0$, then the random walk is nonstationary in the mean.

A random walk is a nonstationary model.

Forecasting

How can we forecast a series of observations, $y_1, \ldots, y_T$, that has been identified as a realization of a random walk model? The technique we use is to forecast the *differences*, or *changes*, in the series and then sum the forecast differences to get the forecast series. This technique is tractable because, by the definition of a random walk model, the differences can be represented using a white noise process, a process that we know how to forecast.

Consider y_{T+l}, the value of the series l lead time units into the future. Let $c_t = y_t - y_{t-1}$ represent the differences in the series, so that

$$\begin{aligned} y_{T+l} &= y_{T+l-1} + c_{T+l} = (y_{T+l-2} + c_{T+l-1}) + c_{T+l} = \cdots \\ &= y_T + c_{T+1} + \cdots + c_{T+l}. \end{aligned}$$

We interpret y_{T+l} to be the current value of the series, y_T, plus the partial sum of future differences.

To forecast y_{T+l}, because at time T we know y_T, we need only forecast the changes $\{c_{T+1}, \ldots, c_{T+l}\}$. Because a forecast of a future value of a white noise process is just the average of the process, the forecast of c_{T+k} is $\bar{c}$ for $k = 1, 2, \ldots, l$. Putting these together, the forecast of y_{T+l} is $y_T + l\bar{c}$. For example, for $l = 1$, we interpret the forecast of the next value of the series to be the current value of the series plus the average change of the series.

Using similar ideas, we have that an approximate 95% prediction interval for y_{T+l} is

$$y_T + l\bar{c} \pm 2 s_c \sqrt{l},$$

where s_c is the standard deviation computed using the changes $c_2, c_3, \ldots, c_T$. Note that the width of the prediction interval, $4 s_c \sqrt{l}$, grows as the lead time l grows. This increasing width simply reflects our diminishing ability to predict into the future.

For example, we rolled the dice $T = 50$ times and we want to forecast y_{60}, our sum of capital after 60 rolls. At time 50, it turned out that our sum of money available was $y_{50} = \$93$. Starting with $y_0 = \$100$, the average change was $\bar{c} = -7/50 = -\$0.14$, with standard deviation $s_c = \$2.703$. Thus, the forecast at time 60 is $93 + 10(-.14) = 91.6$. The corresponding 95% prediction interval is

$$91.6 \pm 2\,(2.703)\,\sqrt{10} = 91.6 \pm 17.1 = (74.5,\ 108.7)\,.$$

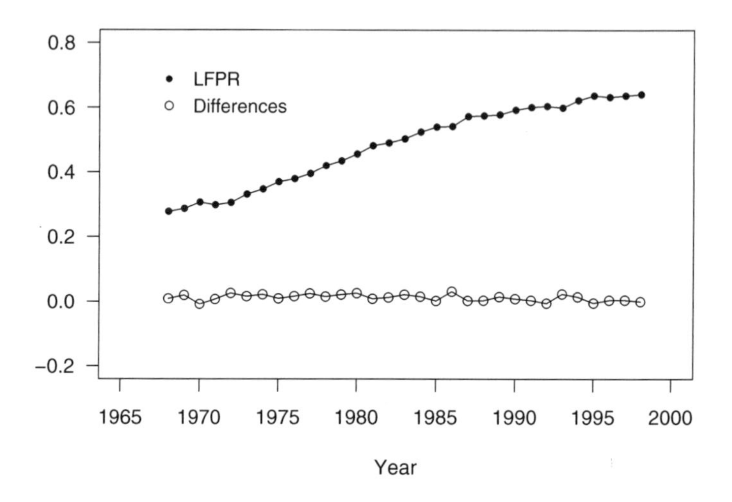

Figure 7.7 Labor force participation rates for females aged 20–44, living in a household with a spouse present and at least one child under six years of age. The plot of the series shows a rapid increase over time. Also shown are the differences that are level.

Ⓡ EMPIRICAL *Filename is "LaborForcePR"*

Example: Labor Force Participation Rates. Labor force participation rate (LFPR) forecasts, coupled with forecasts of the population, provide us with a picture of a nation's future workforce. This picture provides insights to the future workings of the overall economy, and thus LFPR projections are of interest to a number of government agencies. In the United States, LFPRs are projected by the Social Security Administration, the Bureau of Labor Statistics, the Congressional Budget Office, and the Office of Management and Budget. In the context of Social Security, policy makers use labor force projections to evaluate proposals for reforming the Social Security system and to assess its future financial solvency.

A labor force participation rate is the civilian labor force divided by the civilian noninstitutional population. These data are compiled by the Bureau of Labor Statistics. For illustration purposes, let us look at a specific demographic cell and show how to forecast it – forecasts of other cells can be found in Fullerton (1999) and Frees (2006). Specifically, we examine 1968–98 for females, aged 20–44, living in a household with a spouse present and at least one child younger than six years of age. Figure 7.7 shows the rapid increase in LFPR for this group over $T = 31$ years.

To forecast the LFPR with a random walk, we begin with our most recent observation, $LFPR_{31} = 0.6407$. We denote the change in LFPR by c_t, so that $c_t = LFPR_t - LFPR_{t-1}$. It turns out that the average change is $\overline{c} = 0.0121$ with standard deviation $s_c = 0.0101$. Thus, using a random walk model, an approximate 95% prediction interval for the l-step forecast is

$$0.6407 + 0.0121l \pm 0.0202\sqrt{l}.$$

Figure 7.8 illustrates prediction intervals for 1999 through 2002, inclusive.

Identifying Stationarity

We have seen how to do useful things, like forecasting, with random walk models. But how do we identify a series as a realization from a random walk? We know

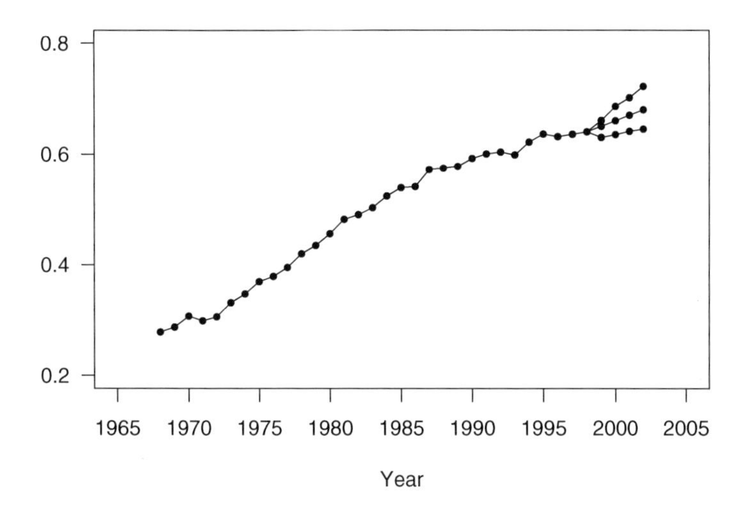

Figure 7.8 Time series plot of labor force participation rates with forecast values for 1999–2002. The middle series represent the point forecasts. The upper and lower series represent the upper and lower 95% forecast intervals. Data for 1968–98 represent actual values.

that the random walk is a special kind of nonstationary model, and so the first step is to examine a series and decide whether it is stationary.

Stationarity quantifies the stability of a process. A process that is strictly stationary has the same distribution over time, so we should be able to take successive samples of modest size and show that they have approximately the same distribution. For weak stationary, the mean and variance are stable over time, so if one takes successive samples of modest size, then we expect the mean level and the variance to be roughly similar. To illustrate, when examining time series plots, if you look at the first five, the next five, the following five, and so forth, successive samples, you should observe approximately the same levels of averages and standard deviations.

A control chart is a time series plot with superimposed reference lines called control limits. *It is used is to detect nonstationarity in a time series.*

In quality management applications, this approach is quantified by looking at *control charts*. A control chart is a useful graphical device for detecting the lack of stationarity in a time series. The basic idea is to superimpose reference lines called *control limits* on a time series plot of the data. These reference lines help us visually detect trends in the data and identify unusual points. The mechanics behind control limits are straightforward. For a given series of observations, calculate the series mean and standard deviation, $\overline{y}$ and s_y. Define the upper control limit by $UCL = \overline{y} + 3s_y$ and the lower control limit by $LCL = \overline{y} - 3s_y$. Time series plots with these superimposed control limits are known as *control charts*.

Sometimes the adjective *retrospective* is associated with this type of control chart. This adjective reminds the user that averages and standard deviations are based on all the available data. In contrast, when the control chart is used as an ongoing management tool for detecting whether an industrial process is "out of control," a *prospective control chart* may be more suitable. Here, prospective merely means using only an early portion of the process, that is, "in control," to compute the control limits.

A control chart that helps us examine the stability of the mean is the *Xbar* chart. An *Xbar* chart is created by combining successive observations of modest size, taking an average over this group, and then creating a control chart for the group averages. By taking averages over groups, the variability associated with each point on the chart is smaller than for a control chart for individual observations. This allows the data analyst to get a clearer picture of any patterns that may be evident in the mean of the series.

A control chart that helps us examine the stability of the variability is the R chart. As with the $Xbar$ chart, we begin by forming successive groups of modest size. With the R chart, for each group we compute the range, which is the largest minus the smallest observation, and then we create a control chart for the group ranges. The range is a measure of variability that is simple to compute, an important advantage in manufacturing applications.

Identifying Random Walks

Suppose that you suspect that a series is nonstationary. How do you identify the fact that these are realizations of a random walk model? Recall that the expected value of a random walk, $\mathrm{E}\, y_t = y_0 + t\mu_c$, suggests that such a series follows a linear trend in time. The variance of a random walk, $\mathrm{Var}\, y_t = t\sigma_c^2$, suggests that the variability of a series gets larger as time t gets large. First, a control chart can help us to detect these patterns, whether they are of a linear trend in time, increasing variability, or both.

If the original data follows a random walk model, then the differenced series follows a white noise process model.

Second, if the original data follows a random walk model, then the differenced series follows a white noise process model. If a random walk model is a candidate model, you should examine the differences of the series. In this case, the time series plot of the differences should be a stationary, white noise process that displays no apparent patterns. Control charts can help us to detect this lack of patterns.

Third, compare the standard deviations of the original series and the differenced series. We expect the standard deviation of the original series to be greater than the standard deviation of the differenced series. Thus, if the series can be represented by a random walk, we expect a substantial reduction in the standard deviation when taking differences.

Example: Labor Force Participation Rates, Continued. In Figure 7.7, the series displays a clear upward trend, whereas the differences show no apparent trends over time. Further, when computing differences of each series, it turned out that

$$0.1197 = SD(series) > SD(differences) = 0.0101.$$

Thus, it seems reasonable to tentatively use a random walk as a model of the labor force participation rate series.

In Chapter 8, we will discuss two additional identification devices. These are scatter plots of the series versus a lagged version of the series and the corresponding summary statistics called *autocorrelations*.

Random Walk versus Linear Trend in Time Models

The labor force participation rate example can be represented using either a random walk or a linear trend in time model. These two models are more closely related to each other than is evident at first glance. To see this relationship, recall that the linear trend in time model can be written as

$$y_t = \beta_0 + \beta_1 t + \varepsilon_t, \tag{7.8}$$

where $\{\varepsilon_t\}$ is a white noise process. If $\{y_t\}$ is a random walk, then it can be modeled as a partial sum as in equation (7.7). We can also decompose the white noise process into a mean μ_c plus another white noise process; that is, $c_t = \mu_c + \varepsilon_t$. Combining these two ideas, a random walk model can be written as

$$y_t = y_0 + \mu_c t + u_t, \tag{7.9}$$

where $u_t = \sum_{j=1}^{t} \varepsilon_j$. Comparing equations (7.8) and (7.9), we see that the two models are similar in that the deterministic portion is a linear function of time. The difference is in the error component. The error component for the linear trend in time model is a stationary, white noise process. The error component for the random walk model is nonstationary because it is the partial sum of white noise processes. That is, the error component is also a random walk. Many introductory treatments of the random walk model focus on the "fair game" example and ignore the drift term μ_c. This is unfortunate because the comparison between the random walk model and the linear trend in time model is not as clear when the parameter μ_c is equal to zero.

7.5 Filtering to Achieve Stationarity

A *filter* is a procedure for reducing observations to white noise. In regression, we accomplished this by simply subtracting the regression function from the observations, that is, $y_i - (\beta_0 + \beta_1 x_{1i} + \cdots + \beta_k x_{ki}) = \varepsilon_i$. Transformation of the data is another device for filtering that we introduced in Chapter 1 when analyzing cross-sectional data. We encountered another example of a filter in Section 7.3. There, by taking differences of observations, we reduced a random walk series to a white noise process.

A filter is a procedure for reducing observations to white noise.

An important theme of this text is to use an iterative approach for fitting models to data. In particular, in this chapter, we discuss techniques for reducing a sequence of observations to a stationary series. By definition, a stationary series is stable and hence is far easier to forecast than an unstable series. This stage, sometimes known as *preprocessing* the data, generally accounts for the most important sources of trends in the data. The next chapter will present models that account for subtler trends in the data.

Transformations

When analyzing longitudinal data, transformation is an important tool used to filter a dataset. Specifically, using a logarithmic transformation tends to shrink "spread out" data. This feature gives us an alternative method to deal with a process where the variability appears to grow with time. Recall that the first option discussed is to posit a random walk model and examine differences of the data. Alternatively, one may take a logarithmic transform that helps to reduce increasing variance through time.

Further, from the random walk discussion, we know that if both the series variance and log series variance increase over time, the differences of the log transform might handle this increasing variability. Differences of natural logarithms are particularly pleasing because they can be interpreted as *proportional changes*. To see this, define $pchange_t = (y_t / y_{t-1}) - 1$. Then,

$$\ln y_t - \ln y_{t-1} = \ln \left(\frac{y_t}{y_{t-1}} \right) = \ln(1 + pchange_t) \approx pchange_t .$$

Here, we use the Taylor series approximation $\ln(1 + x) \approx x$ that is appropriate for small values of $|x|$.

Example: Standard and Poor's Composite Quarterly Index. An important task of a financial analyst is to quantify costs associated with future cash flows. We consider here funds invested in a standard measure of overall market performance, the Standard and Poor's (S&P) 500 Composite Index. The goal is to forecast the performance of the portfolio for discounting of cash flows.

In particular, we examine the S&P Composite Quarterly Index for the years 1936 to 2007, inclusive. By today's standards, this period may not be the most representative because the Depression of the 1930s is included. The motivation to analyze these data is from the Institute of Actuaries "Report of the Maturity Guarantees Working Party" (1980), which analyzed the series from 1936 to 1977, inclusive. This paper studied the long-term behavior of investment returns from an actuarial viewpoint. We complement that work by showing how graphical techniques can suggest a useful transformation for reducing the data to a stationary process.

The data are shown in Figure 7.9. From the original index values in the upper-left-hand panel, we see that the mean level and variability increase with time. This pattern clearly indicates that the series is nonstationary.

From our discussions in Sections 7.2 and 7.3, a candidate model that has these properties is the random walk. However, the time series plot of the differences, in upper-right-hand panel of Figure 7.9, still indicates a pattern of variability increasing with time. The differences are not a white noise process so the random walk is not a suitable model for the S&P 500 Index.

An alternative transformation is to consider logarithmic values of the series. The time series plot of logged values, presented in lower-left-hand panel of Figure 7.9, indicates the the mean level of the series increases over time and is not level. Thus, the logarithmic index is not stationary.

Yet another approach is to examine differences of the logarithmic series. This is especially desirable when looking at indices, or "breadbaskets," because the

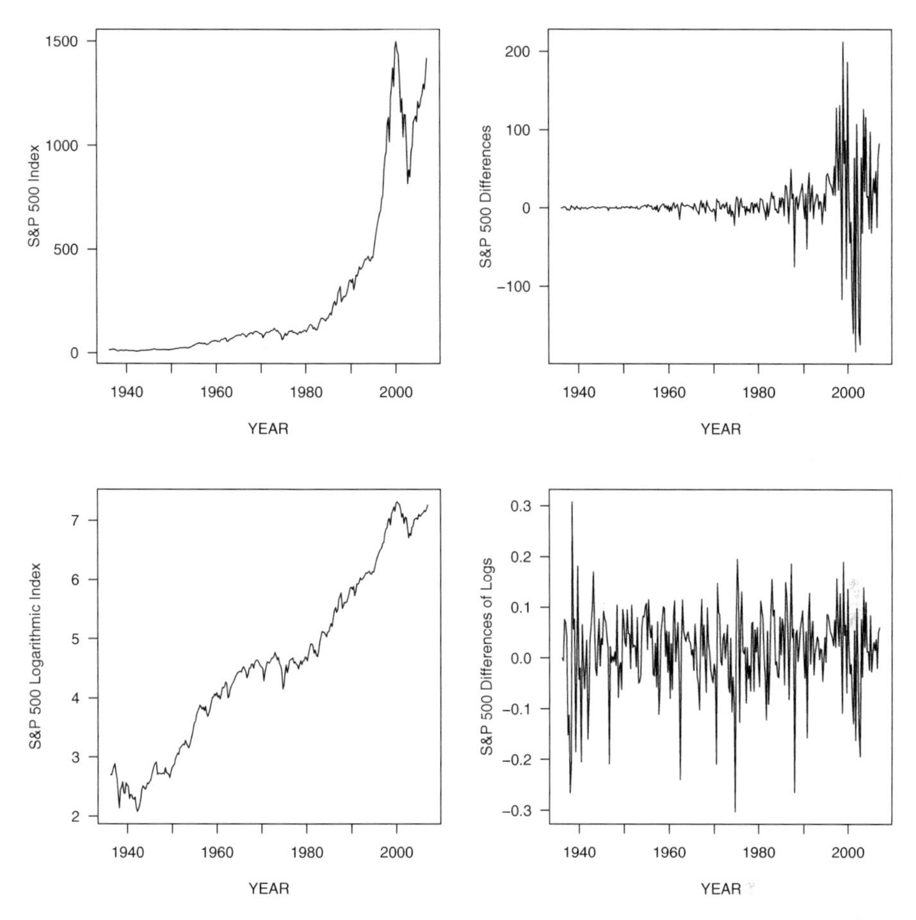

Figure 7.9 Time series plots of the S&P 500 Index. The upper-left-hand panel shows the original series that is nonstationary in the mean and in the variability. The upper-right-hand panel shows the differences in the series that is nonstationary in the variability. The lower-left-hand panel shows the logarithmic index that is nonstationary in the mean. The lower-right-hand panel shows the differences of the logarithmic index that appears to be stationary in the mean and in the variability.

difference of logarithms can be interpreted as proportional changes. From the final time series plot, in the lower-right-hand panel of Figure 7.9, we see that there are fewer discernible patterns in the transformed series, the difference of logs. This transformed series seems to be stationary. It is interesting to note that there seems to be a higher level of volatility at the beginning of the series. This type of changing volatility is more difficult to model and has recently been the subject of considerable attention in the financial economics literature (see, e.g., Hardy, 2003).

7.6 Forecast Evaluation

Judging the accuracy of forecasts is important when modeling time series data. In this section, we present forecast evaluation techniques that

- Help detect recent unanticipated trends or patterns in the data.
- Are useful for comparing different forecasting methods.
- Provide an intuitive and easy-to-explain method for evaluating the accuracy of forecasts.

In the first five sections of Chapter 7, we presented several techniques for detecting patterns in residuals from a fitted model. Measures that summarize the distribution of residuals are called *goodness-of-fit statistics*. As we saw in our study of cross-sectional models, by fitting several different models to a dataset, we introduce the possibility of overfitting the data. To address this concern, we will use *out-of-sample validation* techniques, similar to those introduced in Section 6.5.

To perform an out-of-sample validation of a proposed model, ideally one would develop the model on a dataset and then corroborate the model's usefulness on a second, independent dataset. Because two such ideal datasets are rarely available, in practice we can split a dataset into two subsamples, a *model development subsample* and a *validation subsample*. For longitudinal data, the practice is to use the beginning part of the series, the first T_1 observations, to develop one or more candidate models. The latter part of the series, the last $T_2 = T - T_1$ observations, is used to evaluate the forecasts. For example, we might have 10 years of monthly data so that $T = 120$. It would be reasonable to use the first eight years of data to develop a model and the last two years of data for validation, yielding $T_1 = 96$ and $T_2 = 24$.

Thus, observations $y_1, \ldots, y_{T_1}$ are used to develop a model. From these T_1 observations, we can determine the parameters of the candidate model. Using the fitted model, we can determine fitted values for the model validation subsample for $t = T_1 + 1, T_1 + 2, \ldots, T_1 + T_2$. Taking the difference between the actual and fitted values yield one-step forecast residuals, denoted by $e_t = y_t - \widehat{y}_t$. These forecast residuals are the basic quantities that we will use to evaluate and compare forecasting techniques.

To compare models, we use a four-step process similar to that described in Section 6.5, described as follows.

Out-of-Sample Validation Process

1. Divide the sample of size T into two subsamples, a model development subsample ($t = 1, \ldots, T_1$) and a model validation subsample ($t = T_1 + 1, \ldots, T_1 + T_2$).
2. Using the model development subsample, fit a candidate model to the dataset $t = 1, \ldots, T_1$.
3. Using the model created in Step 2 and the dependent variables up to and including $t - 1$, forecast the dependent variable $\widehat{y}_t$, where $t = T_1 + 1, \ldots, T_1 + T_2$.
4. Use actual observations and the fitted values computed in Step 3 to compute one-step forecast residuals, $e_t = y_t - \widehat{y}_t$, for the model validation subsample. Summarize these residuals with one or more comparison statistics, described subsequently.

Repeat Steps 2 through 4 for each of the candidate models. Choose the model with the smallest set of comparison statistics.

Out-of-sample validation can be used to compare the accuracy of forecasts from virtually any forecasting model. As we saw in Section 6.5, we are not limited to comparisons where one model is a subset of another, where the competing models use the same units for the response, and so on.

There are several statistics that are commonly used to compare forecasts.

Commonly Used Statistics for Comparing Forecasts

1. The *mean error statistic*, defined by
$$ME = \frac{1}{T_2} \sum_{t=T_1+1}^{T_1+T_2} e_t.$$
This statistic measures recent trends that are not anticipated by the model.
2. The *mean percentage error*, defined by
$$MPE = \frac{100}{T_2} \sum_{t=T_1+1}^{T_1+T_2} \frac{e_t}{y_t}.$$
This statistic is also a measure of trend, but examines error relative to the actual value.
3. The *mean square error*, defined by
$$MSE = \frac{1}{T_2} \sum_{t=T_1+1}^{T_1+T_2} e_t^2.$$
This statistic can detect more patterns than *ME*. It is the same as the cross-sectional *SSPE* statistic, except for the division by T_2.
4. The *mean absolute error*, defined by
$$MAE = \frac{1}{T_2} \sum_{t=T_1+1}^{T_1+T_2} |e_t|.$$
Like *MSE*, this statistic can detect more than trend patterns than *ME*. The units of *MAE* are the same as the dependent variable.
5. The *mean absolute percentage error*, defined by
$$MAPE = \frac{100}{T_2} \sum_{t=T_1+1}^{T_1+T_2} \left|\frac{e_t}{y_t}\right|.$$
Like *MAE*, this statistic can detect more than trend patterns. Like *MPE*, it examines error relative to the actual value.

Example: Labor Force Participation Rates, Continued. We can use out-of-sample validation measures to compare two models for the LFPRs: the linear trend in time model and the random walk model. For this illustration, we examined the

Table 7.2
Out-of-Sample Forecast Comparison

	ME	*MPE*	*MSE*	*MAE*	*MAPE*
Linear trend in time model	−0.0488	−0.0766	0.0026	0.0488	0.0766
Random walk model	−0.0007	0.0012	0.0001	0.0115	0.0180

labor rates for years 1968 through 1994, inclusive. This corresponds to $T_1 = 27$ observations defined in step 1. Data were subsequently gathered on rates for 1995 through 1998, inclusive, corresponding to $T_2 = 4$ for out-of-sample validation. For step 2, we fit each model using $t = 1, \ldots, 27$, earlier in this chapter. For step 3, the one-step forecasts are

$$\widehat{y}_t = 0.2574 + 0.0145t$$

and

$$\widehat{y}_t = y_{t-1} + 0.0132$$

for the linear trend in time and the random walk models, respectively. For step 4, Table 7.2 summarizes the forecast comparison statistics. Given these statistics, the choice of the model is clearly the random walk.

7.7 Further Reading and References

For many years, actuaries in North America were introduced to time series analysis by Miller and Wichern (1977), Abraham and Ledolter (1983), and Pindyck and Rubinfeld (1991). A more recent book-length introduction is that of Diebold (2004), which contains a brief introduction to regime-switching models.

Because of the difficulties regarding their specification and limited forecasting use, we do not explore causal models further in this text. For more details on causal models, the interested reader is referred to Pindyck and Rubinfeld (1991).

Chapter References

"Report of the Maturity Guarantees Working Party" (1980). *Journal of the Institute of Actuaries* 107, 103–213.

Abraham, Bovas, and Johannes Ledolter (1983). *Statistical Methods for Forecasting*. John Wiley & Sons, New York.

Diebold, Francis X. (2004). *Elements of Forecasting*, 3rd ed. Thompson South-Western, Mason, Ohio.

Frees, Edward W. (2006). Forecasting of labor force participation rates. *Journal of Official Statistics* 22, no. 3, 453–85.

Fullerton, Howard N., Jr. (1999). Labor force projections to 2008: Steady growth and changing composition. *Monthly Labor Review* (November), 19–32.

Granger, Clive W. J., and P. Newbold (1974). Spurious regressions in econometrics. *Journal of Econometrics* 2, 111–20.

Hardy, Mary (2001). A regime-switching model of long-term stock returns. *North American Actuarial Journal* 5, no. 2, 41–53.

Hardy, Mary (2003). *Investment Guarantees: Modeling and Risk Management for Equity-Linked Life Insurance*. John Wiley & Sons, New York.
Miller, Robert B., and Dean W. Wichern (1977). *Intermediate Business Statistics: Analysis of Variance, Regression and Time Series*. Holt, Rinehart, and Winston, New York.
Pindyck, R.S., and D.L. Rubinfeld (1991). *Econometric Models and Economic Forecasts*, 3rd ed. McGraw-Hill, New York.

7.8 Exercises

7.1. Consider a random walk $\{y_t\}$ as the partial sum of a white noise process $\{c_t\}$ with mean E $c_t = \mu_c$ and variance Var $c_t = \sigma_c^2$. Use equation (7.7) to show that:
 a. E $y_t = y_0 + t\mu_c$, where y_0 is the initial value, and
 b. Var $y_t = t\sigma_c^2$.

7.2. Consider a random walk $\{y_t\}$ as the partial sum of a white noise process $\{c_t\}$.
 a. Show that the l-step forecast error is $y_{T+l} - \widehat{y_{T+l}} = \sum_{j=1}^{l}(c_{T+j} - \bar{c})$.
 b. Show that the approximate variance of the l-step forecast error is $l\sigma_c^2$.

7.3. **Euro Exchange Rates**. The exchange rate that we consider is the amount of euros that one can purchase for one U.S. dollar. We have $T = 699$ daily observations from the period April 1, 2005, through January 8, 2008. The data were obtained from the Federal Reserve (H10 report). The data are based on noon buying rates in New York from a sample of market participants and they represent rates set for cable transfers payable in the listed currencies. These are also the exchange rates required by the Securities and Exchange Commission for the integrated disclosure system for foreign private issuers.

Ⓡ **Empirical**
Filename is "EuroExchange"

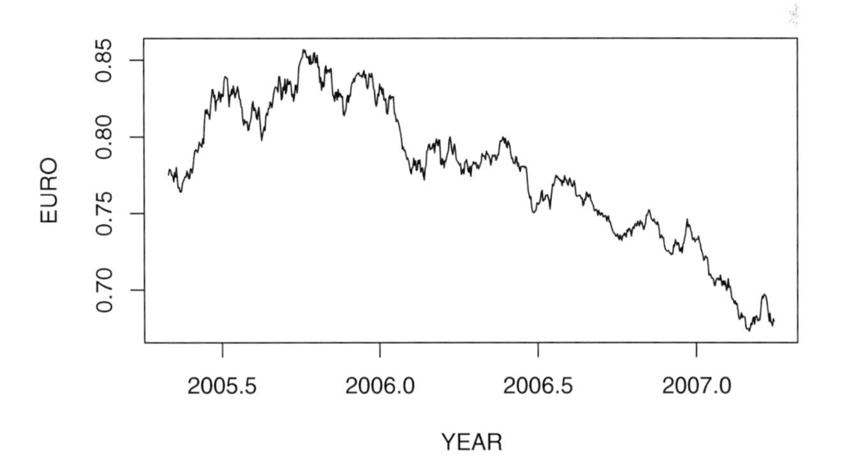

Figure 7.10 Time series plot of the Euro exchange rate.

 a. Figure 7.10 is a time series plot of the Euro exchange rate.
  a(i). Define the concept of a stationary time series.
  a(ii). Is the EURO series stationary? Use your definition in part a(i) to justify your response.

b. After an inspection of Figure 7.10 in part (a), you decide to fit a quadratic trend model of the data. Figure 7.11 superimposes the fitted value on a plot of the series.
   b(i). Cite several basic regression statistics that summarize the quality of the fit.
   b(ii). Briefly describe any residual patterns that you observe in Figure 7.11.
   b(iii). Here, TIME varies from $1, 2, \ldots, 699$. Using this model, calculate the three-step forecast corresponding to TIME = 702.

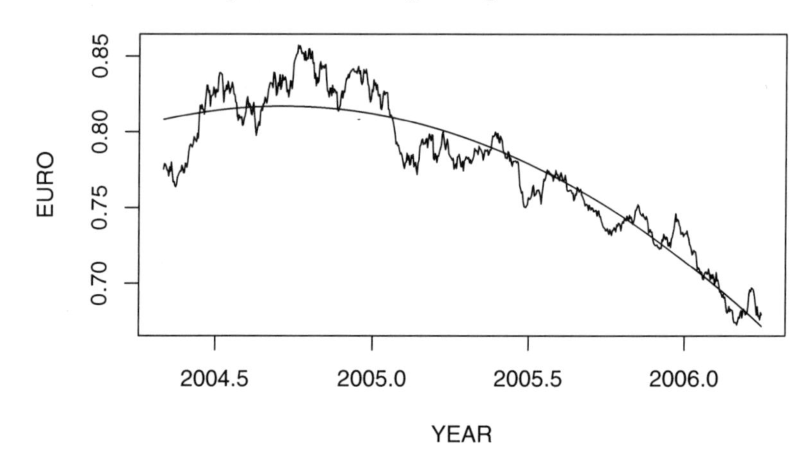

Figure 7.11 Quadratic fitted curve superimposed on the euro exchange rate.

c. To investigate a different approach, DIFFEURO, calculate the difference of EURO. You decide to model DIFFEURO as a white noise process.
   c(i). What is the name for the corresponding model of EURO?
   c(ii). The most recent value of EURO is $EURO_{699} = 0.6795$. Using the model identified in part c(i), provide a three-step forecast corresponding to TIME = 702.
   c(iii). Using the model identified in part c(i) and the point forecast in part c(ii), provide the corresponding 95% prediction interval for $EURO_{702}$.

8

Autocorrelations and Autoregressive Models

Chapter Preview. This chapter continues our study of time series data. Chapter 7 introduced techniques for determining major patterns that provide a good first step for forecasting. Chapter 8 provides techniques for detecting subtle trends in time and models to accommodate these trends. These techniques detect and model relationships between the current and past values of a series using regression concepts.

8.1 Autocorrelations

Application: Inflation Bond Returns

Ⓡ **Empirical** *Filename is "InflationBond"*

To motivate the introduction of methods in this chapter, we work in the context of the inflation bond return series. Beginning in January 2003, the U.S. Treasury Department established an inflation bond index that summarizes the returns on long-term bonds offered by the Treasury Department that are inflation indexed. For a Treasury inflation-protected security (TIPS), the principal of the bond is indexed by the (three-month-lagged) value of the (non-seasonally-adjusted) consumer price index. The bond then pays a semiannual coupon at a rate determined at auction when the bond is issued. The index that we examine is the unweighted average of bid yields for all TIPS with remaining terms to maturity of 10 or more years.

Monthly values of the index from January 2003 through March 2007 are considered, for a total of $T = 51$ returns. A time series plot of the data is presented in Figure 8.1. This plot suggests that the series is stationary, and so it is useful to examine the distribution of the series through summary statistics that appear in Table 8.1.

Our goal is detect patterns in the data and provide models to represent these patterns. Although Figure 8.1 shows a stationary series with no major tendencies, a few subtle patterns are evident. Beginning in mid-2003 and in the beginning of 2004, we see large increases followed by a series of declines in the index. Beginning in 2005, a pattern of increase with some cyclical behavior seems to be occurring. Although it is not clear what economic phenomenon these patterns represent, they are not what we would expect to see with a white noise process. For a white noise process, a series may increase or decrease randomly from one period to the next, producing a nonsmooth, "jagged" series over time.

To help understand these patterns, Figure 8.2 presents a scatter plot of the series (y_t) versus its lagged value (y_{t-1}). Because this is a crucial step for understanding

Table 8.1 Summary Statistics of the Inflation Bond Index

Variable	Mean	Median	Standard Deviation	Minimum	Maximum
INDEX	2.245	2.26	0.259	1.77	2.80

Source: U.S. Treasury.

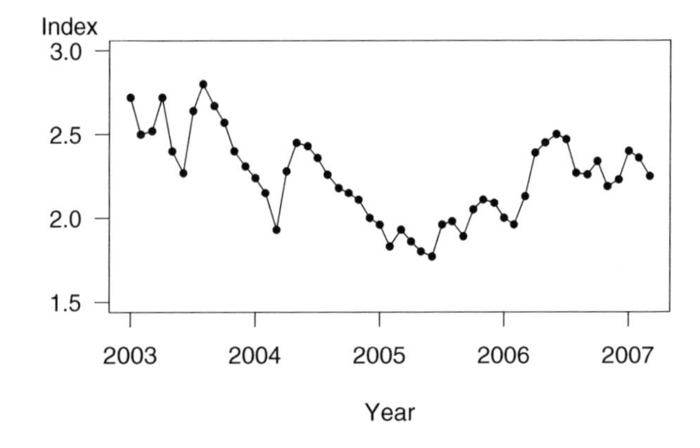

Figure 8.1 Time series plot of the inflation bond index. Monthly values over January 2003 to March 2007, inclusive.

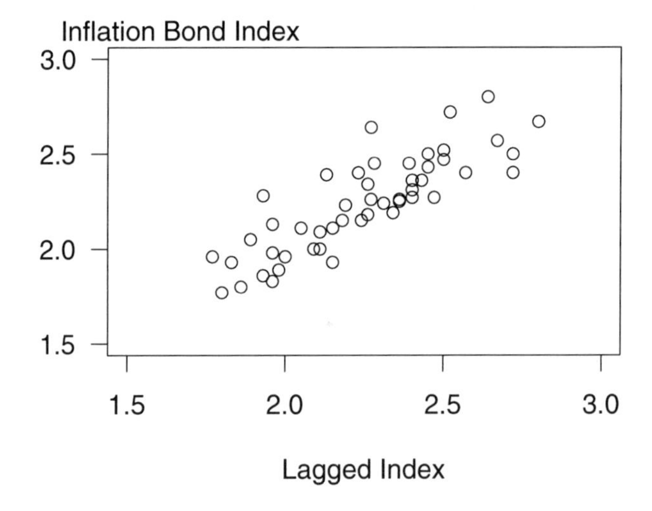

Figure 8.2 Inflation bond versus lagged value. This scatter plot reveals a linear relationship between the index and its lagged value.

this chapter, Table 8.2 presents a small subset of the data so that you can see exactly what each point on the scatter plot represents. Figure 8.2 shows a strong relationship between y_t and y_{t-1}; we will model this relationship in the next section.

Autocorrelations

Scatter plots are useful because they graphically display nonlinear, as well as linear, relationships between two variables. As we established in Chapter 2, correlations can be used to measure the linear relation between two variables.

t	1	2	3	4	5
Index (y_t)	2.72	2.50	2.52	2.72	2.40
Lagged Index (y_{t-1})	*	2.72	2.50	2.52	2.72

Table 8.2 Index and Lagged Index for the First Five of $T = 51$ Values

Lag k	1	2	3	4	5
Autocorrelation r_k	0.814	0.632	0.561	0.447	0.267

Table 8.3 Autocorrelations for the Inflation Bond Series

Recall that, when dealing with cross-sectional data, we summarized relations between $\{y_t\}$ and $\{x_t\}$ using the correlation statistic

$$r = \frac{1}{(T-1)s_x s_y} \sum_{t=1}^{T} (x_t - \overline{x})(y_t - \overline{y}).$$

We now mimic this statistic using the series $\{y_{t-1}\}$ in place of $\{x_t\}$. With this replacement, use $\overline{y}$ in place of $\overline{x}$, and for the denominator, use s_y in place of s_x. With this last substitution, we have $(T-1)s_y^2 = \sum_{t=1}^{T}(y_t - \overline{y})^2$. Our resulting correlation statistic is

$$r_1 = \frac{\sum_{t=2}^{T} (y_{t-1} - \overline{y})(y_t - \overline{y})}{\sum_{t=1}^{T}(y_t - \overline{y})^2}.$$

This statistic is referred to as an *autocorrelation*, that is, a correlation of the series on itself. This statistic summarizes the linear relationship between $\{y_t\}$ and $\{y_{t-1}\}$, that is, observations that are one time unit apart. It will also be useful to summarize the linear relationship between observations that are k time units apart, $\{y_t\}$ and $\{y_{t-k}\}$, as follows.

Definition. The *lag k autocorrelation statistic* is

$$r_k = \frac{\sum_{t=k+1}^{T} (y_{t-k} - \overline{y})(y_t - \overline{y})}{\sum_{t=1}^{T}(y_t - \overline{y})^2}, \quad k = 1, 2, \ldots$$

Properties of autocorrelations are similar to correlations. As with the usual correlation statistic r, the denominator, $\sum_{t=1}^{T}(y_t - \overline{y})^2$, is always nonnegative and hence does not change the sign of the numerator. We use this rescaling device so that r_k always lies within the interval $[-1, 1]$. Thus, when we interpret r_k, a value near -1, 0, and 1, means, respectively, a strong negative, near null, or strong positive relationship between y_t and y_{t-k}. If there is a positive relationship between y_t and y_{t-1}, then $r_1 > 0$, and the process is said to be *positively autocorrelated*. For example, Table 8.3 shows the first five autocorrelations of the inflation bond series. These autocorrelations indicate that there is a positive relationship between adjacent observations.

8.2 Autoregressive Models of Order One

Model Definition and Properties

In Figure 8.2, we noted the strong relationship between the immediate past and current values of the inflation bond index. This suggests using y_{t-1} to explain y_t in a regression model. Using previous values of a series to predict current values of a series is termed, not surprisingly, an *autoregression*. When only the immediate past is used as a predictor, we use the following model.

> **Definition.** The *autoregressive model of order 1*, denoted by $AR(1)$, is written as
>
> $$y_t = \beta_0 + \beta_1 y_{t-1} + \varepsilon_t, \qquad t = 2, \ldots, T, \tag{8.1}$$
>
> where $\{\varepsilon_t\}$ is a white noise process such that $\mathrm{Cov}(\varepsilon_{t+k}, y_t) = 0$ for $k > 0$, and β_0, and β_1 are unknown parameters.

In the $AR(1)$ model, the parameter β_0 may be any fixed constant. However, the parameter β_1 is restricted to be between -1 and 1. By making this restriction, it can be established that the $AR(1)$ series $\{y_t\}$ is stationary. Note that if $\beta_1 = 1$, then the model is a random-walk and hence is nonstationary. This is because, if $\beta_1 = 1$, then equation (8.1) may be rewritten as

$$y_t - y_{t-1} = \beta_0 + \varepsilon_t.$$

If the difference of a series forms a white noise process, then the series itself must be a random walk.

For stationarity in the $AR(1)$ model, we require $|\beta_1| < 1$.

The equation (8.1) is useful in the discussion of model properties. We can view an $AR(1)$ model as a generalization of both a white noise process and a random-walk model. If $\beta_1 = 0$, then equation (8.1) reduces to a white noise process. If $\beta_1 = 1$, then equation (8.1) is a random walk.

A stationary process where there is a linear relationship between y_{t-2} and y_t is said to be *autoregressive of order 2*, and similarly for higher-order processes. Discussion of higher-order processes is in Section 8.5.

Model Selection

When examining the data, how does one recognize that an autoregressive model may be a suitable candidate model? First, an autoregressive model is stationary, and thus a control chart is a good device to examine graphically the data to search for stability. Second, adjacent realizations of an $AR(1)$ model should be related; this can be detected visually by a scatter plot of current versus immediate past values of the series. Third, we can recognize an $AR(1)$ model through its autocorrelation structure, as follows.

A useful property of the $AR(1)$ model is that the correlation between points k time units apart turns out to be β_1^k. Stated another way,

$$\rho_k = \text{Corr}(y_t, y_{t-k}) = \frac{\text{Cov}(y_t, y_{t-k})}{\sqrt{\text{Var}(y_t)\text{Var}(y_{t-k})}} = \frac{\text{Cov}(y_t, y_{t-k})}{\sigma_y^2} = \beta_1^k. \qquad (8.2)$$

The first two equalities are definitions and the third is due to the stationarity. You are asked to check the fourth equality in the exercises. Hence, the absolute values of the autocorrelations of an $AR(1)$ process become smaller as the lag time k increases. In fact, they decrease at a geometric rate. We remark that for a white noise process, we have $\beta_1 = 0$, and thus ρ_k should be equal to zero for all lags k.

For a (stationarity) AR(1) model, $\rho_k = \beta_1^k$.

As an aid in model identification, we use the idea of matching the observed autocorrelations r_k to quantities that we expect from the theory, ρ_k. For white noise, the sample autocorrelation coefficient should be approximately zero for each lag k. Even though r_k is algebraically bounded by -1 and 1, the question arises, How large does r_k need to be, in absolute value, to be considered significantly different from zero? The answer to this type of question is given in terms of the statistic's standard error. Under the hypothesis of no autocorrelation, a good approximation to the standard error of the lag k autocorrelation statistic is

$$se(r_k) = \frac{1}{\sqrt{T}}.$$

Our rule of thumb is that if r_k exceeds $2 \times se(r_k)$ in absolute value, it may be considered significantly nonzero. This rule is based on a 5% level of significance.

Example: Inflation Index Bonds, Continued. Is a white noise process model a good candidate for representing this series? The autocorrelations are given in Table 8.2. For a white noise process model, we expect each autocorrelation r_k to be close to zero but note that, for example, $r_1 = 0.814$. Because there are $T = 51$ returns available, the approximate standard error of each autocorrelation is

$$se(r_k) = \frac{1}{\sqrt{51}} = 0.140.$$

Thus, r_1 is $0.814/0.140 = 5.81$ standard errors above zero. Using the normal distribution as a reference base, this difference is significant, implying that a white noise process is not a suitable candidate model.

Is the autoregressive model of order 1 a suitable choice? Well, because $\rho_k = \beta_1^k$, a good estimate of $\beta_1 = \rho_1$ is $r_1 = 0.814$. If this is the case, then under the $AR(1)$ model, another estimate of ρ_k is $(0.814)^k$. Thus, we have two estimates of ρ_k: (i) r_k, an empirical estimate that does not depend on a parametric model and (ii) $(r_1)^k$, which depends on the $AR(1)$ model. To illustrate, see Table 8.4.

Given that the approximate standard error is $se(r_k) = 0.14$, there seems to be a good match between the two sets of autocorrelations. Because of this match, in Section 8.3, we will discuss how to fit the $AR(1)$ model to this set of data.

Table 8.4 Comparison of Empirical Autocorrelations to Estimated under the $AR(1)$ model

Lag k	1	2	3	4	5
Estimated ρ_k under the $AR(1)$ model	0.814	$(0.814)^2 = .66$	$(0.814)^3 = .54$	$(0.814)^4 = .44$	$(0.814)^5 = .36$
Autocorrelation r_k	0.814	0.632	0.561	0.447	0.267

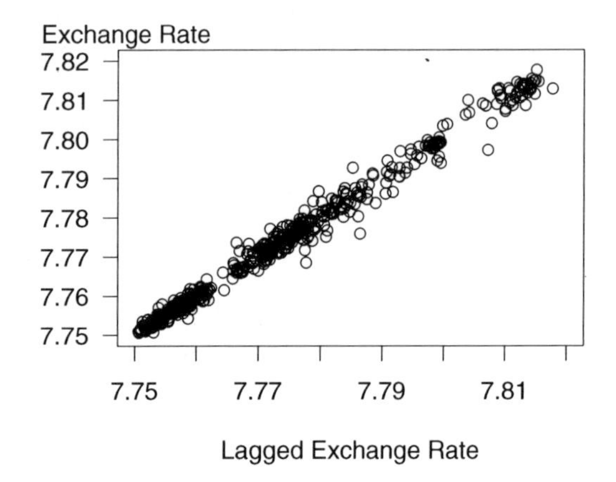

Figure 8.3 Hong Kong daily exchange rates versus lagged values.

Meandering Process

Many processes display the pattern of adjacent points being related to one another. Thinking of a process evolving as a river, Roberts (1991) picturesquely describes such processes as *meandering*. To supplement this intuitive notion, we say that a process is meandering if the lag one autocorrelation of the series is positive. For example, from the plots in Figures 8.1 and 8.2, it seems clear that the inflation bond index is a good example of a meandering series. Indeed, an $AR(1)$ model with a positive slope coefficient is a meandering process.

What about when the slope coefficient approaches one, resulting in a random walk? Consider the Hong Kong exchange rates example in Chapter 7. Although introduced as a quadratic trend in time model, an exercise shows that the series can more appropriately be modeled as a random walk. It seems clear that any point in the process is highly related to each adjacent point in the process. To emphasize this point, Figure 8.3 shows a strong linear relationship between the current and immediate past value of exchange rates. Because of the strong linear relationship in Figure 8.3, we will use the term *meandering process* for a data set that can be modeled using a random walk.

8.3 Estimation and Diagnostic Checking

Having identified a tentative model, the task at hand is to estimate values of β_0 and β_1. In this section, we use the *method of conditional least squares* to determine the estimates, denoted as b_0 and b_1, respectively. This approach is based on the least squares method that was introduced in Section 2.1. Specifically, we use the least squares to find estimates that best fit an observation *conditional* on the previous observation.

Formulas for the conditional least squares estimates are determined from the usual least squares procedures, using the lagged value of y for the explanatory variable. It is easy to see that conditional least squares estimates are closely approximated by

$$b_1 \approx r_1 \qquad \text{and} \qquad b_0 \approx \overline{y}(1 - r_1).$$

Differences between these approximations and the conditional least squares estimates arise because we have no explanatory variable for y_1, the first observation. These differences are typically small in most series and diminish as the series length increases.

Residuals of an $AR(1)$ model are defined as

$$e_t = y_t - (b_0 + b_1 y_{t-1}).$$

As we have seen, patterns in the residuals may reveal ways to improve the model specification. One can use a control chart of the residuals to assess the stationarity and compute the autocorrelation function of residuals to verify the lack of milder patterns through time.

The residuals also play an important role in estimating standard errors associated with model parameter estimates. From equation (8.1), we see that the unobserved errors drive the updating of the new observations. Thus, it makes sense to focus on the variance of the errors, and as in cross-sectional data, we define $\sigma^2 = \sigma_\varepsilon^2 = \text{Var } \varepsilon_t$.

In cross-sectional regression, because the predictor variables were nonstochastic, the variance of the response (σ_y^2) equals the variance of the errors (σ^2). This is not generally true in time series models that use stochastic predictors. For the $AR(1)$ model, taking variances of both sides of equation (8.1) establishes

$$\sigma_y^2(1 - \beta^2) = \sigma^2,$$

so that $\sigma_y^2 > \sigma^2$.

To estimate σ^2, we define

$$s^2 = \frac{1}{T-3} \sum_{t=2}^{T} (e_t - \overline{e})^2. \tag{8.3}$$

In equation (8.3), the first residual, e_1, is not available because y_{t-1} is not available when $t = 1$, and so the number of residuals is $T - 1$. Without the first residual, the average of the residuals is no longer automatically zero and thus is included in the sum of squares. Further, the denominator in the right-hand side of equation (8.3) is still the number of observations minus the number of parameters, keeping in mind the conditions that the "number of observations" is $T - 1$ and the "number of parameters" is two. As in the cross-sectional regression context, we refer to s^2 as the *mean square error* (MSE).

Table 8.5 Residual Autocorrelations from the $AR(1)$ model

Lag k	1	2	3	4	5
Residual Autocorrelation r_k	0.09	−0.33	0.07	0.02	−0.17

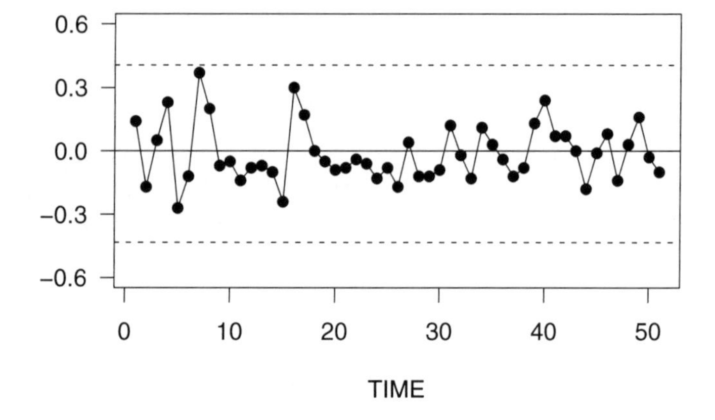

Figure 8.4 Control chart of residuals from an $AR(1)$ fit of the inflation index series. The dashed lines mark the upper and lower control limits that are the mean plus and minus three standard deviations.

Example: Inflation Index Bonds, Continued. The inflation index was fit using an $AR(1)$ model. The estimated equation turns out to be

$$\begin{array}{lccc} \widehat{INDEX}_t & = 0.2923 & + \ 0.8727 & INDEX_{t-1}, \\ \text{standard errors} & (0.0196) & (0.0736) & \end{array}$$

with $s = 0.14$. This is smaller than the standard deviation of the original series (0.259 from Table 8.1), indicating a better fit to the data than with a white noise model. The standard errors, given in parentheses, were computed using the method of conditional least squares. For example, the t-ratio for β_1 is $0.8727/0.0736 = 14.9$, indicating that the immediate past response is an important predictor of the current response.

Residuals were computed as $e_t = INDEX_t - (0.2923 + 0.8727 INDEX_{t-1})$. The control chart of the residuals in Figure 8.4 reveals no apparent patterns. Several autocorrelations of residuals are presented in Table 8.5. With $T = 51$ observations, the approximate standard error is $se(r_k) = 1/\sqrt{51} = 0.14$. The second lag autocorrelation is approximately −2.3 standard errors from zero and the others are smaller, in absolute value. These values are lower than those in Table 8.3, indicating that we have removed some of the temporal patterns with the $AR(1)$ specification. The statistically significant autocorrelation at lag 2 indicates that there is still some potential for model improvement.

8.4 Smoothing and Prediction

Having identified, fit, and checked the identification of the model, we now proceed to basic inference. Recall that by inference we mean the process of using the data

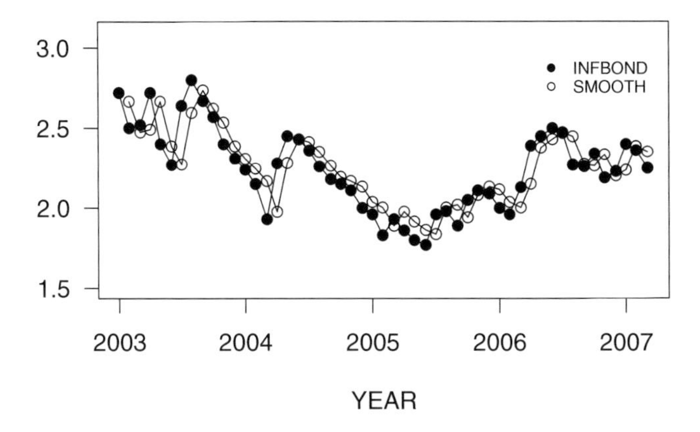

Figure 8.5 Inflation Bond Index with a smoothed series superimposed. The index is given by the open plotting symbols; the smoothed series is represented by the opaque symbols.

set to make statements about the nature of the world. To make statements about the series, analysts often examine the values fitted under the model, called the *smoothed series*. The smoothed series is the estimated expected value of the series given the past. For the $AR(1)$ model, the smoothed series is

$$\widehat{y}_t = b_0 + b_1 y_{t-1}.$$

In Figure 8.5, an open circle represents the actual inflation bond index and an opaque circle represents the corresponding smoothed series. Because the smoothed series is the actual series with the estimated noise component removed, the smoothed series is sometimes interpreted to represent the "real" value of the series.

Typically, the most important application of time series modeling is the forecasting of future values of the series. From equation (8.1), the immediate future value of the series is $y_{T+1} = \beta_0 + \beta_1 y_T + \varepsilon_{T+1}$. Because the series $\{\varepsilon_t\}$ is random, a natural forecast of ε_{T+1} is its mean, zero. Thus, if the estimates b_0 and b_1 are close to the true parameters β_0 and β_1, then a desirable estimate of the series at time $T+1$ is $\widehat{y}_{T+1} = b_0 + b_1 y_T$. Similarly, one can recursively compute an estimate for the series k time points in the future, y_{T+k}.

Definition. The k-step ahead forecast of y_{T+k} for an AR (1) model is recursively determined by

$$\widehat{y}_{T+k} = b_0 + b_1 \widehat{y}_{T+k-1}. \tag{8.4}$$

This is sometimes known as the *chain rule of forecasting*.

To get an idea of the error in using $\widehat{y}_{T+1}$ to predict y_{T+1}, assume for a moment that the error in using b_0 and b_1 to estimate β_0 and β_1 is negligible. With this assumption, the forecast error is

$$y_{T+1} - \widehat{y}_{T+1} = \beta_0 + \beta_1 y_t + \varepsilon_{T+1} - (b_0 + b_1 y_t) \approx \varepsilon_{T+1}.$$

Thus, the variance of this forecast error is approximately σ^2. Similarly, it can be shown that the approximate variance of the forecast error $y_{T+k} - \widehat{y}_{T+k}$ is

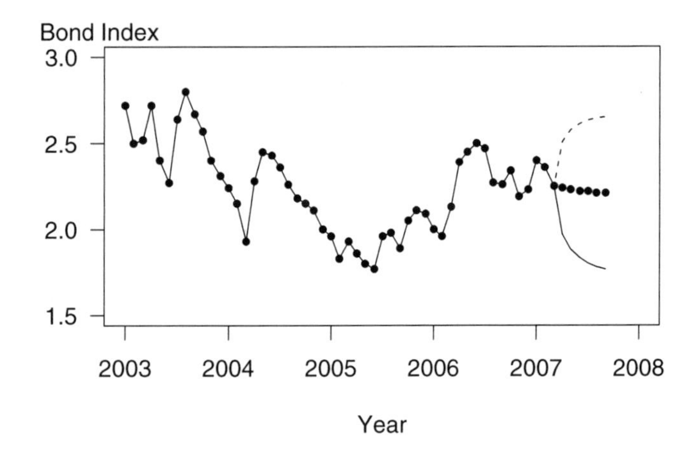

Figure 8.6 Forecast intervals for the inflation bond series.

$\sigma^2(1+\beta_1^2 \cdots + \beta_1^{2(k-1)})$. From this variance calculation and the approximate normality, we have the following prediction interval.

Definition. The k-step ahead forecast interval of y_{T+k} for an $AR(1)$ model is

$$\widehat{y}_{T+k} \pm (t - value)\, s\sqrt{1 + b_1^2 + \cdots + b_1^{2(k-1)}}.$$

Here, the t-value is a percentile from the t-curve using $df = T - 3$ degrees of freedom. The percentile is $1 - (\text{prediction level})/2$.

For example, for 95% prediction intervals, we would have t-value ≈ 2. Thus, one- and two-step 95% prediction intervals are:

$$\begin{array}{ll} \text{one-step:} & \widehat{y}_{T+1} \pm 2s \\ \text{two-step:} & \widehat{y}_{T+2} \pm 2s(1+b_1^2)^{1/2}. \end{array}$$

Figure 8.6 illustrates forecasts of the inflation bond index. The forecast intervals widen as the number of steps into the future increases; this reflects our increasing uncertainty as we forecast further into the future.

8.5 Box-Jenkins Modeling and Forecasting

Sections 8.1 through 8.4 introduced the $AR(1)$ model, including model properties, identification methods, and forecasting. We now introduce a broader class of models known as *autoregressive integrated moving average (ARIMA) models*, attributable to George Box and Gwilym Jenkins (see Box, Jenkins, and Reinsel, 1994).

8.5.1 Models

AR(p) Models

The autoregressive model of order 1 allows us to relate the current behavior of an observation directly to its immediate past value. Moreover, in some applications, there are also important effects of observations that are more distant in the past than simply the immediate preceding observation. To quantify this, we have already introduced the lag k autocorrelation ρ_k that captures the linear relationship between y_t and y_{t-k}. To incorporate this feature into a forecasting framework, we have the *autoregressive model of order p*, denoted by $AR(p)$. The model equation is

$$y_t = \beta_0 + \beta_1 y_{t-1} + \cdots + \beta_p y_{t-p} + \varepsilon_t, \qquad t = p+1, \ldots, T, \tag{8.5}$$

where $\{\varepsilon_t\}$ is a white noise process such that $\text{Cov}(\varepsilon_{t+k}, y_t) = 0$ for $k > 0$, and $\beta_0, \beta_1, \ldots, \beta_p$ are unknown parameters.

As a convention, when data analysts specify an $AR(p)$ model, they include not only y_{t-p} as a predictor variable but also the intervening lags, $y_{t-1}, \ldots, y_{t-p+1}$. The exceptions to this convention are the seasonal autoregressive models, which will be introduced in Section 9.4. Also by convention, the $AR(p)$ is a model of a stationary, stochastic process. Thus, certain restrictions on the parameters $\beta_1, \ldots, \beta_p$ are necessary to ensure (weak) stationarity. These restrictions are developed in the following subsection.

Backshift Notation

The *backshift*, or *backward-shift, operator* B is defined by $\text{B}y_t = y_{t-1}$. The notation B^k means apply the operator k times; that is,

$$\text{B}^k\ y_t = \text{BB} \cdots \text{B}\ y_t = \text{B}^{k-1}\ y_{t-1} = \cdots = y_{t-k}.$$

This operator is linear in the sense that $\text{B}(a_1 y_t + a_2 y_{t-1}) = a_1 y_{t-1} + a_2 y_{t-2}$, where a_1 and a_2 are constants. Thus, we can express the $AR(p)$ model as

$$\begin{aligned} \beta_0 + \varepsilon_t &= y_t - \left(\beta_1 y_{t-1} + \cdots + \beta_p y_{t-p}\right) \\ &= \left(1 - \beta_1 \text{B} - \cdots - \beta_p \text{B}^p\right) y_t = \Phi(\text{B})\, y_t. \end{aligned}$$

If x is a scalar, then $\Phi(x) = 1 - \beta_1 x - \cdots - \beta_p x^p$ is a pth-order polynomial in x. Thus, there exist p roots of the equation $\Phi(x) = 0$. These roots, say, $g_1, \ldots, g_p$, may or may not be complex numbers. It can be shown (see Box, Jenkins, and Reinsel, 1994) that, for stationarity, all roots lie strictly outside the unit circle. To illustrate, for $p = 1$, we have $\Phi(x) = 1 - \beta_1 x$. The root of this equation is $g_1 = \beta_1^{-1}$. Thus, we require $|g_1| > 1$, or $|\beta_1| < 1$, for stationarity.

MA(q) Models

One interpretation of the model $y_t = \beta_0 + \varepsilon_t$ is that the disturbance ε_t perturbs the measure of the "true," expected value of y_t. Similarly, we can consider the model $y_t = \beta_0 + \varepsilon_t - \theta_1 \varepsilon_{t-1}$, where $\theta_1 \varepsilon_{t-1}$ is the perturbation from the previous time period. Extending this line of thought, we introduce the *moving average model of order q*, denoted by $MA(q)$. The model equation is

$$y_t = \beta_0 + \varepsilon_t - \theta_1 \varepsilon_{t-1} - \cdots - \theta_q \varepsilon_{t-q}, \tag{8.6}$$

where the process $\{\varepsilon_t\}$ is a white noise process such that $\text{Cov}(\varepsilon_{t+k}, y_t) = 0$ for $k > 0$ and $\beta_0, \theta_1, \ldots, \theta_q$ are unknown parameters.

With equation (8.6) it is easy to see that $\mathrm{Cov}(y_{t+k}, y_t) = 0$ for $k > q$. Thus, $\rho_k = 0$ for $k > q$. Unlike the $AR(p)$ model, the $MA(q)$ process is stationary for any finite values of the parameters $\beta_0, \theta_1, \ldots, \theta_q$. It is convenient to write the $MA(q)$ using backshift notation, as follows:

$$y_t - \beta_0 = \left(1 - \theta_1 \mathrm{B} - \ldots - \theta_q \mathrm{B}^q\right) \varepsilon_t = \Theta\,(\mathrm{B})\, \varepsilon_t .$$

As with $\Phi\,(x)$, if x is a scalar, then $\Theta\,(x) = 1 - \theta_1 x - \cdots - \theta_q x^q$ is a qth-order polynomial in x. It is unfortunate that the phrase "moving average" is used for the model defined by equation (8.6) and the estimate defined in Section 9.2. We will attempt to clarify the usage as it arises.

ARMA and ARIMA Models

Combining the $AR(p)$ and the $MA(q)$ models yields the *autoregressive moving average model* of order p and q, or $ARMA(p, q)$,

$$y_t - \beta_1 y_{t-1} - \cdots - \beta_p y_{t-p} = \beta_0 + \varepsilon_t - \theta_1 \varepsilon_{t-1} - \cdots - \theta_q \varepsilon_{t-q}, \qquad (8.7)$$

which can be represented as

$$\Phi\,(\mathrm{B})\, y_t = \beta_0 + \Theta\,(\mathrm{B})\, \varepsilon_t . \qquad (8.8)$$

In many applications, the data require differencing to exhibit stationarity. We assume that the data are differenced d times to yield

$$\begin{aligned} w_t &= (1 - \mathrm{B})^d\, y_t = (1 - \mathrm{B})^{d-1} (y_t - y_{t-1}) \\ &= (1 - \mathrm{B})^{d-2} (y_t - y_{t-1} - (y_{t-1} - y_{t-2})) = \cdots \end{aligned} \qquad (8.9)$$

In practice, d is typically 0, 1, or 2. With this, the *autoregressive integrated moving average model* of order (p, d, q), denoted by $ARIMA(p, d, q)$, is

$$\Phi\,(\mathrm{B})\, w_t = \beta_0 + \Theta\,(\mathrm{B})\, \varepsilon_t . \qquad (8.10)$$

Often, β_0 is zero for $d > 0$.

Several procedures are available for estimating model parameters including maximum likelihood estimation, and conditional and unconditional least squares estimation. In most cases, these procedures require iterative fitting procedures. See Abraham and Ledolter (1983) for further information.

Example: Forecasting Mortality Rates. To quantify values in life insurance and annuities, actuaries need forecasts of age-specific mortality rates. Since its publication, the method proposed by Lee and Carter (1992) has proved a popular method to forecast mortality. For example, Li and Chan (2007) used these methods to produce forecasts of 1921–2000 Canadian population rates and 1900–2000 U.S. population rates. They showed how to modify the basic methodology to incorporate atypical events including wars and pandemic events such as influenza and pneumonia.

The Lee-Carter method is usually based on central death rates at age x at time t, denoted by $m_{x,t}$. The model equation is

$$m_{x,t} = \alpha_x + \beta_x \kappa_t + \varepsilon_{x,t}. \tag{8.11}$$

Here, the intercept (α_x) and slope (β_x) depend only on age x, not on time t. The parameter κ_t captures the important time effects (except for those in the disturbance term $\varepsilon_{x,t}$).

At first glance, the Lee-Carter model appears to be a linear regression with one explanatory variable. However, the term κ_t is not observed and so different techniques are required for model estimation. Different algorithms are available, including the singular value decomposition proposed by Lee and Carter (1992), the principal components approach and a Poisson regression model; see Li and Chan (2007) for references.

The time-varying term κ_t is typically represented using an $ARIMA$ model. Li and Chan found that a random walk (with adjustments for unusual events) was a suitable model for Canadian and U.S. rates (with different coefficients), reinforcing the findings of Lee and Carter.

8.5.2 Forecasting

Optimal Point Forecasts

Similar to forecasts that were introduced in Section 8.4, it is common to provide forecasts that are estimates of conditional expectations of the predictive distribution. Specifically, assume that we have available a realization of $\{y_1, y_2, \ldots, y_T\}$ and want to forecast y_{T+l}, the value of the series l lead time units in the future. If the parameters of the process were known, then we would use $\mathrm{E}(y_{T+l}|y_T, y_{T-1}, y_{T-2}, \ldots)$, that is, the conditional expectation of y_{T+l} given the value of the series up to and including time T. We use the notation E_T for this conditional expectation.

To illustrate, taking $t = T + l$ and applying E_T to both sides of equation (8.7) yields

$$\begin{aligned} &y_T(l) - \beta_1 y_T(l-1) - \cdots - \beta_p y_T(l-p) \\ &\quad = \beta_0 + \mathrm{E}_T(\varepsilon_{T+l} - \theta_1 \varepsilon_{T+l-1} - \cdots - \theta_q \varepsilon_{T+l-q}), \end{aligned} \tag{8.12}$$

using the notation $y_T(k) = \mathrm{E}_T\,(y_{T+k})$. For $k \leq 0$, $\mathrm{E}_T\,(y_{T+k}) = y_{T+k}$, as the value of y_{T+k} is known at time T. Further, $\mathrm{E}_T\,(\varepsilon_{T+k}) = 0$ for $k > 0$, as disturbance terms in the future are assumed to be uncorrelated with current and past values of the series. Thus, equation (8.12) provides the basis of the *chain rule of forecasting*, where we recursively provide forecasts at lead time l based on prior forecasts and realizations of the series. To implement equation (8.12), we substitute estimates for parameters and residuals for disturbance terms.

Special Case – MA(1) Model. We have already seen the forecasting chain rule for the $AR(1)$ model in Section 8.4. For the $MA(1)$ model, note that for

$l \geq 2$, we have $y_T(l) = \mathrm{E}_T(y_{T+l}) = \mathrm{E}_T(\beta_0 + \varepsilon_{T+l} - \theta_1 \varepsilon_{T+l-1}) = \beta_0$, because ε_{T+l} and ε_{T+l-1} are in the future at time T. For $l = 1$, we have $y_T(1) = \mathrm{E}_T(\beta_0 + \varepsilon_{T+1} - \theta_1 \varepsilon_T) = \beta_0 - \theta_1 \mathrm{E}_T(\varepsilon_T)$. Typically, one would estimate the term $\mathrm{E}_T(\varepsilon_T)$ using the residual at time T.

ψ-Coefficient Representation

Any *ARIMA*(p, d, q) model can be expressed as

$$y_t = \beta_0^* + \varepsilon_t + \psi_1 \varepsilon_{t-1} + \psi_2 \varepsilon_{t-2} + \cdots = \beta_0^* + \sum_{k=0}^{\infty} \psi_k \varepsilon_{t-k},$$

called the *ψ-coefficient representation*. That is, the current value of a process can be expressed as a constant plus a linear combination of the current and previous disturbances. Values of $\{\psi_k\}$ depend on the linear parameters of the *ARIMA* process and can be determined via straightforward recursive substitution. To illustrate, for the $AR(1)$ model, we have

$$\begin{aligned} y_t &= \beta_0 + \varepsilon_t + \beta_1 y_{t-1} = \beta_0 + \varepsilon_t + \beta_1 (\beta_0 + \varepsilon_{t-1} + \beta_1 y_{t-2}) = \cdots \\ &= \frac{\beta_0}{1-\beta_1} + \varepsilon_t + \beta_1 \varepsilon_{t-1} + \beta_1^2 \varepsilon_{t-2} + \cdots = \frac{\beta_0}{1-\beta_1} + \sum_{k=0}^{\infty} \beta_1^k \varepsilon_{t-k}. \end{aligned}$$

That is, $\psi_k = \beta_1^k$.

Forecast Interval

Using the ψ-coefficient representation, we can express the conditional expectation of y_{T+l} as

$$\mathrm{E}_T(y_{T+l}) = \beta_0^* + \sum_{k=0}^{\infty} \psi_k \mathrm{E}_T(\varepsilon_{T+l-k}) = \beta_0^* + \sum_{k=l}^{\infty} \psi_k \mathrm{E}_T(\varepsilon_{T+l-k}).$$

This is because, at time T, the errors $\varepsilon_T, \varepsilon_{T-1}, \ldots,$ have been determined by the realization of the process. However, the errors $\varepsilon_{T+1}, \ldots, \varepsilon_{T+l}$ have not been realized and hence have conditional expectation zero. Thus, the l-step forecast error is

$$\begin{aligned} y_{T+l} - \mathrm{E}_T(y_{T+l}) &= \beta_0^* + \sum_{k=0}^{\infty} \psi_k \varepsilon_{T+l-k} - \left(\beta_0^* + \sum_{k=l}^{\infty} \psi_k \mathrm{E}_T(\varepsilon_{T+l-k}) \right) \\ &= \sum_{k=0}^{l-1} \psi_k \varepsilon_{T+l-k}. \end{aligned}$$

We focus on the variability of the forecasts errors. That is, straightforward calculations yield $\mathrm{Var}(y_{T+l} - \mathrm{E}_T(y_{T+l})) = \sigma^2 \sum_{k=1}^{l-1} \psi_k^2$. Thus, assuming normality of the errors, a $100(1-\alpha)\%$ forecast interval for y_{T+l} is

$$\widehat{y}_{T+l} \pm (t - value) s \sqrt{\sum_{k=0}^{l-1} \widehat{\psi}_k^2}.$$

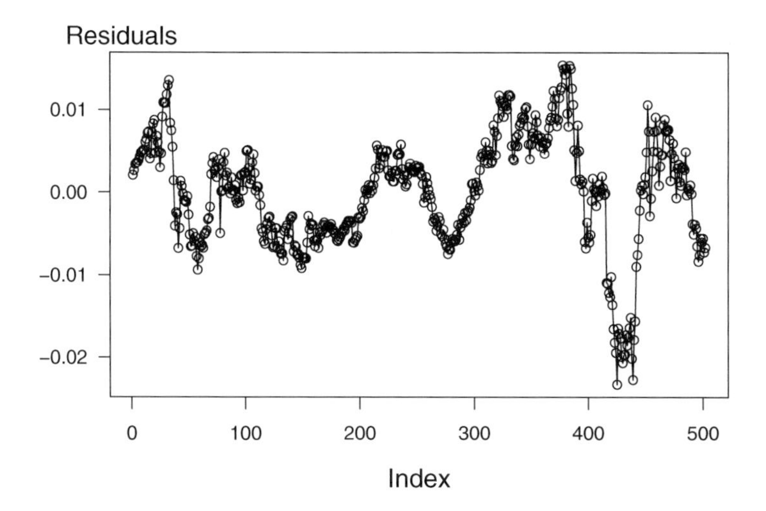

Figure 8.7 Residuals from a quadratic trend in time model of the Hong Kong exchange rates.

where t-value is the $(1-\alpha/2)^{th}$ percentile from a t-distribution with $df = T - (\textit{number of linear parameters})$. If y_t is an $ARIMA(p,d,q)$ process, then ψ_k is a function of $\beta_1,\ldots,\beta_p,\theta_1,\ldots,\theta_q$ and the number of linear parameters is $1+p+q$.

8.6 Application: Hong Kong Exchange Rates

Ⓡ **Empirical** *Filename is "HKExchange"*

Section 7.2 introduced the Hong Kong exchange rate series, based on $T = 502$ daily observations for the period April 1, 2005 through Mary 31, 2007. A quadratic trend was fit to the model that produced an $R^2 = 86.2\%$ with a residual standard deviation of $s = 0.0068$. We now show how to improve on this fit using *ARIMA* modeling.

To begin, Figure 8.7 shows a time series plot of residuals from the quadratic trend time model. This plot displays a meandering pattern, suggesting that there is information in the residuals that can be exploited.

Further evidence of these patterns is in the table of autocorrelations in Table 8.6. Here, we see large residual autocorrelations that do not decrease quickly as the lag k increases. A similar pattern is also evident for the original series, EXHKUS. This confirms the nonstationarity that we observed in Section 7.2.

As an alternative transform, we differenced the series, producing DIFFHKUS. This differenced series has a standard deviation of $s_{DIFF} = 0.0020$, suggesting that it is more stable than the original series or the residuals from the quadratic trend in time model. Table 8.6 presents the autocorrelations from the differenced series, indicating mild patterns. However, these autocorrelations are still significantly different from zero. For $T = 501$ differences, we may use as an approximate standard error for autocorrelations $1/\sqrt{501} \approx 0.0447$. With this, we see that the lag 2 autocorrelation is $0.151/0.0447 \approx 3.38$ standard errors below zero, which is statistically significant. This suggests introducing another model to take advantage of the information in the time series patterns.

Table 8.6 Autocorrelations of Hong Kong Exchange Rates

Lag	1	2	3	4	5	6	7	8	9	10
Residuals from the Quadratic Model	0.958	0.910	0.876	0.847	0.819	0.783	0.748	0.711	0.677	0.636
EXHKUS (Original Series)	0.988	0.975	0.963	0.952	0.942	0.930	0.919	0.907	0.895	0.882
DIFFHKUS	0.078	−0.151	−0.038	−0.001	0.095	−0.005	0.051	−0.012	0.084	−0.001

Model Selection and Partial Autocorrelations

For stationary autoregressive models, $|\rho_k|$ becomes small as the lag k increases.

For all stationary autoregressive models, it can be shown that the absolute values of the autocorrelations become small as the lag k increases. In the case that the autocorrelations decrease approximately like a geometric series, an $AR(1)$ model may be identified. Unfortunately, for other types of autoregressive series, the rules of thumb for identifying the series from the autocorrelations become more cloudy. One device that is useful for identifying the order of an autoregressive series is the *partial autocorrelation function.*

Just like autocorrelations, we now define a *partial autocorrelation* at a specific lag k. Consider the model equation

$$y_t = \beta_{0,k} + \beta_{1,k} y_{t-1} + \cdots + \beta_{k,k} y_{t-k} + \varepsilon_t.$$

Here, $\{\varepsilon_t\}$ is a stationary error that may or may not be a white noise process. The second subscript on the β's, k, is there to remind us that the value of each β may change when the order of the model, k, changes. With this model specification, we can interpret $\beta_{k,k}$ as the correlation between y_t and y_{t-k} after the effects of the intervening variables, $y_{t-1}, \ldots, y_{t-k+1}$, have been removed. This is the same idea as the partial correlation coefficient, introduced in Section 4.4. Estimates of partial correlation coefficients, $b_{k,k}$, can then be calculated using conditional least squares or other techniques. As with other correlations, we may use $1/\sqrt{T}$ as an approximate standard error for detecting significant differences from zero.

A lag k partial autocorrelation is the correlation between y_t and y_{t-k}, controlling for the effects of the intervening variables, $y_{t-1}, \ldots, y_{t-k+1}$.

Partial autocorrelations are used in model identification in the following way. First calculate the first several estimates, $b_{1,1}$, $b_{2,2}$, $b_{3,3}$, and so on. Then, choose the order of the autoregressive model to be the largest k so that the estimate $b_{k,k}$ is significantly different from zero.

To see how this applies in the Hong Kong exchange rate example, recall that the approximate standard error for correlations is $1/\sqrt{501} \approx 0.0447$. Table 8.7 provides the first 10 partial autocorrelations for the rates and for their differences. Using twice the standard error as our cutoff rule, we see that the second partial autocorrelation of the differences exceeds $2 \times 0.0447 = 0.0894$ in absolute value. This would suggest using an $AR(2)$ as a tentative first model choice. Alternatively, the reader may wish to argue that because the fifth and ninth partial autocorrelations are also statistically significant, suggesting a more complex

Table 8.7 Partial Autocorrelations of EXHKUS and DIFFHKUS

Lag	1	2	3	4	5	6	7	8	9	10
EXHKUS	0.988	−0.034	0.051	0.019	−0.001	−0.023	0.010	−0.047	−0.013	−0.049
DIFFHKUS	0.078	−0.158	−0.013	−0.021	0.092	−0.026	0.085	−0.027	0.117	−0.036

$AR(5)$ or $AR(9)$ would be more appropriate. The philosophy is to "use the simplest model possible, but no simpler." We prefer to employ simpler models and thus fit these first and then test to determine whether they capture the important aspects of the data.

Finally, you may be interested to see what happens to partial autocorrelations calculated on a nonstationary series. Table 8.7 provides partial autocorrelations for the original series (EXHKUS). Note how large the first partial autocorrelation is. That is, yet another way of identifying a series as nonstationary is to examine the partial autocorrelation function and look for a large lag 1 partial autocorrelation.

Another way to identify a series as nonstationary is to examine the partial autocorrelation function and look for a large lag one partial autocorrelation.

Residual Checking

Having identified and fit a model, residual checking is still an important part of determining a model's validity. For the $ARMA(p, q)$ model, we compute fitted values as

$$\widehat{y}_t = b_0 + b_1 y_{t-1} + \cdots + b_p y_{t-p} - \widehat{\theta}_1 e_{t-1} - \cdots - \widehat{\theta}_q e_{t-q}. \tag{8.13}$$

Here, $\widehat{\theta}_1, \ldots, \widehat{\theta}_q$ are estimates of $\theta_1, \ldots, \theta_q$. The residuals may be computed in the usual fashion, that is, as $e_t = y_t - \widehat{y}_t$. Without further approximations, note that the initial residuals are missing because fitted values before time $t = \max(p, q)$ cannot be calculated using equation (8.13). To check for patterns, use the devices described in Section 8.3, such as the control chart to check for stationarity and the autocorrelation function to check for lagged variable relationships.

Residual Autocorrelation

Residuals from the fitted model should resemble white noise and, hence, display few discernible patterns. In particular, we expect $r_k(e)$, the lag k autocorrelation of residuals, to be approximately zero. To assess this, we have that $se\,(r_k(e)) \approx 1/\sqrt{T}$. More precisely, MacLeod (1977, 1978) has given approximations for a broad class of $ARMA$ models. It turns out that the $1/\sqrt{T}$ can be improved for small values of k. (These improved values can be seen in the output of most statistical packages.) The improvement depends on the model that is being fit. To illustrate, suppose that an $AR(1)$ model with autoregressive parameter β_1 is fit to the data. Then, the approximate standard error of the lag 1 residual autocorrelation is $|\beta_1|/\sqrt{T}$. This standard error can be much smaller than $1/\sqrt{T}$, depending on the value of β_1.

Testing Several Lags

To test whether there is significant residual autocorrelation at a specific lag k, we use $r_k(e)/se\,(r_k(e))$. Further, to check whether residuals resemble a white noise process, we might test whether $r_k(e)$ is close to zero for several values of k. To test whether the first K residual autocorrelation are zero, use the Box and Pierce (1970) chi-square statistic

$$Q_{BP} = T \sum_{k=1}^{K} r_k\,(e)^2 .$$

Here, K is an integer that is user specified. If there is no real autocorrelation, then we expect Q_{BP} to be small; more precisely, Box and Pierce showed that Q_{BP} follows an approximate χ^2 distribution with $df = K -$ (*number of linear parameters*). For an $ARMA(p, q)$ model, the number of linear parameters is $1 + p + q$. Another widely used statistic is

$$Q_{LB} = T(T+2) \sum_{k=1}^{K} \frac{r_k\,(e)^2}{T-k},$$

Appendix A3.2 provides additional details about the chi-square distribution, including a graph and percentiles.

due to Ljung and Box (1978). This statistic performs better in small samples than does the BP statistic. Under the hypothesis of no residual autocorrelation, Q_{LB} follows the same χ^2 distribution as Q_{BP}. Thus, for each statistic, we reject H_0: no residual autocorrelation if the statistic exceeds chi-value, a $1 - \alpha$ percentile from a χ^2 distribution. A convenient rule of thumb is to use chi-value $= 1.5\ df$.

Example: Hong Kong exchange rates, Continued. Two models were fit, the $ARIMA(2,1,0)$ and the $ARIMA(0,1,2)$; these are the $AR(2)$ and $MA(2)$ models after taking differences. Using $\{y_t\}$ for the differences, the estimated $AR(2)$ model is

$$\begin{array}{lcccccc} \widehat{y_t} & = 0.0000317 & + & 0.0900\ y_{t-1} & - & 0.158\ y_{t-2}, \\ t\text{-statistic} & [0.37] & & [2.03] & & [-3.57] \end{array}$$

with a residual standard error of $s = 0.00193$. The estimated $MA(2)$ is:

$$\begin{array}{lcccccc} \widehat{y_t} & = 0.0000297 & - & 0.0920\ e_{t-1} & + & 0.162\ e_{t-2}, \\ t\text{-statistic} & [0.37] & & [-2.08] & & [3.66] \end{array}$$

with the same residual standard error of $s = 0.00193$. These statistics indicate that the models are roughly comparable. The Ljung-Box statistic in Table 8.8 also indicates a great deal of similarity for the models.

The fitted $MA(2)$ and $AR(2)$ models are roughly similar. We present the $AR(2)$ model for forecasting only because autoregressive models are typically easier to interpret. Figure 8.8 summarizes the predictions, calculated for ten days. Note the widening forecast intervals, typical of forecasts for nonstationary series.

Model	Lag K 2	4	6	8	10
$AR(2)$	0.0050	0.5705	6.3572	10.4746	16.3565
$MA(2)$	0.0146	0.2900	6.6661	11.3655	17.7326

Table 8.8 Ljung-Box Statistics Q_{LB} for Hong Kong Exchange Rate Models

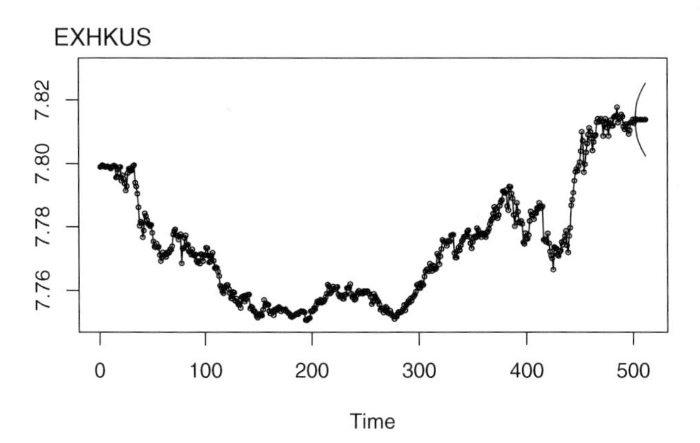

Figure 8.8 Ten-day forecasts and forecast intervals of the Hong Kong exchange rates. Forecasts are based on the ARIMA(2,1,0) model.

8.7 Further Reading and References

The classic book-length introduction to Box-Jenkins time series is Box, Jenkins and Reinsel (1994).

Chapter References

Abraham, Bovas, and Johannes Ledolter (1983). *Statistical Methods for Forecasting*. John Wiley & Sons, New York.

Box, George E. P., Gwilym M. Jenkins, and Gregory C. Reinsel (1994). *Time Series Analysis: Forecasting and Control*, 3rd ed. Prentice-Hall, Englewood Cliffs, New Jersey.

Box, George E. P., and D. A. Pierce (1970). Distribution of residual autocorrelations in autoregressive moving average time series models. *Journal of the American Statistical Association* 65, 1509–26.

Chan, Wai-Sum, and Siu-Hang Li (2007). The Lee-Carter model for forecasting mortality, revisited. *North American Actuarial Journal* 11, no. 1, 68–89.

Lee, Ronald D., and Lawrence R. Carter (1992). Modelling and forecasting U.S. mortality. *Journal of the American Statistical Association* 87, 659–71.

Ljung, G. M., and George E. P. Box (1978). On a measure of lack of fit in time series models. *Biometrika* 65, 297–303.

MacLeod, A. I. (1977). Improved Box-Jenkins estimators. *Biometrika* 64, 531–4.

MacLeod, A. I. (1978). On the distribution of residual autocorrelations in Box-Jenkins models. *Journal of the Royal Statistical Society B* 40, 296–302.

Miller, Robert B., and Dean W. Wichern (1977). *Intermediate Business Statistics: Analysis of Variance, Regression and Time Series*. Holt, Rinehart and Winston, New York.

Roberts, Harry V. (1991). *Data Analysis for Managers with MINITAB*. Scientific Press, South San Francisco, California.

8.8 Exercises

8.1. A mutual fund has provided investment yield rates for five consecutive years as follows:

Year	1	2	3	4	5
Yield	0.09	0.08	0.09	0.12	−0.03

Determine r_1 and r_2, the lag 1 and lag 2 autocorrelation coefficients.

8.2. The *Durbin-Watson* statistic is designed to detect autocorrelation and is defined by

$$DW = \frac{\sum_{t=2}^{T}(y_t - y_{t-1})^2}{\sum_{t=1}^{T}(y_t - \bar{y})^2}.$$

a. Derive the approximate relationship between DW and the lag 1 autocorrelation coefficient r_1.
b. Suppose that $r_1 = 0.4$. What is the approximate value of DW?

8.3. Consider the Chapter 2 linear regression model formulas with y_{t-1} in place of x_t, for $t = 2, \ldots, T$.
a. Provide an exact expression for b_1.
b. Provide an exact expression for b_0.
c. Show that $b_0 \approx \bar{y}(1 - r_1)$.

8.4. Begin with the $AR(1)$ model as in equation (8.1).
a. Take variances of each side of equation (8.1) to show that $\sigma_y^2(1 - \beta_1^2) = \sigma^2$, where $\sigma_y^2 = \text{Var } y_t$ and $\sigma^2 = \text{Var } \varepsilon_t$.
b. Show that $\text{Cov}(y_t, y_{t-1}) = \beta_1 \sigma_y^2$.
c. Show that $\text{Cov}(y_t, y_{t-k}) = \beta_1^k \sigma_y^2$.
d. Use part (c) to establish equation (8.2).

8.5. Consider forecasting with the $AR(1)$ model.
a. Use the forecasting chain rule in equation (8.4) to show

$$y_{T+k} - \widehat{y}_{T+k} \approx \varepsilon_{T+k} + \beta_1 \varepsilon_{T+k-1} + \cdots + \beta_1^{k-1} \varepsilon_{T+1}.$$

b. From part (a), show that the approximate variance of the forecast error is $\sigma^2 \sum_{l=0}^{k-1} \beta_1^{2l}$.

Ⓡ **Empirical**
Filename is "SP500Daily"

8.6. These data consist of the 503 daily returns for the calendar years 2005 and 2006 of the S&P value-weighted index. (The data file contains additional years – this exercise uses only 2005 and 2006 data.) Each year, there are about 250 days on which the exchange is open and stocks were traded – on weekends and holidays it is closed. There are several indices to measure the market's overall performance. The *value-weighted index* is created by assuming that the amount invested in each stock is proportional to its market capitalization. Here, the market capitalization is simply the beginning price per share times the number of outstanding shares. An alternative is the *equally weighted index*, created by taking a simple average of the closing, or last, price of stocks that form the S&P on that trading day.

Financial economic theory states that if the market were predictable, many investors would attempt to take advantage of these predictions, thus forcing unpredictability. For example, suppose a statistical model reliably predicted mutual fund A to increase twofold over the next 18 months. Then, the no-arbitrage principle in financial economics states that several alert investors, armed with information from the statistical model, would bid to buy mutual fund A, thus causing the price to increase because demand is increasing. These alert investors would continue to purchase until the price of mutual fund A rose to the point where the return was equivalent to other investment opportunities in the same risk class. Thus, any advantages produced by the statistical model would disappear rapidly, thus eliminating this advantage.

Thus, financial economic theory states that for liquid markets such as stocks represented through the S&P index there should be no detectable patterns, resulting in a white noise process. In practice, it has been found that the costs of buying and selling equities (called *transactions costs*) are great enough so as to prevent us from taking advantage of these slight tendencies in the swings of the market. This illustrates a point known as *statistically significant but not practically important*. This is not to suggest that statistics is not practical (heaven forbid!). Instead, statistics in and of itself does not explicitly recognize factors, such as economic, psychological, and so on, that may be extremely important in any given situation. It is up to the analyst to interpret the statistical analysis in light of these factors.

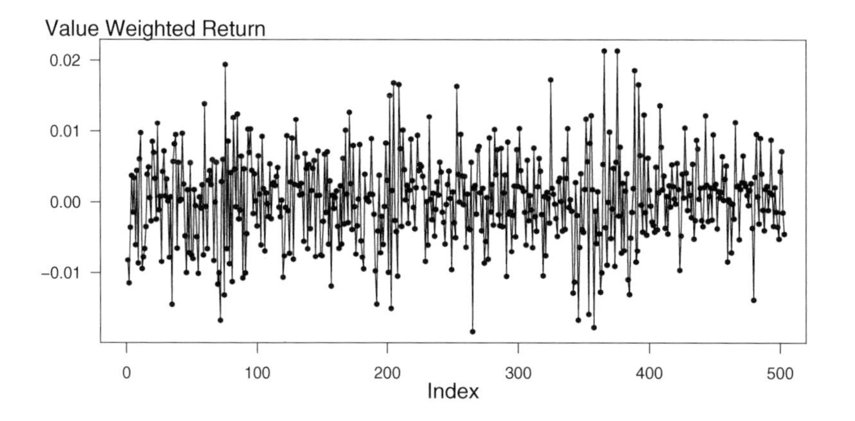

Figure 8.9 Time series plot of the S & P daily market return, 2005–6.

a. The time series plot in Figure 8.9 gives a preliminary idea of the characteristics of the sequence. Comment on the stationarity of the sequence.

b. Calculate summary statistics of the sequence. Suppose that you assume a white noise model for the the sequence. Compute one-, two-, and three-step-ahead forecasts for the daily returns for the first three trading days of 2007.
c. Calculate the autocorrelations for the lags 1 through 10. Do you detect any autocorrelations that are statistically significantly different from zero?

9

Forecasting and Time Series Models

Chapter Preview. This chapter introduces two popular smoothing techniques, moving (running) averages and exponential smoothing, for forecasting. These techniques are simple to explain and easily interpretable. They can be also expressed as regression models, where the technique of weighted least squares is used to compute parameter estimates. Seasonality is then presented, followed by a discussion of two more advanced time series topics, unit root testing, and volatility (*ARCH/GARCH*) models.

9.1 Smoothing with Moving Averages

Smoothing a time series with a moving, or running, average, is a time-tested procedure. This technique continues to be used by many data analysts because of its ease of computation and resulting ease of interpretation. As we discuss subsequently, this estimator can also be motivated as a weighted least squares (WLS) estimator. Thus, the estimator enjoys certain theoretical properties.

The basic *moving, or running, average estimate* is defined by

$$\widehat{s}_t = \frac{y_t + y_{t-1} + \cdots + y_{t-k+1}}{k}, \tag{9.1}$$

where k is the *running average length*. The choice of k depends on the amount of smoothing desired. The larger the value of k, the smoother is the estimate $\widehat{s}_t$ because more averaging is done. The choice $k = 1$ corresponds to no smoothing.

Ⓡ **Empirical** *Filename is "MedCPISmooth"*

Application: Medical Component of the CPI

The consumer price index (CPI) is a breadbasket of goods and services whose price is measured in the United States by the Bureau of Labor Statistics. By measuring this breadbasket periodically, consumers get an idea of the steady increase in prices over time, which, among other things, serves as a proxy for inflation. The CPI is composed of many components, reflecting the relative importance of each component to the overall economy. Here, we study the medical component of the CPI, the fastest-growing part of the overall breadbasket since 1967. The data we consider are quarterly values of the medical component of the CPI (MCPI) over a 60-year period from 1947 to the first quarter of 2007, inclusive. Over this period, the index rose from 13.3 to 346.0. This represents a

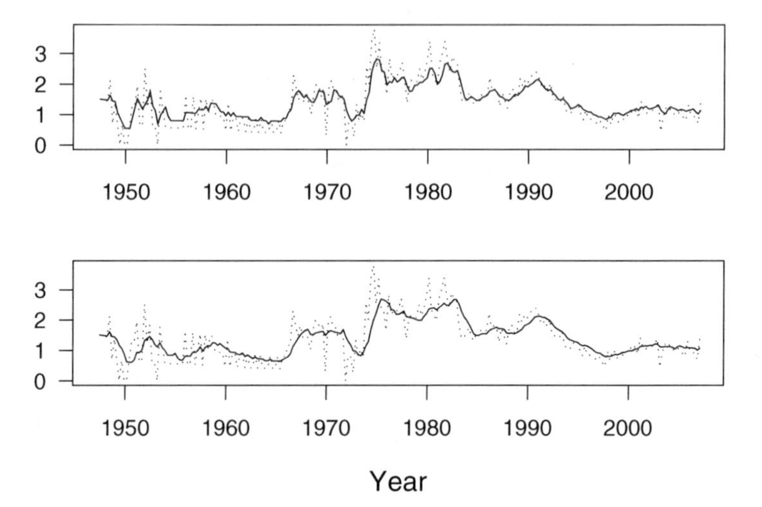

Figure 9.1 Quarterly percentage changes in the medical component of the consumer price index. For both panels, the dashed line is the index. For the upper panel, the solid line is the smoothed version with $k = 4$. For the lower panel, the solid line is the smoothed version with $k = 8$. *Source:* Bureau of Labor Statistics.

26-fold increase over the 60-year period, which translates into a 1.36% quarterly increase.

Figure 9.1 is a time series plot of quarterly percentage changes in MCPI. Note that we have already switched from the nonstationary index to percentage changes. (The index is nonstationary because it exhibits such a tremendous growth over the period considered.) To illustrate the effect of the choice of k, consider the two panels of Figure 9.1. In the upper panel of Figure 9.1, the smoothed series with $k = 4$ is superimposed on the actual series. The lower panel is the corresponding graph with $k = 8$. The fitted values in the lower panel are less jagged than those in upper panel. This helps us to identify graphically the real trends in the series. The danger in choosing too large a value of k is that we may oversmooth the data and lose sight of the real trends.

To forecast the series, reexpress equation (9.1) recursively to get

$$\widehat{s}_t = \frac{y_t + y_{t-1} + \cdots + y_{t-k+1}}{k} = \frac{y_t + k\widehat{s}_{t-1} - y_{t-k}}{k} = \widehat{s}_{t-1} + \frac{y_t - y_{t-k}}{k}. \tag{9.2}$$

If there are no trends in the data, then the second term on the right-hand side, $(y_t - y_{t-k})/k$, may be ignored in practice. This yields the forecasting equation $\widehat{y}_{T+l} = \widehat{s}_T$ for forecasts l lead time units into the future.

Several variants of running averages are available in the literature. For example, suppose that a series can be expressed as $y_t = \beta_0 + \beta_1 t + \varepsilon_t$, a linear trend in time model. This can be handled through the following *double smoothing* procedure:

1. Create a smoothed series using equation (9.1), that is, $\widehat{s}_t^{(1)} = (y_t + \cdots + y_{t-k+1})/k$.
2. Create a doubly smoothed series by using equation (9.1) and treating the smoothed series created in step (1) as input. That is, $\widehat{s}_t^{(2)} = (\widehat{s}_t^{(1)} + \cdots + \widehat{s}_{t-k+1}^{(1)})/k$.

It is easy to check that this procedure smooths out the effect of a linear trend in time. The estimate of the trend is $b_{1,T} = 2\left(\widehat{s}_T^{(1)} - \widehat{s}_T^{(2)}\right)/(k-1)$. The resulting forecasts are $\widehat{y}_{T+l} = \widehat{s}_T + b_{1,T}\, l$ for forecasts l lead time units into the future.

Weighted Least Squares

An important feature of moving, or running, averages is that they can be expressed as weighted least squares (WLS) estimates. WLS estimation was introduced in Section 5.7.3. You will find additional broad discussion in Section 15.1.1. Recall that WLS estimates are minimizers of a weighted sum of squares. The WLS procedure is to find the values of $b_0^*, \ldots, b_k^*$ that minimize

WLS estimation was introduced in Section 5.7.3 and is further described in Section 15.1.1.

$$WSS_T\left(b_0^*, \ldots, b_k^*\right) = \sum_{t=1}^{T} w_t \left(y_t - \left(b_0^* + b_1^* x_{t1}, \ldots, b_k^* x_{tk}\right)\right)^2. \qquad (9.3)$$

Here, WSS_T is the weighted sum of squares at time T.

To arrive at the moving, or running, average estimate, we use the model $y_t = \beta_0 + \varepsilon_t$ with the choice of weights $w_t = 1$ for $t = T-k+1, \ldots, T$, and $w_t = 0$ for $t < T-k+1$. Thus, the problem of minimizing WSS_T in equation (9.3) reduces to finding b_0^* that minimizes $\sum_{t=T-k+1}^{T}\left(y_t - b_0^*\right)^2$. The value of b_0^* that this expression is $b_0 = \widehat{s}_T$, which is the running average of length k.

This model, together with this choice of weights, is called a *locally constant mean model*. Under a *globally constant mean model*, equal weights are used and the least squares estimate of β_0 is the overall average, $\overline{y}$. Under the locally constant mean model, we give equal weight to observations within k time units of the evaluation time T and zero weight to other observations. Although it is intuitively appealing to give more weight to more recent observations, the notion of an abrupt cutoff at a somewhat arbitrarily chosen k is not appealing. This criticism is addressed using exponential smoothing, introduced in the following section.

9.2 Exponential Smoothing

Exponential smoothing estimates are weighted averages of past values of a series, where the weights are given by a series that becomes exponentially small. To illustrate, think of w as a weight number that is between zero and one, and consider the weighted average

$$\frac{y_t + w y_{t-1} + w^2 y_{t-2} + w^3 y_{t-3} + \cdots}{1/(1-w)}.$$

This is a weighted average because the weights $w^k(1-w)$ sum to one; that is, a geometric series expansion yields $\sum_{k=0}^{\infty} w^k = 1/(1-w)$.

Because observations are not available in the infinite past, we use the truncated version

$$\widehat{s}_t = \frac{y_t + w y_{t-1} + \cdots + w^{t-1} y_1 + \cdots + w^t y_0}{1/(1-w)} \qquad (9.4)$$

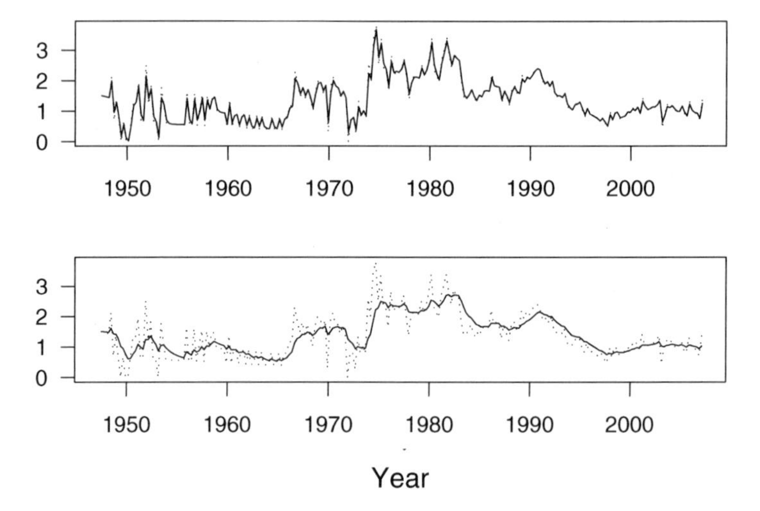

Figure 9.2 Medical component of the consumer price index with smoothing. For both panels, the dashed line is the index. For the upper panel, the solid line is the smoothed version with $w = 0.2$. For the lower panel, the solid line is the smoothed version with $w = 0.8$.

to define the *exponential smoothed estimate* of the series. Here, y_0 is the starting value of the series and is often chosen to be either zero, y_1, or the average value of the series, $\overline{y}$. Like running average estimates, the smoothed estimates in equation (9.4) provide greater weights to more recent observations as compared to observations far in the past with respect to time t. Unlike running averages, the weight function is smooth.

The definition of exponential smoothing estimates in equation (9.4) appears complex. However, as with running averages in equation (9.2), we can reexpress equation (9.4) recursively to yield

$$\widehat{s}_t = \widehat{s}_{t-1} + (1 - w)(y_t - \widehat{s}_{t-1}) = (1 - w)y_t + w\widehat{s}_{t-1}. \tag{9.5}$$

The expression of the smoothed estimates in equation (9.5) is easier to compute than the definition in equation (9.4).

Equation (9.5) also provides insights into the role of w as the smoothing parameter. For example, on the one hand, as w gets close to zero, $\widehat{s}_t$ gets close to y_t. This indicates that little smoothing has taken place. On the other hand, as w gets close to one, there is little effect of y_t on $\widehat{s}_t$. This indicates that a substantial amount of smoothing has taken place because the current fitted value is almost entirely composed of past observations.

Example: Medical Component of the CPI, Continued. To illustrate the effect of the choice of the smoothing parameter, consider the two panels of Figure 9.2. These are time series plots of the quarterly index of the medical component of the CPI. In the upper panel, the smoothed series with $w = 0.2$ is superimposed on the actual series. The lower panel is the corresponding graph with $w = 0.8$. From these figures, we can see that the larger is w, the smoother are our fitted values.

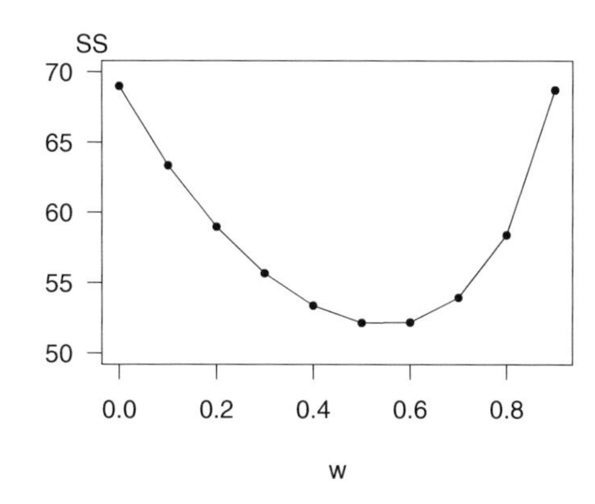

Figure 9.3 Sum of squared one-step prediction errors. Plot of the sum of squared prediction errors $SS(w)$ as a function of the exponential smoothing parameter w.

Equation (9.5) also suggests using the relation $\widehat{y}_{T+l} = \widehat{s}_T$ for our forecast of y_{T+l}, that is, the series at l lead units in the future. Forecasts provide not only a way of predicting the future but also a way of assessing the fit. At time $t-1$, our "forecast" of y_t is $\widehat{s}_{t-1}$. The difference is called the *one-step prediction error.*

To assess the degree of fit, we use the sum of squared one-step prediction errors

$$SS(w) = \sum_{t=1}^{T} (y_t - \widehat{s}_{t-1})^2 . \tag{9.6}$$

An important thing to note is that this sum of squares is a function of the smoothing parameter, w. This then provides a criterion for choosing the smoothing parameter: choose the w that minimizes $SS(w)$. Traditionally, analysts have recommended that w lie within the interval (.70, .95), without providing an objective criterion for the choice. Although minimizing $SS(w)$ provides an objective criterion, it is also computationally intensive. In absence of a sophisticated numerical routine, this minimization is typically accomplished by calculating $SS(w)$ at a number of choices of w and choosing the w that provides the smallest value of $SS(w)$.

To illustrate the choice of the exponential smoothing parameter w, we return to the medical CPI example. Figure 9.3 summarizes the calculation of $SS(w)$ for various values of w. For this dataset, it appears a choice of $w \approx 0.50$ minimizes $SS(w)$.

As with running averages, the presence of a linear trend in time, $T_t = \beta_0 + \beta_1 t$, can be handled through the following double smoothing procedure:

1. Create a smoothed series using equation (9.5), that is, $\widehat{s}_t^{(1)} = (1-w)y_t + w\widehat{s}_{t-1}^{(1)}$.
2. Create a doubly smoothed series by using equation (9.5) and treating the smoothed series created in step (1) as input. That is, $\widehat{s}_t^{(2)} = (1-w)\widehat{s}_t^{(1)} + w\widehat{s}_{t-1}^{(2)}$.

The estimate of the trend is $b_{1,T} = ((1-w)/w)(\widehat{s}_T^{(1)} - \widehat{s}_T^{(2)})$. The forecasts are given by $\widehat{y}_{T+l} = b_{0,T} + b_{1,T}\ l$, where the estimate of the intercept is

$b_{0,T} = 2\widehat{s}_T^{(1)} - \widehat{s}_T^{(2)}$. We will also show how to use exponential smoothing for data with seasonal patterns in Section 9.3.

Weighted Least Squares

As with running averages, an important feature of exponentially smoothed estimates is that they can be expressed as WLS estimates. To see this, for the model $y_t = \beta_0 + \varepsilon_t$, the general weighted sum of squares in equation (9.3) reduces to

$$WSS_T\left(b_0^*\right) = \sum_{t=1}^{T} w_t \left(y_t - b_0^*\right)^2 .$$

The value of b_0^* that minimizes $WSS_T\left(b_0^*\right)$ is $b_0 = \left(\sum_{t=1}^{T} w_t y_t\right)/\left(\sum_{t=1}^{T} w_t\right)$. With the choice $w_t = w^{T-t}$, we have $b_0 \approx \widehat{s}_T$, where there is equality except for the minor issue of the starting value. Thus, exponential smoothing estimates are WLS estimates. Further, because of the choice of the form of the weights, exponential smoothing estimates are also called *discounted least squares estimates*. Here, $w_t = w^{T-t}$ is a discounting function that one might use in considering the time value of money.

9.3 Seasonal Time Series Models

Seasonal patterns appear in many time series that arise in the study of business and economics. Models of seasonality are predominantly used to address patterns that arise as the result of an identifiable, physical phenomenon. For example, seasonal weather patterns affect people's health and, in turn, the demand for prescription drugs. These same seasonal models may be used to model longer cyclical behavior.

There is a variety of techniques available for handling seasonal patterns, including fixed seasonal effects, seasonal autoregressive models, and seasonal exponential smoothing methods. We address each of these techniques in the subsequent sections.

Fixed Seasonal Effects

Recall that, in equations (7.1) and (7.2), we used S_t to represent the seasonal effects under additive and multiplicative decomposition models, respectively. A *fixed seasonal effects model* represents S_t as a function of time t. The two most important examples are the seasonal binary and trigonometric functions. The trends in voting example in Section 7.2 showed how to use a seasonal binary variable and the cost of prescription drugs example here will demonstrate the use of trigonometric functions. The qualifier "fixed effects" means that relationships are constant over time. In contrast, both exponential smoothing and autoregression techniques provide us with methods that adapt to recent events and allow for trends that change over time.

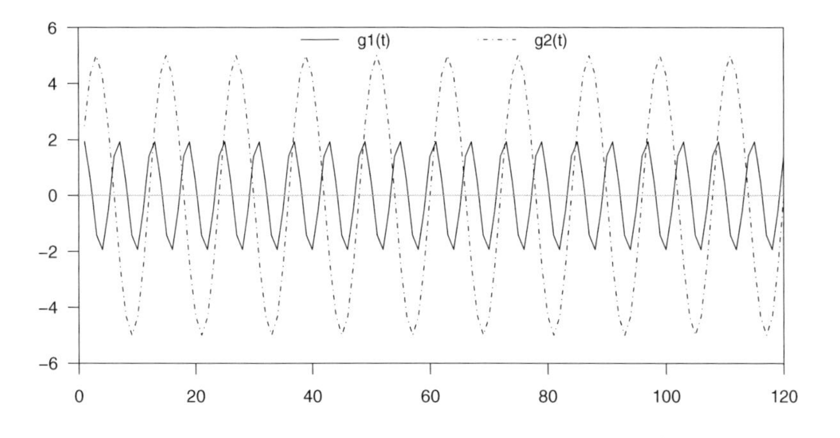

Figure 9.4 Plot of two trigonometric functions. Here, $g_1(t)$ has amplitude $a_1 = 5$, frequency $f_1 = 2\pi/12$, and phase shift $b_1 = 0$. Further, $g_2(t)$ has amplitude $a_2 = 2$, frequency $f_2 = 4\pi/12$, and phase shift $b_2 = \pi/4$.

A large class of seasonal patterns can be represented using trigonometric functions. Consider the function

$$g(t) = a \sin(ft + b),$$

where a is the amplitude (the largest value of the curve), f is the frequency (the number of cycles that occurs in the interval $(0, 2\pi)$), and b is the phase shift. Because of a basic identity, $\sin(x + y) = \sin x \cos y + \sin y \cos x$, we can write

$$g(t) = \beta_1 \sin(ft) + \beta_2 \cos(ft),$$

where $\beta_1 = a \cos b$ and $\beta_2 = a \sin b$. For a time series with *seasonal base SB*, we can represent a wide variety of seasonal patterns using

$$S_t = \sum_{i=1}^{m} a_i \sin(f_i t + b_i) = \sum_{i=1}^{m} \{\beta_{1i} \sin(f_i t) + \beta_{2i} \cos(f_i t)\}, \qquad (9.7)$$

with $f_i = 2\pi i/SB$. To illustrate, the complex function shown in Figure 9.5 was constructed as the sum of the $(m =)$ 2 simpler trigonometric functions that are shown in Figure 9.4.

Consider the model $y_t = \beta_0 + S_t + \varepsilon_t$, where S_t is specified in equation (9.7). Because $\sin(f_i t)$ and $\cos(f_i t)$ are functions of time, they can be treated as known explanatory variables. Thus, the model

$$y_t = \beta_0 + \sum_{i=1}^{m} \{\beta_{1i} \sin(f_i t) + \beta_{2i} \cos(f_i t)\} + \varepsilon_t$$

is a multiple linear regression model with $k = 2m$ explanatory variables. This model can be estimated using standard statistical regression software. Further, we can use our variable selection techniques to choose m, the number of trigonometric functions. We note that m is at most $SB/2$, for SB even. Otherwise, we would have perfect collinearity because of the periodicity of the sine function. The following example demonstrates how to choose m.

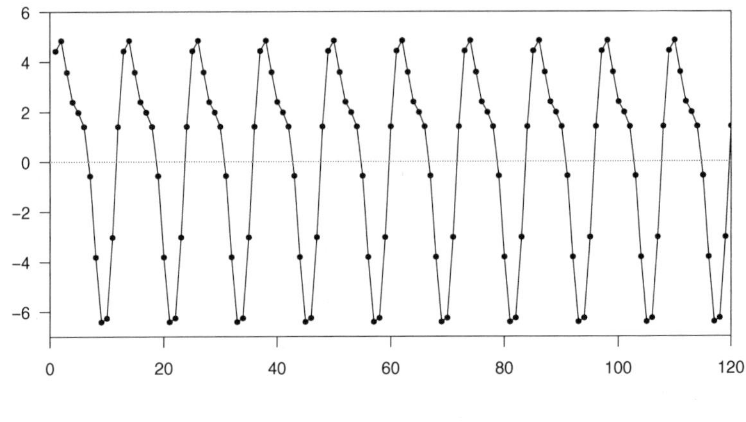

Figure 9.5 Plot of sum of two trigonometric functions in Figure 9.4.

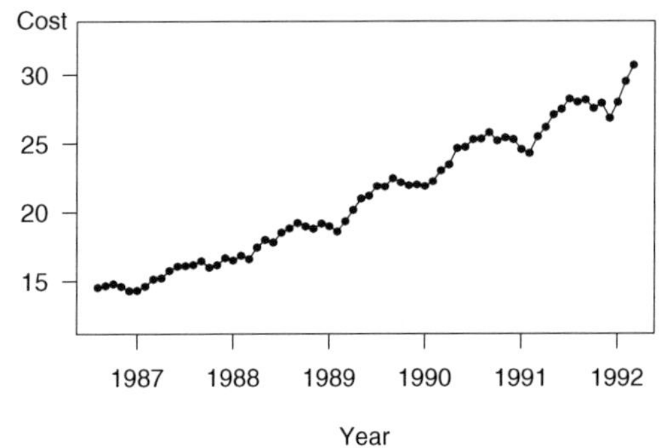

Figure 9.6 Time series plot of cost per prescription claim of the State of New Jersey's prescription drug plan.

Ⓡ **EMPIRICAL** *Filename is "PrescriptionDrug"*

Example: Cost of Prescription Drugs. We consider a series from the State of New Jersey's prescription drug program, the cost per prescription claim. This monthly series is available over the period August 1986 through March 1992, inclusive.

Figure 9.6 shows that the series is clearly nonstationary, in that cost per prescription claims are increasing over time. There are a variety of ways of handling this trend. One may begin with a linear trend in time and include lag claims to handle autocorrelations. For this series, a good approach to the modeling turns out to be to consider the percentage changes in the cost per claim series. Figure 9.7 is a time series plot of the percent changes. In this figure, we see that many of the trends that were evident in Figure 9.6 have been filtered out.

Figure 9.7 displays some mild seasonal patterns in the data. A close inspection of the data reveals higher percentage increases in the spring and lower increases in the fall months. A trigonometric function using $m = 1$ was fit to the data; the fitted model is

$$\begin{array}{rcccc} \widehat{y}_t = & 1.2217 & -1.6956\sin(2\pi t/12) & +0.6536\cos(2\pi t/12) & \\ \text{standard error} & (0.2325) & (0.3269) & (0.3298) & , \\ t\text{-statistic} & [5.25] & [-5.08] & [1.98] & \end{array}$$

with $s = 1.897$ and $R^2 = 31.5\%$. This model reveals some important seasonal patterns. The explanatory variables are statistically significant and an F-test establishes the significance of the model. Figure 9.8 shows the data with fitted

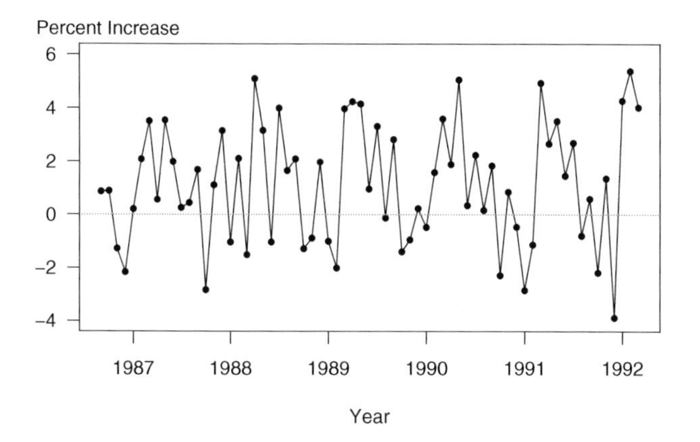

Figure 9.7 Monthly percentage changes of the cost per prescription claim.

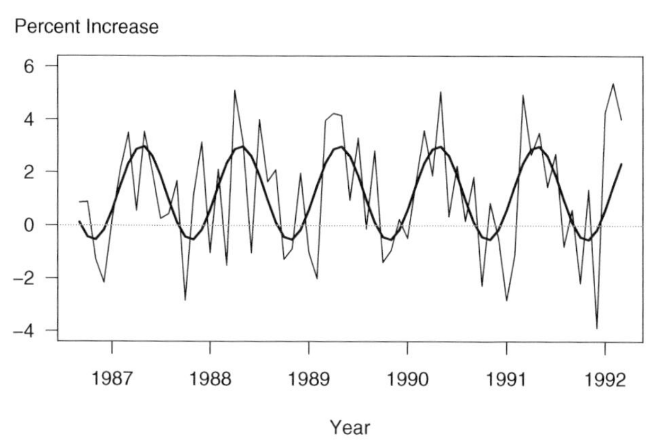

Figure 9.8 Monthly percentage changes of the cost per prescription claim. Fitted values from the seasonal trigonometric model have been superimposed.

values from the model superimposed. These superimposed fitted values help to detect visually the seasonal patterns.

Examination of the residuals from this fitted model revealed few further patterns. In addition, the model using $m = 2$ was fit to the data, resulting in $R^2 = 33.6$ percent. We can decide whether to use $m = 1$ or 2 by considering the model

$$y_t = \beta_0 + \sum_{i=1}^{2} \{\beta_{1i} \sin(f_i t) + \beta_{2i} \cos(f_i t)\} + \varepsilon_t$$

and testing $H_0 : \beta_{12} = \beta_{22} = 0$. Using the partial F-test, with $n = 67, k = p = 2$, we have

$$F\text{-}ratio = \frac{(0.336 - 0.315)/2}{(1.000 - 0.336)/62} = 0.98.$$

With $df_1 = p = 2$ and $df_2 = n - (k + p + 1) = 62$, the 95th percentile of the F-distribution is F-value $= 3.15$. Because $F\text{-}ratio < F\text{-}value$, we cannot reject H_0 and conclude that $m = 1$ is the preferred choice.

Finally, it is also of interest to see how our model of the transformed data works with our original data, in units of cost per prescription claim. Fitted values of percentage increases were converted back to fitted values of cost per claim. Figure 9.9 shows the original data with fitted values superimposed. This figure establishes the strong relationship between the actual and fitted series.

Table 9.1 Autocorrelations of Cost per Prescription Claims

k	1	2	3	4	5	6	7	8	9
r_k	0.08	0.10	−0.12	−0.11	−0.32	−0.33	−0.29	0.07	0.08
k	10	11	12	13	14	15	16	17	18
r_k	0.25	0.24	0.31	−0.01	0.14	−0.10	−0.08	−0.25	−0.18

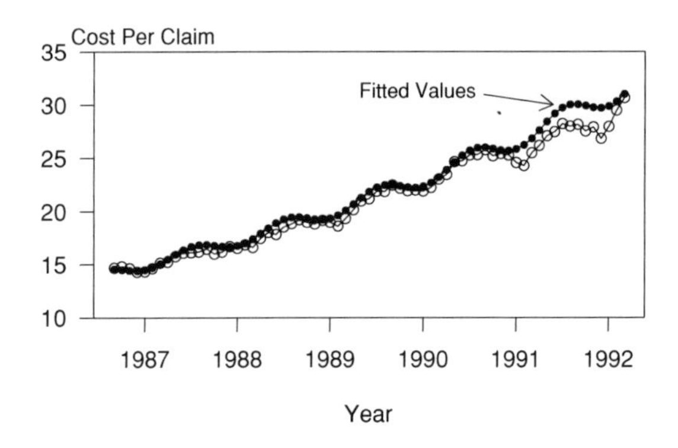

Figure 9.9 Monthly percentage changes of the cost per prescription Claim. Fitted values from the seasonal trigonometric model have been superimposed.

Seasonal Autoregressive Models

In Chapter 8, we examined patterns through time using autocorrelations of the form ρ_k, the correlation between y_t and y_{t-k}. We constructed representations of these temporal patterns using autoregressive models, regression models with lagged responses as explanatory variables. Seasonal time patterns can be handled similarly. We define the *seasonal autoregressive model of order P, SAR(P)*, as

$$y_t = \beta_0 + \beta_1 y_{t-SB} + \beta_2 y_{t-2SB} + \cdots + \beta_P y_{t-PSB} + \varepsilon_t, \qquad (9.8)$$

where SB is the seasonal base of under consideration. For example, using $SB = 12$, a seasonal model of order 1, $SAR(1)$, is

$$y_t = \beta_0 + \beta_1 y_{t-12} + \varepsilon_t.$$

Unlike the $AR(12)$ model defined in Chapter 9, for the $SAR(1)$ model we have omitted $y_{t-1}, y_{t-2}, \ldots, y_{t-11}$ as explanatory variables, though retained y_{t-12}. As in Chapter 8, choice of the order of the model is accomplished by examining the autocorrelation structure and using an iterative model fitting strategy. Similarly, the choice of seasonality SB is based on an examination of the data. We refer the interested reader to Abraham and Ledolter (1983).

Example: Cost of Prescription Drugs, Continued. Table 9.1 presents autocorrelations for the percentage increase in cost per claim of prescription drugs. There are $T = 67$ observations for this dataset, resulting in approximate standard error of $se(r_k) = 1/\sqrt{67} \approx 0.122$. Thus, autocorrelations at and around lags 6, 12, and 18 appear to be significantly different from zero. This suggests using $SB = 6$.

Further examination of the data suggested a *SAR*(2) model. The resulting fitted model is:

$\widehat{y}_t =$	1.2191	$-0.2867 y_{t-6}$	$+0.3120 y_{t-12}$
standard error	(0.4064)	(0.1502)	(0.1489)
t-statistic	[3.00]	[−1.91]	[2.09]

with $s = 2.156$. This model was fit using conditional least squares. Note that because we are using y_{t-12} as an explanatory variable, the first residual that can be estimated is 13. That is, we lose twelve observations when lagging by twelve when using least squares estimates.

Seasonal Exponential Smoothing

An exponential smoothing method that has enjoyed considerable popularity among forecasters is the Holt-Winter additive seasonal model. Although it is difficult to express forecasts from this model as weighted least squares estimates, the model does appear to work well in practice.

Holt (1957) introduced the following generalization of the double exponential smoothing method. Let w_1 and w_2 be smoothing parameters and calculate recursively the parameter estimates:

$$b_{0,t} = (1 - w_1) y_t + w_1 (b_{0,t-1} + b_{1,t-1})$$
$$b_{1,t} = (1 - w_2)(b_{0,t} - b_{0,t-1}) + w_2 b_{1,t-1}.$$

These estimates can be used to forecast the linear trend model, $y_t = \beta_0 + \beta_1 t + \varepsilon_t$. The forecasts are $\widehat{y}_{T+l} = b_{0,T} + b_{1,T}\, l$. With the choice $w_1 = w_2 = 2w/(1 + w)$, the Holt procedure can be shown to produce the same estimates as the double exponential smoothing estimates described in Section 9.2. Because there are two smoothing parameters, the Holt procedure is a generalization of the doubly exponentially smoothed procedure. With two parameters, we need not use the same smoothing constants for the level (β_0) and the trend (β_1) components. This extra flexibility has found appeal with some data analysts.

Winters (1960) extended the Holt procedure to accommodate seasonal trends. Specifically, the *Holt-Winter seasonal additive model* is

$$y_t = \beta_0 + \beta_1 t + S_t + \varepsilon_t,$$

where $S_t = S_{t-SB}$, $S_1 + S_2 + \cdots + S_{SB} = 0$, and SB is the seasonal base. We now employ three smoothing parameters: one for the level, w_1; one for the trend, w_2; and one for the seasonality, w_3. The parameter estimates for this model are determined recursively using:

$$b_{0,t} = (1 - w_1)\left(y_t - \widehat{S}_{t-SB}\right) + w_1 (b_{0,t-1} + b_{1,t-1})$$
$$b_{1,t} = (1 - w_2)(b_{0,t} - b_{0,t-1}) + w_2 b_{1,t-1}$$
$$\widehat{S}_t = (1 - w_3)\left(y_t - b_{0,t}\right) + w_3 \widehat{S}_{t-SB}.$$

With these parameter estimates, forecasts are determined using:

$$\widehat{y}_{T+l} = b_{0,T} + b_{1,T}\, l + \widehat{S}_T(l),$$

where $\widehat{S}_T(l) = \widehat{S}_{T+l}$ for $l = 1, 2, \ldots, SB$, $\widehat{S}_T(l) = \widehat{S}_{T+l-SB}$ for $l = SB + 1, \ldots, 2SB$, and so on.

To compute the recursive estimates, we must decide on (1) initial starting values and (2) a choice of smoothing parameters. To determine initial starting values, we recommend fitting a regression equation to the first portion of the data. The regression equation will include a linear trend in time, $\beta_0 + \beta_1 t$, and $SB - 1$ binary variables for seasonal variation. Thus, only $SB + 1$ observations are required to determine initial estimates $b_{0,0}, b_{1,0}, y_{1-SB}, y_{2-SB}, \ldots, y_0$.

Choosing the three smoothing parameters is more difficult. Analysts have found it difficult to choose three parameters using an objective criterion, such as the minimization of the sum of squared one-step prediction errors, as in Section 9.2. Part of the difficulty stems from the nonlinearity of the minimization, resulting in prohibitive computational time. Another part of the difficulty is that functions such as the sum of squared one-step prediction errors often turn out to be relatively insensitive to the choice of parameters. Analysts have instead relied on rules of thumb to guide the choice of smoothing parameters. In particular, because seasonal effects may take several years to develop, a lower value of w_3 is recommended (resulting in more smoothing). Cryer and Miller (1994) recommend $w_1 = w_2 = 0.9$ and $w_3 = 0.6$.

9.4 Unit Root Tests

We have now seen two competing models that handle nonstationarity with a mean trend: the linear trend in time model and the random walk model. Section 7.6 illustrated how we can choose between these two models on a out-of-sample basis. For a selection procedure based on in-sample data, consider the model

$$y_t = \mu_0 + \phi(y_{t-1} - \mu_0) + \mu_1\,(\phi + (1 - \phi)t) + \varepsilon_t.$$

When $\phi = 1$, this reduces to a random walk model with $y_t = \mu_1 + y_{t-1} + \varepsilon_t$. When $\phi < 1$ and $\mu_1 = 0$, this reduces to an $AR(1)$ model, $y_t = \beta_0 + \phi y_{t-1} + \varepsilon_t$, with $\beta_0 = \mu_0\,(1 - \phi)$. When $\phi = 0$, this reduces to a linear trend in time model with $y_t = \mu_0 + \mu_1 t + \varepsilon_t$.

Running a model where the left-hand-side variable is potentially a random walk is problematic. Hence, it is customary to use least squares on the model

$$y_t - y_{t-1} = \beta_0 + (\phi - 1)\, y_{t-1} + \beta_1 t + \varepsilon_t, \qquad (9.9)$$

where we interpret $\beta_0 = \mu_0\,(1 - \phi) + \phi\mu_1$ and $\beta_1 = \mu_1\,(1 - \phi)$. From this regression, let t_{DF} be the t-statistic associated with the y_{t-1} variable. We wish to use the t-statistic to test the null hypothesis that $H_0 : \phi = 1$ versus the one-sided alternative that $H_a : \phi < 1$. Because $\{y_{t-1}\}$ is a random-walk process under the null hypothesis, the distribution of t_{DF} does not follow the usual t-distribution

Table 9.2 Dickey-Fuller Test Statistics with Critical Values

	Without Trend		With Trend	
Lag (p)	t_{DF}	10% Critical Value	t_{DF}	10% Critical Value
	−1.614	−2.624	−0.266	−3.228
1	−1.816	−2.625	−0.037	−3.230
2	−1.736	−2.626	0.421	−3.233

but rather follows a special distribution (see Dickey and Fuller, 1979). This distribution has been tabulated has been programmed in several statistical packages (see Fuller, 1996).

Example: Labor Force Participation Rates, Continued. We illustrate the performance of the Dickey-Fuller tests on the labor force participation rates introduced in Chapter 7. There, we established that the series was clearly nonstationary and that out-of-sample forecasting showed the random walk to be preferred when compared to the linear trend in time model.

Table 9.2 summarizes the test. Both without ($\mu_1 = 0$) and with ($\mu_1 \neq 0$) the trend line, the t-statistic (t_{DF}) is statistically insignificant (compared to the 10% critical value). This provides evidence that the random walk is the preferred model choice.

One criticism of the Dickey-Fuller test is that the disturbance term in equation (9.9) is presumed to be serially uncorrelated. To protect against this, a commonly used alternative is the *augmented Dickey-Fuller* test statistic. This is the t-statistic associated with the y_{t-1} variable using ordinary least squares on the following equation

$$y_t - y_{t-1} = \beta_0 + (\phi - 1)\, y_{t-1} + \beta_1 t + \sum_{j=1}^{p} \phi_j \left(y_{t-j} - y_{t-j-1}\right) + \varepsilon_t. \quad (9.10)$$

In this equation, we have augmented the disturbance term by autoregressive terms in the differences $\{y_{t-j} - y_{t-j-1}\}$. The idea is that these terms serve to capture serial correlation in the disturbance term. Research has not reached consensus on how to choose the number of lags (p) – in most applications, analysts provide results of the test statistic for a number of choices of lags and hope that conclusions reached are qualitatively similar. This is certainly the case for the labor force participation rates as demonstrated in Table 9.2. Here, we see that for each lag choice, the random walk null hypothesis can not be rejected.

9.5 ARCH/GARCH Models

To this point, we have focused on forecasting the level of the series – that is, the conditional mean. However, there are important applications, notably in the study of finance, where forecasting the variability is important. To illustrate, the

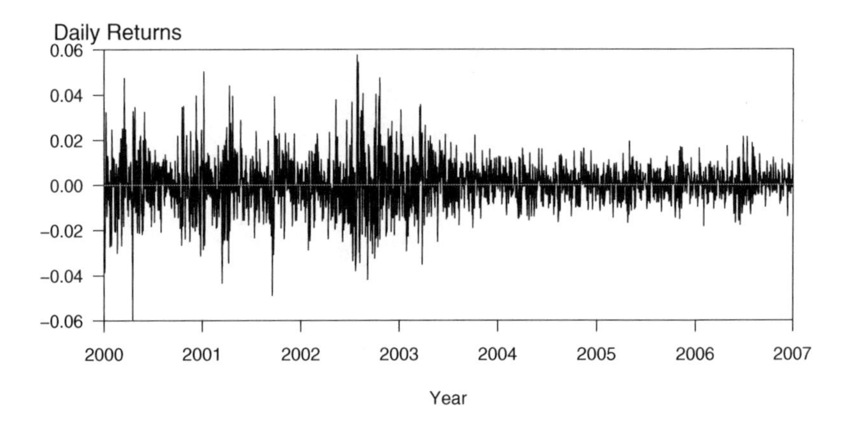

Figure 9.10 Time series plot of daily S&P returns, 2000–6, inclusive.

variance plays a key role in option pricing, such as when using the Black-Scholes formula.

Many financial time series exhibit *volatility clustering*, that is, periods of high volatility (large changes in the series) followed by periods of low volatility. To illustrate, consider the following.

Ⓡ **Empirical** *Filename is "SP500Daily"*

Example: S&P 500 Daily Returns. Figure 9.10 provides a time series plot of daily returns from the S&P 500 over the period 2000–6, inclusive. Here, we see the early part of the series; prior to January 2003 is more volatile than the latter part of the series. Except for the changing volatility, the series appears to be stationary, without dramatic increases or decreases.

The concept of variability changing over time seems at odds with our notions of stationarity. This is because a condition for weak stationary is that the series has a constant variance. The surprising thing is that we can allow for changing variances by conditioning on the past and still retain a weakly stationary model. To see this mathematically, we use the notation Ω_t to denote the information set, the collection of knowledge about the process up to and including time t. For a weakly stationary series, we may denote this as $\Omega_t = \{\varepsilon_t, \varepsilon_{t-1}, \ldots\}$. We allow the variance to depend on time t by conditioning on the past,

$$\sigma_t^2 = \mathrm{Var}_{t-1}(\varepsilon_t) = \mathrm{E}\left([\varepsilon_t - \mathrm{E}(\varepsilon_t|\Omega_{t-1})]^2 \,|\Omega_{t-1}\right).$$

We now present several parametric models of σ_t^2 that allows us to quantify and forecast this changing volatility.

ARCH Model

The *autoregressive changing heteroscedasticity model of order p, ARCH(p)*, is due to Engle (1982). We now assume that the distribution of ε_t given Ω_{t-1} is normally distributed with mean zero and variance σ_t^2. We further assume that the conditional variance is determined recursively by

$$\sigma_t^2 = w + \gamma_1\varepsilon_{t-1}^2 + \cdots + \gamma_p\varepsilon_{t-p}^2 = w + \gamma(B)\varepsilon_t^2,$$

where $\gamma(x) = \gamma_1 x + \cdots + \gamma_p x^p$. Here, $w > 0$ is the "long-run" volatility parameter and $\gamma_1, \ldots, \gamma_p$ are coefficients such that $\gamma_j \geq 0$ and $\gamma(1) = \sum_{j=1}^{p} \gamma_j < 1$.

In the case that $p = 1$, we can see that a large change to the series ε_{t-1}^2 can induce a large conditional variance σ_t^2. Higher orders of p help capture longer-term effects. Thus, this model is intuitively appealing to analysts. Interestingly, Engle provided additional mild conditions to ensure that $\{\varepsilon_t\}$ is weakly stationarity. Thus, despite having a changing *conditional* variance, the *unconditional* variance remains constant over time.

GARCH Model

The *generalized ARCH model of order p, GARCH(p, q)*, complements the *ARCH* model in the same way that the moving average complements the autoregressive model. As with the *ARCH* model, we assume that the distribution of ε_t given Ω_{t-1} is normally distributed with mean zero and variance σ_t^2. The conditional variance is determined recursively by

$$\sigma_t^2 - \delta_1 \sigma_{t-1}^2 + - \cdots - \delta_q \sigma_{t-q}^2 = w + \gamma_1 \varepsilon_{t-1}^2 + \cdots + \gamma_p \varepsilon_{t-p}^2,$$

or $\sigma_t^2 = w + \gamma(B)\varepsilon_t^2 + \delta(B)\sigma_t^2$, where $\delta(x) = \delta_1 x + \cdots + \delta_q x^q$. In addition to the *ARCH*(p) requirements, we also need $\delta_j \geq 0$ and $\gamma(1) + \delta(1) < 1$.

As it turns out, the *GARCH*(p, q) is also a weakly stationary model, with mean zero and (unconditional) variance $\text{Var}\, \varepsilon_t = w/(1 - \gamma(1) - \delta(1))$.

Example: S&P 500 Daily Returns, Continued. After an examination of the data (details not given here), an $MA(2)$ model was fit to the series with *GARCH*(1, 1) errors. Specifically, if y_t denotes the daily return from the S&P series, for $t = 1, \ldots, 1759$, we fit the model

$$y_t = \beta_0 + \varepsilon_t - \theta_1 \varepsilon_{t-1} - \theta_2 \varepsilon_{t-2},$$

where the conditional variance is determined recursively by

$$\sigma_t^2 - \delta_1 \sigma_{t-1}^2 = w + \gamma_1 \varepsilon_{t-1}^2.$$

The fitted model appears in Table 9.3. Here, the statistical package we used employs maximum likelihood to determine the estimated parameters as well as the standard errors needed for the t-statistics. The t-statistics show that all parameter estimates, except θ_1, are statistically significant. As discussed in Chapter 8, the convention is to retain lower-order coefficients, such as θ_1, if higher-order coefficients like θ_2 are significant. Note from Table 9.3 that the sum of the *ARCH* coefficient (δ_1) and the *GARCH* coefficient (γ_1) are nearly one, with *GARCH* coefficient substantially larger than the *ARCH* coefficient. This phenomenon is also reported by Diebold (2004), who states that it is commonly found in studies of financial asset returns.

Table 9.3 S&P 500 Daily Returns Model Fit

Parameter	Estimate	t-Statistic
β_0	0.0004616	2.51
θ_1	−0.0391526	−1.49
θ_2	−0.0612666	−2.51
δ_1	0.0667424	6.97
γ_1	0.9288311	93.55
w	5.61×10^{-7}	2.30
Log likelihood	5,658.852	

9.6 Further Reading and References

For other variations of the running average method and exponential smoothing, see Abraham and Ledolter (1983).

For a more detailed treatment of unit root tests, refer to Diebold (2004) or Fuller (1996).

Chapter References

Abraham, Bovas, and Ledolter, Johannes (1983). *Statistical Methods for Forecasting*. John Wiley & Sons, New York.

Cryer, Jon D., and Robert B. Miller (1994). *Statistics for Business: Data Analysis and Modelling*. PWS-Kent, Boston.

Dickey, D. A., and Wayne A. Fuller (1979). Distribution of the estimators for autoregressive time series with a unit root. *Journal of the American Statistical Association* 74, 427–31.

Diebold, Francis X. (2004). *Elements of Forecasting*, 3rd ed. Thomson, South-Western, Mason, Ohio.

Engle, R. F. (1982). Autoregressive conditional heteroscedasticity with estimates of UK inflation. *Econometrica* 50, 987–1007.

Fuller, Wayne A. (1996). *Introduction to Statistical Time Series*, 2nd ed. John Wiley & Sons, New York.

Holt, C. C. (1957). Forecasting trends and seasonals by exponenetially weighted moving averages. *O.N.R. Memorandum*, No. 52, Carnegie Institute of Technology.

Winters, P. R. (1960). Forecasting sales by exponentially weighted moving averages. *Management Science* 6, 324–42.

10

Longitudinal and Panel Data Models

Chapter Preview. Longitudinal data, also known as panel data, are composed of a cross-section of subjects that we observe repeatedly over time. Longitudinal data enable us to study cross-sectional and dynamic patterns simultaneously; this chapter describes several techniques for visualizing longitudinal data. Two types of models are introduced, fixed and random effects models. This chapter shows how to estimate fixed effects models using categorical explanatory variables. Estimation for random effects models is deferred to a later chapter; this chapter describes when and how to use these models.

10.1 What Are Longitudinal and Panel Data?

In Chapters 1–6, we studied cross-sectional regression techniques that enabled us to predict a dependent variable y using explanatory variables x. For many problems, the best predictor is a value from the preceding period; the times series methods we studied in Chapters 7–9 use the history of a dependent variable for prediction. For example, an actuary seeking to predict insurance claims for a small business will often find that the previous year's claims are the best predictor. However, a limitation of time series methods is that they are based on having available many observations over time (typically 30 or more). When studying annual claims from a business, a long time series is rarely available; either businesses do not have the data, or if they do, it is unreasonable to use the same stochastic model for today's claims as for those 30 years ago. We would like a model that enables us to use information about company characteristics; explanatory variables such as industry, number of employees, age, and sex composition, and so forth; and *recent* claims history. That is, we need a model that combines cross-sectional regression explanatory variables with time series lagged dependent variables as predictors.

Longitudinal data analysis represents a marriage of regression and time series analysis. Longitudinal data are composed of a cross-section of subjects that we observe repeatedly, over time. Unlike regression data, with longitudinal data, we observe subjects over time. By observing a cross-section repeatedly, analysts can make better assessments of regression relationships with a longitudinal data design compared to a regression design. Unlike time series data, with longitudinal data we observe many subjects. By observing time series behavior over many subjects, we can make informed assessments of temporal patterns even when

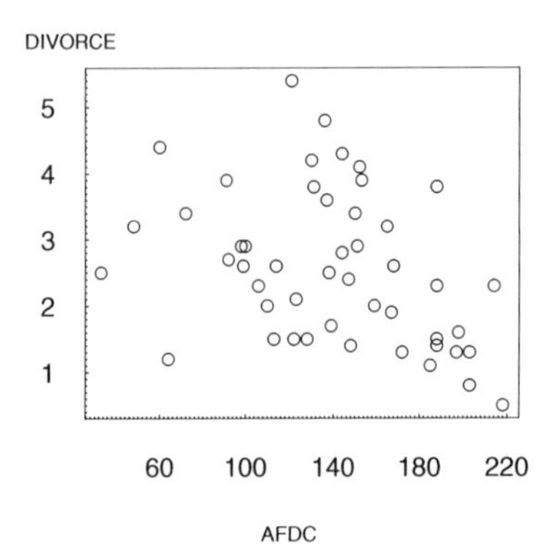

Figure 10.1 Plot of 1965 divorce versus AFDC payments.

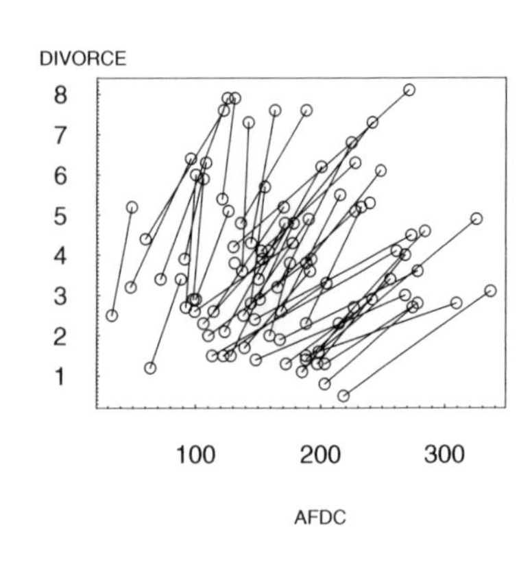

Figure 10.2 Plot of divorce versus AFDC payments, 1965 and 1975.

only a short (time) series is available. Time patterns are also known as *dynamic*. With longitudinal data, we can study cross-sectional and dynamic patterns simultaneously.

With longitudinal data, we can study cross-sectional and dynamic patterns simultaneously.

The descriptor *panel data* comes from surveys of individuals. In this context, a "panel" is a group of individuals surveyed repeatedly over time. We use the terms "longitudinal data" and "panel data" interchangeably, although, for simplicity, we often use only the former term.

As we saw in our Chapter 6 discussion of omitted variables, any new variable can alter our impressions and models of the relationship between y and an x. This is also true of lagged dependent variables. The following example demonstrates that the introduction of a lagged dependent variable can dramatically impact a cross-sectional regression relationship.

Example: Divorce Rates. Figure 10.1 shows the 1965 divorce rates versus Aid to Families with Dependent Children (AFDC) payments for the 50 states. For this example, each state represents an observational unit, the divorce rate is the dependent variable of interest, and the level of AFDC payment represents a variable that may contribute information to our understanding of divorce rates.

Figure 10.1 shows a negative relation; the corresponding correlation coefficient is −0.37. Some argue that this negative relation is counterintuitive in that one would expect a positive relation between welfare payments (AFDC) and divorce rates; states with desirable cultural climates enjoy both low divorce rates and low welfare payments. Others argue that this negative relationship is intuitively plausible; wealthy states can afford high welfare payments and produce economic and cultural climates conducive to low divorce rates. Because the data are observational, it is not appropriate to argue for a causal relationship between welfare payments and divorce rates without additional economic or sociological theory.

Another plot, not displayed here, shows a similar negative relation for 1975; the corresponding correlation is −0.425.

Figure 10.2 shows both the 1965 and 1975 data; a line connects the two observations within each state. The line represents a change over time (dynamic), not a cross-sectional relationship. Each line displays a positive relationship; that is, as welfare payments increase, so do divorce rates *for each state*. Again, we do not infer directions of causality from this display. The point is that the dynamic relation between divorce and welfare payments *within a state* differs dramatically from the cross-sectional relationship *between states.*

Models of longitudinal data are sometimes differentiated from regression and time series through their "double subscripts." We use the subscript i to denote the unit of observation, or *subject*, and t to denote time. To this end, define y_{it} to be the dependent variable for the ith subject during the tth time period. A longitudinal dataset consists of observations of the ith subject over $t = 1, \ldots, T_i$ time periods, for each of $i = 1, \ldots, n$ subjects. Thus, we observe:

$$\begin{array}{rl} \text{first subject} & \{y_{11}, \ldots, y_{1T_1}\} \\ \text{second subject} & \{y_{21}, \ldots, y_{2T_2}\}. \\ \vdots & \qquad \vdots \\ n\text{th subject} & \{y_{n1}, \ldots, y_{nT_n}\} \end{array}$$

In the divorce example, most states have $T_i = 2$ observations and are depicted graphically in Figure 10.2 by a line connecting the two observations. Some states have only $T_i = 1$ observation and are depicted graphically by an open-circle plotting symbol. For many data sets, it is useful to let the number of observations depend on the subject; T_i denotes the number of observations for the ith subject. This situation is known as the *unbalanced data* case. In other datasets, each subject has the same number of observations; this is known as the *balanced data* case.

The applications that we consider are based on many cross-sectional units and only a few time series replications. That is, we consider applications where n is large relative to $T = \max(T_1, \ldots, T_n)$, the maximal number of time periods. Readers will certainly encounter important applications where the reverse is true, $T > n$, or where $n \approx T$.

10.2 Visualizing Longitudinal and Panel Data

To see some ways to visualize longitudinal data, we explore the following example.

Ⓡ **Empirical** *Filename is "Medicare"*

Example: Medicare Hospital Costs. We consider $T = 6$ years, 1990–5, of data for inpatient hospital charges that are covered by the Medicare program. The data were obtained from the Health Care Financing Administration. To illustrate, in 1995, the total covered charges were \$157.8 billions for 12 million discharges. For this analysis, we use state as the subject, or risk class. Here, we consider $n = 54$ states that include the 50 states in the Union, the District of Columbia, the Virgin Islands, Puerto Rico, and an unspecified "other" category. The dependent variable

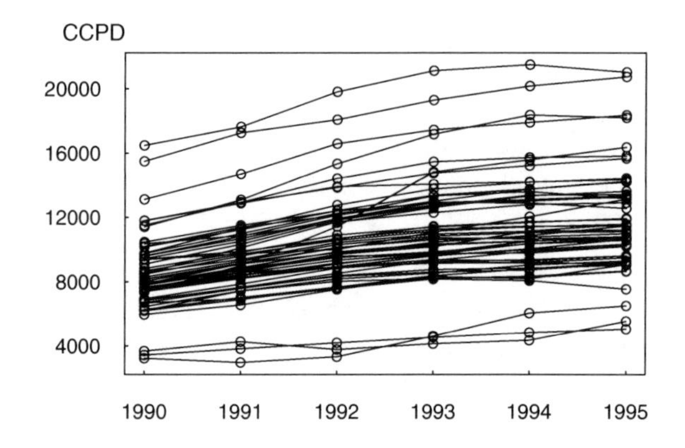

Figure 10.3 Multiple time series plot of CCPD. Covered claims per discharge (CCPD) are plotted over $T = 6$ years, 1990–5. The line segments connect states; thus, we see that CCPD increases for almost every state over time.

of interest is the severity component, covered claims per discharge, which we label as CCPD. The variable CCPD is of interest to actuaries because the Medicare program reimburses hospitals on a per-stay basis. Also, many managed care plans reimburse hospitals on a per-stay basis. Because CCPD varies over state and time, both the state and time (YEAR $= 1, \ldots, 6$) are potentially important explanatory variables. We do not assume a priori that frequency is independent of severity. Thus, number of discharges, NUM_DSCHG, is another potential explanatory variable. We also investigate the importance of another component of hospital utilization, AVE_DAYS, defined to be the average hospital stay per discharge in days.

Figure 10.3 illustrates the *multiple time series plot*. Here, we see not only that are overall claims increasing but also that claims increase for each state. Different levels of hospital costs among states are also apparent; we call this feature *heterogeneity*. Figure 10.3 indicates that there is greater variability among states than over time.

Figure 10.4 is a variation of a scatter plot with symbols. This is a plot of CCPD versus number of discharges. One could use different plotting symbols each state; instead, we connect observations within a state over time. This plot shows a positive overall relationship between CCPD and the number of discharges. Like CCPD, we see substantial state variation on different numbers of discharges. Also like CCPD, the number of discharges increases over time, so that, for each state, there is a positive relationship between CCPD and number of discharges. The slope is higher for those states with smaller number of discharges. This plot also suggests that the number of discharges lagged by one year is an important predictor of CCPD.

Trellis Plot

A technique for graphical display that has recently become popular in the statistical literature is a *trellis plot*. This graphical technique takes its name from a trellis, which is a structure of open latticework. When viewing a house or garden, one typically thinks of a trellis as used to support creeping plants such as vines. We will use this lattice structure and refer to a trellis plot as consisting of one or more panels arranged in a rectangular array. Graphs that contain multiple versions of

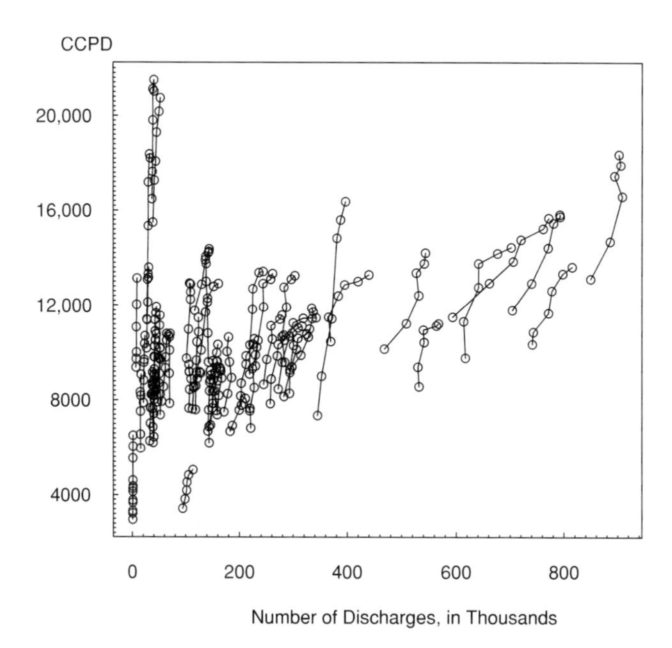

Figure 10.4 Scatter plot of CCPD versus number of discharges. The line segments connect observations within a state over 1990–5. We see a substantial state variation of numbers of discharges. There is a positive relationship between CCPD and number of discharges for each state. Slopes are higher for those states with smaller number of discharges.

a basic graphical form, each version portraying a variation of the basic theme, promote comparisons and assessments of change. By repeating a basic graphical form, we promote the process of communication.

Tufte (1997) states that using small multiples in graphical displays achieves the same desirable effects as using parallel structure in writing. Parallel structure in writing is successful because it allows readers to identify a sentence relationship only once and then focus on the meaning of each individual sentence element, such as a word, phrase, or clause. Parallel structure helps achieve economy of expression and draw together related ideas for comparison and contrast. Similarly, small multiples in graphs allow us to visualize complex relationships across different groups and over time. See guideline 5 in Section 21.3 for further discussion.

Figure 10.5 illustrates the use of small multiples. In each panel, the plot portrayed is identical except that it is based on a different state; this use of parallel structure allows us to demonstrate the increasing covered claims per discharge (CCPD) for each state. Moreover, by organizing the states by average CCPD, we can see the overall level of CCPD for each state as well as variations in the slope (rate of increase).

10.3 Basic Fixed Effects Models

Data

As described in Section 10.1, we let y_{it} denote the dependent variable of the ith subject at the tth time point. Associated with each dependent variable is a set

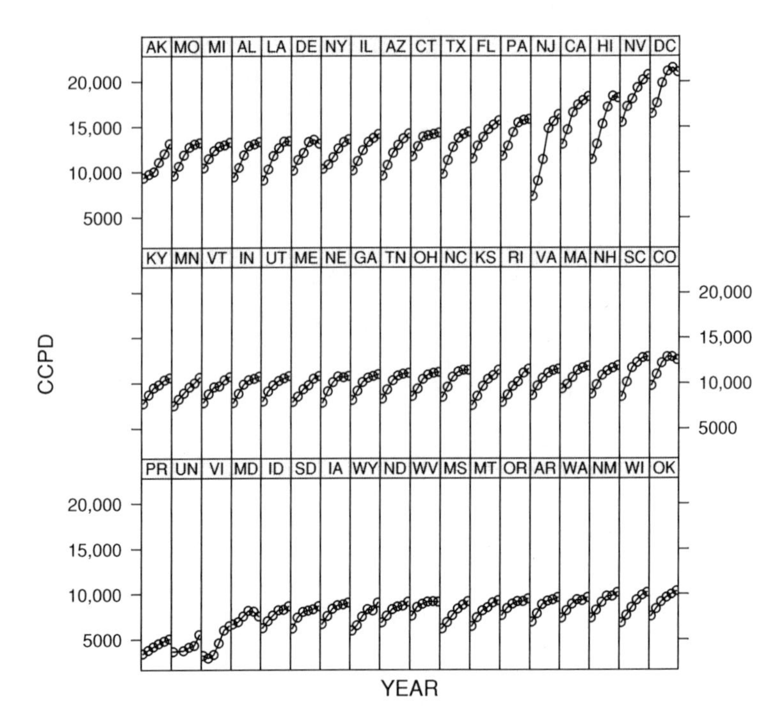

Figure 10.5 Trellis plot of CCPD versus year. Each of the 54 panels represents a plot of CCPD versus YEAR, 1990–5 (the horizontal axis is suppressed). The increase for New Jersey (NJ) is unusually large.

of explanatory variables. For the state hospital costs example, these explanatory variables include the number of discharged patients and the average hospital stay per discharge. In general, we assume there are k explanatory variables $x_{it,1}, x_{it,2}, \ldots, x_{it,k}$ that may vary by subject i and time t. We achieve a more compact notational form by expressing the k explanatory variables as a $k \times 1$ column vector

$$\mathbf{x}_{it} = \begin{pmatrix} x_{it,1} \\ x_{it,2} \\ \vdots \\ x_{it,k} \end{pmatrix}.$$

With this notation, the data for the ith subject consists of:

$$\begin{pmatrix} x_{i1,1}, x_{i1,2}, \ldots, x_{i1,k}, y_{i1} \\ \vdots \\ x_{iT_i,1}, x_{iT_i,2}, \ldots, x_{iT_i,k}, y_{iT_i} \end{pmatrix} = \begin{pmatrix} \mathbf{x}'_{i1}, y_{i1} \\ \vdots \\ \mathbf{x}'_{iT_i}, y_{iT_i} \end{pmatrix}.$$

Model

A basic (and useful) longitudinal data model is a special case of the multiple linear regression model introduced in Section 3.2. We use the modeling assumptions from Section 3.2.3 with the regression function

$$\begin{aligned} \mathrm{E}\ y_{it} &= \alpha_i + \beta_1 x_{it,1} + \beta_2 x_{it,2} + \cdots + \beta_k x_{it,k} \\ &= \alpha_i + \mathbf{x}_{it}^{\prime} \boldsymbol{\beta}, \qquad t = 1, \ldots, T_i, \ \ i = 1, \ldots, n. \end{aligned} \tag{10.1}$$

This is the *basic fixed effects model.*

The parameters $\{\beta_j\}$ are common to each subject and are called *global*, or *population*, parameters. The parameters $\{\alpha_i\}$ vary by subject and are known as *individual*, or *subject-specific*, parameters. In many applications, the population parameters capture broad relationships of interest and hence are the parameters of interest. The subject-specific parameters account for the different features of subjects, not broad population patterns. Hence, they are often of secondary interest and are called *nuisance* parameters. In Section 10.5, we will discuss the case where $\{\alpha_i\}$ are random variables. To distinguish from this case, this section treats $\{\alpha_i\}$ as nonstochastic parameters that are called fixed effects.

The subject-specific parameters help to control for differences, or heterogeneity among subjects. The estimators of these parameters use information in the repeated measurements on a subject. Conversely, the parameters $\{\alpha_i\}$ are nonestimable in cross-sectional regression models without repeated observations. That is, with $T_i = 1$, the model $y_{it} = \alpha_i + \beta_1 x_{it,1} + \beta_2 x_{it,2} + \cdots + \beta_k x_{it,k} + \varepsilon_{it}$ has more parameters $(n + k)$ than observations (n); thus, we cannot identify all the parameters. Typically, the disturbance term ε_{it} includes the information in α_i in cross-sectional regression models. An important advantage of longitudinal data models when compared to cross-sectional regression models is the ability to separate the effects of $\{\alpha_i\}$ from the disturbance terms $\{\varepsilon_{it}\}$. By separating out subject-specific effects, our estimates of the variability become more precise and we achieve more accurate inferences.

Estimation

Estimation of the basic fixed effects model follows directly from the least squares methods. The key insight is that the heterogeneity parameters $\{\alpha_i\}$ simply represent a *factor*, that is, a categorical variable that describes the unit of observation. With this, least squares estimation follows directly with the details given in Section 4.4 and the supporting appendices.

The heterogeneity parameters $\{\alpha_i\}$ can be represented by a factor, *that is, a categorical variable that describes the unit.*

As described in Chapter 4, one can replace categorical variables with an appropriate set of binary variables. For this reason, panel data estimators are sometimes known as *least squares dummy variable model* estimators. However, as we have seen in Chapter 4, be careful with the statistical routines. For some applications, the number of subjects can easily run into the thousands. Creating this many binary variables is computationally cumbersome. When you identify a variable as categorical, statistical packages typically use more computationally efficient recursive procedures (described in Section 4.7.2).

In the basic fixed effects model, coefficients associated with time-constant variables cannot be estimated.

The heterogeneity factor $\{\alpha_i\}$ does not depend on time. Because of this, it is easy to establish that regression coefficients associated with time-constant variables cannot be estimated using the basic fixed effects model. In other words,

Table 10.1 Coefficients and Summary Statistics from Three Models

	Regression Model 1		Regression Model 2		Basic Fixed Effects Model	
	Coefficient	t-Statistic	Coefficient	t-Statistic	Coefficient	t-Statistic
NUM_DCHG	4.70	6.49	4.66	6.44	10.75	4.18
YEAR	744.15	7.96	733.27	7.79	710.88	26.51
AVE_DAYS	325.16	3.85	308.47	3.58	361.29	6.23
YEARNJ			299.93	1.01	1,262.46	9.82
s	2,731.90		2,731.78		529.45	
R^2 (%)	28.6		28.8		99.8	
R_a^2 (%)	27.9		27.9		99.8	

time-constant variables are perfectly collinear with the heterogeneity factor. Because of this limitation, analysts often prefer to design their studies to use the competing random effects model that we will describe in Section 10.5.

Example: Medicare Hospital Costs, Continued. We compare the fit of the basic fixed effects model to ordinary regression models. Model 1 of Table 10.1 shows the fit of an ordinary regression model using number of discharges (NUM_DCHG), YEAR and average hospital stay (AVE_DAYS). Judging by the large t-statistics, each variable is statistically significant. The intercept term is not printed.

Figure 10.5 suggests that New Jersey has an unusually large increase. Thus, an interaction term, YEARNJ, was created that equals YEAR if the observation is from New Jersey and zero otherwise. This variable is incorporated in Model 2, where it does not appear to be significant.

Table 10.1 also shows the fit of a basic fixed effects model with these explanatory variables. In the table, the 54 subject-specific coefficients are not reported. In this model, each variable is statistically significantly, including the interaction term. Most striking is the improvement in the overall fit. The residual standard deviation (s) decrease from 2,731 to 530 and the coefficient of determination (R^2) increased from 29% to 99.8%.

10.4 Extended Fixed Effects Models

Analysis of Covariance Models

In the basic fixed effects model, no special relationships between subjects and time periods are assumed. By interchanging the roles of i and t, we can consider the regression function

$$\mathrm{E}\, y_{it} = \lambda_t + \mathbf{x}_{it}'\boldsymbol{\beta}.$$

Both this regression function and the one in equation (10.1) are based on traditional one-way analysis of covariance models introduced in Section 4.4. For this reason, the basic fixed effects model is also called the *one-way fixed effects model*. By using binary (dummy) variables for the time dimension, we can incorporate time-specific parameters into the population parameters. In this way, it is straightforward to consider the regression function

$$\mathrm{E}\ y_{it} = \alpha_i + \lambda_t + \mathbf{x}_{it}'\boldsymbol{\beta},$$

known as the *two-way fixed effects model.*

Example: Urban Wages. Glaeser and Maré (2001) investigated the effects of determinants on wages, with the goal of understanding why workers in cities earn more than their nonurban counterparts. They examined two-way fixed effects models using data from the National Longitudinal Survey of Youth (NLSY); they also used data from the Panel Study of Income Dynamics (PSID) to assess the robustness of their results to another sample. For the NLSY data, they examined $n = 5{,}405$ male heads of households over the years 1983–93, consisting of a total of $N = 40{,}194$ observations. The dependent variable was logarithmic hourly wage. The primary explanatory variable of interest was a three-level categorical variable that measures the size of the city in which workers reside. To capture this variable, two binary (dummy) variables were used: (1) a variable to indicate whether the worker resides in a large city (with more than a half million residents), a "dense metropolitan area," and (2) a variable to indicate whether the worker resides in a metropolitan area that does not contain a large city, a "nondense metropolitan area." The reference level is nonmetropolitan area. Several other control variables were included to capture effects of a worker's experience, occupation, education, and race. When including time dummy variables, there were $k = 30$ explanatory variables in the reported regressions.

Variable Coefficients Models

In the Medicare hospital costs example, we introduced an interaction variable to represent the unusually high increases in New Jersey costs. However, an examination of Figure 10.5 suggests that many other states are also "unusual." Extending this line of thought, we might want to allow each state to have its own rate of increase, corresponding to the increased hospital charges for that state. We could consider a regression function of the form

$$\mathrm{E}\ CCPD_{it} = \alpha_i + \beta_1 (NUM\_DCHG)_{it} + \beta_{2i} (YEAR)_t + \beta_3 (AVE\_DAYS)_{it}, \tag{10.2}$$

where the slope associated with YEAR is allowed to vary with state i.

Extending this line of thought, we write the regression function for a *variable coefficients* fixed effects model as

$$\mathrm{E}\ y_{it} = \mathbf{x}_{it}'\boldsymbol{\beta}_i.$$

With this notation, we may allow any or all of the variables to be associated with subject-specific coefficients. For simplicity, the subject-specific intercept is now included in the regression coefficient vector $\boldsymbol{\beta}_i$.

Example: Medicare Hospital Costs, Continued. The regression function in equation (10.2) was fit to the data. Not surprisingly, it resulted in excellent fit in the sense that the coefficient of determination is $R^2 = 99.915\%$ and the adjusted version is $R_a^2 = 99.987\%$. However, compared to the basic fixed effects model, there are an additional 52 parameters, a slope for each state (54 states to begin with, minus 1 for the 'population' term and minus 1 for New Jersey already included). Are the extra terms helpful? One way of analyzing this is through the general linear hypothesis test introduced in Section 4.2.2. In this context, the variable coefficients model represents the "full" equation and the basic fixed effects model is our "reduced" equation. From equation (4.4), the test statistic is

$$F\text{-ratio} = \frac{(0.99915 - 0.99809)/52}{(1 - 0.99915)/213} = 5.11.$$

Comparing this to the F-distribution with $df_1 = 52$ and $df_2 = 213$, we see that the associated p-value is less than 0.0001, indicating strong statistical significance. Thus, this is an indication that the variable slope model is preferred when compared to the basic fixed effects model.

Models with Serial Correlation

In longitudinal data, subjects are measured repeatedly over time. For some applications, time trends represent a minor portion of the overall variation. In these cases, one can adjust for their presence by calculating standard errors of regression coefficients robustly, similar to the Section 5.7.2 discussion. However, for other applications, gaining a good understanding of time trends is vital. One such application that is important in actuarial science is prediction; for example, recall the Section 10.1 discussion of an actuary predicting insurance claims for a small business.

We have seen in Chapters 7–9 some basic ways to incorporate time trends, through linear trends in time (e.g., the YEAR term in the Medicare hospital costs example) or dummy variables in time (another type of one-way fixed effects model). Another possibility is to use a lagged dependent variable as a predictor. However, this is known to have some unexpected negative consequences for the basic fixed effects model (see, e.g., the discussion in Hsiao, 2003, section 4.2; or Frees, 2004, section 6.3).

Instead, it is customary to examine the serial correlation structure of the disturbance term $\varepsilon_{it} = y_{it} - \mathrm{E}\ y_{it}$. For example, a common specification is to use an autocorrelation of order 1, $AR(1)$, structure, such as

$$\varepsilon_{it} = \rho_{\varepsilon}\varepsilon_{i,t-1} + \eta_{it},$$

where $\{\eta_{it}\}$ is a set of disturbance random variables and ρ_ε is the autocorrelation parameter. In many longitudinal data sets, the small number of time measurements (T) would inhibit calculation of the correlation coefficient ρ_ε using traditional methods such as those introduced in Chapter 8. However, with longitudinal data, we have many replications (n) of these short time series; intuitively, the replications provide the information needed to provide reliable estimates of the autoregressive parameter.

10.5 Random Effects Models

Suppose that you are interested in studying the behavior of subjects that are randomly selected from a population. For example, you might wish to predict insurance claims for a small business, using characteristics of the business and past claims history. Here, the set of small businesses may be randomly selected from a larger database. In contrast, the Section 10.3 Medicare example dealt with a fixed set of subjects. That is, it is difficult to think of the 54 states as a subset from some "superpopulation" of states. For both situations, it is natural to use subject-specific parameters, $\{\alpha_i\}$, to represent the heterogeneity among subjects. Unlike Section 10.3, we now discuss situations in which it is more reasonable to represent $\{\alpha_i\}$ as random variables instead of fixed, yet unknown, parameters. By arguing that $\{\alpha_i\}$ are draws from a distribution, we will be able to make inferences about subjects in a population that are not included in the sample.

Basic Random Effects Model

The *basic random effects* model equation is

$$y_{it} = \alpha_i + \mathbf{x}_{it}'\boldsymbol{\beta} + \varepsilon_{it}, \quad t = 1, \ldots, T_i, \quad i = 1, \ldots, n. \tag{10.3}$$

This notation is similar to the basic fixed effects model. However, now the term α_i is assumed to be a random variable, not a fixed, unknown parameter. The term α_i is known as a *random effect. Mixed effects* models are ones that include random and fixed effects. Because equation (10.3) includes random effects (α_i) and fixed effects ($\mathbf{x}_{it}$), the basic random effects model is a special case of the *mixed linear model.* The general mixed linear model is introduced in Section 15.1.

To complete the specification, we assume that $\{\alpha_i\}$ are identically and independently distributed with mean zero and variance σ_α^2. Further, we assume that $\{\alpha_i\}$ are independent of the disturbance random variables, ε_{it}. Note that because $\mathrm{E}\ \alpha_i = 0$, it is customary to include a constant within the vector $\mathbf{x}_{it}$. This was not true of the fixed effects models in Section 10.3, where we did not center the subject-specific terms about 0.

Linear combinations of the form $\mathbf{x}_{it}'\boldsymbol{\beta}$ quantify the effect of known variables that may affect the dependent variable. Additional variables that are either unimportant or unobservable constitute the "error term." In equation (10.3), we may think of a regression model $y_{it} = \mathbf{x}_{it}'\boldsymbol{\beta} + \eta_{it}$, where the error term η_{it} is decomposed into two components so that $\eta_{it} = \alpha_i + \varepsilon_{it}$. The term α_i represents the time-constant portion, whereas ε_{it} represents the remaining portion. To identify the model parameters, we assume that the two terms are independent. In the econometrics literature, this is known as the *error components* model; in the biological sciences, is is known as the *random intercepts* model.

Estimation

Estimation of the random effects model does not follows directly from least squares as with the fixed effects models. This is because the observations are no longer independent as a result of the random effects terms. Instead, an extension of least squares known as *generalized least squares* is used to account for this dependency. Generalized least squares, often denoted by the acronym *GLS*, is a type of weighted least squares. Because random effects models are special cases of mixed linear models, we introduce GLS estimation in this broader framework in Section 15.1.

To see the dependency among observations, consider the covariance between the first two observations of the ith subject. Basic calculations show the following:

$$\begin{aligned}
\mathrm{Cov}(y_{i1}, y_{i2}) &= \mathrm{Cov}(\alpha_i + \mathbf{x}_{i1}'\boldsymbol{\beta} + \varepsilon_{i1}, \alpha_i + \mathbf{x}_{i2}'\boldsymbol{\beta} + \varepsilon_{i2}) \\
&= \mathrm{Cov}(\alpha_i + \varepsilon_{i1}, \alpha_i + \varepsilon_{i2}) \\
&= \mathrm{Cov}(\alpha_i, \alpha_i) + \mathrm{Cov}(\alpha_i, \varepsilon_{i2}) + \mathrm{Cov}(\varepsilon_{i1}, \alpha_i) + \mathrm{Cov}(\varepsilon_{i1}, \varepsilon_{i2}) \\
&= \mathrm{Cov}(\alpha_i, \alpha_i) = \sigma_\alpha^2.
\end{aligned}$$

The systematic terms $\mathbf{x}'\boldsymbol{\beta}$ drop out of the covariance calculation because they are nonrandom. Further, the covariance terms involving ε are zero because of the assumed independence. This calculation shows that the covariance between any two observations from the same subject is σ_α^2. Similar calculations show that the variance of an observation is $\sigma_\alpha^2 + \sigma_\varepsilon^2$. Thus, the correlation between observations within a subject is $\sigma_\alpha^2/(\sigma_\alpha^2 + \sigma_\varepsilon^2)$. This quantity is known as the *intraclass correlation*, a commonly reported measure of dependence in random effects studies.

Example: Group Term Life. Frees, Young, and Luo (2001) analyzed claims data provided by an insurer of credit unions. The data contains claims and exposure information from 88 Florida credit unions for the years 1993–6. These are "life savings" claims from a contract between the credit union and its members that provides a death benefit based on the member's savings deposited in the credit union. Actuaries typically price life insurance coverage with knowledge of an insured's age and sex, as well as other explanatory variables such as occupation. However, for the data from small groups, often only a minimal amount of information is available to understand claims behavior.

Of the $88 \times 4 = 352$ potential observations, 27 were not available because these credit unions had zero coverage in that year (and thus were excluded). Thus, the data were unbalanced. The dependent variable is the annual total claims from the life savings contract, in logarithmic units. The explanatory variables were annual coverage, in logarithmic units, and YEAR, a time trend.

A fit of the basic random effects model showed that both year and the annual coverage had positive and strongly statistically significant coefficients. That is, the typical amount of claims increased over the period studied and claims increased as coverage increased, other things being equal. There were strong credit union effects, as well. For example, the estimated intraclass correlation was 0.703, also suggesting strong dependence among observations.

Extended Random Effects Models

As with fixed effects, random effects models can be easily extended to incorporated variable coefficients and serial correlations. For example, Frees et al. (2001) considered the model equation

$$y_{it} = \alpha_{1i} + \alpha_{2i}\text{LNCoverage}_{it} + \beta_1 + \beta_2\text{YEAR}_t + \beta_3\text{LNCoverage}_{it} + \varepsilon_{it}, \tag{10.4}$$

where LNCoverage_{it} is the logarithmic life savings coverage. As with the basic model, it is customary to use a mean zero for the random effects. Thus, the overall intercept is β_1, and α_{1i} represents credit union deviations. Further, the overall or global slope associated with LNCoverage is β_3 and α_{2i} represents credit union deviations. Put another way, the slope corresponding to LNCoverage for the ith credit is $\beta_3 + \alpha_{2i}$.

More generally, the *variable coefficients random effects model* equation can be written as

$$y_{it} = \mathbf{x}_{it}'\boldsymbol{\beta} + \mathbf{z}_{it}'\boldsymbol{\alpha}_i + \varepsilon_{it}. \tag{10.5}$$

As with the fixed effects variable coefficients model, we may allow any or all of the variables to be associated with subject-specific coefficients. The convention used in the literature is to specific fixed effects through the systematic component $\mathbf{x}_{it}'\boldsymbol{\beta}$ and random effects through the component $\mathbf{z}_{it}'\boldsymbol{\alpha}_i$. Here, the vector $\mathbf{z}_{it}$ is typically equal to or a subset of $\mathbf{x}_{it}$, though it need not be so. With this notation, we now have a vector of random effects $\boldsymbol{\alpha}_i$ that are subject specific. To reduce to our basic model, one needs only to choose $\boldsymbol{\alpha}_i$ to be a scalar (a 1×1 vector) and $\mathbf{z}_{it} \equiv 1$. The example in equation (10.4) results from choosing $\boldsymbol{\alpha}_i = (\alpha_{1i}, \alpha_{2i})'$ and $\mathbf{z}_{it} = (1, \text{LNCoverage}_{it})'$.

As with fixed effects models, one can readily incorporate models of serial correlation into random effects models by specifying a correlations structure for $\varepsilon_{i1}, \ldots, \varepsilon_{iT}$. This feature is readily available in statistical packages and is described fully in the references in Section 10.6.

10.6 Further Reading and References

Longitudinal and panel data models are widely used. To illustrate, an index of business and economic journals, ABI/INFORM, lists 685 articles in 2004 and 2005 that use panel data methods. Another index of scientific journals, the ISI Web of Science, lists 1,137 articles in 2004 and 2005 that use longitudinal data methods. A book-length introduction to longitudinal and panel data that emphasizes business and social science applications is Frees (2004). Diggle et al. (2002) provide an introduction from a biomedical perspective. Hsiao (2003) provides a classic introduction from an econometric perspective.

Actuaries are particularly interested in predictions resulting from longitudinal data. These predictions can form the basis for updating insurance prices. This topic is discussed in Chapter 18 on credibility and bonus-malus factors.

Chapter References

Diggle, Peter J., Patrick Heagarty, Kung-Yee Liang, and Scott L. Zeger (2002). *Analysis of Longitudinal Data*, 2nd ed. Oxford University Press, London.

Frees, Edward W. (2004). *Longitudinal and Panel Data: Analysis and Applications in the Social Sciences.* Cambridge University Press, New York.

Frees, Edward W., Virginia R. Young, and Yu Luo (2001). Case studies using panel data models. *North American Actuarial Journal* 5, no. 4, 24–42.

Glaeser, E. L., and D. C. Maré (2001). Cities and skills. *Journal of Labor Economics* 19, 316–42.

Hsiao, Cheng (2003). *Analysis of Panel Data*, 2nd ed. Cambridge University Press, New York.

Tufte, Edward R. (1997). *Visual Explanations*. Graphics Press, Cheshire, Connecticut.

Part III

Topics in Nonlinear Regression

11

Categorical Dependent Variables

Chapter Preview. A model with a categorical dependent variable allows one to predict whether an observation is a member of a distinct group or category. Binary variables represent an important special case; they can indicate whether an event of interest has occurred. In actuarial and financial applications, the event may be whether a claim occurs, whether a person purchases insurance, whether a person retires or a firm becomes insolvent. This chapter introduces logistic regression and probit models of binary dependent variables. Categorical variables may also represent more than two groups, known as *multicategory* outcomes. Multicategory variables can be unordered or ordered, depending on whether it makes sense to rank the variable outcomes. For unordered outcomes, known as *nominal* variables, the chapter introduces generalized logits and multinomial logit models. For ordered outcomes, known as *ordinal* variables, the chapter introduces cumulative logit and probit models.

11.1 Binary Dependent Variables

We have already introduced binary variables as a special type of discrete variable that can be used to indicate whether a subject has a characteristic of interest, such as sex for a person or ownership of a captive insurance company for a firm. Binary variables also describe whether an event of interest, such as an accident, has occurred. A model with a binary dependent variable allows one to predict whether an event has occurred or a subject has a characteristic of interest.

Example: MEPS Expenditures. Section 11.4 will describe an extensive database from the Medical Expenditure Panel Survey (MEPS) on hospitalization utilization and expenditures. For these data, we will consider

$$y_i = \begin{cases} 1 & i\text{th person was hospitalized during the sample period} \\ 0 & \text{otherwise} \end{cases}.$$

There are $n = 2{,}000$ persons in this sample, distributed as:

Table 11.1 Hospitalization by Sex

		Male	Female
Not hospitalized	$y = 0$	902 (95.3%)	941 (89.3%)
Hospitalized	$y = 1$	44 (4.7%)	113 (10.7%)
Total		946	1,054

Table 11.1 suggests that sex has an important influence on whether someone becomes hospitalized.

Like the linear regression techniques introduced in prior chapters, we are interested in using characteristics of a person, such as their age, sex, education, income, and prior health status, to help explain the dependent variable y. Unlike the prior chapters, now the dependent variable is discrete and not even approximately normally distributed. In limited circumstances, linear regression can be used with binary dependent variables – this application is known as a *linear probability model*.

Linear Probability Models

To introduce some of the complexities encountered with binary dependent variables, denote the probability that the response equals 1 by $\pi_i = \Pr(y_i = 1)$. A binary random variable has a *Bernoulli distribution*. Thus, we may interpret the mean response as the probability that the response equals one; that is, $\mathrm{E}\, y_i = 0 \times \Pr(y_i = 0) + 1 \times \Pr(y_i = 1) = \pi_i$. Further, the variance is related to the mean through the expression $\mathrm{Var}\, y_i = \pi_i(1 - \pi_i)$.

Linear probability models enjoy convenient parameter interpretations.

We begin by considering a linear model of the form

$$y_i = \mathbf{x}_i'\boldsymbol{\beta} + \varepsilon_i,$$

known as a *linear probability model*. Assuming that $\mathrm{E}\, \varepsilon_i = 0$, we have that $\mathrm{E}\, y_i = \mathbf{x}_i'\boldsymbol{\beta} = \pi_i$. Because y_i has a Bernoulli distribution, $\mathrm{Var}\, y_i = \mathbf{x}_i'\boldsymbol{\beta}(1 - \mathbf{x}_i'\boldsymbol{\beta})$. Linear probability models are used because of the ease of parameter interpretations. For large datasets, the computational simplicity of ordinary least squares estimators is attractive when compared to some complex alternative nonlinear models introduced later in this chapter. As described in Chapter 3, ordinary least squares estimators for $\boldsymbol{\beta}$ have desirable properties. It is straightforward to check that the estimators are consistent and asymptotically normal under mild conditions on the explanatory variables $\{\mathbf{x}_i\}$. However, linear probability models have several drawbacks that are serious for many applications.

Drawbacks of the Linear Probability Model

- *Fitted values can be poor.* The expected response is a probability and thus must vary between 0 and 1. However, the linear combination, $\mathbf{x}_i'\boldsymbol{\beta}$, can vary between negative and positive infinity. This mismatch implies, for example, that fitted values may be unreasonable.

- *Heteroscedasticity.* Linear models assume homoscedasticity (constant variance), yet the variance of the response depends on the mean that varies over observations. The problem of varying variability is known as *heteroscedasticity*.
- *Residual analysis is meaningless.* The response must be either a 0 or 1, although the regression models typically regard distribution of the error term as continuous. This mismatch implies, for example, that the usual residual analysis in regression modeling is meaningless.

To handle the heteroscedasticity problem, a (two-stage) weighted least squares procedure is possible. In the first stage, one uses ordinary least squares to compute estimates of $\boldsymbol{\beta}$. With this estimate, an estimated variance for each subject can be computed using the relation Var $y_i = \mathbf{x}_i'\boldsymbol{\beta}(1 - \mathbf{x}_i'\boldsymbol{\beta})$. At the second stage, a weighted least squares is performed using the inverse of the estimated variances as weights to arrive at new estimates of $\boldsymbol{\beta}$. It is possible to iterate this procedure, although studies have shown that there are few advantages in doing so (see Carroll and Ruppert, 1988). Alternatively, one can use ordinary least squares estimators of $\boldsymbol{\beta}$ with standard errors that are robust to heteroscedasticity (see Section 5.7.2).

11.2 Logistic and Probit Regression Models

11.2.1 Using Nonlinear Functions of Explanatory Variables

To circumvent the drawbacks of linear probability models, we consider alternative models in which we express the expectation of the response as a function of explanatory variables, $\pi_i = \pi(\mathbf{x}_i'\boldsymbol{\beta}) = \Pr(y_i = 1|\mathbf{x}_i)$. We focus on two special cases of the function $\pi(\cdot)$:

- $\pi(z) = \frac{1}{1+\exp(-z)} = \frac{e^z}{1+e^z}$, the logit case
- $\pi(z) = \Phi(z)$, the probit case

Here, $\Phi(\cdot)$ is the standard normal distribution function. The choice of the identity function (a special kind of linear function), $\pi(z) = z$, yields the linear probability model. In contrast, π is nonlinear for both the logit and probit cases. These two functions are similar in that they are almost linearly related over the interval $0.1 \leq p \leq 0.9$. Thus, to a large extent, the function choice is dependent on the preferences of the analyst. Figure 11.1 compares the logit and probit functions showing that it will be difficult to distinguish between the two specifications with most datasets.

The inverse of the function, π^{-1}, specifies the form of the probability that is linear in the explanatory variables, that is, $\pi^{-1}(\pi_i) = \mathbf{x}_i'\boldsymbol{\beta}$. In Chapter 13, we refer to this inverse as the *link function*.

Table 11.2 Characteristics Used in Some Credit-Scoring Procedures

Characteristics	Potential Values
Time at present address	0–1, 1–2, 3–4, 5+ years
Home status	Owner, tenant, other
Postal code	Band A, B, C, D, E
Telephone	Yes, no
Applicant's annual income	£(0–10,000), £(10,000–20,000) £(20,000+)
Credit card	Yes, no
Type of bank account	Check and/or savings, none
Age	18–25, 26–40, 41–55, 55+ years
County court judgements	Number
Type of occupation	Coded
Purpose of loan	Coded
Marital status	Married, divorced, single, widow, other
Time with bank	Years
Time with employer	Years

Source: Hand and Henley (1997).

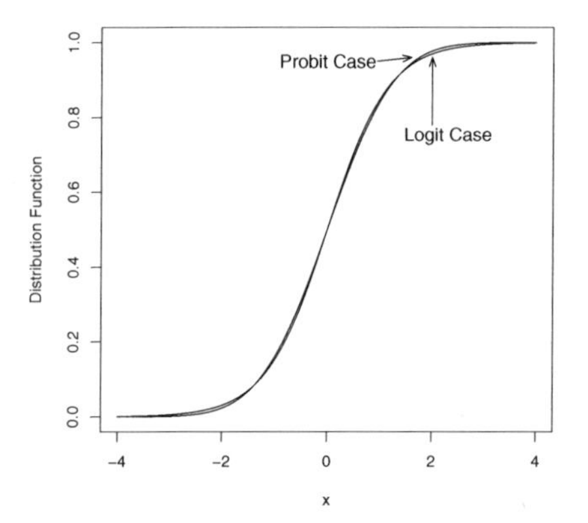

Figure 11.1 Comparison of logit and probit (standard normal) distribution functions.

Example: Credit Scoring. Banks, credit bureaus, and other financial institutions develop credit scores for individuals that are used to predict the likelihood that the borrower will repay current and future debts. Individuals who do not meet stipulated repayment schedules in a loan agreement are said to be in default. A credit score, then, is a predicted probability of being in default, with the credit application providing the explanatory variables used in developing the credit score. The choice of explanatory variables depends on the purpose of the application; credit scoring is used for issuing credit cards for making small consumer purchases as well as for mortgage applications for multimillion-dollar houses. In Table 11.2, Hand and Henley (1997) provide a list of typical characteristics that are used in credit scoring.

With credit application information and default experience, a logistic regression model can be used to fit the probability of default with credit scores resulting from fitted values. Wiginton (1980) provides an early application of logistic regression to consumer credit scoring. At that time, other statistical methods known as *discriminant analysis* were at the cutting edge of quantitative scoring methodologies. In their review article, Hand and Henley (1997) discuss other competitors to logistic regression including machine learning systems and neural networks. As noted by Hand and Henley, there is no uniformly "best" method. Regression techniques are important in their own right because of their widespread usage and because they can provide a platform for learning about newer methods.

Regression techniques are important because of their widespread usage and because they can provide a platform for learning about newer methods.

Credit scores provide estimates of the likelihood of defaulting on loans, but issuers of credit are also interested in the amount and timing of debt repayment. For example, a "good" risk may repay a credit balance so promptly that little profit is earned by the lender. Further, a "poor" mortgage risk may default on a loan so late in the duration of the contract that a sufficient profit was earned by the lender. See Gourieroux and Jasiak (2007) for a broad discussion of how credit modeling can be used to assess the riskiness and profitability of loans.

11.2.2 Threshold Interpretation

Both the logit and the probit cases can be interpreted as follows. Suppose that there exists an *underlying* linear model, $y_i^* = \mathbf{x}_i'\boldsymbol{\beta} + \varepsilon_i^*$. Here, we do not observe the response y_i^*, yet we interpret it to be the propensity to possess a characteristic. For example, we might think about the financial strength of an insurance company as a measure of its propensity to become insolvent (no longer capable of meeting its financial obligations). Under the threshold interpretation, we do not observe the propensity, but we do observe when the propensity crosses a threshold. It is customary to assume that this threshold is 0, for simplicity. Thus, we observe

$$y_i = \begin{cases} 0 & y_i^* \le 0 \\ 1 & y_i^* > 0 \end{cases}.$$

To see how the logit case is derived from the threshold model, assume a logistic distribution function for the disturbances, so that

$$\Pr(\varepsilon_i^* \le a) = \frac{1}{1+\exp(-a)}.$$

Like the normal distribution, one can verify by calculating the density that the logistic distribution is symmetric about zero. Thus, $-\varepsilon_i^*$ has the same distribution as ε_i^*, and so

$$\pi_i = \Pr(y_i = 1|\mathbf{x}_i) = \Pr(y_i^* > 0) = \Pr(\varepsilon_i^* \le \mathbf{x}_i'\beta) = \frac{1}{1+\exp(-\mathbf{x}_i'\boldsymbol{\beta})} = \pi(\mathbf{x}_i'\boldsymbol{\beta}).$$

This establishes the threshold interpretation for the logit case. The development for the probit case is similar and is omitted.

11.2.3 Random Utility Interpretation

Both the logit and probit cases are also justified by appealing to the following random utility interpretation of the model. In some economic applications, individuals select one of two choices. Here, preferences among choices are indexed by an unobserved utility function; individuals select the choice that provides the greater utility.

For the ith subject, we use the notation u_i for this utility function. We model the utility (U) as a function of an underlying value (V) plus random noise (ε); that is, $U_{ij} = u_i(V_{ij} + \varepsilon_{ij})$, where j may be 1 or 2, corresponding to the choice. To illustrate, we assume that the individual chooses the category corresponding to $j = 1$ if $U_{i1} > U_{i2}$ and denote this choice as $y_i = 1$. Assuming that u_i is a strictly increasing function, we have

$$\begin{aligned}\Pr(y_i = 1) &= \Pr(U_{i2} < U_{i1}) = \Pr\left(u_i(V_{i2} + \varepsilon_{i2}) < u_i(V_{i1} + \varepsilon_{i1})\right) \\ &= \Pr(\varepsilon_{i2} - \varepsilon_{i1} < V_{i1} - V_{i2}).\end{aligned}$$

To parameterize the problem, assume that the value V is an unknown linear combination of explanatory variables. Specifically, we take $V_{i2} = 0$ and $V_{i1} = \mathbf{x}_i'\boldsymbol{\beta}$. We may take the difference in the errors, $\varepsilon_{i2} - \varepsilon_{i1}$, as normal or logistic, corresponding to the probit and the logit cases, respectively. The logistic distribution is satisfied if the errors are assumed to have an *extreme-value*, or *Gumbel*, distribution (see, e.g., Amemiya, 1985).

11.2.4 Logistic Regression

An advantage of the logit case is that it permits closed-form expressions, unlike the normal distribution function. *Logistic regression* is another phrase used to describe the logit case.

Using $p = \pi(z) = (1 + \mathrm{e}^{-z})^{-1}$, the inverse of π is calculated as $z = \pi^{-1}(p) = \ln(p/(1-p))$. To simplify future presentations, we define

$$\text{logit}(p) = \ln\left(\frac{p}{1-p}\right)$$

to be the *logit function*. With a logistic regression model, we represent the linear combination of explanatory variables as the logit of the success probability; that is, $\mathbf{x}_i'\boldsymbol{\beta} = \text{logit}(\pi_i)$.

Odds Interpretation

When the response y is binary, knowing only $p = \Pr(y = 1)$ summarizes the entire distribution. In some applications, a simple transformation of p has an important interpretation. The lead example of this is the *odds*, given by $p/(1-p)$. For example, suppose that y indicates whether a horse wins a race and that p is the probability of the horse winning. If $p = 0.25$, then the odds of the horse winning are $0.25/(1.00-0.25)=0.3333$. We might say that the odds of winning are 0.3333 to 1, or 1 to 3. Equivalently, we say that the probability of not winning is $1-p=0.75$, so that the odds of the horse not winning is $0.75/(1-0.75)=3$ and the odds against the horse are 3 to 1.

Odds have a useful interpretation from a betting standpoint. Suppose that we are playing a fair game and that we place a bet of \$1 with one to three odds. If the horse wins, then we get our \$1 back plus winnings of \$3. If the horse loses, then we lose our bet of \$1. It is a fair game in the sense that the expected value of the game is zero because we win \$3 with probability $p=0.25$ and lose \$1 with probability $1-p=0.75$. From an economic standpoint, the odds provide the important numbers (bet of \$1 and winnings of \$3), not the probabilities. Of course, if we know p, then we can always calculate the odds. Similarly, if we know the odds, we can always calculate the probability p.

The logit is the logarithmic odds function, also known as the *log odds*.

Odds Ratio Interpretation

To interpret the regression coefficients in the logistic regression model, $\boldsymbol{\beta}=(\beta_0,\ldots,\beta_k)'$, we begin by assuming that jth explanatory variable, x_{ij}, is either zero or one. Then, with the notation $\mathbf{x}_i=(x_{i0},\ldots,x_{ij},\ldots,x_{ik})'$, we may interpret

$$\begin{aligned}\beta_j &= (x_{i0},\ldots,1,\ldots,x_{ik})'\boldsymbol{\beta}-(x_{i0},\ldots,0,\ldots,x_{ik})'\boldsymbol{\beta}\\ &= \ln\left(\frac{\Pr(y_i=1|x_{ij}=1)}{1-\Pr(y_i=1|x_{ij}=1)}\right)-\ln\left(\frac{\Pr(y_i=1|x_{ij}=0)}{1-\Pr(y_i=1|x_{ij}=0)}\right).\end{aligned}$$

Thus,

$$e^{\beta_j}=\frac{\Pr(y_i=1|x_{ij}=1)/(1-\Pr(y_i=1|x_{ij}=1))}{\Pr(y_i=1|x_{ij}=0)/(1-\Pr(y_i=1|x_{ij}=0))}.$$

This shows that e^{β_j} can be expressed as the ratio of two odds, known as the *odds ratio*. That is, the numerator of this expression is the odds when $x_{ij}=1$, whereas the denominator is the odds when $x_{ij}=0$. Thus, we can say that the odds when $x_{ij}=1$ are $\exp(\beta_j)$ times as large as the odds when $x_{ij}=0$. To illustrate, suppose $\beta_j=0.693$, so that $\exp(\beta_j)=2$. From this, we say that the odds (for $y=1$) are twice as great for $x_{ij}=1$ as for $x_{ij}=0$.

Similarly, assuming that jth explanatory variable is continuous (differentiable), we have

$$\begin{aligned}\beta_j &= \frac{\partial}{\partial x_{ij}}\mathbf{x}_i'\boldsymbol{\beta}=\frac{\partial}{\partial x_{ij}}\ln\left(\frac{\Pr(y_i=1|x_{ij})}{1-\Pr(y_i=1|x_{ij})}\right)\\ &= \frac{\frac{\partial}{\partial x_{ij}}\Pr(y_i=1|x_{ij})/(1-\Pr(y_i=1|x_{ij}))}{\Pr(y_i=1|x_{ij})/(1-\Pr(y_i=1|x_{ij}))}.\end{aligned}\tag{11.1}$$

Thus, we can interpret β_j as the proportional change in the odds ratio, known as an *elasticity* in economics.

We can interpret β_j as the proportional change in the odds ratio.

Example: MEPS Expenditures, Continued. Table 11.1 shows that the percentage of women who were hospitalized is 10.7%; alternatively, the odds of women being hospitalized are $0.107/(1 - 0.107) = 0.120$. For men, the percentage is 4.7%, so that the odds are 0.0493. The odds ratio is $0.120/0.0493 = 2.434$; women are more than twice as likely to be hospitalized as men.

From a logistic regression fit (described in Section 11.4), the coefficient associated with sex is 0.733. Given this model, we say that women are $\exp(0.733) = 2.081$ times as likely as men to be hospitalized. The regression estimate of the odds ratio controls for additional variables (e.g., age, education) compared to the basic calculation based on raw frequencies.

11.3 Inference for Logistic and Probit Regression Models

11.3.1 Parameter Estimation

The customary method of estimation for logistic and probit models is *maximum likelihood*, described in further detail in Section 11.9. To provide intuition, we outline the ideas in the context of binary dependent variable regression models.

The *likelihood* is the observed value of the probability function. For a single observation, the likelihood is

$$\begin{cases} 1 - \pi_i & \text{if } y_i = 0 \\ \pi_i & \text{if } y_i = 1 \end{cases}.$$

The objective of maximum likelihood estimation is to find the parameter values that produce the largest likelihood. Finding the maximum of the logarithmic function yields the same solution as finding the maximum of the corresponding function. Because it is generally computationally simpler, we consider the logarithmic (or log-) likelihood, written as

$$\begin{cases} \ln(1 - \pi_i) & \text{if } y_i = 0 \\ \ln \pi_i & \text{if } y_i = 1 \end{cases}. \tag{11.2}$$

The log-likelihood is viewed as a function of the parameters, with the data held fixed. In contrast, the joint probability mass function is viewed as a function of the realized data, with the parameters held fixed.

More compactly, the log-likelihood of a single observation is

$$y_i \ln \pi(\mathbf{x}_i'\boldsymbol{\beta}) + (1 - y_i) \ln\left(1 - \pi(\mathbf{x}_i'\boldsymbol{\beta})\right),$$

where $\pi_i = \pi(\mathbf{x}_i'\boldsymbol{\beta})$. Assuming independence among observations, the likelihood of the dataset is a product of likelihoods of each observation. Taking logarithms, the log-likelihood of the dataset is the sum of log-likelihoods of single observations.

The log-likelihood of the dataset is

$$L(\boldsymbol{\beta}) = \sum_{i=1}^{n} \left\{ y_i \ln \pi(\mathbf{x}_i'\boldsymbol{\beta}) + (1 - y_i) \ln \left(1 - \pi(\mathbf{x}_i'\boldsymbol{\beta})\right) \right\}. \qquad (11.3)$$

The log-likelihood is viewed as a function of the parameters, with the data held fixed. In contrast, the joint probability mass function is viewed as a function of the realized data, with the parameters held fixed.

The *method of maximum likelihood* involves finding the values of $\boldsymbol{\beta}$ that maximize the log-likelihood. The customary method of finding the maximum is taking partial derivatives with respect to the parameters of interest and finding roots of the resulting equations. In this case, taking partial derivatives with respect to $\boldsymbol{\beta}$ yields the *score equations*

$$\frac{\partial}{\partial \boldsymbol{\beta}} L(\boldsymbol{\beta}) = \sum_{i=1}^{n} \mathbf{x}_i \left(y_i - \pi(\mathbf{x}_i'\boldsymbol{\beta})\right) \frac{\pi'(\mathbf{x}_i'\boldsymbol{\beta})}{\pi(\mathbf{x}_i'\boldsymbol{\beta})(1 - \pi(\mathbf{x}_i'\boldsymbol{\beta}))} = \mathbf{0}, \qquad (11.4)$$

where π' is the derivative of π. The solution of these equations, denoted as $\mathbf{b}_{MLE}$, is the maximum likelihood estimator. For the logit function the score equations reduce to

$$\frac{\partial}{\partial \boldsymbol{\beta}} L(\boldsymbol{\beta}) = \sum_{i=1}^{n} \mathbf{x}_i \left(y_i - \pi(\mathbf{x}_i'\boldsymbol{\beta})\right) = \mathbf{0}, \qquad (11.5)$$

where $\pi(z) = 1/(1 + \exp(-z))$.

11.3.2 Additional Inference

An estimator of the large sample variance of $\boldsymbol{\beta}$ may be calculated taking partial derivatives of the score equations. Specifically, the term

$$\mathbf{I}(\boldsymbol{\beta}) = -\mathrm{E}\left(\frac{\partial^2}{\partial \boldsymbol{\beta}\, \partial \boldsymbol{\beta}'} L(\boldsymbol{\beta})\right)$$

is the *information matrix*. As a special case, using the logit function and equation (11.5), straightforward calculations show that the information matrix is

$$\mathbf{I}(\boldsymbol{\beta}) = \sum_{i=1}^{n} \sigma_i^2 \mathbf{x}_i \mathbf{x}_i',$$

where $\sigma_i^2 = \pi(\mathbf{x}_i'\boldsymbol{\beta})(1 - \pi(\mathbf{x}_i'\boldsymbol{\beta}))$. The square root of the $(j+1)st$ diagonal element of this matrix evaluated at $\boldsymbol{\beta} = \mathbf{b}_{MLE}$ yields the standard error for $b_{j,MLE}$, denoted as $se(b_{j,MLE})$.

To assess the overall model fit, it is customary to cite *likelihood ratio test statistics* in nonlinear regression models. To test the overall model adequacy

$H_0 : \boldsymbol{\beta} = \mathbf{0}$, we use the statistic

$$LRT = 2 \times (L(\mathbf{b}_{MLE}) - L_0),$$

where L_0 is the maximized log-likelihood with only an intercept term. Under the null hypothesis H_0, this statistic has a chi-square distribution with k degrees of freedom. Section 11.9.3 describes likelihood ratio test statistics in greater technical detail.

As described in Section 11.9, measures of goodness of fit can be difficult to interpret in nonlinear models. One measure is the so-called *max-scaled* R^2, defined as $R^2_{ms} = R^2 / R^2_{max}$, where

$$R^2 = 1 - \left(\frac{\exp(L_0/n)}{\exp(L(\mathbf{b}_{MLE})/n)} \right),$$

and $R^2_{max} = 1 - \exp(L_0/n)^2$. Here, L_0/n represents the average value of this log-likelihood.

Another measure is *pseudo-*R^2

$$\frac{L(\mathbf{b}_{MLE}) - L_0}{L_{max} - L_0},$$

where L_0 and L_{max} is the log-likelihood based on only an intercept and on the maximum achievable, respectively. Like the coefficient of determination, the pseudo-R^2 takes on values between zero and one, with larger values indicating a better fit to the data. Other versions of the pseudo-R^2 are available in the literature; see, for example, Cameron and Trivedi (1998). An advantage of this pseudo-R^2 measure is its link to hypothesis testing of regression coefficients.

Example: Job Security. Valletta (1999) studied declining job security using the Panel Survey of Income Dynamics (PSID) database. We consider here one of the regression models presented by Valletta, based on a sample of male heads of households that consists of $n = 24{,}168$ observations over the years 1976–92, inclusive. The PSID survey records reasons why men left their most recent employment, including plant closures, "quit," and changed jobs for other reasons. However, Valletta focused on dismissals ("laid off" or "fired") because involuntary separations are associated with job insecurity.

Table 11.3 presents a probit regression model run by Valletta (1999), using dismissals as the dependent variable. In addition to the explanatory variables listed in Table 11.3, other variables controlled for consisted of education, marital status, number of children, race, years of full-time work experience and its square, union membership, government employment, logarithmic wage, the U.S. employment rate, and location as measured through the Metropolitan Statistical Area residence. In Table 11.3, tenure is years employed at the current firm.

Table 11.3 Dismissal Probit Regression Estimates

Variable	Parameter Estimate	Standard Error
Tenure	−0.084	0.010
Time Trend	−0.002	0.005
Tenure*(Time Trend)	0.003	0.001
Change in Logarithmic Sector Employment	0.094	0.057
Tenure*(Change in Logarithmic Sector Employment)	−0.020	0.009
−2 Log-Likelihood	7,027.8	
Pseudo-R^2	0.097	

Further, sector employment was measured by examining the Consumer Price Survey employment in 387 sectors of the economy, based on 43 industry categories and nine regions of the country.

On the one hand, the tenure coefficient reveals that more experienced workers are less likely to be dismissed. On the other hand, the coefficient associated with the interaction between tenure and time trend reveals an increasing dismissal rate for experienced workers.

The interpretation of the sector employment coefficients is also of interest. With an average tenure of about 7.8 years in the sample, we see that the low-tenure men are relatively unaffected by changes in sector employment. However, for more experienced men, there is an increasing probability of dismissal associated with sectors of the economy where growth declines.

11.4 Application: Medical Expenditures

This section considers data from the Medical Expenditure Panel Survey (MEPS), conducted by the U.S. Agency of Health Research and Quality. The MEPS is a probability survey that provides nationally representative estimates of health-care use, expenditures, sources of payment, and insurance coverage for the U.S. civilian population. This survey collects detailed information on individuals and each medical-care episode by type of services, including physician office visits, hospital emergency room visits, hospital outpatient visits, hospital inpatient stays, all other medical provider visits, and use of prescribed medicines. This detailed information allows one to develop models of health-care utilization to predict future expenditures. We consider MEPS data from the first panel of 2003 and take a random sample of $n = 2{,}000$ individuals between the ages of 18 and 65.

Dependent Variable

Ⓡ **Empirical** *Filename is "HealthExpend"*

Our dependent variable is an indicator of positive expenditures for inpatient admissions. For MEPS, inpatient admissions include persons who were admitted to a hospital and stayed overnight. In contrast, outpatient events include hospital outpatient department visits, office-based provider visits, and emergency room visits excluding dental services. (Dental services, compared to other types of health-care services, are more predictable and occur on a more regular basis.) Hospital stays with the same date of admission and discharge, known as "zero-night stays," were included in outpatient counts and expenditures. Payments associated with emergency room visits that immediately preceded an inpatient stay were included in the inpatient expenditures. Prescribed medicines that can be linked to hospital admissions were included in inpatient expenditures (not in outpatient utilization).

Explanatory Variables

Explanatory variables that can help explain health-care utilization are categorized as demographic, geographic, health status, education, and economic factors. Demographic factors include age, sex, and ethnicity. As persons age, the rate at which their health deteriorates increases; as a result, age has an increasing impact on the demand for health care. Sex and ethnicity can be treated as proxies for inherited health and social habits in maintaining health. For a geographic factor, we use region to proxy the accessibility of health-care services and the overall economic or regional impact on residents' health-care behavior.

The demand for medical services is thought to be influenced by individuals' health status and education. In MEPS, self-rated physical health, mental health, and any functional or activity related limitations during the sample period are used as proxies for health status. Education tends to have ambiguous impact on the demand for health-care services. One theory is that more educated persons are more aware of health risks, thus being more active in maintaining their health; as a result, educated persons may be less prone to severe diseases leading to hospital admissions. Another theory is that less educated persons have greater exposure to health risks and, through exposure, develop a greater tolerance for certain types of risks. In MEPS, education is proxied by degrees received and categorized into three different levels: lower than high school, high school, and college or above education.

Economic covariates include income and insurance coverage. A measure of income in MEPS is income relative to the poverty line. This approach is appropriate because it summarizes effects of different levels of income on health-care utilization in constant dollars. Insurance coverage is also an important variable in explaining health-care utilization. One issue with health insurance coverage is that it reduces the out-of-pocket prices paid by insureds and thus induces moral hazard. Research associated with the Rand Health Insurance Experiment empirically suggested that cost-sharing effects from insurance coverage will affect primarily the number of medical contacts rather than the intensity of each contact. This motivated our introduction of a binary variable that takes the value of one if a person had any public or private health insurance for at least one month and zero otherwise.

Table 11.4 Percentage of Positive Expenditures by Explanatory Variable

Category	Variable	Description	Percentage of Data	Percentage Positive Expend
Demography	AGE	Age in years between 18 to 65 (mean: 39.0)		
	GENDER	1 if female	52.7	10.7
		0 if male	47.3	4.7
Ethnicity	ASIAN	1 if Asian	4.3	4.7
	BLACK	1 if Black	14.8	10.5
	NATIVE	1 if Native	1.1	13.6
	WHITE	Reference level	79.9	7.5
Region	NORTHEAST	1 if Northeast	14.3	10.1
	MIDWEST	1 if Midwest	19.7	8.7
	SOUTH	1 if South	38.2	8.4
	WEST	Reference level	27.9	5.4
Education	COLLEGE	1 if college or higher degree	27.2	6.8
	HIGHSCHOOL	1 if high school degree	43.3	7.9
	Reference level is lower than high school degree		29.5	8.8
Self-rated physical health	POOR	1 if poor	3.8	36.0
	FAIR	1 if fair	9.9	8.1
	GOOD	1 if good	29.9	8.2
	VGOOD	1 if very good	31.1	6.3
	Reference level is excellent health		25.4	5.1
Self-rated mental health	MNHPOOR	1 if poor or fair	7.5	16.8
		0 if good to excellent mental health	92.6	7.1
Any activity limitation	ANYLIMIT	1 if any functional/activity limitation	22.3	14.6
		0 if otherwise	77.7	5.9
Income compared to poverty line	HINCOME	1 if high income	31.6	5.4
	MINCOME	1 if middle income	29.9	7.0
	LINCOME	1 if low income	15.8	8.3
	NPOOR	1 if near poor	5.8	9.5
	Reference level is poor/negative		17.0	13.0
Insurance coverage	INSURE	1 if covered by public/private health insurance in any month of 2003	77.8	9.2
		0 if have no health insurance in 2003	22.3	3.1
Total			100.0	7.9

Summary Statistics

Table 11.4 describes these explanatory variables and provides summary statistics that suggest their effects on the probability of positive inpatient expenditures. For example, we see that women had a higher overall utilization than men. Specifically, 10.7% of women had a positive expenditure during the year, compared to only 4.7% for men. Similarly, utilizations vary by other covariates, suggesting their importance as predictors of expenditures.

Table 11.5 summarizes the fit of several binary regression models. Fits are reported under the "Full Model" column for all variables using the logit function. The t-ratios for many of the explanatory variables exceed two in absolute value,

Table 11.5 Comparison of Binary Regression Models

	Logistic				Probit	
	Full Model		Reduced Model		Reduced Model	
Effect	Parameter Estimate	*t*-Ratio	Parameter Estimate	*t*-Ratio	Parameter Estimate	*t*-Ratio
Intercept	−4.239	−8.982	−4.278	−10.094	−2.281	−11.432
AGE	−0.001	−0.180				
GENDER	0.733	3.812	0.732	3.806	0.395	4.178
ASIAN	−0.219	−0.411	−0.219	−0.412	−0.108	−0.427
BLACK	−0.001	−0.003	0.004	0.019	0.009	0.073
NATIVE	0.610	0.926	0.612	0.930	0.285	0.780
NORTHEAST	0.609	2.112	0.604	2.098	0.281	1.950
MIDWEST	0.524	1.904	0.517	1.883	0.237	1.754
SOUTH	0.339	1.376	0.328	1.342	0.130	1.085
COLLEGE	0.068	0.255	0.070	0.263	0.049	0.362
HIGHSCHOOL	0.004	0.017	0.009	0.041	0.003	0.030
POOR	1.712	4.385	1.652	4.575	0.939	4.805
FAIR	0.136	0.375	0.109	0.306	0.079	0.450
GOOD	0.376	1.429	0.368	1.405	0.182	1.412
VGOOD	0.178	0.667	0.174	0.655	0.094	0.728
MNHPOOR	−0.113	−0.369				
ANYLIMIT	0.564	2.680	0.545	2.704	0.311	3.022
HINCOME	−0.921	−3.101	−0.919	−3.162	−0.470	−3.224
MINCOME	−0.609	−2.315	−0.604	−2.317	−0.314	−2.345
LINCOME	−0.411	−1.453	−0.408	−1.449	−0.241	−1.633
NPOOR	−0.201	−0.528	−0.204	−0.534	−0.146	−0.721
INSURE	1.234	4.047	1.227	4.031	0.579	4.147
Log-Likelihood	−488.69		−488.78		−486.98	
AIC	1,021.38		1,017.56		1,013.96	

suggesting that they are useful predictors. From an inspection of these t-ratios, one might consider a more parsimonious model by removing statistically insignificant variables. Table 11.5 shows a "Reduced Model," where the age and mental health status variables have been removed. To assess their joint significance, we can compute a likelihood ratio test statistic as twice the change in the log-likelihood. This turns out to be only $2 \times (-488.78 - (-488.69)) = 0.36$. Comparing this to a chi-square distribution with $df = 2$ degrees of freedom results in a p-value of 0.835, indicating that the additional parameters for age and mental health status are not statistically significant. Table 11.5 also provides probit model fits. Here, we see that the results are similar to the logit model fits, according to sign of the coefficients and their significance, suggesting that, for this application, there is little difference in the two specifications.

11.5 Nominal Dependent Variables

We now consider a response that is an unordered categorical variable, also known as a *nominal* dependent variable. We assume that the dependent variable y may take on values $1, 2, \ldots, c$, corresponding to c categories. When $c > 2$, we refer to the data as *multicategory,* also known as *polychotomous* or *polytomous*.

In many applications, the response categories correspond to an attribute possessed or choices made by individuals, households, or firms. Some applications include the following:

- Employment choice, such as Valletta (1999)
- Mode of transportation, such as the classic work by McFadden (1978)
- Type of health insurance, as in Browne and Frees (2007)

For an observation from subject i, denote the probability of choosing the jth category as $\pi_{ij} = \Pr(y_i = j)$, so that $\pi_{i1} + \cdots + \pi_{ic} = 1$. In general, we will model these probabilities as a (known) function of parameters and use maximum likelihood estimation for statistical inference. Let y_{ij} be a binary variable that is 1 if $y_i = j$. Extending equation (11.2) to c categories, the likelihood for the ith subject is:

$$\prod_{j=1}^{c} \left(\pi_{i,j}\right)^{y_{i,j}} = \begin{cases} \pi_{i,1} & \text{if } y_i = 1 \\ \pi_{i,2} & \text{if } y_i = 2 \\ \vdots & \vdots \\ \pi_{i,c} & \text{if } y_i = c \end{cases}.$$

Thus, assuming independence among observations, the total log-likelihood is

$$L = \sum_{i=1}^{n} \sum_{j=1}^{c} y_{i,j} \ln \pi_{i,j}.$$

With this framework, standard maximum likelihood estimation is available (Section 11.9). Thus, our main task is to specify an appropriate form for π.

11.5.1 Generalized Logit

Like standard linear regression, generalized logit models employ linear combinations of explanatory variables of the form:

$$V_{i,j} = \mathbf{x}_i' \boldsymbol{\beta}_j. \tag{11.6}$$

Because the dependent variables are not numerical, we cannot model the response y as a linear combination of explanatory variables plus an error. Instead, we use the probabilities

$$\Pr(y_i = j) = \pi_{i,j} = \frac{\exp(V_{i,j})}{\sum_{k=1}^{c} \exp(V_{i,k})}. \tag{11.7}$$

Note here that $\boldsymbol{\beta}_j$ is the corresponding vector of parameters that may depend on the alternative j, whereas the explanatory variables $\mathbf{x}_i$ do not. So that probabilities sum to one, a convenient normalization for this model is $\boldsymbol{\beta}_c = \mathbf{0}$. With this normalization and the special case of $c = 2$, the generalized logit reduces to the logit model introduced in Section 11.2.

Parameter Interpretations

We now describe an interpretation of coefficients in generalized logit models, similar to the logistic model. From equations (11.6) and (11.7), we have

$$\ln \frac{\Pr(y_i = j)}{\Pr(y_i = c)} = V_{i,j} - V_{i,c} = \mathbf{x}_i' \boldsymbol{\beta}_j.$$

The left-hand side of this equation is interpreted to be the logarithmic odds of choosing choice j compared to choice c. Thus, we may interpret $\boldsymbol{\beta}_j$ as the proportional change in the odds ratio.

Generalized logits have an interesting *nested* structure that we will explore briefly in Section 11.5.3. That is, it is easy to check that, conditional on not choosing the first category, the form of $\Pr(y_i = j | y_i \neq 1)$ has a generalized logit form in equation (11.7). Further, if j and h are different alternatives, we note that

$$\Pr(y_i = j | y_i = j \text{ or } y_i = h) = \frac{\Pr(y_i = j)}{\Pr(y_i = j) + \Pr(y_i = h)} = \frac{\exp(V_{i,j})}{\exp(V_{i,j}) + \exp(V_{i,h})}$$
$$= \frac{1}{1 + \exp(\mathbf{x}_i'(\boldsymbol{\beta}_h - \boldsymbol{\beta}_j))}.$$

This has a logit form that was introduced in Section 11.2.

Special Case, Intercept-Only Model. To develop intuition, we now consider the model with only intercepts. Thus, let $\mathbf{x}_i = 1$ and $\boldsymbol{\beta}_j = \beta_{0,j} = \alpha_j$. With the convention $\alpha_c = 0$, we have

$$\Pr(y_i = j) = \pi_{i,j} = \frac{e^{\alpha_j}}{e^{\alpha_1} + e^{\alpha_2} + \cdots + e^{\alpha_{c-1}} + 1}$$

and

$$\ln \frac{\Pr(y_i = j)}{\Pr(y_i = c)} = \alpha_j.$$

From the second relation, we may interpret the jth intercept α_j to be the logarithmic odds of choosing alternative j compared to alternative c.

Example: Job Security, Continued. This is a continuation of the Section 11.2 example on the determinants of job turnover, based on the work of Valletta (1999). The first analysis of this data considered only the binary dependent variable dismissal as this outcome is the main source of job insecurity. Valetta (1999) also presented results from a generalized logit model, his primary motivation being that the economic theory describing turnover implies that other reasons for leaving a job may affect dismissal probabilities.

For the generalized logit model, the response variable has $c = 5$ categories: dismissal, left job because of plant closures, "quit," changed jobs for other reasons, and no change in employment. The "no change in employment" category is the omitted one in Table 11.6. The explanatory variables of the generalized logit are the same as the probit regression; the estimates summarized in Table 11.1 are reproduced here for convenience.

Table 11.6 Turnover Generalized Logit and Probit Regression Estimates

Variable	Probit Regression Model (Dismissal)	Generalized Logit Model Dismissal	Plant Closed	Other Reason	Quit
Tenure	−0.084	−0.221	−0.086	−0.068	−0.127
	(0.010)	(0.025)	(0.019)	(0.020)	(0.012)
Time Trend	−0.002	−0.008	−0.024	0.011	−0.022
	(0.005)	(0.011)	(0.016)	(0.013)	(0.007)
Tenure (Time Trend)	0.003	0.008	0.004	−0.005	0.006
	(0.001)	(0.002)	(0.001)	(0.002)	(0.001)
Change in Logarithmic Sector Employment	0.094	0.286	0.459	−0.022	0.333
	(0.057)	(0.123)	(0.189)	(0.158)	(0.082)
Tenure × (Change in Logarithmic Sector Employment)	−0.020	−0.061	−0.053	−0.005	−0.027
	(0.009)	(0.023)	(0.025)	(0.025)	(0.012)

Notes: Standard errors in parentheses. Omitted category is no change in employment for the generalized logit. Other variables controlled for consist of education, marital status, number of children, race, years of full-time work experience and its square, union membership, government employment, logarithmic wage, the U.S. employment rate, and location.

Table 11.6 shows that turnover declines as tenure increases. To illustrate, consider a typical man in the 1992 sample where we have time $= 16$ and focus on dismissal probabilities. For this value of time, the coefficient associated with tenure for dismissal is $-0.221 + 16\ (0.008) = -0.093$ (due to the interaction term). From this, we interpret an additional year of tenure to imply that the dismissal probability is $\exp(-0.093) = 91\%$ of what it would be otherwise, representing a decline of 9%.

Table 11.6 also shows that the generalized coefficients associated with dismissal are similar to the probit fits.

The standard errors are also qualitatively similar, although higher for the generalized logits when compared to the probit model. In particular, we again see that the coefficient associated with the interaction between tenure and time trend reveals an increasing dismissal rate for experienced workers. The same is true for the rate of quitting.

11.5.2 Multinomial Logit

Similar to equation (11.6), an alternative, linear combination of explanatory variables is

$$V_{i,j} = \mathbf{x}_{i,j}' \boldsymbol{\beta}, \tag{11.8}$$

where $\mathbf{x}_{i,j}$ is a vector of explanatory variables that depends on the jth alternative, whereas the parameters $\boldsymbol{\beta}$ do not. Using the expressions in equations (11.7) and (11.8) forms the basis of the *multinomial logit* model, also known as the *conditional logit* model (McFadden, 1974). With this specification, the total log-likelihood is

$$L = \sum_{i=1}^{n} \sum_{j=1}^{c} y_{i,j} \ln \pi_{i,j} = \sum_{i=1}^{n} \left[\sum_{j=1}^{c} y_{i,j} \mathbf{x}_{i,j}' \boldsymbol{\beta} - \ln \left(\sum_{k=1}^{c} \exp(\mathbf{x}_{i,k}' \boldsymbol{\beta}) \right) \right].$$

This straightforward expression for the likelihood enables maximum likelihood inference to be easily performed.

The generalized logit model is a special case of the multinomial logit model. To see this, consider explanatory variables $\mathbf{x}_i$ and parameters $\boldsymbol{\beta}_j$, each of dimension $k \times 1$. Define

$$\mathbf{x}_{i,j} = \begin{pmatrix} \mathbf{0} \\ \vdots \\ \mathbf{0} \\ \mathbf{x}_i \\ \mathbf{0} \\ \vdots \\ \mathbf{0} \end{pmatrix} \quad \text{and} \quad \boldsymbol{\beta} = \begin{pmatrix} \boldsymbol{\beta}_1 \\ \boldsymbol{\beta}_2 \\ \vdots \\ \boldsymbol{\beta}_c \end{pmatrix}.$$

Specifically, $\mathbf{x}_{i,j}$ is defined as $j - 1$ zero vectors (each of dimension $k \times 1$), followed by $\mathbf{x}_i$ and then followed by $c - j$ zero vectors. With this specification, we have $\mathbf{x}_{i,j}' \boldsymbol{\beta} = \mathbf{x}_i' \boldsymbol{\beta}_j$. Thus, a statistical package that performs multinomial logit estimation can also perform generalized logit estimation through the appropriate coding of explanatory variables and parameters. Another consequence of this connection is that some authors use the descriptor multinomial logit when referring to the generalized logit model.

The generalized logit model is a special case of the multinomial logit model.

Moreover, through similar coding schemes, multinomial logit models can also handle linear combinations of the form

$$V_i = \mathbf{x}_{i,1,j}' \boldsymbol{\beta} + \mathbf{x}_{i,2}' \boldsymbol{\beta}_j.$$

Here, $\mathbf{x}_{i,1,j}$ are explanatory variables that depend on the alternative, whereas $\mathbf{x}_{i,2}$ do not. Similarly, $\boldsymbol{\beta}_j$ are parameters that depend on the alternative, whereas $\boldsymbol{\beta}$ do not. This type of linear combination is the basis of a *mixed logit model*. As with conditional logits, it is customary to choose one set of parameters as the baseline and specify $\boldsymbol{\beta}_c = \mathbf{0}$ to avoid redundancies.

To interpret parameters for the multinomial logit model, we may compare alternatives h and k using equations (11.7) and (11.8) to get

$$\ln \frac{\Pr(y_i = h)}{\Pr(y_i = k)} = (\mathbf{x}_{i,h} - \mathbf{x}_{i,k})' \boldsymbol{\beta}.$$

Thus, we may interpret β_j as the proportional change in the odds ratio, where the change is the value of the jth explanatory variable, moving from the kth to the hth alternative.

With equation (11.7), note that $\pi_{i,1}/\pi_{i,2} = \exp(V_{i,1})/\exp(V_{i,2})$. This ratio does not depend on the underlying values of the other alternatives, $V_{i,j}$, for $j = 3, \ldots, c$. This feature, called the *independence of irrelevant alternatives*, can be a drawback of the multinomial logit model for some applications.

Table 11.7 Percentages of Health Coverage by Law Variable

Disability Law in Effect	Number	Uninsured	Non-group	Government	Group	Odds – Comparing Group to Uninsured	Odds Ratio
No	82,246	20.1	12.2	8.4	59.3	2.946	
Yes	4,229	19.9	10.1	12.5	57.6	2.895	0.983
Total	86,475	20.1	12.1	8.6	59.2		

Example: Choice of Health Insurance. To illustrate, Browne and Frees (2007) examined $c = 4$ health insurance choices, consisting of the following:

$y = 1$: an individual covered by group insurance
$y = 2$: an individual covered by private, non-group insurance
$y = 3$: an individual covered by government but not private insurance
$y = 4$: an individual not covered by health insurance

Their data on health insurance coverage came from the March supplement of the Current Population Survey (CPS), conducted by the Bureau of Labor Statistics. Browne and Frees (2007) analyzed approximately 10,800 single-person households per year, covering 1988–95, yielding $n = 86{,}475$ observations. They examined whether underwriting restrictions, laws passed to prohibit insurers from discrimination, facilitate or discourage consumption of health insurance. They focused on disability laws that prohibited insurers from using physical impairment (disability) as an underwriting criterion.

Table 11.7 suggests that disability laws have little effect on the average health insurance purchasing behavior. To illustrate, for individuals surveyed with disability laws in effect, 57.6% purchased group health compared to 59.3% of those where restrictions were not in effect. Similarly, 19.9% were uninsured when disability restrictions were in effect, compared to 20.1% when they were not. In terms of odds, when disability restrictions were in effect, the odds of purchasing group health insurance compared to becoming uninsured are 57.6/19.9 = 2.895. When disability restrictions were not in effect, the odds are 2.946. The odds ratio, 2.895/2.946 = 0.983, indicates that there is little change in the odds when comparing whether disability restrictions were in effect.

In contrast, Table 11.8 suggests that disability laws may have important effects on the average health insurance purchasing behavior of selected subgroups of the sample. Table 11.8 shows the percentage uninsured and odds of purchasing group insurance (compared to being uninsured) for selected subgroups. To illustrate, for disabled individuals, the odds of purchasing group insurance are 1.329 times higher when disability restrictions are in effect. Table 11.7 suggests that disability restrictions have no effect; this may be true when looking at the entire sample. However, by examining subgroups, Table 11.8 shows that we may see important effects associated with legal underwriting restrictions that are not evident when looking at averages over the whole sample.

Table 11.8 Odds of Health Coverage by Law and Physical Impairment

Selected Subgroups	Disability Law in Effect	Number	Percentage Group	Percentage Uninsured	Odds – Comparing Group to Uninsured	Odds Ratio
Nondisabled	No	72,150	64.2	20.5	3.134	
Nondisabled	Yes	3,649	63.4	21.2	2.985	0.952
Disabled	No	10,096	24.5	17.6	1.391	
Disabled	Yes	580	21.0	11.4	1.848	1.329

Table 11.9 Odds Ratios from Multinomial Logit Regression Model

Variable	Group versus Uninsured	Nongroup versus Uninsured	Government versus Uninsured	Group versus Nongroup	Group versus Government	Nongroup versus Government
Law × Nondisabled	0.825	1.053	1.010	0.784	0.818	1.043
p-Value	0.001	0.452	0.900	0.001	0.023	0.677
Law × Disabled	1.419	0.953	1.664	1.490	0.854	0.573
p-Value	0.062	0.789	0.001	0.079	0.441	0.001

Notes: The regression includes 150 (=50 × 3) state-specific effects, several continuous variables (age, education, and income, as well as higher-order terms), and categorical variables (such as race and year).

There are many ways of picking subgroups of interest. With a large dataset of $n = 86{,}475$ observations, one could probably pick subgroups to confirm almost any hypothesis. Further, there is a concern that the CPS data may not provide a representative sample of state populations. Thus, it is customary to use regression techniques to control for explanatory variables, such as physical impairment.

Table 11.9 reports the main results from a multinomial logit model with many control variables included. A dummy variable for each of 50 states was included (the District of Columbia is a "state" in this dataset, so we need $51 - 1 = 50$ dummy variables). These variables were suggested in the literature and are further described in Browne and Frees (2007). They include an individual's sex, marital status; race; education; whether or not self-employed; and whether an individual worked full-time, part-time, or not at all.

In Table 11.9, "Law" refers to the binary variable that is one if a legal restriction was in effect and "Disabled" is a binary variable that is one if an individual is physically impaired. Thus, the interaction "Law × Disabled" reports the effect of a legal restriction on a physically impaired individual. The interpretation is similar to Table 11.8. Specifically, we interpret the coefficient 1.419 to mean that disabled individuals are 41.9% more likely to purchase group health insurance than to purchase no insurance, when the disability underwriting restriction is in effect. Similarly, nondisabled individuals are 21.2% $(=1/0.825 - 1)$ less likely to purchase group health insurance than to purchase no insurance, when the disability underwriting restriction is in effect. This result suggests that the nondisabled are more likely to be uninsured as a result of prohibitions on the use of disability status as an underwriting criteria. Overall, the results are statistically significant, confirming that this legal restriction does have an impact on the consumption of health insurance.

11.5.3 Nested Logit

To mitigate the problem of independence of irrelevant alternatives in multinomial logits, we now introduce a type of hierarchical model known as a *nested logit* model. To interpret the nested logit model, in the first stage, one chooses an alternative (say, the first alternative) with probability

$$\pi_{i,1} = \Pr(y_i = 1) = \frac{\exp(V_{i,1})}{\exp(V_{i,1}) + \left[\sum_{k=2}^{c} \exp(V_{i,k}/\rho)\right]^{\rho}}. \tag{11.9}$$

Then, conditional on not choosing the first alternative, the probability of choosing any one of the other alternatives follows a multinomial logit model with probabilities

$$\frac{\pi_{i,j}}{1-\pi_{i,1}} = \Pr(y_i = j | y_i \neq 1) = \frac{\exp(V_{i,j}/\rho)}{\sum_{k=2}^{c} \exp(V_{i,k}/\rho)}, \quad j = 2, \ldots, c. \tag{11.10}$$

In equations (11.9) and (11.10), the parameter ρ measures the association among the choices $j = 2, \ldots, c$. The value of $\rho = 1$ reduces to the multinomial logit model that we interpret to mean independence of irrelevant alternatives. We also interpret $\text{Prob}(y_i = 1)$ to be a weighted average of values from the first choice and the others. Conditional on not choosing the first category, the form of $\Pr(y_i = j | y_i \neq 1)$ in equation (11.10) has the same form as the multinomial logit.

The advantage of the nested logit is that it generalizes the multinomial logit model in a way such that we no longer have the problem of independence of irrelevant alternatives. A disadvantage, pointed out by McFadden (1981), is that only one choice is observed; thus, we do not know which category belongs in the first stage of the nesting without additional theory regarding choice behavior. Nonetheless, the nested logit generalizes the multinomial logit by allowing alternative "dependence" structures. That is, one may view the nested logit as a robust alternative to the multinomial logit and examine each one of the categories in the first stage of the nesting.

11.6 Ordinal Dependent Variables

We now consider a response that is an ordered categorical variable, also known as an *ordinal* dependent variable. To illustrate, any type of survey response where you score your impression on a seven-point scale ranging from "very dissatisfied" to "very satisfied" is an example of an ordinal variable.

Example: Health Plan Choice. Pauly and Herring (2007) examined $c = 4$ choices of health-care plan types:

$y = 1$: a health maintenance organization (HMO)
$y = 2$: a point of service (POS) plan
$y = 3$: a preferred provider organization (PPO)
$y = 4$: a fee for service (FFS) plan

A FFS plan is the least restrictive, allowing enrollees to see health-care providers (e.g., primary care physicians) for a fee that reflects the cost of services rendered. The PPO plan is the next least restrictive; this plan uses FFS payments but enrollees generally must choose from a list of preferred providers. Pauly and Herring (2007) took POS and HMO plans to be the third and fourth least restrictive, respectively. An HMO often uses capitation (a flat rate per person) to reimburse providers, restricting enrollees to a network of providers. In contrast, a POS plan gives enrollees the option to see providers outside of the HMO network (for an additional fee).

11.6.1 Cumulative Logit

Models of ordinal dependent variables are based on cumulative probabilities of the form

$$\Pr(y \le j) = \pi_1 + \cdots + \pi_j, \quad j = 1, \ldots, c.$$

In this section, we use *cumulative logits*

$$\text{logit}\,(\Pr(y \le j)) = \ln\left(\frac{\Pr(y \le j)}{1 - \Pr(y \le j)}\right) = \ln\left(\frac{\pi_1 + \cdots + \pi_j}{\pi_{j+1} + \cdots + \pi_c}\right). \tag{11.11}$$

The simplest cumulative logit model is

$$\text{logit}\,(\Pr(y \le j)) = \alpha_j,$$

which does not use any explanatory variables. The cut-point parameters α_j are nondecreasing so that $\alpha_1 \le \alpha_2 \le \ldots \le \alpha_c$, reflecting the cumulative nature of the distribution function $\Pr(y \le j)$.

The *proportional odds model* incorporates explanatory variables. With this model, cumulative logits are expressed as

$$\text{logit}\,(\Pr(y \le j)) = \alpha_j + \mathbf{x}_i'\boldsymbol{\beta}. \tag{11.12}$$

This model provides parameter interpretations similar to those for logistic regression described in Section 11.2.4. For example, if the variable x_1 is continuous, then as in equation (11.1) we have

$$\beta_1 = \frac{\partial}{\partial x_{i1}}\left(\alpha_j + \mathbf{x}_i'\boldsymbol{\beta}\right) = \frac{\frac{\partial}{\partial x_{i1}} \Pr(y_i \le j|\mathbf{x}_i)/\left(1 - \Pr(y_i \le j|\mathbf{x}_i)\right)}{\Pr(y_i \le j|\mathbf{x}_i)/\left(1 - \Pr(y_i \le j|\mathbf{x}_i)\right)}.$$

Table 11.10 Cumulative Logit Model of Health Plan Choice

Variable	Odds Ratio	Variable	Odds Ratio
Age	0.992***	Hispanic	1.735***
Female	1.064***	Risk taker	0.967
Family size	0.985	Smoker	1.055***
Family income	0.963***	Fair/poor health	1.056
Education	1.006	α_1	0.769***
Asian	1.180***	α_2	1.406***
African-American	1.643***	α_3	12.089***
Maximum-rescaled R^2		0.102	

Source: Pauly and Herring (2007). *** indicates that the associated p-values are less than 0.01. *Note:* For race, Caucasian is the omitted variable.

Thus, we may interpret β_1 as the proportional change in the cumulative odds ratio.

Example: Health Plan Choice, Continued. Pauly and Herring used data from the 1996–7 and 1998–9 Community Tracking Study's Household Surveys (CTS-HS) to study the demand for health insurance. This is a nationally representative survey containing more than 60,000 individuals per period. As one measure of demand, Pauly and Herring examined health plan choice, reasoning that individuals who chose (through employment or association membership) less restrictive plans sought greater protection for health care. (They also looked at other measures, including the number of restrictions placed on plans and the amount of cost sharing.) Table 11.10 provides determinants of health plan choice based on $n = 34{,}486$ individuals who had group health insurance, aged 18–64, without public insurance. Pauly and Herring also compared these results to those who had individual health insurance to understand the differences in determinants between these two markets.

To interpret the odds ratios in Table 11.10, we first note that the cut-point estimates, corresponding to α_1, α_2, and α_3, increase as choices become less restrictive, as anticipated. For sex, we see that the estimated odds for women are 1.064 times those of men in the direction of choosing a less restrictive health plan. Controlling for other variables, women are more likely to choose less restrictive plans than are men. Similarly, younger, less wealthy, non-Caucasians, and smokers are more likely to choose less restrictive plans. Coefficients associated with family size, education, risk taking, and self-reported health were not statistically significant in this fitted model.

11.6.2 Cumulative Probit

As in Section 11.2.2 for logistic regression, cumulative logit models have a threshold interpretation. Specifically, let y_i^* be a latent, unobserved, random variable on which we base the observed dependent variable as

$$y_i = \begin{cases} 1 & y_i^* \leq \alpha_1 \\ 2 & \alpha_1 < y_i^* \leq \alpha_2 \\ \vdots & \vdots \\ c-1 & \alpha_{c-2} < y_i^* \leq \alpha_{c-1} \\ c & \alpha_{c-1} < y_i^* \end{cases}.$$

If $y_i^* - \mathbf{x}_i'\boldsymbol{\beta}$ has a logistic distribution, then

$$\Pr(y_i^* - \mathbf{x}_i'\boldsymbol{\beta} \leq a) = \frac{1}{1+\exp(-a)},$$

and thus

$$\Pr(y_i \leq j) = \Pr(y_i^* \leq \alpha_j) = \frac{1}{1+\exp\left(-(\alpha_j - \mathbf{x}_i'\boldsymbol{\beta})\right)}.$$

Applying the logit transform to both sides yields equation (11.12).

Alternatively, assume that $y_i^* - \mathbf{x}_i'\boldsymbol{\beta}$ has a standard normal distribution. Then,

$$\Pr(y_i \leq j) = \Pr(y_i^* \leq \alpha_j) = \Phi\left(\alpha_j - \mathbf{x}_i'\boldsymbol{\beta}\right).$$

This is the *cumulative probit* model. As with binary variable models, the cumulative probit gives results that are similar to the cumulative logit model.

11.7 Further Reading and References

Regression models of binary variables are used extensively. For more detailed introductions, see Hosmer and Lemshow (1989) or Agresti (1996). You may also want to examine more rigorous treatments such as those in Agresti (1990) and Cameron and Trivedi (1998). The work by Agresti (1990, 1996) discuss multicategory dependent variables, as does the advanced econometrics treatment in Amemiya (1985).

Chapter References

Agresti, Alan (1990). *Categorical Data Analysis*. Wiley, New York.

Agresti, Alan (1996). *An Introduction to Categorical Data Analysis*. Wiley, New York.

Amemiya, Takeshi (1985). *Advanced Econometrics*. Harvard University Press, Cambridge, Massachusetts.

Browne, Mark J., and Edward W. Frees (2007). Prohibitions on health insurance underwriting. University of Wisconsin working paper.

Cameron, A. Colin, and Pravin K. Trivedi (1998) *Regression Analysis of Count Data*. Cambridge University Press, Cambridge.

Carroll, Raymond J., and Ruppert, David (1988). *Transformation and Weighting in Regression*. Chapman-Hall, New York.

Gourieroux, Christian and Joann Jasiak (2007). *The Econometrics of Individual Risk*. Princeton University Press, Princeton, New Jersey.

Hand, D. J., and W. E. Henley (1997). Statistical classification methods in consumer credit scoring: A review. *Journal of the Royal Statistical Society A*, 160, no. 3, 523–41.

Hosmer, David W., and Stanley Lemeshow (1989). *Applied Logistic Regression*. Wiley, New York.

Pauly, Mark V., and Bradley Herring (2007). The demand for health insurance in the group setting: Can you always get what you want? *Journal of Risk and Insurance* 74, 115–40.

Smith, Richard M., and Phyllis Schumacher (2006). Academic attributes of college freshmen that lead to success in actuarial studies in a business college. *Journal of Education for Business* 81, no. 5, 256–60.

Valletta, R. G. (1999). Declining job security. *Journal of Labor Economics* 17, S170–S197.

Wiginton, John C. (1980). A note on the comparison of logit and discriminant models of consumer credit behavior. *Journal of Financial and Quantitative Analysis* 15, no. 3, 757–70.

11.8 Exercises

11.1. **Similarity of Logit and Probit**. Suppose that the random variable y^* has a logit distribution function, $\Pr(y^* \leq y) = \mathrm{F}(y) = e^y/(1+e^y)$.

a. Calculate the corresponding probability density function.

b. Use the probability density function to compute the mean (μ_y).

c. Compute the corresponding standard deviation (σ_y).

d. Define the rescaled random variable $y^{**} = \frac{y^*-\mu_y}{\sigma_y}$. Determine the probability density function for y^{**}.

e. Plot the probability density function in part (d). Overlay this plot with a plot of a standard normal probability density function. (This provides a density function version of the distribution function plots in Figure 11.1.)

11.2. **Threshold Interpretation of the Probit Regression Model**. Consider an underlying linear model, $y_i^* = \mathbf{x}_i'\boldsymbol{\beta} + \epsilon_i^*$, where ϵ_i^* is normally distributed with mean zero and variance σ^2. Define $y_i = \mathrm{I}(y_i^* > 0)$, where $\mathrm{I}(\cdot)$ is the indicator function. Show that $\pi_i = \Pr(y_i = 1|\mathbf{x}_i) = \Phi(\mathbf{x}_i'\beta/\sigma)$, where $\Phi(\cdot)$ is the standard normal distribution function.

11.3. **Random Utility Interpretation of the Logistic Regression Model**. Under the random utility interpretation, an individual with utility $U_{ij} = u_i(V_{ij} + \epsilon_{ij})$, where j may be 1 or 2, selects category corresponding to $j = 1$ with probability

$$\pi_i = \Pr(y_i = 1) = \Pr(U_{i2} < U_{i1})$$
$$= \Pr(\epsilon_{i2} - \epsilon_{i1} < V_{i1} - V_{i2}).$$

As in Section 11.2.3, we take $V_{i2} = 0$ and $V_{i1} = \mathbf{x}_i'\boldsymbol{\beta}$. Further suppose that the errors are from an extreme value distribution of the form

$$\Pr(\epsilon_{ij} < a) = \exp(-e^{-a}).$$

Show that the choice probability π_i has a logit form. That is, show

$$\pi_i = \frac{1}{1+\exp(-\mathbf{x}_i'\boldsymbol{\beta})}.$$

11.4. **Two Populations.**

a. Begin with one population and assume that $y_1, \ldots, y_n$ is an i.i.d. sample from a Bernoulli distribution with mean π. Show that the maximum likelihood estimator of π is $\overline{y}$.

b. Now consider two populations. Suppose that $y_1, \ldots, y_{n_1}$ is an i.i.d. sample from a Bernoulli distribution with mean π_1 and that $y_{n_1+1}, \ldots, y_{n_1+n_2}$ is an i.i.d. sample from a Bernoulli distribution with mean π_2, where the samples are independent of one another.

b(i). Show that the maximum likelihood estimator of $\pi_2 - \pi_1$ is $\overline{y}_2 - \overline{y}_1$.

b(ii). Determine the variance of the estimator in part b(i).

c. Now express the two population problem in a regression context using one explanatory variable. Specifically, suppose that x_i only takes on the values of zero and one. Of the n observations, n_1 take on the value $x = 0$. These n_1 observations have an average y value of $\overline{y}_1$. The remaining $n_2 = n - n_1$ observations have value $x = 1$ and an average y value of $\overline{y}_2$. Using the logit case, let $b_{0,MLE}$ and $b_{1,MLE}$ represent the maximum likelihood estimators of β_0 and β_1, respectively.

c(i). Show that the maximum likelihood estimators satisfy the equations

$$\overline{y}_1 = \pi\left(b_{0,MLE}\right)$$

and

$$\overline{y}_2 = \pi\left(b_{0,MLE} + b_{1,MLE}\right).$$

c(ii). Use part c(i) to show that the maximum likelihood estimator for β_1 is $\pi^{-1}(\overline{y}_2) - \pi^{-1}(\overline{y}_1)$.

c(iii). With the notation $\pi_1 = \pi(\beta_0)$ and $\pi_2 = \pi(\beta_0 + \beta_1)$, confirm that the information matrix can be expressed as

$$\mathbf{I}(\beta_0, \beta_1) = n_1\pi_1(1-\pi_1)\begin{pmatrix} 1 & 0 \\ 0 & 0 \end{pmatrix} + n_2\pi_2(1-\pi_2)\begin{pmatrix} 1 & 1 \\ 1 & 1 \end{pmatrix}.$$

c(iv). Use the information matrix to determine the large sample variance of the maximum likelihood estimator for β_1.

11.5. **Fitted Values.** Let $\widehat{y}_i = \pi\left(\mathbf{x}_i'\mathbf{b}_{MLE}\right)$ denote the ith fitted value for the logit function. Assume that an intercept is used in the model so that one of the explanatory variables x is a constant equal to one. Show that the average response is equal to the average fitted value; that is, show $\overline{y} = n^{-1}\sum_{i=1}^{n}\widehat{y}_i$.

11.6. Beginning with the score equations (11.4), verify the expression for the logit case in equation (11.5).

11.7. **Information Matrix**

a. Beginning with the score function for the logit case in equation (11.5), show that the information matrix can be expressed as

$$\mathbf{I}(\boldsymbol{\beta}) = \sum_{i=1}^{n} \sigma_i^2 \mathbf{x}_i \mathbf{x}_i',$$

where $\sigma_i^2 = \pi(\mathbf{x}_i'\boldsymbol{\beta})(1 - \pi(\mathbf{x}_i'\boldsymbol{\beta}))$.

b. Beginning with the general score function in equation (11.4), determine the information matrix.

Ⓡ **Empirical** *Filename is "AutoBI"*

11.8. **Automobile Injury Insurance Claims.** Refer to the description in Exercise 1.5.

We consider $n = 1{,}340$ bodily injury liability claims from a single state using a 2002 survey conducted by the Insurance Research Council (IRC). The IRC is a division of the American Institute for Chartered Property Casualty Underwriters and the Insurance Institute of America. The survey asked participating companies to report claims closed with payment during a designated two-week period. In this exercise, we are interested in understanding the characteristics of the claimants who choose to be presented by an attorney when settling their claim. Variable descriptions are given Table 11.11.

Table 11.11 Bodily Injury Claims

Variable	Description
ATTORNEY	Whether the claimant is represented by an attorney (=1 if yes and =2 if no)
CLMAGE	Claimant's age
CLMSEX	Claimant's sex (=1 if male and =2 if female)
MARITAL	Claimant's marital status (=1 if married, =2 if single, =3 if widowed, and =4 if divorced/separated)
SEATBELT	Whether the claimant was wearing a seatbelt/child restraint (=1 if yes, =2 if no, and =3 if not applicable)
CLMINSUR	Whether the driver of the claimant's vehicle was uninsured (=1 if yes, =2 if no, and =3 if not applicable)
LOSS	The claimant's total economic loss (in thousands)

a. *Summary Statistics.*
   i. Calculate histograms and summary statistics of continuous explanatory variables CLMAGE and LOSS. On the basis of these results, create a logarithm version of LOSS, say, lnLOSS.
   ii. Examine the means of CLMAGE, LOSS and lnLOSS by level of ATTORNEY. Do these statistics suggest that the continuous variables differ by ATTORNEY?
   iii. Create tables of counts (or percentages) of ATTORNEY by level of CLMSEX, MARITAL, SEATBELT, and CLMINSUR. Do these statistics suggest that the categorial variables differ by ATTORNEY?

iv. Identify the number of missing values for each explanatory variable.

b. *Logistic Regression Models.*

i. Run a logistic regression model using only the explanatory variable CLMSEX. Is it an important factor in determining the use of an attorney? Provide an interpretation in terms of the odds of using an attorney.

ii. Run a logistic regression model using the explanatory variables CLMAGE, CLMSEX, MARITAL, SEATBELT, and CLMINSUR. Which variables appear to be statistically significant?

iii. For the model in part (ii), who uses attorneys more, men or women? Provide an interpretation in terms of the odds of using an attorney for the variable CLMSEX.

iv. Run a logistic regression model using the explanatory variables CLMAGE, CLMSEX, MARITAL, SEATBELT, CLMINSUR, LOSS, and lnLOSS. Decide which of the two loss measures is more important and rerun the model using only one of the variables. In this model, is the measure of losses a statistically significant variable?

v. Run your model from part (iv) but omit the variable CLMAGE. Describe differences between this model fit and that in part (iv), focusing on statistically significant variables and number of observations used in the model fit.

vi. Consider a single male claimant who is age 32. Assume that the claimant was wearing a seat belt, that the driver was insured and the total economic loss is \$5,000. For the model in part (iv), what is the estimate of the probability of using an attorney?

c. *Probit Regression.* Repeat part b(v) using probit regression models but interpret only the sign of the regression coefficients.

Ⓡ **EMPIRICAL**
Filename is "HKHorse"

11.9. **Hong Kong Horse Racing.** The race track is a fascinating example of financial market dynamics at work. Let's go to the track and make a wager. Suppose that, from a field of 10 horses, we simply want to pick a winner. In the context of regression, we will let y be the response variable indicating whether a horse wins ($y = 1$) or not ($y = 0$). From racing forms, newspapers and so on, there are many explanatory variables that are publicly available that might help us predict the outcome for y. Some candidate variables may include the age of the horse, recent track performance of the horse and jockey, pedigree of the horse, and so on. These variables are assessed by the investors present at the race, the betting crowd. Like many financial markets, it turns out that one of the most useful explanatory variable is the crowd's overall assessment of the horse's abilities. These assessments are not made from a survey of the crowd but rather from the wagers placed. Information about the crowd's wagers is available on a large sign at the race called the *tote board.* The tote board provides the odds of each horse winning a race. Table 11.12 is a hypothetical tote board for a race of 10 horses.

Table 11.12 Hypothetical Tote Board

Horse	1	2	3	4	5	6	7	8	9	10
Posted Odds	1-1	79-1	7-1	3-1	15-1	7-1	49-1	49-1	19-1	79-1

The odds that appear on the tote board have been adjusted to provide a "track take." That is, for every dollar that has been wagered, $\$T$ goes to the track for sponsoring the race and $\$(1-T)$ goes to the winning bettors. Typical track takes are in the neighborhood of 20%, or $T = 0.20$.

We can readily convert the odds on the tote board to the crowd's assessment of the probabilities of winning. To illustrate this, Table 11.13 shows hypothetical bets to win which resulted in the displayed information on the hypothetical tote board in Table 11.12.

Table 11.13 Hypothetical Bets

Horse	1	2	3	4	5	6	7	8	9	10	Total
Bets to Win	8,000	200	2,000	4,000	1,000	3,000	400	400	800	200	20,000
Probability	0.40	0.01	0.10	0.20	0.05	0.15	0.02	0.02	0.04	0.02	1.000
Posted Odds	1-1	79-1	7-1	3-1	15-1	7-1	49-1	49-1	19-1	79-1	

For this hypothetical race, \$20,000 was bet to win. Because \$8,000 of this \$20,000 was bet on the first horse, interpret the ratio $8{,}000/20{,}000 = 0.40$ as the crowd's assessment of the probability to win. The theoretical odds are calculated as $0.4/(1-0.4) = 2/3$, or a 0.67 bet wins \$1. However, the theoretical odds assume a fair game with no track take. To adjust for the fact that only $\$(1 - \text{T})$ are available to the winner, the posted odds for this horse would be $0.4/(1 - T - 0.4) = 1$, if $T = 0.20$. For this case, it now takes a \$1 bet to win \$1. We then have the relationship *adjusted odds* $= x/(1 - T - x)$, where x is the crowd's assessment of the probability of winning.

Before the start of the race, the tote board provides us with adjusted odds that can readily be converted into x, the crowd's assessment of winning. We use this measure to help us to predict y, the event of the horse actually winning the race.

We consider data from 925 races run in Hong Kong from September 1981 through September 1989. In each race, there were ten horses, one of which was randomly selected to be in the sample. In the data, use FINISH $= y$ to be the indicator of a horse winning a race and WIN $= x$ to be the crowd's a priori probability assessment of a horse winning a race.

a. A statistically naive colleague would like to double the sample size by picking two horses from each race instead of randomly selecting one horse from a field of ten.

i. Describe the relationship between the dependent variables of the two horses selected.
ii. Say how this violates the regression model assumptions.

b. Calculate the average FINISH and summary statistics for WIN. Note that the standard deviation of FINISH is greater than that of WIN, even though the sample means are about the same. For the variable FINISH, what is the relationship between the sample mean and standard deviation?
c. Calculate summary statistics of WIN by level of FINISH. Note that the sample mean is larger for horses that won (FINISH = 1) than for those that lost (FINISH = 0). Interpret this result.
d. Estimate a linear probability model, using WIN to predict FINISH.
   i. Is WIN a statistically significant predictor of FINISH?
   ii. How well does this model fit the data using the usual goodness-of-fit statistic?
   iii. For this estimated model, is it possible for the fitted values to lie outside the interval [0, 1]? Note, by definition, that the x-variable WIN must lie within the interval [0, 1].
e. Estimate a logistic regression model, using WIN to predict FINISH. Is WIN a statistically significant predictor of FINISH?
f. Compare the fitted values from the models in parts (d) and (e)
   i. For each model, provide fitted values at WIN = 0, 0.01, 0.05, 0.10, and 1.0.
   ii. Plot fitted values from the linear probability model versus fitted values from the logistic regression model.
g. Interpret WIN as the crowd's prior probability assessment of the probability of a horse winning a race. The fitted values, FINISH, is your new estimate of the probability of a horse winning a race, based on the crowd's assessment.
   i. Plot the difference FINISH – WIN versus WIN.
   ii. Discuss a betting strategy that you might employ based on the difference, FINISH – WIN.

Ⓡ **Empirical**
Filename is "TermLife"

11.10. **Demand for Term Life Insurance.** We continue our study of term life insurance demand from Chapters 3 and 4. Specifically, we examine the 2004 Survey of Consumer Finances (SCF), a nationally representative sample that contains extensive information on assets, liabilities, income, and demographic characteristics of those sampled (potential U.S. customers). We now return to the original sample of $n = 500$ families with positive incomes to study whether a family purchases term life insurance. From our sample, it turns out that 225 did not (FACEPOS = 0), whereas 275 did purchase term life insurance (FACEPOS = 1).

a. Summary Statistics. Provide a table of means of explanatory variables by level of the dependent variable FACEPOS. Interpret what we learn from this table.

b. Linear Probability Model. Fit a linear probability model using FACEPOS as the dependent variable and LINCOME, EDUCATION, AGE, and GENDER as continuous explanatory variables, together with the factor MARSTAT.

   b(i). Briefly define a linear probability model.

   b(ii). Comment on the quality of the fitted model.

   b(iii). What are the three main drawbacks of the linear probability model?

c. Logistic Regression Model. Fit a logistic regression model using the same set of explanatory variables.

   c(i). Identify which variables appear to be statistically significant. In your identification, describe the basis for your conclusions.

   c(ii). Which measure summarizes the goodness of fit?

d. Reduced Logistic Regression Model. Define MARSTAT1 to be a binary variable that indicates MARSTAT $= 1$. Fit a second logistic regression model using LINCOME, EDUCATION, and MARSTAT1.

   d(i). Compare the two models using a likelihood ratio test. State your null and alternative hypotheses, decision-making criterion, and decision-making rule.

   d(ii). Who is more likely to purchase term life insurance, married or nonmarried? Provide an interpretation in terms of the odds of purchasing term life insurance for the variable MARSTAT1.

   d(iii). Consider a married male who is age 54. Assume that this person has 13 years of education, annual wages of $70,000, and lives in a household composed of four people. For this model, what is the estimate of the probability of purchasing term life insurance?

11.11. **Success in Actuarial Studies**. Much like the medical and legal fields, members of the actuarial profession face interesting problems and are generally well compensated for their efforts in resolving these problems. Also like the medical and legal professions, the educational barriers to becoming an actuary are challenging, which limits entrance into the field.

To advise students on whether they have the potential to meet the demands of this intellectually challenging field, Smith and Schumacher (2006) studied attributes of students in a business college. Specifically, they examined $n = 185$ freshman at Bryant University in Rhode Island who had begun their college careers in 1995–2001. The dependent variable of interest was whether they graduated with an actuarial concentration–for these students the first step to becoming a professional actuary. Of these, 77 graduated with an actuarial concentration and the other 108 dropped the concentration (at Bryant, most transferred to other concentrations, though some left the university).

Smith and Schumacher (2006) reported the effects of four early assessment mechanisms as well as GENDER, a control variable. The assessment mechanisms were PLACE%, performance on a mathematics placement exam administered just prior to the freshman year; MSAT and VSAT, mathematics (M) and verbal (V) portions of the Scholastic Aptitude Test (SAT); and RANK, high school rank given as a proportion (with closer to one being better). Table 11.14 shows that students who eventually graduated with an actuarial concentration performed higher on these early assessment mechanisms than did actuarial dropouts.

A logistic regression was fit to the data with the results reported in Table 11.14.

a. To get a sense of which variables are statistically significance, calculate t-ratios for each variable. For each variable, state whether it is statistically significant.
b. To get a sense of the relative impact of the assessment mechanisms, use the coefficients in Table 11.14 to compute estimated success probabilities for the following combination of variables. In your calculations, assume that GENDER = 1.
   b(i). Assume PLACE% = 0.80, MSAT = 680, VSAT = 570 and RANK = 0.90.
   b(ii). Assume PLACE% = 0.60, MSAT = 680, VSAT = 570 and RANK = 0.90.
   b(iii). Assume PLACE% = 0.80, MSAT = 620, VSAT = 570 and RANK = 0.90.
   b(iv). Assume PLACE% = 0.80, MSAT = 680, VSAT = 540 and RANK = 0.90.
   b(v). Assume PLACE% = 0.80, MSAT = 680, VSAT = 570 and RANK = 0.70.

Table 11.14 Summary Statistics and Logistic Regression Fits for Predicting Actuarial Graduation

	Average for Actuarial		Logistic Regression	
Variable	Graduates	Dropouts	Estimate	Std. Error
Intercept	—	—	−12.094	2.575
GENDER	—	—	0.256	0.407
PLACE%	0.83	0.64	4.336	1.657
MSAT	679.25	624.25	0.008	0.004
VSAT	572.20	544.25	−0.002	0.003
RANK	0.88	0.76	4.442	1.836

11.12. **Case Control.** Consider the following "case-control" sample selection method for binary dependent variables. Intuitively, if we are working with a problem in which the event of interest is rare, we want to make sure that we sample a sufficient number of events so that our estimation procedures are reliable.

Suppose that we have a large database consisting of $\{y_i, \mathbf{x}_i\}$, $i = 1, \ldots, N$ observations. (For insurance company records, N could easily be 10 million or more.) We want to make sure to get plenty of $y_i = 1$

(corresponding to claims or cases) in our sample, plus a sample of $y_i = 0$ (corresponding to nonclaims or controls). Thus, we split the dataset into two subsets. For the first subset, consisting of observations with $y_i = 1$, we take a random sample with probability τ_1. Similarly, for the second subset, consisting of observations with $y_i = 0$, we take a random sample with probability τ_0. For example, in practice we might use $\tau_1 = 1$ and $\tau_0 = 0.10$, corresponding to taking all of the claims and a 10% sample of nonclaims – thus, τ_1 and τ_1 are considered known to the analyst.

a. Let $\{r_i = 1\}$ denote the event that the observation is selected to be part of the analysis. Determine $\Pr(y_i = 1, r_i = 1)$, $\Pr(y_i = 0, r_i = 1)$ and $\Pr(r_i = 1)$ in terms of τ_0, τ_1, and $\pi_i = \Pr(y_i = 1)$.
b. Using the calculations in part (a), determine the conditional probability $\Pr(y_i = 1 | r_i = 1)$.
c. Now assume that π_i has a logistic form ($\pi(z) = \exp(z)/(1 + \exp(z))$ and $\pi_i = \pi(\mathbf{x}_i' \boldsymbol{\beta})$). Rewrite your answer part (b) using this logistic form.
d. Write the likelihood of the observed y_i's (conditional on $r_i = 1, i = 1, \ldots, n$). Show how we can interpret this as the usual logistic regression likelihood with the exception that the intercept has changed. Specify the new intercept in terms of the original intercept, τ_0 and τ_1.

11.9 Technical Supplements – Likelihood-Based Inference

Begin with random variables $(y_1, \ldots, y_n)' = \mathbf{y}$ whose joint distribution is known up to a vector of parameters $\boldsymbol{\theta}$. In regression applications, $\boldsymbol{\theta}$ consists of the regression coefficients, $\boldsymbol{\beta}$, and possibly a scale parameter σ^2 as well as additional parameters. This joint probability density function is denoted as $\mathrm{f}(\mathbf{y}; \boldsymbol{\theta})$. The function may also be a probability mass function for discrete random variables or a mixture distribution for random variables that have discrete and continuous components. In each case, we can use the same notation, $\mathrm{f}(\mathbf{y}; \boldsymbol{\theta})$, and call it the *likelihood function*. The likelihood is a function of the parameters with the data ($\mathbf{y}$) fixed rather than a function of the data with the parameters ($\boldsymbol{\theta}$) fixed.

It is customary to work with the logarithmic version of the likelihood function and thus we define the *log-likelihood function* to be

$$L(\boldsymbol{\theta}) = L(\mathbf{y}; \boldsymbol{\theta}) = \ln \mathrm{f}(\mathbf{y}; \boldsymbol{\theta}),$$

evaluated at a realization of $\mathbf{y}$. In part, this is because we often work with the important special case in which the random variables $y_1, \ldots, y_n$ are independent. In this case, the joint density function can be expressed as a product of the marginal density functions and, by taking logarithms, we can work with sums. Even when not dealing with independent random variables, as with time series data, it is often computationally more convenient to work with log-likelihoods than with the original likelihood function.

11.9.1 Properties of Likelihood Functions

Two basic properties of likelihood functions are:

$$\mathrm{E}\left(\frac{\partial}{\partial \boldsymbol{\theta}} L(\boldsymbol{\theta})\right) = \mathbf{0} \tag{11.1}$$

and

$$\mathrm{E}\left(\frac{\partial^2}{\partial \boldsymbol{\theta} \partial \boldsymbol{\theta}'} L(\boldsymbol{\theta})\right) + \mathrm{E}\left(\frac{\partial L(\boldsymbol{\theta})}{\partial \boldsymbol{\theta}} \frac{\partial L(\boldsymbol{\theta})}{\partial \boldsymbol{\theta}'}\right) = \mathbf{0}. \tag{11.2}$$

The derivative of the log-likelihood function, $\partial L(\boldsymbol{\theta})/\partial \boldsymbol{\theta}$, is called the *score function*. Equation (11.1) shows that the score function has mean zero. To see this, under suitable regularity conditions, we have

$$\mathrm{E}\left(\frac{\partial}{\partial \boldsymbol{\theta}} L(\boldsymbol{\theta})\right) = \mathrm{E}\left(\frac{\frac{\partial}{\partial \boldsymbol{\theta}} \mathrm{f}(\mathbf{y};\boldsymbol{\theta})}{\mathrm{f}(\mathbf{y};\boldsymbol{\theta})}\right) = \int \frac{\partial}{\partial \boldsymbol{\theta}} \mathrm{f}(\mathbf{y};\boldsymbol{\theta}) d\mathbf{y} = \frac{\partial}{\partial \boldsymbol{\theta}} \int \mathrm{f}(\mathbf{y};\boldsymbol{\theta}) d\mathbf{y}$$
$$= \frac{\partial}{\partial \boldsymbol{\theta}} 1 = \mathbf{0}.$$

For convenience, this demonstration assumes a density for $\mathrm{f}(\cdot)$; extensions to mass and mixtures distributions are straightforward. The proof of equation (11.2) is similar and is omitted. To establish equation (11.1), we implicitly used "suitable regularity conditions" to allow the interchange of the derivative and integral sign. To be more precise, an analyst working with a specific type of distribution can use this information to check that the interchange of the derivative and integral sign is valid.

Using equation (11.2), we can define the *information matrix*

$$\mathbf{I}(\boldsymbol{\theta}) = \mathrm{E}\left(\frac{\partial L(\boldsymbol{\theta})}{\partial \boldsymbol{\theta}} \frac{\partial L(\boldsymbol{\theta})}{\partial \boldsymbol{\theta}'}\right) = -\mathrm{E}\left(\frac{\partial^2}{\partial \boldsymbol{\theta} \partial \boldsymbol{\theta}'} L(\boldsymbol{\theta})\right). \tag{11.3}$$

This quantity is used extensively in the study of large sample properties of likelihood functions.

The information matrix appears in the large sample distribution of the score function. Specifically, under broad conditions, we have that $\partial L(\boldsymbol{\theta})/\partial \boldsymbol{\theta}$ has a large sample normal distribution with mean $\mathbf{0}$ and variance $\mathbf{I}(\boldsymbol{\theta})$. To illustrate, suppose that the random variables are independent so that the score function can be written as

$$\frac{\partial}{\partial \boldsymbol{\theta}} L(\boldsymbol{\theta}) = \frac{\partial}{\partial \boldsymbol{\theta}} \ln \prod_{i=1}^{n} \mathrm{f}(y_i;\boldsymbol{\theta}) = \sum_{i=1}^{n} \frac{\partial}{\partial \boldsymbol{\theta}} \ln \mathrm{f}(y_i;\boldsymbol{\theta}).$$

The score function is the sum of mean zero random variables because of equation (11.1); central limit theorems are widely available to ensure that sums of independent random variables have large sample normal distributions (see Section 1.4 for an example). Further, if the random variables are identical, then from equation (11.3) we can see that the second moment of $\partial \ln \mathrm{f}(y_i;\boldsymbol{\theta})/\partial \boldsymbol{\theta}$ is the information matrix, yielding the result.

11.9.2 Maximum Likelihood Estimators

Maximum likelihood estimators are values of the parameters $\boldsymbol{\theta}$ that are "most likely" to have been produced by the data. The value of $\boldsymbol{\theta}$, say, $\boldsymbol{\theta}_{MLE}$, that maximizes $\mathrm{f}(\mathbf{y};\boldsymbol{\theta})$ is called the *maximum likelihood estimator*. Because $\ln(\cdot)$ is a one-to-one function, we can also determine $\boldsymbol{\theta}_{MLE}$ by maximizing the log-likelihood function, $L(\boldsymbol{\theta})$.

Under broad conditions, we have that $\boldsymbol{\theta}_{MLE}$ has a large sample normal distribution with mean $\boldsymbol{\theta}$ and variance $(\mathbf{I}(\boldsymbol{\theta}))^{-1}$. This is a critical result on which much of estimation and hypothesis testing is based. To underscore this result, we examine the special case of "normal-based" regression.

Special Case: Regression with Normal Distributions. Suppose that $y_1, \ldots, y_n$ are independent and normally distributed, with mean $\mathrm{E}\, y_i = \mu_i = \mathbf{x}_i'\boldsymbol{\beta}$ and variance σ^2. The parameters can be summarized as $\boldsymbol{\theta} = \left(\boldsymbol{\beta}', \sigma^2\right)'$. Recall from equation (1.1) that the normal probability density function is

$$\mathrm{f}(y;\mu_i,\sigma^2) = \frac{1}{\sigma\sqrt{2\pi}} \exp\left(-\frac{1}{2\sigma^2}(y-\mu_i)^2\right).$$

With this, the two components of the score function are

$$\frac{\partial}{\partial \boldsymbol{\beta}} L(\boldsymbol{\theta}) = \sum_{i=1}^{n} \frac{\partial}{\partial \boldsymbol{\beta}} \ln \mathrm{f}(y_i;\mathbf{x}_i'\boldsymbol{\beta},\sigma^2) = -\frac{1}{2\sigma^2}\sum_{i=1}^{n}\frac{\partial}{\partial \boldsymbol{\beta}}\left(y_i - \mathbf{x}_i'\boldsymbol{\beta}\right)^2$$
$$= -\frac{(-2)}{2\sigma^2}\sum_{i=1}^{n}\left(y_i - \mathbf{x}_i'\boldsymbol{\beta}\right)\mathbf{x}_i$$

and

$$\frac{\partial}{\partial \sigma^2} L(\boldsymbol{\theta}) = \sum_{i=1}^{n} \frac{\partial}{\partial \sigma^2} \ln \mathrm{f}(y_i;\mathbf{x}_i'\boldsymbol{\beta},\sigma^2) = -\frac{n}{2\sigma^2} + \frac{1}{2\sigma^4}\sum_{i=1}^{n}\left(y_i - \mathbf{x}_i'\boldsymbol{\beta}\right)^2.$$

Setting these equations to zero and solving yields the maximum likelihood estimators

$$\boldsymbol{\beta}_{MLE} = \left(\sum_{i=1}^{n}\mathbf{x}_i\mathbf{x}_i'\right)^{-1}\sum_{i=1}^{n}\mathbf{x}_i y_i = \mathbf{b}$$

and

$$\sigma^2_{MLE} = \frac{1}{n}\sum_{i=1}^{n}\left(y_i - \mathbf{x}_i'\mathbf{b}\right)^2 = \frac{n-(k+1)}{n}s^2.$$

Thus, the maximum likelihood estimator of $\boldsymbol{\beta}$ is equal to the usual least squares estimator. The maximum likelihood estimator of σ^2 is a scalar multiple of the usual least squares estimator. The least squares estimators s^2 is unbiased, whereas as σ^2_{MLE} is only approximately unbiased in large samples.

The information matrix is

$$\mathbf{I}(\boldsymbol{\theta}) = -\mathrm{E}\begin{pmatrix} \frac{\partial^2}{\partial \boldsymbol{\beta}\, \partial \boldsymbol{\beta}'} L(\boldsymbol{\theta}) & \frac{\partial^2}{\partial \boldsymbol{\beta}\, \partial \sigma^2} L(\boldsymbol{\theta}) \\ \frac{\partial^2}{\partial \sigma^2 \partial \boldsymbol{\beta}'} L(\boldsymbol{\theta}) & \frac{\partial^2}{\partial \sigma^2 \partial \sigma^2} L(\boldsymbol{\theta}) \end{pmatrix} = \begin{pmatrix} \frac{1}{\sigma^2}\sum_{i=1}^n \mathbf{x}_i \mathbf{x}_i' & 0 \\ 0 & \frac{n}{2\sigma^4} \end{pmatrix}.$$

Thus, $\boldsymbol{\beta}_{MLE} = \mathbf{b}$ has a large sample normal distribution with mean $\boldsymbol{\beta}$ and variance-covariance matrix $\sigma^2 \left(\sum_{i=1}^n \mathbf{x}_i \mathbf{x}_i'\right)^{-1}$, as seen previously. Moreover, σ^2_{MLE} has a large sample normal distribution with mean σ^2 and variance $2\sigma^4/n$.

Maximum likelihood is a general estimation technique that can be applied in many statistical settings, not just regression and time series applications. It can be applied broadly and enjoys certain optimality properties. We have already cited the result that maximum likelihood estimators typically have a large sample normal distribution. Moreover, maximum likelihood estimators are the most efficient in the following sense. Suppose that $\widehat{\boldsymbol{\theta}}$ is an alternative unbiased estimator. The Cramer-Rao theorem states, under mild regularity conditions, for all vectors $\mathbf{c}$, that $\mathrm{Var}\ \mathbf{c}'\boldsymbol{\theta}_{MLE} \leq \mathrm{Var}\ \mathbf{c}'\widehat{\boldsymbol{\theta}}$, for sufficiently large n.

We also note that $2\left(L(\boldsymbol{\theta}_{MLE}) - L(\boldsymbol{\theta})\right)$ has a chi-square distribution with degrees of freedom equal to the dimension of $\boldsymbol{\theta}$.

In a few applications, such as the regression case with a normal distribution, maximum likelihood estimators can be computed analytically as a closed-form expression. Typically, this can be done by finding roots of the first derivative of the function. However, in general, maximum likelihood estimators cannot be calculated with closed-form expressions and are determined iteratively. Two general procedures are widely used:

1. *Newton-Raphson* uses the iterative algorithm

$$\boldsymbol{\theta}_{NEW} = \boldsymbol{\theta}_{OLD} - \left.\left\{\left(\frac{\partial^2 L}{\partial \boldsymbol{\theta} \partial \boldsymbol{\theta}'}\right)^{-1} \frac{\partial L}{\partial \boldsymbol{\theta}}\right\}\right|_{\boldsymbol{\theta}=\boldsymbol{\theta}_{OLD}}. \tag{11.4}$$

2. *Fisher scoring* uses the iterative algorithm

$$\boldsymbol{\theta}_{NEW} = \boldsymbol{\theta}_{OLD} + \mathbf{I}(\boldsymbol{\theta}_{OLD})^{-1} \left.\left\{\frac{\partial L}{\partial \boldsymbol{\theta}}\right\}\right|_{\boldsymbol{\theta}=\boldsymbol{\theta}_{OLD}}, \tag{11.5}$$

where $\mathbf{I}(\boldsymbol{\theta})$ is the information matrix.

11.9.3 Hypothesis Tests

We consider testing the null hypothesis $H_0 : h(\boldsymbol{\theta}) = \mathbf{d}$, where $\mathbf{d}$ is a known vector of dimension $r \times 1$ and $\mathrm{h}(\cdot)$ is known and differentiable. This testing framework encompasses the general linear hypothesis introduced in Chapter 4 as a special case.

There are three general approaches for testing hypotheses, called the *likelihood ratio*, *Wald*, and *Rao* tests. The Wald approach evaluates a function of the likelihood at $\boldsymbol{\theta}_{MLE}$. The likelihood ratio approach uses $\boldsymbol{\theta}_{MLE}$ and $\boldsymbol{\theta}_{Reduced}$. Here, $\boldsymbol{\theta}_{Reduced}$ is the value of $\boldsymbol{\theta}$ that maximizes $L(\boldsymbol{\theta}_{Reduced})$ under the constraint that $h(\boldsymbol{\theta}) = \mathbf{d}$. The Rao approach also uses $\boldsymbol{\theta}_{Reduced}$ but determines it by maximizing $L(\boldsymbol{\theta}) - \boldsymbol{\lambda}'(h(\boldsymbol{\theta}) - \mathbf{d})$, where $\boldsymbol{\lambda}$ is a vector of Lagrange multipliers. Hence, Rao's test is also called the *Lagrange multiplier test*.

The test statistics associated with the three approaches are:

1. $LRT = 2 \times \{L(\boldsymbol{\theta}_{MLE}) - L(\boldsymbol{\theta}_{Reduced})\}$.
2. Wald: $TS_W(\boldsymbol{\theta}_{MLE})$, where

$$TS_W(\boldsymbol{\theta}) = (h(\boldsymbol{\theta}) - \mathbf{d})' \left\{ \frac{\partial}{\partial \boldsymbol{\theta}} h(\boldsymbol{\theta})' \left(-\mathbf{I}(\boldsymbol{\theta})\right)^{-1} \frac{\partial}{\partial \boldsymbol{\theta}} h(\boldsymbol{\theta}) \right\}^{-1} (h(\boldsymbol{\theta}) - \mathbf{d}).$$

3. Rao: $TS_R(\boldsymbol{\theta}_{Reduced})$, where $TS_R(\boldsymbol{\theta}) = \frac{\partial}{\partial \boldsymbol{\theta}} L(\boldsymbol{\theta}) \left(-\mathbf{I}(\boldsymbol{\theta})\right)^{-1} \frac{\partial}{\partial \boldsymbol{\theta}} L(\boldsymbol{\theta})'$.

Under broad conditions, all three test statistics have large sample chi-square distributions with r degrees of freedom under H_0. All three methods work well when the number of parameters is finite dimensional and the null hypothesis specifies that $\boldsymbol{\theta}$ is on the interior of the parameter space.

The main advantage of the Wald statistic is that it requires only computation of $\boldsymbol{\theta}_{MLE}$ and not $\boldsymbol{\theta}_{Reduced}$. In contrast, the main advantage of the Rao statistic is that it requires only computation of $\boldsymbol{\theta}_{Reduced}$ and not $\boldsymbol{\theta}_{MLE}$. In many applications, computation of $\boldsymbol{\theta}_{MLE}$ is onerous. The likelihood ratio test is a direct extension of the partial F-test introduced in Chapter 4 – it allows one to directly compare nested models, a helpful technique in applications.

11.9.4 Information Criteria

Likelihood ratio tests are useful for choosing between two models that are *nested*, that is, where one model is a subset of the other. How do we compare models when they are not nested? One way is to use the following information criteria.

The distance between two probability distributions given by probability density functions g and $f_{\boldsymbol{\theta}}$ can be summarized by

$$\mathrm{KL}(g, f_{\boldsymbol{\theta}}) = \mathrm{E}_g \ln \frac{g(y)}{f_{\boldsymbol{\theta}}(y)}.$$

This is the *Kullback-Leibler distance*. Here, we have indexed f by a vector of parameters $\boldsymbol{\theta}$. If we let the density function g be fixed at a hypothesized value, say, $f_{\boldsymbol{\theta}_0}$, then minimizing $\mathrm{KL}(f_{\boldsymbol{\theta}_0}, f_{\boldsymbol{\theta}})$ is equivalent to maximizing the log-likelihood.

However, maximizing the likelihood does not impose sufficient structure on the problem because we know that we can always make the likelihood greater by introducing additional parameters. Thus, Akaike in 1974 showed that a reasonable alternative is to minimize

$$AIC = -2 \times L(\boldsymbol{\theta}_{MLE}) + 2 \times (\textit{number of parameters}),$$

known as *Akaike's information criterion*. Here, the additional term $2\times$ (*number of parameters*) is a penalty for the complexity of the model. With this penalty, one cannot improve on the fit simply by introducing additional parameters. This statistic can be used when comparing several alternative models that are not necessarily nested. One picks the model that minimizes AIC. If the models under consideration have the same number of parameters, this is equivalent to choosing the model that maximizes the log-likelihood.

We remark that this definition is not uniformly adopted in the literature. For example, in time series analysis, the AIC is rescaled by the number of parameters. Other versions that provide finite sample corrections are also available in the literature.

In 1978, Schwarz derived an alternative criterion using Bayesian methods. His measure is known as the *Bayesian information criterion*, defined as

$$BIC = -2 \times L(\boldsymbol{\theta}_{MLE}) + (number\ of\ parameters) \\ \times \ln(number\ of\ observations).$$

This measure gives greater weight to the number of parameters. That is, other things being equal, BIC will suggest a more parsimonious model than AIC.

Like the adjusted coefficient of determination R_a^2 that we have introduced in the regression literature, both AIC and BIC provide measures of fit with a penalty for model complexity. In normal linear regression models, Section 5.6 pointed out that minimizing AIC is equivalent to minimizing $n \ln s^2 + k$. Another linear regression statistic that balances the goodness of fit and complexity of the model is Mallows C_p statistic. For p candidate variables in the model, this is defined as $C_p = (Error\ SS)_p / s^2 - (n - 2p)$. See, for example, Cameron and Trivedi (1998) for references and further discussion of information criteria.

12

Count Dependent Variables

Chapter Preview. In this chapter, the dependent variable y is a count, taking on values of zero, one, and two, and so on, which describes a number of events. Count dependent variables form the basis of actuarial models of claims *frequency*. In other applications, a count dependent variable may be the number of accidents, the number of people retiring, or the number of firms becoming insolvent.

The chapter introduces Poisson regression, a model that includes explanatory variables with a Poisson distribution for counts. This fundamental model handles many datasets of interest to actuaries. However, with the Poisson distribution, the mean equals the variance, a limitation suggesting the need for more general distributions such as the negative binomial. Even the two-parameter negative binomial can fail to capture some important features, motivating the need for even more complex models such as the zero-inflated and latent variable models introduced in this chapter.

12.1 Poisson Regression

12.1.1 Poisson Distribution

A count random variable y is one that has outcomes on the nonnegative integers, $j = 0, 1, 2, \ldots$ The Poisson is a fundamental distribution used for counts that has probability mass function

$$\Pr(y = j) = \frac{\mu^j}{j!} e^{-\mu}, \quad j = 0, 1, 2, \ldots. \tag{12.1}$$

It can be shown that $\mathrm{E}\, y = \sum_{j=0}^{\infty} j \Pr(y = j) = \mu$, so we may interpret the parameter μ to be the mean of the distribution. Similarly, one can show that $\mathrm{Var}\, y = \mu$, so that the mean equals the variance for this distribution.

An early application (Bortkiewicz, 1898) was based on using the Poisson distribution to represent the annual number of deaths in the Prussian army due to "mule kicks." The distribution is still widely used as a model of the number of accidents, such as injuries in an industrial environment (for workers' compensation coverage) and property damages in automobile insurance.

Ⓡ **Empirical** *Filename is "SingaporeAuto"*

Example: Singapore Automobile Data. These data are from a 1993 portfolio of $n = 7{,}483$ automobile insurance policies from a major insurance company

Table 12.1 Comparison of Observed to Fitted Counts Based on Singapore Automobile Data

Count (j)	Observed (n_j)	Fitted Counts Using the Poisson Distribution ($n\widehat{p}_j$)
0	6,996	6,977.86
1	455	487.70
2	28	17.04
3	4	0.40
4	0	0.01
Total	7,483	7,483.00

in Singapore. The data will be described further in Section 12.2. Table 12.1 provides the distribution of the number of accidents. The dependent variable is the number of automobile accidents per policyholder. For this dataset, it turns out that the maximum number of accidents in a year was three. There were on average $\overline{y} = 0.06989$ accidents per person.

Table 12.1 also provides fitted counts that were computed using the maximum likelihood estimator of μ. Specifically, from equation (12.1), we can write the mass function as $\mathrm{f}(y, \mu) = \mu^y e^{-\mu}/y!$, and so the log-likelihood is

$$L(\mu) = \sum_{i=1}^{n} \ln \mathrm{f}(y_i, \mu) = \sum_{i=1}^{n} (-\mu + y_i \ln \mu - \ln y_i!) . \qquad (12.2)$$

It is straightforward to show that the log-likelihood has a maximum at $\widehat{\mu} = \overline{y}$, the average claims count. Estimated probabilities, using equation (12.1) and $\widehat{\mu} = \overline{y}$, are denoted as $\widehat{p}_j$. We used these estimated probabilities in Table 12.1 when computing the fitted counts with $n = 7{,}483$.

To compare observed and fitted counts, a widely used goodness of fit statistic is *Pearson's chi-square statistic*, given by

$$\sum_j \frac{\left(n_j - n\widehat{p}_j\right)^2}{n\widehat{p}_j} . \qquad (12.3)$$

Under the null hypothesis that the Poisson distribution is a correct model, this statistic has a large sample chi-square distribution where the degrees of freedom is the number of cells minus one minus the number of estimated parameters. For the Singapore data in Table 12.1, this is $df = 5 - 1 - 1 = 3$. It turns out the statistic is 41.98, indicating that this basic Poisson model is inadequate.

12.1.2 Regression Model

To extend the basic Poisson model, we first allow the mean to vary by a known amount called an *exposure* E_i, so that

$$\mathrm{E}\, y_i = E_i \times \mu .$$

To motivate this specification, recall that sums of independent Poisson random variables also have a Poisson distribution so that it is sensible to think of exposures as large positive numbers. Thus, it is common to model the number of accidents per thousand vehicles or the number of homicides per million population. Further, we also consider instances in which the units of exposure may be fractions. To illustrate, for our Singapore data, E_i represents the fraction of the year that a policyholder had insurance coverage. The logic behind this is that the expected number of accidents is directly proportional to the length of coverage. (This can also be motivated by a probabilistic framework based on collections of Poisson distributed random variables known as *Poisson processes;* see, e.g., Klugman et al., 2008).

More generally, we wish to allow the mean to vary according to information contained in other explanatory variables. For the Poisson, it is customary to specify

$$\mathrm{E}\ y_i = \mu_i = \exp\left(\mathbf{x}_i'\boldsymbol{\beta}\right).$$

Using the exponential function to map the systematic component $\mathbf{x}_i'\boldsymbol{\beta}$ into the mean ensures that E y_i will remain positive. Assuming the linearity of the regression coefficients allows for easy interpretation. Specifically, because

$$\frac{\partial \mathrm{E}\ y_i}{\partial x_{ij}} \times \frac{1}{\mathrm{E}\ y_i} = \beta_j,$$

we may interpret β_j to be the proportional change in the mean per unit change in x_{ij}. The function that connects the mean to the systematic component is known as the *logarithmic link function*; that is, $\ln \mu_i = \mathbf{x}_i'\boldsymbol{\beta}$.

With a logarithmic link function, we may interpret β_j to be the proportional change in the mean per unit change in x_j.

To incorporate exposures, one can always specify one of the explanatory variables to be ln E_i and restrict the corresponding regression coefficient to be 1. This term is known as an *offset*. With this convention, the link function is

$$\ln \mu_i = \ln E_i + \mathbf{x}_i'\boldsymbol{\beta}. \tag{12.4}$$

Example: California Automobile Accidents. Weber (1971) provided the first application of Poisson regression to automobile accident frequencies in his study of California driving records. In one model, Weber examined the number of automobile accidents during 1963 of nearly 87,000 male drivers. His explanatory variables consisted of the following:

- $x_1 =$ the natural logarithm of the traffic density index of the county in which the driver resides
- $x_2 = 5/(age - 13)$
- $x_3 =$ the number of countable convictions incurred during years 1961–2
- $x_4 =$ the number of accident involvements incurred during years 1961–2
- $x_5 =$ the number of noncountable convictions incurred during years 1961–2.

Interestingly, in this early application, Weber achieved a satisfactory fit representing the mean as a linear combination of explanatory variables (E $y_i = \mathbf{x}_i'\boldsymbol{\beta}$), not the exponentiated version as in equation (12.4) that is now commonly fit.

12.1.3 Estimation

Maximum likelihood is the usual estimation technique for Poisson regression models. Using the logarithmic link function in equation (12.4), the log-likelihood is given by

$$L(\boldsymbol{\beta}) = \sum_{i=1}^{n} (-\mu_i + y_i \ln \mu_i - \ln y_i!)$$

$$= \sum_{i=1}^{n} \left(-E_i \exp\left(\mathbf{x}_i'\boldsymbol{\beta}\right) + y_i \left(\ln E_i + \mathbf{x}_i'\boldsymbol{\beta}\right) - \ln y_i!\right).$$

Setting the *score function* equal to zero yields

$$\left.\frac{\partial}{\partial \boldsymbol{\beta}} \mathrm{L}(\boldsymbol{\beta})\right|_{\beta=\mathbf{b}} = \sum_{i=1}^{n} \left(y_i - E_i \exp\left(\mathbf{x}_i'\mathbf{b}\right)\right) \mathbf{x}_i = \sum_{i=1}^{n} (y_i - \widehat{\mu}_i)\, \mathbf{x}_i = \mathbf{0}, \qquad (12.5)$$

where $\widehat{\mu}_i = E_i \exp\left(\mathbf{x}_i'\mathbf{b}\right)$. Solving this equation (numerically) yields $\mathbf{b}$, the maximum likelihood estimator of $\boldsymbol{\beta}$. From equation (12.5), we see that if a row of $\mathbf{x}_i$ is constant (corresponding to a constant intercept regression term), then the sum of residuals $y_i - \widehat{\mu}_i$ is zero.

In Poisson regression with an intercept, the sum, and hence the average, of residuals is zero.

Taking second derivatives yields the *information matrix*,

$$\mathbf{I}(\boldsymbol{\beta}) = -\mathrm{E} \frac{\partial^2}{\partial \boldsymbol{\beta} \partial \boldsymbol{\beta}'} \mathrm{L}(\boldsymbol{\beta}) = \sum_{i=1}^{n} E_i \exp\left(\mathbf{x}_i'\boldsymbol{\beta}\right) \mathbf{x}_i \mathbf{x}_i' = \sum_{i=1}^{n} \mu_i \mathbf{x}_i \mathbf{x}_i'.$$

Standard maximum likelihood estimation theory (Section 11.9.2) shows that the asymptotic variance-covariance matrix of $\mathbf{b}$ is

$$\widehat{\mathrm{Var}\, \mathbf{b}} = \left(\sum_{i=1}^{n} \widehat{\mu}_i \mathbf{x}_i \mathbf{x}_i' \right)^{-1}.$$

The square root of the jth diagonal element of $\widehat{\mathrm{Var}\, \mathbf{b}}$ yields the standard error for b_j, which we denote as $se(b_j)$.

Example: Medical Malpractice Insurance. Physicians make errors and may be sued by parties harmed by these errors. Like many professionals, it is common for physicians to carry insurance coverage that mitigates the financial consequences of malpractice lawsuits.

Because insurers wish to accurately price this type of coverage, it seems natural to ask what type of physicians are likely to submit medical malpractice claims.

Table 12.2 Regression Coefficients of Medical Malpractice Poisson Regression Model

Explanatory Variables	Coef-ficient	Standard Error	Explanatory Variables	Coef-ficient	Standard Error
Intercept	−1.634	0.254	MSA: Miami Dade-Broward	0.377	0.094
Log Years Licensed	−0.392	0.054	MSA: Other	0.012	0.084
Female	−0.432	0.082	Speciality		
Patient Volume	0.643	0.045	Anesthesiology	0.944	0.099
(Patient Volume)^2	−0.066	0.008	Emergency Medicine	0.583	0.105
Per Capita Education	−0.015	0.006	Internal Medicine	0.428	0.066
Per Capita Income	0.047	0.011	Obstetrics-Gynecology	1.226	0.070
Regional Variables			Otorhinolaryngology	1.063	0.109
Second Circuit	0.066	0.072	Pediatrics	0.385	0.089
Third Circuit	0.103	0.088	Radiology	0.478	0.099
Fourth Circuit	0.214	0.098	Surgery	1.410	0.061
Fifth Circuit	0.287	0.069	Other Specialties	0.011	0.076

Fournier and McInnes (2001) examined a sample of $n = 9{,}059$ Florida physicians using data from the Florida Medical Professional Liability Insurance Claims File. The authors examined closed claims in years 1985–9 for physicians who were licensed before 1981, thus omitting claims for newly licensed physicians. Medical malpractice claims can take a long time to be resolved ("settled"); in their study, Fournier and McInnes found that 2% of claims were still not settled after five years of the malpractice event. Thus, they chose an early period (1985–9) to allow the experience to mature. The authors also ignored minor claims by only considering claims that exceeded $100.

Table 12.2 provides fitted Poisson regression coefficients along with standard errors that appear in Fournier and McInnes (2001). The table shows that physicians' practice area, region, practice size, and physicians' personal characteristics (experience and sex) to be important determinants of the number of medical malpractice suits. For example, we may interpret the coefficient associated with sex to say that men are expected to have exp(0.432) = 1.540 times as many claims as women.

12.1.4 Additional Inference

In Poisson regression models, we anticipate *heteroscedastic* dependent variables because of the relation Var $y_i = \mu_i$. This characteristic means that ordinary residuals $y_i - \widehat{\mu}_i$ are of less use, so that it is more common to examine *Pearson residuals*, defined as

$$r_i = \frac{y_i - \widehat{\mu}_i}{\sqrt{\widehat{\mu}_i}}.$$

By construction, Pearson residuals are approximately homoscedastic. Plots of Pearson residuals can be used to identify unusual observations or to detect whether additional variables of interest can be used to improve the model specification.

Pearson residuals can also be used to calculate a Pearson goodness-of-fit statistic,

$$\sum_{i=1}^{n} r_i^2 = \sum_{i=1}^{n} \frac{(y_i - \widehat{\mu}_i)^2}{\widehat{\mu}_i}. \tag{12.6}$$

This statistic is an overall measure of how well the model fits the data. If the model is specified correctly, then this statistic should be approximately $n - (k + 1)$. In general, Pearson goodness-of-fit statistics take the form $\sum (O - E)^2 / E$, where O is some observed quantity and E is the corresponding estimated (expected) value based on a model. The statistic in equation (12.6) is computed at the observation level whereas the statistic in equation (12.3) was computed summarizing information over cells.

In linear regression, the coefficient of determination R^2 is a widely accepted goodness-of-fit measure. In nonlinear regression such as for binary and count dependent variables, this is not true. Information statistics, such as *Akaike's information criterion,*

$$AIC = -2L(\mathbf{b}) + 2(k + 1),$$

represents a type of statistic useful for goodness of fit that is broadly defined over a large range of models. Models with smaller values of AIC fit better, and are preferred.

As noted in Section 12.1.3, t-statistics are regularly used for testing the significance of individual regression coefficients. For testing collections of regression coefficients, it is customary to use the *likelihood ratio test.* The likelihood ratio test is a well-known procedure for testing the null hypothesis $H_0 : \mathrm{h}(\boldsymbol{\beta}) = \mathbf{d}$, where $\mathbf{d}$ is a known vector of dimension $r \times 1$ and $\mathrm{h}(\cdot)$ is known and differentiable function. This approach uses $\mathbf{b}$ and $\mathbf{b}_{\text{Reduced}}$, where $\mathbf{b}_{\text{Reduced}}$ is the value of $\boldsymbol{\beta}$ that maximizes $L(\boldsymbol{\beta})$ under the restriction that $\mathrm{h}(\boldsymbol{\beta}) = \mathbf{d}$. One computes the test statistic

$$LRT = 2\left(L(\mathbf{b}) - L(\mathbf{b}_{\text{Reduced}})\right). \tag{12.7}$$

Under the null hypothesis H_0, the test statistic LRT has an asymptotic chi-square distribution with r degrees of freedom. Thus, large values of LRT suggest that the null hypothesis is not valid.

12.2 Application: Singapore Automobile Insurance

Frees and Valdez (2008) investigate hierarchical models of Singapore driving experience. Here we examine in detail a subset of their data, focusing on 1993 counts of automobile accidents. The purpose of the analysis is to understand the impact of vehicle and driver characteristics on accident experience. These relationships provide a foundation for an actuary working in *ratemaking*, that is, setting the price of insurance coverages.

Table 12.3 Description of Covariates

Covariate	Description
Vehicle Type	The type of vehicle being insured, either automobile (A) or other (O).
Vehicle Age	The age of the vehicle, in years, grouped into six categories.
Gender	The policyholder's sex, either male or female
Age	The age of the policyholder, in years, grouped into seven categories.
NCD	No claims discount. This is based on the previous accident record of the policyholder. The higher the discount, the better is the prior accident record.

Table 12.4 Effect of Vehicle Characteristics on Claims

	Count = 0	Count = 1	Count = 2	Count = 3	Totals
Vehicle Type					
Other	3,441	184	13	3	3,641
	(94.5)	(95.1)	(0.4)	(0.1)	(48.7)
Automobile	3,555	271	15	1	3,842
	(92.5)	(7.1)	(0.4)	(0.0)	(51.3)
Vehicle Age (in years)					
0–2	4,069	313	20	4	4,406
	(92.4)	(7.1)	(0.5)	(0.1)	(50.8)
3–5	708	59	4		771
	(91.8)	(7.7)	(0.5)		(10.3)
6–10	872	49	3		924
	(94.4)	(5.3)	(0.3)		(12.3)
11–15	1,133	30	1		1,164
	(97.3)	(2.6)	(0.1)		(15.6)
16 and older	214	4			218
	(98.2)	(1.8)			(2.9)
Totals	6,996	455	28	4	7,483

Note: Numbers in parentheses are percentages.

The data are from the General Insurance Association of Singapore, an organization consisting of general (property and casualty) insurers in Singapore (see the organization's Web site at www.gia.org.sg). From this database, several characteristics were available to explain automobile accident frequency. These characteristics include vehicle variables, such as type and age, as well as person level variables, such as age, sex, and prior driving experience. Table 12.3 summarizes these characteristics.

Table 12.4 shows the effects of vehicle characteristics on claim count. The "Automobile" category has lower overall claims experience. The "Other" category consists primarily of (commercial) goods vehicles, as well as weekend and hire cars. The vehicle age shows nonlinear effects of the age of the vehicle. Here, we see low claims for new cars with initially increasing accident frequency over time. However, for vehicles in operation for long periods of time, the accident

Table 12.5 Effect of Personal Characteristics on Claims. Based on Sample with Auto = 1.

	Count = 0		
	Number	Percentage	Total
Gender			
Female	654	93.4	700
Male	2,901	92.3	3,142
Age Category			
22–25	131	92.9	141
26–35	1,354	91.7	1,476
36–45	1,412	93.2	1,515
46–55	503	93.8	536
56–65	140	89.2	157
66 and over	15	88.2	17
No Claims Discount			
0	889	89.6	992
10	433	91.2	475
20	361	92.8	389
30	344	93.5	368
40	291	94.8	307
50	1,237	94.4	1,311
Total	3,555	92.5	3,842

frequencies are relatively low. There are also some important interaction effects between vehicle type and age that are not shown here. Nonetheless, Table 12.4 clearly suggests the importance of these two variables on claim frequencies.

Table 12.5 shows the effects of person-level characteristics, sex, age, and no claims discount on the frequency distribution. Person-level characteristics were largely unavailable for commercial use vehicles, and so Table 12.5 presents summary statistics for only those observations having automobile coverage with the requisite sex and age information. When we restricted consideration to (private use) automobiles, relatively few policies did not contain sex and age information.

Table 12.5 suggests that driving experience was roughly similar between men and women. This company insured very few young drivers, so the young male driver category that typically has extremely high accident rates in most automobiles studies is less important for these data. Nonetheless, Table 12.5 suggests strong age effects, with older drivers having better driver experience. Table 12.5 also demonstrates the importance of the no claims discounts (NCD). As anticipated, drivers with better previous driving records who enjoy a higher NCD have fewer accidents.

As part of the examination process, we investigated interaction terms among the covariates and nonlinear specifications. However, Table 12.6 summarizes a simpler fitted Poisson model with only additive effects. Table 12.6 shows that both vehicle age and no claims discount are important categories in that the t-ratios for

Table 12.6 Parameter Estimates from a Fitted Poisson Model

Variable	Parameter Estimate	t-Ratio	Variable	Parameter Estimate	t-Ratio
Intercept	−3.306	−6.602	(Auto = 1) × No Claims Discount*		
Auto	−0.667	−1.869	0	0.729	4.704
Female	−0.173	−1.115	10	0.528	2.732
			20	0.293	1.326
(Auto = 1) × Age Category*			30	0.260	1.152
22–25	0.747	0.961	40	−0.095	−0.342
26–35	0.489	1.251			
36–45	−0.057	−0.161	Vehicle Age (in years)*		
46–55	0.124	0.385	0–2	1.674	3.276
56–65	0.165	0.523	3–5	1.504	2.917
			6–10	1.081	2.084
			11–15	0.362	0.682

Note: The omitted reference levels are "66 and over" for age, "50" for no claims discount, and "16 and over" for vehicle age.

many of the coefficients are statistically significant. The overall log-likelihood for this model is $L(\mathbf{b}) = -1{,}776.730$.

Omitted reference levels are given in the footnote of Table 12.6 to help interpret the parameters. For example, for $NCD = 0$, we expect that a poor driver with $NCD = 0$ will have exp(0.729) = 2.07 times as many accidents as a comparable excellent driver with $NCD = 50$. In the same vein, we expect that a poor driver with $NCD = 0$ will have exp(0.729 − 0.293) = 1.55 times as many accidents as a comparable average driver with $NCD = 30$.

For a more parsimonious model, one might consider removing the automobile, sex, and age variables. Removing these seven variables results in a model with a log-likelihood of $L\left(\mathbf{b}_{\text{Reduced}}\right) = -1{,}779.420$. To understand whether this is a significant reduction, we can compute a likelihood ratio statistic (equation 12.7),

$$LRT = 2 \times (-1{,}776.730 - (-1{,}779.420)) = 5.379.$$

Comparing this to a chi-square distribution with $df = 7$ degrees of freedom, the statistic p-value $= \Pr\left(\chi_7^2 > 5.379\right) = 0.618$ indicates that these variables are not statistically significant. Nonetheless, for purposes of further model development, we retained automobile, sex, and age as it is customary to include these variables in ratemaking models.

As described in Section 12.1.4, there are several ways of assessing a model's overall goodness of fit. Table 12.7 compares several fitted models, providing fitted values for each response level and summarizing the overall fit with Pearson chi-square goodness-of-fit statistics. The left portion of the table repeats the baseline information that appeared in Table 12.1, for convenience. To begin, first note that, even without covariates, the inclusion of the offset, exposures, dramatically improves the fit of the model. This is intuitively appealing; as a driver has more insurance coverage during a year, he or she is more likely to be

Table 12.7 Comparison of Fitted Frequency Models

Count	Observed	Without Exposures/ No Covariates	With Exposures No Covariates	Poisson	Negative Binomial
0	6,996	6,977.86	6,983.05	6,986.94	6,996.04
1	455	487.70	477.67	470.30	453.40
2	28	17.04	21.52	24.63	31.09
3	4	0.40	0.73	1.09	2.28
4	0	0.01	0.02	0.04	0.18
Pearson Goodness of Fit		41.98	17.62	8.77	1.79

in an accident covered under the insurance contract. Table 12.7 also shows the improvement in the overall fit when including the fitted model summarized in Table 12.6. When compared to a chi-square distribution, the statistic p-value $=$ $\Pr\left(\chi_4^2 > 8.77\right) = 0.067$ suggests agreement between the data and the fitted value. However, this model specification can be improved – the following section introduces a negative binomial model that proves an even better fit for this dataset.

12.3 Overdispersion and Negative Binomial Models

Although simplicity is a virtue of the Poisson regression model, its form can also be too restrictive. In particular, the requirement that the mean equal the variance, known as *equidispersion*, is not satisfied for many datasets of interest. If the variance exceeds the mean, then the data are said to be *overdispersed.* A less common case occurs when the variance is less than the mean, known as *underdispersion.*

Adjusting Standard Errors for Data Not Equidispersed

To mitigate this concern, a common specification is to assume that

$$\mathrm{Var}\ y_i = \phi \mu_i, \tag{12.8}$$

where $\phi > 0$ is a parameter to accommodate the potential over- or underdispersion. As suggested by equation (12.5), consistent estimation of $\boldsymbol{\beta}$ requires only that the mean function be specified correctly, not that the equidispersion or Poisson distribution assumptions hold. This feature also holds for linear regression. Because of this, the estimator **b** is sometimes referred to as a *quasi-likelihood estimator*. With this estimator, we may compute estimated means $\widehat{\mu}_i$ and then estimate ϕ as

$$\widehat{\phi} = \frac{1}{n-(k+1)} \sum_{i=1}^{n} \frac{(y_i - \widehat{\mu}_i)^2}{\widehat{\mu}_i}. \tag{12.9}$$

Standard errors are then based on

$$\widehat{\mathrm{Var}\, \mathbf{b}} = \left(\widehat{\phi} \sum_{i=1}^{n} \widehat{\mu}_i \mathbf{x}_i \mathbf{x}_i' \right)^{-1}.$$

A drawback of equation (12.8) is that one assumes the variance of each observation is a constant multiple of its mean. For datasets where this assumption is in doubt, it is common to use a *robust standard error*, computed as the square root of the diagonal element of

$$\mathrm{Var}\, \mathbf{b} = \left(\sum_{i=1}^{n} \mu_i \mathbf{x}_i \mathbf{x}_i' \right)^{-1} \left(\sum_{i=1}^{n} (y_i - \mu_i)^2 \mathbf{x}_i \mathbf{x}_i' \right) \left(\sum_{i=1}^{n} \mu_i \mathbf{x}_i \mathbf{x}_i' \right)^{-1},$$

evaluated at $\widehat{\mu}_i$. Here, the idea is that $(y_i - \mu_i)^2$ is an unbiased estimator of Var y_i, regardless of the form. Although $(y_i - \mu_i)^2$ is a poor estimator of Var y_i for each observation i, the weighted sum $\sum_i (y_i - \mu_i)^2 \mathbf{x}_i \mathbf{x}_i'$ is a reliable estimator of $\sum_i (\mathrm{Var}\, y_i) \mathbf{x}_i \mathbf{x}_i'$.

For the quasi-likelihood estimator, the estimation strategy assumes only a correct specification of the mean and uses a more robust specification of the variance than implied by the Poisson distribution. The advantage and disadvantage of this estimator is that it is not linked to a full distribution. This assumption makes it difficult, for example, if the interest is in estimating the probability of zero counts. An alternative approach is to assume a more flexible parametric model that permits a wider range of dispersion.

Negative Binomial

A widely used model for counts is the *negative binomial*, with probability mass function

$$\Pr(y = j) = \binom{j + r - 1}{r - 1} p^r (1 - p)^j, \qquad (12.10)$$

where r and p are parameters of the model. To help interpret the parameters of the model, straightforward calculations show that $\mathrm{E}\, y = r(1 - p)/p$ and $\mathrm{Var}\, y = r(1 - p)/p^2$.

The negative binomial has several important advantages when compared to the Poisson distribution. First, because there are two parameters describing the negative binomial distribution, it has greater flexibility for fitting data. Second, it can be shown that the Poisson is a limiting case of the negative binomial (by allowing $p \to 1$ and $r \to 0$ such that $rp \to \lambda$). In this sense, the Poisson is nested within the negative binomial distribution. Third, one can show that negative binomial distribution arises from a mixture of the Poisson variables. For example, think about the Singapore dataset with each driver having his or her own value of λ. Conditional on λ, assume that the driver's accident distribution has a Poisson distribution with parameter λ. Further assume that the distribution of λ's can be described as a gamma distribution. Then, it can be shown that the overall accident counts have a negative binomial distribution. See, for example,

Klugman, Panjer, and Willmot (2008). Such "mixture" interpretations are helpful in explaining results to consumers of actuarial analyses.

For regression modeling, the p parameter varies by subject i. It is customary to reparameterize the model and use a log-link function such that $\sigma = 1/r$ and that p_i related to the mean through $\mu_i = r(1-p_i)/p_i = \exp(\mathbf{x}_i'\boldsymbol{\beta})$. Because the negative binomial is a probability frequency distribution, there is no difficulty in estimating features of this distribution, such as the probability of zero counts, after a regression fit. This is in contrast to the quasi-likelihood estimation of a Poisson model with an ad hoc specification of the variance summarized in equation (12.9).

Example: Singapore Automobile Data, Continued. The negative binomial distribution was fit to the Section 12.2 Singapore data using the set of covariates summarized in Table 12.6. The resulting log-likelihood was $L_{NegBin}(\mathbf{b}) = -1{,}774.494$; this is larger than the Poisson likelihood fit $L_{Poisson}(\mathbf{b}) = -1{,}776.730$ because of an additional parameter. The usual likelihood ratio test is not formally appropriate because the models are only nested in a limiting sense. It is more useful to compare the goodness-of-fit statistics given in Table 12.7. Here, we see that the negative binomial is a better fit than the Poisson (with the same systematic components). A chi-square test of whether the negative binomial with covariates is suitable yields p-value $= \Pr\left(\chi_4^2 > 1.79\right) = 0.774$, suggesting strong agreement between the observed data and fitted values. We interpret the findings of Table 12.7 to mean that the negative binomial distribution well captures the heterogeneity in the accident frequency distribution.

12.4 Other Count Models

Actuaries are familiar with a host of frequency models; see, for example, Klugman, Panjer, and Willmot (2008). In principle, each frequency model could be used in a regression context by simply incorporating a systematic component, $\mathbf{x}'\boldsymbol{\beta}$, into one or more model parameters. However, analysts have found that four variations of the basic models perform well in fitting models to data and provide an intuitive platform for interpreting model results.

12.4.1 Zero-Inflated Models

For many datasets, a troublesome aspect is the "excess" number of zeros, relative to a specified model. For example, this could occur in automobile claims data because insureds are reluctant to report claims, fearing that a reported claim will result in higher future insurance premiums. Thus, we have a higher than anticipated number of zeros because of the nonreporting of claims.

A zero-inflated model represents the claims number y_i as a mixture of a point mass at zero and another claims frequency distribution, say, $g_i(j)$ (which is typically Poisson or negative binomial). (We might interpret the point mass as the tendency of nonreporting.) The probability of getting the point mass would be modeled by a binary count model such as, for example, the logit model

$$\pi_i = \frac{\exp\left(\mathbf{x}_i'\boldsymbol{\beta}_1\right)}{1+\exp\left(\mathbf{x}_i'\boldsymbol{\beta}_1\right)}.$$

As a consequence of the mixture assumption, the zero-inflated count distribution can be written as

$$\Pr(y_i = j) = \begin{cases} \pi_i + (1-\pi_i)g_i(0) & j=0 \\ (1-\pi_i)g_i(j) & j=1,2,\ldots \end{cases}. \tag{12.11}$$

From equation (12.11), we see that zeros could arise from either the point mass or the other claims frequency distribution.

To see the effects of a zero-inflated model, suppose that g_i follows a Poisson distribution with mean μ_i. Then, easy calculations show that

$$\mathrm{E}\ y_i = (1-\pi_i)\mu_i$$

and

$$\mathrm{Var}\ y_i = \pi_i\mu_i + \pi_i\mu_i^2(1-\pi_i).$$

Thus, for the zero-inflated Poisson, the variance always exceeds the mean, thus accommodating overdispersion relative to the Poisson model.

Example: Automobile Insurance. Yip and Yau (2005) examine a portfolio of $n =$ 2,812 automobile policies available from SAS Institute Inc. Explanatory variables include age, sex, marital status, annual income, job category, and education level of the policyholder. For this dataset, they found that several zero-inflated count models accommodated well the presence of extra zeros.

12.4.2 Hurdle Models

A hurdle model provides another mechanism to modify basic count distributions to represent situations with an excess number of zeros. Hurdle models can be motivated by sequential decision-making processes confronted by individuals. For example, in health-care choice, we can think about an individual's decision to seek health care as an initial process. Conditional on having sought health care $\{y \geq 1\}$, the amount of health care is a decision made by a health-care provider (e.g., a physician or hospital), thus representing a different process. One needs to pass the first "hurdle" (the decision to seek health care) to address the second (the amount of health care). An appeal of the hurdle model is its connection to the "principal-agent" model, where the provider (agent) decides on the amount after initial contact by the insured (principal) is made. As another example, in property and casualty insurance, the decision process an insured uses for reporting the initial claim may differ from that used for reporting subsequent claims.

To represent hurdle models, let π_i represent the probability that $\{y_i = 0\}$ used for the first decision and suppose that g_i represents the count distribution that will be used for the second decision. We define the probability mass function as

$$\Pr(y_i = j) = \begin{cases} \pi_i & j = 0 \\ k_i g_i(j) & j = 1, 2, \ldots \end{cases}. \tag{12.12}$$

where $k_i = (1 - \pi_i)/(1 - g_i(0))$. As with zero-inflated models, a logit model might be suitable for representing π_i.

To see the effects of a hurdle model, suppose that g_i follows a Poisson distribution with mean μ_i. Then, easy calculations show that

$$\mathrm{E}\ y_i = k_i \mu_i$$

and

$$\mathrm{Var}\ y_i = k_i \mu_i + k_i \mu_i^2 (1 - k_i).$$

Because k_i may be larger or smaller than 1, this model allows for both under- and overdispersion relative to the Poisson model.

The hurdle model is a special case of the two-part model described in Chapter 16. There, we will see that for two-part models, the amount of health care utilized may be a continuous as well as a count variable. An appeal of two-part models is that parameters for each hurdle or part can be analyzed separately. Specifically, the log-likelihood for the ith subject can be written as

$$\ln[\Pr(y_i = j)] = [\mathrm{I}(j = 0)\ln \pi_i + \mathrm{I}(j \geq 1)\ln(1 - \pi_i)] \\ + \mathrm{I}(j \geq 1)\ln \frac{g_i(j)}{(1 - g_i(0))}.$$

The terms in the square brackets on the right-hand side correspond to the likelihood for a binary count model. The latter terms correspond to a count model with zeros removed (known as a truncated model). If the parameters for the two pieces are different ("separable"), then the maximization may be done separately for each part.

12.4.3 Heterogeneity Models

In a heterogeneity model, one allows one or more model parameters to vary randomly. The motivation is that the random parameters capture unobserved features of a subject. For example, suppose that α_i represents a random parameter and that y_i given α_i has conditional mean $\exp\left(\alpha_i + \mathbf{x}_i'\boldsymbol{\beta}\right)$. We interpret α_i, called a *heterogeneity component,* to represent unobserved subject characteristics that contribute linearly to the systematic component $\mathbf{x}_i'\boldsymbol{\beta}$.

To see the effects of the heterogeneity component on the count distribution, basic calculations show that

$$\mathrm{E}\ y_i = \exp\left(\mathbf{x}_i'\boldsymbol{\beta}\right) = \mu_i$$

and

$$\mathrm{Var}\, y_i = \mu_i + \mu_i^2 \mathrm{Var}\,(e^{\alpha_i}),$$

where we typically assume that $\mathrm{E}\,(e^{\alpha_i}) = 1$ for parameter identification. Thus, heterogeneity models readily accommodate overdispersion in datasets.

It is common to assume that the count distribution is Poisson conditional on α_i. There are several choices for the distribution of α_i, the two most common being the log-gamma and the lognormal. For the former, one first assumes that $\exp(\alpha_i)$ has a gamma distribution, implying that $\exp\left(\alpha_i + \mathbf{x}_i'\boldsymbol{\beta}\right)$ also has a gamma distribution. Recall that we have already noted in Section 12.3 that using a gamma mixing distribution for Poisson counts results in a negative binomial distribution. Thus, this choice provides another motivation for the popularity of the negative binomial as the choice of the count distribution. For the latter, assuming that an observed quantity such as $\exp(\alpha_i)$ has a normal distribution is quite common in applied data analysis. Although there are no closed-form analytic expressions for the resulting marginal count distribution, there are several software packages that readily lend themselves to easing computational difficulties.

The heterogeneity component is particularly useful in repeated samples, where it can be used to model clustering of observations. Observations from different clusters tend to be dissimilar compared to observations within a cluster, a feature known as *heterogeneity*. The similarity of observations within a cluster can be captured by a common term α_i. Different heterogeneity terms from observations from different clusters can capture the heterogeneity. For an introduction to modeling from repeated sampling, see Chapter 10.

Example: Spanish Third-Party Automobile Liability Insurance. Boucher, Denuit, and Guillén (2007) analyzed a portfolio of $n = 548{,}830$ automobile contracts from a major insurance company operating in Spain. Claims were for third-party automobile liability, so that in the event of an automobile accident, the amount that the insured is liable for nonproperty damages to other parties is covered under the insurance contract. For these data, the average claims frequency was approximately 6.9%. Explanatory variables include age, sex, driving location, driving experience, engine size, and policy type. The paper considers a wide variety of zero-inflated, hurdle, and heterogeneity models, showing that each was a substantial improvement over the basic Poisson model.

12.4.4 Latent Class Models

In most datasets, it is easy to think about classifications of subjects that the analyst would like to make to promote homogeneity among observations. Some examples include

- "Healthy" and "ill" people when examining health-care expenditures

- Automobile drivers who are likely to file a claim in the event of an accident compared to those who are reluctant to do so
- Physicians who are "low" risks compared to "high" risks when examining medical malpractice insurance coverage.

For many datasets of interests, such obvious classification information is not available and are said to be unobserved, or *latent*. A *latent class* model still employs this classification idea but treats it as an unknown discrete random variable. Thus, like Sections 12.4.1–12.4.3, we use mixture models to modify basic count distributions but now assume that the mixture is a discrete random variable that we interpret to be the latent class.

To be specific, assume that we have two classes, "low risk" and "high risk," with probability π_L that a subject belongs to the low-risk class. Then, we can write the probability mass function as

$$\Pr(y_i = j) = \pi_L \Pr(y_i = j; L) + (1 - \pi_L) \Pr(y_i = j; H), \tag{12.13}$$

where $\Pr(y_i = j; L)$ and $\Pr(y_i = j; H)$ are the probability mass functions for the low and high risks, respectively.

This model is intuitively pleasing in that it corresponds to an analyst's perception of the behavior of the world. It is flexible in the sense that the model readily accommodates under- and overdispersion, long tails, and bimodal distributions. However, this flexibility also leads to difficulty regarding computational issues. There is a possibility of multiple local maxima when estimating via maximum likelihood. Convergence can be slow compared to other methods described in Sections 12.4.1–12.4.3.

Nonetheless, latent class models have proved fruitful in applications of interest to actuaries.

Example: Rand Health Insurance Experiment. Deb and Trivedi (2002) find strong evidence that a latent class model performs well when compared to the hurdle model. They examined counts of utilization of health-care expenditures for the Rand Health Insurance Experiment, a dataset that has extensively analyzed in the health economics literature. They interpreted $\Pr(y_i = j; L)$ to be a distribution of infrequent health-care users and $\Pr(y_i = j; H)$ to be a distribution of frequent health-care users. Each distribution was based on a negative binomial distribution, with different parameters for each class. They found statistically significant differences for their four insurance variables, two coinsurance variables, a variable indicating whether there was an individual deductible and a variable describing the maximum limit reimbursed. Because subjects were randomly assigned to insurance plans (very unusual), the effects of insurance variables on health-care utilization are particularly interesting from a policy standpoint, as are differences among low- and high-use subjects. For their data, they estimated that approximately 20% were in the high-use class.

12.5 Further Reading and References

The Poisson distribution was derived by Poisson (1837) as a limiting case of the binomial distribution. Greenwood and Yule (1920) derived the negative binomial distribution as a mixture of a Poisson with a gamma distribution. Interestingly, one example of the 1920 paper was to use the Poisson distribution as a model of accidents, with the mean as a gamma random variable, reflecting the variation of workers in a population. Greenwood and Yule referred to this as individuals subject to "repeated accidents" that other authors have dubbed as "accident proneness."

The first application of Poisson regression is due to Cochran (1940) in the context of ANOVA modeling and to Jorgensen (1961) in the context of multiple linear regression. As described in Section 12.2, Weber (1971) gives the first application to automobile accidents.

This chapter focuses on insurance and risk management applications of count models. For those interested in automobiles, there is a related literature on studies of motor vehicle crash process, see for example, Lord, Washington, and Ivan (2005). For applications in other areas of social science and additional model development, we refer to Cameron and Trivedi (1998).

Chapter References

Bortkiewicz, L. von (1898). *Das Gesetz de Kleinen Zahlen*. Teubner, Leipzig.

Boucher, Jean-Philippe, Michel Denuit, and Montserratt Guillén (2007). Risk classification for claim counts: A comparative analysis of various zero-inflated mixed Poisson and hurdle models. *North American Actuarial Journal* 11, no. 4, 110–131.

Cameron, A. Colin, and Pravin K. Trivedi. (1998) *Regression Analysis of Count Data*. Cambridge University Press, Cambridge.

Cochran, W. G. (1940). The analysis of variance when experimental errors follow the Poisson or binomial law. *Annals of Mathematical Statistics* 11, 335–47.

Deb, Partha, and Pravin K. Trivedi (2002). The structure of demand for health care: latent class versus two-part models. *Journal of Health Economics* 21, 601–25.

Fournier, Gary M., and Melayne Morgan McInnes (2001). The case of experience rating in medical malpractice insurance: An empirical evaluation. *Journal of Risk and Insurance* 68, 255–76.

Frees, Edward W., and Emiliano Valdez (2008). Hierarchical insurance claims modeling. *Journal of the American Statistical Association* 103, 1457–69.

Greenwood, M., and G. U. Yule (1920). An inquiry into the nature of frequency distributions representative of multiple happenings with particular reference to the occurrence of multiple attacks of disease or of repeated accidents. *Journal of the Royal Statistical Society* 83, 255–79.

Jones, Andrew M. (2000). Health econometrics. In *Handbook of Health Economics, Volume 1*, Antonio. J. Culyer and Joseph.P. Newhouse, eds., 265–344. Elsevier, Amsterdam.

Jorgensen, Dale W. (1961). Multiple regression analysis of a Poisson process. *Journal of the American Statistical Association* 56, 235–45.

Lord, Dominique, Simon P. Washington, and John N. Ivan (2005). Poisson, Poisson-gamma and zero-inflated regression models of motor vehicle crashes: Balancing statistical theory and fit. *Accident Analysis and Prevention* 37, 35–46.

Klugman, Stuart A., Harry H. Panjer, and Gordon E. Willmot (2008). *Loss Models: From Data to Decisions*. John Wiley & Sons, Hoboken, New Jersey.

Purcaru, Oana, and Michel Denuit (2003). Dependence in dynamic claim frequency credibility models. *ASTIN Bulletin* 33, no. 1, 23–40.

Weber, Donald C. (1971). Accident rate potential: An application of multiple regression analysis of a Poisson process. *Journal of the American Statistical Association* 66, 285–88.

Yip, Karen C. H., and Kelvin K.W. Yau (2005). On modeling claim frequency data in general insurance with extra zeros. *Insurance: Mathematics and Economics* 36(2) 153–63.

12.6 Exercises

12.1. Show that the log-likelihood in equation (12.2) has a maximum at $\widehat{\mu} = \overline{y}$.

12.2. For the data in Table 12.1, confirm that the Pearson statistic in equation (12.3) is 41.98.

12.3. **Poisson Residuals.** Consider a Poisson regression. Let $e_i = y_i - \widehat{\mu}_i$ denote the ith ordinary residual. Assume that an intercept is used in the model so that one of the explanatory variables x is a constant equal to one.

a. Show that the average ordinary residual is 0.

b. Show that the correlation between the ordinary residuals and each explanatory variable is zero.

12.4. **Negative Binomial Distribution.**

a. Assume that $y_1, \ldots, y_n$ are i.i.d. with a negative binomial distribution with parameters r and p. Determine the maximum likelihood estimators.

b. Use the sampling mechanism in part (a) but with parameters $\sigma = 1/r$ and μ where $\mu = r(1-p)/p$. Determine the maximum likelihood estimators of σ and μ.

c. Assume that $y_1, \ldots, y_n$ are independent with y_i having a negative binomial distribution with parameters r and p_i, where $\sigma = 1/r$ and p_i satisfies $r(1-p_i)/p_i = \exp(\mathbf{x}_i'\boldsymbol{\beta})(= \mu_i)$. Determine the score function in terms of σ and $\boldsymbol{\beta}$.

Ⓡ **Empirical**
Filename is "HealthExpend"

12.5. **Medical Expenditures Data.** This exercise considers data from the Medical Expenditure Panel Survey (MEPS) described in Exercise 1.1 and Section 11.4. Our dependent variable consists of the number of outpatient (COUNTOP) visits. For MEPS, outpatient events include hospital outpatient department visits, office-based provider visits, and emergency room visits excluding dental services. (Dental services, compared to other types of health care services, are more predictable and occur on a more regular basis.) Hospital stays with the same date of admission and discharge, known as *zero-night stays*, were also included in outpatient counts and expenditures. (Payments associated with emergency room visits that immediately preceded an inpatient stay were included in the inpatient expenditures. Prescribed medicines that can be linked to hospital admissions were included in inpatient expenditures, not outpatient utilization.)

Consider the explanatory variables described in Section 11.4.

a. Provide a table of counts, a histogram, and summary statistics of COUNTOP. Note the shape of the distribution and the relationship between the sample mean and sample variance.

b. Create tables of means of COUNTOP by level of GENDER, ethnicity, region, education, self-rated physical health, self-rated mental health, activity limitation, income, and insurance. Do the tables suggest that the explanatory variables have an impact on COUNTOP?

c. As a baseline, estimate a Poisson model without any explanatory variables and calculate a Pearson's chi-square statistic for goodness of fit (at the individual level).

d. Estimate a Poisson model using the explanatory variables in part (b).

  d(i). Comment briefly on the statistical significance of each variable.

  d(ii). Provide an interpretation for the GENDER coefficient.

  d(iii). Calculate a (individual-level) Pearson's chi-square statistic for goodness of fit. Compare this to the one in part (b). On the basis of this statistic and the statistical significance of coefficients discussed in part d(i), which model do you prefer?

  d(iv). Reestimate the model using the quasi-likelihood estimator of the dispersion parameter. How have your comments in part d(i) changed?

e. Estimate a negative binomial model using the explanatory variables in part (d).

  e(i). Comment briefly on the statistical significance of each variable.

  e(ii). Calculate a (individual-level) Pearson's chi-square statistic for goodness of fit. Compare this to the ones in parts (b) and (d). Which model do you prefer? Also cite the AIC statistic in your comparison.

  e(iii). Reestimate the model, dropping the factor income. Use the likelihood ratio test to say whether income is a statistically significant factor.

f. As a robustness check, estimate a logistic regression model using the explanatory variables in part (d). Do the signs and significance of the coefficients of this model fit give the same interpretation as with the negative binomial model in part (e)?

12.6. **Two Population Poissons.** We can express the two population problem in a regression context using one explanatory variable. Specifically, suppose that x_i only takes on the values of zero and one. Out of the n observations, n_0 take on the value $x = 0$. These n_0 observations have an average y value of $\overline{y}_0$. The remaining $n_1 = n - n_0$ observations have value $x = 1$ and an average y value of $\overline{y}_1$.

Use the Poisson model with the logarithmic link function and systematic component $\mathbf{x}_i'\boldsymbol{\beta} = \beta_0 + \beta_1 x_i$.

i. Determine the maximum likelihood estimators of β_0 and β_1, respectively.

ii. Suppose that $n_0 = 10, n_1 = 90, \overline{y}_0 = 0.20$, and $\overline{y}_1 = 0.05$. Using your results in part a(i), compute the maximum likelihood estimators of β_0 and β_1, respectively.

iii. Determine the information matrix.

13

Generalized Linear Models

Chapter Preview. This chapter describes a unifying framework for the Part I linear model and the binary and count models in Chapters 11 and 12. Generalized linear models, often known by the acronym GLM, represent an important class of nonlinear regression models that have found extensive use in actuarial practice. This unifying framework not only encompasses many models we have seen but also provides a platform for new ones, including gamma regressions for fat-tailed data and "Tweedie" distributions for two-part data.

13.1 Introduction

There are many ways to extend, or generalize, the linear regression model. This chapter introduces an extension that is so widely used that it is known as the generalized linear model, or as the acronym GLM.

Generalized linear models include linear, logistic, and Poisson regressions, all as special cases. One common feature of these models is that in each case we can express the mean response as a function of linear combinations of explanatory variables. In the GLM context, it is customary to use $\mu_i = \mathrm{E}\ y_i$ for the mean response and call $\eta_i = \mathbf{x}_i'\boldsymbol{\beta}$ the *systematic component* of the model. We have seen that we can express the systematic component as

- $\mathbf{x}_i'\boldsymbol{\beta} = \mu_i$, for (normal) linear regression
- $\mathbf{x}_i'\boldsymbol{\beta} = \exp(\mu_i)/(1+\exp(\mu_i))$, for logistic regression
- $\mathbf{x}_i'\boldsymbol{\beta} = \ln(\mu_i)$, for Poisson regression

For GLMs, the systematic component is related to the mean through the expression

$$\eta_i = \mathbf{x}_i'\boldsymbol{\beta} = \mathrm{g}\left(\mu_i\right). \tag{13.1}$$

Here, $\mathrm{g}(\cdot)$ is known and called the *link* function. The inverse of the link function, $\mu_i = \mathrm{g}^{-1}(\mathbf{x}_i'\boldsymbol{\beta})$, is the mean function.

The second common feature involves the distribution of the dependent variables. In Section 13.2, we will introduce the *linear exponential family of distributions*, an extension of the exponential distribution. This family includes the normal, Bernoulli, and Poisson distributions as special cases.

Distribution	Variance Function $v(\mu)$
Normal	1
Bernoulli	$\mu(1-\mu)$
Poisson	μ
Gamma	μ^2
Inverse Gaussian	μ^3

Table 13.1 Variance Functions for Selected Distributions

The third common feature of GLM models is the robustness of inference to the choice of distributions. Although linear regression is motivated by normal distribution theory, we have seen that responses need not be normally distributed for statistical inference procedures to be effective. The Section 3.2 sampling assumptions focus on

- The form of the mean function (assumption F1),
- Nonstochastic or exogenous explanatory variables (F2)
- Constant variance (F3)
- Independence among observations (F4)

The GLM models maintain assumptions F2 and F4 and generalize F1 through the link function. The choice of different distributions allows us to relax F3 by specifying the variance to be a function of the mean, written as Var $y_i = \phi v(\mu_i)$. Table 13.1 shows how the variance depends on the mean for different distributions. As we will see when considering estimation (Section 13.3), it is the choice of the variance function that drives the most important inference properties, not the choice of the distribution.

The choice of the variance function, not the choice of the distribution, drives many inference properties.

By considering regression in the GLM context, we will be able to handle dependent variables that are approximately normally distributed, binary, or representing counts, all within one framework. This will aid our understanding of regression by allowing us to see the big picture and not be so concerned with the details. Further, the generality of GLMs will allow us to introduce new applications, such as gamma regressions that are useful for fat-tailed distributions and the so-called Tweedie distributions for two-part data. Two-part data, where there is a mass at zero and a continuous component, is a topic taken up in Chapter 16. For insurance claims data, the zero represents no claim and the continuous component represents the amount of a claim.

This chapter describes estimation procedures for calibrating GLM models, significance tests and goodness-of-fit statistics for documenting the usefulness of the model, and residuals for assessing the robustness of the model fit. We will see that our earlier work done on linear, binary, and count regression models provides the foundations for the tools needed for the GLM model. Indeed, many are slight variations of tools and concepts developed earlier in this text and we will be able to build on these foundations.

13.2 GLM Model

To specify a GLM, the analyst chooses an underlying response distribution, the topic of Section 13.2.1, and a function that links the mean response to the covariates, the topic of Section 13.2.2.

13.2.1 Linear Exponential Family of Distributions

Definition. The distribution of the *linear exponential family* is

$$\mathrm{f}(y;\theta,\phi)=\exp\left(\frac{y\theta-b(\theta)}{\phi}+S\left(y,\phi\right)\right). \tag{13.2}$$

Here, y is a dependent variable and θ is the parameter of interest. The quantity ϕ is a scale parameter. The term $b(\theta)$ depends only on the parameter θ, not on the dependent variable. The statistic $S\left(y,\phi\right)$ is a function of the dependent variable and the scale parameter, not the parameter θ.

The dependent variable y may be discrete, continuous, or a mixture. Thus, $\mathrm{f}(\cdot)$ may be interpreted to be a density or mass function, depending on the application. Table 13.8 provides several examples, including the normal, binomial, and Poisson distributions. To illustrate, consider a normal distribution with a probability density function of the form

$$\begin{aligned}\mathrm{f}(y;\mu,\sigma^2)&=\frac{1}{\sigma\sqrt{2\pi}}\exp\left(-\frac{(y-\mu)^2}{2\sigma^2}\right)\\&=\exp\left(\frac{(y\mu-\mu^2/2)}{\sigma^2}-\frac{y^2}{2\sigma^2}-\frac{1}{2}\ln\left(2\pi\sigma^2\right)\right).\end{aligned}$$

With the choices $\theta=\mu$, $\phi=\sigma^2$, $b(\theta)=\theta^2/2$, and $S\left(y,\phi\right)=-y^2/(2\phi)-\ln\left(2\pi\phi\right)/2$, we see that the normal probability density function can be expressed as in equation (13.2).

For the distribution in equation (13.2), some straightforward calculations show that

- $\mathrm{E}\ y=b'(\theta)$ and
- $\mathrm{Var}\ y=\phi b''(\theta)$.

For reference, these calculations appear in Section 13.9.2. To illustrate, in the context of the normal distribution example earlier, it is easy to check that $\mathrm{E}\ y=b'(\theta)=\theta=\mu$ and $\mathrm{Var}\ y=\sigma^2 b''(\theta)=\sigma^2$, as anticipated.

In regression modeling situations, the distribution of y_i varies by observation through the subscript i. It is customary to let the distribution family remain constant but allow the parameters to vary by observation through the notation θ_i and ϕ_i. For our applications, the variation of the scale parameter is due to known weight factors. Specifically, when the scale parameter varies by observation, it is according to $\phi_i=\phi/w_i$, that is, a constant divided by a known weight w_i.

Table 13.2 Mean Functions and Canonical Links for Selected Distributions

Distribution	Mean function $b'(\theta)$	Canonical link $g(\mu)$
Normal	θ	μ
Bernoulli	$e^{\theta}/(1+e^{\theta})$	$\text{logit}(\mu)$
Poisson	e^{θ}	$\ln \mu$
Gamma	$-1/\theta$	$-1/\mu$
Inverse Gaussian	$(-2\theta)^{-1/2}$	$-1/(2\mu^2)$

With the relation Var $y_i = \phi_i b''(\theta_i) = \phi b''(\theta_i)/w_i$, we have that a larger weight implies a smaller variance, other things being equal.

13.2.2 Link Functions

In regression situations, we wish to understand the impact of $\eta_i = \mathbf{x}_i'\boldsymbol{\beta}$, the systematic component. As we saw in the prior subsection, we can express the mean of y_i as E $y_i = \mu_i = b'(\theta_i)$. Equation (13.1) serves to "link" the systematic component to μ_i and thus to the parameter θ_i. It is possible to use the identity function for g(·) so that $\mu_i = b'(\theta_i)$. Indeed, this is the usual case in linear regression. However, linear combinations of explanatory variables, $\mathbf{x}_i'\boldsymbol{\beta}$, may vary between negative and positive infinity whereas means are often restricted to smaller range. For example, Poisson means vary between zero and infinity. The link function serves to map the domain of the mean function onto the whole real line.

Special Case: Links for the Bernoulli Distribution. Bernoulli means are probabilities and thus vary between zero and one. For this case, it is useful to choose a link function that maps the unit interval (0, 1) onto the whole real line. The following are three important examples of link functions for the Bernoulli distribution:

- Logit: $g(\mu) = \text{logit}(\mu) = \ln(\mu/(1-\mu))$.
- Probit: $g(\mu) = \Phi^{-1}(\mu)$, where Φ^{-1} is the inverse of the standard normal distribution function.
- Complementary log-log: $g(\mu) = \ln(-\ln(1-\mu))$.

This illustration demonstrates that there may be several link functions that are suitable for a particular distribution. To help with the selection, an intuitively appealing case occurs when the systematic component equals the parameter of interest ($\eta = \theta$). To see this, first recall that $\eta = g(\mu)$ and $\mu = b'(\theta)$, dropping the i subscripts for the moment. Then, it is easy to see that if $g^{-1} = b'$, then $\eta = g(b'(\theta)) = \theta$. The choice of g that is the inverse of $b'(\theta)$ is called the *canonical link*.

The choice of g that is the inverse of $b'(\theta)$ is called the canonical link. With this choice, the systematic component equals the parameter of interest.

Table 13.2 shows the mean function and corresponding canonical link for several important distributions.

Table 13.3 Private Passenger Automobile UK Collision Data

Age Group	Vehicle Use	Average Severity	Claim Count	Age Group	Vehicle Use	Average Severity	Claim Count
17–20	Pleasure	250.48	21	35–39	Pleasure	153.62	151
17–20	DriveShort	274.78	40	35–39	DriveShort	201.67	479
17–20	DriveLong	244.52	23	35–39	DriveLong	238.21	381
17–20	Business	797.80	5	35–39	Business	256.21	166
21–24	Pleasure	213.71	63	40–49	Pleasure	208.59	245
21–24	DriveShort	298.60	171	40–49	DriveShort	202.80	970
21–24	DriveLong	298.13	92	40–49	DriveLong	236.06	719
21–24	Business	362.23	44	40–49	Business	352.49	304
25–29	Pleasure	250.57	140	50–59	Pleasure	207.57	266
25–29	DriveShort	248.56	343	50–59	DriveShort	202.67	859
25–29	DriveLong	297.90	318	50–59	DriveLong	253.63	504
25–29	Business	342.31	129	50–59	Business	340.56	162
30–34	Pleasure	229.09	123	60+	Pleasure	192.00	260
30–34	DriveShort	228.48	448	60+	DriveShort	196.33	578
30–34	DriveLong	293.87	361	60+	DriveLong	259.79	312
30–34	Business	367.46	169	60+	Business	342.58	96

Source: Mildenhall (1999).
Note: "DriveShort" means drive to work but less than 10 miles. "DriveLong" means drive to work but more than 10 miles.

Links relate the mean to the systematic component and to the regression parameters. Because the regression parameters are unknown, it is common to specify the links only up to scale. For example, it is common to specify the inverse Gaussian canonical link as $1/\mu^2$ (instead of $-1/(2\mu^2)$). If necessary, one can always recover the scale when estimating the unknown regression coefficients.

Ⓡ **Empirical** *Filename is "AutoCollision"*

Example: Ratemaking Classification. The process of grouping risks with similar characteristics is known as *risk classification. Ratemaking* is the art of setting premiums, or rates, based on loss experience and exposures of risk classes. For example, Mildenhall (1999) considered 8,942 collision losses from private passenger United Kingdom automobile insurance policies. The data were derived from Nelder and McCullagh (1989, Section 8.4.1) but originated from Baxter, Coutts, and Ross (1980). A typical personal auto rating plan is based on driver and vehicle characteristics. Driver characteristics may include the driver's age, sex, marital status, history (accidents and violations), and good student discount. Vehicle characteristics may include vehicle model type and year, purpose (business/school or pleasure), garage territory, and so forth. We can represent the systematic component as

$$\eta_{ij} = \beta_0 + \alpha_i + \tau_j,$$

where α_i represents the effect of the ith category of driver classification and τ_j the effect of the jth vehicle type. Table 13.3 displays the Mildenhall data for eight driver types (age groups) and four vehicle classes (vehicle use). The average severity is in pounds sterling adjusted for inflation.

In GLM terminology, an *additive rating plan* is based on the identity link function, whereas a *multiplicative plan* is based on a logarithmic link function. Specifically, if we use $\eta_{ij} = \ln(\mu_{ij})$, then we can write the mean as

$$\mu_{ij} = \exp(\beta_0 + \alpha_i + \tau_j) = B \times A_i \times T_j, \tag{13.3}$$

where $B = \exp(\beta_0)$ is a scaling constant, $A_i = \exp(\alpha_i)$ represents driver effects, and $T_j = \exp(\tau_j)$ represents vehicle effects.

13.3 Estimation

This section presents maximum likelihood, the customary form of estimation. To provide intuition, we focus on the simpler case of canonical links. Results for more general links appear in Section 13.9.3.

13.3.1 Maximum Likelihood Estimation for Canonical Links

From equation (13.2) and the independence among observations, the log-likelihood is

$$\ln \mathrm{f}(\mathbf{y}) = \sum_{i=1}^{n} \left\{ \frac{y_i \theta_i - b(\theta_i)}{\phi_i} + S(y_i, \phi_i) \right\}. \tag{13.4}$$

Recall that, for canonical links, we have equality between the distribution's parameter and the systematic component, so that $\theta_i = \eta_i = \mathbf{x}_i' \boldsymbol{\beta}$. Thus, with $\phi_i = \phi / w_i$, the log-likelihood is

$$L(\boldsymbol{\beta}, \phi) = \ln \mathrm{f}(\mathbf{y}) = \sum_{i=1}^{n} \left\{ \frac{y_i \mathbf{x}_i' \boldsymbol{\beta} - b(\mathbf{x}_i' \boldsymbol{\beta})}{\phi / w_i} + S(y_i, \phi / w_i) \right\}. \tag{13.5}$$

Taking the partial derivative with respect to $\boldsymbol{\beta}$ yields the score function

$$\frac{\partial}{\partial \boldsymbol{\beta}} L(\boldsymbol{\beta}, \phi) = \frac{1}{\phi} \sum_{i=1}^{n} \left(y_i - b'(\mathbf{x}_i' \boldsymbol{\beta}) \right) w_i \mathbf{x}_i. \tag{13.6}$$

Because $\mu_i = b'(\theta_i) = b'(\mathbf{x}_i' \boldsymbol{\beta})$, we can solve for the maximum likelihood estimators of $\boldsymbol{\beta}$, $\mathbf{b}_{MLE}$, through the "normal equations"

$$\mathbf{0} = \sum_{i=1}^{n} w_i \left(y_i - \mu_i \right) \mathbf{x}_i. \tag{13.7}$$

There are $k + 1$ equations and $k + 1$ unknowns in the equation (13.7). Typically, the solution is unique and we use the notation $\mathbf{b}_{MLE}$ to denote the solution. One reason for the widespread use of GLM methods is that the maximum likelihood estimators can be computed quickly through a technique known as *iterated reweighted least squares*, described in Section 13.9.4.

Note that, like ordinary linear regression normal equations, we do not need to consider estimation of the variance scale parameter ϕ at this stage. That is, we can first compute $\mathbf{b}_{MLE}$ and, when necessary, estimate ϕ. (The parameter ϕ is known for certain distributions such as the binomial and Poisson and so does not require estimation.)

As described in Section 11.9, maximum likelihood estimators are consistent and have large sample normal distributions under broad conditions. Maximum likelihood inference provides a mechanism for calculating this distribution. From equations (13.6) and the likelihood technical supplement (Section 11.9, equation 11.14), the corresponding information matrix is

$$\mathbf{I}(\mathbf{b}_{MLE}) = \frac{1}{\phi} \sum_{i=1}^{n} w_i \, b''(\mathbf{x}_i' \mathbf{b}_{MLE}) \, \mathbf{x}_i \mathbf{x}_i'. \tag{13.8}$$

The inverse of the information matrix is the large-sample variance-covariance matrix of $\mathbf{b}_{MLE}$. Specifically, the square root of the $(j+1)$st diagonal element of the inverse of this matrix yields the standard error for $b_{j,MLE}$, which we denote as $se(b_{j,MLE})$. Extensions to general links are similar.

Inference for $\mathbf{b}_{MLE}$ is robust to the choice of distributions in the following sense. The solution of the maximum likelihood estimators in equation (13.7) depends only on the mean function; it can be shown that consistency of the estimators depends only on proper choice of this function. Further, the large sample behavior of $\mathbf{b}_{MLE}$ essentially only requires that the mean and variance functions be correctly specified, not the choice of the distribution. (A few additional regularity conditions are required, but these are mild technical requirements.) For example, suppose an analyst chooses a Poisson distribution with a logarithmic link. If the log link is appropriate, then only the equality between the mean and the variance is needed, see Table 13.1. Unlike the usual domain of the Poisson distribution, the dependent variables could be nonintegers or even be negative. Large-sample inference for $\mathbf{b}_{MLE}$ only requires that we choose the mean and variance functions correctly.

Example: Ratemaking Classification, Continued. Using the data in Table 13.3, a log-link function with gamma distribution was fit using claim counts as weights (w_i). Table 13.4 shows estimates of the expected severity using equation (13.3). The averages suggest that young drivers (ages 17–20 and 21–24) have the highest claims. For vehicle use, those driving for pleasure had the lowest and those driving for business had the highest claims.

13.3.2 Overdispersion

For some members of the linear exponential family, such as the Bernoulli and the Poisson distributions, the variance is determined by the mean. In contrast, the

Table 13.4 Estimated Expected Severity for a Multiplicative Rating Plan

Age Group	DriverShort	DriveLong	Pleasure	Business	Average
17–20	322.17	265.56	254.90	419.07	315.42
21–24	320.66	264.31	253.70	417.10	313.94
25–29	297.26	245.02	235.19	386.66	291.03
30–34	284.85	234.80	225.37	370.53	278.89
35–39	229.37	189.06	181.47	298.35	224.56
40–49	248.15	204.54	196.33	322.78	242.95
50–59	251.95	207.67	199.34	327.72	246.67
60+	246.47	203.16	195.00	320.60	241.31
Average	275.11	226.77	217.66	357.85	269.35

normal distribution has a separate scale parameter. When fitting models to data with binary or count dependent variables, it is common to observe that the variance exceeds that anticipated by the fit of the mean parameters. This phenomenon is known as *overdispersion*. Several alternative probabilistic models are available to explain this phenomenon, depending on the application at hand. See Section 12.3 for an example and McCullagh and Nelder (1989) for a more detailed inventory.

Although arriving at a satisfactory probabilistic model is the most desirable route, in many situations analysts are content to postulate an approximate model through the relation

$$\mathrm{Var}\ y_i = \sigma^2 \phi\, b''(\theta_i)/w_i .$$

The parameter ϕ is specified through the choice of the distribution whereas the scale parameter σ^2 allows for extra variability. For example, Table 13.8 shows that by specifying either the Bernoulli or Poisson distribution, we have $\phi = 1$. Although the scale parameter σ^2 allows for extra variability, it may also accommodate situations in which the variability is smaller than specified by the distributional form (although this situation is less common). Finally, note that for some distributions, such as the normal distribution, the extra term is already incorporated in the ϕ parameter and thus serves no useful purpose.

When the additional scale parameter σ^2 is included, it is customary to estimate it by Pearson's chi-square statistic divided by the error degrees of freedom. That is,

$$\widehat{\sigma}^2 = \frac{1}{N-k} \sum_{i=1}^{n} w_i \frac{\left(y_i - b'(\mathbf{x}_i'\mathbf{b}_{MLE})\right)^2}{\phi b''(\mathbf{x}_i'\mathbf{b}_{MLE})}.$$

As with the Poisson distribution in Section 12.3, another way of handling unusual variance patterns is through robust or empirical standard errors. Section 13.9.3 provides additional details.

13.3.3 Goodness-of-Fit Statistics

In linear regression models, the most widely cited goodness-of-fit statistic is the R^2 measure that is based on the decomposition

$$\sum_i (y_i - \overline{y})^2 = \sum_i (y_i - \widehat{y}_i)^2 + \sum_i (\widehat{y}_i - \overline{y})^2 + 2 \times \sum_i (y_i - \widehat{y}_i)(\widehat{y}_i - \overline{y}).$$

In the language of Section 2.3, this decomposition is

$$\textit{Total SS} = \textit{Error SS} + \textit{Regression SS} + 2 \times \textit{Sum of Cross-Products}.$$

R^2 is not a useful statistic in nonlinear models, in part because of the analysis of variance decomposition is no longer valid.

The difficulty with nonlinear models is that the *Sum of Cross-Products* term rarely equals zero. Thus, one gets different statistics when defining R^2 as (*Regression SS/Total SS*) as compared to (1-*Error SS/Total SS*). Section 11.3.2 described some alternative R^2 measures that are sometimes cited in GLM settings.

A widely cited goodness-of-fit measure is the Pearson chi-square statistic that was introduced in Section 12.1.4. In the GLM context, we suppose that $\mathrm{E}\ y_i = \mu_i$, $\mathrm{Var}\ y_i = \phi v(\mu_i)$ is the variance function (as in the Table 13.1 examples) and that $\widehat{\mu}_i$ is an estimator of μ_i. Then, the Pearson chi-square statistic is defined as $\sum_i (y_i - \widehat{\mu}_i)^2 / (\phi v(\widehat{\mu}_i))$. As we have seen for Poisson models of count data, this formulation reduces to the form $\sum_i (y_i - \widehat{\mu}_i)^2 / \widehat{\mu}_i$.

General information criteria, including *AIC* and *BIC*, that were defined in Section 11.9 are also regularly cited in GLM studies.

A goodness-of-fit measure that is specific to GLM modeling is the *deviance statistic*. To define this statistic, we work with the notion a *saturated model*, where there are as many parameters as observations, $\theta_i, i = 1, \ldots, n$. A saturated model provides the best possible fit. With a parameter for each observation, we maximize the likelihood on an observation-by-observation basis. Thus, taking derivatives of logarithmic likelihood from equation (13.2) yields

$$\frac{\partial}{\partial \theta_i} \ln \mathrm{f}(y_i; \theta_i, \phi) = \frac{y_i - b'(\theta_i)}{\phi}.$$

Setting this equal to zero yields the parameter estimate, say, $\theta_{i,SAT}$, as the solution of $y_i = b'(\theta_{i,Sat})$. Letting $\boldsymbol{\theta}_{SAT}$ be the vector of parameters, the likelihood $L(\boldsymbol{\theta}_{SAT})$ is the largest possible value of the log-likelihood. Then, for a generic estimator $\widehat{\boldsymbol{\theta}}$, the *scaled* deviance statistic is defined as

$$\mathrm{D}^*(\widehat{\boldsymbol{\theta}}) = 2 \times \left(L(\boldsymbol{\theta}_{SAT}) - L(\widehat{\boldsymbol{\theta}}) \right).$$

In linear exponential families, one multiplies by the scaling factor ϕ to define the *deviance statistic*, $\mathrm{D}(\widehat{\boldsymbol{\theta}}) = \phi \mathrm{D}^*(\widehat{\boldsymbol{\theta}})$. This multiplication actually removes the variance scaling factor from the definition of the statistic.

It is straightforward to check that the deviance statistic reduces to the following forms for three special cases:

- Normal: $\mathrm{D}(\widehat{\boldsymbol{\mu}}) = \sum_i (y_i - \widehat{\mu}_i)^2$,
- Bernoulli: $\mathrm{D}(\widehat{\boldsymbol{\pi}}) = \sum_i \left\{ y_i \ln \frac{y_i}{\widehat{\pi}_i} + (1 - y_i) \ln \frac{1 - y_i}{1 - \widehat{\pi}_i} \right\}$, and
- Poisson: $\mathrm{D}(\widehat{\boldsymbol{\mu}}) = \sum_i \left\{ y_i \ln \frac{y_i}{\widehat{\mu}_i} + (y_i - \widehat{\mu}_i) \right\}$.

Here, we use the convention that $y \ln y = 0$ when $y = 0$.

Table 13.5 Median Expenditures by Explanatory Variable Based on a Sample of $n = 157$ with Positive Expenditures

Category	Variable	Description	Percentage of Data	Median Expend
	COUNTIP	Number of expenditures (median: 1.0)		
Demography	AGE	Age in years between 18 to 65 (median: 41.0)		
	GENDER	1 if female	72.0	5,546
		0 if male	28.0	7,313
Ethnicity	ASIAN	1 if Asian	2.6	4,003
	BLACK	1 if Black	19.8	6,100
	NATIVE	1 if Native	1.9	2,310
	WHITE	Reference level	75.6	5,695
Region	NORTHEAST	1 if Northeast	18.5	5,833
	MIDWEST	1 if Midwest	21.7	7,999
	SOUTH	1 if South	40.8	5,595
	WEST	Reference level	19.1	4,297
Education	COLLEGE	1 if college or higher degree	23.6	5,611
	HIGHSCHOOL	1 if high school degree	43.3	5,907
	Reference level is lower than high school degree		33.1	5,338
Self-rated physical health	POOR	1 if poor	17.2	10,447
	FAIR	1 if fair	10.2	5,228
	GOOD	1 if good	31.2	5,032
	VGOOD	1 if very good	24.8	5,546
	Reference level is excellent health		16.6	5,277
Self-rated mental health	MPOOR	1 if poor or fair	15.9	6,583
		0 if good to excellent mental health	84.1	5,599
Any activity limitation	ANYLIMIT	1 if any functional/activity limitation	41.4	7,826
		0 if otherwise	58.6	4,746
Income compared to poverty line	Reference level is high income		21.7	7,271
	MINCOME	1 if middle income	26.8	5,851
	LINCOME	1 if low income	16.6	6,909
	NPOOR	1 if near poor	7.0	5,546
	POORNEG	if poor/negative income	28.0	4,097
Insurance coverage	INSURE	1 if covered by public/private health insurance in any month of 2003	91.1	5,943
		0 if have no health insurance in 2003	8.9	2,668
Total			100.0	5,695

13.4 Application: Medical Expenditures

Ⓡ **Empirical** *Filename is "HealthExpend"*

We now return to the Medical Expenditures Panel Survey (MEPS) data introduced in Section 11.4. In that section, we sought to develop a model to understand the event of an inpatient admission to a hospital. In this section, we now wish to model the amount of the expenditure. In actuarial terminology, Section 11.4 considered the "frequency" whereas this section involves the "severity."

Out of the 2,000 randomly sampled observations from year 2003 considered in Section 11.4, only $n = 157$ were admitted to the hospital during the year. Table 13.5 summarizes the data using the same explanatory variables as in Table 11.4.

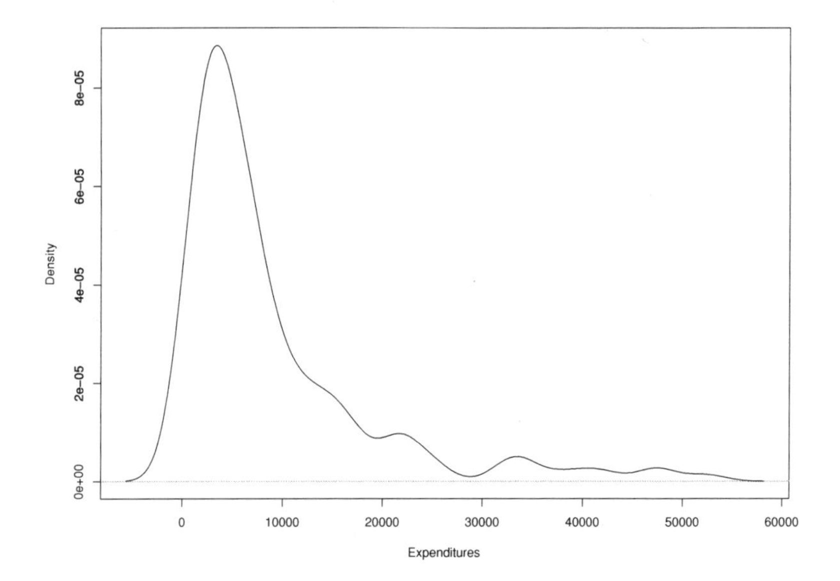

Figure 13.1 Smooth Empirical Histogram of Positive Inpatient Expenditures. The largest expenditure is omitted.

For example, Table 13.5 shows that the sample is 72% female, almost 76% white, and more than 91% insured. The table also shows relatively few expenditures by Asians, Native Americans, and the uninsured in our sample.

Table 13.5 also gives median expenditures by categorical variable. This table suggests that sex, a poor self-rating of physical health, and income that is poor or negative may be important determinants of the amount of medical expenditures.

Table 13.5 uses medians as opposed to means because the distribution of expenditures is skewed to the right. This is evident in Figure 13.1, which provides a smooth histogram (known as a *kernel density estimate*, see Section 15.2) for inpatient expenditures. For skewed distributions, the median often provides a more helpful idea of the center of the distribution than the mean. The distribution is even more skewed than suggested by this figure because the largest expenditure (which is $607,800) is omitted from the graphical display.

A gamma regression model using a logarithmic link was fit to inpatient expenditures using all explanatory variables. The result of this model fit appear in Table 13.6. Here, we see that many of the potentially important determinants of medical expenditures are not statistically significant. This is common in expenditure analysis, where variables help predict the frequency though are not as useful in explaining severity.

Because of collinearity, we have seen in linear models that having too many variables in a fitted model can lead to statistical insignificance of important variables and even cause signs to be reversed. For a simpler model, we removed the Asian, Native American, and uninsured variables because they account for a small subset of our sample. We also used only the POOR variable for self-reported health status and only POORNEG for income, essentially reducing these categorical variables to binary variables. Table 13.6, under the heading "Reduced Model," reports the result of this model fit. This model has almost the

Table 13.6 Comparison of Gamma and Inverse Gaussian Regression Models

	Gamma				Inverse Gaussian	
	Full Model		Reduced Model		Reduced Model	
Effect	Parameter Estimate	*t*-Value	Parameter Estimate	*t*-Value	Parameter Estimate	*t*-Value
Intercept	6.891	13.080	7.859	17.951	6.544	3.024
COUNTIP	0.681	6.155	0.672	5.965	1.263	0.989
AGE	0.021	3.024	0.015	2.439	0.018	0.727
GENDER	−0.228	−1.263	−0.118	−0.648	0.363	0.482
ASIAN	−0.506	−1.029				
BLACK	−0.331	−1.656	−0.258	−1.287	−0.321	−0.577
NATIVE	−1.220	−2.217				
NORTHEAST	−0.372	−1.548	−0.214	−0.890	0.109	0.165
MIDWEST	0.255	1.062	0.448	1.888	0.399	0.654
SOUTH	0.010	0.047	0.108	0.516	0.164	0.319
COLLEGE	−0.413	−1.723	−0.469	−2.108	−0.367	−0.606
HIGHSCHOOL	−0.155	−0.827	−0.210	−1.138	−0.039	−0.078
POOR	−0.003	−0.010	0.167	0.706	0.167	0.258
FAIR	−0.194	−0.641				
GOOD	0.041	0.183				
VGOOD	0.000	0.000				
MNHPOOR	−0.396	−1.634	−0.314	−1.337	−0.378	−0.642
ANYLIMIT	0.010	0.053	0.052	0.266	0.218	0.287
MINCOME	0.114	0.522				
LINCOME	0.536	2.148				
NPOOR	0.453	1.243				
POORNEG	−0.078	−0.308	−0.406	−2.129	−0.356	−0.595
INSURE	0.794	3.068				
Scale	1.409	9.779	1.280	9.854	0.026	17.720
Log-Likelihood	−1,558.67		−1,567.93		−1,669.02	
AIC	3,163.34		3,163.86		3,366.04	

same goodness-of-fit statistic, *AIC*, suggesting that it is a reasonable alternative. Under the reduced model fit, the variables COUNTIP (inpatient count), AGE, COLLEGE, and POORNEG, are statistically significant variables. For these significant variables, the signs are intuitively appealing. For example, the positive coefficient associated with COUNTIP means that as the number of inpatient visits increases, the total expenditures increases, as anticipated.

For another alternative model specification, Table 13.6 also shows a fit of an inverse Gaussian model with a logarithmic link. From the *AIC* statistic, we see that this model does not fit nearly as well as the gamma regression model. All variables are statistically insignificant, making it difficult to interpret this model.

13.5 Residuals

One way of selecting an appropriate model is to fit a tentative specification and analyze ways of improving it. This diagnostic method relies on residuals. In the linear model, we defined residuals as the response minus the corresponding fitted values. In a GLM context, we refer to these as *raw residuals*, denoted as $y_i - \widehat{\mu}_i$. As we have seen, residuals are useful for

- Discovering new covariates or the effects of nonlinear patterns in existing covariates
- Identifying poorly fitting observations
- Quantifying the effects of individual observations on model parameters
- Revealing other model patterns, such as heteroscedasticity or time trends.

As this section emphasizes, residuals are also the building blocks for many goodness-of-fit statistics.

As we saw in Section 11.1, on binary dependent variables, analysis of raw residuals can be meaningless in some nonlinear contexts. Cox and Snell (1968, 1971) introduced a general notion of residuals that is more useful in nonlinear models than the simple raw residuals. For our applications, we assume that y_i has a distribution function $\mathrm{F}(\cdot)$ that is indexed by explanatory variables $\mathbf{x}_i$ and a vector of parameters $\boldsymbol{\theta}$ denoted as $\mathrm{F}(\mathbf{x}_i, \boldsymbol{\theta})$. Here, the distribution function is common to all observations but the distribution varies through the explanatory variables. The vector of parameters $\boldsymbol{\theta}$ includes the regression parameters $\boldsymbol{\beta}$ as well as scale parameters. With knowledge of $\mathrm{F}(\cdot)$, we now assume that a new function $\mathrm{R}(\cdot)$ can be computed such that $\varepsilon_i = \mathrm{R}(y_i; \mathbf{x}_i, \boldsymbol{\theta})$, where ε_i are identically and independently distributed. Here, the function $\mathrm{R}(\cdot)$ depends on explanatory variables $\mathbf{x}_i$ and parameters $\boldsymbol{\theta}$. The Cox-Snell residuals are then $e_i = \mathrm{R}(y_i; \mathbf{x}_i, \widehat{\boldsymbol{\theta}})$. If the model is well specified and the parameter estimates $\widehat{\boldsymbol{\theta}}$ are close to the model parameters $\boldsymbol{\theta}$, then the residuals e_i should be close to i.i.d.

To illustrate, in the linear model we use the residual function $\mathrm{R}(y_i; \mathbf{x}_i, \boldsymbol{\theta}) = y_i - \mathbf{x}_i'\boldsymbol{\beta} = y_i - \mu_i = \varepsilon_i$, resulting in (ordinary) raw residuals. Another choice is to rescale by the standard deviation

$$\mathrm{R}(y_i; \mathbf{x}_i, \boldsymbol{\theta}) = \frac{y_i - \mu_i}{\sqrt{\mathrm{Var}\, y_i}},$$

that yields *Pearson residuals*.

Another type of residual is based on a first transforming the response with a known function $\mathrm{h}(\cdot)$,

$$\mathrm{R}(y_i; \mathbf{x}_i, \boldsymbol{\theta}) = \frac{\mathrm{h}(y_i) - \mathrm{E}\,\mathrm{h}(y_i)}{\sqrt{\mathrm{Var}\,\mathrm{h}(y_i)}}.$$

Choosing the function so that $\mathrm{h}(y_i)$ is approximately normally distributed yields *Anscombe residuals*. Another popular choice is to choose the transform so that the variance is stabilized. Table 13.7 gives transforms for the binomial, Poisson, and gamma distributions.

Table 13.7 Approximate Normality and Variance-Stabilizing Transforms

Distribution	Approximate Normality Transform	Variance-Stabilizing Transform
Binomial(m, p)	$\frac{\mathrm{h}_B(y/m)-[\mathrm{h}_B(p)+(p(1-p))^{-1/3}(2p-1)/(6m)]}{(p(1-p)^{1/6}/\sqrt{m}}$	$\frac{\sin^{-1}(\sqrt{y/m})-\sin^{-1}(\sqrt{p})}{a/(2\sqrt{m})}$
Poisson(λ)	$1.5\lambda^{-1/6}\left[y^{2/3}-(\lambda^{2/3}-\lambda^{-1/3}/9)\right]$	$2(\sqrt{y}-\sqrt{\lambda})$
Gamma(α, γ)	$3\alpha^{1/6}\left[(y/\gamma)^{1/3}-(\alpha^{1/3}-\alpha^{-2/3}/9)\right]$	Not considered

Note: $\mathrm{h}_B(u)=\int_0^u s^{-1/3}(1-s)^{-1/3}ds$ is the incomplete beta function evaluated at (2/3, 2/3), up to a scalar constant. *Source:* Pierce and Schafer (1986).

Another widely used choice in GLM studies are *deviance residuals*, defined as

$$\text{sign}\,(y_i-\widehat{\mu}_i)\sqrt{2\left(\ln \mathrm{f}(y_i;\theta_{i,SAT})-\ln \mathrm{f}(y_i;\widehat{\theta}_i)\right)}.$$

Section 13.3.3 introduced the parameter estimates of the saturated model, $\theta_{i,SAT}$. As discussed by Pierce and Schafer (1986), deviance residuals are very close to Anscombe residuals in many cases – we can check deviance residuals for approximate normality. Further, deviance residuals can be defined readily for any GLM model and are easy to compute.

13.6 Tweedie Distribution

We have seen that the natural exponential family includes continuous distributions, such as the normal and gamma, as well as discrete distributions, such as the binomial and Poisson. It also includes distributions that are mixtures of discrete and continuous components. In insurance claims modeling, the most widely used mixture is the Tweedie (1984) distribution. It has a positive mass at zero representing no claims and a continuous component for positive values representing the amount of a claim.

The Tweedie distribution is defined as a Poisson sum of gamma random variables, known as an *aggregate loss* in actuarial science. Specifically, suppose that N has a Poisson distribution with mean λ, representing the number of claims. Let y_j be an i.i.d. sequence, independent of N, with each y_j having a gamma distribution with parameters α and γ, representing the amount of a claim. Then, $S_N = y_1 + \ldots + y_N$ is Poisson sum of gammas. Section 16.5 will discuss aggregate loss models in further detail. This section focuses on the important special case of the Tweedie distribution.

To understand the mixture aspect of the Tweedie distribution, first note that it is straightforward to compute the probability of zero claims as

$$\Pr(S_N = 0) = \Pr(N = 0) = e^{-\lambda}.$$

The distribution function can be computed using conditional expectations,

$$\Pr(S_N \le y) = e^{-\lambda} + \sum_{n=1}^{\infty} \Pr(N = n)\Pr(S_n \le y), \qquad y \ge 0.$$

Because the sum of i.i.d. gammas is a gamma, S_n (not S_N) has a gamma distribution with parameters $n\alpha$ and γ. Thus, for $y > 0$, the density of the Tweedie distribution is

$$\mathrm{f}_S(y) = \sum_{n=1}^{\infty} e^{-\lambda} \frac{\lambda^n}{n!} \frac{\gamma^{n\alpha}}{\Gamma(n\alpha)} y^{n\alpha-1} e^{-y\gamma}. \tag{13.9}$$

At first glance, this density does not appear to be a member of the linear exponential family given in equation (13.2). To see the relationship, we first calculate the moments using iterated expectations as

$$\mathrm{E}\ S_N = \lambda \frac{\alpha}{\gamma} \quad \text{and} \quad \mathrm{Var}\ S_N = \frac{\lambda\alpha}{\gamma^2}(1+\alpha). \tag{13.10}$$

Now, define three parameters μ, ϕ, and p through the relations

$$\lambda = \frac{\mu^{2-p}}{\phi(2-p)}, \qquad \alpha = \frac{2-p}{p-1}, \qquad \text{and} \qquad \frac{1}{\gamma} = \phi(p-1)\mu^{p-1}.$$

Inserting these new parameters in equation (13.9) yields

$$\mathrm{f}_S(y) = \exp\left[\frac{-1}{\phi}\left(\frac{\mu^{2-p}}{2-p} + \frac{y}{(p-1)\mu^{p-1}}\right) + S(y,\phi)\right]. \tag{13.11}$$

We leave the calculation of $S(y, \phi)$ as an exercise.

Thus, the Tweedie distribution is a member of the linear exponential family. Easy calculations show that

$$\mathrm{E}\ S_N = \mu \quad \text{and} \quad \mathrm{Var}\ S_N = \phi\mu^p, \tag{13.12}$$

where $1 < p < 2$. Examining our variance function Table 13.1, the Tweedie distribution can also be viewed as a choice that is intermediate between the Poisson and the gamma distributions.

13.7 Further Reading and References

A more extensive treatment of GLM can be found in the classic work by McCullagh and Nelder (1989) or the gentler introduction by Dobson (2002). De Jong and Heller (2008) provide a book-length introduction to GLMs with a special insurance emphasis.

The GLMs have enjoyed substantial attention from property and casualty (general) insurance practicing actuaries recently, see Anderson et al. (2004) and Clark and Thayer (2004) for introductions.

Mildenhall (1999) provides a systematic analysis of the connection between GLM models and algorithms for assessing different rating plans. Fu and Wu (2008) provide an updated version, introducing a broader class of iterative algorithms that tie nicely to GLM models.

For further information on the Tweedie distribution, see Jørgensen and de Souza (1994) and Smyth and Jørgensen (2002).

Chapter References

Anderson, Duncan, Sholom Feldblum, Claudine Modlin, Doris Schirmacher, Ernesto Schirmacher, and Neeza Thandi (2004). A practitioner's guide to generalized linear models. *Casualty Actuarial Society 2004 Discussion Papers* 1–116, Casualty Actuarial Society. Arlington, Virginia.

Baxter, L. A., S. M. Coutts, and G. A. F. Ross (1980). Applications of linear models in motor insurance. *Proceedings of the 21st International Congress of Actuaries*, 11–29. Zurich.

Clark, David R., and Charles A. Thayer (2004). A primer on the exponential family of distributions. *Casualty Actuarial Society 2004 Discussion Papers* 117-148, Casualty Actuarial Society. Arlington, Virginia.

Cox, David R., and E. J. Snell (1968). A general definition of residuals. *Journal of the Royal Statistical Society*, Ser. B, 30, 248–75.

Cox, David R., and E. J. Snell (1971). On test statistics computed from residuals. *Biometrika* 71, 589–94.

Dobson, Annette J. (2002). *An Introduction to Generalized Linear Models*, 2nd ed. Chapman & Hall, London.

Fu, Luyang, and Cheng-sheng Peter Wu (2008). General iteration algorithm for classification ratemaking. *Variance* 1, no. 2, 193–213.

de Jong, Piet, and Gillian Z. Heller (2008). *Generalized Linear Models for Insurance Data*. Cambridge University Press, Cambridge.

Jørgensen, Bent, and Marta C. Paes de Souza (1994). Fitting Tweedie's compound Poisson model to insurance claims data. *Scandinavian Actuarial Journal* 1, 69–93.

McCullagh, P., and J. A. Nelder (1989). *Generalized Linear Models*, 2nd ed. Chapman & Hall, London.

Mildenhall, Stephen J. (1999). A systematic relationship between minimum bias and generalized linear models. *Casualty Actuarial Society Proceedings* 86, 393–487, Casualty Actuarial Society. Arlington, Virginia.

Pierce, Donald A., and Daniel W. Schafer (1986). Residuals in generalized linear models. *Journal of the American Statistical Association* 81, 977–86.

Smyth, Gordon K., and Bent Jørgensen (2002). Fitting Tweedie's compound Poisson model to insurance claims data: Dispersion modelling. *Astin Bulletin* 32, no. 1, 143–57.

Tweedie, M. C. K. (1984). An index which distinguishes between some important exponential families. In *Statistics: Applications and New Directions. Proceedings of the Indian Statistical Golden Jubilee International Conference*, (J. K. Ghosh and J. Roy, eds.), 579–604. Indian Statistical Institute, Calcutta.

13.8 Exercises

13.1. Verify the entries in Table 13.8 for the gamma distribution. Specifically:

a. Show that the gamma is a member of the linear exponential family of distributions.

b. Describe the components of the linear exponential family $(\theta, \phi, b(\theta), S(y, \phi))$ in terms of the parameters of the gamma distribution.

c. Show that the mean and variance of the gamma and linear exponential family distributions agree.

13.2. **Ratemaking Classification**. This exercise considers the data described in the Section 13.2.2 ratemaking classification example using data in Table 13.3.

Ⓡ **Empirical** *Filename is "AutoCollision"*

a. Fit a gamma regression model using a log-link function with claim counts as weights (w_i). Use the categorical explanatory variables age group and vehicle use to estimate expected average severity.

b. On the basis of your estimated parameters in part (a), verify the estimated expected severities in Table 13.4.

13.3. Verify that the Tweedie distribution is a member of the linear exponential family of distributions by checking equation (13.9). In particular, provide an expression for $S(y, \phi)$ (note that $S(y, \phi)$ also depends on p but not on μ). You may wish to see Clark and Thayer (2004) to check your work. Further, verify the moments in equation (13.12).

13.9 Technical Supplements – Exponential Family

13.9.1 Linear Exponential Family of Distributions

The distribution of the random variable y may be discrete, continuous or a mixture. Thus, f(·) in equation (13.2) may be interpreted to be a density or mass function, depending on the application. Table 13.8 provides several examples, including the normal, binomial, and Poisson distributions.

13.9.2 Moments

To assess the moments of exponential families, it is convenient to work with the moment generating function. For simplicity, we assume that the random variable y is continuous. Define the moment generating function

$$\begin{aligned} M(s) = \mathrm{E}\, e^{sy} &= \int \exp\left(sy + \frac{y\theta - b(\theta)}{\phi} + S(y,\phi)\right) dy \\ &= \exp\left(\frac{b(\theta + s\phi) - b(\theta)}{\phi}\right) \int \exp\left(\frac{y(\theta + s\phi) - b(\theta + s\phi)}{\phi} + S(y,\phi)\right) dy \\ &= \exp\left(\frac{b(\theta + s\phi) - b(\theta)}{\phi}\right), \end{aligned}$$

because $\int \exp\left(\frac{1}{\phi}\left[y(\theta + s\phi) - b(\theta + s\phi)\right] + S(y,\phi)\right) dy = 1$. With this expression, we can generate the moments. Thus, for the mean, we have

$$\begin{aligned} \mathrm{E}\, y = M'(0) &= \frac{\partial}{\partial s} \exp\left(\frac{b(\theta + s\phi) - b(\theta)}{\phi}\right)\Bigg|_{s=0} \\ &= \left[b'(\theta + s\phi) \exp\left(\frac{b(\theta + s\phi) - b(\theta)}{\phi}\right)\right]_{s=0} = b'(\theta). \end{aligned}$$

Similarly, for the second moment, we have

$$\begin{aligned} M''(s) &= \frac{\partial}{\partial s}\left[b'(\theta + s\phi) \exp\left(\frac{b(\theta + s\phi) - b(\theta)}{\phi}\right)\right] \\ &= \phi b''(\theta + s\phi) \exp\left(\frac{b(\theta + s\phi) - b(\theta)}{\phi}\right) \\ &\quad + (b'(\theta + s\phi))^2 \left(\frac{b(\theta + s\phi) - b(\theta)}{\phi}\right). \end{aligned}$$

This yields $\mathrm{E}y^2 = M''(0) = \phi b''(\theta) + (b'(\theta))^2$ and $\mathrm{Var}\ y = \phi b''(\theta)$.

Table 13.8 Selected Distributions of the One-Parameter Exponential Family

Distribution	Parameters	Density or Mass Function	Components	E y	Var y
General	$\theta,\ \phi$	$\exp\left(\frac{y\theta-b(\theta)}{\phi}+S(y,\phi)\right)$	$\theta,\ \phi, b(\theta), S(y,\phi)$	$b'(\theta)$	$b''(\theta)\phi$
Normal	μ, σ^2	$\frac{1}{\sigma\sqrt{2\pi}}\exp\left(-\frac{(y-\mu)^2}{2\sigma^2}\right)$	$\mu, \sigma^2, \frac{\theta^2}{2}, -\left(\frac{y^2}{2\phi}+\frac{\ln(2\pi\phi)}{2}\right)$	$\theta=\mu$	$\phi=\sigma^2$
Binomial	π	$\binom{n}{y}\pi^y(1-\pi)^{n-y}$	$\ln\left(\frac{\pi}{1-\pi}\right), 1, n\ln(1+e^\theta),$	$n\frac{e^\theta}{1+e^\theta}$	$n\frac{e^\theta}{(1+e^\theta)^2}$
			$\ln\binom{n}{y}$	$=n\pi$	$=n\pi(1-\pi)$
Poisson	λ	$\frac{\lambda^y}{y!}\exp(-\lambda)$	$\ln\lambda, 1, e^\theta, -\ln(y!)$	$e^\theta=\lambda$	$e^\theta=\lambda$
Negative Binomial[a]	r, p	$\frac{\Gamma(y+r)}{y!\Gamma(r)}p^r(1-p)^y$	$\ln(1-p), 1, -r\ln(1-e^\theta),$	$\frac{r(1-p)}{p}$	$\frac{r(1-p)}{p^2}$
			$\ln\left[\frac{\Gamma(y+r)}{y!\Gamma(r)}\right]$	$=\mu$	$=\mu+\mu^2/r$
Gamma	α, γ	$\frac{\gamma^\alpha}{\Gamma(\alpha)}y^{\alpha-1}\exp(-y\gamma)$	$-\frac{\gamma}{\alpha}, \frac{1}{\alpha}, -\ln(-\theta), \phi^{-1}\ln\phi$	$-\frac{1}{\theta}=\frac{\alpha}{\gamma}$	$\frac{\phi}{\theta^2}=\frac{\alpha}{\gamma^2}$
			$-\ln\left(\Gamma(\phi^{-1})\right)+(\phi^{-1}-1)\ln y$		
Inverse Gaussian	μ, λ	$\sqrt{\frac{\lambda}{2\pi y^3}}\exp\left(-\frac{\lambda(y-\mu)^2}{2\mu^2 y}\right)$	$-1/(2\mu^2), 1/\lambda, -\sqrt{-2\theta},$	$(-2\theta)^{-1/2}$	$\phi(-2\theta)^{-3/2}$
			$\theta/(\phi y)-0.5\ln(\phi 2\pi y^3)$	$=\mu$	$=\frac{\mu^3}{\lambda}$
Tweedie	See Section 13.6				

[a] This assumes that the parameter r is fixed but need not be an integer.

13.9.3 Maximum Likelihood Estimation for General Links

For general links, we no longer assume the relationship $\theta_i = \mathbf{x}_i'\boldsymbol{\beta}$ but assume that $\boldsymbol{\beta}$ is related to θ_i through the relations $\mu_i = b'(\theta_i)$ and $\mathbf{x}_i'\boldsymbol{\beta} = g(\mu_i)$. We continue to assume that the scale parameter varies by observation so that $\phi_i = \phi/w_i$, where w_i is a known weight function. Using equation (13.4), we have that the jth element of the score function is

$$\frac{\partial}{\partial \beta_j} \ln \mathrm{f}(\mathbf{y}) = \sum_{i=1}^{n} \frac{\partial \theta_i}{\partial \beta_j} \frac{y_i - \mu_i}{\phi_i},$$

because $b'(\theta_i) = \mu_i$. Now, use the chain rule and the relation $\mathrm{Var}\ y_i = \phi_i b''(\theta_i)$ to get

$$\frac{\partial \mu_i}{\partial \beta_j} = \frac{\partial b'(\theta_i)}{\partial \beta_j} = b''(\theta_i)\frac{\partial \theta_i}{\partial \beta_j} = \frac{\mathrm{Var}\ y_i}{\phi_i}\frac{\partial \theta_i}{\partial \beta_j}.$$

Thus, we have

$$\frac{\partial \theta_i}{\partial \beta_j}\frac{1}{\phi_i} = \frac{\partial \mu_i}{\partial \beta_j}\frac{1}{\mathrm{Var}\ y_i}.$$

This yields

$$\frac{\partial}{\partial \beta_j} \ln \mathrm{f}(\mathbf{y}) = \sum_{i=1}^{n} \frac{\partial \mu_i}{\partial \beta_j} (\mathrm{Var}\ y_i)^{-1} (y_i - \mu_i),$$

which we summarize as

$$\frac{\partial L(\boldsymbol{\beta})}{\partial \boldsymbol{\beta}} = \frac{\partial}{\partial \boldsymbol{\beta}} \ln \mathrm{f}(\mathbf{y}) = \sum_{i=1}^{n} \frac{\partial \mu_i}{\partial \boldsymbol{\beta}} (\mathrm{Var}\ y_i)^{-1} (y_i - \mu_i), \tag{13.13}$$

which is known as the *generalized estimating equations* form.

Solving for roots of the score equation $L(\boldsymbol{\beta}) = \mathbf{0}$ yield maximum likelihood estimates, $\mathbf{b}_{MLE}$. In general, this requires iterative numerical methods. An exception is the following special case.

Special Case: No Covariates. Suppose that $k = 1$ and that $x_i = 1$. Thus, $\beta = \eta = g(\mu)$, and μ does not depend on i. Then, from the relation $\mathrm{Var}\ y_i = \phi b''(\theta)/w_i$ and equation (13.13), the score function is

$$\frac{\partial L(\beta)}{\partial \beta} = \frac{\partial \mu}{\partial \beta}\frac{1}{\phi b''(\theta)}\sum_{i=1}^{n} w_i (y_i - \mu).$$

Setting this equal zero yields $\overline{y}_w = \sum_i w_i y_i / \sum_i w_i = \widehat{\mu}_{MLE} = g^{-1}(b_{MLE})$, or $b_{MLE} = g(\overline{y}_w)$, where $\overline{y}_w$ is a weighted average of y's.

For the information matrix, use the independence among subjects and equation (13.13) to get

$$\begin{aligned}\mathbf{I}(\boldsymbol{\beta}) &= \mathrm{E}\left(\frac{\partial L(\boldsymbol{\beta})}{\partial \boldsymbol{\beta}}\frac{\partial L(\boldsymbol{\beta})}{\partial \boldsymbol{\beta}'}\right) \\ &= \sum_{i=1}^{n} \frac{\partial \mu_i}{\partial \boldsymbol{\beta}} \mathrm{E}\left((\mathrm{Var}\ y_i)^{-1}(y_i-\mu_i)^2(\mathrm{Var}\ y_i)^{-1}\right)\frac{\partial \mu_i}{\partial \boldsymbol{\beta}'} \\ &= \sum_{i=1}^{n} \frac{\partial \mu_i}{\partial \boldsymbol{\beta}}(\mathrm{Var}\ y_i)^{-1}\frac{\partial \mu_i}{\partial \boldsymbol{\beta}'}. \end{aligned} \tag{13.14}$$

Taking the square root of the diagonal elements of the inverse of $\mathbf{I}(\boldsymbol{\beta})$, after inserting parameter estimates in the mean and variance functions, yields *model-based standard errors*. For data where the correct specification of the variance component is in doubt, one replaces Var y_i by an unbiased estimate $(y_i - \mu_i)^2$. The resulting estimator

$$se(b_j) = \sqrt{jth\ diagonal\ element\ of\ [\sum_{i=1}^{n}\frac{\partial \mu_i}{\partial \boldsymbol{\beta}}(y_i-\mu_i)^{-2}\frac{\partial \mu_i}{\partial \boldsymbol{\beta}'}]^{-1}|_{\boldsymbol{\beta}=\mathbf{b}_{MLE}}}$$

is known as the *empirical*, or *robust, standard error*.

13.9.4 Iterated Reweighted Least Squares

To see how general iterative methods work, we use the canonical link so that $\eta_i = \theta_i$. Then, the score function is given in equation (13.6) and the Hessian is in equation (13.8). Using the Newton-Raphson equation (11.15) in Section 11.9 yields

$$\boldsymbol{\beta}_{NEW} = \boldsymbol{\beta}_{OLD} - \left(\sum_{i=1}^{n} w_i b''(\mathbf{x}_i'\boldsymbol{\beta}_{OLD})\mathbf{x}_i\mathbf{x}_i'\right)^{-1}\left(\sum_{i=1}^{n} w_i(y_i - b'(\mathbf{x}_i'\boldsymbol{\beta}_{OLD}))\mathbf{x}_i\right). \tag{13.15}$$

Note that, because the matrix of second derivatives is nonstochastic, the Newton-Raphson is equivalent to the Fisher scoring algorithm. As described in Section 13.3.1, the estimation of $\boldsymbol{\beta}$ does not require knowledge of ϕ.

Continue using the canonical link and define an "adjusted dependent variable"

$$y_i^*(\boldsymbol{\beta}) = \mathbf{x}_i'\boldsymbol{\beta} + \frac{y_i - b'(\mathbf{x}_i'\boldsymbol{\beta})}{b''(\mathbf{x}_i'\boldsymbol{\beta})}.$$

This has variance

$$\mathrm{Var}\ y_i^*(\boldsymbol{\beta}) = \frac{\mathrm{Var}\ y_i}{(b''(\mathbf{x}_i'\boldsymbol{\beta}))^2} = \frac{\phi_i b''(\mathbf{x}_i'\boldsymbol{\beta})}{(b''(\mathbf{x}_i'\boldsymbol{\beta}))^2} = \frac{\phi/w_i}{b''(\mathbf{x}_i'\boldsymbol{\beta})}.$$

Use the new weight as the reciprocal of the variance, $w_i(\boldsymbol{\beta}) = w_i b''(\mathbf{x}_i'\boldsymbol{\beta})/\phi$. Then, with the expression

$$w_i\left(y_i - b'(\mathbf{x}_i'\boldsymbol{\beta})\right) = w_i b''(\mathbf{x}_i'\boldsymbol{\beta})(y_i^*(\boldsymbol{\beta}) - \mathbf{x}_i'\boldsymbol{\beta}) = \phi w_i(\boldsymbol{\beta})(y_i^*(\boldsymbol{\beta}) - \mathbf{x}_i'\boldsymbol{\beta}),$$

from the Newton-Raphson iteration in equation (13.15), we have
$\boldsymbol{\beta}_{NEW}$

$$= \boldsymbol{\beta}_{OLD} - \left(\sum_{i=1}^n w_i b''(\mathbf{x}_i'\boldsymbol{\beta}_{OLD})\mathbf{x}_i\mathbf{x}_i'\right)^{-1}\left(\sum_{i=1}^n w_i(y_i - b'(\mathbf{x}_i'\boldsymbol{\beta}_{OLD}))\mathbf{x}_i\right)$$

$$= \boldsymbol{\beta}_{OLD} - \left(\sum_{i=1}^n \phi w_i(\boldsymbol{\beta}_{OLD})\mathbf{x}_i\mathbf{x}_i'\right)^{-1}\left(\sum_{i=1}^n \phi w_i(\boldsymbol{\beta}_{OLD})(y_i^*(\boldsymbol{\beta}_{OLD}) - \mathbf{x}_i'\boldsymbol{\beta}_{OLD})\mathbf{x}_i\right)$$

$$= \boldsymbol{\beta}_{OLD} - \left(\sum_{i=1}^n w_i(\boldsymbol{\beta}_{OLD})\mathbf{x}_i\mathbf{x}_i'\right)^{-1} \times$$

$$\left(\sum_{i=1}^n w_i(\boldsymbol{\beta}_{OLD})\mathbf{x}_i y_i^*(\boldsymbol{\beta}_{OLD}) - \sum_{i=1}^n w_i(\boldsymbol{\beta}_{OLD})\mathbf{x}_i\mathbf{x}_i'\boldsymbol{\beta}_{OLD})\right)$$

$$= \left(\sum_{i=1}^n w_i(\boldsymbol{\beta}_{OLD})\mathbf{x}_i\mathbf{x}_i'\right)^{-1}\left(\sum_{i=1}^n w_i(\boldsymbol{\beta}_{OLD})\mathbf{x}_i y_i^*(\boldsymbol{\beta}_{OLD})\right).$$

Thus, this provides a method for iteration using weighted least squares. Iterative reweighted least squares is also available for general links using Fisher scoring; see McCullagh and Nelder (1989) for further details.

14

Survival Models

Chapter Preview. This chapter introduces regression where the dependent variable is the time until an event, such as the time until death, the onset of a disease, or the default on a loan. Event times are often limited by sampling procedures and so ideas of censoring and truncation of data are summarized in this chapter. Event times are nonnegative and their distributions are described in terms of survival and hazard functions. Two types of hazard-based regression are considered, a fully parametric accelerated failure time model and a semiparametric proportional hazards models.

14.1 Introduction

In survival models, the dependent variable is the time until an event of interest. The classic example of an event is time until death (the complement of death being survival). Survival models are now widely applied in many scientific disciplines; other examples of events of interest include the onset of Alzheimer's disease (biomedical), time until bankruptcy (economics), and time until divorce (sociology).

Example: Time until Bankruptcy. Shumway (2001) examined the time to bankruptcy for 3,182 firms listed on Compustat Industrial File and the CRSP Daily Stock Return File for the New York Stock Exchange over the period 1962–92. Several explanatory financial variables were examined, including working capital to total assets, retained earnings to total assets, earnings before interest and taxes to total assets, market equity to total liabilities, sales to total assets, net income to total assets, total liabilities to total assets, and current assets to current liabilities. The dataset included 300 bankruptcies from 39,745 firm years.

See also Kim et al. (1995) for a similar study on insurance insolvencies.

A distinguishing feature of survival modeling is that dependent variables are often limited by censoring and truncation.

A distinguishing feature of survival modeling is that it is common for the dependent variable to be observed only in a limited sense. Some events of interest, such as bankruptcy or divorce, never occur for specific subjects (and so can be thought of as taking an infinite time). For other subjects, even when the event time is finite, it may occur after the study period so that the data are (right) *censored*.

That is, complete information about event times may not be available due to the design of the study. Moreover, firms may merge with or be acquired by other firms and individuals may move from a geographical area, leaving the study. Thus, the data may be limited by events that are extraneous to the research question under consideration, represented as *random* censoring. Censoring is a regular feature of survival data; large values of a dependent variable require more time to develop so that they can be more difficult to observe than small values, other things being equal. Other types of limitations may also occur; subjects whose observation depends on the experience of the event of interest are said to be *truncated.* To illustrate, in an investigation of old-age mortality, only those who survive to age 85 are recruited to be a part of the study. Section 14.2 describes censoring and truncation in more detail.

Another distinguishing feature of survival modeling is that the dependent variable is positive valued. Thus, normal curve approximations used in linear regression are less helpful in survival analysis; this chapter introduces alternative regression models. Further, it is customary to interpret models of survival using the *hazard function*, defined as

$$\mathrm{h}(t) = \frac{\textit{probability density function}}{\textit{survival function}} = \frac{\mathrm{f}(t)}{\mathrm{S}(t)},$$

the "instantaneous" probability of an event, conditional on survivorship up to time t. The hazard function goes by many other names: it is known as the *force of mortality* in actuarial science, the *failure rate* in engineering and the *intensity function* in stochastic processes.

Duration dependence describes the relation between the instantaneous event probability and the time spent in a given state.

In economics, hazard functions are used to describe *duration dependence*, the relation between the instantaneous event probability (the density) and the time spent in a given state. *Negative* duration dependence is associated with decreasing hazard rates. For example, the longer the time until a claimant requests a payment from an insured injury, the lower is the probability of making a request. *Positive* duration dependence is associated with increasing hazard rates. For example, old-age human mortality generally displays an increasing hazard rate. The older someone is, the higher the near-term probability of death.

A related quantity of interest is the *cumulative hazard function*, $H(t) = \int_0^t h(s)\,ds$. This quantity can also be expressed as the negative log survival function, and conversely, $\Pr(y > t) = \mathrm{S}(t) = \exp(-H(t))$.

The two most widely used regression models in survival analysis are based on hazard functions. Section 14.3 introduces the *accelerated failure time model*, where one assumes a linear model for the logarithmic time to failure but with an error distribution that need not be approximately normal. Section 14.4 introduces the the *proportional hazards model* due to Cox (1972), where one assumes that the hazard function can be written as the product of a "baseline" hazard and a function of a linear combination of explanatory variables.

With survival data, we observe a cross-section of subjects where time is the dependent variable of interest. As with Chapter 10 on longitudinal and panel data, there are also applications in which we are interested in repeated observations

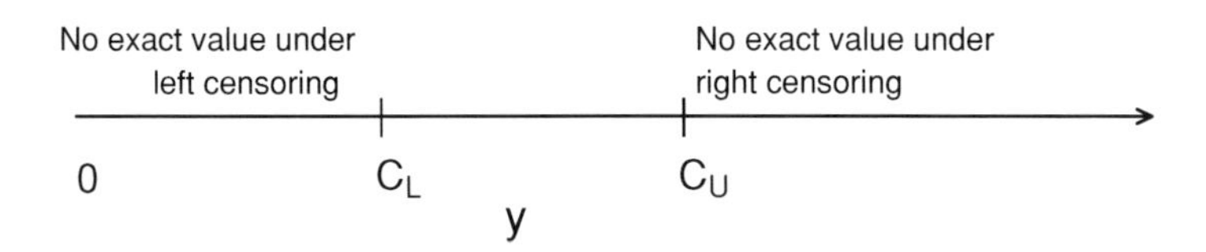

Figure 14.1 Figure illustrating left and right censoring.

for each subject. For example, if you are injured in an accident that is covered by insurance, the payments that arise from this claim can occur repeatedly over time, depending on the time to recovery. Section 14.5 introduces the notion of repeated event times, called *recurrent events*.

14.2 Censoring and Truncation

14.2.1 Definitions and Examples

Two types of limitations encountered in survival data are *censoring* and *truncation*. Censoring and truncation are also common features in other actuarial applications, including the Chapter 16 two-part and Chapter 17 fat-tailed models. Thus, this section describes these concepts in detail.

For censoring, the most common form is *right censoring*, in which we observe the smaller of the "true" dependent variable and a censoring time variable. For example, suppose that we wish to study the time until a new employee leaves the firm and that we have five years of data to conduct our analysis. Then, we observe the smaller of five years and the amount of time that the employee was with the firm. We also observe whether the employee has departed within five years.

Using notation, let y denote the time to the event, such as the amount of time the employee worked with a firm. Let C_U denote the censoring time, such as $C_U = 5$. Then, we observe the variable $y_U^* = \min(y, C_U)$. We also observe whether censoring has occurred. Let $\delta_U = \mathrm{I}(y \geq C_U)$ be a binary variable that is 1 if censoring occurs, $y \geq C_U$, and 0 otherwise. For example, in Figure 14.1, values of y that are greater than the upper censoring limit C_U are not observed – thus, this is often called *right censoring*.

Other common forms of censoring are *left censoring* and *interval censoring*. With left censoring, we observe $y_L^* = \max(y, C_L)$ and $\delta_L = \mathrm{I}(y \leq C_L)$, where C_L is the censoring time. For example, if you are conducting a study and interviewing a person about an event in the past, the subject may recall that the event occurred before C_L but not the exact date.

Common forms of censoring include left, right, and interval censoring.

With interval censoring, there is an interval of time, such as (C_L, C_U), in which y is known to occur but the exact value is not observed. For example, you may be looking at two successive years of annual employee records. People employed in the first year but not the second have left sometime during the year. With an exact departure date, you could compute the amount of time that they were with the firm. Without the departure date, then you know only that they departed sometime during a yearlong interval.

Censoring times may be fixed or random. Fixed and independent random censoring are said to be noninformative.

Censoring times such as C_L and C_U may or may not be stochastic. If the censoring times represent variables known to the analyst, such as the observation

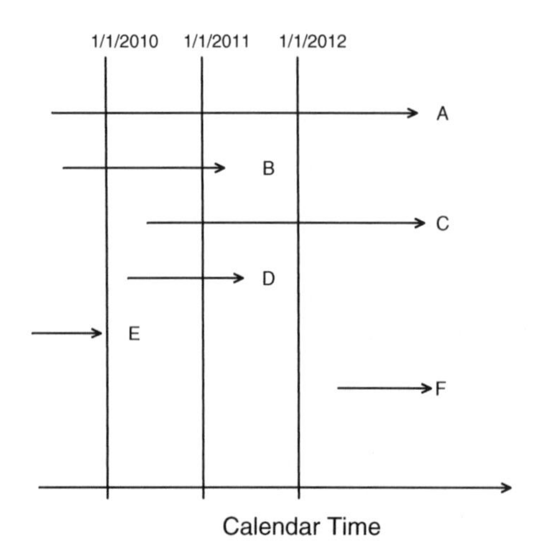

Figure 14.2 Timeline for several subjects in a mortality study.

period, they are said to be *fixed censoring* times. Here, the adjective "fixed" means that the times are known in advance. Censoring times may vary by individual but still be fixed. However, censoring may also be the result of an unforeseen phenomena, such as a firm merger or a subject moving from a geographical area. In this case, the censoring C is represented by a random variable and said to be *random censoring*. If the censoring is fixed or if the random censoring time is independent of the event of interest, then the censoring is said to be *noninformative*. If the censoring times are independent of the event of interest, then we can essentially treat the censoring as fixed. When censoring and the time to event are dependent (known as *informative censoring*), then special models are required. For example, if the event of interest is time until bankruptcy of a firm and the censoring mechanism involves filing adequate financial statements, then there is a potential relationship. Financially weak firms are more likely to become bankrupt and less likely to go through the accounting effort to file adequate statements.

Censored observations are available for study, although in a limited form. In contrast, *truncated* responses are a type of missing data. To illustrate, with *left-truncated* data, if y is less than a threshold (say, C_L), then it is not observed. For *right-truncated* data, if y exceeds a threshold (say, C_U), then it is not observed. To compare truncated and censored observations, consider the following example.

Example: Mortality Study. Suppose that you are conducting a two-year study of mortality of high-risk subjects, beginning January 1, 2010, and finishing January 1, 2012. Figure 14.2 graphically portrays the six types of subjects recruited. For each subject, the beginning of the arrow represents that the the subject was recruited and the arrow end represents the event time. Thus, the arrow represents exposure time.

- Type A: right censored. This subject is alive at the beginning and the end of the study. Because the time of death is not known by the end of the study, it is right censored. Most subjects are type A.
- Type B: complete information is available for a type B subject. The subject is alive at the beginning of the study and the death occurs within the observation period.
- Type C: right censored and left truncated. A type C subject is right censored, in that death occurs after the observation period. However, the subject entered after the start of the study and is said to have a *delayed entry time*. Because the subject would not have been observed had death occurred before entry, it is left truncated.
- Type D: left truncated. A type D subject also has delayed entry. Because death occurs within the observation period, this subject is not right censored.
- Type E: left truncated. A type E subject is not included in the study because death occurs prior to the observation period.
- Type F: right truncated. Similarly, a type F subject is not included because the entry time occurs after the observation period.

14.2.2 Likelihood Inference

Many inference techniques for survival modeling involve likelihood estimation, so it is helpful to understand the implications of censoring and truncation when specifying likelihood functions. For simplicity, we assume fixed censoring times and a continuous time to event y.

To begin, consider the case of right-censored data, where we observe $y_U^* = \min(y, C_U)$ and $\delta_U = \mathrm{I}(y \geq C_U)$. If censoring occurs so that $\delta_U = 1$, then $y \geq C_U$ and the likelihood is $\Pr(y \geq C_U) = \mathrm{S}(C_U)$. If censoring does not occur so that $\delta_U = 0$, then $y < C_U$ and the likelihood is $\mathrm{f}(y)$. Summarizing, we have

$$\begin{aligned} Likelihood &= \begin{cases} \mathrm{f}(y) & \text{if } \delta = 0 \\ \mathrm{S}(C_U) & \text{if } \delta = 1 \end{cases} \\ &= (\mathrm{f}(y))^{1-\delta} (\mathrm{S}(C_U))^{\delta}. \end{aligned}$$

The right-hand expression allows us to present the likelihood more compactly. Now, for an independent sample of size n, $\{(y_{U1}, \delta_1), \ldots, (y_{Un}, \delta_n)\}$, the likelihood is

$$\prod_{i=1}^{n} (\mathrm{f}(y_i))^{1-\delta_i} (\mathrm{S}(C_{Ui}))^{\delta_i} = \prod_{\delta_i=0} \mathrm{f}(y_i) \prod_{\delta_i=1} \mathrm{S}(C_{Ui}),$$

with potential censoring times $\{C_{U1}, \ldots, C_{Un}\}$. Here, the notation "$\prod_{\delta_i=0}$" means take the product over uncensored observations, and similarly for "$\prod_{\delta_i=1}$."

Truncated data are handled in likelihood inference via conditional probabilities. Specifically, we adjust the likelihood contribution by dividing by the probability that the variable was observed. Summarizing, we have the following contributions to the likelihood for six types of outcomes.

Outcome	Likelihood Contribution
Exact value	$\mathrm{f}(y)$
Right censoring	$\mathrm{S}(C_U)$
Left censoring	$1 - \mathrm{S}(C_L)$
Right truncation	$\mathrm{f}(y)/(1 - \mathrm{S}(C_U))$
Left truncation	$\mathrm{f}(y)/\mathrm{S}(C_L)$
Interval censoring	$\mathrm{S}(C_L) - \mathrm{S}(C_U)$

For known event times and censored data, the likelihood is

$$\prod_E \mathrm{f}(y_i) \prod_R \mathrm{S}(C_{Ui}) \prod_L (1 - \mathrm{S}(C_{Li})) \prod_I (\mathrm{S}(C_{Li}) - \mathrm{S}(C_{Ui})),$$

where "$\prod_E$" is the product over observations with *e*xact values (E), and similarly for *r*ight (R), *l*eft (L), and *i*nterval censoring (I).

For right-censored and left-truncated data, the likelihood is

$$\prod_E \frac{\mathrm{f}(y_i)}{\mathrm{S}(C_{Li})} \prod_R \frac{\mathrm{S}(C_{Ui})}{\mathrm{S}(C_{Li})},$$

and similarly for other combinations. To get further insights, consider the following.

Special Case: Exponential Distribution. Consider data that are right censored and left truncated, with dependent variables y_i that are exponentially distributed with mean μ. With these specifications, recall that $\mathrm{f}(y) = \mu^{-1} \exp(-y/\mu)$ and $\mathrm{S}(y) = \exp(-y/\mu)$.

For this special case, the logarithmic likelihood is

$$\begin{aligned} \ln Likelihood &= \sum_E (\ln \mathrm{f}(y_i) - \ln \mathrm{S}(C_{Li})) + \sum_R (\ln \mathrm{S}(C_{Ui}) - \ln \mathrm{S}(C_{Li})) \\ &= \sum_E (-\ln \mu - (y_i - C_{Li})/\mu) - \sum_R (C_{Ui} - C_{Li})/\mu. \end{aligned}$$

To simplify the notation, define $\delta_i = \mathrm{I}(y_i \geq C_{Ui})$ to be a binary variable that indicates right censoring. Let $y_i^{**} = \min(y_i, C_{Ui}) - C_{Li}$ be the amount that the observed variable exceeds the lower truncation limit. With this, the logarithmic likelihood is

$$\ln Likelihood = -\sum_{i=1}^{n} ((1 - \delta_i) \ln \mu + \frac{y_i^{**}}{\mu}). \qquad (14.1)$$

Taking derivatives with respect to the parameter μ and setting it equal to zero yields the maximum likelihood estimator

$$\widehat{\mu} = \frac{1}{n_u} \sum_{i=1}^{n} y_i^{**},$$

where $n_u = \sum_i (1 - \delta_i)$ is the number of uncensored observations.

14.2.3 Product-Limit Estimator

It can be useful to calibrate likelihood methods with nonparametric methods that do not rely on a parametric form of the distribution. The *product-limit estimator* due to Kaplan and Meier (1958) is a well-known estimator of the distribution in the presence of censoring.

To introduce this estimator, we consider the case of right-censored data. Let $t_1 < \cdots < t_c$ be distinct time points at which an event of interest occurs and let d_j be the number of events at time point t_j. Further, define R_j to be the corresponding risk set – this is the number of observations that are active at an instant just prior to t_j. Using notation, the risk set is $R_j = \sum_{i=1}^{n} I\left(y_i \geq t_j\right)$. With this notation, the product-limit estimator of the survival function is

$$\widehat{S}(t) = \begin{cases} 1 & t < t_1 \\ \prod_{t_j \leq t} \left(1 - \frac{d_j}{R_j}\right) & t \geq t_1 \end{cases}. \tag{14.2}$$

To interpret the product-limit estimator, we look to evaluating it at event times. At the first event time t_1, the estimator is $\widehat{S}(t_1) = 1 - d_1/R_1 = (R_1 - d_1)/R_1$, the proportion of nonevents from the risk set R_1. At the second event time t_2, the survival probability conditional on survivorship to time t_1 is $\widehat{S}(t_2)/\widehat{S}(t_1) = 1 - d_2/R_2 = (R_2 - d_2)/R_2$, the proportion of nonevents from the risk set R_2. Similarly, at the jth event time,

$$\frac{\widehat{S}(t_j)}{\widehat{S}(t_{j-1})} = 1 - \frac{d_j}{R_j} = \frac{R_j - d_j}{R_j}.$$

Starting from these conditional probabilities, one can build the survival estimate as

$$\widehat{S}(t_j) = \frac{\widehat{S}(t_j)}{\widehat{S}(t_{j-1})} \times \cdots \times \frac{\widehat{S}(t_2)}{\widehat{S}(t_1)} \times \widehat{S}(t_1),$$

resulting in equation (14.2). In this sense, the estimator is a "product," up to the time "limit." For times between event times, the survival estimate is taken to be a constant.

To see how to use the product-limit estimator, we consider a small dataset of $n = 23$ observations, where 18 are event times and 5 have been right censored. This example is from Miller (1997), where the events correspond to survival for

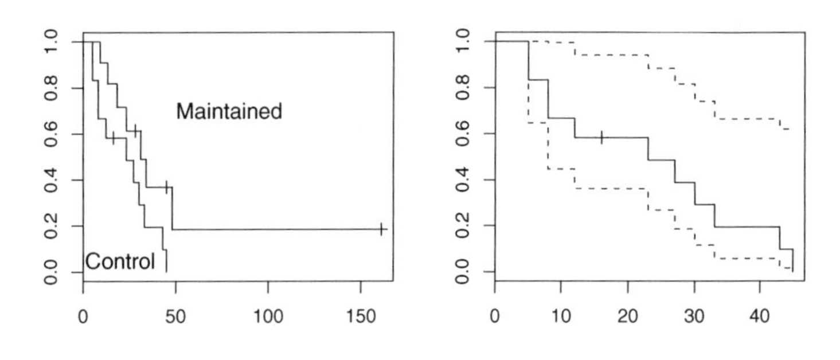

Figure 14.3 Product-limit estimate of the survival functions for two groups. This graph shows that those with maintained chemotherapy treatment have higher estimated survival probabilities than those in the control group.

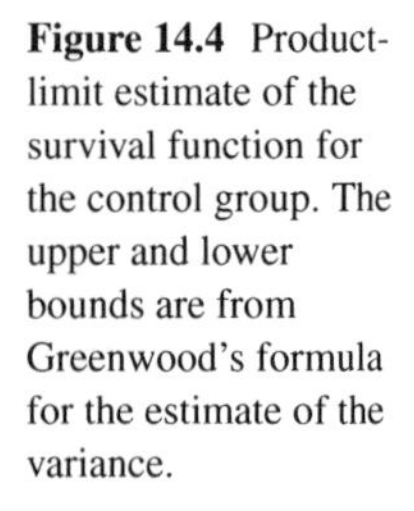

Figure 14.4 Product-limit estimate of the survival function for the control group. The upper and lower bounds are from Greenwood's formula for the estimate of the variance.

patients with acute myelogenous leukemia. After patients received chemotherapy to achieve complete remission, they were randomly allocated to one of two groups, those who received maintenance chemotherapy and those who did not (the control group). The event was time in weeks to relapse from the complete remission state. For those in the maintenance group, the times are 9, 13, 13+, 18, 23, 28+, 31, 34, 45+, 48, 161+. For those in the control group, the times are: 5, 5, 8, 8, 12, 16+, 23, 27, 30, 33, 43 45. Here, the plus sign (+) indicates right censoring of an observation.

Figures 14.3 and 14.4 show the product-limit survival function estimates for the data. Notice the step nature of the function, where drops (from the left, or jumps from the right) correspond to event times. When no censoring is involved, the product-limit estimator reduces to the usual empirical estimator of the survival function.

In the figures, censored observations are depicted with a plus (+) plotting symbol. When the last observation has been censored as in Figure 14.3, there are different methods for defining the survival curve for times exceeding this observation. Analysts making estimates at these times will need to be aware of the option that their statistical package uses. The options are described in standard books on survival analysis, such as Klein and Moschberger (1997).

Figure 14.4 also shows an estimate of the standard error. These are calculated using Greenwood's (1926) formula for an estimate of the variance, given by

$$\widehat{\mathrm{Var}\,(\widehat{S}(t))} = (\widehat{S}(t))^2 \sum_{t_j \le t} \frac{d_j}{R_j(R_j - d_j)}.$$

14.3 Accelerated Failure Time Model

An accelerated failure time (AFT) model can be expressed using a linear regression equation $\ln y_i = \mathbf{x}_i'\boldsymbol{\beta} + \varepsilon_i$, where y_i is the event time. Unlike the usual linear regression model, the AFT model makes a parametric assumption about the disturbance term, such as a normal, extreme-value, or logistic distribution. As we have seen, the normal distribution is the basis for usual linear regression model. Thus, many analysts begin by taking logarithms of event data and using the usual linear regression routines to explore their data.

Table 14.1 Location-Scale Distributions

Standard-Form Survival Distribution	Location-Scale Distribution	Log Location-Scale Distribution
$S_0(t) = \exp(-e^t)$	Extreme value distribution	Weibull
$S_0(t) = 1 - \Phi(t)$	Normal	Lognormal
$S_0(t) = (1 + e^t)^{-1}$	Logistic	Log-logistic

Location-Scale Distributions

To understand the name *accelerated failure*, also known as *accelerated hazards*, it is helpful to review the statistical idea of location-scale families. A parametric *location-scale distribution* is one where the density function is of the form

$$f(t) = \frac{1}{\sigma} f_0\left(\frac{t-\mu}{\sigma}\right),$$

where μ is the location parameter and $\sigma > 0$ is the scale parameter. The case $\mu = 0$ and $\sigma = 1$ corresponds to the *standard form* of the distribution. Within a location-scale distribution, additive shifts, such as going from degrees Kelvin to degrees Centigrade, $^\circ K = {}^\circ C + 273.15$, and scalar multiples, such as going from dollars to thousands of dollars, remain in the same distribution family. As we will see in Chapter 17, a location parameter is a natural place to introduce regression covariates.

A random variable y is said to have a *log location-scale distribution* if $\ln y$ has a location-scale distribution. Suppose that a random variable y_0 has a standard-form location-scale distribution with survival function $S_0(z)$. Then, the survival function of y defined by $\ln y = \mu + \sigma y_0$ can be expressed as

$$\begin{aligned} S(t) = \Pr(y > t) &= \Pr\left(y_0 > \frac{\ln t - \mu}{\sigma}\right) \\ &= S_0\left(\frac{\ln t - \mu}{\sigma}\right) = S_0^*\left(\left(\frac{t}{e^\mu}\right)^{1/\sigma}\right), \end{aligned}$$

where $S_0^*(t) = S_0(\ln t)$ is the survival function of the standard form of $\ln y$. In the context of survival modeling where t represents time, the effect of rescaling by dividing by e^μ can be thought of as "accelerating time" (when $e^\mu < 1$), which is the motivation for the name *accelerated failure time models*. Table 14.1 provides some special cases of widely used location-scale distributions and their log location-scale counterparts.

Inference for AFT Models

To get an idea of the complexities in estimating an AFT model, we return to the exponential distribution. This is a special case of Weibull regression with the scale parameter $\sigma = 1$.

Special Case: Exponential Distribution, Continued. To introduce regression covariates, we let $\mu_i = \exp(\mathbf{x}_i'\boldsymbol{\beta})$. Using the same reasoning as with equation (14.1), the logarithmic likelihood is

$$\begin{aligned}\ln \textit{Likelihood} &= -\sum_{i=1}^{n}((1-\delta_i)\ln \mu_i + \frac{y_i^{**}}{\mu_i})\\ &= -\sum_{i=1}^{n}((1-\delta_i)\mathbf{x}_i'\boldsymbol{\beta} + y_i^{**}\exp(-\mathbf{x}_i'\boldsymbol{\beta})),\end{aligned}$$

where $\delta_i = I(y_i \geq C_{Ui})$ and $y_i^{**} = \min(y_i, C_{Ui}) - C_{Li}$. Taking derivatives with respect to $\boldsymbol{\beta}$ yields the score function

$$\begin{aligned}\frac{\partial \ln \textit{Likelihood}}{\partial \boldsymbol{\beta}} &= -\sum_{i=1}^{n}((1-\delta_i)\mathbf{x}_i - \mathbf{x}_i y_i^{**}\exp(-\mathbf{x}_i'\boldsymbol{\beta}))\\ &= \sum_{i=1}^{n}\mathbf{x}_i\frac{y_i^{**} - \mu_i(1-\delta_i)}{\mu_i}.\end{aligned}$$

This has the form of a generalized estimating equation, introduced in Chapter 13. Although closed-form solutions rarely exist, they can readily be solved by modern statistical packages.

As this special case illustrates, estimation of AFT regression models can be readily addressed through maximum likelihood. Moreover, properties of maximum likelihood estimates are well understood and so we readily have general inference tools, such as estimation and hypothesis testing, available. Standard statistical packages provide output that supports this inference.

14.4 Proportional Hazards Model

The assumption of proportional hazards is defined in Section 14.4.1 and inference techniques are discussed in Section 14.4.2.

14.4.1 Proportional Hazards

In the *proportional hazards* (PH) model due to Cox (1972), one assumes that the hazard function can be written as the product of some "baseline" hazard and a function of a linear combination of explanatory variables. To illustrate, we use

$$h_i(t) = h_0(t)\exp(\mathbf{x}_i'\boldsymbol{\beta}), \tag{14.3}$$

where $h_0(t)$ is the baseline hazard. This is known as a *proportional* hazards model because if one takes the ratio of hazard functions for two sets of covariates, say, $\mathbf{x}_1$ and $\mathbf{x}_2$, one gets

$$\frac{h_1(t|\mathbf{x}_1)}{h_2(t|\mathbf{x}_1)} = \frac{h_0(t)\exp(\mathbf{x}_1'\boldsymbol{\beta})}{h_0(t)\exp(\mathbf{x}_2'\boldsymbol{\beta})} = \exp((\mathbf{x}_1 - \mathbf{x}_2)'\boldsymbol{\beta}).$$

Note that the ratio does not depend on time t.

As we have seen in many regression applications, users are interested in variable effects and not always concerned with other aspects of the model. The reason that the PH model in equation (14.3) has proved so popular that it specifies the explanatory variable effects as a simple function of the linear combination while permitting a flexible baseline component, h_0. Although this baseline is common to all subjects, it need not be specified parametrically as with AFT models. Further, one need not use the "exp" function for the explanatory variables; however, this is the common specification as it ensures that the hazard function will remain nonnegative.

Proportional hazards can be motivated as an extension of exponential regression. Consider a random variable y^* that has an exponential distribution with mean $\mu = \exp(\mathbf{x}'\boldsymbol{\beta})$. Suppose that we observe $y = \mathrm{g}(y^*)$, where $\mathrm{g}(\cdot)$ is unknown except that it is monotonically increasing. Many survival distributions are transformations of the exponential distribution. For example, if $y = \mathrm{g}(y^*) = (y^*)^{\sigma}$, then it is easy to check that y has a Weibull distribution given in Table 14.1. The Weibull distribution is the only AFT model that has proportional hazards (see Lawless, 2003, exercise 6.1).

Straightforward calculations show that the hazard function of y can be expressed as

$$\mathrm{h}_y(t) = \mathrm{g}'(t)/\mu = \mathrm{g}'(t)\exp(-\mathbf{x}'\boldsymbol{\beta});$$

see, for example, Zhou (2001). This has a proportional hazards structure, as in equation (14.3), with $\mathrm{g}'(t)$ serving as the baseline hazard. In the PH model, the baseline hazard is assumed unknown. In contrast, for the Weibull and other AFT regression models, the baseline hazard function is assumed known up to one or two parameters that can be estimated from the data.

14.4.2 Inference

Because of the flexible specification of the baseline component, the usual maximum likelihood estimation techniques are not available to estimate the PH model. To outline the estimation procedure, we let $(y_1, \delta_1), \ldots, (y_n, \delta_n)$ be independent and assume that y_i follows equation (14.3) with regressors $\mathbf{x}_i$. That is, we now drop the asterisk $(*)$ notation and let y_i denote observed values (exact or censored) and use δ_i to be the binary variable that indicates (right) censoring. Further, let H_0 be the cumulative hazard function associated with the baseline function h_0. Recalling the general relationship $\mathrm{S}(t) = \exp(-H(t))$, with equation (14.3) we have $\mathrm{S}(t) = \exp\left(-H_0(t)\exp(\mathbf{x}_i'\boldsymbol{\beta})\right)$.

Starting from the usual likelihood perspective, from Section 14.2.2 the likelihood is

$$L(\boldsymbol{\beta}, h_0) = \prod_{i=1}^{n} \mathrm{f}(y_i)^{1-\delta_i} \mathrm{S}(y_i)^{\delta_i} = \prod_{i=1}^{n} h(y_i)^{1-\delta_i} \exp(-H(y_i))$$

$$= \prod_{i=1}^{n} \left(h_0(t) \exp(\mathbf{x}_i' \boldsymbol{\beta})\right)^{1-\delta_i} \exp\left(-H_0(y_i) \exp(\mathbf{x}_i' \boldsymbol{\beta})\right).$$

Now, parameter estimates that maximize $L(\boldsymbol{\beta}, h_0)$ do not follow the usual properties of maximum likelihood estimation because the baseline hazard h_0 is not specified parametrically. One way to think about this problem is due to Breslow (1974), who showed that a *nonparametric* maximum likelihood estimator of h_0 could be used in $L(\boldsymbol{\beta}, h_0)$. This results in what Cox called a *partial likelihood*,

$$L_P(\boldsymbol{\beta}) = \prod_{i=1}^{n} \left(\frac{\exp(\mathbf{x}_i' \boldsymbol{\beta})}{\sum_{j \in R(y_i)} \exp(\mathbf{x}_j' \boldsymbol{\beta})} \right)^{1-\delta_i}, \qquad (14.4)$$

where $R(t)$ is the risk set at time t. Specifically, this is the set of all $\{y_1, \ldots, y_n\}$ such that $y_i \geq t$, that is, the set of all subjects still under study at time t.

Equation (14.4) is only a "partial" likelihood in that does not use all of the information in $(y_1, \delta_1), \ldots, (y_n, \delta_n)$. For example, from equation (14.4), we see that inference for the regression coefficients depends only on the ranks of the dependent variables $\{y_1, \ldots, y_n\}$, not their actual values.

Nonetheless, equation (14.4) suggests (and it is true) that large sample distribution theory has properties similar to the usual desirable (fully) parametric theory. From a user's perspective, this partial likelihood can be treated as a usual likelihood function. That is, the regression parameters that maximize equation (14.4) are consistent and have a large sample normal distribution with the usual variance estimates (see Section 11.9 for a review). This is mildly surprising because the proportional hazards model is semiparametric; in equation (14.3), the hazard function has a fully parametric component, $\exp(\mathbf{x}_i' \boldsymbol{\beta})$, but also contains a nonparametric baseline hazard, $h_0(t)$. In general, nonparametric models are more flexible than parametric counterparts for model fitting but result in less desirable large sample properties (specifically, slower rates of convergence to an asymptotic distribution).

Example: Credit Scores. Stepanova and Thomas (2002) used proportional hazards models to create credit scores that banks could use to assess the quality of personal loans. Their scores depended on the loan purposes and characteristics of the applicants. Potential customers apply to banks for loans for many purposes, including financing the purchase of a house, car, boat, or musical instrument, for home improvement and car repair, or for funding a wedding and honeymoon; Stepanova and Thomas listed 22 loan purposes. On the loan application, people provided their age, requested loan amount, years with current employer and other personal characteristics; Stepanova and Thomas listed 22 application characteristics used in their analysis.

Their data was from a major U.K. financial institution. It consisted of application information from 50,000 personal loans, with repayment status for each of the first 36 months of the loan. Thus, each loan was (fixed) right censored by the smaller of the length of the study period, 36 months, and the repayment term of the loan that varied from 6 to 60 months.

To create the scores, the authors examined two dependent variables that have negative financial consequences for the loan provider, the time to loan default and time to early repayment. For this study, the definition of default is 3 or more months delinquent. (In principle, one could analyze both times simultaneously in a so-called competing risks framework.) When Stepanova and Thomas used a proportional hazards model with time to loan default as a dependent variable, the time to early repayment was a random right-censoring variable. Conversely, when time to early repayment was the dependent variable, time to loan default was a random right-censoring variable.

Stepanova and Thomas used model estimates of the survival functions at 12 months and at 24 months as their credit scores for both dependent variables. They compared these scores to those from a logistic regression model where, for example, the dependent variable was loan default within 12 months. Neither approach dominated the other; they found situations in which each modeling approach provided better predictors of an individual's credit risk.

An important feature of the proportional hazards model is that it can readily be extended to handle time-varying covariates of the form $\mathbf{x}_i(t)$. In this case, one can write the partial likelihood as

$$L_P(\boldsymbol{\beta}) = \prod_{i=1}^{n} \left(\frac{\exp(\mathbf{x}_i'(y_i)\boldsymbol{\beta})}{\sum_{j \in R(y_i)} \exp(\mathbf{x}_j'(y_j)\boldsymbol{\beta})} \right)^{1-\delta_i}.$$

Using this likelihood is complex, although maximization can be readily accomplished with modern statistical software. Inference is complex because it can be difficult to disentangle observed time-varying covariates $\mathbf{x}_i(t)$ from the unobserved time-varying baseline hazard $\mathrm{h}_0(t)$. See the references in Section 14.6 for more details.

14.5 Recurrent Events

Recurrent events are event times that may occur repeatedly over time for each subject. For the ith subject, we use the notation $y_{i1} < y_{i2} < \cdots < y_{im_i}$ to denote the m_i event times. Among other applications, recurrent events can be used to model

- Warranty claims
- Payments from claims that have been reported to an insurance company

Table 14.2 Warranty Claims Frequency Distribution

Number of Claims	0	1	2	3	4	5+	Total
Number of Cars	13,987	1,243	379	103	34	29	15,775

Source: Cook and Lawless (2007, Table 1.3)

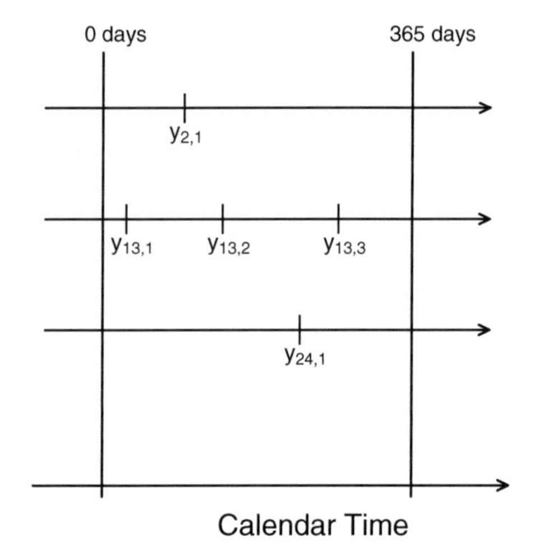

Figure 14.5 Figure illustrating (potentially) repeated warranty claims.

- Claims from events that have been incurred but not yet reported to an insurance company

---

Example: Warranty Automobile Claims. Cook and Lawless (2007) consider a sample of 15,775 automobiles that were sold and under warranty for 365 days. Warranties are guarantees of product reliability issued by the manufacturer. The warranty data are for one vehicle system (e.g., brakes or power train) and cover one year with a 12,000 mile limit on coverage. Table 14.2 summarizes the distribution of the 2,620 claims from this sample.

Table 14.2 shows that there are 1,788 (=15,775 − 13,987) automobiles with at least one ($m_i > 0$) claim. Figure 14.5 represents the structure of event times.

We present models of recurrent events using *counting processes*, specifically employ *Poisson processes*. Although there are alternative approaches, for statistical analyses the counting process concept seems the most fruitful (Cook and Lawless, 2007). Moreover, there is a strong historical connection of Poisson processes with actuarial ruin theory (Klugman, Panjer, and Willmot, 2008, Chapter 11) and so many actuaries are familiar with Poisson processes.

Let $N_i(t)$ be the number of events that have occurred by time t. Using algebra, we can write this as $N_i(t) = \sum_{j \geq 1} \mathrm{I}(y_{ij} \leq t)$. Because $N_i(t)$ varies with a subject's experience, it is a random variable for each fixed t. Further, when viewing the entire evolution of claims, $\{N_i(t), t \geq 0\}$, is known as a *stochastic process*.

A counting process is a special kind of stochastic process that describes counts that are monotonically increasing over time. A Poisson process is a special kind of counting process. If $\{N_i(t), t \geq 0\}$ is a Poisson process, then $N_i(t)$ has a Poisson distribution with mean, say, $\mathrm{h}_i(t)$ for each fixed t. In the counting process literature, it is customary to refer to $\mathrm{h}_i(t)$ as the *intensity function*.

Statistical inference for recurrent events can be conducted using Poisson processes and parametric maximum likelihood techniques. As discussed in Cook and Lawless (2007), the likelihood for the ith subject is based on the conditional probability density of the observed outcomes "m_i events, at times $y_{i1} < \cdots < y_{im_i}$". This yields the likelihood

$$Likelihood_i = \prod_{j=1}^{m_i} \{\mathrm{h}_i(y_{ij})\} \exp\left(- \int_0^{\infty} y_i(s) \mathrm{h}_i(s) ds \right), \qquad (14.5)$$

where $y_i(s)$ is a binary variable to indicate whether the ith subject is observed by time s.

To parameterize the intensity function $\mathrm{h}_i(t)$, we assume that we have available explanatory variables. For example, when examining warranty automobile claims, we might have available the make and model of the vehicle, or driver characteristics such as sex and age at time of purchase. We might have characteristics that are functions of the time of claim, such as the number of miles driven. Thus, we use the notation $\mathbf{x}_i(t)$ to denote the potential dependence of the explanatory variables on the event time.

Similar to equation (14.3), it is customary to write the intensity function as

$$h_i(t) = h_0(t) \exp(\mathbf{x}_i'(t) \boldsymbol{\beta}).$$

where $h_0(t)$ is the baseline intensity. For a full parametric specification, one would specify the baseline intensity in terms of several parameters. Then, one would use the intensity function $h_i(t)$ in the likelihood equation (14.5) that would serve as the basis for likelihood inference. As discussed in Cook and Lawless (2007), semiparametric approaches where the baseline is not fully parametrically specified (e.g., the PH model) are also possible.

14.6 Further Reading and References

As described in Collett (1994), the product-limit estimator had been used since the early part of the twentieth century. Greenwood (1926) established the formula for an approximate variance. The work of Kaplan and Meier (1958) is often associated with the product-limit estimator; they showed that it is a nonparametric maximum likelihood estimator of the survival function.

There are several good sources for further study of survival analysis, particularly in the biomedical sciences; Collett (1994) is a one such source. The text by Klein and Moeschberger (1997) was used for several years as required reading for the North American Society of Actuaries syllabus. Lawless (2003) provides an introduction from an engineering perspective. Lancaster (1990) discusses econometric issues. Hougaard (2000) provides an advanced treatment. Refer to Cook and Lawless (2007) for a book-length treatment of recurrent events.

Chapter References

Breslow, Norman (1974). Covariance analysis of censored survival data. *Biometrics* 30, 89–99.

Collett, D. (1994). *Modelling Survival Data in Medical Research.* Chapman & Hall, London.

Cook, Richard J., and Jerald F. Lawless (2007). *The Statistical Analysis of Recurrent Events.* Springer-Verlag, New York.

Cox, David R. (1972). Regression models and life-tables. *Journal of the Royal Statistical Society*, Ser. B, 34, 187–202.

Gourieroux, Christian, and Joann Jasiak (2007). *The Econometrics of Individual Risk.* Princeton University Press, Princeton, New Jersey.

Greenwood, M. (1926). The errors of sampling of the survivorship tables. *Reports on Public Health and Statistical Subjects*, 33, Appendix, HMSO, London.

Kaplan, E. L., and Meier, P. (1958). Nonparametric estimation from incomplete observations. *Journal of the American Statistical Association* 53, 457–81.

Kim, Yong-Duck, Dan R. Anderson, Terry L. Amburgey, and James C. Hickman (1995). The use of event history analysis to examine insurance insolvencies. *Journal of Risk and Insurance* 62, 94–110.

Klein, John P., and Melvin L. Moeschberger (1997). *Survival Analysis: Techniques for Censored and Truncated Data.* Springer-Verlag, New York.

Klugman, Stuart A., Harry H. Panjer, and Gordon E. Willmot (2008). *Loss Models: From Data to Decisions.* John Wiley & Sons, Hoboken, New Jersey.

Lancaster, Tony (1990). *The Econometric Analysis of Transition Data.* Cambridge University Press, New York.

Lawless, Jerald F. (2003). *Statistical Models and Methods for Lifetime Data, Second Edition.* John Wiley & Sons, New York.

Hougaard, Philip (2000). *Analysis of Multivariate Survival Data.* Springer-Verlag, New York.

Miller, Rupert G. (1997). *Survival Analysis.* John Wiley & Sons, New York.

Shumway, Tyler (2001). Forecasting bankruptcy more accurately: A simple hazard model. *Journal of Business* 74, 101–124.

Stepanova, Maria, and Lyn Thomas (2002). Survival analysis methods for personal loan data. *Operations Research* 50, no. 2, 277–90.

Zhou, Mai (2001). Understanding the Cox regression models with time-changing covariates. *American Statistician* 55, no. 2, 153–55.

15

Miscellaneous Regression Topics

Chapter Preview. This chapter provides a quick tour of several regression topics that an analyst is likely to encounter in different regression contexts. The goal of this chapter is to introduce these topics, provide definitions, and illustrate contexts in which the topics may be applied.

15.1 Mixed Linear Models

Although mixed linear models are an established part of statistical methodology, their use is not as widespread as regression in actuarial and financial applications. Thus, the section introduces this modeling framework, beginning with a widely used special case. After introducing the modeling framework, this section describes estimation of regression coefficients and variance components.

We begin with the *one-way random effects* model, with model equation

$$y_{it} = \mu + \alpha_i + \varepsilon_{it}, \qquad t = 1, \ldots, T_i, \;\; i = 1, \ldots, n. \qquad (15.1)$$

We may use this model to represent repeated observations of subject or group i. The subscript t is used to denote replications that may be over time or multiple group membership (e.g., several employees in a firm). Repeated observations over time were the focus of Chapter 10.

When there is only one observation per group so that $T_i = 1$, the disturbance term represents the unobservable information about the dependent variable. With repeated observations, we have an opportunity to capture unobservable characteristics of the group through the term α_i. Here, α_i is assumed to be a random variable and is known as a *random effect*. Another approach, introduced in Section 4.3, represented α_i as a parameter to be estimated using a categorical explanatory variable.

For this model, μ represents an overall mean, α_i the deviation from the mean due to unobserved group characteristics, and ε_{it} the individual response variation. We assume that $\{\alpha_i\}$ are i.i.d. with mean zero and variance σ_α^2. Further assume that $\{\varepsilon_{it}\}$ are i.i.d. with mean zero and variance σ^2 and are independent of α_i.

One extension of equation (15.1) is the basic random effects model described in Section 10.5, based on model equation

$$y_{it} = \alpha_i + \mathbf{x}_{it}'\boldsymbol{\beta} + \varepsilon_{it}. \qquad (15.2)$$

Mixed effects models are ones that include random as well as fixed effects.

In this extension, the overall mean μ is replaced by the regression function $\mathbf{x}_{it}'\boldsymbol{\beta}$. This model includes random effects (α_i) as well as fixed effects ($\mathbf{x}_{it}$). *Mixed effects* models are ones that include random and fixed effects.

Stacking the model equations in an appropriate fashion yields an expression for the *mixed linear model*

$$\mathbf{y} = \mathbf{Z}\boldsymbol{\alpha} + \mathbf{X}\boldsymbol{\beta} + \boldsymbol{\varepsilon}. \tag{15.3}$$

Here, $\mathbf{y}$ is a $N \times 1$ vector of dependent variables; $\boldsymbol{\varepsilon}$ is a $N \times 1$ vector of errors; $\mathbf{Z}$ and $\mathbf{X}$ are $N \times q$ and $N \times k$ known matrices of explanatory variables, respectively; and $\boldsymbol{\alpha}$ and $\boldsymbol{\beta}$ are $q \times 1$ and $k \times 1$ vectors of unknown parameters. In the mixed linear model, the $\boldsymbol{\beta}$ parameters are fixed (nonstochastic) and the $\boldsymbol{\alpha}$ parameters are random (stochastic).

For the mean structure, we assume $\mathrm{E}(\mathbf{y}|\boldsymbol{\alpha}) = \mathbf{Z}\boldsymbol{\alpha} + \mathbf{X}\boldsymbol{\beta}$ and $\mathrm{E}\,\boldsymbol{\alpha} = \mathbf{0}$, so that $\mathrm{E}\,\mathbf{y} = \mathbf{X}\boldsymbol{\beta}$. For the covariance structure, we assume $\mathrm{Var}(\mathbf{y}|\boldsymbol{\alpha}) = \mathbf{R}$, $\mathrm{Var}\,(\boldsymbol{\alpha}) = \mathbf{D}$, and $\mathrm{Cov}(\boldsymbol{\alpha}, \boldsymbol{\varepsilon}') = \mathbf{0}$. This yields $\mathrm{Var}\,\mathbf{y} = \mathbf{ZDZ}' + \mathbf{R} = \mathbf{V}$. In longitudinal applications, the matrix $\mathbf{R}$ is used to model the intrasubject serial correlation.

The mixed linear model is quite general and includes many models as special cases. For a book-length treatment of mixed linear models, see Pinheiro and Bates (2000). To illustrate, we return to the basic random effects model in equation (15.2). Stacking the replications from the ith group, we may write

$$\mathbf{y}_i = \alpha_i \mathbf{1}_i + \mathbf{X}_i\boldsymbol{\beta} + \boldsymbol{\varepsilon}_i,$$

where $\mathbf{y}_i = (y_{i1}, \ldots, y_{iT_i})'$ is the vector of dependent variables, $\boldsymbol{\varepsilon}_i = (\varepsilon_{i1}, \ldots, \varepsilon_{iT_i})'$ is the corresponding vector of disturbance terms, $\mathbf{X}_i = (\mathbf{x}_{i1}, \ldots, \mathbf{x}_{iT_i})'$ is the $T_i \times k$ matrix of explanatory variables, and $\mathbf{1}_i$ is a $T_i \times 1$ vectors of ones. Stacking the groups $i = 1, \ldots, n$ yields equation (15.3) with $\mathbf{y} = (\mathbf{y}_1', \ldots, \mathbf{y}_n')'$, $\boldsymbol{\varepsilon} = (\boldsymbol{\varepsilon}_1', \ldots, \boldsymbol{\varepsilon}_n')'$, and $\boldsymbol{\alpha} = (\alpha_1, \ldots, \alpha_n)'$,

$$\mathbf{X} = \begin{pmatrix} \mathbf{X}_1 \\ \vdots \\ \mathbf{X}_n \end{pmatrix} \quad \text{and} \quad \mathbf{Z} = \begin{pmatrix} \mathbf{1}_1 & \mathbf{0} & \cdots & \mathbf{0} \\ \mathbf{0} & \mathbf{1}_2 & \cdots & \mathbf{0} \\ \vdots & \vdots & \ddots & \vdots \\ \mathbf{0} & \mathbf{0} & \cdots & \mathbf{1}_n \end{pmatrix}.$$

Estimation of the mixed linear model proceeds in two stages. In the first stage, we estimate the regression coefficients $\boldsymbol{\beta}$, assuming knowledge of the variance-covariance matrix $\mathbf{V}$. Then, in the second stage, components of the variance-covariance matrix $\mathbf{V}$ are estimated.

15.1.1 Weighted Least Squares

In Section 5.7.3, we introduced the notion of *weighted* least squares estimates of the regression coefficients of the form

$$\mathbf{b}_{WLS} = \left(\mathbf{X}'\mathbf{W}\mathbf{X}\right)^{-1}\mathbf{X}'\mathbf{W}\mathbf{y}. \tag{15.4}$$

The $n \times n$ matrix $\mathbf{W}$ was chosen to be of the form $\mathbf{W} = diag(w_i)$ so that the ith diagonal element of $\mathbf{W}$ is a weight w_i. As introduced in Section 5.7.3, this allowed us to fit heteroscedastic regression models.

More generally, we may allow $\mathbf{W}$ to be any (symmetric) matrix (such that $\mathbf{X'WX}$ is invertible). This extension allows us to accommodate other types of dependencies that appear, for example, in mixed linear models. Assuming only that $\mathrm{E}\ \mathbf{y} = \mathbf{X}\boldsymbol{\beta}$ and $\mathrm{Var}\ \mathbf{y} = \mathbf{V}$, it is easy to establish

$$\mathrm{E}\ \mathbf{b}_{WLS} = \boldsymbol{\beta} \tag{15.5}$$

and

$$\mathrm{Var}\ \mathbf{b}_{WLS} = \left(\mathbf{X'WX}\right)^{-1}\left(\mathbf{X'WVWX}\right)\left(\mathbf{X'WX}\right)^{-1}. \tag{15.6}$$

Equation (15.5) indicates that $\mathbf{b}_{WLS}$ is an unbiased estimator of $\boldsymbol{\beta}$. Equation (15.6) is a basic result that is used for statistical inference, including evaluation of standard errors.

The best choice of the weight matrix is the inverse of the variance-covariance matrix so that $\mathbf{W} = \mathbf{V}^{-1}$. This choice results in the *generalized least squares estimator*, commonly denoted by the acronym GLS. The variance is

$$\mathrm{Var}\ \mathbf{b}_{GLS} = \left(\mathbf{X'V}^{-1}\mathbf{X}\right)^{-1}. \tag{15.7}$$

This is best in the sense that it can be shown that $\mathbf{b}_{GLS} = \left(\mathbf{X'V}^{-1}\mathbf{X}\right)^{-1}\mathbf{X'V}^{-1}\mathbf{y}$ has minimum variance among the class of all unbiased estimators of the parameter vector $\boldsymbol{\beta}$. This is property is known as the *Gauss-Markov theorem*, an extension for general variance-covariance matrices $\mathbf{V}$ of the property introduced in Section 3.2.3.

15.1.2 Variance Components Estimation

Generalized least squares estimation assumes that $\mathbf{V}$ is known, at least up to a scalar constant. Of course, it is unlikely that a general $n \times n$ matrix $\mathbf{V}$ could be estimated from n observations. However, estimation of special cases of $\mathbf{V}$ is possible and done routinely. Let $\boldsymbol{\tau}$ denote the vector of parameters that index $\mathbf{V}$; once $\boldsymbol{\tau}$ is known, the matrix $\mathbf{V}$ is fully specified. We call elements of $\boldsymbol{\tau}$ the *variance components*. For example, in our basic regression case, we have $\mathbf{V} = \sigma^2\mathbf{I}$, so that $\boldsymbol{\tau} = \sigma^2$. As another example, in the basic one-way random effects model, the variance structure is described by the variance components $\boldsymbol{\tau} = (\sigma^2, \sigma_\alpha^2)'$.

There are several methods for estimating variance components, some of which are likelihood based and others that use method of moments. These methods are readily available in statistical software. To give readers a feel for the computations involved, we briefly sketch the procedure based on maximum likelihood using normal distributions.

For normally distributed observations $\mathbf{y}$ with mean E $\mathbf{y} = \mathbf{X}\boldsymbol{\beta}$ and Var $\mathbf{y} = \mathbf{V} = \mathbf{V}(\boldsymbol{\tau})$, the logarithmic likelihood is given by

$$L(\boldsymbol{\beta}, \boldsymbol{\tau}) = -\frac{1}{2}\left[N \ln(2\pi) + \ln \det(\mathbf{V}(\boldsymbol{\tau})) + (\mathbf{y} - \mathbf{X}\boldsymbol{\beta})'\mathbf{V}(\boldsymbol{\tau})^{-1}(\mathbf{y} - \mathbf{X}\boldsymbol{\beta})\right]. \tag{15.8}$$

The generalized least squares estimator $\mathbf{b}_{GLS}$ is also the maximum likelihood estimator of $\boldsymbol{\beta}$.

This log-likelihood is to be maximized in terms of the parameters $\boldsymbol{\beta}$ and $\boldsymbol{\tau}$. In the first stage, we hold $\boldsymbol{\tau}$ fixed and maximize equation (15.8) over $\boldsymbol{\beta}$. Pleasant calculations show that $\mathbf{b}_{GLS}$ is, in fact, the maximum likelihood estimator of $\boldsymbol{\beta}$. Putting this into equation (15.8) yields the profile likelihood

$$L_P(\boldsymbol{\tau}) = L(\mathbf{b}_{GLS}, \boldsymbol{\tau}) \propto -\frac{1}{2}\left[\ln \det(\mathbf{V}(\boldsymbol{\tau})) + (\mathbf{y} - \mathbf{X}\mathbf{b}_{GLS})'\mathbf{V}(\boldsymbol{\tau})^{-1}(\mathbf{y} - \mathbf{X}\mathbf{b}_{GLS})\right], \tag{15.9}$$

where we have dropped constants that do not depend on $\boldsymbol{\tau}$. (The symbol $\propto$ means "is proportional to.")

To implement this two-stage procedure, computer software will typically use ordinary least squares (OLS) estimates $\mathbf{b}$ for starting values. Then, in the second stage, estimates of $\boldsymbol{\tau}$ are determined by iterative methods (numerical optimization) by finding the values of $\boldsymbol{\tau}$ that maximize $L(\mathbf{b}, \boldsymbol{\tau})$. These estimates are then used to update the regression coefficient estimates using weighted least squares. This process is continued until convergence is achieved.

There are two advantages to this two-stage procedure. First, by decoupling the regression from the variance component parameters estimation, we can apply any method that we like to the variance components and then "plug-in" the estimates into the regression component (estimated) generalized least squares estimation. Second, we have a closed-form expression for the regression estimates. This is faster computationally than the iterative methods required by general optimization routines.

15.1.3 Best Linear Unbiased Prediction

This section develops *best linear unbiased predictors* (BLUPs) in the context of mixed linear models. We introduce BLUPs as the minimum mean square error predictor of a random variable, w. This development is originally attributable to Goldberger (1962), who coined the phrase "best linear unbiased predictor." The acronym BLUP was first used by Henderson (1973).

The generic goal is to *predict* a random variable w, such that E $w = \boldsymbol{\lambda}'\boldsymbol{\beta}$ and Var $w = \sigma_w^2$. Denote the covariance between w and $\mathbf{y}$ as the $1 \times N$ vector $\mathrm{Cov}(w, \mathbf{y}) = \mathrm{E}\{(w - \mathrm{E}w)(\mathbf{y} - \mathrm{E}\mathbf{y})'\}$. The choice of w, and thus $\boldsymbol{\lambda}$ and σ_w^2, will depend on the application at hand.

Under these assumptions, it can be shown that the BLUP of w is

$$w_{BLUP} = \boldsymbol{\lambda}'\mathbf{b}_{GLS} + \mathrm{Cov}(w, \mathbf{y})\mathbf{V}^{-1}(\mathbf{y} - \mathbf{X}\mathbf{b}_{GLS}). \tag{15.10}$$

The BLUP predictors are optimal, assuming the variance components implicit in $\mathbf{V}$ and $\mathrm{Cov}(w, \mathbf{y})$ are known. Applications of BLUP typically require that the variance components be estimated, as described in Section 15.1.2. BLUPs with estimated variance components are known as *empirical BLUPs*, or EBLUPs.

There are three important types of choice for w:

1. $w = \varepsilon$, resulting in so-called BLUP residuals
2. Random effects, such as $\boldsymbol{\alpha}$
3. Future observations, resulting in optimal forecasts

For the first choice, you will find that BLUP residuals are regularly coded in statistical software packages that fit linear mixed models. For the second choice, by letting w be an arbitrary linear combination of random effects, it can be shown that the BLUP predictor of $\boldsymbol{\alpha}$ is

$$\mathbf{a}_{BLUP} = \mathbf{DZ}'\mathbf{V}^{-1}(\mathbf{y} - \mathbf{X}\mathbf{b}_{GLS}). \tag{15.11}$$

For examples of the third choice, forecasting with linear mixed models, we refer to Frees (2004, Chapters 4 and 8).

To consider an application of equation (15.11), consider the following.

Special Case: One-Way Random Effects Model. Consider the model based on equation (15.1) and suppose that we want to estimate the conditional mean of the ith group, $w = \mu + \alpha_i$. Then, direct calculations (see Frees, 2004, Chapter 4) based on equation (15.11) show that the BLUP is

$$\zeta_i \bar{y}_i + (1 - \zeta_i) m_{\alpha,GLS}, \tag{15.12}$$

with weight $\zeta_i = T_i/(T_i + \sigma^2/\sigma_\alpha^2)$ and *GLS* estimate of μ, $m_{\alpha,GLS} = \sum_i \zeta_i \bar{y}_i / \sum_i \zeta_i$. In Chapter 18, we will interpret ζ_i to be a *credibility factor*.

15.2 Bayesian Regression

With Bayesian statistical models, one views both the model parameters and the data as random variables. In this section, we use a specific type of Bayesian model, the *normal linear hierarchical model* discussed by, for example, Gelman et al. (2004). As with the two-stage sampling scheme described in Section 3.3.1, the hierarchical linear model is one that is specified in stages. Specifically, we consider the following two-level hierarchy:

1. Given the parameters $\boldsymbol{\alpha}$ and $\boldsymbol{\beta}$, the response model is $\mathbf{y} = \mathbf{Z}\boldsymbol{\alpha} + \mathbf{X}\boldsymbol{\beta} + \boldsymbol{\varepsilon}$. This level is an ordinary (fixed) linear model that was introduced in Chapters 3 and 4. Specifically, we assume that the vector of responses $\mathbf{y}$ conditional on $\boldsymbol{\alpha}$ and $\boldsymbol{\beta}$ is normally distributed and that $\mathrm{E}\,(\mathbf{y}|\boldsymbol{\alpha}, \boldsymbol{\beta}) = \mathbf{Z}\boldsymbol{\alpha} + \mathbf{X}\boldsymbol{\beta}$ and Var $(\mathbf{y}|\boldsymbol{\alpha}, \boldsymbol{\beta}) = \mathbf{R}$.
2. Assume that $\boldsymbol{\alpha}$ is distributed normally with mean $\boldsymbol{\mu}_\alpha$ and variance $\mathbf{D}$ and that $\boldsymbol{\beta}$ is distributed normally with mean $\boldsymbol{\mu}_\beta$ and variance $\boldsymbol{\Sigma}_\beta$, each independent of the other.

The technical differences between the mixed linear model and the normal hierarchical linear model are as follows:

- In the mixed linear model, $\boldsymbol{\beta}$ is an unknown, fixed parameter, whereas in the normal hierarchical linear model, $\boldsymbol{\beta}$ is a random vector.
- The mixed linear model is distribution free, whereas distributional assumptions are made in each stage of the normal hierarchical linear model.

Moreover, there are important differences in interpretation. To illustrate, suppose that $\boldsymbol{\beta} = \mathbf{0}$ with probability of one. In the classic non-Bayesian, also known as the *frequentist*, interpretation, we think of the distribution of $\{\boldsymbol{\alpha}\}$ as representing the likelihood of drawing a realization of $\boldsymbol{\alpha}_i$. The likelihood interpretation is most suitable when we have a population of firms or people and each realization is a draw from that population. In contrast, in the Bayesian case, one interprets the distribution of $\{\boldsymbol{\alpha}\}$ as representing the knowledge that one has of this parameter. This distribution may be subjective and allows the analyst a formal mechanism to inject his or her assessments into the model. In this sense, the frequentist interpretation can be regarded as a special case of the Bayesian framework.

The joint distribution of $(\boldsymbol{\alpha}', \boldsymbol{\beta}')'$ is known as the *prior* distribution. To summarize, the joint distribution of $(\boldsymbol{\alpha}', \boldsymbol{\beta}', \mathbf{y}')'$ is

$$\begin{pmatrix} \boldsymbol{\alpha} \\ \boldsymbol{\beta} \\ \mathbf{y} \end{pmatrix} \sim N\left(\begin{pmatrix} \boldsymbol{\mu}_{\alpha} \\ \boldsymbol{\mu}_{\beta} \\ \mathbf{Z}\boldsymbol{\mu}_{\alpha} + \mathbf{X}\boldsymbol{\mu}_{\beta} \end{pmatrix}, \begin{pmatrix} \mathbf{D} & \mathbf{0} & \mathbf{DZ}' \\ \mathbf{0} & \boldsymbol{\Sigma}_{\beta} & \boldsymbol{\Sigma}_{\beta}\mathbf{X}' \\ \mathbf{ZD} & \mathbf{X}\boldsymbol{\Sigma}_{\beta} & \mathbf{V} + \mathbf{X}\boldsymbol{\Sigma}_{\beta}\mathbf{X}' \end{pmatrix} \right), \quad (15.13)$$

where $\mathbf{V} = \mathbf{R} + \mathbf{ZDZ}'$.

The distribution of parameters given the data is known as the *posterior distribution*. To calculate this conditional distribution, we use standard results from multivariate analysis. Specifically, the posterior distribution of $(\boldsymbol{\alpha}', \boldsymbol{\beta}')'$ given $\mathbf{y}$ is normal. It is not hard to verify that the conditional mean is

$$\mathrm{E}\begin{pmatrix} \boldsymbol{\alpha} \\ \boldsymbol{\beta} \end{pmatrix} \mid \mathbf{y} = \begin{pmatrix} \boldsymbol{\mu}_{\alpha} + \mathbf{DZ}'(\mathbf{V} + \mathbf{X}\boldsymbol{\Sigma}_{\beta}\mathbf{X}')^{-1}(\mathbf{y} - \mathbf{Z}\boldsymbol{\mu}_{\alpha} - \mathbf{X}\boldsymbol{\mu}_{\beta}) \\ \boldsymbol{\mu}_{\beta} + \boldsymbol{\Sigma}_{\beta}\mathbf{X}'(\mathbf{V} + \mathbf{X}\boldsymbol{\Sigma}_{\beta}\mathbf{X}')^{-1}(\mathbf{y} - \mathbf{Z}\boldsymbol{\mu}_{\alpha} - \mathbf{X}\boldsymbol{\mu}_{\beta}) \end{pmatrix}. \quad (15.14)$$

Up to this point, the treatment of parameters $\boldsymbol{\alpha}$ and $\boldsymbol{\beta}$ has been symmetric. In some applications, such as with longitudinal data, one typically has more information about the global parameters $\boldsymbol{\beta}$ than about subject-specific parameters $\boldsymbol{\alpha}$. To see how the posterior distribution changes depending on the amount of information available, we consider two extreme cases.

First, consider the case $\boldsymbol{\Sigma}_{\beta} = \mathbf{0}$, so that $\boldsymbol{\beta} = \boldsymbol{\mu}_{\beta}$ with probability one. Intuitively, this means that $\boldsymbol{\beta}$ is precisely known, generally from collateral information. Then, from equation (15.14), we have

$$\mathrm{E}\,(\boldsymbol{\alpha}|\mathbf{y}) = \boldsymbol{\mu}_{\alpha} + \mathbf{DZ}'\mathbf{V}^{-1}(\mathbf{y} - \mathbf{Z}\boldsymbol{\mu}_{\alpha} - \mathbf{X}\boldsymbol{\beta}). \quad (15.15)$$

Assuming that $\boldsymbol{\mu}_{\alpha} = \mathbf{0}$, the best linear unbiased estimator of $\mathrm{E}\,(\boldsymbol{\alpha}|\mathbf{y})$ is

$$\mathbf{a}_{BLUP} = \mathbf{DZ}'\mathbf{V}^{-1}(\mathbf{y} - \mathbf{X}\mathbf{b}_{GLS}).$$

Recall from equation (15.11) that $\mathbf{a}_{BLUP}$ is also the best linear unbiased predictor in the frequentist (non-Bayesian) model framework.

Second, consider the case where $\boldsymbol{\Sigma}_{\beta}{}^{-1} = \mathbf{0}$. In this case, prior information about the parameter $\boldsymbol{\beta}$ is vague; this is known as using a *diffuse* prior. In this case, one can check that

$$\mathrm{E}\left(\boldsymbol{\alpha} \mid \mathbf{y}\right) \rightarrow \mathbf{a}_{BLUP},$$

as $\boldsymbol{\Sigma}_{\beta}{}^{-1} \rightarrow \mathbf{0}$. (See, e.g., Frees, 2004, Section 4.6.)

Thus, it is interesting that in both extreme cases, we arrive at the statistic $\mathbf{a}_{BLUP}$ as a predictor of $\boldsymbol{\alpha}$. This analysis assumes that $\mathbf{D}$ and $\mathbf{R}$ are matrices of fixed parameters. It is also possible to assume distributions for these parameters; typically, independent Wishart distributions are used for $\mathbf{D}^{-1}$ and $\mathbf{R}^{-1}$, as these are conjugate priors. Alternatively, one can estimate $\mathbf{D}$ and $\mathbf{R}$ using methods described in Section 15.1. The general strategy of substituting point estimates for certain parameters in a posterior distribution is called *empirical Bayes estimation*.

To examine intermediate cases, we look to the following special case. Generalizations may be found in the work of Luo, Young, and Frees (2001).

Special Case: One-Way Random Effects Model. We return to the model considered in equation (15.2) and, for simplicity, assume balanced data so that $T_i = T$. The goal is to determine the posterior distributions of the parameters. For illustrative purposes, we focus on the posterior means. Thus, rewriting equation (15.2), the model is

$$y_{it} = \beta + \alpha_i + \varepsilon_{it},$$

where we use the random $\beta \sim N(\mu_\beta, \sigma_\beta^2)$ in lieu of the fixed mean μ. The prior distribution of α_i is independent with $\alpha_i \sim N(0, \sigma_\alpha^2)$.

Using equation (15.14), we obtain the posterior mean of β,

$$\hat{\beta} = \mathrm{E}(\beta \mid \mathbf{y}) = \left(\frac{1}{\sigma_\beta^2} + \frac{nT}{\sigma_\varepsilon^2 + T\sigma_\alpha^2}\right)^{-1} \left(\frac{nT}{\sigma_\varepsilon^2 + T\sigma_\alpha^2}\bar{y} + \frac{\mu}{\sigma_\beta^2}\right), \tag{15.16}$$

after some algebra. Thus, $\hat{\beta}$ is a weighted average of the sample mean, $\bar{y}$, and the prior mean, μ_β. It is easy to see that $\hat{\beta}$ approaches the sample mean as $\sigma_\beta^2 \rightarrow \infty$, that is, as prior information about β becomes "vague." Conversely, $\hat{\beta}$ approaches the prior mean μ_β as $\sigma_\beta^2 \rightarrow 0$, that is, as information about becomes "precise."

Similarly, using equation (15.14), the posterior mean of α_i is

$$\hat{\alpha_i} = \mathrm{E}\left(\alpha_i \mid \mathbf{y}\right) = \zeta\left[(\bar{y}_i - \mu_\beta) - \zeta_\beta(\bar{y} - \mu_\beta)\right]$$

where we have that

$$\zeta = \frac{T\sigma_\alpha^2}{\sigma_\varepsilon^2 + T\sigma_\alpha^2},$$

and define

$$\zeta_\beta = \frac{nT\sigma_\beta^2}{\sigma_\varepsilon^2 + T\sigma_\alpha^2 + nT\sigma_\beta^2}.$$

Note that ζ_β measures the precision of knowledge about β. Specifically, we see that ζ_β approaches one as $\sigma_\beta^2 \to \infty$ and approaches zero as $\sigma_\beta^2 \to 0$.

Combining these two results, we have that

$$\hat{\alpha}_i + \hat{\beta} = (1 - \zeta_\beta)\left[(1-\zeta)\mu_\beta + \zeta\bar{y}_i\right] + \zeta_\beta\left[(1-\zeta)\bar{y} + \zeta\bar{y}_i\right].$$

Thus, if our knowledge of the distribution of β is vague, then $\zeta_\beta = 1$ and the predictor reduces to the expression in equation (15.12) (for balanced data). Conversely, if our knowledge of the distribution of β is precise, then $\zeta_\beta = 0$ and the predictor reduces to the expression given in Chapter 18. With the Bayesian formulation, we may entertain situations in which knowledge is available though imprecise.

To summarize, there are several advantages of the Bayesian approach. First, one can describe the entire distribution of parameters conditional on the data. This allows one, for example, to provide probability statements regarding the likelihood of parameters. Second, this approach allows analysts to blend information known from other sources with the data in a coherent manner. In our development, we assumed that information may be known through the vector of $\boldsymbol{\beta}$ parameters, with their reliability control through the dispersion matrix $\boldsymbol{\Sigma}_\beta$. Values of $\boldsymbol{\Sigma}_\beta = \mathbf{0}$ indicate complete faith in values of $\boldsymbol{\mu}_\beta$, whereas values of $\boldsymbol{\Sigma}_\beta^{-1} = \mathbf{0}$ indicate complete reliance on the data in lieu of prior knowledge.

Third, the Bayesian approach provides for a unified approach for estimating $(\boldsymbol{\alpha}, \boldsymbol{\beta})$. Section 15.1, on non-Bayesian methods, required a separate subsection on variance components estimation. In contrast, in Bayesian methods, all parameters can be treated in a similar fashion. This is convenient for explaining results to consumers of the data analysis. Fourth, Bayesian analysis is particularly useful for forecasting future responses.

15.3 Density Estimation and Scatter plot Smoothing

When exploring a variable or relationships between two variables, one often wishes to get an overall idea of patterns without imposing strong functional relationships. Typically graphical procedures work well because we can comprehend potentially nonlinear relationships more readily and visually than with numerical summaries. This section introduces *(kernel) density estimation* to visualize the distribution of a variable and *scatter plot smoothing* to visualize the relationship between two variables.

To get a quick impression of the distribution of a variable, a histogram is easy to compute and interpret. However, as suggested in Chapter 1, changing the location and size of rectangles that comprise the histogram can give viewers different impression of the distribution. To introduce an alternative, suppose that we have a random sample $y_1, \ldots, y_n$ from a probability density function $\mathrm{f}(\cdot)$. We define the *kernel density estimator* as

$$\hat{\mathrm{f}}(y) = \frac{1}{nb_n} \sum_{i=1}^{n} \mathrm{k}\left(\frac{y - y_i}{b_n}\right),$$

where b_n is a small number called a *bandwidth* and $\mathrm{k}(\cdot)$ is a probability density function called a *kernel*.

To develop intuition, we first consider the case where the kernel $\mathrm{k}(\cdot)$ is a probability density function for a uniform distribution on $(-1, 1)$. For the uniform kernel, the kernel density estimate counts the number of observations y_i that are within b_n units of y, and then expresses the density estimate as the count divided by the sample size times the rectangle width (i.e., the count divided by $n \times 2b_n$). In this way, it can be viewed as a "local" histogram estimator in the sense that the center of the histogram depends on the argument y.

There are several possibilities for the kernel. Some widely used choices are the following:

- The uniform kernel, $\mathrm{k}(u) = \frac{1}{2}$ for $-1 \leq u \leq 1$ and 0 otherwise
- The "Epanechikov" kernel, $\mathrm{k}(u) = \frac{3}{4}(1 - u^2)$ for $-1 \leq u \leq 1$ and 0 otherwise
- The Gaussian kernel, $\mathrm{k}(u) = \phi(u)$ for $-\infty < u < \infty$, the standard normal density function

The Epanechnikov kernel is a smoother version that uses a quadratic polynomial so that discontinuous rectangles are not used. The Gaussian kernel is even more continuous in the sense that the domain is no longer plus or minus b_n but is the whole real line.

Ⓡ **Empirical**
Filename is "WiscNursingHome"

The bandwidth b_n controls the amount of averaging. To see the effects of different bandwidth choices, we consider a dataset on nursing home utilization that will be introduced in Section 17.3.2. Here, we consider occupancy rates, a measure of nursing home utilization. A value of 100 means full occupancy, but because of the way this measure is constructed, it is possible for values to exceed 100. Specifically, there are $n = 349$ occupancy rates that are displayed in Figure 15.1. Both figures use a Gaussian kernel. The left-hand panel is based on a bandwidth of 0.1. This panel appears very ragged; the relatively small bandwidth means that there is little averaging being done. For the outlying points, each spike represents a single observation. In contrast, the right-hand panel is based on a bandwidth of 1.374. In comparison to the left-hand panel, this picture displays a smoother picture, allowing the analyst to search for patterns and not be distracted by jagged edges. From this panel, we can readily see that most of the mass is less than 100%. Moreover, the distribution is left skewed, with values of 100 to 120 being rare.

The bandwidth 1.374 was selected using an automatic procedure built into the software. These automatic procedures choose the bandwidth to find the best

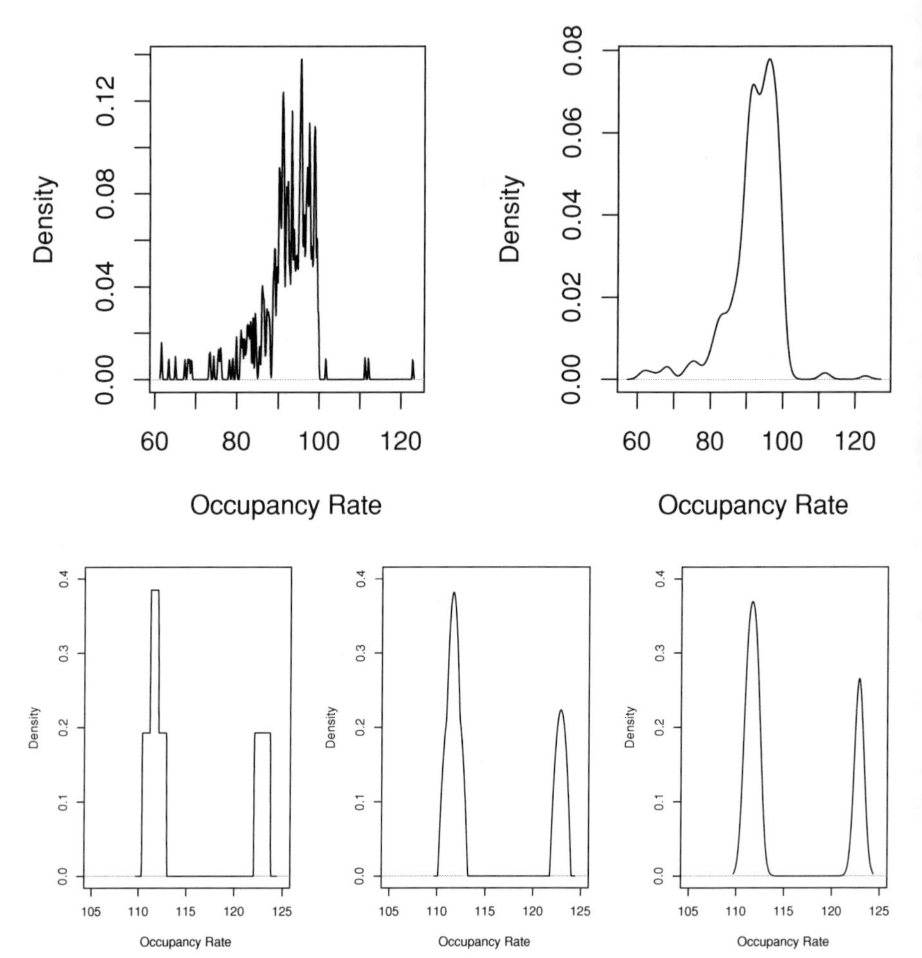

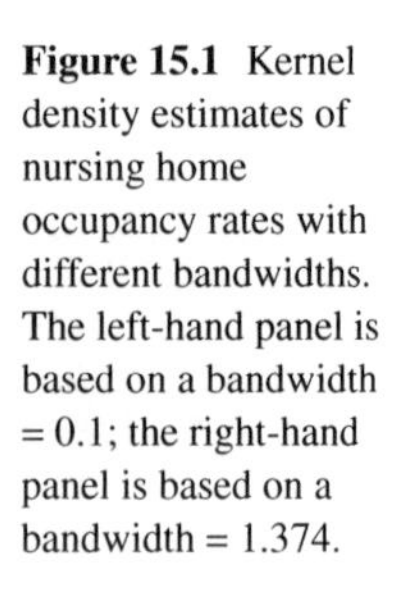

Figure 15.1 Kernel density estimates of nursing home occupancy rates with different bandwidths. The left-hand panel is based on a bandwidth = 0.1; the right-hand panel is based on a bandwidth = 1.374.

Figure 15.2 Kernel density estimates of nursing home occupancy rates with different kernels. From left to right, the panels use the uniform, Epanechnikov, and Gaussian kernels.

trade-off between the accuracy and the smoothness of the estimates. (For this figure, we used the statistical software "R," which has Silverman's procedure built in.)

Kernel density estimates also depend on the choice of the kernel, although this is typically much less important in applications than the choice of the bandwidth. To show the effects of different kernels, we show only the $n = 3$ occupancy rates that exceeded 110 in Figure 15.2. The left-hand panel shows the stacking of rectangular histograms based on the uniform kernel. The smoother Epanechnikov and Gaussian kernels in the middle and right-hand panels are visually indistinguishable. Unless you are working with very small sample sizes, you will usually not need to be concerned about the choice of the kernel. Some analysts prefer the uniform kernel because of its interpretability, some prefer the Gaussian because of its smoothness, and some prefer the Epanechnikov kernel as a reasonable compromise.

Some *scatter plot smoothers*, which show relationships between an x and a y, can also be described in terms of kernel estimation. Specifically, a kernel estimate of the regression function $\mathrm{E}(y|x)$ is

$$\hat{\mathrm{m}}(x) = \frac{\sum_{i=1}^{n} w_{i,x} y_i}{\sum_{i=1}^{n} w_{i,x}}$$

with the local weight $w_{i,x} = \mathrm{k}((x_i - x)/b_n)$. This is the now-classic Nadaraya-Watson estimator (see, e.g., Ruppert, Wand, and Carroll, 2003).

More generally, for a pth-order local polynomial fit, consider finding parameter estimates $\beta_0, \ldots, \beta_p$ that minimize

$$\sum_{i=1}^{n} \left\{ y_i - \beta_0 - \cdots - \beta_p (x_i - x)^p \right\}^2 w_{i,x}. \qquad (15.17)$$

The best value of the intercept β_0 is taken to be the estimate of the regression function $\mathrm{E}(y|x)$. Ruppert, Wand, and Carroll (2003) recommend values of $p = 1$ or 2 for most applications (the choice $p = 0$ yields the Nadaraya-Watson estimator). As a variation, taking $p = 1$ and letting the bandwidth vary so that the number of points used to estimate the regression function is fixed results in the *lowess* estimator (for "local regression") due to Cleveland (see, e.g., Ruppert, Wand, and Carroll, 2003).

As an example, we used the lowess estimator in Figure 6.11 to get a sense of the relationship between the residuals and the riskiness of an industry as measured by INDCOST. As an analyst, you will find that kernel density estimators and scatter plot smoothers are straightforward to use when searching for patterns and developing models.

15.4 Generalized Additive Models

Classic linear models are based on the regression function

$$\mu = \mathrm{E}(y|x_1, \ldots, x_k) = \beta_0 + \sum_{j=1}^{k} \beta_j x_j.$$

With a generalized linear model (GLM), we have seen that we can substantially extend applications through a function that links the mean to the systematic component,

$$\mathrm{g}(\mu) = \beta_0 + \sum_{j=1}^{k} \beta_j x_j,$$

from equation (13.1). As in linear models, the systematic component is linear in the parameters β_j, not necessarily the underlying explanatory variables. For example, we have seen that we can use polynomial functions (e.g., x^2 in Chapter 3), trigonometric functions (e.g., $\sin x$ in Chapter 8), binary and categorizations in Chapter 4 and the "broken-stick" (piecewise linear) representation in Section 3.5.2.

The *generalized additive model* (*GAM*) extends the GLM by allowing each explanatory variable to be replaced by a function that may be nonlinear,

$$\mathrm{g}(\mu) = \beta_0 + \sum_{j=1}^{k} \beta_j \ \mathrm{m}_j(x_j). \tag{15.18}$$

Here, the function $\mathrm{m}_j(\cdot)$ may differ by explanatory variable. Depending on the application, $\mathrm{m}_j(\cdot)$ may include the traditional parametric specifications (e.g., polynomials and categorizations) as well as more flexible nonparametric specifications such as the scatter plot smoothers introduced in Section 15.3.

For example, suppose that we have a large insurance database and want to model the probability of a claim. Then, we might use the model

$$\ln\left(\frac{\pi}{1-\pi}\right) = \beta_0 + \sum_{j=1}^{k} \beta_j x_j + \mathrm{m}(z).$$

The left-hand side is the usual logit link function used in logistic regression, with π being the probability of a claim. For the right-hand side, we might consider a host of rating variables such as territory, sex, and type of vehicle or house (depending on the coverage) that are included in the linear component $\beta_0 + \sum_{j=1}^{k} \beta_j x_j$. The additional variable z is some continuous variable (e.g., age) that we wish to allow for the possibility of nonlinear effects. For the function $\mathrm{m}(z)$, we could use a pth-order polynomial fit with discontinuities at several ages, such as in equation (15.17).

This is known as a *semiparametric* model, in that the systematic component consists of parametric ($\beta_0 + \sum_{j=1}^{k} \beta_j x_j$) as well as nonparametric ($\mathrm{m}(z)$) pieces. Although we do not present the details here, modern statistical software allows the simultaneous estimation of parameters from both components. For example, the statistical software SAS implements generalizes additive models in its PROC GAM procedure as does the software R through the VGAM package.

The specification of the GAM in equation (15.18) is quite general. For a narrower class, the choice of $\mathrm{g}(\cdot)$ as the identity function yields the *additive model*. Although general, the nonparametric forms of $\mathrm{m}_j(\cdot)$ make the model more flexible, yet the additivity allows us to interpret the model in much the same way as before. Readers interested in further information about GAMs will find Ruppert, Wand, and Carroll (2003) and Hastie, Tibshirani, and Freedman (2001) to be useful resources.

15.5 Bootstrapping

The *bootstrap* is a general tool for assessing the distribution of a statistic. We first describe the general procedure and then discuss ways to implement it in a regression context.

Suppose that we have an i.i.d. sample $\{z_1, \ldots, z_n\}$ from a population. From these data, we wish to understand the reliability of a statistic $\mathrm{S}(z_1, \ldots, z_n)$. To calculate a bootstrap distribution, we compute the following two steps:

1. *Bootstrap sample*: Generate an i.i.d sample of size n, $\{z^*_{1r}, \ldots, z^*_{nr}\}$, from $\{z_1, \ldots, z_n\}$.
2. *Bootstrap replication*: Calculate the bootstrap replication, $S^*_r = \mathrm{S}(z^*_{1r}, \ldots, z^*_{nr})$.

Repeat steps (i) and (ii) $r = 1, \ldots, R$ times, where R is a large number of replications. In the first step, the bootstrap sample is randomly drawn from the original sample with replacement. When repeating steps (1) and (2), the bootstrap samples are independent of one another, conditional on the original sample $\{z_1, \ldots, z_n\}$. The resulting *bootstrap distribution*, $\{S^*_1, \ldots, S^*_R\}$, can be used to assess the distribution of the statistic S.

There are three variations of this basic procedure used in regression. In the first variation, we treat $z_i = (y_i, \mathbf{x}_i)$, and use the basic bootstrap procedure. This variation is known as *resampling pairs*.

In the second variation, we treat the regression residuals as the "original" sample and create a bootstrap sample by sampling the residuals. This variation is known as *resampling residuals*. Specifically, consider a generic regression model of the form $y_i = \mathrm{F}(\mathbf{x}_i, \boldsymbol{\theta}, \varepsilon_i)$, where $\boldsymbol{\theta}$ represents a vector of parameters. Suppose that we estimate this model and compute residuals $e_i, i = 1, \ldots, n$. In Section 13.5, we denoted the residuals as $e_i = \mathrm{R}(y_i; \mathbf{x}_i, \widehat{\boldsymbol{\theta}})$, where the function R was determined by the model form and $\widehat{\boldsymbol{\theta}}$ represented the estimated vector of parameters. The residuals may be the raw residuals, Pearson residuals, or some other choice.

Using a bootstrap residual e^*_{jr}, we can create a pseudoresponse:

$$y^*_{jr} = \mathrm{F}(\mathbf{x}_i, \widehat{\boldsymbol{\theta}}, e^*_{jr}).$$

We can then use the set of pseudo-observations $\{(y^*_{1r}, \mathbf{x}_1), \ldots, (y^*_{nr}, \mathbf{x}_n)\}$ to calculate the bootstrap replication S^*_r. As earlier, the resulting bootstrap distribution, $\{S^*_1, \ldots, S^*_R\}$, can be used to assess the distribution of the statistic S.

Comparing these two options, the strengths of the first variation are that it employs fewer assumptions and is simpler to interpret. The limitation is that it uses a *different* set of explanatory variables $\{\mathbf{x}^*_{1r}, \ldots, \mathbf{x}^*_{nr}\}$ in the calculation of each bootstrap replication. Some analysts reason that their inference about the statistic S is conditional on the observed explanatory variables $\{\mathbf{x}_1, \ldots, \mathbf{x}_n\}$ and using a different set attacks a problem that is not of interest. The second variation addresses this, but at a cost of slightly less generality. In this variation, there is a stronger assumption that the analyst has correctly identified the model and that the disturbance process $\varepsilon_i = \mathrm{R}(y_i; \mathbf{x}_i, \boldsymbol{\theta})$ is i.i.d.

The third variation is known as a *parametric bootstrap*. Here, we assume that the disturbances, and hence the original dependent variables, come from a model that is known up to a vector of parameters. For example, suppose that we want the accuracy of a statistic S from a Poisson regression. As described in Chapter 12, we assume that $y_i \sim Poisson(\mu_i)$, where $\mu_i = \exp(\mathbf{x}_i'\boldsymbol{\beta})$. The estimate of the regression parameters is $\mathbf{b}$, and so the estimated mean is $\widehat{\mu}_i = \exp(\mathbf{x}_i'\mathbf{b})$. From this, we can simulate to create a set of pseudoresponses

$$y_{ir}^* \sim Poisson(\widehat{\mu}_i), i = 1, \ldots, n, \quad r = 1, \ldots, R.$$

These pseudoresponses can be used to form the rth bootstrap sample, $\{(y_{1r}^*, \mathbf{x}_1), \ldots, (y_{nr}^*, \mathbf{x}_n)\}$, and from this, the bootstrap replication, S_r^*. Thus, the main difference between the parametric bootstrap and the first two variations is that we simulate from a distribution (the Poisson, in this case), not from an empirical sample. The parametric bootstrap is easy to interpret and explain because the procedure is similar to the usual Monte Carlo simulation (see, e.g., Klugman et al., 2008). The difference is that with the bootstrap, we use the estimated parameters to calibrate the bootstrap sampling distribution, whereas this distribution is assumed known in Monte Carlo simulation.

There are two common ways to summarize the accuracy of the statistic S using the bootstrap distribution, $\{S_1^*, \ldots, S_R^*\}$. The first, a model-free approach, involves using the percentiles of the bootstrap distribution to create a confidence interval for S. For example, we might use the 2.5th and 97.5th percentiles of $\{S_1^*, \ldots, S_R^*\}$ for a 95% confidence interval for S. For the second, one assumes some distribution for S_r^*, typically approximate normality. With this approach, one can estimate a mean and standard deviation to get the usual confidence interval for S.

Special Case: Bootstrapping Loss Reserves. England and Verrall (2002) discuss bootstrapping loss reserves. As we will see in Chapter 19, by assuming that losses follow an overdisperse Poisson, predictions for loss reserves can be obtained by a simple mechanistic procedure known as the *chain-ladder* technique. Parametric bootstrapping is not available for an overdisperse Poisson because, as we saw in Chapter 12, it is not a true probability distribution. Instead, England and Verrall show how to use residual resampling, employing Pearson residuals.

In many instances, computation of the bootstrap replication S_r^* of the statistic can be computationally intensive and require specialized software. However, as pointed out by England and Verrall, the case of loss reserves with an overdisperse Poisson is straightforward. One essentially uses the chain-ladder technique to estimate model parameters and calculate Pearson residuals. Then, one simulates from the residuals and creates pseudoresponses and bootstrap distributions. Because simulation is widely available, the entire procedure can be readily mechanized to work with standard spreadsheet packages, without the need for statistical software. See Appendix 3 of England and Verrall (2002) for additional details on the algorithm.

15.6 Further Reading and References

The formula in equations (15.10) does not account for the uncertainty in variance component estimation. Inflation factors that account for this additional uncertainty have been proposed (Kackar and Harville, 1984), but they tend to be small, at least for data sets commonly encountered in practice. McCulloch and Searle (2001) provide further discussions.

Silverman (1986) is a now-classic introduction to density estimation.

Ruppert, Wand and Carroll (2003) provide an excellent book-length introduction to scatter plot smoothing. Moreover, they provide a complete discussion of spline-based smoothers, an alternative to local polynomial fitting.

Efron and Tibshirani (1991) is a now-classic introduction to the bootstrap.

Chapter References

Efron, Bradley, and Robert Tibshirani (1991). *An Introduction to the Bootstrap.* Chapman & Hall, London.

England, Peter D., and Richard J. Verrall (2002). Stochastic claims reserving in general insurance. *British Actuarial Journal* 8, 443–544.

Frees, Edward W. (2004). *Longitudinal and Panel Data: Analysis and Applications in the Social Sciences.* Cambridge University Press, New York.

Frees, Edward W., Virginia R. Young, and Yu Luo (2001). Case studies using panel data models. *North American Actuarial Journal* 5, no. 4, 24–42.

Gelman, A., J. B. Carlin, H. S. Stern, and D. B. Rubin (2004). *Bayesian Data Analysis*, 2nd ed. Chapman & Hall, New York.

Goldberger, Arthur S. (1962). Best linear unbiased prediction in the generalized linear regression model. *Journal of the American Statistical Association* 57, 369–75.

Hastie, Trevor, Robert Tibshirani, and Jerome Friedman (2001). *The Elements of Statistical Learning: Data Mining, Inference and Prediction.* Springer, New York.

Henderson, C. R. (1973), Sire evaluation and genetic trends. In *Proceedings of the Animal Breeding and Genetics Symposium in Honor of Dr. Jay L. Lush*, 10–41. American Society of Animal Science, American Dairy Science Association, and Poultry Science Association, Champaign, Illinois.

Kackar, R. N., and D. Harville (1984). Approximations for standard errors of estimators of fixed and random effects in mixed linear models. *Journal of the American Statistical Association* 79, 853–62.

Klugman, Stuart A., Harry H. Panjer, and Gordon E. Willmot (2008). *Loss Models: From Data to Decisions.* John Wiley & Sons, Hoboken, New Jersey.

McCulloch, Charles E., and Shayle R. Searle (2001). *Generalized, Linear and Mixed Models.* John Wiley & Sons, New York.

Pinheiro, José C., and Douglas M. Bates (2000). *Mixed-Effects Models in S and S-PLUS.* Springer-Verlag, New York.

Ruppert, David, M. P. Wand, and Raymond J. Carroll (2003). *Semiparametric Regression.* Cambridge University Press, Cambridge.

Silverman, B. W. (1986). *Density Estimation for Statistics and Data Analysis.* Chapman & Hall, London.

Part IV

Actuarial Applications

16

Frequency-Severity Models

Chapter Preview. Many datasets feature dependent variables that have a large proportion of zeros. This chapter introduces a standard econometric tool, known as a *tobit model*, for handling such data. The tobit model is based on observing a left-censored dependent variable, such as sales of a product or claim on a health-care policy, where it is known that the dependent variable cannot be less than zero. Although this standard tool can be useful, many actuarial datasets that feature a large proportion of zeros are better modeled in "two parts," one part for frequency and one part for severity. This chapter introduces two-part models and provides extensions to an *aggregate loss model*, where a unit under study, such as an insurance policy, can result in more than one claim.

16.1 Introduction

Many actuarial datasets come in "two parts:"

1. One part for the frequency, indicating whether a claim has occurred or, more generally, the number of claims
2. One part for the severity, indicating the amount of a claim

Two-part models consist of frequency (number) and severity (amount) components.

In predicting or estimating claims distributions, we often associate the cost of claims with two components: the event of the claim and its amount, if the claim occurs. Actuaries term these the claims *frequency* and *severity components*, respectively. This is the traditional way of decomposing two-part data, where one can consider a zero as arising from a policy without a claim (Bowers et al., 1997, Chapter 2). Because of this decomposition, two-part models are also known as *frequency-severity models*. However, this formulation has been traditionally used without covariates to explain either the frequency or the severity components. In the econometrics literature, Cragg (1971) introduced covariates into the two components, citing an example from fire insurance.

Health-care data also often feature a large proportion of zeros that must be accounted for in the modeling. Zero values can represent an individual's lack of healthcare utilization, no expenditure, or nonparticipation in a program. With respect to health care, Mullahy (1998) cites some prominent areas of potential applicability:

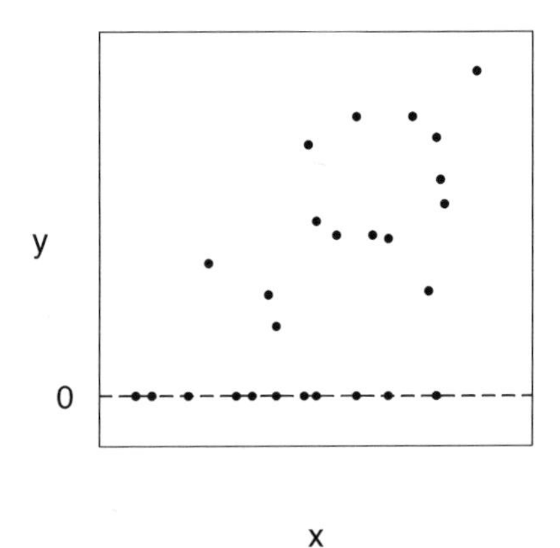

Figure 16.1 When individuals do not purchase insurance, they are recorded as $y = 0$ sales. The sample in this plot represents two subsamples, those who purchased insurance, corresponding to $y > 0$, and those who did not, corresponding to $y = 0$.

- Outcomes research: amount of health-care utilization or expenditures
- Demand for health care: amount of health care sought, such as number of physician visits
- Substance abuse: amount consumed of tobacco, alcohol, and illicit drugs

The two-part aspect can be obscured by a natural way of recording data; enter the amount of the claim when the claim occurs (a positive number) and a zero for no claim. It is easy to overlook a large proportion of zeros, particularly when the analyst is also concerned with many covariates that may help explain a dependent variable. As we will see in this chapter, ignoring the two-part nature can lead to serious bias. To illustrate, recall from Chapter 6 a plot of an individual's income (x) versus the amount of insurance purchased (y) (Figure 6.3). Fitting a single line to these data would misinform users about the effects of x on y.

In contrast, many insurers keep separate data files for frequency and severity. For example, insurers maintain a "policyholder" file that is established when a policy is underwritten. This file records much underwriting information about the insured(s), such as age, sex, and prior claims experience; policy information such as coverage, deductibles, and limitations; and the insurance claims event. A separate file, often known as the claims file, records details of the claim against the insurer, including the amount. (There may also be a "payments" file that records the timing of the payments, though we will not deal with that here.) This recording process makes it natural for insurers to model the frequency and severity as separate processes.

16.2 Tobit Model

One way of modeling a large proportion of zeros is to assume that the dependent variable is (left) censored at zero. This chapter introduces left-censored regression, beginning with the well-known *tobit model*, based on the pioneering work of James Tobin (1958). Subsequently, Goldberger (1964) coined the phrase *tobit model*, acknowledging the work of Tobin and its similarity to the probit model.

A latent variable is not observed by the analyst.

As with probit (and other binary response) models, we use an unobserved, or latent, variable y^* that is assumed to follow a linear regression model of the form

$$y_i^* = \mathbf{x}_i'\boldsymbol{\beta} + \varepsilon_i. \tag{16.1}$$

The responses are censored or "limited" in the sense that we observe $y_i = \max\left(y_i^*, d_i\right)$. The limiting value, d_i, is a known amount. Many applications use $d_i = 0$, corresponding to zero sales or expenses, depending on the application. However, we also might use d_i for the daily expenses claimed for travel reimbursement and allow the reimbursement (e.g., \$50 or \$100) to vary by employee i. Some readers may want to review Section 14.2 for an introduction to censoring.

The model parameters consist of the regression coefficients, $\boldsymbol{\beta}$, and the variability term, $\sigma^2 = \mathrm{Var}\ \varepsilon_i$. With equation (16.1), we interpret the regression coefficients as the marginal change of E y^* per unit change in each explanatory variable. This may be satisfactory in some applications, such as when y^* represents an insurance loss. However, for most applications, users are typically interested in marginal changes in E y, that is, the expected value of the *observed* response.

To interpret these marginal changes, it is customary to adopt the assumption of normality for the latent variable y_i^* (or equivalently for the disturbance ε_i). With this assumption, standard calculations (see Exercise 16.1) show that

$$\mathrm{E}\ y_i = d_i + \Phi\left(\frac{\mathbf{x}_i'\boldsymbol{\beta} - d_i}{\sigma}\right)\left(\mathbf{x}_i'\boldsymbol{\beta} - d_i + \sigma\lambda_i\right), \tag{16.2}$$

where

$$\lambda_i = \frac{\phi\left((\mathbf{x}_i'\boldsymbol{\beta} - d_i)/\sigma\right)}{\Phi\left((\mathbf{x}_i'\boldsymbol{\beta} - d_i)/\sigma\right)}.$$

Here, $\phi(\cdot)$ and $\Phi(\cdot)$ are the standard normal density and distribution functions, respectively. The ratio of a probability density function to a cumulative distribution function is sometimes called an *inverse Mills ratio*. Although complex in appearance, equation (16.2) allows one to readily compute E y. For large values of $(\mathbf{x}_i'\boldsymbol{\beta} - d_i)/\sigma$, we see that λ_i is close to 0 and $\Phi\left((\mathbf{x}_i'\boldsymbol{\beta} - d_i)/\sigma\right)$ is close to 1. We interpret this to mean that, for large values of the systematic component $\mathbf{x}_i'\boldsymbol{\beta}$, the regression function E y_i tends to be linear and the usual interpretations apply. The tobit model specification has the greatest impact on observations close to the limiting value d_i.

The tobit model specification has the greatest impact on observations close to the limiting value d_i.

Equation (16.2) shows that if an analyst ignores the effects of censoring, then the regression function can be quite different from the typical linear regression function, E $y = \mathbf{x}'\boldsymbol{\beta}$, resulting in biased estimates of coefficients. The other tempting path is to exclude limited observations ($y_i = d_i$) from the dataset and again run ordinary regression. However, standard calculations also show that

$$\mathrm{E}\ (y_i | y_i > d_i) = \mathbf{x}_i'\boldsymbol{\beta} + \sigma \frac{\phi\left((\mathbf{x}_i'\boldsymbol{\beta} - d_i)/\sigma\right)}{1 - \Phi\left((\mathbf{x}_i'\boldsymbol{\beta} - d_i)/\sigma\right)}. \tag{16.3}$$

Thus, this procedure also results in biased regression coefficients.

A commonly used method of estimating the tobit model is maximum likelihood. Employing the normality assumption, standard calculations show that the log-likelihood can be expressed as

$$\ln L = \sum_{i:y_i=d_i} \ln\left\{1-\Phi\left(\frac{\mathbf{x}_i'\boldsymbol{\beta}-d_i}{\sigma}\right)\right\}$$
$$-\frac{1}{2}\sum_{i:y_i>d_i}\left\{\ln 2\pi\sigma^2+\frac{(y_i-\mathbf{x}_i'\boldsymbol{\beta})^2}{\sigma^2}\right\}, \qquad (16.4)$$

where $\{i : y_i = d_i\}$ and $\{i : y_i > d_i\}$ mean the sum over the censored and noncensored observations, respectively. Many statistical software packages can readily compute the maximum likelihood estimators, $\mathbf{b}_{MLE}$ and s_{MLE}, as well as corresponding standard errors. Section 11.9 introduces likelihood inference.

For some users, it is convenient to have an algorithm that does not rely on specialized software. A two-stage algorithm attributable to Heckman (1976) fulfills this need. For this algorithm, first subtract d_i from each y_i, so that one may take d_i to be zero without loss of generality. Even for those who wish to use the more efficient maximum likelihood estimators, Heckman's algorithm can be useful in the model exploration stage, as one uses linear regression to help select the appropriate form of the regression equation.

Heckman's Algorithm for Estimating Tobit Model Parameters

(i) For the first stage, define the binary variable

$$r_i = \begin{cases} 1 & \text{if } y_i > 0 \\ 0 & \text{if } y_i = 0 \end{cases},$$

indicating whether the observation is censored. Run a probit regression using r_i as the dependent variable and $\mathbf{x}_i$ as explanatory variables. Call the resulting regression coefficients $\mathbf{g}_{PROBIT}$.

(ii) For each uncensored observation, compute the estimated variable

$$\widehat{\lambda}_i = \frac{\phi\left(\mathbf{x}_i'\mathbf{g}_{PROBIT}\right)}{\Phi\left(\mathbf{x}_i'\mathbf{g}_{PROBIT}\right)},$$

an inverse Mill's ratio. With this, run a regression of y_i on $\mathbf{x}_i$ and $\widehat{\lambda}_i$. Call the resulting regression coefficients $\mathbf{b}_{2SLS}$.

The idea behind this algorithm is that equation (16.1) has the same form as the probit model; thus, consistent estimates of the regression coefficients (up to scale) can be computed. The regression coefficients $\mathbf{b}_{2SLS}$ provide consistent and asymptotically normal estimates of $\boldsymbol{\beta}$. They are, however, inefficient compared to the maximum likelihood estimators, $\mathbf{b}_{MLE}$. Standard calculations (see Exercise 16.1) show that $\mathrm{Var}(y_i | y_i > d_i)$ depends on i (even when d_i is constant). Thus, it is customary to use heteroscedasticity-consistent standard errors for $\mathbf{b}_{2SLS}$.

16.3 Application: Medical Expenditures

This section considers data from the Medical Expenditure Panel Survey (MEPS), which were introduced in Section 11.4. Recall that MEPS is a probability survey that provides nationally representative estimates of health-care use, expenditures, sources of payment, and insurance coverage for the U.S. civilian population. We consider MEPS data from the first panel of 2003 and take a random sample of $n = 2{,}000$ individuals between the ages of 18 and 65. Section 11.4 analyzed the frequency component, trying to understand the determinants that influenced whether people were hospitalized. Section 13.4 analyzed the severity component; given that a person was hospitalized, what are the determinants of medical expenditures? This chapter seeks to unify these two components into a single model of health-care utilization.

Summary Statistics

Ⓡ **Empirical** *Filename is "HealthExpend"*

Table 16.1 reviews these explanatory variables and provides summary statistics that suggest their effects on expenditures of inpatient visits. The second column, "Average Expend," displays the average logarithmic expenditure by explanatory variable, treating no expenditures as a zero (logarithmic) expenditure. This would be the primary variable of interest if one did not decompose the total expenditure into a discrete zero and continuous amount.

Examining this overall average (logarithmic) expenditure, we see that women had greater expenditures than men. In terms of ethnicity, native Americans and Asians had the lowest average expenditures. However, these two ethnic groups accounted for only 5.4% of the total sample size. Regarding regions, it appears that individuals from the West had the lowest average expenditures. In terms of education, more educated persons had lower expenditures. This observation supports the theory that more educated persons take more active roles in keeping their health. When it comes to self-rated health status, poorer physical, mental health, and activity-related limitations led to greater expenditures. Lower-income individuals had greater expenditures and those with insurance coverage had greater average expenditures.

Table 16.1 also describes the effects of explanatory variables on the frequency of utilization and average expenditures for those that used inpatient services. As in Table 11.4, the column "Percentage Positive Expend" gives the percentage of individuals who had some positive expenditure, by explanatory variable. The column "Average of Pos Expend" gives the average (logarithmic) expenditure when there was an expenditure, ignoring the zeros. This is comparable to the median expenditure in Table 13.5 (given in dollars, not log dollars).

A variable's effect on overall expenditures may be positive, negative, or nonsignificant; this effect can be quite different when we decompose expenditures into frequency and amount components.

To illustrate, consider that women had greater average expenditures than men by looking at the "Average Expend" column. Breaking this down into frequency and amount of utilization, we see that women had a greater frequency of utilization, but when they had a positive utilization, the average (logarithmic) expenditure was less than that of men. An examination of Table 16.1 shows that this observation holds true for other explanatory variables. A variable's effect on

Table 16.1 Percentage of Positive Expenditures and Average Logarithmic Expenditure, by Explanatory Variable

Category	Variable	Description	Percentage of Data	Average Expend.	Percentage Positive Expend.	Average of Pos. Expend.
Demography	AGE	Age in years between 18 to 65 (mean: 39.0)				
	GENDER	1 if female	52.7	0.91	10.7	8.53
	GENDER	1 if male	47.3	0.40	4.7	8.66
Ethnicity	ASIAN	1 if Asian	4.3	0.37	4.7	7.98
	BLACK	1 if Black	14.8	0.90	10.5	8.60
	NATIVE	1 if Native	1.1	1.06	13.6	7.79
	WHITE	Reference level	79.9	0.64	7.5	8.59
Region	NORTHEAST	1 if Northeast	14.3	0.83	10.1	8.17
	MIDWEST	1 if Midwest	19.7	0.76	8.7	8.79
	SOUTH	1 if South	38.2	0.72	8.4	8.65
	WEST	Reference level	27.9	0.46	5.4	8.51
Education	COLLEGE	1 if college or higher degree	27.2	0.58	6.8	8.50
	HIGHSCHOOL	1 if high school degree	43.3	0.67	7.9	8.54
	Reference level is lower than high school degree		29.5	0.76	8.8	8.64
Self-rated physical health	POOR	1 if poor	3.8	3.26	36.0	9.07
	FAIR	1 if fair	9.9	0.66	8.1	8.12
	GOOD	1 if good	29.9	0.70	8.2	8.56
	VGOOD	1 if very good	31.1	0.54	6.3	8.64
	Reference level is excellent health		25.4	0.42	5.1	8.22
Self-rated mental health	MNHPOOR	1 if poor or fair	7.5	1.45	16.8	8.67
		0 if good to excellent mental health	92.5	0.61	7.1	8.55
Any activity limitation	ANYLIMIT	1 if any functional or activity limitation	22.3	1.29	14.6	8.85
		0 if otherwise	77.7	0.50	5.9	8.36
Income compared to poverty line	HINCOME	1 if high income	31.6	0.47	5.4	8.73
	MINCOME	1 if middle income	29.9	0.61	7.0	8.75
	LINCOME	1 if low income	15.8	0.73	8.3	8.87
	NPOOR	1 if near poor	5.8	0.78	9.5	8.19
	Reference level is poor/negative		17.0	1.06	13.0	8.18
Insurance coverage	INSURE	1 if covered by public or private health insurance in any month of 2003	77.8	0.80	9.2	8.68
		0 if had no health insurance in 2003	22.3	0.23	3.1	7.43
Total			100.0	0.67	7.9	8.32

Table 16.2 Comparison of OLS, Tobit MLE, and Two-Stage Estimates

	OLS		Tobit MLE		Two-Stage	
Effect	Parameter Estimate	t-Ratio	Parameter Estimate	t-Ratio	Parameter Estimate	t-Ratio[a]
Intercept	−0.123	−0.525	−33.016	−8.233	2.760	0.617
AGE	0.001	0.091	−0.006	−0.118	0.001	0.129
GENDER	0.379	3.711	5.727	4.107	0.271	1.617
ASIAN	−0.115	−0.459	−1.732	−0.480	−0.091	−0.480
BLACK	0.054	0.365	0.025	0.015	0.043	0.262
NATIVE	0.350	0.726	3.745	0.723	0.250	0.445
NORTHEAST	0.283	1.702	3.828	1.849	0.203	1.065
MIDWEST	0.255	1.693	3.459	1.790	0.196	1.143
SOUTH	0.146	1.133	1.805	1.056	0.117	0.937
COLLEGE	−0.014	−0.089	0.628	0.329	−0.024	−0.149
HIGHSCHOOL	−0.027	−0.209	−0.030	−0.019	−0.026	−0.202
POOR	2.297	7.313	13.352	4.436	1.780	1.810
FAIR	−0.001	−0.004	1.354	0.528	−0.014	−0.068
GOOD	0.188	1.346	2.740	1.480	0.143	1.018
VGOOD	0.084	0.622	1.506	0.815	0.063	0.533
MNHPOOR	0.000	−0.001	−0.482	−0.211	−0.011	−0.041
ANYLIMIT	0.415	3.103	4.695	3.000	0.306	1.448
HINCOME	−0.482	−2.716	−6.575	−3.035	−0.338	−1.290
MINCOME	−0.309	−1.868	−4.359	−2.241	−0.210	−0.952
LINCOME	−0.175	−0.976	−3.414	−1.619	−0.099	−0.438
NPOOR	−0.116	−0.478	−2.274	−0.790	−0.065	−0.243
INSURE	0.594	4.486	8.534	4.130	0.455	2.094
Inverse Mill's ratio $\widehat{\lambda}$					−3.616	−0.642
Scale σ^2	4.999		14.738		4.997	

[a] Two-stage t-ratios are calculated using heteroscedasticity-consistent standard errors.

overall expenditures may be positive, negative, or nonsignificant; this effect can be quite different when we decompose expenditures into frequency and amount components.

Table 16.2 compares the ordinary least squares (OLS) regression to maximum likelihood estimates for the tobit model. From this table, we can see that there is a substantial agreement among the t-ratios for these fitted models. This agreement comes from examining the sign (positive or negative) and the magnitude (e.g., exceeding two for statistical significance) of each variable's t-ratio. The regression coefficients also largely agree in sign. However, it is not surprising that the magnitudes of the regression coefficients differ substantially. This is because, from equation (16.2), we can see that the tobit coefficients measure the marginal change of the expected latent variable y^*, not the marginal change of the expected observed variable y, as does OLS.

Table 16.2 also reports the fit using the two-stage Heckman algorithm. The coefficient associated with the inverse Mill's ratio selection correction is statistically insignificant. Thus, there is general agreement between the OLS coefficients and those estimated using the two-stage algorithm. The two-stage t-ratios were calculated using heteroscedasticity-consistent standard errors, described in Section 5.7.2. Here, we see some disagreement between the t-ratios calculated using Heckman's algorithm and the maximum likelihood values calculated using the tobit model. For example, GENDER, POOR, HINCOME, and MINCOME are statistically significant in the tobit model but not in the two-stage algorithm. This is troubling because both techniques yield consistent estimators, providing that the assumptions of the tobit model are valid. Thus, we suspect the validity of the model assumptions for these data; the next section provides an alternative model that turns out to be more suitable for this dataset.

16.4 Two-Part Model

One drawback of the tobit model is its reliance on the normality assumption of the latent response. A second, and more important, drawback is that a single latent variable dictates the magnitude both of the response and of the censoring. As pointed out by Cragg (1971), there are many instances in which the limiting amount represents a choice or activity that is separate from the magnitude. For example, in a population of smokers, zero cigarettes consumed during a week may simply represent a lower bound (or limit) and may be influenced by available time and money. However, in a general population, zero cigarettes consumed during a week can indicate that a person is a nonsmoker, a choice that could be influenced by other lifestyle decisions (where time and money may or may not be relevant). As another example, when studying health-care expenditures, a zero represents a person's choice or decision not to utilize health care during a period. For many studies, the *amount* of health-care expenditure is strongly influenced by a health-care provider (e.g., a physician); the decision to utilize and the amount of health care can involve very different considerations.

In the traditional actuarial literature (see, e.g., Bowers et al. 1997, Chapter 2), the *individual risk model* decomposes a response, typically an insurance claim, into frequency (number) and severity (amount) components. Specifically, let r_i be a binary variable indicating whether the ith subject has an insurance claim and let y_i describe the amount of the claim. Then, the claim is modeled as

$$(claim\ recorded)_i = r_i \times y_i .$$

This is the basis for the two-part model, where we also use explanatory variables to understand the influence of each component.

Definition: Two-Part Model

(i) Use a binary regression model with r_i as the dependent variable and $\mathbf{x}_{1i}$ as the set of explanatory variables. Denote the corresponding set of regression coefficients as $\boldsymbol{\beta}_1$. Typical models include the linear probability, logit, and probit models.

Table 16.3 Comparison of Full and Reduced Two-Part Models

	Full Model				Reduced Model			
	Frequency		Severity		Frequency		Severity	
Effect	Estimate	t-Ratio	Estimate	t-Ratio	Estimate	t-Ratio	Estimate	t-Ratio
Intercept	−2.263	−10.015	6.828	13.336	−2.281	−11.432	6.879	14.403
AGE	−0.001	−0.154	0.012	1.368			0.020	2.437
GENDER	0.395	4.176	−0.104	−0.469	0.395	4.178	−0.102	−0.461
ASIAN	−0.108	−0.429	−0.397	−0.641	−0.108	−0.427	−0.159	−0.259
BLACK	0.008	0.062	0.088	0.362	0.009	0.073	0.017	0.072
NATIVE	0.284	0.778	−0.639	−0.905	0.285	0.780	−1.042	−1.501
NORTHEAST	0.283	1.958	−0.649	−2.035	0.281	1.950	−0.778	−2.422
MIDWEST	0.239	1.765	0.016	0.052	0.237	1.754	−0.005	−0.016
SOUTH	0.132	1.099	−0.078	−0.294	0.130	1.085	−0.022	−0.081
COLLEGE	0.048	0.356	−0.597	−2.066	0.049	0.362	−0.470	−1.743
HIGHSCHOOL	0.002	0.017	−0.415	−1.745	0.003	0.030	−0.256	−1.134
POOR	0.955	4.576	0.597	1.594	0.939	4.805		
FAIR	0.087	0.486	−0.211	−0.527	0.079	0.450		
GOOD	0.184	1.422	0.145	0.502	0.182	1.412		
VGOOD	0.095	0.736	0.373	1.233	0.094	0.728		
MNHPOOR	−0.027	−0.164	−0.176	−0.579			−0.177	−0.640
ANYLIMIT	0.318	2.941	0.235	0.981	0.311	3.022	0.245	1.052
HINCOME	−0.468	−3.131	0.490	1.531	−0.470	−3.224		
MINCOME	−0.314	−2.318	0.472	1.654	−0.314	−2.345		
LINCOME	−0.241	−1.626	0.550	1.812	−0.241	−1.633		
NPOOR	−0.145	−0.716	0.067	0.161	−0.146	−0.721		
INSURE	0.580	4.154	1.293	3.944	0.579	4.147	1.397	4.195
Scale σ^2			1.249				1.333	

(ii) Conditional on $r_i = 1$, specify a regression model with y_i as the dependent variable and $\mathbf{x}_{2i}$ as the set of explanatory variables. Denote the corresponding set of regression coefficients as $\boldsymbol{\beta}_2$. Typical models include the linear and gamma regression models.

Unlike the tobit, in the two-part model, one need not have the same set of explanatory variables influencing the frequency and amount of response. However, there is usually overlap in the sets of explanatory variables, where variables are members of both $\mathbf{x}_1$ and $\mathbf{x}_2$. Typically, one assumes that $\boldsymbol{\beta}_1$ and $\boldsymbol{\beta}_2$ are not related so that the joint likelihood of the data can be separated into two components and run separately, as described earlier.

Example: MEPS Expenditure Data, Continued. Consider the Section 16.3 MEPS expenditure data using a probit model for the frequency and a linear regression model for the severity. Table 16.3 shows the results from using all explanatory variables to understand their influence on the decision to seek health care (frequency) and the amount of health care utilized (severity). Unlike the Table 16.2 Tobit model, the two-part models allow each variable to have a separate influence on frequency and severity. To illustrate, the full model results in Table 16.3 show that COLLEGE has no significant impact on frequency but a strong, positive impact on severity.

Because of the flexibility of the two-part model, one can also reduce the model complexity for each component by removing extraneous variables. Table 16.3 shows a reduced model in which age and mental health status variables have been removed from the frequency component; regional, educational, physical status, and income variables have been removed from the severity component.

Tobit Type II Model

To connect the tobit and two-part models, let us assume that the frequency is represented by a probit model and use

$$r_i^* = \mathbf{x}_{1i}'\boldsymbol{\beta}_1 + \eta_{1i}$$

as the latent tendency to be observed. Define $r_i = \mathrm{I}\left(r_i^* > 0\right)$ to be the binary variable indicating that an amount has been observed. For the severity component, define

$$y_i^* = \mathbf{x}_{2i}'\boldsymbol{\beta}_2 + \eta_{2i}$$

to be the latent amount variable. The "observed" amount is

$$y_i = \begin{cases} y_i^* & \text{if } r_i = 1 \\ 0 & \text{if } r_i = 0 \end{cases}.$$

Because responses are censored, the analyst is aware of the subject i and has covariate information even when $r_i = 0$.

If $\mathbf{x}_{1i} = \mathbf{x}_{2i}$, $\boldsymbol{\beta}_1 = \boldsymbol{\beta}_2$, and $\eta_{1i} = \eta_{2i}$, then this is the tobit framework with $d_i = 0$. If $\boldsymbol{\beta}_1$ and $\boldsymbol{\beta}_2$ are not related and if η_{1i} and η_{2i} are independent, then this is the two-part framework. For the two-part framework, the likelihood of the observed responses $\{r_i, y_i\}$ is given by

$$L = \prod_{i=1}^{n} \left\{(p_i)^{r_i} (1 - p_i)^{1-r_i}\right\} \prod_{r_i=1} \phi\left(\frac{y_i - \mathbf{x}_{2i}'\boldsymbol{\beta}_2}{\sigma_{\eta 2}}\right) \Big/ \sigma_{\eta 2}. \tag{16.5}$$

Here, $p_i = \Pr(r_i = 1) = \Pr\left(\mathbf{x}_{1i}'\boldsymbol{\beta}_1 + \eta_{1i} > 0\right) = 1 - \Phi\left(-\mathbf{x}_{1i}'\boldsymbol{\beta}_1\right) = \Phi\left(\mathbf{x}_{1i}'\boldsymbol{\beta}_1\right)$, assuming $\mathrm{Var}\, \eta_{i1} = 1$. Assuming that $\boldsymbol{\beta}_1$ and $\boldsymbol{\beta}_2$ are not related, one can separately maximize the two pieces of the likelihood function.

In some instances, it is sensible to assume that the frequency and severity components are related. The tobit model considers a perfect relationship (with $\eta_{1i} = \eta_{2i}$), whereas the two-part models assumes independence. For an intermediate model, the tobit type II model allows for a nonzero correlation between η_{1i} and η_{2i}. See Amemiya (1985) for additional details. Hsiao, Kim, and Taylor (1990) provide an application of the tobit type II model to Canadian collision coverage of private passenger automobile experience.

16.5 Aggregate Loss Model

We now consider two-part models where the frequency may exceed one. For example, if we are tracking automobile accidents, a policyholder may have more than one accident in a year. As another example, we may be interested in the claims for a city or a state and expect many claims per government unit.

To establish notation, for each $\{i\}$, the observable responses consist of

- N_i: the number of claims (events)
- y_{ij}, $j = 1, \ldots, N_i$: the amount of each claim (loss)

By convention, the set $\{y_{ij}\}$ is empty when $N_i = 0$. If one uses N_i as a binary variable, then this framework reduces to the two-part setup.

Although we have detailed information on losses per event, the interest often is in *aggregate losses*, $S_i = y_{i1} + \ldots + y_{i,N_i}$. In traditional actuarial modeling, one assumes that the distribution of losses are, conditional on the frequency N_i, identical and independent over replicates j. This representation is known as the *collective risk model*; see, for example, Klugman et al. (2008). We also maintain this assumption.

Data are typically available in two forms:

1. $\{N_i, y_{i1}, \ldots, y_{i,N_i}\}$, so that detailed information about each claim is available. For example, when examining personal automobile claims, losses for each claim are available. Let $\mathbf{y}_i = \left(y_{i1}, \ldots, y_{i,N_i}\right)'$ be the vector of individual losses.
2. $\{N_i, S_i\}$, so that only aggregate losses are available. For example, when examining losses at the city level, only aggregate losses are available.

We are interested in both forms. Because there are multiple responses (events) per subject $\{i\}$, one might approach the analysis using multilevel models as described in, for example, Raudenbush and Bryk (2002). Unlike a multilevel structure, we consider data where the number of events are random that we want to model stochastically and thus use an alternative framework. When only $\{S_i\}$ is available, the Tweedie GLM introduced in Section 13.6 may be used.

To see how to model these data, consider the first data form. Suppressing the $\{i\}$ subscript, we decompose the joint distribution of the dependent variables as:

$$\begin{aligned} \mathrm{f}(N, \mathbf{y}) &= \mathrm{f}(N) \times \mathrm{f}(\mathbf{y}|N) \\ \text{joint} &= \text{frequency} \times \text{conditional severity,} \end{aligned}$$

where $\mathrm{f}(N, \mathbf{y})$ denotes the joint distribution of $(N, \mathbf{y})$. This joint distribution equals the product of the two components:

1. Claims frequency: $\mathrm{f}(N)$ denotes the probability of having N claims.
2. Conditional severity: $\mathrm{f}(\mathbf{y}|N)$ denotes the conditional density of the claim vector $\mathbf{y}$ given N.

We represent the frequency and severity components of the aggregate loss model as follows.

Definition: Aggregate Loss Model I

(i) Use a count regression model with N_i as the dependent variable and $\mathbf{x}_{1i}$ as the set of explanatory variables. Denote the corresponding set of regression coefficients as $\boldsymbol{\beta}_1$. Typical models include the Poisson and negative binomial models.
(ii) Conditional on $N_i > 0$, use a regression model with y_{ij} as the dependent variable and $\mathbf{x}_{2i}$ as the set of explanatory variables. Denote the corresponding set of regression coefficients as $\boldsymbol{\beta}_2$. Typical models include the linear regression, gamma regression, and mixed linear models. For the mixed linear models, one uses a subject-specific intercept to account for heterogeneity among subjects.

To model the second data form, the setup is similar. The count data model in step 1 will not change. However, the regression model in step 2 will use S_i as the dependent variable. Because the dependent variable is the sum over N_i independent replicates, it may be that you need to allow the variability to depend on N_i.

Example: MEPS Expenditure Data, Continued. To get a sense of the empirical observations for claim frequency, we present the overall claim frequency. According to this table, there were a total of 2,000 observations, of which 92.15% did not have any claims. There are a total of 203 ($= 1 \times 130 + 2 \times 19 + 3 \times 2 + 4 \times 3 + 5 \times 2 + 6 \times 0 + 7 \times 1$) claims.

Frequency of Claims

Count	0	1	2	3	4	5	6	7	Total
Number	1,843	130	19	2	3	2	0	1	2,000
Percentage	92.15	6.50	0.95	0.10	0.15	0.10	0.00	0.10	100.00

Table 16.4 summarizes the regression coefficient parameter fits using the negative binomial model. The results are comparable to those for the fitted probit models in Table 16.3, where many of the covariates are statistically significant predictors of claim frequency.

This fitted frequency model is based on $n = 2{,}000$ persons. The Table 16.4 fitted severity models are based on $n_1 + \ldots + n_{2000} = 203$ claims. The gamma regression model is based on a logarithmic link,

$$\mu_i = \exp\left(\mathbf{x}_i' \boldsymbol{\beta}_2\right).$$

Table 16.4 shows that the results from fitting an ordinary regression model are similar to those from fitting the gamma regression model. They are similar in the sense that the sign and statistical significance of coefficients for each

Table 16.4 Aggregate Loss Models

Effect	Negative Binomial Frequency		Ordinary Regression Severity		Gamma Regression Severity	
	Parameter Estimate	t-Ratio	Parameter Estimate	t-Ratio	Parameter Estimate	t-Ratio
Intercept	−4.214	−9.169	7.424	15.514	8.557	20.521
AGE	−0.005	−0.756	−0.006	−0.747	−0.011	−1.971
GENDER	0.617	3.351	−0.385	−1.952	−0.826	−4.780
ASIAN	−0.153	−0.306	−0.340	−0.588	−0.711	−1.396
BLACK	0.144	0.639	0.146	0.686	−0.058	−0.297
NATIVE	0.445	0.634	−0.331	−0.465	−0.512	−0.841
NORTHEAST	0.492	1.683	−0.547	−1.792	−0.418	−1.602
MIDWEST	0.619	2.314	0.303	1.070	0.589	2.234
SOUTH	0.391	1.603	0.108	0.424	0.302	1.318
COLLEGE	0.023	0.089	−0.789	−2.964	−0.826	−3.335
HIGHSCHOOL	−0.085	−0.399	−0.722	−3.396	−0.742	−4.112
POOR	1.927	5.211	0.664	1.964	0.299	0.989
FAIR	0.226	0.627	−0.188	−0.486	0.080	0.240
GOOD	0.385	1.483	0.223	0.802	0.185	0.735
VGOOD	0.348	1.349	0.429	1.511	0.184	0.792
MNHPOOR	−0.177	−0.583	−0.221	−0.816	−0.470	−1.877
ANYLIMIT	0.714	3.499	0.579	2.720	0.792	4.171
HINCOME	−0.622	−2.139	0.723	2.517	0.557	2.290
MINCOME	−0.482	−1.831	0.720	2.768	0.694	3.148
LINCOME	−0.460	−1.611	0.631	2.241	0.889	3.693
NPOOR	−0.465	−1.131	−0.056	−0.135	0.217	0.619
INSURE	1.312	4.207	1.500	4.551	1.380	4.912
Dispersion	2.177		1.314		1.131	

variable are comparable. As discussed in Chapter 13, the advantage of the ordinary regression model is its relatively simplicity in terms of ease of implementation and interpretation. In contrast, the gamma regression model can be a better model for fitting long-tail distributions such as medical expenditures.

16.6 Further Reading and References

Property and Casualty

There is a rich literature on modeling the joint frequency and severity distribution of automobile insurance claims. To distinguish this modeling from classical risk theory applications (see, e.g., Klugman et al., 2008), we focus on cases where explanatory variables, such as policyholder characteristics, are available. There has been substantial interest in statistical modeling of claims frequency, yet the literature on modeling claims severity, especially in conjunction with claims frequency, is less extensive. A possible explanation, noted by Coutts (1984), is that most of the variation in overall claims experience may be attributed to claim frequency (at least when inflation was low). Coutts (1984) also remarks that the first paper to analyze claim frequency and severity separately seems to be that of Kahane and Levy (1975).

Brockman and Wright (1992) provide an early overview of how statistical modeling of claims and severity can be helpful for pricing automobile coverage. For computational convenience, they focused on categorical pricing variables to form cells that could be used with traditional insurance underwriting forms. Renshaw (1994) shows how generalized linear models can be used to analyze both the frequency and severity portions on the basis of individual-policyholder-level data. Hsiao, Kim, and Taylor (1990) note the excess number of zeros in policyholder claims data (because of no claims) and compare tobit, two-part, and simultaneous equation models, building on the work of Weisberg and Tomberlin (1982) and Weisberg, Tomberlin, and Chatterjee (1984). All of these papers use grouped data, not individual level data as in this chapter.

At the individual policyholder level, Frangos and Vrontos (2001) examined a claim frequency and severity model, using negative binomial and Pareto distributions, respectively. They used their statistical model to develop experience rated (bonus-malus) premiums. Pinquet (1997, 1998) provides a more modern statistical approach, not only fitting cross-sectional data but also following policyholders over time. Pinquet was interested in two lines of business, claims at fault and not at fault with respect to a third party. For each line, Pinquet hypothesized a frequency and severity component that were allowed to be correlated to each other. In particular, the claims frequency distribution was assumed to be bivariate Poisson. Severities were modeled using lognormal and gamma distributions.

Health Care

The two-part model became prominent in the health-care literature on adoption by Rand Health Insurance Experiment researchers (Duan et al., 1983, Manning et al., 1987). They used the two-part model to analyze health insurance cost sharing's effect on health-care utilization and expenditures because of the close resemblance of the demand for medical care to the two decision-making processes. That is, the amount of health-care expenditures is largely unaffected by an individual's decision to seek treatment. This is because physicians, as the patients' (principal) agents, would tend to decide the intensity of treatments as suggested by the principal-agent model of Zweifel (1981).

The two-part model has become widely used in the health-care literature despite some criticisms. For example, Maddala (1985) argues that two-part modeling is not appropriate for nonexperimental data because individuals' self-selection into different health insurance plans is an issue. (In the Rand Health Insurance Experiment, the self-selection aspect was not an issue because participants were randomly assigned to health insurance plans.) See Jones (2000) and Mullahy (1998) for overviews.

Two-part models remain attractive in modeling health-care usage because they provide insights into the determinants of initiation and level of health-care usage. Individuals' decision to utilize health care is related primarily to personal characteristics, whereas the cost per user may be more related to characteristics of the health-care provider.

Chapter References

Amemiya, T. (1985). *Advanced Econometrics*. Harvard University Press, Cambridge, Massachusetts.

Bowers, Newton L., Hans U. Gerber, James C. Hickman, Donald A. Jones, and Cecil J. Nesbitt (1997). *Actuarial Mathematics*. Society of Actuaries, Schaumburg, Illinois.

Brockman, M. J., and T. S. Wright. (1992). Statistical motor rating: Making effective use of your data. *Journal of the Institute of Actuaries* 119, 457–543.

Cameron, A. Colin, and Pravin K. Trivedi. (1998) *Regression Analysis of Count Data*. Cambridge University Press, Cambridge.

Coutts, S. M. (1984). Motor insurance rating, an actuarial approach. *Journal of the Institute of Actuaries* 111, 87–148.

Cragg, John G. (1971). Some statistical models for limited dependent variables with application to the demand for durable goods. *Econometrica* 39, no. 5, 829–44.

Duan, Naihua, Willard G. Manning, Carl N. Morris, and Joseph P. Newhouse (1983). A comparison of alternative models for the demand for medical care. *Journal of Business and Economics* 1(2), 115–26.

Frangos, Nicholas E., and Spyridon D. Vrontos (2001). Design of optimal bonus-malus systems with a frequency and a severity component on an individual basis in automobile insurance. *ASTIN Bulletin* 31, no. 1, 1–22.

Goldberger, Arthur S. (1964). *Econometric Theory*. John Wiley & Sons, New York.

Heckman, James J. (1976). The common structure of statistical models of truncation, sample selection and limited dependent variables, and a simple estimator for such models. *Ann. Econ. Soc. Meas.* 5, 475–92.

Hsiao, Cheng, Changseob Kim, and Grant Taylor (1990). A statistical perspective on insurance rate-making. *Journal of Econometrics* 44, 5–24.

Jones, Andrew M. (2000). Health econometrics. In *Handbook of Health Economics, Volume 1*, Antonio J. Culyer and Joseph P. Newhouse, eds., 265–344. Elsevier, Amsterdam.

Kahane, Yehuda, and Haim Levy (1975). Regulation in the insurance industry: Determination of premiums in automobile insurance. *Journal of Risk and Insurance* 42, 117–32.

Klugman, Stuart A., Harry H. Panjer, and Gordon E. Willmot (2008). *Loss Models: From Data to Decisions*. John Wiley & Sons, Hoboken, New Jersey.

Maddala, G. S. (1985). A survey of the literature on selectivity as it pertains to health care markets. *Advances in Health Economics and Health Services Research* 6, 3–18.

Manning, Willard G., Joseph P. Newhouse, Naihua Duan, Emmett B. Keeler, Arleen Leibowitz and M. Susan Marquis (1987). Health insurance and the demand for medical care: Evidence from a randomized experiment. *American Economic Review* 77(3), 251–277.

Mullahy, John (1998). Much ado about two: Reconsidering retransformation and the two-part model in health econometrics. *Journal of Health Economics* 17, 247–81.

Pinquet, Jean (1997). Allowance for cost of claims in bonus-malus systems. *ASTIN Bulletin* 27, no. 1, 33–57.

Pinquet, Jean (1998). Designing optimal bonus-malus systems from different types of claims. *ASTIN Bulletin* 28, no. 2, 205–29.

Raudenbush, Steven W., and Anthony S. Bryk (2002). *Hierarchical Linear Models: Applications and Data Analysis Methods*, 2nd ed. London: Sage.

Tobin, James (1958). Estimation of relationships for limited dependent variables. *Econometrica* 26, 24–36.

Weisberg, Herbert I., and Thomas J. Tomberlin (1982). A statistical perspective on actuarial methods for estimating pure premiums from cross-classified data. *Journal of Risk and Insurance* 49, 539–63.

Weisberg, Herbert I., Thomas J. Tomberlin, and Sangit Chatterjee (1984). Predicting insurance losses under cross-classification: A comparison of alternative approaches. *Journal of Business and Economic Statistics* 2, no. 2, 170–8.

Zweifel, P. (1981). Supplier-induced demand in a model of physician behavior. In *Health, Economics and Health Economics*, J. van der Gaag and M. Perlman, eds., 245–67. North-Holland, Amsterdam.

16.7 Exercises

16.1. Assume that y is normally distributed with mean μ and variance σ^2. Let $\phi(\cdot)$ and $\Phi(\cdot)$ be the standard normal density and distribution functions, respectively. Define $\mathrm{h}(d) = \phi(d)\,/\,(1 - \Phi(d))$, a hazard rate. Let d be a known constant and $d_s = (d - \mu)/\sigma$ be the standardized version.
   a. Determine the density of y, conditional on $\{y > d\}$.
   b. Show that $\mathrm{E}\,(y|y > d) = \mu + \sigma \mathrm{h}(d_s)$.
   c. Show that $\mathrm{E}\,(y|y \leq d) = \mu - \sigma\phi(d_s)\,/\,\Phi(d_s)$.
   d. Show that $\mathrm{Var}\,(y|y > d) = \sigma^2\,(1 - \delta\,(d_s))$, where $\delta\,(d) = \mathrm{h}\,(d)\,(\mathrm{h}\,(d) - d)$.
   e. Show that $\mathrm{E}\ \max\,(y, d) = (\mu + \sigma \mathrm{h}\,(d_s))\,(1 - \Phi(d_s)) + d\Phi(d_s)$.
   f. Show that $\mathrm{E}\ \min\,(y, d) = \mu + d - ((\mu + \sigma \mathrm{h}\,(d_s))\,(1 - \Phi(d_s)) + d\Phi(d_s))$.

16.2. Verify the log-likelihood in equation (16.4) for the Tobit model.

16.3. Verify the likelihood in equation (16.5) for the two-part model.

16.4. Derive the likelihood for the tobit type II model. Show that your likelihood reduces to equation (16.5) in the case of uncorrelated disturbance terms.

17

Fat-Tailed Regression Models

Chapter Preview. When modeling financial quantities, we are just as interested in the extreme values as in the center of the distribution; extreme values can represent the most unusual claims, profits, or sales. Actuaries often encounter situations where the data exhibit *fat tails*, or cases in which extreme values in the data are more likely to occur than in normally distributed data. Traditional regression focuses on the center of the distribution and downplays extreme values. In contrast, the focus of this chapter is on the entire distribution. This chapter surveys four techniques for regression analysis of fat-tailed data: transformation, generalized linear models, more general distributions, and quantile regression.

17.1 Introduction

Actuaries often encounter situations in which the data exhibit *fat tails*, meaning that extreme values in the data are more likely to occur than in normally distributed data. These distributions can be described as "fat," "heavy," "thick" or "long" as compared to the normal distribution. (Section 17.3.1 will be more precise on what constitutes fat tailed.) In finance, for example, the asset pricing theories assume normally distributed asset returns. Empirical distributions of the returns of financial assets, however, suggest fat-tailed distributions rather than normal distributions, as assumed in the pricing theories (see, e.g., Rachev, Menn, and Fabozzi, 2005). In health care, fat-tailed data are also common. For example, outcomes of interest such as the number of inpatient days or inpatient expenditures are typically right skewed and heavy tailed as a result of a few high-cost patients (Basu, Manning, and Mullahy, 2004). Actuaries also regularly analyze fat-tailed data in non-life insurance (Klugman, Panjer, and Willmot, 2008).

As with any other dataset, the outcome of interest may be related to other factors, and so regression analysis is of interest. However, employing the usual regression routines without addressing the fat-tailed nature can lead to serious difficulties:

- Regression coefficients can be expressed as weighted sums of dependent variables. Thus, coefficients may be unduly influenced by extreme observations.
- Because the distribution is fat tailed, the usual rules of thumb for approximate (large-sample) normality of parameter estimates no longer apply. Thus,

for example, the standard t-ratios and p-values associated with regression estimates may no longer be meaningful indicators of statistical significance.
- The usual regression routines minimize a squared error loss function. For some problems, we are more concerned with errors in one direction (either low or high), not a symmetric function.
- Large values in the dataset may be the most important in a financial sense, for example, an extremely high expenditure when examining medical costs. Although atypical, this is not an observation that we want to neglect or downweight simply because it does not fit the usual normal-based regression model.

This chapter describes four basic approaches for handling fat-tailed regression data:

1. Transformations of the dependent variable
2. Generalized linear models
3. Models using more flexible positive random variable distributions, such as the generalized gamma
4. Quantile regression models

Sections 17.2–17.5 address each in turn.

Another area of statistics that is devoted to the analysis of tail behavior is known as *extreme-value statistics*. This area concerns modeling tail behavior, largely at the expense of ignoring the rest of the distribution. In contrast, traditional regression focuses on the center of the distribution and downplays extreme values. For financial quantities, we are just as interested in the extremes as the center of the distribution; extreme values can represent the most unusual claims, profits, or sales. The focus of this chapter is on the entire distribution. Regression modeling within extreme-value statistics is a topic that has only begun to receive serious attention; Section 17.6 provides a brief survey.

17.2 Transformations

As we have seen throughout this text, the most commonly used approach for handling fat-tailed data is to simply transform the dependent variable. As a matter of routine, analysts take a logarithmic transformation of y and then use ordinary least squares on $\ln(y)$. Even though this technique is not always appropriate, it has proved effective for a surprisingly large number of datasets. This section summarizes what we have learned about transformations and provides some additional tools that can be helpful in certain applications.

Section 1.3 introduced the idea of transformations and showed how (power) transforms could be used to symmetrize a distribution. Power transforms, such as y^λ, can "pull in" extreme values so that any observation will not have an undue effect on parameter estimates. Moreover, the usual rules of thumb for approximate (large-sample) normality of parameter estimates are more likely to apply when data are approximately symmetric (as compared to skewed data).

However, there are three major drawbacks to transformation. The first is that it can be difficult to interpret the resulting regression coefficients. One of the main reasons that we introduced the natural logarithmic transform in Section 3.2.2 was its ability to provide interpretations of the regression coefficients as proportional changes. Other transformations may not enjoy this intuitively appealing interpretation.

The second drawback, introduced in Section 5.7.4, is that a transformation also affects other aspects of the model, such as heteroscedasticity. For example, if the original model is a multiplicative (heteroscedastic) model of the form $y_i = (\mathrm{E} y_i)\, \varepsilon_i$, then a logarithmic transform means that the new model is

$$\ln y_i = \ln \mathrm{E}(y_i) + \ln \varepsilon_i .$$

Often, the ability to stabilize the variance is viewed as a positive aspect of transformations. However, the point is that a transformation affects both the symmetry and the heteroscedasticity, when only one aspect may be viewed as troublesome.

The third drawback of transforming the dependent variable is that the analyst is implicitly optimizing on the transformed scale. This has been viewed negatively by some scholars. As noted by Manning (1998, p. 285), "No one is interested in log model results per se. Congress does not appropriate log dollars."

"No one is interested in log model results per se. Congress does not appropriate log dollars" Manning (1998).

Our discussions of transformation refers to functions of the dependent variables. As we have seen in Section 3.5, it is common to transform explanatory variables. The adjective "linear" in the phrase "multiple linear regression" refers to linear combinations of parameters – the explanatory variables themselves may be highly nonlinear.

Another technique that we have used implicitly throughout the text for handling fat-tailed data is known as *rescaling*. In rescaling, one divides the dependent variable by an explanatory variable so that the resulting ratio is more comparable among observations. For example, in Section 6.5, we used property and casualty premiums and uninsured losses divided by assets as the dependent variable. Although the numerator, a proxy for annual expenditures associated with insurable events, is the key measure of interest, it is common to standardize by company size (as measured by assets).

Many transformations are special cases of the Box-Cox family of transforms, introduced in Section 1.3. Recall that this family is given as

$$y^{(\lambda)} = \begin{cases} \frac{y^{\lambda}-1}{\lambda} & \text{if } \lambda \neq 0 \\ \ln y & \text{if } \lambda = 0 \end{cases},$$

where λ is the transformation parameter (typically $\lambda = 1$, 1/2, 0, or -1). When data are nonpositive, it is common to add a constant to each observation so that all observations are positive prior to transformation. For example, the transform $\ln(1 + y)$ accommodates the presence of zeros. One can also multiply by a constant so that the approximate original units are retained. For example, the

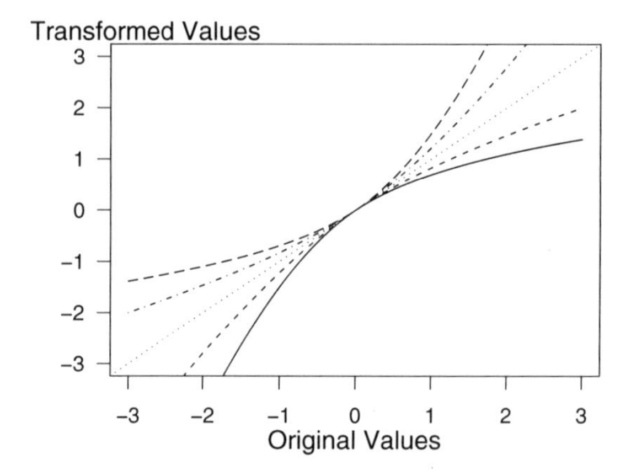

Figure 17.1 Yeo-Johnson transformations. From bottom to top, the curves correspond to $\lambda = 0, 0.5, 1, 1.5$, and 2.

transform $100 \ln(1 + y/100)$ may be applied to percentage data where negative percentages sometimes appear. For the binomial, Poisson and gamma distributions, we also showed how to use power transforms for approximate normality and variance stabilization in Section 13.5.

Alternatively, for handling nonpositive data, an easy-to-use modification is the *signed-log transformation*, given by $\text{sign}(y) \ln(|y| + 1)$. This is a special case of the family introduced by John and Draper (1980):

$$y^{(\lambda)} = \begin{cases} \text{sign}(y) \left\{ (|y| + 1)^{\lambda} - 1 \right\} / \lambda, & \lambda \neq 0 \\ \text{sign}(y) \ln(|y| + 1), & \lambda = 0 \end{cases}.$$

A drawback of the John and Draper family is that its derivative is not continuous at zero meaning that there can be abrupt discontinuities for observations around zero. To address this, Yeo and Johnson (2000) recommend the following extension of the Box-Cox family,

$$y^{(\lambda)} = \begin{cases} \frac{(1+y)^{\lambda}-1}{\lambda} & y \geq 0, \lambda \neq 0 \\ \ln(1 + y) & y \geq 0, \lambda = 0 \\ -\frac{(1+|y|)^{2-\lambda}-1}{2-\lambda} & y < 0, \lambda \neq 2 \\ -\ln(1 + |y|) & y < 0, \lambda = 2 \end{cases}.$$

For nonnegative values of y, this transform is the same as the Box-Cox family with the use of $1 + y$ instead of y to accommodate zeros. For negative values, the power λ is replaced by $2 - \lambda$, so that a right-skewed distribution remains right skewed after the change sign. Figure 17.1 displays this function for several values of λ.

Both the John and Draper and the Yeo and Johnson families are based on power transforms. An alternative family, attributable to Burbidge and Magee (1988), is a modification of the inverse hyperbolic sine transformation. This family is given by

$$y^{(\lambda)} = \sinh^{-1}(\lambda y)/\lambda.$$

17.3 Generalized Linear Models

As introduced in Chapter 13, the generalized linear model (GLM) method has become popular in financial and actuarial statistics. An advantage of this methodology is the ability to fit distributions with tails heavier than the normal distribution. In particular, GLM methods are based on the exponential family that includes the normal, gamma, and inverse Gaussian distributions. As we will see in Section 17.3.1, it is customary to think of the gamma distribution as having intermediate tails and the inverse Gaussian as having heavier tails than the thin-tailed normal distribution.

The idea of a GLM is to map a linear systematic component $\mathbf{x}'\boldsymbol{\beta}$ into the mean of the variable of interest through a known function. Thus, GLMs provide a natural way to include covariates into the modeling. With a GLM, the variance is not required to be constant as in the linear model but is a function of the mean. Once the distribution family and link function have been specified, estimation of GLM regression coefficients depends only on the mean and thus is robust to some model distribution misspecifications. This is both a strength and weakness of the GLM approach. Although more flexible than the linear model, this approach does not handle many of the long-tail distributions traditionally used for modeling insurance data. Thus, in Section 17.4, we will present more flexible distributions.

17.3.1 What Is Fat Tailed?

Many analysts begin discussions of tail heaviness through skewness and kurtosis coefficients. *Skewness* measures the lack of symmetry, or lopsidedness, of a distribution. It is typically quantified by the third standardized moment, $\mathrm{E}(y-\mathrm{E}\,y)^3/(\mathrm{Var}\,y)^{3/2}$. *Kurtosis* measures tail heaviness, or its converse, "peakedness." It is typically quantified by the fourth standardized moment minus 3, $\mathrm{E}(y-\mathrm{E}\,y)^4/(\mathrm{Var}\,y)^2-3$. The "minus 3" is to center discussions around the normal distribution; that is, for a normal distribution, one can check that $\mathrm{E}(y-\mathrm{E}\,y)^4/(\mathrm{Var}\,y)^2=3$. Distributions with positive kurtosis are called *leptokurtic*, whereas those with negative kurtosis are called *platykurtic*. These definitions focus heavily on the normal that has traditionally been viewed as the benchmark distribution.

For many actuarial and financial applications, the normal distribution is not an appropriate starting point and so we seek other definitions of *fat tail*. In addition to moments, the size of the tail can be measured using a density (or mass, for discrete distributions) function, the survival function, or a conditional moment. Typically, the measure would be used to compare one distribution to another.

For example, comparing the right tails of the normal to a gamma density function, we have

$$\begin{aligned}\frac{\mathrm{f}_{normal}(y)}{\mathrm{f}_{gamma}(y)} &= \frac{\sqrt{2\pi\sigma^2}\exp\left(-(y-\mu)^2/(2\sigma^2)\right)}{\left[\lambda^{\alpha}\Gamma(\alpha)\right]^{-1}y^{\alpha-1}\exp(-y/\lambda)}\\ &= C_1\exp\left(-(\alpha-1)\ln y+y/\lambda-(y-\mu)/(2\sigma^2)\right)\to 0,\end{aligned}$$

as $y\to\infty$, indicating that the gamma has a heavier, or fatter, tail than the normal.

Both the normal and the gamma are members of the exponential family of distributions. For comparison with another member of this family, the inverse Gaussian distribution, consider

$$\frac{\mathrm{f}_{gamma}(y)}{\mathrm{f}_{invGaussian}(y)} = \frac{[\lambda^{\alpha}\Gamma(\alpha)]^{-1} y^{\alpha-1}\exp(-y/\lambda)}{\sqrt{\theta/(2\pi y^3)}\exp\left(-\theta(y-\mu)^2/(2y\mu^2)\right)}$$

$$= C_2\exp\left((\alpha+1/2)\ln y - y/\lambda + \theta(y-\mu)^2/(2y\mu^2)\right).$$

As $y \to \infty$, this ratio tends to zero for $\theta/(2\mu^2) < \lambda$, indicating that the inverse Gaussian can have a heavier tail than the gamma.

For a distribution that is not a member of the exponential family, consider the Pareto distribution. Similar calculations show

$$\frac{\mathrm{f}_{gamma}(y)}{\mathrm{f}_{Pareto}(y)} = \frac{[\lambda^{\alpha}\Gamma(\alpha)]^{-1} y^{\alpha-1}\exp(-y/\lambda)}{\alpha\theta^{-\alpha}(y+\theta)^{-\alpha-1}}$$

$$= C_3\exp((\alpha-1)\ln y - y/\lambda + (\alpha+1)\ln(y+\theta)) \to 0,$$

as $y \to \infty$, indicating that the Pareto has a heavier tail than the gamma.

The ratio of densities is an easily interpretable measure for comparing the tail heaviness of two distributions. Because densities and survival functions have a limiting value of zero, by L'Hôpital's rule, the ratio of survival functions is equivalent to the ratio of densities. That is,

$$\lim_{y\to\infty}\frac{\mathrm{S}_1(y)}{\mathrm{S}_2(y)} = \lim_{y\to\infty}\frac{\mathrm{S}_1'(y)}{\mathrm{S}_2'(y)} = \lim_{y\to\infty}\frac{\mathrm{f}_1(y)}{\mathrm{f}_2(y)}.$$

This provides another motivation for using this measure.

17.3.2 Application: Wisconsin Nursing Homes

Ⓡ **Empirical** *Filename is "WiscNursingHome"*

Nursing home financing has drawn the attention of policymakers and researchers for the past several decades. With aging populations and increasing life expectancies, expenditures on nursing homes and demands of long-term care are expected to increase in the future. In this section, we analyze the data of 349 nursing facilities in the state of Wisconsin in the cost report year 2001.

The state of Wisconsin Medicaid program funds nursing home care for individuals qualifying on the basis of need and financial status. Most, but not all, nursing homes in Wisconsin are certified to provide Medicaid-funded care. Those that do not accept Medicaid are generally paid directly by the resident or the resident's insurer.

Similarly, most but not all nursing facilities are certified to provide Medicare-funded care. Medicare provides postacute care for 100 days following a related hospitalization. Medicare does not fund care provided by intermediate-care facilities to individuals with developmental disabilities. As part of the conditions for participation, Medicare-certified nursing homes must file an annual cost report to the Wisconsin Department of Health and Family Services that summarizes the

Table 17.1 Nursing Home Descriptive Statistics

Variable	Description	Percentage	Median TPY
TPY	Total person years (median 81.89)		
Continuous Explanatory Variables			
NumBed	Number of beds (median 90)		
SqrFoot	Nursing home net square footage (in thousands, median 40.25)		
Categorical Explanatory Variables		Percentage	Median TPY
POPID	Nursing home identification number		
SelfIns	Self-funding of insurance		
	Yes	62.8	88.40
	No	37.2	67.84
MCert	Medicare certified		
	Yes	90.5	84.06
	No	9.5	53.38
Organizational Structure	Pro (for profit)	51.9	77.23
	TaxExempt (tax exempt)	37.5	81.13
	Govt (governmental unit)	10.6	106.70
Location	Urban	53.3	91.55
	Rural	46.7	74.12

volume and cost of care provided to all of its residents, Medicare-funded and otherwise.

Nursing homes are owned and operated by a variety of entities, including the state, counties, municipalities, for-profit businesses, and tax-exempt organizations. Private firms often own several nursing homes. Periodically, facilities may change ownership and, less frequently, ownership type.

Typically, utilization of nursing home care is measured in patient days. Facilities bill the fiscal intermediary at the end of each month for total patient days incurred in the month, itemized by resident and level of care. Projections of patient days by facility and level of care play a key role in the annual process of updating facility rate schedules. Rosenberg et al. (2007) provide additional discussion.

Summarizing the Data

After examining the data, we found some minor variations in the number of days that a facility was open, primarily because of openings and closing of facilities. Thus, to make utilization more comparable among facilities, we examine TPY, defined as the total number of patient days divided by the number of days the facility was open; this has a median value of 81.99 per facility.

Table 17.1 describes the variables that will be used to explain the distribution of TPY. More than half of the facilities have self-funding of insurance. Approximately 90.5% of the facilities are Medicare certified. Regarding the organizational structure, about half (51.9%) are run on a for-profit basis, and about one-third (37.5%) are organized as tax exempt; the remainder are governmental

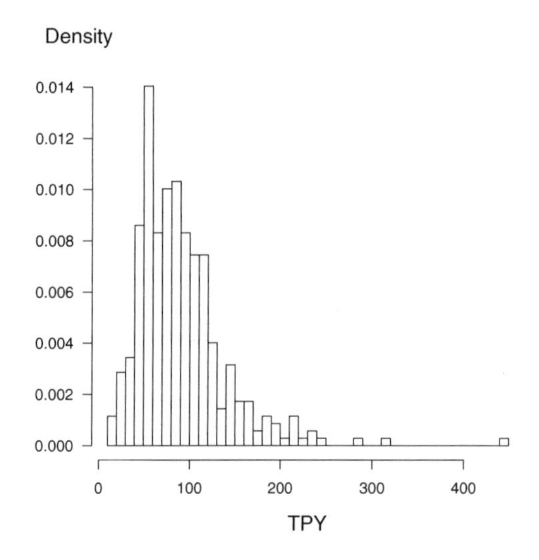

Figure 17.2 Histogram of TPY. This plot demonstrates the right skewness of the distribution.

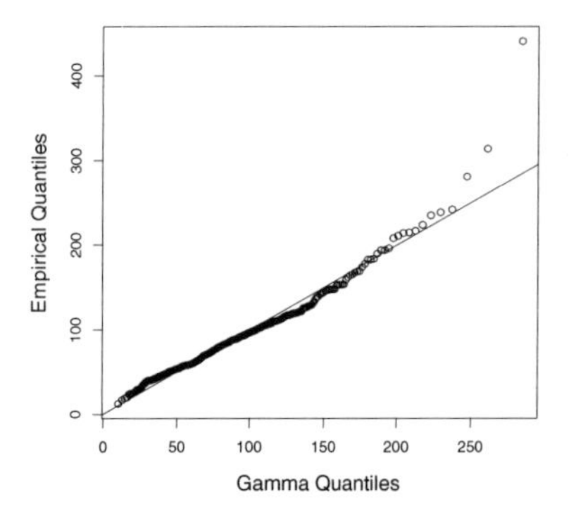

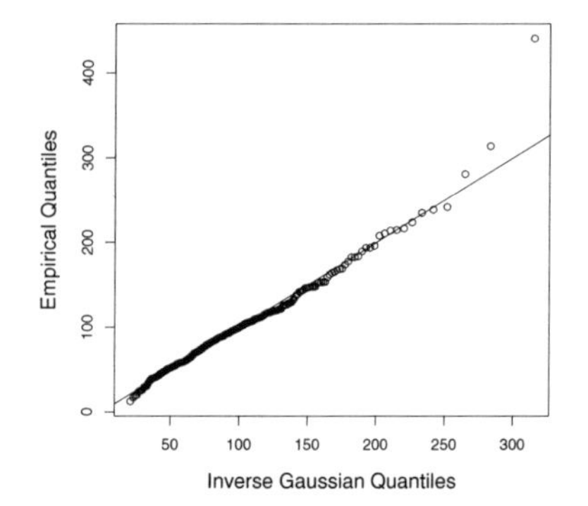

Figure 17.3 The qq plots of TPY for the gamma and inverse Gaussian distributions.

organizations. The tax-exempt facilities have the highest median occupancy rates. Slightly more than half of the facilities are located in an urban environment (53.3%).

Fitting Generalized Linear Models

Figure 17.2 shows the distribution of the dependent variable TPY. From this figure, we see clear evidence of the right skewness of the distribution. One option would be to take a transform as described in Section 17.2. Rosenberg et al. (2007) explore this option using a logarithmic transformation.

Another option is to directly fit a skewed distribution to the data. Figure 17.3 presents the qq plots of the gamma and inverse Gaussian distributions. The data fall fairly close to the line in both panels, meaning that both models are reasonable choices. The normal qq plot, not shown here, indicates that the normal regression model is not a reasonable fit.

We fit the generalized linear models using the gamma and inverse Gaussian distributions. In both models, we choose the logarithmic link function. The linear

Table 17.2 Fitted Nursing Home Generalized Linear Models

	Gamma		Inverse Gaussian	
Variables	Estimate	t-Ratio	Estimate	t-Ratio
Intercept	−0.159	−3.75	−0.196	−4.42
ln(NumBed)	0.996	66.46	1.023	65.08
ln(SqrFoot)	0.026	2.07	0.003	0.19
SelfIns	0.006	0.75	0.003	0.27
MCert	−0.008	−0.55	−0.008	−0.57
Pro	0.004	0.29	0.007	0.36
TaxExempt	0.018	1.28	0.021	1.12
Urban	−0.011	−1.25	−0.006	−0.64
Scale	165.64		0.0112	
Goodness-of-fit statistics				
Log-likelihood	−1,131.24		−1,218.15	
AIC	2,280.47		2,454.31	
BIC	2,315.17		2,489.00	

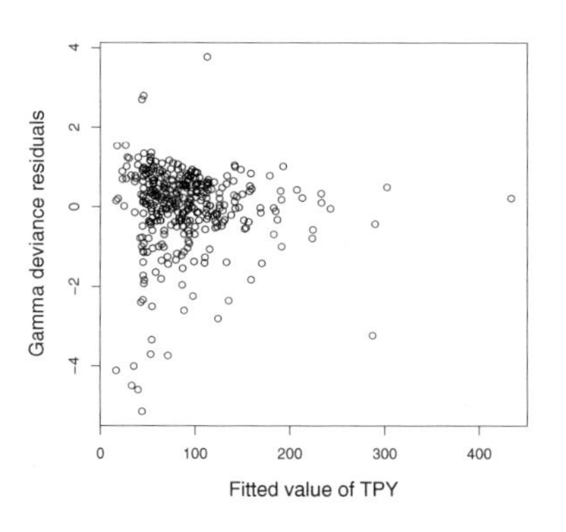

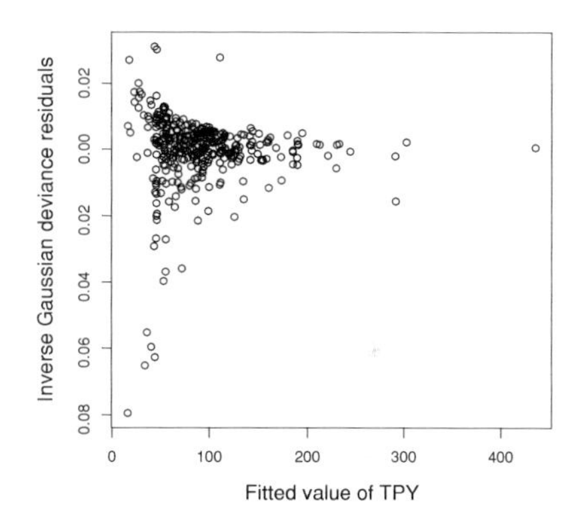

Figure 17.4 Plots of deviance residuals versus fitted values for the gamma and inverse Gaussian models.

systematic component that is common to each model is

$$\eta = \beta_0 + \beta_1 \ln(\text{NumBed}) + \beta_2 \ln(\text{SqrFoot}) + \beta_3 \text{Pro} \qquad (17.1)$$
$$+\beta_4 \text{TaxExempt} + \beta_5 \text{SelfIns} + \beta_6 \text{MCert} + \beta_7 \text{Urban}.$$

Table 17.2 summarizes the parameter estimates of the models. By comparing the BIC statistics, or the AIC and log-likelihood in that the number of estimated parameters and the sample size in both models are identical, we find that the gamma model performs better than the inverse Gaussian. As anticipated, the coefficient for the size variable NumBed is positive and significant. The only other variable that is statistically significant is the SqrFoot variable, and this only in the gamma model.

Figure 17.4 presents the plots of deviance residuals against the fitted value of TPY for the gamma and inverse Gaussian models. No patterns are found in the plots, supporting the position that the models are reasonable fits to the data.

17.4 Generalized Distributions

Another approach for handling fat-tailed regression data is to use parametric distributions, such as those from survival modeling. Although survival analysis focuses on censored data, the methods can certainly be applied to complete data. In Section 14.3, we introduced an accelerated failure time (AFT) model. The AFT is a log location-scale model, so that $\ln(y)$ follows a parametric location-scale density distribution in the form $\mathrm{f}(y) = \mathrm{f}_0\left((y-\mu)/\sigma\right)/\sigma$, where μ and $\sigma > 0$ are location and scale parameters, and f_0 is the standard form of the distribution. The Weibull, lognormal, and log-logistic distributions are commonly used lifetime distributions that are special cases of the AFT framework.

For fitting fat-tailed distributions of interest in actuarial science, we consider the following minor variation, and examine distributions from the relation

$$\ln y = \mu + \sigma \ln y_0. \tag{17.2}$$

As before, the distribution associated with y_0 is a standard one, and we are interested in the distribution of the random variable y. Two important special cases are the generalized gamma and the generalized beta of the second kind. These distributions have been used extensively in modeling insurance data; see, for example, Klugman et al. (2008), though most applications have not utilized regression covariates.

The *generalized gamma distribution* is obtained when y_0 has a gamma distribution with shape parameter α and scale parameter 1. When including limiting distributions (e.g., allowing coefficients to become arbitrarily large), it includes the exponential, Weibull, gamma, and lognormal distributions as special cases. Therefore, it can be used to discriminate between the alternate models. The generalized gamma distribution is also known as the transformed gamma distribution (Klugman et al., 2008).

When y_0 has a distribution that is the ratio of two gammas, then y is said to have a *generalized beta of the second kind distribution,* commonly known by the acronym *GB2*. Specifically, we assume that $y_0 = Gamma_1/Gamma_2$, where $Gamma_i$ has a gamma distribution with shape parameter α_i and scale parameter 1, $i = 1, 2$, and that $Gamma_1$ and $Gamma_2$ are independent. Thus, the GB2 family has four parameters (α_1, α_2, μ, and σ) compared to the three-parameter generalized gamma distribution. When including limiting distributions, the GB2 encompasses the generalized gamma (by allowing $\alpha_2 \to \infty$) and hence the exponential, Weibull, and so forth. It also encompasses the Burr type 12 (by allowing $\alpha_1 = 1$), as well as other families of interest, including the Pareto distributions.

The distribution of y from equation (17.2) contains location parameter μ, scale parameter σ, and additional parameters that describe the distribution of y_0. In principle, one could allow for any distribution parameter to be a function of the covariates. However, following this principle would lead to a large number of parameters; this typically yields computational difficulties and problems of interpretations. To limit the number of parameters, it is customary to assume that the parameters from y_0 do not depend on covariates. It is natural to allow the location parameter to be a linear function of covariates so that $\mu = \mu(\mathbf{x}) = \mathbf{x}'\boldsymbol{\beta}$. One may also allow the scale parameter σ to depend on $\mathbf{x}$. For σ positive, a common specification is $\sigma = \sigma(\mathbf{x}) = \exp(\mathbf{x}'\boldsymbol{\beta}_\sigma)$, where $\boldsymbol{\beta}_\sigma$ are regression coefficients associated with the scale parameter. Other parameters are typically held fixed.

The interpretability of parameters is one reason to hold the scale and other nonlocation parameters fixed. By doing this, it is straightforward to show that the regression function is of the form

$$\mathrm{E}(y|\mathbf{x}) = C \exp(\mu(\mathbf{x})) = C\, e^{\mathbf{x}'\boldsymbol{\beta}},$$

where the constant C is a function of other (nonlocation) model parameters. Thus, one can interpret the regression coefficients in terms of a proportional change (an *elasticity* in economics). That is, $\partial\left[\ln \mathrm{E}(y)\right]/\partial x_k = \beta_k$.

Another reason for holding nonlocation parameters fixed is the ease of computing sensible residuals and using these residuals to assist with model selection. Specifically, with equation (17.2), one can compute residuals of the form

$$r_i = \frac{\ln y_i - \widehat{\mu}_i}{\widehat{\sigma}},$$

where $\widehat{\mu}_i$ and $\widehat{\sigma}$ are maximum likelihood estimates. For large datasets, we may assume little estimation error so that $r_i \approx (\ln y_i - \mu_i)/\sigma$, and the quantity on the right-hand side has a known distribution.

To illustrate, consider the case when y follows a GB2 distribution. In this case,

$$y_0 = \frac{Gamma_1}{Gamma_2} = \frac{\alpha_1}{\alpha_2} \times \frac{Gamma_1/(2\alpha_1)}{Gamma_2/(2\alpha_2)} = \frac{\alpha_1}{\alpha_2} \times F,$$

where F has an F-distribution with numerator and denominator degrees of freedom $df_1 = 2\alpha_1$ and $df_2 = 2\alpha_2$. Then, $\exp(r_i) \approx (\alpha_1/\alpha_2)F_i$, so that the exponentiated residuals should have an approximate F-distribution (up to a scale parameter). This fact allows us to compute quantile-quantile (*qq*) plots to assess model adequacy graphically.

To illustrate, we consider a few insurance related examples that use fat-tailed regression models. McDonald and Butler (1990) discussed regression models, including those commonly used as well as the GB2 and generalized gamma distribution. They applied the model to the duration of poverty spells and found that the GB2 improved model fitting significantly over the lognormal. Beirlant et al. (1998) proposed two Burr regression models and applied them to portfolio segmentation for fire insurance. The Burr is a an extension of the Pareto distribution, though still a special case of the GB2. Manning, Basu, and Mullahy (2005) applied the generalized gamma distribution to inpatient expenditures using the data from a study of hospitals conducted at the University of Chicago.

Because the regression model is fully parametric, maximum likelihood is generally the estimation method of choice. If y follows a GB2 distribution, straightforward calculations show that its density can be expressed as

$$f(y;\mu,\sigma,\alpha_1,\alpha_2) = \frac{[\exp(z)]^{\alpha_1}}{y|\sigma|B(\alpha_1,\alpha_2)[1+\exp(z)]^{\alpha_1+\alpha_2}}, \tag{17.3}$$

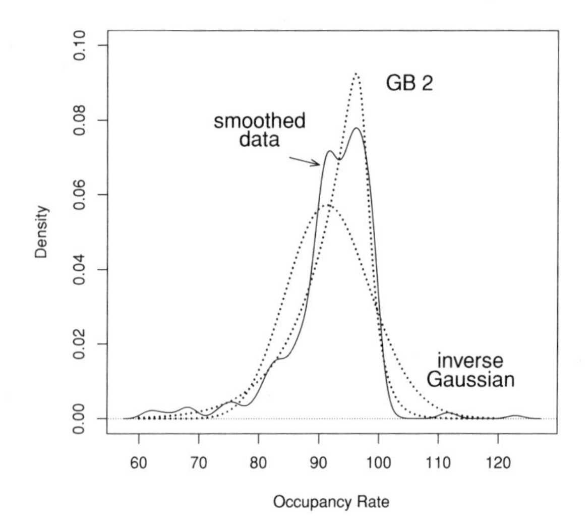

Figure 17.5 Nursing Home Densities. The empirical version, based on a kernel density estimate, is compared to fitted GB2 and inverse gaussian densities.

where $z = (\ln y - \mu)/\sigma$ and $\mathrm{B}(\cdot,\cdot)$ is the beta function, defined as $\mathrm{B}(\alpha_1, \alpha_2) = \Gamma(\alpha_1)\Gamma(\alpha_2)/\Gamma(\alpha_1 + \alpha_2)$. This density can be used directly in likelihood routines of many statistical packages. As described in Section 11.9, the method of maximum likelihood automatically provides

- Standard errors for the parameter estimates
- Methods of model selection via likelihood ratio testing
- Goodness-of-fit statistics such as *AIC* and *BIC*.

Application: Wisconsin Nursing Homes

In the fitted generalized linear models summarized in Table 17.2, we saw that the coefficients associated with ln(NumBed) were close to one. This suggests identifying ln(NumBed) as an *offset variable*, that is, forcing the coefficient associated with ln(NumBed) to be 1. For another modeling strategy, it also suggests rescaling the dependent variable by NumBed. This is sensible because we used a logarithmic link function so that the expected value of TPY is proportional to NumBed. Pursuing this approach, we now define the annual occupancy rate (Rate) to be

$$\text{Occupancy Rate} = \frac{\text{Total Patient Days}}{\text{Number of Beds} \times \text{Days Open}} \times 100. \qquad (17.4)$$

This new dependent variable is easy to interpret – it measures the percentage of beds being used on any given day. Occupancy rates were calculated using the average number of licensed beds in a cost report year rather than the number of licensed beds on a specific day. This gives rise to a few occupancy rates of greater than 100.

One difficulty of using occupancy rates is that its distribution cannot reasonably be approximated by a member of the exponential family. Figure 17.5 shows a smoothed histogram of the rate variable (using a kernel smoother); this

Table 17.3 Wisconsin Nursing Home Generalized Models Fits

	Generalized Gamma		GB2	
Variables	Estimate	t-Ratio	Estimate	t-Ratio
Intercept	4.522	78.15	4.584	198.47
ln(NumBed)	−0.027	−2.06	−0.010	−1.17
ln(SqrFoot)	0.031	2.89	0.010	1.28
SelfIns	0.003	0.44	−0.001	−0.25
MCert	−0.010	−0.81	−0.010	−1.30
Pro	−0.021	−1.46	−0.002	−0.20
TaxExempt	−0.007	−0.48	0.015	1.66
Urban	−0.014	−1.78	−0.003	−0.60
	Estimate	Std. Error	Estimate	Std. Error
$\ln \sigma$	−2.406	0.131	−5.553	1.716
$\ln \alpha_1$	0.655	0.236	−2.906	1.752
$\ln \alpha_2$			−1.696	1.750
Log-likelihood	−1,148.135		−1,098.723	
AIC	2,316.270		2,219.446	
BIC	2,319.296		2,223.822	

distribution is *left* skewed. Superimposed on it with the dotted line is the inverse Gaussian distribution, where the parameters were fit without covariates, using method of moments. The gamma and normal distributions are very close to the inverse Gaussian and hence not shown here. In contrast, the fitted (also without covariates) GB2 distribution shown in Figure 17.5 captures important parts of the distribution; in particular, it captures the peakedness and left skewness.

The GB2 distribution was fit using maximum likelihood with the same covariates as in Table 17.2. Specifically, we used location parameter $\mu = \exp(\eta)$, where η is specified in equation (17.1). As is customary in likelihood estimation, we reparameterized the scale and two shape parameters, σ, α_1, and α_2, to be transformed on the log scale so that they could range over the whole real line. In this way, we avoided boundary problems that could arise when trying to fit models with negative parameter values. Table 17.3 summarizes the fitted model. Unfortunately, for this fitted model, none of the explanatory variables turned out to be statistically significant. (Recall that we rescaled by number of beds, a very important explanatory variable.)

To further assess the model fit, Figure 17.6 shows residuals from this fitted model. For these figures, residuals are computed using $r_i = (\ln y_i - \widehat{\mu}_i)/\widehat{\sigma}$. The left-hand panel shows the residuals versus fitted values $(\exp(\widehat{\mu}_i))$, no apparent patterns are evident in this display. The right-hand panel is a qq plot of residuals, where the reference distributions is the logarithmic F-distribution (plus a constant) described earlier. This figure shows some discrepancies for smaller values of nursing homes. Because of this, Table 17.3 also reports fits from the generalized gamma model. This fit is more pleasing in the sense that two of the explanatory variable are statistically significant. However, from the goodness-of-fit statistics, we see that the GB2 is a better-fitting model. Note that the goodness-of-fit

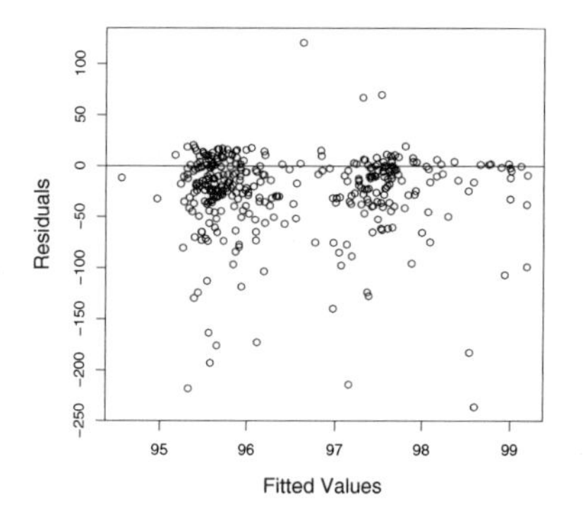

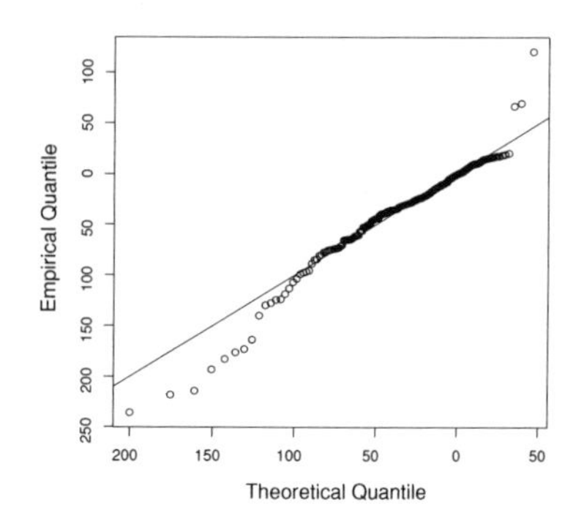

Figure 17.6 Residual analysis of the GB2 model. The left-hand panel is a plot of residuals versus fitted values. The right-hand panel is a *qq* plot of residuals.

statistics for the generalized gamma model are not directly comparable with the gamma regression fits in Table 17.2; this is only because the dependent variable differs by the scale variable NumBeds.

17.5 Quantile Regression

Quantile regression is an extension of median regression, so it is helpful to introduce this concept first.

In median regression, one finds the set of regression coefficients $\boldsymbol{\beta}$ that minimizes

$$\sum_{i=1}^{n} |y_i - \mathbf{x}_i'\boldsymbol{\beta}|.$$

That is, we simply replace the usual squared loss function with an absolute value function. Although we do not go into the details here, finding these optimal coefficients is a simple optimization problem in nonlinear programming that can be readily implemented in statistical software.

Because this procedure uses the absolute value as the loss function, median regression is also known as *LAD*, for *least absolution deviations*, as compared to *OLS* (for ordinary least squares). The adjective *median* comes from the special case in which there are no regressors, so that $\mathbf{x}$ is a scalar 1. In this case the minimization problem reduces to finding an intercept β_0 that minimizes

$$\sum_{i=1}^{n} |y_i - \beta_0|.$$

The solution to this problem is the *median* of $\{y_1, \ldots, y_n\}$.

Suppose that you would also like to find the 25th, 75th, or some other percentile of $\{y_1, \ldots, y_n\}$. One can also use this optimization procedure to find any percentile, or *quantile*. Let τ be a fraction between 0 and 1. Then, the τth sample quantile of $\{y_1, \ldots, y_n\}$ is the value of β_0 that minimizes

$$\sum_{i=1}^{n} \rho_\tau(y_i - \beta_0).$$

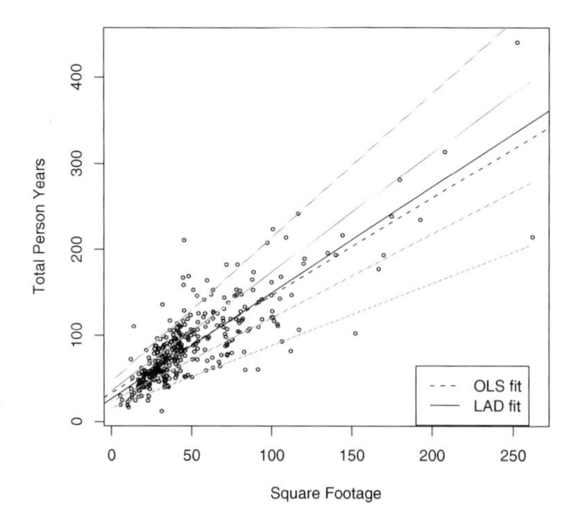

Figure 17.7 Quantile regression fits of square footage on total person years. Superimposed are fits from mean (OLS) and median (LAD) regressions, indicated in the legend. Also superimposed with gray lines are quantile regression fits – from bottom to top, the fits correspond to $\tau = 0.05, 0.25, 0.75$, and 0.95.

Here, $\rho_\tau(u) = u(\tau - \mathrm{I}(u \leq 0))$ is called a *check function* and $\mathrm{I}(\cdot)$ is the indicator function.

Extending this procedure, in quantile regression one finds the set of regression coefficients $\boldsymbol{\beta}$ that minimizes

$$\sum_{i=1}^{n} \rho_\tau(y_i - \mathbf{x}_i'\boldsymbol{\beta}).$$

The estimated regression coefficients depend on the fraction τ, so we use the notation $\widehat{\boldsymbol{\beta}}(\tau)$ to emphasize this dependence. The quantity $\mathbf{x}_i'\widehat{\boldsymbol{\beta}}(\tau)$ represents the τth quantile of the distribution of y_i for the explanatory vector $\mathbf{x}_i$. To illustrate, for $\tau = 0.5$, $\mathbf{x}_i'\widehat{\boldsymbol{\beta}}(0.5)$ represents the estimated *median* of the distribution of y_i. In contrast, the OLS fitted value $\mathbf{x}_i'\mathbf{b}$ represents the estimated *mean* of the distribution of y_i.

Example: Wisconsin Nursing Homes, Continued. To illustrate quantile regression techniques, we fit a regression of square footage (SqrFoot) on total person years (TPY). Figure 17.7 shows the relationship between the two variables, with mean (OLS) and median (LAD) fitted lines superimposed. Unlike the original TPY distribution that is skewed, for each value of SqrFoot we can see little difference between the mean and median values. This suggests that the conditional distribution of TPY given SqrFoot is not skewed.

Figure 17.7 also shows the fitted lines that result from fitting quantile regressions at four additional values of $\tau = 0.05, 0.25, 0.75$, and 0.95. These fits are indicated by the gray lines. At each value of SqrFoot, we can visually get a sense of the 5th, 25th, 50th, 75th, and 95th percentiles of the distribution of TPY. Although classic ordinary least squares also provides this, the classic recipes generally assume homoscedasticity. From 17.7, we see that the distribution of y seems to widen as SqrFoot increases, suggesting a heteroscedastic relationship.

Quantile regressions perform well in situations when ordinary least squares requires careful attention to be used with confidence. As demonstrated in the Wisconsin nursing home example, quantile regression handles skewed distributions and heteroscedasticity readily. Just as ordinary quantiles are relatively robust to unusual observations, quantile regression estimates are much less sensitive to outlying observations than the usual regression routines.

17.6 Extreme Value Models

Extreme value models focus on the extremes, the "tip of the iceberg," such as the highest temperature over a month, the fastest time to run a kilometer, or the lowest return from the stock market. Some extreme value models are motivated by maximal statistics. Suppose that we consider annual chief executive officer (CEO) compensation in a country, $y_1, y_2, \ldots$. Then, $M = \max(y_1, \ldots, y_n)$ represents the compensation of the highest-paid CEO during the year. If values y were observed, then we could use some mild additional assumptions (e.g., independence) to make inferences about the the distribution of M. However, in many cases, only M is directly observed, forcing us to base inference procedures on extreme observations M. As a variation, we might have observations for the top 20 CEOs – not the entire population. This variation uses inference based on the 20 largest-order statistics; see, for example, Coles (2003, Section 3.5.2).

Modeling M is often based on the *generalized extreme value*, or GEV, distribution, defined by the distribution function

$$\Pr(M \leq x) = \exp\left[-(1+\gamma z)^{-1/\gamma}\right], \tag{17.5}$$

where $z = (x - \mu)/\sigma$. This is a location-scale model, with location and scale parameters μ and σ, respectively. In the standard case where $\mu = 0$ and $\sigma = 1$, allowing $\gamma \to \infty$ means that $\Pr(M \leq x) \to \exp\left[-e^{-x}\right]$, the classical extreme-value distribution. Thus, the parameter γ provides the generalization of this classical distribution.

Beirlant, Goegebeur, Segers, and Teugels (2004) discuss ways in which one could introduce regression covariates into the GEV distribution, essentially by allowing each parameter to depend on covariates. Estimation is done via maximum likelihood. In their inference procedures, the focus is on the behavior of the extreme quantiles (conditional on the covariates).

Another approach to modeling extreme values is to focus on data that must be large to be included in the sample.

Example: Large Medical Claims. Cebrián, Denuit, and Lambert (2003) analyzed 75,789 large group medical insurance claims from 1991. To be included in the database, claims must exceed \$25,000. Thus, the data are left truncated at \$25,000. The interest in their study was to interpret the long-tailed distribution in terms of covariates age and sex.

The *peaks-over-threshold* approach to modeling extreme values is motivated by left-truncated data where the truncation point, or "threshold," is large. To be included in the dataset, the observations must exceed a large threshold that we refer to as a "peak." Following our Section 14.2 discussion on truncation, if C_L is the left truncation point, then the distribution function of $y - C_L$, given that $y > C_L$, is $1 - \Pr(y - C_L > x | y > C_L) = 1 - (1 - F_y(C_L + x))/(1 - F_y(C_L))$. Instead of modeling the distribution of y, F_y, directly as in prior sections, one assumes that it can be directly approximated by a *generalized Pareto distribution*. That is, we assume

$$\Pr(y - C_L \leq x | y > C_L) \approx 1 - (1 + \frac{z}{\theta})^{-\theta}, \tag{17.6}$$

where $z = x/\sigma$, σ is a scale parameter, $x \geq 0$ if $\theta \geq 0$, and $0 \leq x \leq -\theta$ if $\theta < 0$. Here, the right-hand side of equation (17.6) is the generalized Pareto distribution. The usual Pareto distribution restricts θ to be positive; this specification allows for negative values of θ. Allowing $\theta \to 0$ means that $1 - (1 + z/\theta)^{-\theta} \to 1 - e^{-x/\sigma}$, the exponential distribution.

Example: Large Medical Claims, Continued. To incorporate age and sex covariates, Cebrián, Denuit, and Lambert (2003) categorized the variables, allowed parameters to vary by category, and estimated each category in isolation of the others. Alternative, more efficient, approaches are described in Chapter 7 of Beirlant et al. (2004).

17.7 Further Reading and References

The literature on long-tailed claims modeling is actively developing. A standard reference is Klugman, Panjer, and Willmot (2008). Kleiber and Kotz (2003) provide an excellent survey of the univariate literature, with many historical references. Carroll and Ruppert (1988) provide extensive discussions of transformations in regression modeling.

This chapter has emphasized the GB2 distribution with its many special cases. Venter (2007) discusses extensions of the generalized linear model, focusing on loss reserving applications. Balasooriya and Low (2008) provide recent applications to insurance claims modeling, although without any regression covariates. Another approach is to use a skewed elliptical (e.g., normal, t-) distribution. Bali and Theodossiou (2008) provide a recent application, showing how to use such distributions in time series modeling of stock returns.

Koenker (2005) is an excellent book-length introduction to quantile regression. Yu, Lu, and Stander (2003) provide an accessible shorter introduction.

Coles (2003) and Beirlant et al. (2004) are two excellent book-length introductions to extreme value statistics.

Chapter References

Balasooriya, Uditha, and Chan-Kee Low (2008). Modeling insurance claims with extreme observations: Transformed kernel density and generalized lambda distribution. *North American Actuarial Journal* 11, no. 2, 129–42.

Bali, Turan G., and Panayiotis Theodossiou (2008). Risk measurement of alternative distribution functions. *Journal of Risk and Insurance* 75, no. 2, 411–37.

Beirlant, Jan, Yuir Goegebeur, Johan Segers, and Jozef Teugels (2004). *Statistics of Extremes*. Wiley, New York.

Beirlant, Jan, Yuir Goegebeur, Robert Verlaak, and Petra Vynckier (1998). Burr regression and portfolio segmentation. *Insurance: Mathematics and Economics* 23, 231–50.

Burbidge, J. B., L. Magee, and A. L. Robb (1988). Alternative transformations to handle extreme values of the dependent variable. *Journal of the American Statistical Association* 83, 123–7.

Carroll, Raymond, and David Ruppert (1988). *Transformation and Weighting in Regression*. Chapman-Hall, New York.

Cebrián, Ana C., Michel Denuit, and Philippe Lambert (2003). Generalized Pareto fit to the Society of Actuaries' large claims database. *North American Actuarial Journal* 7, no. 3, 18–36.

Coles, Stuart (2003). *An Introduction to Statistical Modeling of Extreme Values*. Springer, New York.

Cummins, J. David, Georges Dionne, James B. McDonald, and B. Michael Pritchett (1990). Applications of the GB2 family of distributions in modeling insurance loss processes. *Insurance: Mathematics and Economics* 9, 257–72.

John, J. A., and Norman R. Draper (1980). An alternative family of transformations. *Applied Statistics* 29, no. 2, 190–7.

Kleiber, Christian, and Samuel Kotz (2003). *Statistical Size Distributions in Economics and Actuarial Sciences*. John Wiley & Sons, New York.

Klugman, Stuart A., Harry H. Panjer, and Gordon E. Willmot (2008). *Loss Models: From Data to Decisions*. John Wiley & Sons, Hoboken, New Jersey.

Koenker, Roger (2005). *Quantile Regression*. Cambridge University Press, New York.

Manning, Willard G. (1998). The logged dependent variable, heteroscedasticity, and the retransformation problem. *Journal of Health Economics* 17, 283–95.

Manning, Willard G., Anirban Basu, and John Mullahy (2005). Generalized modeling approaches to risk adjustment of skewed outcomes data. *Journal of Health Economics* 24, 465–88.

McDonald, James B., and Richard J. Butler (1990). Regression models for positive random variables. *Journal of Econometrics* 43, 227–51.

Rachev, Svetiozar T., Christian Menn, and Frank Fabozzi (2005). *Fat-Tailed and Skewed Asset Return Distributions: Implications for Risk Management, Portfolio Selection, and Option Pricing*. Wiley, New York.

Rosenberg, Marjorie A., Edward W. Frees, Jiafeng Sun, Paul Johnson, and James M. Robinson (2007). Predictive modeling with longitudinal data: A case study of Wisconsin nursing homes. *North American Actuarial Journal* 11, no. 3, 54–69.

Sun, Jiafeng, Edward W. Frees, and Marjorie A. Rosenberg (2008). Heavy-tailed longitudinal data modeling using copulas. *Insurance: Mathematics and Economics* 42, no. 2, 817–30.

Venter, Gary (2007). Generalized linear models beyond the exponential family with loss reserve applications. *Astin Bulletin: Journal of the International Actuarial Association* 37, no. 2, 345–64.

Yeo, In-Kwon, and Richard A. Johnson (2000). A new family of power transformations to improve normality or symmetry. *Biometrika* 87, 954–9.

Yu, Keming, Zudi Lu, and Julian Stander (2003). Quantile regression: Applications and current research areas. *Journal of the Royal Statistical Society Series D (The Statistician)* 52, no. 3, 331–50.

17.8 Exercises

17.1. Quantiles and Simulation. Use equation (17.2) to establish the following distributional relationships that are helpful for calculating quantiles.

a. Assume that $y_0 = \alpha_1 F/\alpha_2$, where F has an F-distribution with numerator and denominator degrees of freedom $df_1 = 2\alpha_1$ and $df_2 = 2\alpha_2$. Show that y has a GB2 distribution.

b. Assume that $y_0 = B/(1-B)$, where B has a beta distribution with parameters α_1 and α_2. Show that y has a GB2 distribution.

c. Describe how to use parts (a) and (b) for calculating quantiles.

d. Describe how to use parts (a) and (b) for simulation.

17.2. Consider a GB2 probability density function given in equation (17.3).

a. Reparameterize the distribution by defining the new parameter $\theta = e^{\mu}$. Show that the density can be expressed as

$$\mathrm{f}_{GB2}(y;\theta,\sigma,\alpha_1,\alpha_2) = \frac{\Gamma(\alpha_1+\alpha_2)}{\Gamma(\alpha_1)\Gamma(\alpha_2)} \frac{(y/\theta)^{\alpha_2/\sigma}}{\sigma y\left[1+(y/\theta)^{1/\sigma}\right]^{\alpha_1+\alpha_2}}.$$

b. Using part (a), show that

$$\lim_{\alpha_2\to\infty} \mathrm{f}_{GB2}(y;\theta\alpha_2^{\sigma},\sigma,\alpha_1,\alpha_2) = \frac{1}{\sigma y \Gamma(\alpha_1)} (y/\theta)^{\alpha_1/\sigma} \exp\left(-(y/\theta)^{1/\sigma}\right)$$
$$= \mathrm{f}_{GG}(y;\theta,\sigma,\alpha_1),$$

a generalized gamma density.

c. Using part (a), show that

$$\mathrm{f}_{GB2}(y;\theta,\sigma,1,\alpha_2) = \frac{\alpha_2 (y/\theta)^{\alpha_2/\sigma}}{\sigma y\left[1+(y/\theta)^{1/\sigma}\right]^{1+\alpha_2}} = \mathrm{f}_{Burr}(y;\theta,\sigma,\alpha_2),$$

a Burr type 12 density.

17.3. Recall that the density of a gamma distribution with shape parameter α and scale parameter θ has a density given by $\mathrm{f}(y) = [\theta^{\alpha}\Gamma(\alpha)]^{-1} y^{\alpha-1}\exp(-y/\theta)$ and kth moment given by $\mathrm{E}(y^k) = \theta^k \Gamma(\alpha+k)/\Gamma(\alpha)$, for $k > -\alpha$.

a. For the GB2 distribution, show that

$$\mathrm{E}(y) = e^{\mu}\frac{\Gamma(\alpha_1+\sigma)\Gamma(\alpha_2-\sigma)}{\Gamma(\alpha_1)\Gamma(\alpha_2)}.$$

b. For the generalized gamma distribution, show that

$$\mathrm{E}(y) = e^{\mu}\Gamma(\alpha_1+\sigma)/\Gamma(\alpha_1).$$

c. Calculate the moments of a Burr type 12 density.

18

Credibility and Bonus-Malus

Chapter Preview. This chapter introduces regression applications of pricing in credibility and bonus-malus experience rating systems. Experience rating systems are formal methods for including claims experience into renewal premiums of short-term contracts, such automobile, health, and workers' compensation. This chapter provides brief introductions to credibility and bonus-malus, emphasizing their relationship with regression methods.

18.1 Risk Classification and Experience Rating

Risk classification is a key ingredient of insurance pricing. Insurers sell coverage at prices that are sufficient to cover anticipated claims, administrative expenses, and an expected profit to compensate for the cost of capital necessary to support the sale of the coverage. In many countries and lines of business, the insurance market is mature and highly competitive. This strong competition induces insurers to classify risks they underwrite to receive fair premiums for the risk undertaken. This classification is based on *known* characteristics of the insured, the person, or firm seeking the insurance coverage.

Insurers classify risks on the basis of known characteristics of the insured, the person or firm seeking coverage.

For example, suppose that you are working for a company that insures small businesses for time lost because of employees injured on the job. Consider pricing this insurance product for two businesses that are identical with respect to number of employees, location, age and sex distribution, and so forth, except that one company is a management consulting firm and the other is a construction firm. From experience, you expect the management consulting firm to have a lower claims level than the construction firm and you need to price accordingly. If you do not, another insurance company will offer a lower insurance price to the consulting firm and take this potential customer away, leaving your company with only the more costly construction business.

Competition among insurers leads to charging premiums according to observable characteristics, known as *risk classification*. In the context of regression modeling, we can think of this as modeling the claims distributions in terms of explanatory variables.

Many pricing situations are based on a relationship between the insurer and the insured that develops over time. These relationships allow insurers to base prices on *unobservable* characteristics of the insured by taking into account the prior claims experience of the insured. Modifying premiums with claims history is known as *experience rating*, also sometimes referred to as *merit rating*.

Experience rating methods are either applied retrospectively or prospectively. With retrospective methods, a "refund" of a portion of the premium is provided to the insured in the event of favorable (to the insurer) experience. Retrospective premiums are common in life insurance arrangements (where insureds earned dividends in the United States and bonuses in the United Kingdom). In property and casualty insurance, prospective methods are more common, where favorable insured experience is rewarded with a lower renewal premium.

In this chapter, we discuss two prospective methods that are well suited for regression modeling: credibility and bonus-malus. Bonus-malus methods are used extensively in Asia and Europe, though almost exclusively with automobile insurance. As we will see in Section 18.4, the idea is to use claims experience to modify the classification of an insured. Credibility methods, introduced in Section 18.2, are more broadly applied in terms of lines of business and geography.

18.2 Credibility

Credibility is a technique for pricing insurance coverages that is widely used by health, group term life, and property and casualty actuaries. In the United States, the standards are described under the Actuarial Standard of Practice Number 25, published by the Actuarial Standards Board of the American Academy of Actuaries (see www.actuary.org). Further, several insurance laws and regulation require the use of credibility.

The theory of credibility has been called a "cornerstone" of the field of actuarial science (Hickman and Heacox, 1999). The basic idea is to use claims experience and additional information to develop a pricing formula, such as through the relation

$$\textit{New Premium} = \zeta \times \textit{Claims Experience} + (1 - \zeta) \times \textit{Old Premium}. \quad (18.1)$$

Here, ζ (the Greek letter "zeta") is known as the *credibility factor*; values generally lie between zero and one. The case $\zeta = 1$ is known as *full credibility*, where claims experience is used solely to determine the premium. The case $\zeta = 0$ can be thought of as "no credibility," where claims experience is ignored and external information is used as the sole basis for pricing.

To keep this chapter self-contained, we begin by introducing some basic credibility concepts. Section 18.2.1 reviews classic concepts of credibility, including when to use it and linear pricing formulas. Section 18.2.2 describes the modern version of credibility by introducing a formal probabilistic model that can be used for updating insurance prices. Section 18.3 discusses the link with regression modeling.

18.2.1 Limited Fluctuation Credibility

Credibility has a long history in actuarial science, with fundamental contributions dating back to Mowbray (1914). Subsequently, Whitney (1918) introduced the intuitively appealing concept of using a weighted average of (1) claims from the risk class and (2) claims over all risk classes to predict future expected claims.

Standards for Full Credibility

The title of Mowbray's paper was "How Extensive a Payroll Exposure Is Necessary to Give a Dependable Pure Premium?" It is still the first question that an analyst needs to confront: when do I need to use credibility estimators? To get a better handle on this question, consider the following situation.

Example: Dental Costs. Suppose that you are pricing dental insurance coverage for a small employer. For males aged 18–25, the employer provides the following experience:

Year	2007	2008	2009
Number	8	12	10
Average Dental Cost	500	400	900

The "manual rate," available from a tabulation of a much larger set of data, is \$700 per employee. Ignoring inflation and expenses, what would you use to anticipate dental costs in 2010? The manual rate? The average of the available data? Or some combination?

Mowbray wanted to distinguish between situations when large employers with substantial information would use their own experience and when small employers with limited experience would use external sources, so-called manual rates. In statistical terminology, we can think about forming an estimator from an employer's experience of the true mean costs. We are asking whether the distribution of the estimator is sufficiently close to the mean to be reliable. Of course, "sufficiently close" is the tricky part, so let us look at a more concrete situation.

The simplest setup is to assume that you have claims $y_1, \ldots, y_n$ that are identically and independently distributed (i.i.d.) with mean μ and variance σ^2. As a standard for full credibility, we could require that n be large enough so that

$$\Pr((1-r)\mu \leq \bar{y} \leq (1+r)\mu) \geq p, \tag{18.2}$$

where r and p are given constants. For example, if $r = 0.05$ and $p = 0.9$, then we wish to have at least 90% chance of being within 5% of the mean.

Using normal approximations, it is straightforward to show that sufficient for equation (18.2) is

$$n \geq \left(\frac{\Phi^{-1}(\frac{p+1}{2})\sigma}{r\mu} \right)^2 . \tag{18.3}$$

We define n_F, the number of observations required for full credibility, to be the smallest value of n that satisfies equation (18.3).

Example: Dental Costs, Continued. From the table, average costs are $\bar{y} = (500 \times 8 + 400 \times 12 + 900 \times 10)/30 = 593.33$. Suppose that we have available an estimate of the standard deviation $\sigma \approx \widehat{\sigma} = 200$. Using $p = 0.90$, the 90th percentile of the normal distribution is $\Phi^{-1}(.95) = 1.645$. With $r = 0.05$, the approximate sample size required is

$$\left(\frac{\Phi^{-1}(.95)\widehat{\sigma}}{r\bar{y}}\right)^2 = \left(\frac{1.645 \times 200}{0.05 \times 593.33}\right)^2 = 122.99,$$

or $n_F = 123$. With a sample of size 30, we do not have enough observations for full credibility.

The standards for full credibility given in equations (18.2) and (18.3) are based on approximate normality. It is easy to construct similar rules for other distributions such as binomial and Poisson count data or mixtures of distributions for aggregate losses. See Klugman, Panjer, and Willmot (2008) for more details.

Partial Credibility

Actuaries do not always work with massive datasets. You may work with the experience from a small employer or association and not have sufficient experience to meet the full credibility standard. Or, you may be working with a large employer but have decided to decompose your data into small, homogeneous subsets. For example, if you are working with dental claims, you may wish to create several small groups based on age and sex.

For smaller groups that do not meet the full credibility threshold, Witney (1918) proposed using a weighted average of the group's claims experience and a manual rate. Assuming approximate normality, the expression for partial credibility is

$$\textit{New Premium} = Z \times \bar{y} + (1 - Z) \times \textit{Manual Premium}, \tag{18.4}$$

where Z is the credibility factor, defined as

$$Z = \min\{1, \sqrt{\frac{n}{n_F}}\}. \tag{18.5}$$

Here, n is the sample size and n_F is the number of observations required for full credibility.

Example: Dental Costs, Continued. From prior work, the standard for full credibility is $n_F = 123$. Thus, the credibility factor is $\min\{1, \sqrt{\frac{30}{123}}\} = 0.494$. With this, the partial credibility premium is

$$\textit{New Premium} = 0.494 \times 593.33 + (1 - 0.494) \times 700 = 647.31.$$

One line of justification for the partial credibility formulas in equations (18.4) and (18.5) is given in the exercises. From equation (18.5), we see that the credibility factor Z is bounded by 0 and 1; as the sample size n and hence the experience becomes larger, Z tends to 1. This means that larger groups are more "credible." As the credibility factor Z increases, a greater weight is placed on the group's experience ($\bar{y}$). As Z decreases, more weight is placed on the manual premium, the rate that is developed externally on the basis of the group's characteristics.

18.2.2 Greatest Accuracy Credibility

Credibility theory was used for more than 50 years in insurance pricing before it was placed on a firm mathematical foundation by Bühlmann (1967). To introduce this framework, sometimes known as *greatest accuracy credibility*, let us begin with the assumption that we have a sample of claims $y_1, \ldots, y_n$ from a small group and that we want to estimate the mean for this group. Although the sample average $\bar{y}$ is certainly a sensible estimator, the sample size may be too small to rely exclusively on $\bar{y}$. We also suppose that have an external estimate of overall mean claims, M, that we think of as a "manual premium." The question is whether we can combine the two estimates, $\bar{y}$ and M, to provide an estimator that is superior to either alternative.

Bühlmann hypothesized the existence of unobserved characteristics of the group that we denote as α; he referred to these as *structure variables*. Although unobserved, these characteristics are common to all observations from the group. For dental claims, the structure variables may include the water quality where the group is located, the number of dentists to provide preventative care in the area, the educational level of the group, and so forth. Thus, we assume, conditional on α, that $\{y_1, \ldots, y_n\}$ are a random sample from an unknown population and hence are i.i.d. For notation, we will let $\mathrm{E}(y|\alpha)$ denote the conditional expected claims and $\mathrm{Var}(y|\alpha)$ be the corresponding conditional variance. Our goal is to determine a sensible "estimator" of $\mathrm{E}(y|\alpha)$.

Although unobserved, we can learn something about the characteristics α from repeated observations of claims. For each group, the (conditional) mean and variance functions are $\mathrm{E}(y|\alpha)$ and $\mathrm{Var}(y|\alpha)$, respectively. The expectation over all groups of the variance functions is $\mathrm{E}\,\mathrm{Var}(y|\alpha)$. Similarly, the variance of conditional expectations is $\mathrm{Var}\,\mathrm{E}(y|\alpha)$.

With these quantities in hand, we are able to give Bühlmann's credibility premium.

$$\textit{New Premium} = \zeta \times \bar{y} + (1 - \zeta) \times M, \tag{18.6}$$

where ζ is the credibility factor, defined as

$$\zeta = \frac{n}{n + \textit{Ratio}}, \quad \text{with} \quad \textit{Ratio} = \frac{\mathrm{E}\,\mathrm{Var}(y|\alpha)}{\mathrm{Var}\,\mathrm{E}(y|\alpha)}. \tag{18.7}$$

The credibility formula in equation (18.6) is the same as the classic partial credibility formula in equation (18.4), with the credibility factor ζ in place of Z. Thus, it shares the same intuitively pleasing expression as a weighted average.

Further, both credibility factors lie in the interval (0, 1) and both increase to one as the sample size n increases.

Example: Dental Costs, Continued. From prior work, we have that an estimate of the conditional mean is 593.33. Use similar calculations to show that the estimated conditional variance is 48,622.22.

Now assume that there are three additional groups with conditional means and variances given as follows:

Group	Unobserved Variable	Conditional Mean $\mathrm{E}(y\mid\alpha)$	Conditional Variance $\mathrm{Var}(y\mid\alpha)$	Probability $\Pr(\alpha)$
1	α_1	593.33	48,622.22	0.20
2	α_2	625.00	50,000.00	0.30
3	α_3	800.00	70,000.00	0.25
4	α_4	400.00	40,000.00	0.25

We assume that probability of being a member of a group is given as $\Pr(\alpha)$. For example, this may be determined by taking proportions of the number of members in each group.

With this information, it is straightforward to calculate the expected conditional variance:

$$\begin{aligned}\mathrm{E}\,\mathrm{Var}(y\mid\alpha) &= 0.2(48622.22) + 0.3(50000) + .25(70000) + .25(40000) \\ &= 52,224.44.\end{aligned}$$

To calculate the variance of conditional expectations, one can begin with the overall expectation

$$\mathrm{E}\,\mathrm{E}(y\mid\alpha) = 0.2(593.33) + 0.3(625) + .25(800) + .25(400) = 606.166,$$

and then use a similar procedure to calculate the expected value of the conditional second moment, $\mathrm{E}(\mathrm{E}(y\mid\alpha))^2 = 387,595.6$. With these two pieces, the variance of conditional expectations is $\mathrm{E}\,\mathrm{Var}(y\mid\alpha) = 387,595.6 - 606.166^2 = 20,158$.

This yields the $Ratio = 52,224.44/20,158 = 2.591$ and thus the credibility factor $\zeta = \frac{30}{30+2.591} = 0.9205$. With this, the credibility premium is

$$New\ Premium = 0.9205 \times 593.33 + (1 - 0.9205) \times 700 = 601.81.$$

To see how to use the credibility formula using alternative distributions, consider the following.

Example: Credibility with Count Data. Suppose that the number of claims each year for an individual insured has a Poisson distribution. The expected annual claim frequency of the entire population of insureds is uniformly distributed over the interval (0, 1). An individual's expected claims frequency is constant through time.

Consider a particular insured that had three claims during the prior three years.

Under these assumptions, we have the claims for an individual y with latent characteristics α that are Poisson distributed with conditional mean α and conditional variance α. The distribution of α is uniform on the interval (0, 1), so easy calculations show that

$$\mathrm{E}\ \mathrm{Var}(y|\alpha) = \mathrm{E}\ \alpha = 0.5 \quad \text{and} \quad \mathrm{Var}\ \mathrm{E}(y|\alpha) = \mathrm{Var}\ \alpha = 1/12 = 0.08333.$$

Thus, with $n = 3$, the credibility factor is

$$\zeta = \frac{3}{3 + 0.5/0.08333} = 0.3333.$$

With $\bar{y} = 3/3 = 1$ and the overall mean $\mathrm{E}\ \mathrm{E}(y|\alpha) = 0.5$ as the manual premium, the credibility premium is

$$\textit{New Premium} = 0.3333 \times 1 + (1 - 0.3333) \times 0.5 = 0.6667.$$

More formally, the optimality of the credibility estimator is based on the following.

Property. Assume, conditional on α, that $\{y_1, \ldots, y_n\}$ are identically and independently distributed with conditional mean and variance $\mathrm{E}(y|\alpha)$ and and $\mathrm{Var}(y|\alpha)$, respectively. Suppose that we want to estimate $\mathrm{E}\ (y_{n+1}|\alpha)$. Then, the credibility premium given in equations (18.6) and (18.7) has the smallest variance within the class of all linear unbiased predictors.

This property indicates that the credibility premium has "greatest accuracy" in the sense that it has minimum variance among linear unbiased predictors. As we have seen, it is couched in terms of means and variances that can be applied to many distributions; unlike the partial credibility premium, there is no assumption of normality. It is a fundamental result in that it is based on (conditionally) i.i.d. observations. Not surprisingly, it is easy to modify this basic result to allow for different exposures for observations, trends in times, and so forth.

The property is silent on how one would estimate quantities associated with the distribution of α. To do this, we will introduce a more detailed sampling scheme that will allow us to incorporate regression methods. Although not the only way of sampling, this framework will allow us to introduce many variations of interest and will help to interpret credibility in a natural way.

18.3 Credibility and Regression

By expressing credibility in the framework of regression models, actuaries can realize several benefits:

- Regression models provide a wide variety of models from which to choose.
- Standard statistical software makes analyzing data relatively easy.
- Actuaries have another method for explaining the ratemaking process.
- Actuaries can use graphical and diagnostic tools to select a model and assess its usefulness.

18.3.1 One-Way Random Effects Model

Assume that we are interested in pricing for n groups and that for each of the $i = 1, \ldots, n$ groups, we have claims experience $y_{it}, t = 1, \ldots, T$. Although this is the longitudinal data setup introduced in Chapter 10, we need not assume that claims evolve over time; the t subscripts may represent different members of a group. To begin, we assume that we do not have an explanatory variables. Claims experience follows

$$y_{it} = \mu + \alpha_i + \varepsilon_{it}, \qquad t = 1, \ldots, T, i = 1, \ldots, n, \tag{18.8}$$

where μ represents an overall claim average; α_i, the unobserved group characteristics; and ε_{it}, the individual claim variation. We assume that $\{\alpha_i\}$ are i.i.d. with mean zero and variance σ_α^2. Further assume that $\{\varepsilon_{it}\}$ are i.i.d. with mean zero and variance σ^2 and are independent of α_i. These are the assumptions of a basic one-way random effects model described in Section 10.5.

It seems reasonable to use the quantity $\mu + \alpha_i$ to predict a new claim from the ith group. For the model in equation (18.8), it seems intuitively plausible that $\bar{y}$ is a desirable estimator of μ and that $\bar{y}_i - \bar{y}$, is a desirable "estimator" of α_i. Thus, $\bar{y}_i$ is a desirable predictor of $\mu + \alpha_i$. More generally, consider predictors of $\mu + \alpha_i$ that are linear combinations of $\bar{y}_i$ and $\bar{y}$, that is, $c_1 \bar{y}_i + c_2 \bar{y}$, for constants c_1 and c_2. To retain the unbiasedness, we use $c_2 = 1 - c_1$. Some basic calculations show that the best value of c_1 that minimizes $\mathrm{E}\,(c_1 \bar{y}_i + (1 - c_1)\bar{y} - (\mu + \alpha_i))^2$ is

$$c_1 = \frac{T}{T + \sigma^2/\sigma_\alpha^2} = \zeta,$$

the credibility factor. This yields the *shrinkage estimator*, or predictor, of $\mu + \alpha_i$, defined as

$$\bar{y}_{i,s} = \zeta \bar{y}_i + (1 - \zeta)\bar{y}. \tag{18.9}$$

The shrinkage estimator is equivalent to credibility premium when we see that

$$\mathrm{Var}\,(\mathrm{E}(y_{it}|\alpha_i)) = \mathrm{Var}\,(\mathrm{E}(\mu + \alpha_i)) = \sigma_\alpha^2$$

and

$$\mathrm{E}\,(\mathrm{Var}(y_{it}|\alpha_i)) = \mathrm{E}\left(\mathrm{E}(\sigma^2)\right) = \sigma^2,$$

so that $Ratio = \sigma^2/\sigma_\alpha^2$. Thus, the one-way random effects model is sometimes referred to as the *balanced Bühlmann model*. This shrinkage estimator is also a best linear unbiased predictor (BLUP), introduced in Section 15.1.3.

Example: Visualizing Shrinkage. Consider the following illustrative data:

Group i	Replications t 1	2	3	4	Group Average ($\bar{y}_i$)
1	14	12	10	12	12
2	9	16	15	12	13
3	8	10	7	7	8

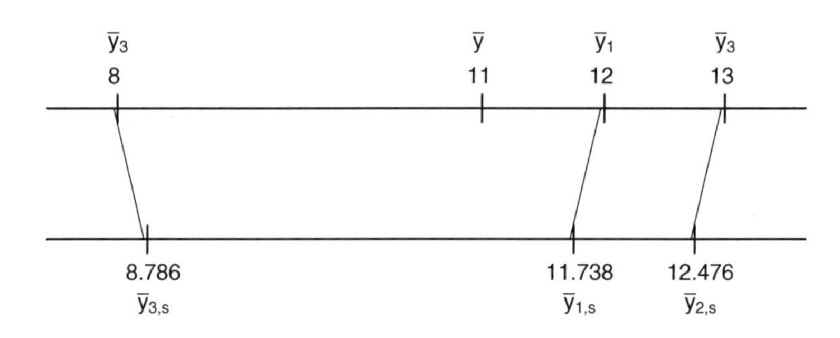

Figure 18.1 Comparison of group-specific means to shrinkage estimates. For an illustrative dataset, group-specific and overall means are graphed on the upper scale. The corresponding shrinkage estimates are graphed on the lower scale. This figure shows the shrinkage aspect of models with random effects.

That is, we have $n = 3$ groups, each of which has $T = 4$ observations. The sample mean is $\bar{y} = 8$, and group-specific sample means are $\bar{y}_1 = 12$, $\bar{y}_2 = 13$ and $\bar{y}_3 = 8$. We now fit the one-way random effects ANOVA model in equation (18.8) using maximum likelihood estimation assuming normality. Standard statistical software shows that the estimates of σ^2 and σ_α^2 are 4.889 and 3.444, respectively. It follows that the estimated ζ factor is 0.738. Using equation (18.9), the corresponding predictions for the subjects are 11.738, 12.476, and 8.786, respectively.

Figure 18.1 compares group-specific means to the corresponding predictions. Here, we see less spread in the predictions compared to the group-specific means; each group's estimate is "shrunk" to the overall mean, $\bar{y}$. These are the best predictors, assuming that α_i are random. In contrast, the subject-specific means are the best predictors assuming α_i are deterministic. Thus, this shrinkage effect is a consequence of the random effects specification.

Under the one-way random effects model, we have that $\bar{y}_i$ is an unbiased predictor of $\mu + \alpha_i$ in the sense that $\mathrm{E}\,(\bar{y}_i - (\mu + \alpha_i)) = 0$. However, $\bar{y}_i$ is inefficient in the sense that the shrinkage estimator, $\bar{y}_{i,s}$, has a smaller mean square error than $\bar{y}_i$. Intuitively, because $\bar{y}_{i,s}$ is a linear combination of $\bar{y}_i$ and $\bar{y}$, we say that has been shrunk toward the estimator $\bar{y}$. Further, because of the additional information in $\bar{y}_{i,s}$, it is customary to interpret a shrinkage estimator as "borrowing strength" from the estimator of the overall mean.

Note that the shrinkage estimator reduces to the fixed effects estimator $\bar{y}_i$ when the credibility factor, ζ, becomes 1. It is easy to see that $\zeta \rightarrow 1$ as either (i) $T \rightarrow \infty$ or (ii) $\sigma^2/\sigma_\alpha^2 \rightarrow 0$. That is, the best predictor approaches the group mean as either (i) the number of observations per group becomes large or (ii) the variability among groups becomes large relative to the response variability. In actuarial language, either case supports the idea that the information from the ith group is becoming more "credible."

18.3.2 Longitudinal Models

As we have seen the context of the one-way random effects model with balanced replications, Bühlmann's greatest accuracy credibility estimator is equivalent to the best linear unbiased predictor introduced in Section 15.1.3. This is also true when considering a more general longitudinal sampling setup introduced in

Chapter 10 (see, e.g., Frees, Young, and Luo, 1999). By expressing the credibility problem in terms of a regression-based sampling scheme, we can use well-known regression techniques to estimate parameters and to predict unknown quantities. We now consider a longitudinal model that handles many special cases of interest in actuarial practice.

Assume that claims experience follows

$$y_{it} = M_i + \mathbf{x}_{it}'\boldsymbol{\beta} + \mathbf{z}_{it}'\boldsymbol{\alpha}_i + \varepsilon_{it}, \qquad t = 1, \ldots, T_i, i = 1, \ldots, n. \tag{18.10}$$

Let us consider each model component in turn:

- M_i represents the manual premium that is assumed to be known. In the language of generalized linear models, M_i is an *offset variable.* In estimating the regression model, one simply uses $y_{it}^* = y_{it} - M_i$ as the dependent variable. If the manual premium is not available, then take $M_i = 0$.
- $\mathbf{x}_{it}'\boldsymbol{\beta}$ is the usual linear combination of explanatory variables. These can be used to adjust the manual premium. For example, you may have a larger than typical proportion of men (or women) in your group and want to allow for the sex of group members. This can be done by using a binary sex variable as a predictor and thinking of the regression coefficient as the amount to adjust the manual premium. In a similar fashion, you could use age, experience, or other characteristics of group members to adjust manual premiums. Explanatory variables may also be used to describe the group, not just the members. For example, you may wish to include explanatory variables that give information about location of place of employment (e.g., urban versus rural) to adjust manual premiums.
- $\mathbf{z}_{it}'\boldsymbol{\alpha}_i$ represents a linear combination of random effects that can be written as $\mathbf{z}_{it}'\boldsymbol{\alpha}_i = z_{it,1}\alpha_{i,1} + \cdots + z_{it,q}\alpha_{i,q}$. Often, there is only a single random intercept so that $q = 1$, $z_{it} = 1$ and $\mathbf{z}_{it}'\boldsymbol{\alpha}_i = \alpha_{i1} = \alpha_i$. The random effects have mean zero but a nonzero mean can be incorporated using $\mathbf{x}_{it}'\boldsymbol{\beta}$. For example, we might use time t as an explanatory variable and define $\mathbf{x}_{it} = \mathbf{z}_{it} = (1\ t)'$. Then, equation (18.10) reduces to $y_{it} = \beta_0 + \alpha_{i1} + (\beta_1 + \alpha_{i2}) \times t + \varepsilon_{it}$, a model due to Hachemeister (1975).
- ε_{it} is the mean zero disturbance term. In many applications, a weight is attached to it in the sense that Var $\varepsilon_{it} = \sigma^2/w_{it}$. Here, the weight w_{it} is known and accounts for an exposure such as the amount of insurance premium, number of employees, size of payroll, number of insured vehicles, and so forth. The introduction of weights was proposed by Bühlmann and Straub in 1970. The disturbance terms are typically assumed independent among groups (over i) but, in some applications, may incorporate time patterns, such as $AR(1)$ (autoregressive of order 1).

Ⓡ **Empirical** *Filename is "WorkersComp"*

Example: Workers' Compensation. We consider workers' compensation insurance, examining losses due to permanent, partial disability claims. The data are from Klugman (1992), who explored Bayesian model representations, and are originally from the National Council on Compensation Insurance. We consider $n = 121$ occupation classes over $T = 7$ years. To protect the data sources, further information on the occupation classes and years are not available. We summarize the analysis in Frees, Young, and Luo (2001).

The response variable of interest is the pure premium (PP), defined to be losses due to permanent, partial disability per dollar of PAYROLL. The variable PP is of interest to actuaries because workers' compensation rates are determined and quoted per unit of payroll. The exposure measure, PAYROLL, is a potential explanatory variable. Other explanatory variables are YEAR (=1, ..., 7) and occupation class.

Among other representations, Frees, Young, and Luo (2001) considered the model Bühlmann-Straub model,

$$\ln(PP)_{it} = \beta_0 + \alpha_{i1} + \varepsilon_{it}, \tag{18.11}$$

the Hachemeister model

$$\ln(PP)_{it} = \beta_0 + \alpha_{i1} + (\beta_1 + \alpha_{i2}) YEAR_t + \varepsilon_{it}, \tag{18.12}$$

and an intermediate version

$$\ln(PP)_{it} = \beta_0 + \alpha_{i1} + \alpha_{i2} YEAR_t + \varepsilon_{it}. \tag{18.13}$$

In all three cases, the weights are given by $w_{it} = PAYROLL_{it}$. These models are all special cases of the general model in equation (18.10), with $y_{it} = \ln(PP)_{it}$ and $M_i = 0$.

Parameter estimation and related statistical inference, including prediction, for the mixed linear regression model in equation (18.10) have been well investigated. The literature is summarized briefly in Section 15.1. From Section 15.1.3, the best linear unbiased predictor of $\mathrm{E}(y_{it}|\boldsymbol{\alpha})$ is of the form

$$M_i + \mathbf{x}_{it}' \mathbf{b}_{GLS} + \mathbf{z}_{it}' \mathbf{a}_{BLUP,i}, \tag{18.14}$$

where $\mathbf{b}_{GLS}$ is the generalized least squares estimator of $\boldsymbol{\beta}$, and the general expression for $\mathbf{a}_{BLUP,i}$ is given in equation (15.11). This is a general credibility estimator that can be readily calculated using statistical packages.

Special Case: Bühlmann-Straub Model. For the Bühlmann-Straub model, the credibility factor is

$$\zeta_{i,w} = \frac{WT_i}{WT_i + \sigma^2/\sigma_\alpha^2},$$

where WT_i is the sum of weights for the ith group, $WT_i = \sum_{t=1}^{T_i} w_{it}$.

Using equation (18.14) in the Bühlmann-Straub model, it is easy to check that the prediction for the ith group is

$$\zeta_{i,w} \bar{y}_{i,w} + (1 - \zeta_{i,w}) \bar{y}_w,$$

Table 18.1 Workers' Compensation Model Fits

Parameter	Bühlmann	Bühlmann-Straub Eq. (18.11)	Hachemeister Eq. (18.12)	Eq. (18.13)
β_0	–4.3665	–4.4003	–4.3805	–4.4036
(t-statistic)	(–50.38)	(–51.47)	(–44.38)	(–51.90)
β_1			–0.00446	
(t-statistic)			(–0.47)	
$\sigma_{\alpha,1}$	0.9106	0.8865	0.9634	0.9594
$\sigma_{\alpha,2}$			0.0452	0.0446
σ	0.5871	42.4379	41.3386	41.3582
AIC	1,715.924	1,571.391	1,567.769	1,565.977

Table 18.2 Workers' Compensation Predictions

Occupational Class	Fixed Effects	Bühlmann	Bühlmann-Straub	Hachemeister	Eq. (18.13)
1	2.981	2.842	2.834	2.736	2.785
2	1.941	1.895	1.875	1.773	1.803
3	1.129	1.137	1.135	1.124	1.139
4	0.795	0.816	0.765	0.682	0.692
5	1.129	1.137	1.129	1.062	1.079

where

$$\bar{y}_{i,w} = \frac{\sum_{t=1}^{T_i} w_{it} y_{it}}{W T_i} \quad \text{and} \quad \bar{y}_w = \frac{\sum_{t=1}^{T_i} \zeta_{i,w} \bar{y}_{i,w}}{\sum_{t=1}^{T_i} \zeta_{i,w}}$$

are the ith weighted group mean and the overall weighted mean, respectively. This reduces to the balanced Bühlmann predictor by taking weights identically equal to 1. See, for example, Frees (2004, Section 4.7) for further details.

Example: Workers' Compensation, Continued. We estimated the models in equations (18.11)–(18.13) and the unweighted Bühlmann model using maximum likelihood. Table 18.1 summarizes the results. This table suggests that the annual trend factor is not statistically significant, at least for conditional means. The annual trend that varies by occupational class does seem to be helpful. The information criterion AIC suggests that the intermediate model given in equation (18.13) provides the best fit to the data.

Table 18.2 illustrates the resulting credibility predictions for the first five occupational classes. Here, for each method, after making the prediction, we exponentiated the result and multiplied by 100, so that these are the number of cents of predicted losses per dollar of payroll. Also included are predictions for the "fixed effects" model that amount to taking the average over occupational class for the seven-year time span. Predictions for the Hachemeister and equation (18.13) models were made for year 8.

Table 18.2 shows substantial agreement between the Bühlmann and Bühlmann-Straub predictions, indicating that payroll weighting is less important for this data set. There is also substantial agreement between the Hachemeister and equation (18.13) model predictions, indicating that the overall time trend is less important. Comparing the two sets of predictions indicates that a time trend that varies by occupational class does make a difference.

18.4 Bonus-Malus

Bonus-malus methods of experience rating are used extensively in automobile insurance pricing in Europe and Asia. To understand this type of experience rating, let us first consider pricing based on observable characteristics. In automobile insurance, these include driver characteristics (e.g., age, sex), vehicle characteristics (e.g., car type, whether used for work), and territory characteristics (e.g., county of residence). Use of only these characteristics for pricing results in an *a priori premium*. In the United States and Canada, this is the primary basis of the premium; experience rating enters in a limited fashion in the form of premium surcharges for at-fault accidents and moving traffic violations.

Bonus-malus systems (BMS) provide a more detailed integration of claims experience into the pricing. Typically, a BMS classifies policyholders into one of several ordered categories. A policyholder enters the system in a specific category. In the following year, policyholders with an accident-free year are awarded a "bonus" and moved up a category. Policyholders that experience at-fault accidents during the year receive a "malus" and moved down a specified number of categories. The category that one resides in dictates the *bonus-malus factor*, or *BMF*. The *BMF* times the a priori premium is known as the *a posteriori premium*.

To illustrate, Lemaire (1998) gives an example of a Brazilian system that is summarized in Table 18.3. In this system, one begins in class 7, paying 100% of premiums dictated by the insurer's a priori premium. In the following year, if the policyholder experiences an accident-free year, he or she pays only 90% of the a priori premium. Otherwise, the premium is 100% of the a priori premium.

As described in Lemaire (1998), the Brazilian system is simple compared to others (the Belgian system has 23 classes). Insurers operating in countries with detailed bonus-malus systems do not require the extensive a priori rating variable compared to the United States and Canada. This typically means fewer underwriting expenses and thus a less costly insurance system. Moreover, many argue that it is fairer to policyholders in the sense that those with poor claims experience bear the burden of higher premiums and one is not penalized simply because of sex or other rating variables that are outside of the policyholder's control. See Lemaire (1995) for a broad discussion of institutional, regulatory, and ethical issues involving bonus-malus systems.

Table 18.3 Brazilian Bonus-Malus System

Class Before	BMF	Class After: 0 Claims	1 Claims	2 Claims	3 Claims	4 Claims	5 Claims	≥ 6 Claims
7	100	6	7	7	7	7	7	7
6	90	5	7	7	7	7	7	7
5	85	4	6	7	7	7	7	7
4	80	3	5	6	7	7	7	7
3	75	2	4	5	6	7	7	7
2	70	1	3	4	5	6	7	7
1	65	1	2	3	4	5	6	7

Source: Lemaire (1998).

Throughout this book, we have seen how to use regression techniques to compute a priori premiums. Dionne and Vanasse (1992) pointed out the advantages of using a regression framework to calculate bonus-malus factors. Essentially, they used latent variable to represent a policyholder's unobserved tendencies to become involved in an accident (e.g., aggressiveness, swiftness of reflexes) with regression count models to compute a posterior premiums. See Denuit et al. (2007) for a recent overview of this developing area.

18.5 Further Reading and References

See Norberg (1986) for an early account relating credibility theory to the framework of mixed linear models. The treatment here follows that of Frees, Young, and Luo (1999).

By demonstrating that many important credibility models can be viewed in a (linear) longitudinal data framework, we restrict our consideration to certain types of credibility models. Specifically, the longitudinal data models accommodate only unobserved risks that are additive. This chapter does not address models of nonlinear random effects that have been investigated in the actuarial literature; see, for example, Taylor (1977) and Norberg (1980). Taylor (1977) allowed insurance claims to be possibly infinite dimensional using Hilbert space theory and established credibility formulas in this general context. Norberg (1980) considered the more concrete but still general context of multivariate claims and established the relationship between credibility and statistical empirical Bayes estimation. As described in Section 18.4, Denuit et al. (2007) provide a recent overview of nonlinear longitudinal claim count models.

To account for the entire distribution of claims, a common approach used in credibility is to adopt a Bayesian perspective. Keffer (1929) initially suggested using a Bayesian perspective for experience rating in the context of group life insurance. Subsequently, Bailey (1945, 1950) showed how to derive the linear credibility form from a Bayesian perspective as the mean of a predictive distribution. Several authors have provided useful extensions of this paradigm. Jewell

(1980) extended Bailey's results to a broader class of distributions, the exponential family, with conjugate prior distributions for the structure variables.

Chapter References

Bailey, Arthur (1945). A generalized theory of credibility. *Proceedings of the Casualty Actuarial Society* 32, 13–20.

Bailey, Arthur (1950). Credibility procedures: LaPlace's generalization of Bayes' rule and the combination of collateral knowledge with observed data. *Proceedings of the Casualty Actuarial Society Society* 37, 7–23.

Bühlmann, Hans (1967). Experience rating and credibility. *ASTIN Bulletin* 4, 199–207.

Bühlmann, Hans, and E. Straub (1970). Glaubwürdigkeit für schadensätze. *Mitteilungen der Vereinigung Schweizerischer Versicherungs-Mathematiker* 70, 111–33.

Dionne, George, and C. Vanasse (1992). Automobile insurance ratemaking in the presence of asymmetrical information. *Journal of Applied Econometrics* 7, 149–65.

Denuit, Michel, Xavier Marechal, Sandra Pitrebois, and Jean-Francois Walhin (2007). *Actuarial Modelling of Claim Counts: Risk Classification, Credibility and Bonus-Malus Systems.* Wiley, New York.

Frees, Edward W., Virginia R. Young, and Yu Luo (1999). A longitudinal data analysis interpretation of credibility models. *Insurance: Mathematics and Economics* 24, 229–47.

Frees, Edward W., Virginia R. Young, and Yu Luo (2001). Case studies using panel data models. *North American Actuarial Journal* 5, no. 4, 24–42.

Frees, Edward W. (2004). *Longitudinal and Panel Data: Analysis and Applications in the Social Sciences.* Cambridge University Press, New York.

Hachemeister, Charles A. (1975). Credibility for regression models with applications to trend. In *Credibility: Theory and Applications*, Paul M. Kahn, ed., 129–63. Academic Press, New York.

Keffer, R (1929). An experience rating formula. *Transactions of the Actuarial Society of America* 30, 130–39.

Klugman, Stuart A. (1992). *Bayesian Statistics in Actuarial Science.* Kluwer, Boston.

Klugman, Stuart A., Harry H. Panjer, and Gordon E. Willmot (2008). *Loss Models: From Data to Decisions.* John Wiley & Sons, Hoboken, New Jersey.

Lemaire, Jean (1995). *Bonus-Malus Systems in Automobile Insurance.* Kluwer, Boston.

Lemaire, Jean (1998). Bonus-malus systems: The European and Asian approach to merit-rating. *North American Actuarial Journal* 2, no. 1, 26–47.

Mowbray, Albert H. (1914). How extensive a payroll exposure is necessary to give a dependable pure premium. *Proceedings of the Casualty Actuarial Society* 1, 24–30.

Norberg, Ragnar (1980). Empirical Bayes credibility. *Scandinavian Actuarial Journal*, 177–94.

Norberg, Ragnar (1986). Hierarchical credibility: Analysis of a random effect linear model with nested classification. *Scandinavian Actuarial Journal* 204–22.

Taylor, Greg C. (1977). Abstract credibility. *Scandinavian Actuarial Journal* 149–68.

Whitney, Albert W. (1918). The theory of experience rating. *Proceedings of the Casualty Actuarial Society* 4, 274–92.

19

Claims Triangles

Chapter Preview. This chapter introduces a classic actuarial reserving problem that is encountered extensively in property and casualty as well as health insurance. The data are presented in a triangular format to emphasize their longitudinal and censored nature. This chapter explains how such data arise naturally and introduces regression methods to address the actuarial reserving problem.

19.1 Introduction

In many types of insurance, little time elapses between the event of a claim, notification to an insurance company, and payment to beneficiaries. For example, in life insurance notification and benefit payments typically occurs within two weeks of an insured's death. However, for other lines of insurance, times from claim occurrence to the final payment can be much longer, taking months and even years. To introduce this situation, this section describes the evolution of a claim; introduces summary measures used by insurers, and then describes the prevailing deterministic method for forecasting claims, the chain-ladder method.

19.1.1 Claims Evolution

For example, suppose that you become injured in an automobile accident covered by insurance. It can take months for the injury to heal and all of the medical care payments to become known and paid by the insurance company. Moreover, disputes may arise among you, other parties to the accident, your insurer, and insurer(s) of other parties, thus lengthening the time until claims are settled and paid. When claims take a long time to develop, an insurer's claim obligations may be incurred in one accounting period, but not paid until a later accounting period. In the example of your accident, the insurer is aware that a claim has occurred and may have even made some payments in the current accounting period. Future payment amounts are unknown by the end of the current accounting period, but the insurer wishes to make an accurate forecast of future obligations to set aside a fair amount of money of future obligations, known as a *reserve*. The insurer's objective is to use current claim information to predict the timing and amount of future claim payments.

The insurer's objective is to use current claim information to predict the timing and amount of future claim payments.

To set terminology, it is helpful to follow the time line of a claim as it develops. In Figure 19.1, the claim occurs at time t_1 and the insuring company is notified

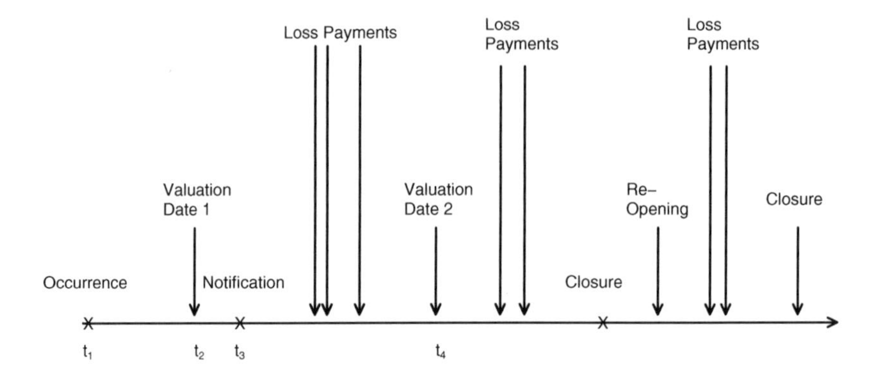

Figure 19.1 Time line of claim development.

at time t_3. There can be a long gap between occurrence and notification such that a *valuation date* (t_2) may occur within this gap. Here, t_2 is the time when the obligations are valued, which is typically but not always at the end of a company financial reporting period. In this case, the claim is said to be *incurred but not reported* at this valuation date.

After claim notification, there may one or more loss payments. Not all of the payments may be made by the next valuation date (t_4). As the claim develops, eventually the company deems its financial obligations on the claim to be resolved and declares the claim closed. However, it is possible that new facts arise and the claim must be reopened, giving rise to additional loss payments prior to being closed again.

19.1.2 Claims Triangles

Insurers do not typically model the evolution of each claim and then sum them (as is done on a policy basis in life insurance). Instead, portfolios of claims are summarized at each valuation date; it is these summaries on which forecasts of outstanding claims are based.

Specifically, let i represent the year in which a claim has been incurred and j represent the number of years from incurral to the time when the payment is made, the development (or delay) year. Thus, y_{ij} represents the sum of payment amounts in the ith incurral and jth development year. Here, *year* refers to the accounting period – in subsequent examples, you will see that we often use month, quarter, or other fixed period. Table 19.1 shows the information that actuaries typically confront.

The term *claims triangle* is evident from Table 19.1. We observe data in the upper-left-hand triangle, $y_{ij}, i = 1, \ldots, 5, j = 1, \ldots, 6 - i$. The goal is to complete the triangle, that is, to forecast values in the lower-right-hand triangle ($y_{ij}, i = 2, \ldots, 5, j = 7 - i, \ldots, 5$). For example, the most recent incurral year is $i = 5$, for which we have only one year of claims experience, y_{51}. The other values, $y_{5j}, j = 2, \ldots, 5$, are unknown at the valuation date. In some situations, it is also of interest to forecast claims in development years six and beyond.

Incurral Year (i)	Development Year (j) 1	2	3	4	5
1	y_{11}	y_{12}	y_{13}	y_{14}	y_{15}
2	y_{21}	y_{22}	y_{23}	y_{24}	.
3	y_{31}	y_{32}	y_{33}	.	.
4	y_{41}	y_{42}	.	.	.
5	y_{51}	.	.	.	.

Table 19.1 Classic Claims Run-Off Triangle

Claims triangle data are both longitudinal and censored.

Table 19.1 underscores that data are both longitudinal and censored. This is the key point regarding statistical modeling assumptions. In applications, the observations can vary significantly depending on the type and purpose. Each element of the triangle may represent actual payments by the insurance company, known as *incremental payments*, or the cumulative sum of payments since development. For some lines of business, estimates of the outstanding payments are available on a claim-by-claim basis, known as *case estimates*. Here, triangle elements may represent incurred claims in lieu of paid claims. That is, an incurred claim is the paid amount plus the reserve. Because case estimates are revised as new information about a claim comes in, it is not uncommon for incurred payments to be negative. Further, even paid incremental payments can be negative because of reinsurance or recovery from other parties of claims amounts. (Many insurers will write checks to pay claimants quickly on the occurrence of an accident and later be reimbursed by another party responsible for the accident.) Claims triangle data can also be in the form of the number of claim notifications. This type of data is particularly useful for estimating incurred but not reported reserves.

Example: Singapore Property Damage. Table 19.2 reports incremental payments from a portfolio of automobile policies for a Singapore property and casualty (general) insurer. Here, payments are for third-party property damage from comprehensive auto insurance policies. All payments have been deflated using a Singaporean consumer price index, so they are in constant dollars. The data are for policies with coverages from 1997–2001, inclusive. Table 19.2 also provides the premiums for these policies (in thousands of Singaporean dollars) to provide a sense of the insurer's increasing exposure to potential claim obligations.

Ⓡ EMPIRICAL
Filename is "SingaporeProperty"

19.1.3 Chain-Ladder Method

To introduce the basic chain-ladder method, we continue to work in the context of the Singapore property damage example. Table 19.3 shows the same payments as Table 19.2 but in cumulative rather than incremental form. Let $S_{ij} = y_{i1} + \cdots + y_{ij}$ denote cumulative claims.

Table 19.2 Singapore Incremental Payments

Incurral Year	Premium (in thousands)	Development Year 1	2	3	4	5
1997	32,691	1,188,675	2,257,909	695,237	166,812	92,129
1998	33,425	1,235,402	3,250,013	649,928	211,344	.
1999	34,849	2,209,850	3,718,695	818,367	.	.
2000	37,011	2,662,546	3,487,034	.	.	.
2001	40,152	2,457,265	.	.	.	.

Table 19.3 Singapore Cumulative Payments with Chain-Ladder Estimates

Incurral Year	Premium	Development Year 1	2	3	4	5	Reserve	Ultimate Loss Ratio (%)
1997	32,691	1,188,675	3,446,584	4,141,821	4,308,633	4,400,762		13.5
1998	33,425	1,235,402	4,485,415	5,135,343	5,346,687	**5,461,012**	114,325	16.3
1999	34,849	2,209,850	5,928,544	6,746,912	**7,021,930**	**7,172,075**	425,163	20.6
2000	37,011	2,662,546	6,149,580	**7,109,486**	**7,399,283**	**7,557,497**	1,407,917	20.4
2001	40,152	2,457,265	**6,738,898**	**7,790,792**	**8,108,361**	**8,281,737**	5,824,471	20.6
Total Reserve							7,771,877	
Chain-Ladder Factors			2.742	1.156	1.041	1.021		

The chain-ladder factor for the jth development year is calculated by taking the ratio of the sum of claims over all incurral years for the jth development year divided by the sum of the same incurral year payments for the $j-1st$ development year. Using notation, we have

$$CL_j = \frac{\sum_{i=1}^{6-j} S_{ij}}{\sum_{i=1}^{6-j} S_{i,j-1}}.$$

For example, $CL_5 = 4{,}400{,}762/4{,}308{,}633 = 1.021$ and $CL_4 = (5{,}346{,}687 + 4{,}308{,}633)/(5{,}135{,}343 + 4{,}141{,}821) = 1.041$.

Bold numbers are forecasts calculated recursively using the chain-ladder factors and $\widehat{S}_{i,j} = CL_j \times \widehat{S}_{i,j-1}$. The recursion starts when the value of the cumulative payment is known so that $\widehat{S}_{ij} = S_{ij}$. For example, for incurral year 2, $\widehat{S}_{25} = CL_5 \times S_{24} = (1.021)(5{,}346{,}687) = 5{,}461{,}012$. For incurral year 3, $\widehat{S}_{34} = CL_4 \times S_{33} = (1.041)(6{,}746{,}912) = 7{,}021{,}930$ and $\widehat{S}_{35} = CL_5 \times \widehat{S}_{34} = (1.021)(7{,}021{,}930) = 7{,}172{,}075$. Alternatively, one can go directly to the last development year and use $\widehat{S}_{35} = CL_5 \times \widehat{S}_{34} = CL_5 \times CL_4 \times S_{33} = (1.041)(1.021)(6{,}746{,}912) = 7{,}172{,}075$.

In Table 19.3, the reserve is the ultimate forecast amount minus the most recent cumulative paid claim. The ultimate loss ratio is the ratio of the forecast of cumulative claims in the last development year (5) to premiums paid (expressed as a percentage – recall that premiums are in thousands).

Incurral Year	Development Year								
	1	2	3	4	5	6	7	8	9
1993	14,695	205,515	118,686	416,286	93,544	185,043	37,750	0	14,086
1994	153,615	467,722	645,513	421,896	146,576	96,470	27,765	38,017	.
1995	24,741	547,862	754,475	417,573	156,596	55,155	36,984	.	.
1996	68,630	188,627	492,306	179,249	34,062	443,436	.	.	.
1997	29,177	364,672	437,507	385,571	529,319	.	.	.	.
1998	40,430	241,809	678,541	528,026	.	.	.	.	.
1999	45,125	372,935	704,168	.	.	.	.	.	.
2000	21,695	158,005	.	.	.	.	.	.	.
2001	6,626	.	.	.	.	.	.	.	.

Table 19.4 Singapore Incremental Injury Payments, 1993–2001

19.2 Regression Using Functions of Time as Explanatory Variables

The chain-ladder method is an important tool that actuaries use extensively when forecasting claims. It is typically presented as deterministic, as in Section 19.1.3. Alternative stochastic models for claim forecasting have two primary advantages.

- By explicitly modeling the distribution of claims, estimates for the uncertainty of the reserve forecasts can be made.
- There are many variations of chain-ladder techniques available because they are applied in many different situations. As we have seen in Chapter 5, stochastic methods feature a disciplined way of model selection that can help determine the most appropriate model for a given set of data.

As we will see, one need not make a choice between using the chain-ladder method and using a stochastic model. The Section 19.2.3 overdisperse Poisson model and the Section 19.3 Mack model both yield point forecasts that are equal to chain-ladder forecasts.

19.2.1 Lognormal Model

Our starting point is the lognormal model for incremental claims. That is, we consider a two-factor model of the form

$$\ln y_{ij} = \mu + \alpha_i + \tau_j + \varepsilon_{ij}, \tag{19.1}$$

where $\{\alpha_i\}$ are parameters for the incurral year factor and $\{\tau_j\}$ are parameters for the development year factor. A regression model with two factors was introduced in Section 4.4. Recall that we require constraints on the factor parameters for estimability, such as $\sum_i \alpha_i = 0$ and $\sum_j \tau_j = 0$. Assuming normality of $\{\varepsilon_{ij}\}$ gives rise to the lognormal specification for the incremental claims y_{ij}.

Example: Singapore Third-Party Injury. Table 19.4 reports payments from a portfolio of automobile policies for a Singapore property and casualty (general) insurer. Payments, deflated for inflation, are for third-party injury from

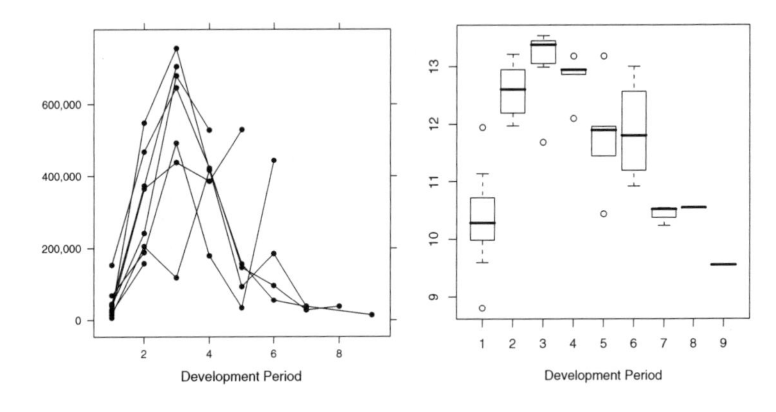

Figure 19.2 Singapore incremental injury payments. The left-hand panel shows payments by development year with each line connecting payments from the same incurral year. The right-hand panel shows the distribution of logarithmic payments for each development year.

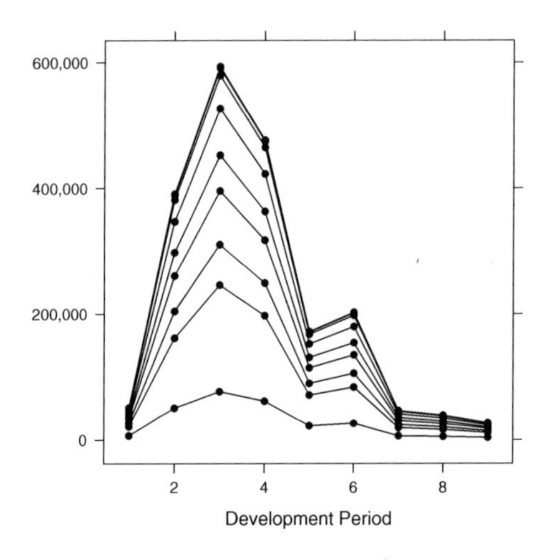

Figure 19.3 Fitted values for the Singapore incremental injury payments. These estimates are based on the two-factor lognormal model.

Ⓡ **Empirical** *Filename is "SingaporeInjury"*

comprehensive insurance policies. The data are for policies with coverages from 1993–2001, inclusive.

In automobile insurance, it typically takes longer to settle and pay injury than property damage claims. Thus, the number of development years, the runoff, is longer in Table 19.4 than in Table 19.2. Table 19.4 also shows a lack of stability of injury payments that Figure 19.2 helps us visualize. The left-hand panel shows trend lines by development for each incurral year. The right-hand panel presents box plots of logarithmic payments for each development year. This display shows that payments tend to rise for the first two development periods ($j = 1, 2$), reach a peak at the third period ($j = 3$), and decline thereafter.

The lognormal model based on equation (19.1) was fit to these data. Not surprisingly, both the factors incurral and development year were statistically significant. The coefficient of determination from the fit is $R^2 = 73.3\%$. A *qq* plot (not presented here) showed reasonable agreement with the assumption of normality. Fitted values from the model, after exponentiation to convert back to dollars, appear in Figure 19.3. This figure seems to capture the payment patterns

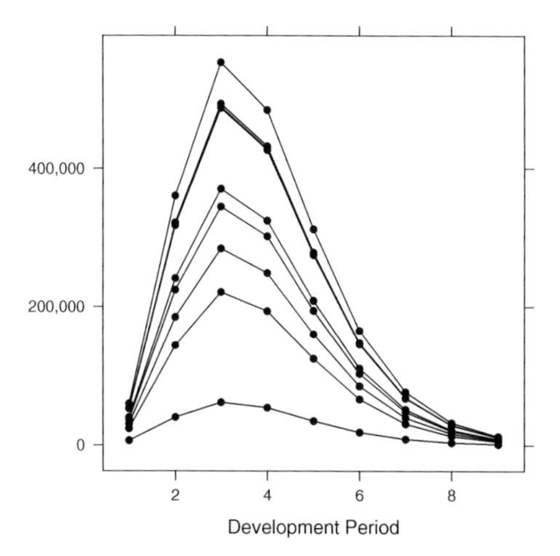

Figure 19.4 Fitted values from the reduced Hoerl model in equation (19.3).

that appear in the left-hand panel of Figure 19.2. Note that the fitted values for the unobserved portion of the triangle are forecasts.

19.2.2 Hoerl Curve

The systematic component of equation (19.1) can be easily modified. One possibility is the so-called "Hoerl curve," leading to the model equation

$$\ln y_{ij} = \mu + \alpha_i + \beta_i \ln(j) + \gamma_i \times j + \varepsilon_{ij}. \tag{19.2}$$

An advantage of treating development time j as a continuous covariate is that extrapolation is possible beyond the range of development times observed. As a variation, England and Verrall (2002) suggest allowing the first few development years to have their own levels and imposing the same runoff pattern for all incurral years ($\beta_i = \beta, \gamma_i = \gamma$).

Example: Singapore Third-Party Injury, Continued. The basic model from equation (19.2) fit well, and the coefficient of determination is $R^2 = 87.8\%$. We also examined a simpler model based on the equation

$$\ln y_{ij} = \mu + \alpha_i + \beta \ln(j) + \gamma \times j + \varepsilon_{ij}. \tag{19.3}$$

This simpler model did not fit the data as well as the more complete Hoerl model from equation (19.2), having $R^2 = 78.6\%$. However, a partial F-test established that the additional parameters were not statistically significant and so the simpler model in equation (19.3) is preferred.

On the basis of the simpler model, fitted values are displayed in Figure 19.4. This figure displays the geometrically declining fitted values beginning in the fourth development period.

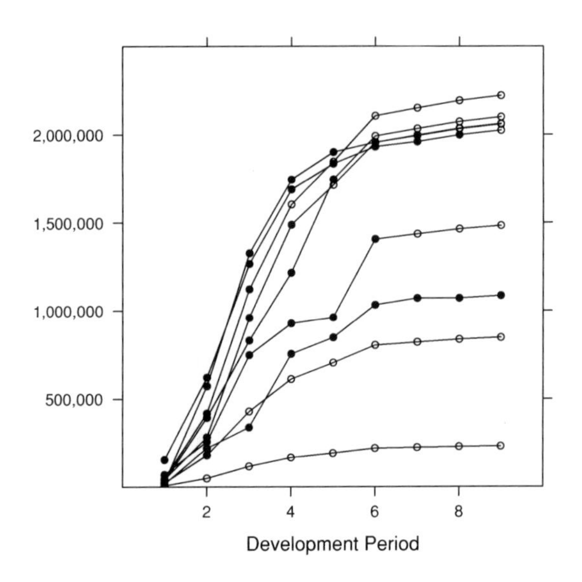

Figure 19.5 Actual and forecast values for the Singapore cumulative injury payments. Actual values are denoted with an opaque plotting symbol. Chain-ladder forecasts, from an overdisperse Poisson model, are denoted with an open plotting symbol.

19.2.3 Poisson Models

A drawback of the lognormal model is that the predictions produced by it do not replicate the traditional chain-ladder estimates. This section introduces the overdisperse Poisson model that does have this desirable feature.

To begin, from equation (19.1), we may write

$$\mathrm{E}\ y_{ij} = \exp(\eta_{i,j})\mathrm{E}\ e^{\varepsilon},$$

where the systematic component is $\eta_{i,j} = \mu + \alpha_i + \tau_j$. This is a model with a logarithmic link function (i.e., $\ln \mathrm{E}\ y = \eta$). Instead of using the lognormal distribution for y, this section assumes that y follows an overdisperse Poisson with variance function

$$\mathrm{Var}\ y_{ij} = \phi \exp(\eta_{i,j}).$$

Note that we have absorbed the scalar $\mathrm{E}\ e^{\varepsilon}$ into the overdispersion parameter ϕ.

We introduced overdisperse Poisson models in Section 12.3. Thus, this model can be estimated with standard statistical software and, as with the lognormal model, forecasts readily produced. It can be shown that the forecasts produced by the overdisperse Poisson are equivalent to the deterministic chain-ladder forecasts. See, for example, Taylor (2000) or Wüthrich and Merz (2008) for a proof. Not only does this give us a mechanism to quantify the uncertainty associated with chain-ladder forecasts, we can also use standard statistical software to compute these estimates.

Example: Singapore Third-Party Injury, Continued. The overdisperse Poisson model was fit to the Singapore injury payments data. Standard statistical software was use to compute parameter estimates, using techniques described in Section 12.3. Figure 19.5 summarizes the forecasts from these models. This figure shows cumulative, not incremental, payments, marked with opaque plotting symbols in the figure. Forecasts of incremental payments were produced and then summed to get the cumulative payment forecasts in Figure 19.5; these are marked with the open plotting symbols. The reader is invited to check that these forecasts are identical to those produced by the deterministic chain ladder method up to the eighth development year. Here, the value of zero for the first incurral year causes small differences between the Poisson model and chain ladder forecasts.

19.3 Using Past Developments

As with autoregressive models, one can use prior history to forecast payments. What is prior history? In the claim triangle setup in Table 19.1, there are *two* dimensions of time: incurral year and development year. Most models focus on using prior development (j) for forecasting. By focusing on prior development experience, we allow ourselves the flexibility to our model cumulative (S_{ij}) or incremental (y_{ij}) payments. As we learned in our Chapters 7 and 8 study of time series, it is useful to be able to model both a series (S_{ij}) and its changes ($y_{ij} = S_{ij} - S_{i,j-1}$).

19.3.1 Mack Model

The model put forward by Mack (1993) specifies the first two conditional moments of cumulative payments and uses generalized least squares to fit the model. Under this stochastic specification, traditional chain-ladder forecasts are produced.

Specifically, we assume

$$\mathrm{E}\left(S_{i,j}|S_{i,j-1}\right) = \nu_j S_{i,j-1} \tag{19.4}$$

and

$$\mathrm{Var}\left(S_{i,j}|S_{i,j-1}\right) = \sigma_j^2, S_{i,j-1}, \tag{19.5}$$

where ν_j and σ_j^2 are model parameters.

The mean parameters, ν_j, are determined through generalized least squares by minimizing the quantity

$$Q = \sum_{j=2}^{n} \sum_{i=1}^{n+1-i} \frac{(S_{ij} - \nu_j S_{i,j-1})^2}{\sigma_j^2 S_{i,j-1}}.$$

Taking derivatives with respect to the parameters ν_j and setting them equal to zero yields

$$\frac{\partial}{\partial \nu_j} Q = \sum_{i=1}^{n+1-i} \frac{(-2)(S_{ij} - \nu_j S_{i,j-1})}{\sigma_j^2 S_{i,j-1}} = 0.$$

The solution of this equation yields

$$\hat{\nu}_j = \frac{\sum_{i=1}^{n+1-i} S_{ij}}{\sum_{i=1}^{n+1-i} S_{i,j-1}},$$

the chain-ladder factor. Here, the caret on ν_j indicates that $\hat{\nu}_j$ is an estimator determined by the data.

With these parameter estimates, one can use equation (19.4) to produce fitted values that equal chain-ladder estimates. Moreover, one can estimate the scale parameters σ_j^2 and then use equation (19.5) to quantify the uncertainty of the estimates. See England and Verrall (2002) or Wüthrich and Merz (2008) for details on the scale parameter estimation.

The strength and limitation of this model is that it employs assumptions only about the first two conditional moments. It is a strength in the sense that we need not worry about whether the underlying distribution is close to lognormal or Poisson. Thus, it is sometimes referred to as a nonparametric model. It is a limitation in the sense that measures of uncertainty in equation (19.5) are related to the second moment that uses a squared error loss function. For insurance claims, data are typically skewed so that the variance is not a good scale measure. Moreover, in loss reserving, we want to know whether reserves are too high or too low; using a measure of uncertainty that reports only absolute deviations does not provide the actuary with the type of information needed.

19.3.2 Distributional Models

Models supplementing the moment assumptions in equations (19.4) and (19.5) with distributional assumptions on payments have been proposed in the literature.

For example, Verrall proposed using the negative binomial as a distribution for $\{y_{ij}\}$ with conditional moments

$$\mathrm{E}\big(y_{i,j}|S_{i,j-1}\big) = (\nu_j - 1)S_{i,j-1}$$

and

$$\mathrm{Var}\big(y_{i,j}|S_{i,j-1}\big) = \phi\nu_j(\nu_j - 1)S_{i,j-1},$$

where ϕ and ν_j are model parameters. Note that the conditional mean assumption is the same as equation (19.4) because of the relation $\mathrm{E}\big(S_{i,j}|S_{i,j-1}\big) = S_{i,j-1} + \mathrm{E}\big(y_{i,j}|S_{i,j-1}\big)$. Similarly, we can express the variance of cumulative payments as $\mathrm{Var}\big(S_{i,j}|S_{i,j-1}\big) = \mathrm{Var}\big(y_{i,j}|S_{i,j-1}\big)$. Thus, the conditional variance assumption is as in equation (19.5) with parameters $\sigma_j^2 = \phi\nu_j(\nu_j - 1)$. This model can be easily implemented using generalized linear model software by specifying the negative binomial model with logarithmic link function

$$\ln \mu_{ij} = \ln \lambda_j + \ln S_{i,j-1},$$

where $\mu_{ij} = \mathrm{E}\big(y_{i,j}|S_{i,j-1}\big)$. See England and Verrall (2002, Section 7.3) for further discussion.

Another distributional model, suggested by England and Verrall (2002, Section 2.5), is to use the moments in equations (19.4) and (19.5) but to specify a normal distribution for payments $\{y_{ij}\}$. This can be useful as an approximation to the negative binomial model, particularly in the case when estimates of ν_j are less than one, indicating that the conditional variance is misspecified.

As emphasized at the beginning of Section 19.2, the important strengths of the distributional models are that they allow analysts to quantify the uncertainty of the forecasts and provide disciplined mechanisms for model selection.

19.4 Further Reading and References

Two book-length introductions to claim triangles are Taylor (2000) and Wüthrich and Merz (2008). Practicing actuaries will find the review article by England and Verrall (2002) helpful.

When there is instability in the runoff patterns in the early years, a method developed by Bornheuetter and Ferguson (1972) can be useful in that it allows the actuary to incorporate an external assessment of ultimate paid claims. As with the chain ladder, this technique can be expressed in terms of stochastic modeling via Bayesian modeling. See, for example, the discussion by England and Verrall (2002), Wüthrich and Merz (2008), and de Alba (2006).

An emerging problem is developing reserves when one triangle is correlated with another. Such correlation might be expected among lines of insurance business. Braun (2004) provides an introduction to this topic.

Another emerging area is developing reserve based on individual claims runoff patterns, such as described in Section 19.1.1. Antonio et al. (2006) provides an introduction to this topic, where they use mixed linear models to develop reserves for claims that have been reported but not yet settled. See also Section 14.5 on recurrent event theory.

Chapter References

Antonio, Katrien, Jan Beirlant, Tom Hoedemakers, and Robert Verlaak (2006). Lognormal mixed models for reported claim reserves. *North American Actuarial Journal* 10, no. 1, 30–48.

Bornheuetter, Ronald L., and Ronald E. Ferguson (1972). The actuary and IBNR. *Proceedings of the Casualty Actuarial Society* 59, 181–95.

Braun, Christian (2004). The prediction error of the chain ladder method applied to correlated run-off triangles. *ASTIN Bulletin* 34, no. 2, 399–423.

de Alba, Enrique (2006). Claims reserving when there are negative values in the runoff triangle: Bayesian analysis using the three-parameter log-normal distribution. *North American Actuarial Journal* 10, no. 3, 45–59.

England, Peter D., and Richard J. Verrall (2002). Stochastic claims reserving in general insurance. *British Actuarial Journal* 8, 443–544.

Gamage, Jinadasa, Jed Linfield, Krzysztof Ostaszewski, and Steven Siegel (2007). Statistical methods for health actuaries – IBNR estimates: An introduction. Society of Actuaries Working Paper, Schaumburg, Illinois.

Mack, Thomas (1993). Distribution-free calculation of the standard error of chain-ladder reserve estimates. *ASTIN Bulletin* 23, no. 2, 213–25.

Mack, Thomas (1994). Measuring the variability of chain-ladder reserve estimates. *Casualty Actuarial Society*, Spring Forum, Arlington, Virginia.

Taylor, Greg (2000). *Loss Reserving: An Actuarial Perspective*. Kluwer Academic Publishers, Boston.

Wacek, Michael G. (2007). The path of the ultimate loss ratio estimate. *Variance* 1, no. 2, 173–92.

Wüthrich, Mario V., and Michael Merz (2008). *Stochastic Claims Reserving Methods in Insurance*. Wiley, New York.

19.5 Exercises

19.1. The data in Table 19.5 originate from the 1991 edition of the Historical Loss Development Study, published by the Reinsurance Association of American. These data have been widely used to illustrate triangle methods, beginning with Mack (1994) and later by England and Verrall (2002). These data are from automatic facultative reinsurance business in general liability (excluding asbestos and environmental) coverages. (Under a facultative basis, each risk is underwritten by the reinsurer on its own merits.) Table 19.5 reports incremental incurred losses from 1981–90, in thousands of U.S. dollars.

Table 19.5 Loss Development Study (1991) Facultative Reinsurance

Year	1	2	3	4	5	6	7	8	9	10
1	5,012	3,257	2,638	898	1,734	2,642	1,828	599	54	172
2	106	4,179	1,111	5,270	3,116	1,817	−103	673	535	
3	3,410	5,582	4,881	2,268	2,594	3,479	649	603		
4	5,655	5,900	4,211	5,500	2,159	2,658	984			
5	1,092	8,473	6,271	6,333	3,786	225				
6	1,513	4,932	5,257	1,233	2,917					
7	557	3,463	6,926	1,368						
8	1,351	5,596	6,165							
9	3,133	2,262								
10	2,063									

Ⓡ **Empirical** *Filename is "ReinsGenLiab"*

a. Begin by calculating the deterministic chain-ladder factors. Note the element in the second origin and seventh development year is negative. You may want to first convert the incremental to cumulative payments. Use these factors to complete the triangle.
b. Use your work in part (a) to calculate the reserve estimate.
c. Remove the observation in the second origin and seventh development year. Fit a lognormal model to the remaining data. Comment on the statistical significance of each factor and the goodness of fit.
d. Fit the Hoerl model to the data in part (c). Produce a graph of fitted values.
e. Fit the overdisperse Poisson model to the data in part (c). Check the proximity of the fitted values to the chain-ladder values produced in part (a).

19.2. The data in Table 19.6 is an excerpt from Braun (2004), based on the 2001 Historical Loss Development Study, published by the Reinsurance Association of American. The larger data (available in the file "ReinsGL2004") contains data for years 1987–2000, inclusive.

Repeat parts (a)–(e) of Exercise 19.1 for these data.

Ⓡ **EMPIRICAL** *Filename is "ReinsGL2004"*

Table 19.6 Reinsurance General Liability

Year	1	2	3	4	5	6
1995	97,518	343,218	575,441	769,017	934,103	1,019,303
1996	173,686	459,416	722,336	955,335	1,141,750	
1997	139,821	436,958	809,926	1,174,196		
1998	154,965	528,080	1,032,684			
1999	196,124	772,971				
2000	204,325					

19.3. The data in Table 19.7 are from Wacek (2007). The data represent industry aggregates for private passenger auto liability/medical coverages from year 2004, in millions of dollars. They are based on insurance company annual statements, specifically, Schedule P, Part 3B. The elements of the triangle represent cumulative net payments, including defense and cost containment expenses.

Repeat parts (a)–(e) of Exercise 19.1 for these data.

Ⓡ **EMPIRICAL** *Filename is "AutoIndustry"*

Table 19.7 2004 U.S. Insurance Industry Aggregates for Private Passenger Auto Liability and Medical

Year	1	2	3	4	5	6	7	8	9	10
1995	17,674	32,062	38,619	42,035	43,829	44,723	45,162	45,375	45,483	45,540
1996	18,315	32,791	39,271	42,933	44,950	45,917	46,392	46,600	46,753	
1997	18,606	32,942	39,634	43,411	45,428	46,357	46,681	46,921		
1998	18,816	33,667	40,575	44,446	46,476	47,350	47,809			
1999	20,649	36,515	43,724	47,684	49,753	50,716				
2000	22,327	39,312	46,848	51,065	53,242					
2001	23,141	40,527	48,284	52,661						
2002	24,301	42,168	50,356							
2003	24,210	41,640								
2004	24,468									

19.4. The data in Table 19.8 are from Gamage et al. (2007). These data are for 36 months of medical-care payments, from January 2001 through December 2003, inclusive. These are payments for medical-care coverage with no deductible or coinsurance. There were relatively low copayments, such as \$10 per office visit. The payments exclude prescription drugs that typically have a shorter payment pattern than other medical claims.

Repeat parts (a)–(e) of Exercise 19.1 for these data.

Ⓡ **EMPIRICAL** *Filename is "MedicalCare"*

Table 19.8 Monthly Medical Care Payments, 2001–3

Date	Members	Month	1	2	3	4	5	6	7	8	9	10	11	12	13
Jan-01	11,154	1	180	436,082	933,353	116,978	42,681	41,459	5,088	22,566	4,751	3,281	−188	1,464	1,697
Feb-01	11,118	2	5,162	940,722	561,967	21,694	171,659	11,008	19,088	5,213	4,337	7,844	2,973	4,061	10,236
Mar-01	11,070	3	42,263	844,293	720,302	94,634	182,077	32,216	12,937	22,815	1,754	4,695	1,326	758	2,177
Apr-01	11,069	4	20,781	762,302	394,625	78,043	157,950	46,173	126,254	4,839	337	1,573	9,573	1,947	5,937
May-01	11,130	5	20,346	772,404	392,330	315,888	39,197	21,360	8,721	5,452	16,627	2,118	4,119	5,666	−1,977
Jun-01	11,174	6	20,491	831,793	738,087	65,526	27,768	12,185	1,493	11,265	1,805	29,278	13,020	2,967	−83
Jul-01	11,180	7	37,954	1,126,675	360,514	89,317	40,126	16,576	16,701	2,444	8,266	11,310	8,006	1,403	3,124
Aug-01	11,420	8	138,558	806,362	589,304	273,117	36,912	16,831	19,941	13,310	8,619	4,679	3,094	4,609	236
Sep-01	11,400	9	28,332	954,543	246,571	205,528	60,060	15,198	42,208	17,568	1,686	9,897	3,367	2,062	421
Oct-01	11,456	10	104,160	704,796	565,939	323,789	45,307	32,518	26,227	7,976	3,364	992	33,963	2,200	1,293
Nov-01	11,444	11	40,747	927,158	425,794	146,145	66,663	31,214	12,808	15,859	374	3,079	412	937	1,875
Dec-01	11,555	12	10,861	847,338	272,165	134,798	71,804	27,800	17,917	3,930	2,794	846	1,962	1,879	16,060
Jan-02	11,705	13	77,938	896,195	544,372	173,606	41,595	4,209	16,473	6,000	−66	−1,881	−4,054	84,233	4,921
Feb-02	11,823	14	38,041	1,035,439	438,153	115,587	12,489	22,260	13,203	6,395	2,056	−3,323	33,397	3,479	−1,625
Mar-02	11,753	15	39,410	1,022,024	255,002	169,881	35,230	40,307	21,067	5,378	5,508	17,606	−24,320	1,298	1,362
Apr-02	11,654	16	68,253	1,414,379	317,110	91,880	53,970	10,888	3,171	11,660	20,861	1,033	−21,670	2,634	149
May-02	11,703	17	124,824	1,053,972	516,876	145,954	25,171	12,609	7,704	29,633	4,555	6,203	3,872	1,116	666
Jun-02	11,580	18	49,725	1,119,099	533,444	80,182	32,203	23,205	18,807	7,944	4,152	−910	3,664	608	528
Jul-02	11,577	19	44,317	1,297,335	385,789	141,155	150,726	35,075	16,176	8,070	67	14,217	2,326	7,091	687
Aug-02	11,655	20	134,152	1,111,151	493,175	101,439	46,657	22,824	12,818	3,781	1,265	2,467	−62,165	247	−8,689
Sep-02	11,735	21	29,968	1,382,043	178,587	71,030	25,708	15,068	3,145	−4,058	−1,920	4,984	−1,523	−3,539	−478
Oct-02	11,889	22	210,377	999,963	528,880	201,410	58,003	26,174	−9,371	2,017	9,795	6,688	−40	453	−73
Nov-02	11,951	23	56,654	1,206,370	376,504	56,322	19,591	12,055	21,077	11,573	4,039	822	6,612	−9,678	715
Dec-02	12,132	24	89,181	1,240,938	279,553	57,164	75,344	12,665	71,741	9,049	1,298	12,164	19,616	−4,604	−3,184
Jan-03	12,227	25	131,568	1,301,927	716,180	150,253	110,031	78,148	4,610	19,855	18,448	14,432	119	2,748	
Feb-03	12,201	26	76,262	1,130,312	692,736	174,283	38,891	41,811	8,834	18,123	4,268	−291	2,119		
Mar-03	12,130	27	159,575	1,313,809	704,116	68,412	30,185	64,402	19,229	−3,021	3,220	1,994			
Apr-03	11,986	28	76,313	1,505,842	437,084	50,872	116,723	18,160	10,975	12,664	8,805				
May-03	11,927	29	104,028	1,667,823	360,676	153,274	37,529	34,840	17,479	9,374					
Jun-03	11,814	30	79,688	1,235,573	776,240	65,303	18,723	10,779	10,615						
Jul-03	11,787	31	76,395	1,689,354	442,965	234,171	36,806	22,351							
Aug-03	11,689	32	110,460	1,492,980	589,184	93,366	180,095								
Sep-03	11,731	33	196,687	2,011,979	313,416	166,839									
Oct-03	11,843	34	268,365	1,027,925	897,097										
Nov-03	11,902	35	58,510	1,225,307											
Dec-03	11,844	36	96,378												

20

Report Writing: Communicating Data Analysis Results

Chapter Preview. Statistical reports should be accessible to different types of readers. Such reports inform managers who desire broad overviews in nontechnical language and analysts who require technical details to replicate the study. This chapter summarizes methods of writing and organizing statistical reports. To illustrate, we will consider a report of claims from third-party automobile insurance.

20.1 Overview

The last relationship has been explored, the last parameter has been estimated, the last forecast has been made, and now you are ready to share the results of your statistical analysis with the world. The medium of communication can come in many forms: you may simply recommend to a client to "buy low, sell high" or you may give an oral presentation to your peers. Most likely, however, you will need to summarize your findings in a written report.

Communicating technical information is difficult for a variety of reasons. First, in most data analyses, there is no one "right" answer that the author is trying to communicate to the reader. To establish a right answer, one need only position the pros and cons of an issue and weigh their relative merits. In statistical reports, the author is trying to communicate data features and the relationship of the data to more general patterns, a much more complex task. Second, most reports written are directed at a primary client or audience. In contrast, statistical reports are often read by many different readers whose knowledge of statistical concepts varies extensively; it is important to take into consideration the characteristics of this heterogeneous readership when judging the pace and order in which the material is presented. This is particularly difficult when a writer can only guess as to the secondary audience. Third, authors of statistical reports need to have a broad and deep knowledge base, including a good understanding of underlying substantive issues, knowledge of statistical concepts, and language skills. Drawing on these different skill sets can be challenging. Even for a generally effective writer, any confusion in the analysis is inevitably reflected in the report.

Provide enough details of the study so that the analysis can be independently replicated with access to the original data.

Communication of data analysis results can be a brief oral recommendation to a client or a 500-page dissertation. However, a 10- to 20-page report summarizing the main conclusions and outlining the details of the analysis suffices for most business purposes. A key aspect of such a report is to provide the reader with an understanding of the salient features of the data. Enough details of the study should be provided so that the analysis could be independently replicated with access to the original data.

20.2 Methods for Communicating Data

To allow readers to interpret numerical information effectively, data should be presented using a combination of words, numbers, and graphs that reveal the data's complexity. Thus, the creators of data presentations must draw on background skills from several areas, including

- An understanding of the underlying substantive area
- A knowledge of the related statistical concepts
- An appreciation of design attributes of data presentations
- An understanding of the characteristics of the intended audience

This balanced background is vital if the purpose of the data presentation is to inform. If the purpose is to enliven the data (because data are inherently "boring") or to attract attention, then the design attributes may take on a more prominent role. Conversely, some creators with strong quantitative skills take great pains to simplify data presentations to reach a broad audience. By not using the appropriate design attributes, they reveal only part of the numerical information and hide the true story of their data. To quote Albert Einstein, "You should make your models as simple as possible, but no simpler."

This section presents the basic elements and rules for constructing successful data presentations. To this end, we discuss three modes of presenting numerical information: (1) within text data, (2) tabular data, and (3) data graphics. These three modes are ordered roughly in the complexity of data that they are designed to present, from the within-text data mode that is most useful for portraying the simplest types of data to the data graphics mode that is able to convey numerical information from extremely large sets of data.

Within-Text Data

Within-text data simply means numerical quantities that are cited within the usual sentence structure. For example:

> The price of Vigoro stock today is $36.50 per share, a record high.

When presenting data in text, you will have to decide whether to use figures or spell out a particular number. There are several guidelines for choosing between figures and words, although generally for business writing you will use words if this choice results in a concise statement. Some of the important guidelines include the following:

- Spell out whole numbers from one to ninety-nine.
- Use figures for fractional numbers.
- Spell out round numbers that are approximations.

- Spell out numbers that begin a sentence.
- Use figures in sentences that contain several numbers.

For example:

There are forty-three students in my class.
With 0.2267 U.S. dollars, I can buy one Swedish kroner.
There are about forty-three thousand students at this university.
Three thousand four hundred fifty-six people voted for me.
Those boys are 3, 4, 6, and 7 years old.

Text flows linearly; this makes it difficult for the reader to make comparisons of data within a sentence. When lists of numbers become long or important comparisons are to be made, a useful device for presenting data is the *within-text table*, also called the *semitabular* form. For example:

For 2005, net premiums by major line of business written by property and casualty insurers in billions of U.S. dollars, were

Private passenger auto, 159.57
Homeowners multiple peril, 53.01
Workers' compensation, 39.73
Other lines, 175.09.

Tables

When the list of numbers is longer, the tabular form, or *table*, is the preferred choice for presenting data. The basic elements of a table are identified in Table 20.1.

These basic elements are as follows:

1. *Title.* A short description of the data, placed above or to the side of the table. For longer documents, provide a table number for easy reference within the main body of the text. The title may be supplemented by additional remarks, thus forming a *caption*.

Table 20.1 Summary Statistics of Stock Liquidity Variables

Column Headings →; Rule →; Stub →; Title →; Body ←

	Mean	Median	Standard Deviation	Minimum	Maximum
VOLUME	13.423	11.556	10.632	0.658	64.572
AVGT	5.441	4.284	3.853	0.590	20.772
NTRAN	6436	5071	5310	999	36420
PRICE	38.80	34.37	21.37	9.12	122.37
SHARE	94.7	53.8	115.1	6.7	783.1
VALUE	4.116	2.065	8.157	0.115	75.437
DEB_EQ	2.697	1.105	6.509	0.185	53.628

Source: Francis Emory Fitch Inc., Standard & Poor's Compustat, and University of Chicago's Center for Research on Security Prices.

2. *Column Headings*. Brief indications of the material in the columns.
3. *Stub*. The left-hand vertical column, which often provides identifying information for individual row items.
4. *Body*. The other vertical columns of the table.
5. *Rules*. Lines that separate the table into its various components.
6. *Source*. Provides the origin of the data.

As with the semitabular form, tables can be designed to enhance comparisons between numbers. Unlike the semitabular form, tables are separate from the main body of the text. Because they are separate, tables should be self-contained so that the reader can draw information from the table with little reference to the text. The title should draw attention to the important features of the table. The layout should guide the reader's eye and facilitate comparisons. Table 20.1 illustrates the application of some basic rules for constructing "user friendly" tables. These rules include the following:

1. For titles and other headers, STRINGS OF CAPITALS ARE DIFFICULT TO READ, so keep these to a minimum.
2. Reduce the physical size of a table so that the eye does not have to travel as far as it might otherwise; use single spacing and reduce the type size.
3. Use columns for figures to be compared rather than rows; columns are easier to compare, though this makes documents longer.
4. Use row and column averages and totals to provide focus. This allows readers to make comparisons.
5. When possible, order rows and/or columns by size to facilitate comparisons. Generally, ordering by alphabetical listing of categories contributes little to an understanding complex datasets.
6. Use combinations of spacing and horizontal and vertical rules to facilitate comparisons. Horizontal rules are useful for separating major categories; vertical rules should be used sparingly. White space between columns serves to separate categories; closely spaced pairs of columns encourage comparison.
7. Use tinting and different type size and attributes to draw attention to figures. Use of tint is also effective for breaking up the monotonous appearance of a large table.
8. The first time that the data are displayed, provide the source.

Graphs

For portraying large, complex datasets, or data for which the actual numerical values are less important than the relations to be established, graphical representations of data are useful. Figure 20.1 describes some of the basic elements of a *graph*, also known as a *chart, illustration*, or *figure*. These include:

1. *Title* and *caption*. As with a table, these provide short descriptions of the main features of the figure. Long captions may be used to describe

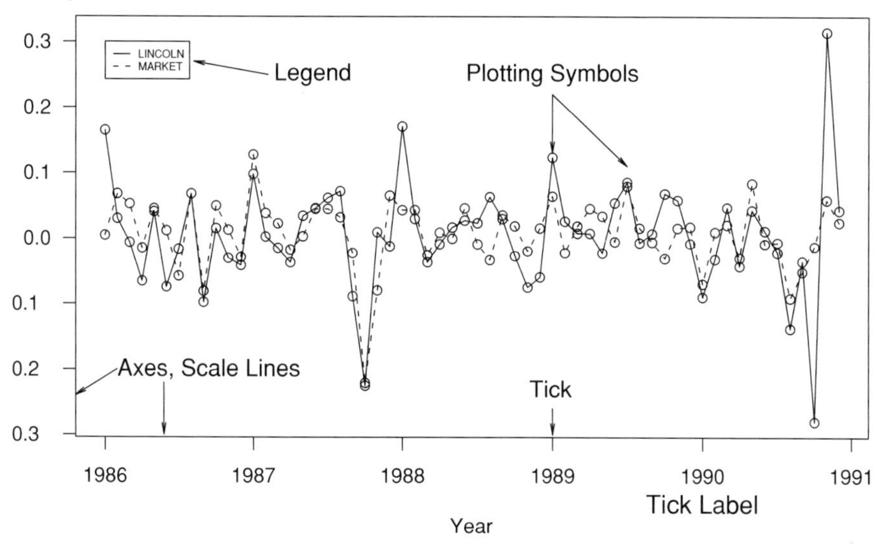

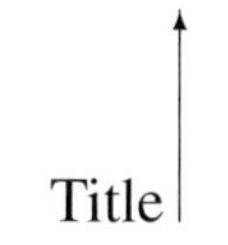

Figure 20.1 Time series plot of returns from the Lincoln National Corporation and the market. There are 60 monthly returns over the period January, 1986–December, 1990.

everything that is being graphed, to draw attention to the important features, and to describe the conclusions to be drawn from the data. Include the source of the data here or on a separate line immediately below the graph.

2. *Scale lines (Axes)* and *scale labels*. Choose the scales so that the data fill up as much of the data region as possible. Do not insist that zero be included; assume that the viewer will look at the range of the scales and understand them.
3. *Ticks* and *tick labels*. Choose the range of the ticks to include almost all of the data. Three to ten tick marks are generally sufficient. When possible put the tick outside of the data region, so that it does not interfere with the data.
4. *Plotting Symbols*. Use different plotting symbols to encode different levels of a variable. Plotting symbols should be chosen so that they are easy to identify, for example, "O" for one and "T" for two. However, be sure that plotting symbols are easy to distinguish; for example, it can be difficult to distinguish "F" and "E".
5. *Legend (keys)*. These are small textual displays that help identify certain aspects of the data. Do not let these displays interfere with the data or clutter the graph.

As with tables, graphs are separate from the main body of the text and thus should be self-contained. Especially with long documents, tables and graphs may contain a separate story line, thus providing a look at the main message of the document in a different way than the main body of the text. Cleveland (1994) and Tufte (1990) provide several tips to make graphs more "user friendly."

1. Make lines as thin as possible. Thin lines distract the eye less from the data than do thicker lines. However, make the lines thick enough so that the image will not degrade under reproduction.

2. Try to use as few lines as possible. Again, several lines distract the eye from the data, which carries the information. Try to avoid grid lines if possible. If you must use grid lines, a light ink, such as a gray or half-tone is the preferred choice.
3. Spell out words and avoid abbreviations. Rarely is the space saved worth the potential confusion that the shortened version may cause the viewer.
4. Use a type that includes both capital and small letters.
5. Place graphs on the same page as the text that discusses the graph.
6. Make words run from left to right, not vertically.
7. Use the substance of the data to suggest the shape and size of the graph. For time series graphs, make the graph twice as wide as tall. For scatter plots, make the graph equally wide as tall. If a graph displays an important message, make the graph large.

Of course, for most graphs it will be impossible to follow all these pieces of advice simultaneously. To illustrate, if we spell out the scale label on a left-hand vertical axis and make it run from left to right, then we cut into the vertical scale. This forces us to reduce the size of the graph, perhaps at the expense of reducing the message.

A graph is a powerful tool for summarizing and presenting numerical information. Graphs can be used to break up long documents; they can provoke and maintain reader interest. Further, graphs can reveal aspects of the data that other methods cannot.

20.3 How to Organize

Writing experts agree that results should be reported in an organized fashion with some logical flow, though there is no consensus as to how this goal should be achieved. Every story has a beginning and an end, usually with an interesting path connecting the two endpoints. There are many types of paths, or methods of development, that connect the beginning and the end. For general technical writing, the method of development may be organized chronologically, spatially, by order of importance, general to specific or specific to general, by cause and effect, or by any other logical development of the issues. This section presents one method of organization for statistical report writing that has achieved desirable results in a number of different circumstances, including the 10- to 20-page report described previously. This format, though not appropriate for all situations, serves as a workable framework on which to base your first statistical report.

The broad outline of the recommended format is as follows:

1. Title and abstract
2. Introduction
3. Data characteristics
4. Model selection and interpretation
5. Summary and concluding remarks
6. References and appendix

Sections (1) and (2) serve as the preparatory material, designed to orient the reader. Sections (3) and (4) form the main body of the report while Sections (5) and (6) are parts of the ending.

Title and Abstract

If your report is disseminated widely (as you hope), here is some disappointing news. A vast majority of your intended audience gets no further than the title and the abstract. Even for readers who carefully read your report, they will usually carry in their memory the impressions left by the title and abstract unless they are experts in the subject that you are reporting on (and most readers are not). Choose the title of your report carefully. It should be concise and to the point. Do not include deadwood (phrases like *The Study of, An Analysis of*) but do not be too brief, for example, by using only one word titles. In addition to being concise, the title should be comprehensible, complete, and correct.

The abstract is a one- to two-paragraph summary of your investigation; 75–200 words is a reasonable rule of thumb. The language should be nontechnical, as you are trying to reach as broad an audience as possible. The abstract should summarize the main findings of your report. Be sure to respond to such questions as. What problem was studied? How was it studied? and What were the findings? Because you are summarizing not only your results but also your report, it is generally easiest to write the abstract last.

The language of the abstract should be nontechnical.

Introduction

As with the general report, the introduction should be partitioned into three sections: orientation material, key aspects of the report, and a plan.

To begin the orientation material, reintroduce the problem at the level of technicality that you wish to use in the report. It may or may not be more technical than the statement of the problem in the abstract. The introduction sets the pace, or the speed at which new ideas are introduced, in the report. Throughout the report, be consistent in the pace. To clearly identify the nature of the problem, in some instances, a short literature review is appropriate. The literature review cites other reports that provide insights on related aspects of the same problem. This helps crystallize the new features of your report.

As part of the key aspects of the report, identify the source and nature of the data used in your study. Make sure that the manner in which your dataset can address the stated problem is apparent. Give an indication of the class of modeling techniques that you intend to use. Is the purpose behind this model selection clear (e.g., understanding versus forecasting)?

At this point, things can get a bit complex for many readers. It is a good idea to provide an outline of the remainder of the report at the close of the introduction. This provides a map to guide the reader through the complex arguments of the report. Further, many readers will be interested only in specific aspects of the report and, with the outline, will be able to "fast-forward" to the sections that interest them most.

It is also useful to describe the data without reference to a specific model.

Data Characteristics

In a data analysis project, the job is to summarize the data and use the summary information to make inferences about the state of the world. Much of this summarization is done through statistics that are used to estimate model parameters. However, it is also useful to describe the data without reference to a specific model for at least two reasons. First, by using basic summary measures of the data, you can appeal to a larger audience than if you had restricted your considerations to a specific statistical model. Indeed, with a carefully constructed graphical summary device, you should be able to reach virtually any reader who is interested in the subject material. Conversely, familiarity with statistical models requires a certain amount of mathematical sophistication, and you may or may not want to restrict your audience at this stage of the report. Second, constructing statistics that are immediately connected to specific models leaves you open to the criticism that your model selection is incorrect. For most reports, the selection of a model is an unavoidable juncture in the process of inference, but you need not do it at this relatively early stage of your report.

In the data characteristics section, identify the nature of data. For example, be sure to identify the component variables, and state whether the data are longitudinal versus cross-sectional, observational versus experimental, and so forth. Present any basic summary statistics that would help the reader develop an overall understanding of the data. It is a good idea to include about two graphs. Use scatter plots to emphasize primary relationships in cross-sectional data and time series plots to indicate the most important longitudinal trends. The graphs, and concomitant summary statistics, should not only underscore the most important relationships but also serve to identify unusual points that are worthy of special consideration. Carefully choose the statistics and graphical summaries that you present in this section. Do not overwhelm the reader with a plethora of numbers. The details presented in this section should foreshadow the development of the model in the subsequent section. Other salient features of the data can appear in the appendix.

Model Selection and Interpretation

This section is the heart and soul of your report. The results reported in this section generally took the longest to achieve. However, the length of the section need not be in proportion to the time it took you to accomplish the analysis. Remember, you are trying to spare readers the anguish that you went through in arriving at your conclusions. However, at the same time, you want to convince readers of the thoughtfulness of your recommendations. Here is an outline for the model selection and interpretation section that incorporates the following key elements:

1. An outline of the section
2. A statement of the recommended model
3. An interpretation of the model, parameter estimates, and any broad implications of the model

4. The basic justifications of the model
5. An outline of a thought process that would lead up to this model
6. A discussion of alternative models.

In this section, develop your ideas by discussing the general issues first and specific details later. Use subsections (1)–(3) to address the broad, general concerns that a nontechnical manager or client may have. Additional details can be provided in subsections (4)–(6) to address the concerns of the technically inclined reader. In this way, the outline is designed to accommodate the needs of these two types of readers. More details of each subsection are described in the following.

You are again confronted with the conflicting goals of wanting as large an audience as possible and yet needing to address the concerns of technical reviewers. Start this all-important section with an outline of things to come. That will enable the reader to pick and choose. Indeed, many readers will want only to examine your recommended model and the corresponding interpretations and will assume that your justifications are reliable. So, after providing the outline, immediately provide a *statement of the recommended model* in no uncertain terms. Now, it may not be clear at all from the dataset that your recommended model is superior to alternative models and, if that is the case, just say so. However, be sure to state, without ambiguity, what you considered the best. Do not let the confusion that arises from several competing models representing the data equally well drift over into your statement of a model.

The statement of a model is often in statistical terminology, a language used to express model ideas precisely. Immediately follow the statement of the recommended model with the concomitant *interpretations*. The interpretations should be done using nontechnical language. In addition to discussing the overall form of the model, the parameter estimates may provide an indication of the strength of any relationships that you have discovered. Often, a model is easily digested by the reader when discussed in terms of the resulting implications of a model, such as a confidence or prediction interval. Although only one aspect of the model, a single implication may be important to many readers.

It is a good idea to discuss briefly some of the technical *justifications of the model* in the main body of the report. This is to convince the reader that you know what you are doing. Thus, to defend your selection of a model, cite some of the basic justifications such as t-statistics, coefficient of determination, residual standard deviation, and so forth in the main body and include more detailed arguments in the appendix. To further convince the reader that you have seriously thought about the problem, include a brief description of a *thought process* that would lead one from the data to your proposed model. Do *not* describe to the reader all the pitfalls that you encountered on the way. Describe instead a clean process that ties the model to the data, with as little fuss as possible.

As mentioned, in data analysis there is rarely, if ever, a "right" answer. To convince the reader that you have thought about the problem deeply, it is a good idea to mention *alternative models*. This will show that you considered the problem from more than one perspective and are aware that careful, thoughtful individuals may arrive at different conclusions. However, in the end, you still need to give your recommended model and stand by your recommendation. You will sharpen your arguments by discussing a close competitor and comparing it with your recommended model.

Summary and Concluding Remarks

This final section may may serve as a springboard for questions and suggestions about future investigations.

This section should rehash the results of the report in a concise fashion, in different words than the abstract. The language may or may not be more technical than the abstract, depending on the tone you set in the introduction. Refer to the key questions posed when you began the study and tie these to the results. This section may look back over the analysis or serve as a springboard for questions and suggestions about future investigations. Include ideas that you have about future investigations, keeping in mind costs and other considerations that may be involved in collecting further information.

References and Appendix

The appendix may contain many auxiliary figures and analyses. The reader will not give the appendix the same level of attention as the main body of the report. However, the appendix is a useful place to include many crucial details for the technically inclined reader and important features that are not critical to the main recommendations of your report. Because the level of technical content here is generally higher than in the main body of the report, it is important that each portion of the appendix be clearly identified, especially with respect to its relation to the main body of the report.

20.4 Further Suggestions for Report Writing

1. Be as brief as you can but still include all important details. On the one hand, the key aspects of several regression outputs can often be summarized in one table. Often a number of graphs can be summarized in one sentence. On the other hand, recognize the value of a well-constructed graph or table for conveying important information.
2. Keep your readership in mind when writing your report. Explain what you now understand about the problem, with little emphasis on how you happened to get there. Give practical interpretations of results, in language that the client will be comfortable with.
3. Outline, outline. Develop your ideas in a logical, step-by-step fashion. It is *vital* that there is a logical flow to the report. Start with a broad outline that specifies the basic layout of the report. Then make a more detailed outline, listing each issue that you will discuss in each section. You retain literary freedom only by imposing structure on your reporting.
4. Simplicity, simplicity, simplicity. Emphasize your primary ideas in simple language. Replace complex words with simpler words if the meaning

remains the same. Avoid the use of clichés and trite language. Although technical language can be used, avoid the use of jargon or slang. Statistical jargon, such as "Let $x_1, x_2, \ldots$ be i.i.d. random variables" is rarely necessary.

5. Include important summary tables and graphs in the body of the report. Label all figures and tables so each is understandable when viewed alone.
6. Use one or more appendices to provide supporting details. Graphs of secondary importance, such as residuals plots, and statistical software output, such as regression fits, can be included in an appendix. Include enough detail so that another analyst, with access to the data, could replicate your work. Provide a strong link between the primary ideas that are described in the main body of the report and the supporting material in the appendix.

20.5 Case Study: Swedish Automobile Claims

® **EMPIRICAL** *Filename is "SwedishMotor Insurance"*

Determinants of Swedish Automobile Claims

Abstract

Automobile ratemaking depends on an actuary's ability to estimate the probability of a claim and, in the event of a claim, the likely amount. This study examines a classic Swedish dataset of third-party automobile insurance claims. Poisson and gamma regression models were fit to the frequency and severity portions, respectively. Distance driven by a vehicle, geographic area, recent driver claims experience, and the type of automobile are found to be important determinants of claim frequency. Only geographic area and automobile type turn out to be important determinants of claim severity. Although the experience is dated, the techniques used and the importance of these determinants give helpful insights into current experience.

What problem was studied? How was it studied? What were the findings?

Section 1. Introduction

Actuaries seek to establish premiums that are fair to consumers in the sense that each policyholder pays according to his or her own expected claims. These expected claims are based on policyholder characteristics that may include age, sex, and driving experience. Motivation for this rating principle is not entirely altruistic; an actuary understands that rate mispricing can lead to serious adverse financial consequences for the insuring company. For example, if rates are too high relative to the marketplace, then the company is unlikely to gain sufficient market share. Conversely, if rates are too low relative to actual experience, then premiums received will be unlikely to cover claims and related expenses.

Begin with some orientation material.

Setting appropriate rates is important in automobile insurance that indemnifies policyholders and other parties in the event of an automobile accident. For a short-term coverage like automobile insurance, claims resulting from policies are quickly realized and the actuary can calibrate the rating formula to actual experience.

For many analysts, data on insurance claims can be difficult to access. Insurers want to protect the privacy of their customers and so do not want to share data. For some insurers, data are not stored in an electronic format that is convenient for statistical analyses; it can be expensive to access data even though it is available to the insurer. Perhaps most important, insurers are reluctant to release data to the public because they fear that disseminating proprietary information will help their competitors in pricing wars.

When describing the key aspects of the report, include sources of data.

Because of this lack of up-to-date automobile data, this study examines a classic Swedish dataset of third-party automobile insurance claims that occurred in 1977. Third-party claims involve payments to someone other than the policyholder and the insurance company, typically someone injured as a result of an automobile accident. Although the experience is dated, the regression techniques used in this report work equally well with current experience. Further, the determinants of claims investigated, such as vehicle use and driver experience, are likely to be important in today's driving world.

Provide a plan for the remainder of the paper.

The outline of the remainder of this report is as follows. In Section 2, I present the most important characteristics of the data. To summarize these characteristics, Section 3 features a discussion of a model to represent the data. Concluding remarks can be found in Section 4, and many of the details of the analysis are in the appendix.

Section 2. Data Characteristics

The data were compiled by the Swedish Committee on the Analysis of Risk Premium in Motor Insurance, summarized in Hallin and Ingenbleek (1983) and Andrews and Herzberg (1985). The data are cross-sectional, describing third-party automobile insurance claims for the year 1977.

Identify the nature of the data.

The outcomes of interest are the number of claims (the frequency) and sum of payments (the severity), in Swedish kroners. Outcomes are based on five categories of distance driven by a vehicle, broken down by seven geographic zones, seven categories of recent driver claims experience (captured by the "bonus") and 9 types of automobile. Even though there are 2,205 potential distance, zone, experience, and type combinations ($5 \times 7 \times 7 \times 9 = 2{,}205$), only $n = 2{,}182$ were realized in the 1977 dataset. For each combination, in addition to outcomes of interest, we have available the number of policyholder years as a measure of exposure. A policyholder year is the fraction of the year that the policyholder has a contract with the issuing company. More detailed explanations of these variables are available in Appendix A2.

Use selected plots and statistics to emphasize the primary trends. Do not refer to a statistical model in this section.

In these data, there were 113,171 claims from 2,383,170 policyholder years, for a 4.75% claims rate. From these claims, a total of 560,790,681 kroners were paid, for an average of 4,955 per claim. For reference, in June 1977, one Swedish kroner could be exchanged for 0.2267 U.S. dollars.

Table 20.2 provides more details on the outcomes of interest. This table is organized by the $n = 2{,}182$ distance, zone, experience, and type combinations. For example, the combination with the greatest exposure (127,687.27 policyholder years) comes from those driving a minimal amount in rural areas of southern Sweden, having at least six accident-free years and driving a car that is not one of the basic eight types (kilometers = 1, zone = 4, bonus = 7, and make = 9; see

Table 20.2 Swedish Automobile Summary Statistics

Variable	Mean	Median	Standard Deviation	Minimum	Maximum
Policyholder Years	1,092.20	81.53	5,661.16	0.01	127,687.27
Claims	51.87	5.00	201.71	0.00	3,338.00
Payments	257,008	27,404	1,017,283	0	18,245,026
Average Claim Number (per Policyholder Year)	0.069	0.051	0.086	0.000	1.667
Average Payment (per Claim)	5,206.05	4,375.00	4,524.56	72.00	31,442.00

Note: Distributions are based on $n = 2,182$ distance, zone, experience, and type combinations.
Source: Hallin and Ingenbleek (1983).

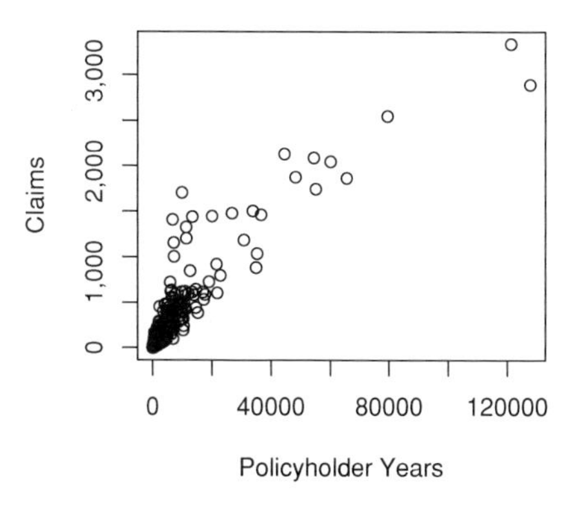

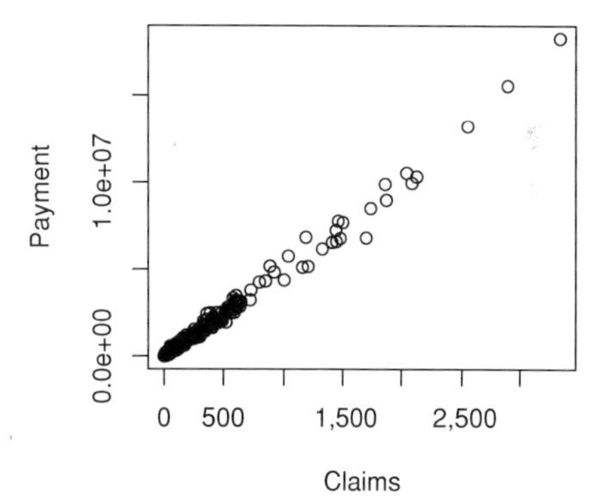

Figure 20.2 Scatter Plots of Claims versus Policyholder Years and Payments versus Claims.

Appendix A2). This combination had 2,894 claims with payments of 15,540,162 kroners. Further, there were 385 combinations that had zero claims.

Table 20.2 also shows the distribution of the average claim number per insured. Not surprisingly, the largest average claim number occurred in a combination where there was only a single claim with a small number (0.6) of policyholder years. Because I will use policyholder years as a weight in the Section 3 analysis, this type of aberrant behavior will be automatically downweighted; thus, no special techniques are required to deal with it. For the largest average payment, there are 27 combinations with a single claim of 31,442 (and one combination with two claims of 31,442). This apparently represents some type of policy limit imposed that I do not have documentation on. I ignore this feature in the analysis.

Figure 20.2 shows the relationships between the outcomes of interest and exposure bases. For the number of claims, I use policyholder years as the exposure basis. It is clear that the number of insurance claims increases with exposure. Further, the payment amounts increase with the claims number in a linear fashion.

To understand the explanatory variable effects on frequency, Figure 20.3 presents box plots of the average claim number per insured versus each rating variable. To visualize the relationships, three combinations in which the average

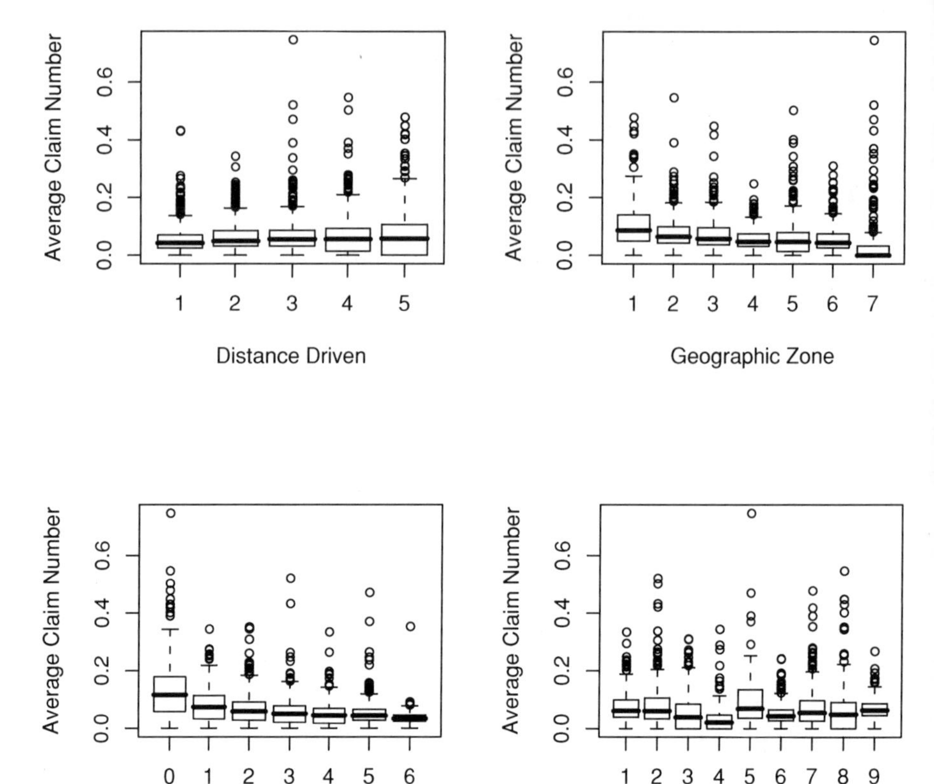

Figure 20.3 Box plots of frequency by distance driven, geographic zone, accident-free years and make of automobile.

claim exceeds 1.0 have been omitted. This figure shows lower frequencies associated with lower driving distances, nonurban locations, and a higher number of accident-free years. The automobile type also appears to have a strong impact on claim frequency.

For severity, Figure 20.4 presents box plots of the average payment per claim versus each rating variable. Here, effects of the explanatory variables are not as pronounced as they are for frequency. The upper-right-hand panel shows that the average severity is much smaller for zone = 7. This corresponds to Gotland, a county and municipality of Sweden that occupies the largest island in the Baltic Sea. Figure 20.4 also suggests some variation based on the type of automobile.

Section 3. Model Selection and Interpretation

Section 2 established that there are real patterns between claims frequency and severity and the rating variables, despite the great variability in the variables. This section summarizes these patterns using regression modeling. Following the statement of the model and its interpretation, this section describes features of the data that drove the selection of the recommended model.

Start with a statement of your recommended model.

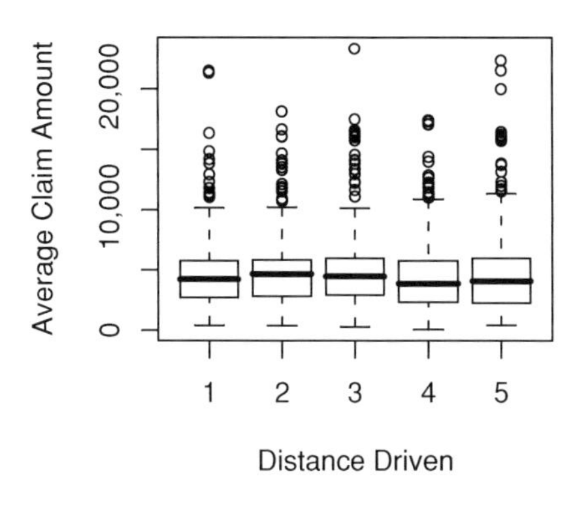

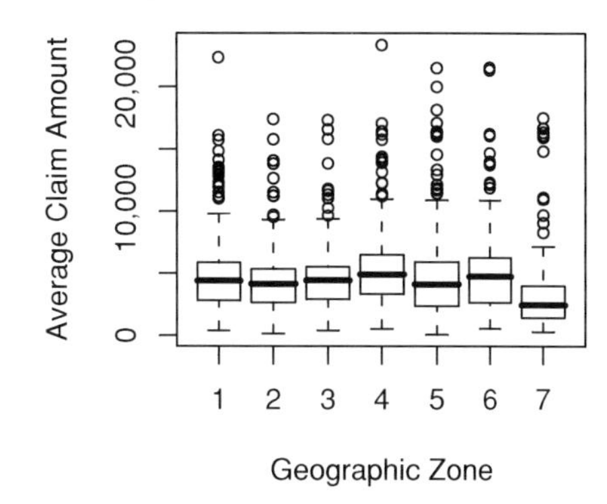

Figure 20.4 Box plots of severity by distance driven, geographic zone, accident-free years and make of automobile.

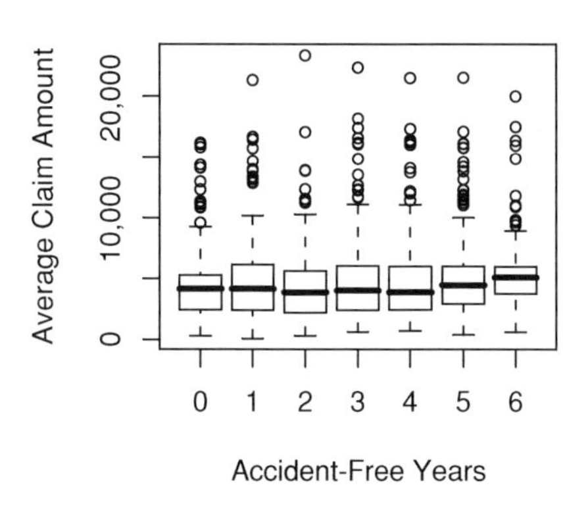

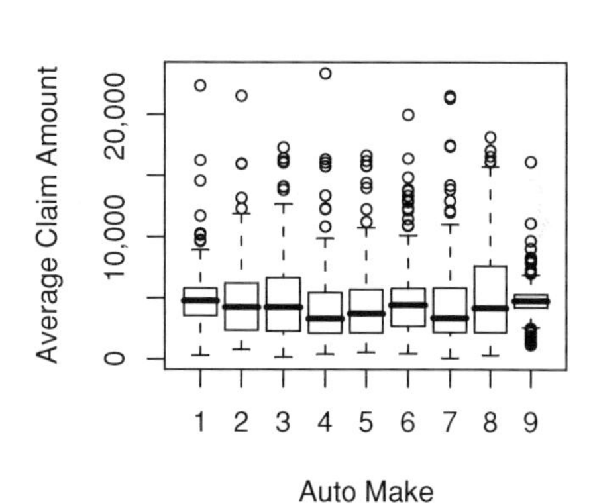

As a result of this study, I recommend a Poisson regression model using a logarithmic link function for the frequency portion. The systematic component includes the rating factors distance, zone, experience, and type as additive categorical variables, as well as an offset term in logarithmic number of insureds.

Interpret the model; discuss variables, coefficients and broad implications of the model.

This model was fit using maximum likelihood, with the coefficients appearing in Table 20.3; more details appear in Appendix A4. Here, the base categories correspond to the first level of each factor. To illustrate, consider a driver living in Stockholm (zone = 1) who drives between 1 and 15,000 kilometers per year (kilometers = 2), has had an accident within the past year (bonus = 1) and is driving car type make = 6. Then, from Table 20.3, the systematic component is $-1.813 + 0.213 - 0.336 = -1.936$. For a typical policy from this combination, we would estimate a Poisson number of claims with mean $\exp(-1.936) = 0.144$. For example, the probability of no claims in a year is $\exp(-0.144) = 0.866$. In 1977, there were 354.4 policyholder years in this combination, for an expected number of claims of $354.4 \times 0.144 = 51.03$. It turned out that there were only 48 claims in this combination in 1977.

For the severity portion, I recommend a gamma regression model using a logarithmic link function. The systematic component consists of the rating factors zone and type as additive categorical variables, as well as an offset term in

Table 20.3 Poisson Regression Model Fit

Variable	Coefficient	t-Ratio	Variable	Coefficient	t-Ratio
Intercept	−1.813	−131.78	Bonus = 2	−0.479	−39.61
Kilometers = 2	0.213	28.25	Bonus = 3	−0.693	−51.32
Kilometers = 3	0.320	36.97	Bonus = 4	−0.827	−56.73
Kilometers = 4	0.405	33.57	Bonus = 5	−0.926	−66.27
Kilometers = 5	0.576	44.89	Bonus = 6	−0.993	−85.43
Zone = 2	−0.238	−25.08	Bonus = 7	−1.327	−152.84
Zone = 3	−0.386	−39.96	Make = 2	0.076	3.59
Zone = 4	−0.582	−67.24	Make = 3	−0.247	−9.86
Zone = 5	−0.326	−22.45	Make = 4	−0.654	−27.02
Zone = 6	−0.526	−44.31	Make = 5	0.155	7.66
Zone = 7	−0.731	−17.96	Make = 6	−0.336	−19.31
			Make = 7	−0.056	−2.40
			Make = 8	−0.044	−1.39
			Make = 9	−0.068	−6.84

Table 20.4 Gamma Regression Model Fit

Variable	Coefficient	t-Ratio	Variable	Coefficient	t-Ratio
Intercept	8.388	76.72	Make = 2	−0.050	−0.44
Zone = 2	−0.061	−0.64	Make = 3	0.253	2.22
Zone = 3	0.153	1.60	Make = 4	0.049	0.43
Zone = 4	0.092	0.94	Make = 5	0.097	0.85
Zone = 5	0.197	2.12	Make = 6	0.108	0.92
Zone = 6	0.242	2.58	Make = 7	−0.020	−0.18
Zone = 7	0.106	0.98	Make = 8	0.326	2.90
			Make = 9	−0.064	−0.42
Dispersion	0.483				

logarithmic number of claims. Further, the square root of the claims number was used as a weighting variable to give greater weight to those combinations with a greater number of claims.

This model was fit using maximum likelihood, with the coefficients appearing in Table 20.4; more details appear in Appendix A6. Consider again the illustrative driver living in Stockholm (zone = 1) who drives between 1 and 15,000 kilometers per year (kilometers = 2), has had an accident within the past year (bonus = 1), and is driving car type make = 6. For this person, the systematic component is $8.388 + 0.108 = 8.496$. Thus, the expected claims under the model are $\exp(8.496) = 4{,}895$. For comparison, the average 1977 payment was 3,467 for this combination and 4,955 per claim for all combinations.

Discussion of the Frequency Model

What are some of the basic justifications of the model?

Both models provided a reasonable fit to the available data. For the frequency portion, the t-ratios in Table 20.3 associated with each coefficient exceed

Model	Pearson	Weighted Pearson
Poisson without Covariates	44,639	653.49
Final Poisson Model	3,003	6.41
Negative Binomial Model	3,077	9.03

Table 20.5 Pearson Goodness of Fit for Three Frequency Models

three in absolute value, which indicates strong statistical significance. Moreover, Appendix A5 demonstrates that each categorical factor is strongly statistically significant.

Provide strong links between the main body of the report and the appendix.

There were no other major patterns between the residuals from the final fitted model and the explanatory variables. Figure A1 displays a histogram of the deviance residuals, indicating approximate normality, a sign that the data are in congruence with model assumptions.

A number of competing frequency models were considered. Table 20.5 lists two others, a Poisson model without covariates and a negative binomial model with the same covariates as the recommended Poisson model. This table shows that the recommended model is best among these three alternatives, given the Pearson goodness-of-fit statistic and a version weighted by exposure. Recall that the Pearson fit statistic is of the form $\sum(O - E)^2/E$, comparing observed (O) to data expected under the model fit (E). The weighted version summarizes $\sum w(O - E)^2/E$, where our weights are policyholder years in units of 100,000. In each case, we prefer models with smaller statistics. Table 20.5 shows that the recommended model is the clear choice among the three competitors.

Is there a thought process that leads us to conclude that the model is a useful one?

In developing the final model, the first decision made was to use the Poisson distribution for counts. This is in accord with accepted practice and because a histogram of claims numbers (not displayed here) showed a skewed Poisson-like distribution.

Covariates displayed important features that could affect the frequency, as shown in Section 2 and Appendix A3.

A good way to justify your recommended model is to compare it to one or more alternatives.

In addition to the Poisson and negative binomial models, I also fit a quasi-Poisson model with an extra parameter for dispersion. Although this seemed to be useful, ultimately I chose not to recommend this variation because the rate-making goal is to fit expected values. All rating factors were statistically significant with and without the extra dispersion factor, and so the extra parameter added only complexity to the model. Hence, I elected not to include this term.

Discussion of the Severity Model

For the severity model, the categorical factors zone and make are statistically significant, as is shown in Appendix A7. Although not displayed here, residuals from this model were well behaved. Deviance residuals were approximately normally distributed. Residuals, when rescaled by the square root of the claims number were approximately homoscedastic. There were no apparent relations with explanatory variables.

This complex model was specified after a long examination of the data. Given the evident relations between payments and number of claims in Figure 20.2, the first step was to examine the distribution of payments per claim. This distribution was skewed, and so an attempt to fit logarithmic payments per claim was made. After fitting explanatory variables to this dependent variable, residuals from the model fitting were heteroscedastic. These were weighted by the square root of the claims number and achieved approximate homoscedasticity. Unfortunately, as seen in Appendix Figure A2, the fit is still poor in the lower tails of the distribution.

A similar process was then undertaken using the gamma distribution with a log-link function, with payments as the response and logarithmic claims number as the offset. Again, I established the need for the square root of the claims number as a weighting factor. The process began with all four explanatory variables but distance and accident-free years were dropped because of their lack of statistical significance. I also created a binary variable "Safe" to indicate that a driver had six or more accident-free years (based on my examination of Figure 20.4). However, this was not statistically significant and so was not included in the final model specification.

Section 4. Summary and Concluding Remarks

Although insurance claims vary significantly, it is possible to establish important determinants of claims number and payments. The recommended regression models conclude that insurance outcomes can be explained in terms of the distance driven by a vehicle, geographic area, recent driver claims experience, and type of automobile. Separate models were developed for the frequency and severity of claims. In part, this was motivated by the evidence that fewer variables seem to influence payment amounts compared to claims number.

Rehash the results in a concise fashion. Discuss shortcomings and potential extensions of the work.

This study was based on 113,171 claims from 2,383,170 policyholder years, for a total of 560,790,681 kroners. This is a large dataset that allows us to develop complex statistical models. The grouped form of the data allows us to work with only $n = 2{,}182$ cells, which is relatively low by today's standards. Ungrouped data would allow us to consider additional explanatory variables. One might conjecture about any number of additional variables that could be included; age, sex, and good student discount are some good candidates. I note that the article by Hallin and Ingenbleek (1983) considered vehicle age – this variable was not included in my database because analysts responsible for the data publication considered it an insignificant determinant of insurance claims.

Further, my analysis of data is based on 1977 experience of Swedish drivers. The lessons learned from this report may or may not transfer to modern drivers. Nonetheless, the techniques explored in this report should be immediately applicable with the appropriate set of modern experience.

Appendix

Appendix Table of Contents

A table of contents, or outline, is useful for long appendices.

A1. References

Include references, detailed data analysis and other materials of lesser importance in the appendices.

Andrews, D. F., and A. M. Herzberg (1985). *A Collection from Many Fields for the Student and Research Worker*. Springer, New York.

Hallin, Marc, and Jean-Fran cois Ingenbleek (1983). The Swedish automobile portfolio in 1977: A statistical study. *Scandinavian Actuarial Journal* 1983, 49–64.

A2. Variable Definitions

Table A.1 Variable Definitions

Name	Description
Kilometres	Kilometers traveled per year 1: <1,000 2: 1,000–15,000 3: 15,000–20,000 4: 20,000–25,000 5: >25,000
Zone	Geographic zone 1: Stockholm, Göteborg, Malmö with surroundings 2: Other large cities with surroundings 3: Smaller cities with surroundings in southern Sweden 4: Rural areas in southern Sweden 5: Smaller cities with surroundings in northern Sweden 6: Rural areas in northern Sweden 7: Gotland
Bonus	No claims bonus. Equal to the number of years, plus one, since the last claim.
Make	1–8 represent eight different common car models. All other models are combined in class 9.
Exposure	Amount of policyholder years
Claims	Number of claims
Payment	Total value of payments in Swedish kroner

A3. Basic Summary Statistics for Frequency

Table A.2 Averages of Claims per Insured by Rating Factor

```
Kilometre
     1         2       3       4       5
0.0561    0.0651  0.0718  0.0705  0.0827
Zone
     1         2       3       4       5       6       7
0.1036    0.0795  0.0722  0.0575  0.0626  0.0569  0.0504
Bonus
     1         2       3       4       5       6       7
0.1291    0.0792  0.0676  0.0659  0.0550  0.0524  0.0364
Make
     1         2       3       4       5       6       7       8       9
0.0761    0.0802  0.0576  0.0333  0.0919  0.0543  0.0838  0.0729  0.0712
```

A4. Final Fitted Frequency Regression Model: R Output

```
Call: glm(formula = Claims ~ factor(Kilometres) + factor(Zone) +
factor(Bonus) + factor(Make), family = poisson(link = log),
offset = log(Insured))

Deviance Residuals:
   Min       1Q  Median       3Q      Max
-6.985  -0.863  -0.172   0.600    6.401

Coefficients:
                    Estimate Std. Error z value Pr(>|z|)
(Intercept)         -1.81284    0.01376 -131.78  < 2e-16 ***
factor(Kilometres)2  0.21259    0.00752   28.25  < 2e-16 ***
factor(Kilometres)3  0.32023    0.00866   36.97  < 2e-16 ***
factor(Kilometres)4  0.40466    0.01205   33.57  < 2e-16 ***
factor(Kilometres)5  0.57595    0.01283   44.89  < 2e-16 ***
factor(Zone)2       -0.23817    0.00950  -25.08  < 2e-16 ***
factor(Zone)3       -0.38639    0.00967  -39.96  < 2e-16 ***
factor(Zone)4       -0.58190    0.00865  -67.24  < 2e-16 ***
factor(Zone)5       -0.32613    0.01453  -22.45  < 2e-16 ***
factor(Zone)6       -0.52623    0.01188  -44.31  < 2e-16 ***
factor(Zone)7       -0.73100    0.04070  -17.96  < 2e-16 ***
factor(Bonus)2      -0.47899    0.01209  -39.61  < 2e-16 ***
factor(Bonus)3      -0.69317    0.01351  -51.32  < 2e-16 ***
factor(Bonus)4      -0.82740    0.01458  -56.73  < 2e-16 ***
factor(Bonus)5      -0.92563    0.01397  -66.27  < 2e-16 ***
```

(continued)

A4. (Continued)

```
factor(Bonus)6      -0.99346    0.01163  -85.43  < 2e-16 ***
factor(Bonus)7      -1.32741    0.00868 -152.84  < 2e-16 ***
factor(Make)2        0.07624    0.02124    3.59  0.00033 ***
factor(Make)3       -0.24741    0.02509   -9.86  < 2e-16 ***
factor(Make)4       -0.65352    0.02419  -27.02  < 2e-16 ***
factor(Make)5        0.15492    0.02023    7.66  1.9e-14 ***
factor(Make)6       -0.33558    0.01738  -19.31  < 2e-16 ***
factor(Make)7       -0.05594    0.02334   -2.40  0.01655 *
factor(Make)8       -0.04393    0.03160   -1.39  0.16449
factor(Make)9       -0.06805    0.00996   -6.84  8.2e-12 ***
---
Signif. codes:  0 *** .001 ** .01 * 0.05 . 0.1   1

(Dispersion parameter for poisson family taken to be 1)

    Null deviance: 34070.6  on 2181  degrees of freedom
Residual deviance:  2966.1  on 2157  degrees of freedom AIC: 10654
```

A5. Checking Significance of Factors in the Final Fitted Frequency Regression Model: R Output

```
Analysis of Deviance Table

Terms added sequentially (first to last)
           Df Deviance Resid. Df Resid. Dev P(>|Chi|)
NULL                        2181     34071
factor(Kilometres)  4    1476   2177     32594  2.0e-318
factor(Zone)        6    6097   2171     26498         0
factor(Bonus)       6   22041   2165      4457         0
factor(Make)        8    1491   2157      2966  1.4e-316
```

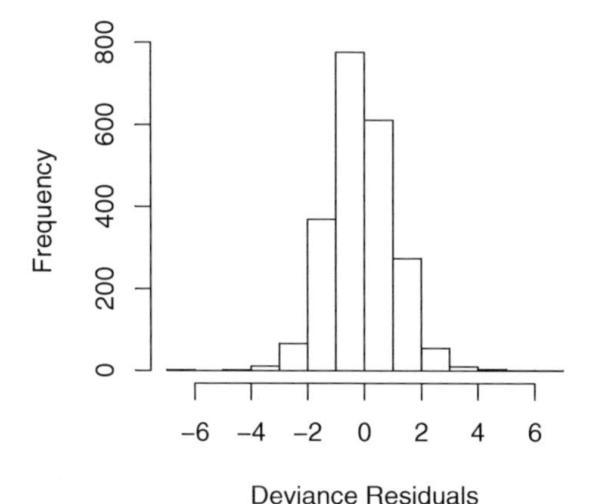

Figure A1 Histogram of deviance residuals from the final frequency model

A6. Final Fitted Severity Regression Model: R Output

```
Call:
glm(formula = Payment ~ factor(Zone) + factor(Make),
    family = Gamma (link = log), weights = Weight,
    offset = log(Claims))

Deviance Residuals:
     Min        1Q    Median        3Q       Max
-2.56968  -0.39928  -0.06305   0.07179   2.81822

Coefficients:
               Estimate Std. Error t value Pr(>|t|)
(Intercept)     8.38767  0.10933  76.722  < 2e-16 ***
factor(Zone)2 -0.06099  0.09515  -0.641  0.52156
factor(Zone)3  0.15290  0.09573   1.597  0.11041
factor(Zone)4  0.09223  0.09781   0.943  0.34583
factor(Zone)5  0.19729  0.09313   2.119  0.03427 *
factor(Zone)6  0.24205  0.09377   2.581  0.00992 **
factor(Zone)7  0.10566  0.10804   0.978  0.32825
factor(Make)2 -0.04963  0.11306  -0.439  0.66071
factor(Make)3  0.25309  0.11404   2.219  0.02660 *
factor(Make)4  0.04948  0.11634   0.425  0.67067
factor(Make)5  0.09725  0.11419   0.852  0.39454
factor(Make)6  0.10781  0.11658   0.925  0.35517
factor(Make)7 -0.02040  0.11313  -0.180  0.85692
factor(Make)8  0.32623  0.11247   2.900  0.00377 **
factor(Make)9 -0.06377  0.15061  -0.423  0.67205
---
Signif. codes:  0 *** 0.001 ** 0.01 * 0.05 . 0.1  1

(Dispersion parameter for Gamma family taken
to be 0.4830309)

    Null deviance: 617.32 on 1796 degrees of freedom
Residual deviance: 596.79 on 1782 degrees of freedom
AIC: 16082
```

A7. Checking Significance of Factors in the Final Fitted Severity Regression Model: R Output

```
Analysis of Deviance Table

Terms added sequentially (first to last)

              Df Deviance Resid. Df Resid. Dev P(>|Chi|)
NULL                           1796     617.32
factor(Zone)   6     8.06      1790     609.26      0.01
factor(Make)   8    12.47      1782     596.79  0.001130
```

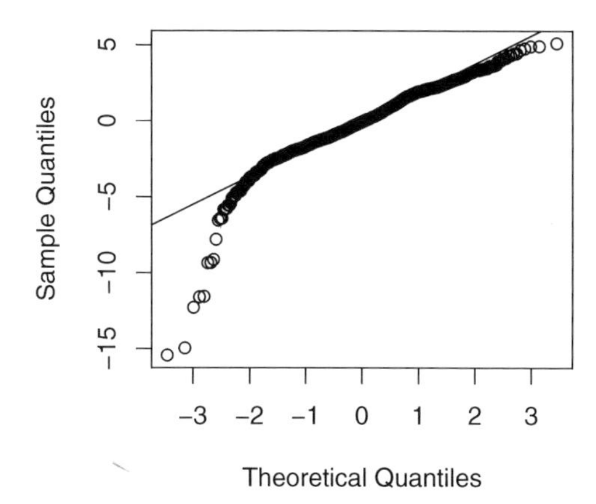

Figure A2 A qq plot of weighted residuals from a lognormal model. The dependent variable is average severity per claim. Weights are the square root of the number of claims. The poor fit in the tails suggests using an alternative to the lognormal model.

20.6 Further Reading and References

You can find further discussion of guidelines for presenting within-text data in *The Chicago Manual of Style*, a well-known reference for preparing and editing written copy.

You can find further discussion of guidelines for presenting tabular data in Ehrenberg (1977) and Tufte (1983).

Miller (2005) is a book-length introduction to writing statistical reports with an emphasis on regression methods.

Chapter References

The Chicago Manual of Style, 15th ed. (2003). University of Chicago Press, Chicago.

Cleveland, William S. (1994). *The Elements of Graphing Data.* Wadsworth, Monterey, California.

Ehrenberg, A. S. C. (1977). Rudiments of numeracy. *Journal of the Royal Statistical Society A* 140, 277–97.

Miller, Jane E. (2005). *The Chicago Guide to Writing about Multivariate Analysis.* University of Chicago Press, Chicago.

Tufte, Edward R. (1983). *The Visual Display of Quantitative Information*. Graphics Press, Cheshire, Connecticut.
Tufte, Edward R. (1990). *Envisioning Information*. Graphics Press, Cheshire, Connecticut.

20.7 Exercises

Ⓡ **Empirical**
Filename is "CeoCompensation"

20.1. Determinants of CEO Compensation. Chief executive officer (CEO) compensation varies significantly from firm to firm. For this exercise, you will report on a sample of firms from a survey by *Forbes* magazine to establish important patterns in the compensation of CEOs. Specifically, introduce a regression model that explains CEO salaries in terms of the firm's sales and the CEO's length of experience, education level, and ownership stake in the firm. Among other things, the model should show that larger firms tend to pay CEOs more and that, somewhat surprisingly, CEOs with more education earn less than otherwise comparable CEOs. In addition to establishing important influences on CEO compensation, the model should be used to predict CEO compensation for salary negotiation purposes.

21

Designing Effective Graphs

Chapter Preview.[†] Actuaries, like other business professionals, communicate quantitative ideas graphically. Because the process of reading, or decoding, graphs is more complex than reading text, graphs are vulnerable to abuse. To underscore this vulnerability, we give several examples of commonly encountered graphs that mislead and hide information. To help creators design more effective graphs and to help viewers recognize misleading graphs, this chapter summarizes guidelines for designing graphs that show important numerical information. When designing graphs, creators should

(1) Avoid chartjunk.
(2) Use small multiples to promote comparisons and assess change.
(3) Use complex graphs to portray complex patterns.
(4) Relate graph size to information content.
(5) Use graphical forms that promote comparisons.
(6) Integrate graphs and text.
(7) Demonstrate an important message.
(8) Know the audience.

Some of the guidelines for designing effective graphs, such as (6), (7), and (8), are drawn directly from principles for effective writing. Others, such as guidelines (3), (4), and (5), come from cognitive psychology, the science of perception. Guidelines (1) and (2) have roots both in effective writing and in graphical perception. For example, the writing principle of brevity demonstrates how eliminating pseudo three-dimensional perspectives and other forms of chartjunk improve graphs. As another example, the writing principle of parallel structure suggests using small multiple variations of a basic graphical form to visualize complex relationships across different groups and over time.

To underscore the scientific aspect of graphical perception, we examine the process of communicating with a graph, beginning with a sender's interpretation of data and ending with a receiver's interpretation of the graph. In keeping with scientific tradition, this chapter discusses several studies in the literature on the effectiveness of graphs.

We conclude that the actuarial profession has many opportunities to improve its practice, thus making communication more efficient and precise.

† This chapter is based on "Designing Effective Graphs," by Edward W. Frees and Robert B. Miller, 1990, *North American Actuarial Journal*, vol. 2, no. 2, 53–70. Published by the Society of Actuaries, reprinted with permission.

21.1 Introduction

Like other business professionals, actuaries communicate ideas orally and in writing, as well as through presentations, which are interactive forms of communication that encompass oral and written messages. Actuaries, as well as other financial analysts, communicate ideas with important quantitative components. Writers express quantitative ideas as (1) numbers within paragraphs, (2) numbers within tabular forms, (3) functional relationships such as equations, and (4) data or equations as graphs.

Graphs are a simple yet powerful medium for written communication of quantitative ideas. Graphs can present a large amount of data in a small space, express important relationships among quantities, compare different sets of data, and describe data, thus providing a coherent picture of complex systems. Graphs do more than merely state an idea; they demonstrate it.

Graphs are powerful because they are flexible, but flexibility can be a disadvantage because of the potential for abuse. Well-accepted references dealing with methods of quantitative data presentation mitigate opportunities for abuse. The 2003 *Chicago Manual of Style*, a standard reference, discusses presentation of in-text data, and Ehrenberg (1977) and Tufte (1983) discuss presentation of tabular data. In contrast, we focus on data presentation through *graphical* displays.

This chapter seeks to improve actuarial practice as it relates to graphical displays. We intend to: (1) demonstrate the importance of graphical displays, (2) provide guidelines to improve graphical practice, and (3) introduce some of the scientific underpinnings of good graphical practice. The agenda is ambitious, yet the goal of this chapter is to provide practicing actuaries with the basic tools to become critical consumers and effective producers of graphs. We also hope that readers will adopt our enthusiasm and explore the graphical design literature on their own.

"... sixty-three words that could change the world."

An important theme of this chapter is that principles of vigorous writing can and should be applied to the practice of making effective graphs. The *Elements of Style* (Strunk and White 1979, p. xiv) summarizes vigorous writing:

> Vigorous writing is concise. A sentence should contain no unnecessary words, a paragraph no unnecessary sentences, for the same reason that a drawing should have no unnecessary lines and a machine no unnecessary parts. This requires not that the writer make all his sentences short, or that he avoid all detail and treat his subjects only in outline, but that every word tell.

Chartjunk is any unnecessary appendage in a graph.

White attributes this quotation to William Strunk. White calls it "a short, valuable essay on the nature and beauty of brevity – sixty-three words that could change the world." Brevity is especially important when making effective graphs. This was also understood by Strunk; he said "a drawing should contain no unnecessary lines." We use the term *chartjunk*, introduced by Tufte (1983), for any unnecessary appendage in a graph.

Vigorous writing principles other than brevity also apply to the practice of making effective graphs. As with writing, effective graphs are the result of repeated revising and editing. Poorly designed graphs can and do hide information and mislead. Fancy or pretentious graphs are distracting when simpler graphs suffice.

Although the principles of effective writing are valuable, they are not sufficient for producing effective graphs. Writing is processed in a serial manner, word by word, sentence by sentence, with a beginning and an ending. The process of reading, or *decoding*, a graph is nonlinear and more complex. The additional complexities mean that even authors who follow effective writing practices may produce ineffective graphs. Often the form of written prose is the sole determinant of its value, whereas in graphics the communication process plays the dominant role. We assume that readers are familiar with effective writing forms. Thus, we first review the communication process in which a graph plays a crucial role.

To underscore the importance of effective graphical design, Section 21.2 provides several illustrations of graphs that hide information and are misleading; the defects illustrated are more serious drawbacks than mere chartjunk. The Section 21.2 illustrations motivate the need for additional guidelines and methods for constructing effective graphs.

Section 21.3 introduces eight important guidelines for creating and viewing graphs. Although the guidelines do not provide a panacea for all graphical defects, they do provide business professionals such as actuaries with a key checklist for creating effective graphs. The guidelines are organized so that the first two, on chartjunk and the use of multiples, are based on both effective writing and graphical perception perspectives. Guidelines 3, 4, and 5 are related primarily to the graphical perception literature, whereas guidelines 6, 7, and 8 are based primarily on effective-writing principles.

As with effective writing, questions of style enter into the discussion of what is and what is not an effective graph. Many style decisions are based on accepted practices without a firm scientific foundation. However, the process of perceiving graphs has been the subject of inquiry in several scientific disciplines, including psychophysics, cognitive psychology, and computational visions (Cleveland 1995, Chapter 4). Section 21.4 illustrates some types of experimental evidence for determining an effective graphical form on the basis of both the receiver and the graph itself as units of study. Section 21.4 also illustrates how such mainstays of business publications as bar charts and pie charts are poor communicators of numerical information.

Sections 21.5 and 21.6 contain concluding remarks and descriptions of some resources for actuaries who want to learn more about designing effective graphs.

Most readers are removed from the detailed data summarized by a graph. Several difficulties and misconceptions can arise because of the distance between the original data and a viewer's interpretation of the graph. Figure 21.1 illustrates the challenge of communicating with a graph. The sender (and creator) of the graph has a message derived from an interpretation of data. Although a few graphs

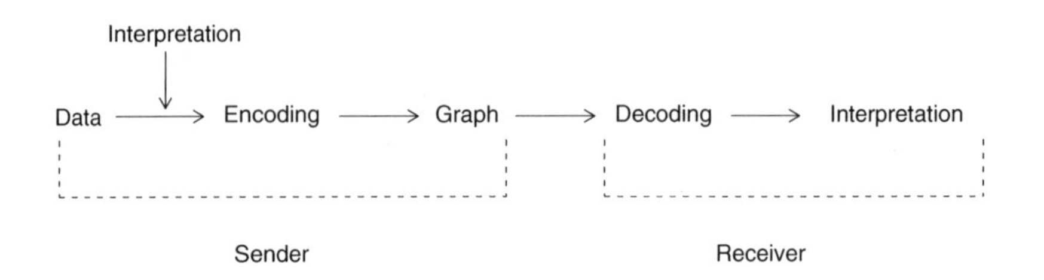

Figure 21.1 Flow chart of the process of communicating with a graph. The graph is a crucial intermediary in the process of communicating data interpretation to the receiver.

communicate raw data, the primary purpose of most graphs is to communicate the sender's interpretation. The message the sender intends is encoded in a graph and passed on to the receiver.

In general, the receiver is party to neither the exact interpretation intended by the sender nor the raw data. Thus, the receiver must decode the graph and develop an interpretation of its message. Two issues arise:

1. Whether the interpretation constructed by the receiver is congruent to the interpretation of the sender
2. Whether the receiver's interpretation is consistent with and supported by the data

The first issue depends on the skill with which the sender constructs the graph and the skill with which the receiver decodes it. A poorly constructed graph can hide or distort the sender's message. A graph that is hard to read can discourage the receiver from spending the time necessary to decode the message correctly. The receiver can ignore or misinterpret a graph that is not constructed with care.

The second issue depends not only on the skills mentioned earlier but also on the skill with which the sender draws meaning from the data. How carefully does the sender document the process of interpretation? Is this communicated to the receiver? Is the receiver capable of assessing the extent to which the graph is a credible summary of the data? Failure at any of these points could result in the receiver ignoring or misinterpreting the graph.

This chapter assumes that the graphs included in business communications are the subject of scrutiny by serious readers. Graphs that appear quickly on the television screen, a flip chart, or presentation package are designed to attract attention and to entertain the viewer. Design, rather than information, considerations dominate these media. We focus instead on graphs that are part of professional writing and are designed to inform. As with effective writing, we assume that, in creating graphs, "one must believe – in the truth and worth of the scrawl, in the ability of the reader to receive and decode the message" (Strunk and White 1979, p. 84). We now turn to examples of graphs that mislead.

21.2 Graphic Design Choices Make a Difference

As noted by Schmid (1992), the ancient proverb "One picture is worth ten thousand words," when applied to graphs might well read, "One picture *can*

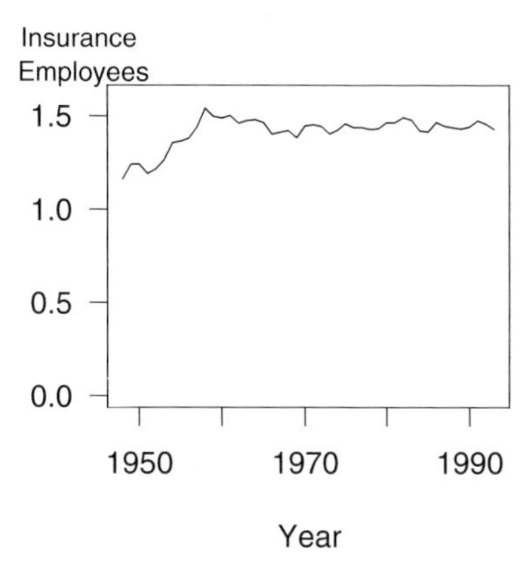

(a) A stable insurance industry

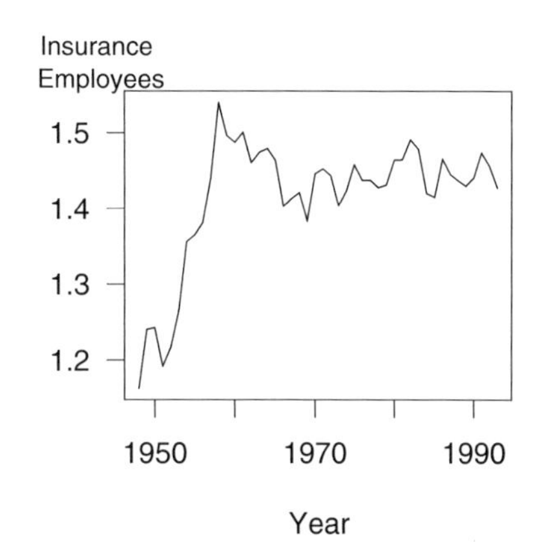

(b) The insurance industry workforce increased dramatically in the 1950s

Figure 21.2 Annual insurance employees, 1948–93. "Insurance employees" is the percentage of full-time-equivalent employees who are working for insurance carriers. Allowing the data to determine the scale ranges reveals interesting aspects of the data.

be worth ten thousand words *or figures*." Graphic potential is not easily realized. Because of their flexibility, graphs too easily render visual displays of quantitative information that are uninformative, confusing, or even misleading.

Examples 21.2.1–21.2.5 illustrate five different types of deceptive graphs. In each case, the data were not altered and different dimensions of the data were not portrayed. The common theme of the examples is that, by altering only the data scales, the creator can alter dramatically a viewer's interpretation.

Example 21.2.1: Including Zero to Compress Data. Figure 21.2 shows a time series of the percentage of full-time equivalent workers employed in the insurance industry. The annual data, 1948–93, are from the National Income and Product Accounts produced by the Bureau of Labor Statistics. The left-hand panel, Figure 21.2(a), provides the impression of a stable employment environment for the insurance industry. Including zero on the vertical axis produces this seeming stability. By doing this, most of the graph is devoted to white space that does not show the variability in the data. In contrast, the right-hand panel, Figure 21.2(b), uses the data to set the range on the axes. This panel clearly shows the large employment increases in the years following the Korean War, circa 1952. It also allows the reader to see the employment declines that the insurance industry has suffered in the most recent three years.

This example is similar to a popular illustration from Huff's well-known *How to Lie with Statistics* (Huff, 1954). The point is that motivation external to the data, such as including zero on an axis, can invite us to alter the data scale and change a viewer's interpretation of the data. As Example 21.2.2 shows, creators of graphs can also alter a viewer's interpretation by changing both scales of a two-dimensional graph.

Figure 21.3 Cost-effectiveness of a firm's risk management practices versus firm size. The data represented in each figure are the same. However, the wider scales in panel (b) suggest that the data are more highly correlated.

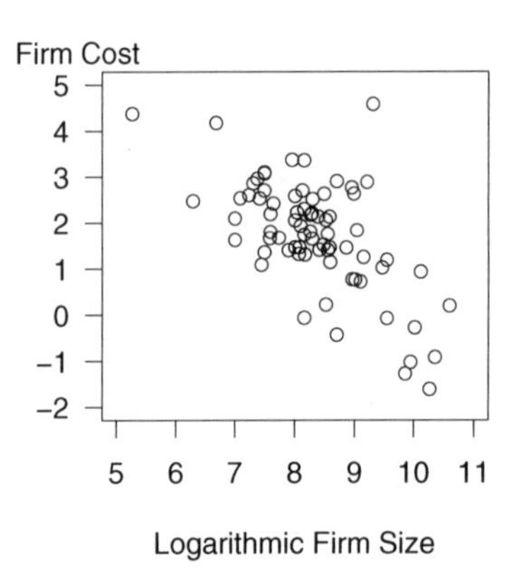

(a) The data in this figure appear less correlated.

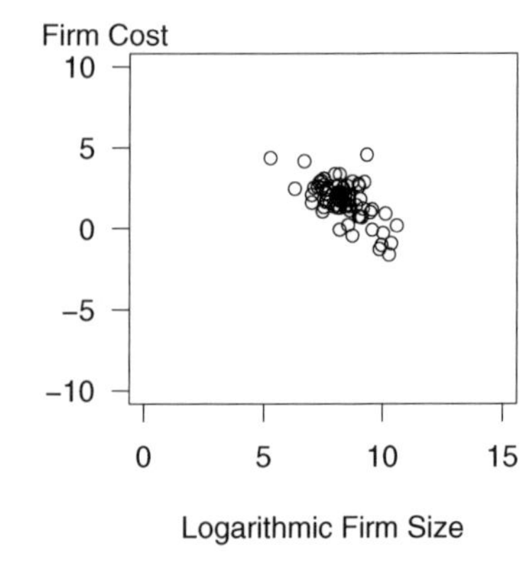

(b) The data in this figure appear more correlated.

Ⓡ **EMPIRICAL** *Filename is "RiskSurvey"*

Example 21.2.2: Perception of Correlation. Figure 21.3 relates risk management cost-effectiveness to firm size. The data are from a survey of 73 risk managers of large, U.S.-based, international firms that was originally reported in Schmit and Roth (1990). The data are analyzed in Section 6.5. Here, the measure of risk management cost-effectiveness, firm cost, is defined to be the logarithm of the firm's total property and casualty premiums and uninsured losses as a percentage of total assets. The firm size measure is total assets in logarithmic units.

The left-hand panel, Figure 21.3(a), shows a negative relationship between firm costs and firm size, as anticipated by Schmit and Roth. The correlation coefficient between the two variables is -0.64. The data are in a small center portion of Figure 21.3(b) as compared to the left-hand panel, Figure 21.3(a). Figure 21.3(a) uses the data to determine the axes and thus shows more patterns in the data. As Cleveland, Diaconis, and McGill (1982) show, the scaling makes the data in the right-hand panel appear more correlated than in the left-hand panel.

Change of scales can also alter the viewer's perception of trend in time series data, as illustrated in Example 21.2.3.

Example 21.2.3: Transforming to a Logarithmic Scale. Figure 21.4 exhibits a time series of the U.S. credit insurance market over 1950 to 1989. These data are analyzed in Frees (1996) and are originally from the *Life Insurance Fact Book* (1990). When the amount of insurance is examined on a linear scale in Figure 21.4(a), the credit insurance market appears to be expanding rapidly. However, Figure 21.4(b) shows that, when examined on a logarithmic scale, the market is leveling off. As discussed in Section 3.2.2, changes on a logarithmic scale can be interpreted as proportional changes. Thus, Figure 21.4(a) shows the market is increasing rapidly, and Figure 21.4(b) shows that the rate of increase is leveling off. These messages are not contradictory, but viewers must interpret each graph critically to understand the intended message.

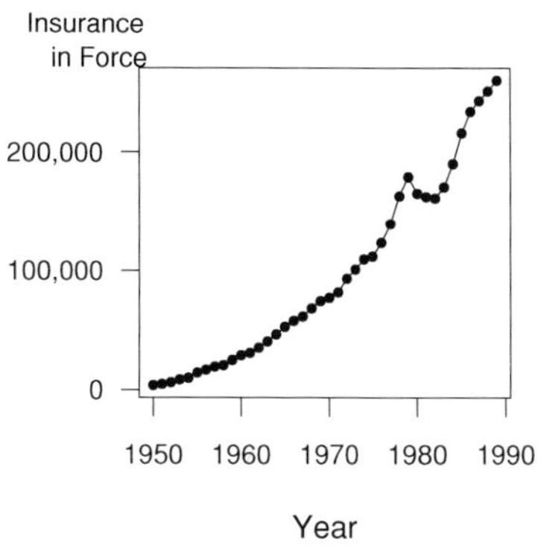

(a) U.S. credit life insurance market exploding

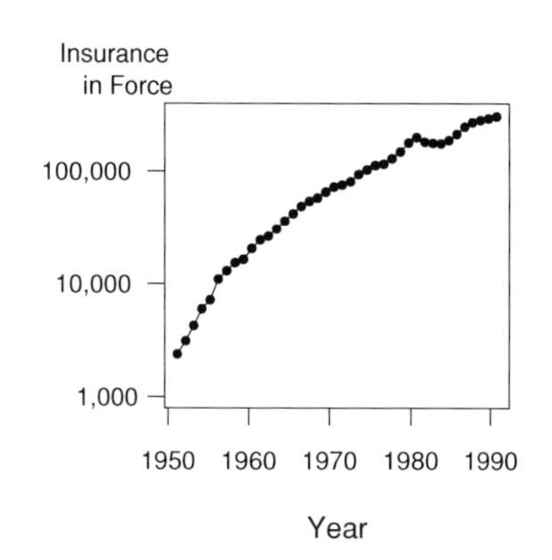

(b) U.S. credit life insurance market leveling off

Figure 21.4 Annual U.S. credit life insurance in force, 1950–89. Different vertical scales give different impressions of the rate of growth over time.

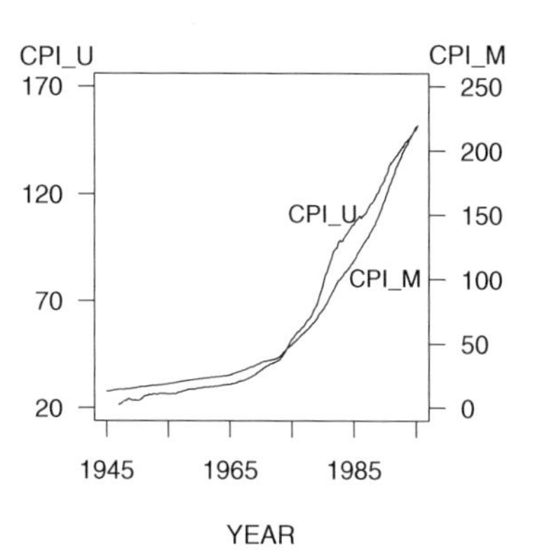

(a) Overall CPI is similar to the medical component of the CPI

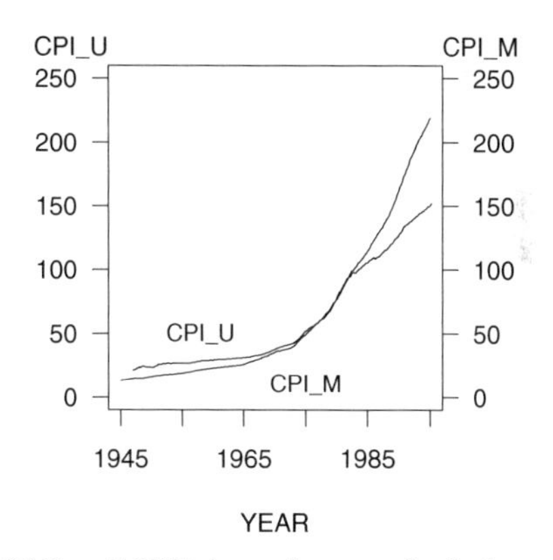

(b) Overall CPI is increasing more slowly than the medical component of the CPI

Figure 21.5 Monthly values of the overall CPI and the medical component of the CPI, January 1947 April 1995. Different scale ranges alter the appearances of relative growth of the two series.

Example 21.2.4: Double Y-Axes. Figure 21.5 displays two measures of inflation that are produced by the Bureau of Labor Statistics. On the left-hand axes are CPI_U, the consumer price index for urban consumers. On the right-hand axes are CPI_M, the consumer price index for medical components of the overall index. Each series consists of monthly values from January 1947 through April 1995.

The left-hand panel, Figure 21.5(a), suggests that the CPI_U and the CPI_M begin and end in approximately the same position, thus implying that they have increased at about the same rate over the period. The creator could argue that each index measures the value of a standard bundle of goods, thus justifying the argument for using a different scale for each series.

The right-hand panel, Figure 21.5(b), provides a more useful representation of the data by using the same scale for each series. Here, CPI_M begins lower than CPI_U and ends higher. That is, the medical component index has increased more quickly than the index of prices for urban consumers. Other patterns are also evident in Figure 21.5: each series increased at roughly the same rate over 1979 to 1983 and CPI_M increased much more quickly from 1983 to 1994 than from 1948 to 1979.

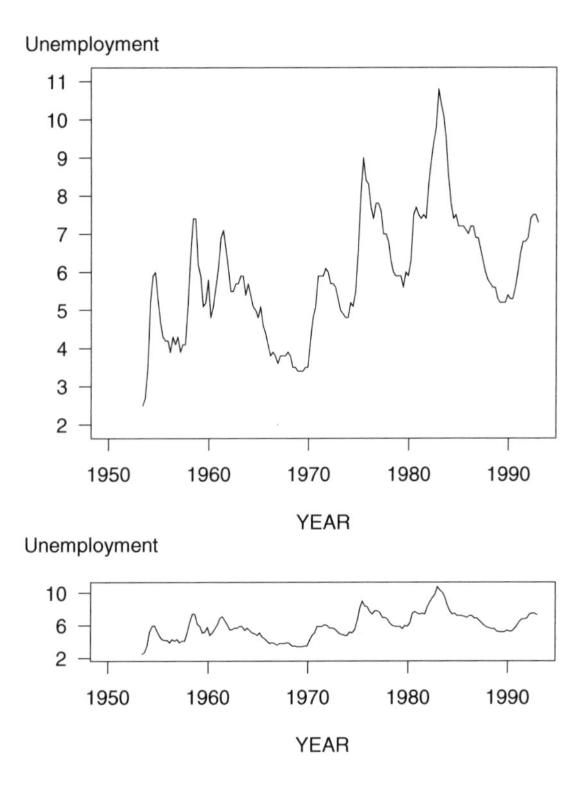

Figure 21.6 Time series plot of quarterly values of the U.S. unemployment rate, 1953–92. The lower panel displays a feature that is not evident in the upper panel; unemployment declines more slowly than it rises.

Example 21.2.5: Aspect Ratio. Figure 21.6 shows a time series plot of the monthly unemployment rate, from April 1953 through December 1992. The unemployment rate is the percentage of unemployed civilian labor force, seasonally adjusted. It is part of the Household Survey produced by the Bureau of Labor Statistics, Department of Labor. This series was analyzed in Frees et al. (1997). The top panel of Figure 21.6 shows that the unemployment rate averaged 5.9% with a peak of 10.8% in the fourth quarter of 1982 and a minimum of 2.7% in the third quarter of 1953.

A figure's aspect ratio is defined to be the height of the data frame divided by its width.

The two panels in Figure 21.6 differ only in their shape, not in the scaling of either variable or in the relative amount of space that the data take within the figure frame. To differentiate the two shapes, we can use the concept of a figure's *aspect ratio*, defined as the height of the data frame divided by its width (some sources use the reciprocal of this value for the aspect ratio). The data frame is simply a rectangle whose height and width just allow the graph *of the data* to fit inside. To illustrate, in the upper panel in Figure 21.6, the length of the vertical side is equal to the length of the horizontal side. In the lower panel, the vertical side is only 25% of the horizontal side.

Both panels show that the unemployment series oscillated widely over this 39-year period. The lower panel, however, displays a feature that is not apparent in the upper panel; the rise to the peak of an unemployment cycle is steeper than the descent from the peak. Within each unemployment cycle, the percentage of workers unemployed tends to rise quickly to a maximum and then to fall gradually to a minimum. This behavior is surprisingly regular over the almost-39-year period displayed in the plot.

Different aspect ratios can leave substantially different impressions on the eye, as Figure 21.6 illustrates. Thus, the aspect ratio can be chosen to emphasize different features of the data.

21.3 Design Guidelines

Understanding the issues illustrated in Section 21.2 can help actuaries and other business professionals create and interpret graphs. This section presents eight guidelines for designing effective graphs. One of the main points is that current practice is not in accord with these guidelines. Thus, we anticipate that not all of our readers will find the demonstrations of the guidelines visually appealing, but, as stated in Section 21.1, many of the guidelines are based on a scientific foundation outlined in Section 21.4. "Intuition" is something we learn and cultivate; progress in science does not always conform to intuition. It was widely believed at one time that the earth was flat and that the sun revolved about the earth. The demonstrations of this section may or may not be immediately intuitive, but they are logical conclusions from the design guidelines advocated here.

Guideline 1: Avoid Chartjunk

In Section 21.1, we defined chartjunk to be any unnecessary appendage in a graph. Creators of graphs who use chartjunk lower their credibility with serious receivers. Even when senders convey a correct interpretation accompanied by chartjunk, they ask receivers to process and properly ignore the chartjunk. If chartjunk is part of the default, or easily used, options of a software package, then the sender can clutter a graph or even make a graph misleading, simply by punching a button.

Avoid pictures that are not worth 10,000 words.

Senders who avoid chartjunk increase their credibility. They ask receivers to look only at meaningful characters and marks. Senders may have to spend considerable time with their software to make effective graphs, but the respect and attention of their receivers reward them. Another way to avoid chartjunk is not to use a graph at all if a few words will do. If the message in a graph can be summarized in a few words, then the graph is not needed. Avoid pictures that are not worth 10,000 words!

Avoiding chartjunk is based in part on the concept of brevity in vigorous writing principles. From the graphical perception viewpoint, avoiding chartjunk reduces the noise when communicating between the graph's sender and its receiver. Thus, this guideline is important because it has roots in both writing and perception principles.

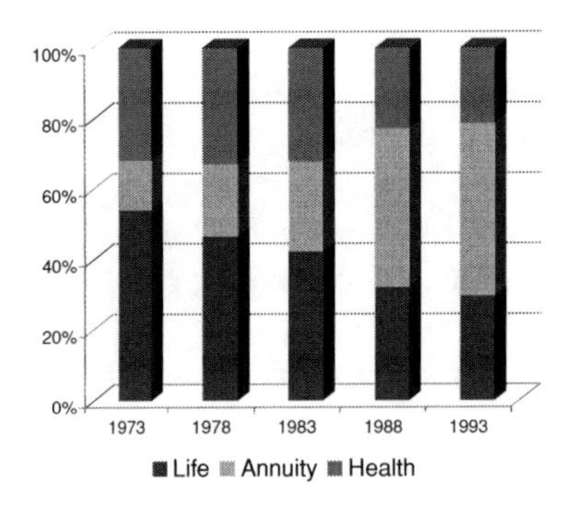

(a) The three-dimensional stacked bar chart is a poor graphical form for making comparisons over time and across lines of business.

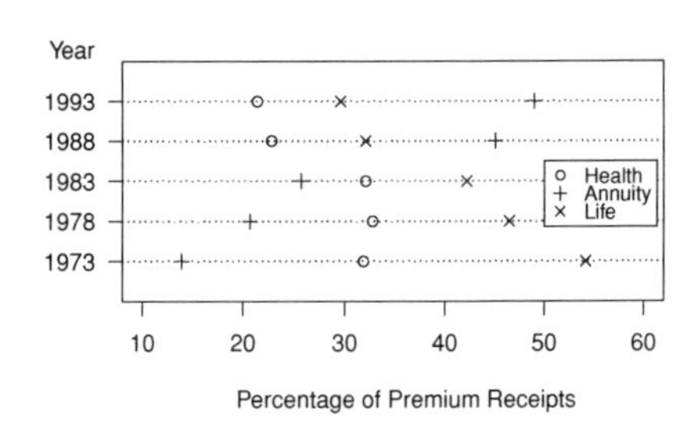

(b) The dot plot allows for direct comparison over time and across lines of business.

Figure 21.7 Distribution of premium receipts, 1973–93. The excessive chartjunk of (a) hides the large change in distribution types between 1983 and 1988.

Example 21.3.1: Premium Receipts of Life Insurance Companies. Figure 21.7(a) is an adaptation of a graph on page 69 of the *Life Insurance Fact Book* (1994). The graph reports ten bits of information: five years and two percentages for each year (a third percentage is found by subtraction). A three-dimensional box represents each percentage, and each box displays different shadings to represent the three lines of business: health, annuity, and life. These figures could be reported compactly in a small table. However, because a graph may help the receiver appreciate trends in the figures, the graph's simplicity should reflect the simplicity of the information available in the figures. In particular, a small plotting symbol suffices to report a percentage. A three-dimensional, shaded box is hardly called for. It is interesting that the three-dimensional box was an "innovation" in 1994. Earlier editions of the *Fact Book* used two-dimensional boxes. The volume of chartjunk took a big jump in 1994.

Figure 21.7(b) is a *dot plot*, discussed by Cleveland (1994). Different plotting symbols show the different lines of business. The tick marks on the lower horizontal axes help us estimate the percentages, and the light, dotted grid lines help us scan across the graph to the plotting symbols of interest. The major shifts, and the approximate magnitudes of the shifts, that happened between 1983 and 1988 are clear here.

Guideline 2: Use Small Multiples to Promote Comparisons and Assess Change

Statistical thinking is directed toward comparing measurements of different entities and assessing the change of a measurement over time or some other unit of measurement. Graphical displays are inherently limited when portraying comparisons or assessing changes because they are static, two-dimensional media. Graphs that contain multiple versions of a basic graphical form, each version portraying a variation of the basic theme, promote comparisons and assessments of change. By repeating a basic graphical form, we promote the process of communication.

Tufte (1997) states that using *small multiples* in graphical displays achieves the same desirable effects as using parallel structure in writing. Parallel structure in writing is successful because it allows readers to identify a sentence relationship only once and then focus on the meaning of each individual sentence element, such as a word, phrase, or clause. Parallel structure helps achieve economy of expression and draw together related ideas for comparison and contrast. Similarly, small multiples in graphs allow us to visualize complex relationships across different groups and over time.

Small multiples in graphs allow us to visualize complex relationships across different groups and over time.

The Section 21.2 figures illustrated the use of small multiples. In each figure, the two plots portrayed were identical except for the change in scale; this use of parallel structure allowed us to demonstrate the importance of scaling when interpreting graphs. Example 21.3.2 illustrates another application of small multiples in graphical displays, Cleveland's (1993) multiway dot plot.

Example 21.3.2: Relative Importance of Risk Source. Figure 21.8, called a *multiway dot plot*, demonstrates conclusions reached by using a model introduced in Frees (1998) concerning the relative importance of risk sources in a block of short-term insurance contracts. The risk sources are the stochastic interest environment, the frequency of claims (mortality), and the possibility of a catastrophic event (disaster) occurring. The relative importance of these three risk sources is considered by letting two parameters of interest vary. These parameters are the expected year until disaster and, in the event of disaster, the expected proportion (probability) of policyholders that will succumb to disaster.

Figure 21.8 shows that when no policyholders succumb to disaster ($q = 0$), then the frequency component, mortality, dominates the other risk sources. At the opposite extreme, when all policyholders succumb to disaster ($q = 1$), then the disaster component dominates the other risk factors. This is true even when the expected time until disaster is 500 years! For the intermediate cases, when either the expected proportion of policyholders succumbing to disaster increases or the expected year until disaster decreases, the importance of the disaster component increases at the expense of the mortality component. Because of the short-term nature of the contract considered, the interest component does not play an important role in Figure 21.8.

This story of relative importance could not be told using analytic expressions because of the complexity of the underlying models. The story behind Figure 21.8 could be told, however, using tabular displays. The advantage of Figure 21.8 is that it allows the viewer to make comparisons over three different risk sources when two parameters of interest vary. Although such comparisons are possible with tabular displays, graphical displays are more effective.

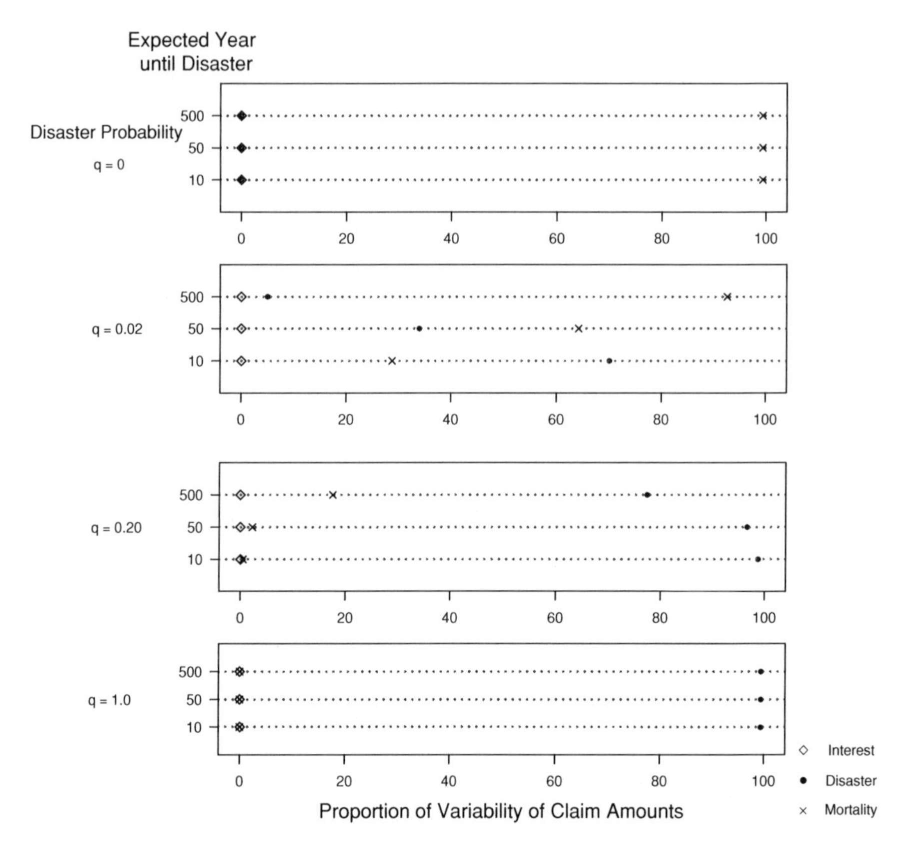

Figure 21.8 The relative importance of risk sources. This complex graph allows us to visualize differences over sources of risk (interest, disaster, and mortality), expected year until disaster, and probability of disaster. The multiway dot plot demonstrates how quickly the importance of the disaster component increases as the probability of disaster increases.

Guideline 3: Use Complex Graphs to Portray Complex Patterns

Many authors believe that a graph should be simple and immediately understood by the viewer. Simple graphs are desirable because they can deliver their message to a broad audience and can be shown quickly and digested immediately. Although this notion may be appropriate for popular writing, for professional writing, the concept of instant understanding is limiting in that it precludes the notion that graphs demonstrate complex ideas. Complex patterns should be portrayed as simply as possible, though the patterns themselves should not be unnecessarily simplified.

One way for a graph to represent complex patterns is for some of its basic elements to serve more than one purpose. Tufte (1983) called such elements *multifunctioning*. For example, we can use plotting symbols to represent not only elements corresponding to the horizontal and vertical scales but also a level of a categorical variable.

Ⓡ **Empirical**
Filename is "WiscHospCosts"

Example 21.3.3: Frequency and Severity of Hospital Costs. Figure 21.9 displays the relationship between average hospital costs and frequency of hospital usage. These data for the year 1989 were obtained from the Office of Health Care Information, Wisconsin's Department of Health and Human Services, and are further analyzed in Section 4.4. The data represent averages over the state of

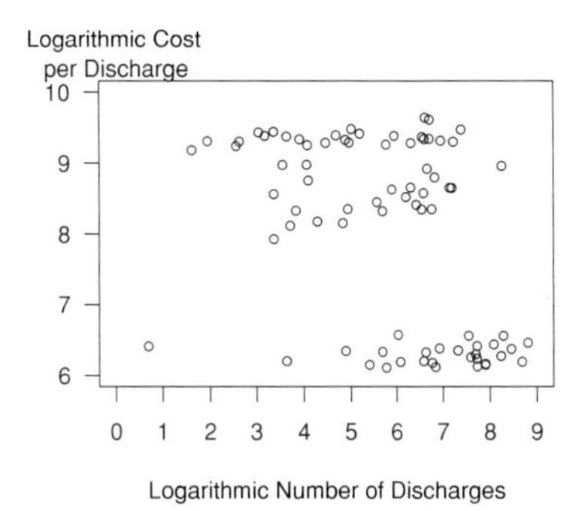

(a) With the exception of one outlying observation in the lower-left-hand region, there appears to be a significant, negative relationship between cost and number of hospital discharges.

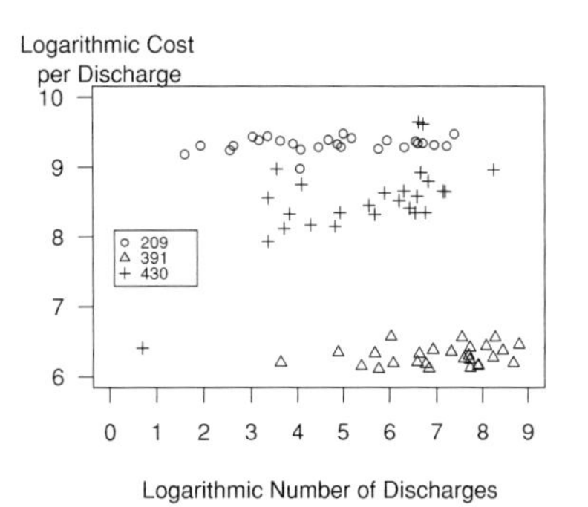

(b) By introducing the DRG codes, we see a small positive relationship between cost and number of hospital discharges within each group.

Figure 21.9 Logarithmic cost per discharge versus the logarithmic number of discharges. By adding a plotting symbol code for the level of DRG, the three distinct groups are evident. The three DRGs, 209, 391, and 430, represent major joint and limb reattachment, normal newborns, and psychoses, respectively.

Wisconsin, broken down by nine health service areas, three types of providers (fee for service, health maintenance organization, and other) and three types of diagnosis-related groups (DRGs). The three DRGs, numbers 209, 391, and 430, represent major joint and limb reattachment, normal newborns, and psychoses, respectively. Each plotting symbol in Figure 21.9 represents a combination of health service area, type of payer, and type of DRG. The horizontal axis provides the number of patients admitted in 1989 for each combination, in natural logarithmic units. The vertical scale provides the average hospital cost per discharge for each combination, in natural logarithmic units.

The story in the left-hand panel, Figure 21.9(a), is one of increased economies of scale. That is, combinations of health service areas, type of payer, and DRG that have a larger number of patients, measured by discharges, have lower costs. A substantial negative relationship is evident in Figure 21.9(a); the correlation coefficient is −0.43. This is true despite the aberrant point in the lower-left-hand region of Figure 21.9(a). The aberrant point is less important economically than the others; it represents a combination with only two discharges. When the point is removed, the correlation becomes −0.50, thus representing an even stronger negative relationship.

Despite its simplicity, Figure 21.9(a) hides an important relationship. The right-hand panel, Figure 21.9(b), is a redrawing of Figure 21.9(a) that includes different plotting symbols for different DRGs. Here, the story is the opposite to the one of increased economies of scale. For combinations representing major joint and limb reattachments and normal newborns, the relationship between frequency and cost is fairly flat. For these DRGs, there are few economies of scale. For the psychoses DRG, number 430, Figure 21.9(b) shows a small positive relationship between frequency and cost, even discounting for the combination with only two patients discharged.

The two panels illustrate a phenomenon in statistics referred to as *Simpson's paradox*, or a problem of *aggregation of data*. See Section 4.4 for further discussion. The important point for this chapter is that sometimes simple graphs are misleading. Complex graphs may take more time for viewers to interpret, but they more effectively summarize complex relationships.

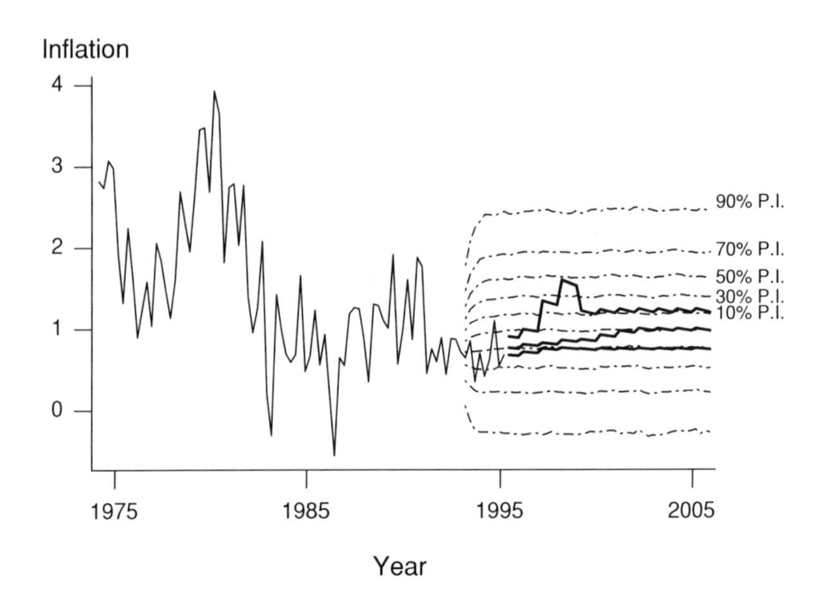

Figure 21.10 Comparison of stochastic prediction intervals to held-out actual experience and to Social Security assumptions. The thin solid lines represent actual inflation rates, and the thick solid lines represent projections by Social Security experts. The dotted lines represent prediction intervals generated by a stochastic time series model. This complex graph allows viewers to make comparisons based on approximately 600 points.

Guideline 4: Relate Graph Size to Information Content

"How large should the graph be?" is an important question. The bounds on size are clear. Graphs should not be so small that they are not clearly legible, particularly on reproduction that degrades an image, nor should they be so large that they exceed a page. With large graphs, it is difficult to compare elements in the graph, thus defeating a primary purpose of graphs.

The data density of a graph is the number of data entries per unit area of the graph.

Within these bounds, a graph should be proportional to the amount of information that it contains. To discuss the proportion of information content, Tufte (1983) introduced the *data density of a graph*, the number of data entries per unit area of the graph. For comparing graph size and information, the data density is a quantity to be maximized, either by increasing the number of data entries or by reducing the size of the graph. By examining this density over a number of popular publications, Tufte concluded that most graphs could be effectively shrunk.

For example, Figure 21.7(a) is a chart with a low data density. This chart represents only fifteen numbers. With an area of approximately nine square inches, the graph's data density is roughly 15/9. For comparison, Figure 21.10 shows approximately 600 numbers. Although Figure 21.10's area is about twice as large as that of Figure 21.7(a), the data density is much larger in Figure 21.10 than in Figure 21.7(a).

Example 21.3.4: Inflation Rate Forecasts. Figure 21.10 is a complex graph that contains much information about a complex subject, forecasting the inflation rate (CPI) for projections of Social Security funds (Frees et al., 1997). The graph shows actual experience of quarterly inflation rates up through the first quarter of 1995. Experience up through 1992 was used to fit a time series model described in Frees et al. (1997), and this model was used to generate prediction intervals (PIs) of the inflation rate. These prediction intervals can be compared to held-out experience that was not used to fit the model (1993–5) as well as to projections of inflation by Social Security experts. The thick lines represent high-, intermediate-, and low-cost inflation projections as determined by Social Security experts.

Figure 21.10 is complex and may not be immediately understood by the viewer. However, almost every stroke in the data region represents numerical information. Although complex, Figure 21.10 allows the viewer to compare 20 years of experience to a ten-year forecast, recent held-out experience to forecasts, and expert projections to forecasts generated by a time series model. The graph's complexity reflects the complexity of forecasting inflation rates; this complexity is not due to unnecessary elements that distract viewers and make them more "interested" in the graph.

Guideline 5: Use Graphical Forms That Promote Comparisons

Creators of graphs often face the choice of several graphical forms to represent a feature of the data. As we describe in Guideline 8, the receiver's knowledge of graphical forms can influence the choice. Graphical perception is also an important determinant. In Section 21.4, we discuss this issue in detail. We include it here as part of the guidelines for completeness.

Guideline 6: Integrate Graphs and Text

Data graphics should be carefully integrated with text, tables, and other graphs. A legend summarizes the graph and its main message, but the surrounding text develops the theme leading up to the message and discusses its impact. Although "a picture is worth ten thousand words," a graph needs supporting text. Tufte (1983) encourages readers and writers to think of data graphics as paragraphs and to treat them as such.

Data graphics can be complemented by a tabular presentation of data: graphics can highlight relationships among the data, and tables can present precise numerical descriptions of the data. The two modes are complementary. A good writing device is to place a graphical display in the main body of the report and to reinforce the graph with a tabular display in an appendix.

The American Statistical Association, in its *Style Guide* for journal publications, reminds us that a detailed legend is helpful when interpreting graphs. The *Style Guide* recommends that a legend describe a graph, draw attention to the graph's important features, and explain this importance.

Guideline 7: Demonstrate an Important Message

Detailed legends and graphs should reinforce messages developed in the main body of the text. To illustrate, when considering ways to portray a complex dataset, choose a graphical form that highlights an important message. All too often, creators of graphs display data features that are not part of the theme being developed.

Cleveland (1994) recommends that we "put major conclusions in a graphical form." In regression data analysis, major conclusions are about patterns in the data that are summarized using models. Usually major conclusions are best presented graphically. Graphs display a large amount of information that is retained by the viewer because it is visualized. Graphs communicate patterns directly to a viewer, without using an equation to represent the patterns. In this way, a wider audience can be reached than if the presentation relies solely on a model-based interpretation of the data. Further, patterns suggested by a graph reinforce those represented by a model, and vice versa. Thus the two tools, graphs and models, reinforce and strengthen each other.

"The greatest value of a picture is when it forces us to notice what we never expected to see." Tukey (1977)

Tukey (1977) states, "The greatest value of a picture is when it forces us to notice what we never expected to see." Unexpected phenomena are usually memorable events; viewers of graphs remember these results, which makes them powerful. In writing this chapter, we did not expect the results of Figure 21.6. This figure demonstrates that unemployment rises much more quickly than it declines; it is a powerful example of the use of aspect ratios.

Guideline 8: Know Your Audience

A basic precept of effective writing, familiarity with one's audience, is also valid for designing effective graphs. As stated previously, our primary motivation in developing guidelines is to encourage the precise and concise communication of quantitative ideas to a scientific audience using a written medium. As discussed in Section 21.4, the graphical form is subservient to the real role of the graphical display, *communicating* quantitative ideas of the creator to the viewer of a graph. If the audience does not have an understanding of the graphical form, then the form will hinder the communication flow rather than aid it. Thus, each of the seven guidelines already discussed can be modified or even ignored on occasion, depending on the audience for the graph. To illustrate, in Example 21.3.1, we argued that the dot plot was superior to the three-dimensional stacked bar chart. As another example, in Section 21.4, we argue that pie charts are ineffective communicators of information based on the science of cognitive perception. However, for some audiences, creators of graphs will prefer the less effective forms because of the level of audience familiarity. We hope that practice will eventually shift from these ineffective modes of communication. Still, it is important to recognize the background of the audience of the graph. We recommend that creators of graphs not so much swim against the tide of poor graphic design as bend their course toward more effective modes of communication.

21.4 Empirical Foundations for Guidelines

This section consists of two different scientific aspects of graphical studies: science of perception and surveys of graphical practice.

1. Position along a common scale
2. Position along identical, nonaligned scales
3. Length
4. Angles and slopes
5. Area
6. Volume
7. Color and density

Table A.1 Basic Graphical Perception Tasks

This chapter does not include a number of graphical forms that are mainstays in business publications and the popular press, such as pie charts, pictographs, and stacked bar charts. In fact, we have shown stacked bar charts in Section 21.3 only as an example of how *not* to draw figures. Why are these widely used graphical forms not adopted in an chapter emphasizing data graphics? The reasons lie in how graphical forms communicate information and how we perceive graphical information. We demonstrate that, given how we perceive information, pie and stacked bar charts are poor communicators of numerical information.

As described in Section 21.1, data graphics encode information, and we, as viewers, decode this information when viewing a graph. The efficiency of this transmission can be considered in the context of cognitive psychology, the science of perception. This discipline provides a framework for distinguishing among different types of information processing that we do when decoding graphs. Identifying different types of information processing will help us decide what are effective, and ineffective, graphical forms.

21.4.1 Viewers as Units of Study

Table 21.1 is an ordered list of basic graphical perception tasks, according to Cleveland (1994). Here, the ordering begins with a set of tasks that is least difficult for a viewer to perform and ends with a set that is most difficult. Thus, for example, judging position along a common scale is the least difficult for viewers and judging relative shadings of colors and density (the amount of ink) is the most difficult.

To understand the relative difficulty of the tasks, Cleveland and McGill (1984) performed a series of tests on many experimental subjects. To illustrate, Figure 21.11 presents a series of tests that are analogous to the first five tasks. Cleveland and McGill summarized the performance of the experimental subjects by calculating the accuracy with which the subjects performed each set of tasks. Through these measures of relative accuracy and arguments from cognitive psychology, Cleveland and McGill developed the ordering presented in Table 21.1.

This chapter does not discuss the use of color because of the complexities of coding and decoding it effectively. We refer interested readers to Cleveland (1994, Section 3.13) and Tufte (1990, Chapter 5) for further information.

The ordered list of graphical perception tasks can help the creator choose the appropriate graphical form to portray a dataset. When confronted with a choice

Figure 21.11
Experiments in judgments about graphical perception.

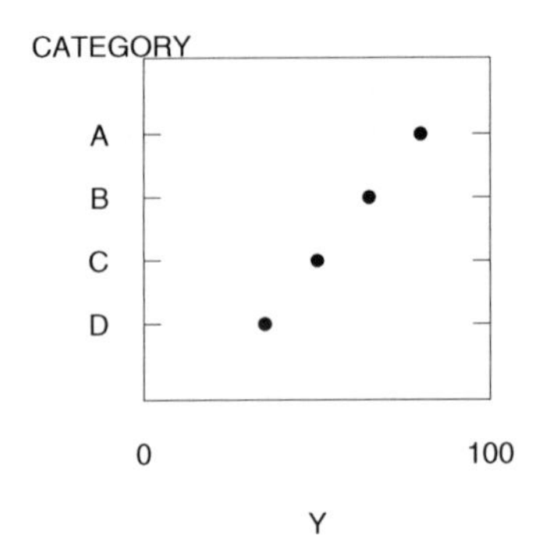

(a) Experiment to judge position along a common scale. Assess the relative values of A, B, C, and D along this 100-point scale.

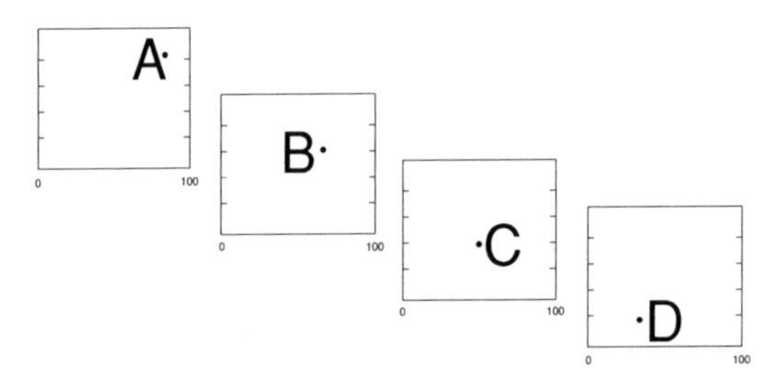

(b) Experiment to judge position along identical, nonaligned scales. Assess the relative values of A, B, C, and D on a common 100-point scale.

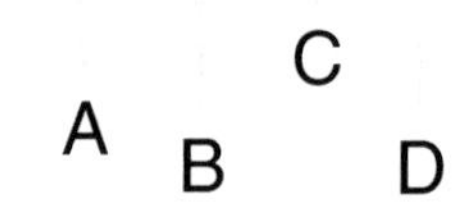

(c) Experiment to understand length judgments. Suppose that line A is 100 units long. Assess the relative lengths of lines B, C, and D.

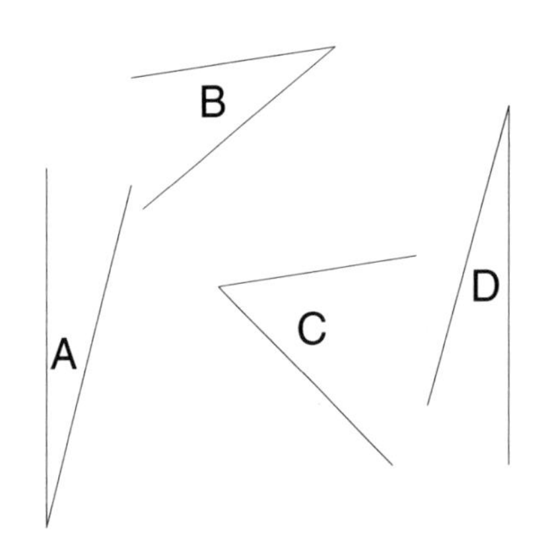

(d) Experiment to understand angle judgments. Suppose that angle A is 100 units. Assess the relative values of angles B, C and D.

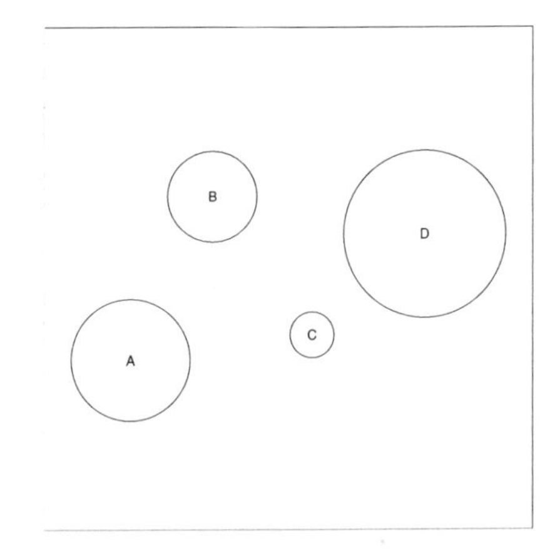

(e) Experiment to understand area judgments. Suppose that circle A has area 100 units. Assess the relative areas of circles B, C, and D.

of two graphical forms, a creator should select the form that is least difficult for the viewer. Other things being equal, a task that can be performed with little difficulty by the viewer means that information can be transmitted more reliably. To illustrate, we discuss two examples in which Table 21.1 can help you decide on the appropriate graphical form for portraying a dataset.

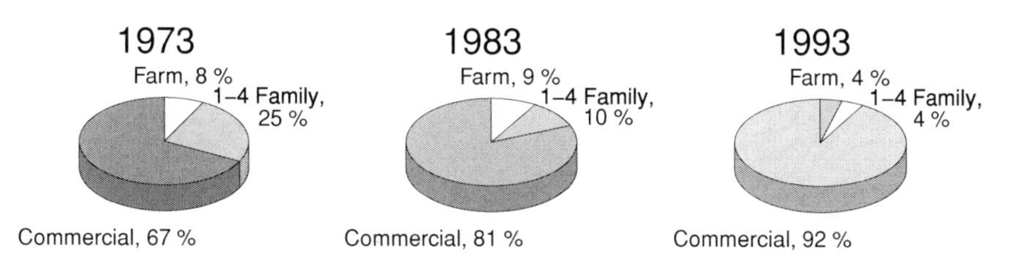

Figure 21.12 Distribution of mortgages for the years 1973, 1983, and 1993. The three-dimensional pie chart is a poor graphical form for making comparisons over time and across types of mortgages.

Example 21.4.1: Distribution of Premium Income. The first example demonstrates some shortcomings of the stacked bar chart. For this discussion, we return to Example 21.3.1. Figure 21.7(a) is a three-dimensional stacked bar chart. We have already discussed the substantial amount of chartjunk in this figure. Even without the useless pseudo – third dimension, the stacked bar chart requires the viewer to make length judgments to understand, for example, the distribution of annuity receipts over time. In contrast, the dot plot in Figure 21.7(b) requires the viewer to make comparisons only according to positions along a common scale. As described in Table 21.1, the latter is an easier task, resulting in more reliable information for the viewer. Thus, we conclude that the dot plot is preferred to the stacked bar chart.

Example 21.4.2: Distribution of Mortgages. Our second example demonstrates the inadequacy of pie charts. Figure 21.12 is an adaptation of the figure on page 100 of the *Life Insurance Fact Book* (1994). It reports, for the years 1973, 1983, and 1993, commercial, one- to four-family, and farm mortgages as percentages of total mortgages. Pie charts make comparisons difficult. For example, the graph makes it difficult to detect whether farm mortgages are more prevalent than one- to four-family mortgages in 1983, or whether farm mortgage percentages increased or decreased from 1973 to 1983. The comparison of percentages across years is a linear operation, yet the pie charts require us to decode angles, a difficult task according to the ordering in Table 21.1. As with Example 21.3.1, the charts in Figure 21.12 make things worse by reporting in three dimensions; these figures not only require us to decode volumes but also add substantially to the chartjunk in the graphic. *Only nine numbers are reported in this graphic*, three years and two percentages in each year. (The third percentage can be computed by subtraction.)

If a graphic is needed, then the dot plot in Figure 21.13 is more than sufficient. Here, comparisons are made according to positions along a common scale, a task easier than comparing angles. Pie charts require us to make comparisons using angles, which are more difficult and less reliable than comparisons using other graphical forms.

Although Figure 21.13 is a more effective graph than Figure 21.12, for these data we recommend a tabular display (Table 21.2), which allows for clear comparisons across mortgage types and across years. Further, more detailed information

Table A.2
Commercial, One- to Four-Family, and Farm Mortgages as Percentages of Total Mortgages for 1973, 1983, and 1993

	Year		
Mortgage Type	1973	1983	1993
Commercial	67.5	81.3	91.7
1–4 Family	24.9	10.1	4.1
Farm	7.6	8.6	4.2

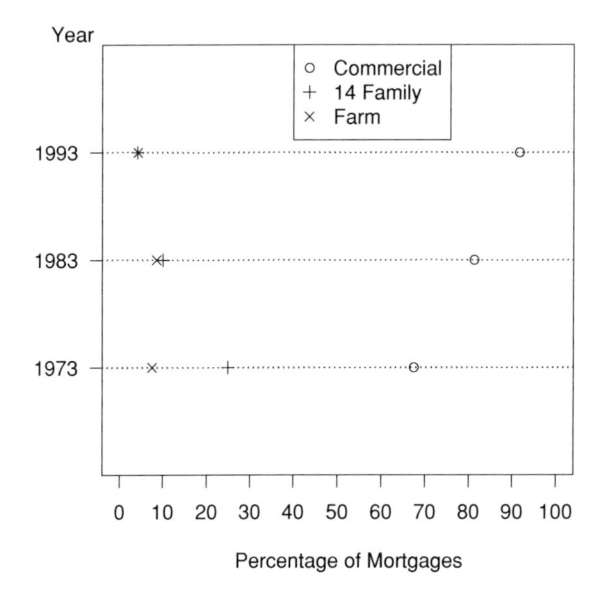

Figure 21.13
Commercial, one- to four-family, and farm mortgages as percentages of total mortgages for 1973, 1983, and 1993. A negative aspect of this graph is the overlap of the one- to four-family and farm plotting symbols in 1983 and 1993.

about mortgage percentages is available in Table 21.2 than in Figure 21.12 or 21.13. Of course, we can always superimpose the actual percentages, as is often done with pie charts, as illustrated in Figure 21.12. Our response to this approach is to question the worth of the entire graph. As with writing, each stroke should offer new information; let creators of graphs make each stroke tell!

21.4.2 Graphs as Units of Study

Surveys of graphical practice in professional publications provide an important database for assessing prevalence of good and bad practice and changes in practice over time. Tufte (1983) discusses a survey of approximately 4,000 graphs randomly selected from 15 news publications for the years 1974 to 1980. The graphs were assessed for "sophistication," defined as presentation of relationship between variables, excluding time series or maps. Cleveland and McGill (1985) report a similar survey of scientific publications, assessing the prevalence of graphical errors.

Harbert (1995) assessed every graph and table in the 1993 issues of four psychology journals on 34 measures of quality. The measures of quality were gleaned

Table A.3 Factors Affecting Assessment of Graphic Quality, Harbert Study

Variables with Positive Coefficients	Variables with Negative Coefficients
Data-ink ratio	Proportion of page used by graphic
Comparisons made easy	Vertical labels on Y-axis
Sufficient data to make a rich graphic	Abbreviations used
	Optical art used
	Comparisons using areas or volumes

from the current research literature on graphic quality. They were converted into a checklist, and a checklist was filled out for each graph and table in the selected psychology journals. Harbert's study yielded data on 439 graphs and tables. We summarize the analysis of the 212 graphs.

Harbert assigned letter grades to the graphics: A, AB, B, BC, C, CD, D, DF, and F. These grades reflected her overall evaluation of the graphs as communicators of statistical information. The grades were converted to numerical values: 4.0, 3.5, 3.0, 2.5, 2.0, 1.5, 1.0, 0.5, and 0.0. The numerical values were the dependent variable in a regression. The independent variables were the 34 measures of quality, suitably coded. The purpose of the study was to determine which factors were statistically significant predictors of the grades assigned by an expert evaluator of graphics. By trial and error, Harbert selected a multiple linear regression equation in which all the predictors were statistically significant (5% level) and no other predictors achieved this level of significance when added to the equation. Table 21.3 shows the variables included in the regression equation ($R^2 = 0.612$).

Data-ink ratio was defined by Tufte (1983, p. 93) as the "proportion of the graphics ink devoted to the nonredundant display of data-information" or equivalently as "1.0 minus the proportion of a graphic that can be erased without loss of data-information." The data-ink ratio is more readily calculated than the data density measure defined in Section 21.3 of this paper. Optical art is decoration that does not tell the viewer anything new.

One variable that had been anticipated as very significant was data density, which is difficult and time-consuming to measure. An important finding of the study was that the easier-to-measure data-ink ratio and proportion of page variables were sufficient to predict the grades. A quotation from Harbert's thesis sums up the finding: "The highest grades were given to those graphics that take up small proportions of the page, have a large data-ink ratio, make comparisons easy, have enough data points, have horizontally printed labels, do not have abbreviations, do not have optical art, and do not use volume or 3-D comparisons" (Harbert 1995, p. 56).

As a small follow-up study to Harbert's work, we examined each of the 19 non-table graphics in the *Life Insurance Fact Book* (1994), assessing them on seven negative factors. Table A.4 shows the percentage of graphs that displayed each of the negative factors.

Table A.4
Percentage of Graphs Displaying Negative Factors in Life Insurance Fact Book 1994

Negative Factor	Percentage of Graphics
Use of 3-D bars	79
Grid lines too dense	79
Making comparison of time series values hard	37
Use of stacked bars	37
Growth displayed poorly	32
Use of lines that are wider than need be	16
Use of pies	5

Our review suggests that every graphic could have been reduced by 50% to 75% without loss of clarity. This observation is in keeping with Harbert's finding about the proportion-of-page variable. In a word, the graphs in the *Life Insurance Fact Book* could be produced much more ably. Doing so would improve the quality of communication and would potentially increase the respect with which knowledgeable professionals in other fields view the insurance industry.

We hope that other investigators will engage in further study of graphic practice in actuarial publications. By using data from such studies, the profession can improve its practice, making communications efficient and precise.

21.5 Concluding Remarks

The Society of Actuaries' motto is a quotation of Ruskin: "The work of science is to substitute facts for appearances and demonstrations for impressions." Armed with the guidelines outlined in this paper and discussed further in the references, actuaries can be leaders in presenting data graphically, thus substituting demonstrations for impressions. Surveys of recent actuarial literature should be the basis for assessing current practice. Editors and referees of professional publications can be especially influential in bringing about a rapid improvement in standards of practice. Moreover, actuaries can recommend and use statistics textbooks that pay attention to graphic quality.

Because actuaries read material that contains graphs, they are consumers. They should become tough customers! All too often the defaults in spreadsheet and statistical graphics software become the norm. Actuaries should not allow the choices made by software programmers to drive graphic quality or standards. Although it is easy to create graphs using defaults in the graphics software, the resulting graphs are seldom fully satisfactory. If a graph is not worth doing well, let's leave it out of our publications.

21.6 Further Reading and References

In addition to the references listed, other resources are available to actuaries interested in improving their graphic design skills. Like the Society of Actuaries, another professional organization, the American Statistical Association (ASA), has special interest sections. In particular, the ASA now has a section on statistical graphics. Interested actuaries can join ASA and that section to get the newsletter *Statistical Computing & Graphics*. This publication has examples of excellent graphical practice in the context of scientific discovery and application.

The technical *Journal of Computational and Graphical Statistics* contains more in-depth information on effective graphs. We also recommend accessing and using the *ASA Style Guide* at www.amstat.org/publications/style-guide.html as an aid to effective communication of quantitative ideas.

Chapter References

American Council of Life Insurance. Various years. *Life Insurance Fact Book*. American Council of Life Insurance, Washington, D.C.

Cleveland, William S. (1994). *The Elements of Graphing Data*. Wadsworth, Monterey, California.

Cleveland, William S. (1993). *Visualizing Data*. Hobart Press, Summit, New Jersey.

Cleveland, William S., P. Diaconis, and R. McGill (1982). Variables on scatter plots look more highly correlated when the scales are increased. *Science* 216, 1138–41.

Cleveland, William S., and R. McGill (1984). Graphical perception: Theory, experimentation, and application to the development of graphical methods. *Journal of the American Statistical Association* 79, 531–554.

Cleveland, William S., and R. McGill (1985). Graphical perception and graphical methods for analyzing and presenting scientific data. *Science* 229, 828–33.

Ehrenberg, A.S.C. (1977). Rudiments of Numeracy. *Journal of the Royal Statistical Society A* 140, 277–97.

Frees, Edward W. (1996). *Data Analysis Using Regression Models*. Prentice Hall, Englewood Cliffs, New Jersey.

Frees, Edward W. (1998). Relative importance of risk sources in insurance systems. *North American Actuarial Journal* 2, no. 2, 34–51.

Frees, Edward W., Yueh C., Kung, Marjorie A., Rosenberg, Virginia R., Young, and Siu-Wai Lai (1997). Forecasting Social Security Assumptions. *North American Actuarial Journal* 1, no. 3, 49–82.

Harbert, D. (1995). The Quality of Graphics in 1993 Psychology Journals, Senior honors thesis, University of Wisconsin–Madison.

Huff, D. (1954). *How to Lie with Statistics*. Norton, New York.

Schmid, C. F. (1992). *Statistical Graphics: Design Principles and Practices*. Krieger Publishing, Malabar, Florida.

Schmit, Joan T., and K. Roth (1990). Cost effectiveness of risk management practices. *Journal of Risk and Insurance* 57, 455–70.

Strunk, W., and E. B. White (1979). *The Elements of Style*, 3rd ed. Macmillan, New York.

Tufte, Edward R. (1983). T*he Visual Display of Quantitative Information*. Graphics Press, Cheshire, Connecticut.

Tufte, Edward R. (1990). *Envisioning Information*. Graphics Press, Cheshire, Connecticut.

Tufte, Edward R. (1997). *Visual Explanations*. Graphics Press, Cheshire, Connecticut.

Tukey, John (1977). *Exploratory Data Analysis*. Addison-Wesley, Reading, Massachusetts.

University of Chicago Press (2003). *The Chicago Manual of Style*, 15th ed. University of Chicago Press, Chicago.

Brief Answers to Selected Exercises

Chapter 1

1.1 a(i). Mean = 12,840, and median = 5,695.

a(ii). Standard deviation = 48,836.7 = 3.8 times the mean. The data appear to be skewed.

b. The plots are not presented here. When viewing them, the distribution appears to be skewed to the right.

c(i). The plots are not presented here. When viewing them, although the distribution has moved toward symmetry, it is still quite lopsided.

c(ii). The plots are not presented here. When viewing them, the distribution appears to be much more symmetric.

d. Mean = 1,854.0, median = 625.7, and standard deviation = 3,864.3. A similar pattern holds true for outpatient as for inpatient.

1.2 Part 1. a. Descriptive Statistics for the 2000 data

	Min.	1st Quartile	Median	Mean	3rd Quartile	Max.	Standard Deviation
TPY	11.57	56.72	80.54	88.79	108.60	314.70	46.10
NUMBED	18.00	60.25	90.00	97.08	118.8	320.00	48.99
SQRFOOT	5.64	28.64	39.22	50.14	65.49	262.00	34.50

b. The plots are not presented here. When viewing them, the histogram appears to be skewed to the right but only mildly.

c. The plots are not presented here. When viewing them, both the histogram and the *qq* plot suggest that the transformed distribution is close to a normal distribution.

Part 2. a. Descriptive Statistics for the 2001 data

	Min.	1st Quartile	Median	Mean	3rd Quartile	Max.	Standard Deviation
TPY	12.31	56.89	81.13	89.71	109.90	440.70	49.05
NUMBED	18.00	60.00	90.00	97.33	119.00	457.00	51.97
SQRFOOT	5.64	28.68	40.26	50.37	63.49	262.00	35.56

c. Both the histogram and the qq plot (not presented here) suggest that the transformed distribution is close to the normal distribution.

1.5 a. Mean = 5.953, and median = 2.331.

b. The plots are not presented here. When viewing them, the histogram appears to be skewed to the right. The qq plot indicates a serious departure from normality.

c(i). For ATTORNEY = 1, we have mean = 9.863 and median = 3.417. For ATTORNEY = 2, we have mean = 1.865 and median = 0.986. This suggests that the losses associated with attorney involvement (ATTORNEY = 1) are higher than when an attorney is not involved (ATTORNEY = 1).

1.7 a. The plots are not presented here. When viewing them, the histogram appears to be skewed to the left. The qq plot indicates a serious departure from normality.

b. The plots are not presented here. When viewing them, the transformation does little to symmetrize the distribution.

Chapter 2

2.1 $r = 0.5491$, $b_0 = 4.2054$, and $b_1 = 0.1279$.

2.3 a.

$$
\begin{aligned}
0 &\le \frac{1}{n-1}\sum_{i=1}^{n}\left(a\frac{x_i-\overline{x}}{s_x} - c\frac{y_i-\overline{y}}{s_y}\right)^2 \\
&= \frac{1}{n-1}\sum_{i=1}^{n}\left[a^2\frac{(x_i-\overline{x})^2}{s_x^2} - 2ac\frac{(x_i-\overline{x})(y_i-\overline{y})}{s_xs_y} + c^2\frac{(y_i-\overline{y})^2}{s_y^2}\right] \\
&= a^2\frac{1}{s_x^2}\frac{1}{n-1}\sum_{i=1}^{n}(x_i-\overline{x})^2 - 2ac\frac{1}{s_xs_y}\frac{1}{n-1}\sum_{i=1}^{n}(x_i-\overline{x})(y_i-\overline{y}) \\
&\quad + c^2\frac{1}{s_y^2}\frac{1}{n-1}\sum_{i=1}^{n}(y_i-\overline{y})^2 \\
&= a^2\frac{1}{s_x^2}s_x^2 - 2acr + c^2\frac{1}{s_y^2}s_y^2 \\
&= a^2 + c^2 - 2acr.
\end{aligned}
$$

b. From part (a), we have $a^2 + c^2 - 2acr \ge 0$. So,

$$
\begin{aligned}
a^2 + c^2 - 2ac + 2ac &\ge 2acr \\
(a-c)^2 &\ge 2acr - 2ac \\
(a-c)^2 &\ge 2ac(r-1).
\end{aligned}
$$

c. Using the result in part (b) and taking $a = c$, we can get $2a^2(r-1) \leq 0$. Also $a^2 \geq 0$, so $r - 1 \leq 0$. Thus, $r \leq 1$.
d. Using the result in part (b) and taking $a = -c$, we can get $-2a^2(r-1) \leq 4a^2$. Also $-2a^2 \leq 0$, so $r - 1 \geq -2$. Thus, $r \geq -1$.
e. If all of the data lie on a straight line that goes through the upper-left- and lower-right-hand quadrants, then $r = -1$. If all of the data lie on a straight line that goes through the lower-left- and upper-right-hand quadrants, then $r = 1$.

2.5 a.

$$b_1 = r\frac{s_y}{s_x} = \frac{1}{(n-1)s_x^2}\sum_{i=1}^{n}(x_i - \overline{x})(y_i - \overline{y})$$

$$= \frac{1}{\sum_{i=1}^{n}(x_i - \overline{x})^2}\sum_{i=1}^{n}\left[\frac{y_i - \overline{y}}{x_i - \overline{x}}(x_i - \overline{x})^2\right]$$

$$= \frac{\sum_{i=1}^{n} weight_i \; slope_i}{\sum_{i=1}^{n} weight_i},$$

where

$$slope_i = \frac{y_i - \overline{y}}{x_i - \overline{x}} \quad \text{and} \quad weight_i = (x_i - \overline{x})^2 .$$

b. $slope_1 = -1.5$, and $weight_1 = 4$.

2.7 a. For the model in this exercise, the least squares estimate of β_1 is the b_1 that minimizes the sum of squares $SS(b_1^*) = \sum_{i=1}^{n}\left(y_i - b_1^* x_i\right)^2$. So, taking derivative with respect to b_1^*, we have

$$\frac{\partial}{\partial b_1^*}SS(b_1^*) = \sum_{i=1}^{n}(-2x_i)\left(y_i - b_1^* x_i\right).$$

Setting this quantity equal to zero and canceling constant terms yields

$$\sum_{i=1}^{n}\left(x_i y_i - b_1^* x_i^2\right) = 0.$$

So, we get the conclusion

$$b_1 = \frac{\sum_{i=1}^{n} x_i y_i}{\sum_{i=1}^{n} x_i^2}.$$

b. From the problem, we have $x_i = z_i^2$. Using the result of part (a), we can reach the conclusion that

$$b_1 = \frac{\sum_{i=1}^{n} z_i^2 y_i}{\sum_{i=1}^{n} z_i^4}.$$

2.10 a(i). Correlation $= 0.9372$

a(ii). Table of correlations

	TPY	NUMBED	SQRFOOT
TPY	1.0000	0.9791	0.8244
NUMBED	0.9791	1.0000	0.8192
SQRFOOT	0.8244	0.8192	1.0000

a(iii). Correlation $= 0.9791$. Correlations are unaffected by scale changes.

b. The plots are not presented here. When viewing them, there is a strong linear relationship between NUMBED and TPY. The linear relationship of SQRFOOT and TPY is not as strong as that of NUMBED and TPY.

c(i). $b_1 = 0.92142$, t-ratio $= 91.346$, and $R^2 = 0.9586$.

c(ii). $R^2 = 0.6797$. The model using NUMBED is preferred.

c(iii). $b_1 = 1.01231$, t-ratio $= 81.235$, and $R^2 = 0.9483$.

c(iv). $b_1 = 0.68737$, t-ratio $= 27.25$, and $R^2 = 0.6765$.

Part 2: $b_1 = 0.932384$, t-ratio $= 120.393$, and $R^2 = 0.9762$. The pattern is similar to the cost report for year 2000.

2.11 $\hat{e}_1 = -23$.

2.13 a.

$$\hat{y}_i - \overline{y} = (b_0 + b_1 x_i) - \overline{y} = (\overline{y} - b_1 \overline{x} + b_1 x_i) - \overline{y} = b_1 (x_i - \overline{x}).$$

b.

$$\sum_{i=1}^{n} (y_i - \overline{y})^2 = \sum_{i=1}^{n} (b_1 (x_i - \overline{x}))^2 = b_1^2 \sum_{i=1}^{n} (x_i - \overline{x})^2 = b_1^2 s_x^2 (n-1).$$

c.

$$R^2 = \frac{Regression\ SS}{Total\ SS} = \frac{b_1^2 s_x^2 (n-1)}{\sum_{i=1}^{n} (y_i - \overline{y})^2} = \frac{b_1^2 s_x^2 (n-1)}{s_y^2 (n-1)} = \frac{b_1^2 s_x^2}{s_y^2}.$$

2.15 a. From the definition of the correlation coefficient and Exercise 2.8(b), we have

$$r(y,x)(n-1) s_y s_x = \sum_{i=1}^{n} (y_i - \overline{y})(x_i - \overline{x}) = \sum_{i=1}^{n} y_i x_i - n \overline{x}\overline{y}.$$

If either $\overline{y} = 0$, $\overline{x} = 0$ or both $\overline{x}$ and $\overline{y} = 0$, then $r(y,x)(n-1) s_y s_x = \sum_{i=1}^{n} y_i x_i$. Therefore, $r(y,x) = 0$ implies $\sum_{i=1}^{n} y_i x_i = 0$ and vice versa.

b.

$$\sum_{i=1}^{n} x_i e_i = \sum_{i=1}^{n} x_i(y_i - (\overline{y} + b_1(x_i - \overline{x})))$$
$$= \sum_{i=1}^{n} x_i(y_i - \overline{y}) - b_1 \sum_{i=1}^{n} x_i(x_i - \overline{x})$$
$$= \sum_{i=1}^{n} x_i b_1(x_i - \overline{x}) - b_1 \sum_{i=1}^{n} x_i(x_i - \overline{x}) = 0.$$

c.

$$\sum_{i=1}^{n} \widehat{y}_i e_i = \sum_{i=1}^{n} (\overline{y} + b_1(x_i - \overline{x}))e_i$$
$$= \overline{y} \sum_{i=1}^{n} e_i + b_1 \sum_{i=1}^{n} ((x_i - \overline{x}))e_i = 0.$$

2.17 When $n = 100$, $k = 1$, $Error\ SS = [n - (k+1)]s^2 = 98s^2$,
a. $e_{10}^2/(Error\ SS) = (8s)^2/(98s^2) = 65.31\%$.
b. $e_{10}^2/(Error\ SS) = (4s)^2/(98s^2) = 16.33\%$.
When $n = 20$, $k = 1$, $Error\ SS = [n - (k+1)]s^2 = 18s^2$.
c. $e_{10}^2/(Error\ SS) = (4s)^2/(18s^2) = 88.89\%$.

2.20 a. Correlation = 0.9830
Descriptive Statistics
b. $R^2 = 0.9664$, $b_1 = 1.01923$, and $t(b_1) = 100.73$.
c(i). The degrees of freedom is $df = 355 - (1+1) = 353$. The corresponding t-value is 1.96. Because the t-statistic $t(b_1) = 100.73 > 1.9667$, we reject H_0 in favor of the alternative.
c(ii). The t-statistic is $t\text{-ratio} = (b_1 - 1)/se(b_1) = (1.01923 - 1)/0.01012 = 1.9002$. Because t-ratio < 1.9667, we do not reject H_0 in favor of the alternative.
c(iii). The corresponding t-value is 1.645. The t-statistic is t-ratio = 1.9002. We reject H_0 in favor of the alternative.
c(iv). The corresponding t-value is -1.645. The t-statistic is t-ratio = 1.9002. We do not reject H_0 in favor of the alternative.
d(i). A point estimate is 2.0384.
d(ii). 95% C.I. for slope b_1 is $1.0192 \pm 1.9667 \times 0.0101 = (0.9993, 1.0391)$. A 95% C.I. for expected change of LOGTPY is $(0.9993 \times 2, 1.0391 \times 2) = (1.9987, 2.0781)$
d(iii). A 99% C.I. is $(2 \times (1.0192 - 2.5898 \times 0.0101), 2 \times (1.0192 + 2.5898 \times 0.0101)) = (1.9861, 2.0907)$
e(i). $\widehat{y} = -0.1747 + 1.0192 \times \ln 100 = 4.519037$.

e(ii). The standard error of the prediction is

$$se(pred) = s\sqrt{1 + \frac{1}{n} + \frac{(x^* - \overline{x})^2}{(n-1)s_x^2}}$$

$$= 0.09373\sqrt{1 + \frac{1}{355} + \frac{(\ln(100) - 4.4573)^2}{(355-1)0.4924^2}} = 0.0938.$$

	Min.	1st Quartile	Median	Mean	3rd Quartile	Max.	Standard Deviation
LOGTPY	2.51	4.04	4.40	4.37	4.70	6.09	0.51
LOGNUMBED	2.89	4.09	4.50	4.46	4.78	6.13	0.49

e(iii). The 95% prediction interval at x^* is

$$\widehat{y}^* \pm t_{n-2,1-\alpha/2}\, se(pred) = 4.519037 \pm 1.9667(0.0938)$$
$$= (4.3344, 4.7034).$$

e(iv). The point prediction is $e^{4.519037} = 91.747$.
The prediction interval is $(e^{4.334405} = 76.280, e^{4.703668} = 110.351)$.

e(v). The prediction interval is $(e^{4.364214} = 78.588, e^{4.673859} = 107.110)$.

2.22 a. Fitted US *LIFEEXP* $= 83.7381 - 5.2735 \times 2.0 = 73.1911$.

b. A 95% prediction interval for the life expectancy in Dominica is

$$\widehat{y}_* \pm t_{n-2,1-\alpha/2}\, se(pred) = 73.1911 \pm (1.96)(6.642)$$
$$= (60.173, 86.209)$$

c.

$$e_i = y_i - \widehat{y}_i = y_i - (b_0 + b_1 x_i) = 72.5 - (83.7381 - 5.2735 \times 1.7)$$
$$= -2.273.$$

This residual is 2.273/6.615 = 0.3436 multiples of s below zero.

d. Test $H_0 : \beta_1 = -6.0$ versus $H_a : \beta_1 > -6.0$ at the 5% level of significance using t-value = 1.645. The calculated t-statistics $= \frac{-5.2735-(-6)}{0.2887} =$ 2.5165, which is ≥ 1.645. Hence, we reject H_0 in favor of the alternative. The corresponding p-value is 0.00637.

Chapter 3

3.1 a. $R_a^2 = 1 - s/s_y^2 = 1 - (50)^2/(100)^2 = 1 - 1/4 = 0.75$.

b. *Total SS* $= (n-1)s_y^2 = 99(100)^2 = 990{,}000$ and
Error SS $= (n-(k+1))s^2 = (100-(3+1))(50)^2 = 240{,}000$.

Source	SS	df	MS	F
Regression	750,000	3	250,000	100
Error	240,000	96	2,500	
Total	990,000	99		

c. $R^2 = (Regression\ SS)/(Total\ SS) = 750{,}000/990{,}000 = 75.76\%$.

3.3 a. $\mathbf{y} = (0\ 1\ 5\ 8)'$, $\mathbf{X} = \begin{pmatrix} 1 & -1 & 0 \\ 1 & 2 & 0 \\ 1 & 4 & 1 \\ 1 & 6 & 1 \end{pmatrix}$.

b. $\hat{y}_3 = \mathbf{x}_3'\mathbf{b} = (1\ 4\ 1)\begin{pmatrix} 0.15 \\ 0.692 \\ 2.88 \end{pmatrix} = 5.798$.

c. $se(b_2) = s\sqrt{3rd\ diagonal\ element\ of\ (\mathbf{X}'\mathbf{X})^{-1}} = 1.373\sqrt{4.11538} = 2.785$.

d. $t(b_1) = b_1/se(b_1) = 0.15/(1.373 \times \sqrt{0.15385}) = 0.279$.

3.6 a. The regression coefficient is -0.1846, meaning that when public education expenditures increase by 1% of gross domestic product, life expectancy is expected to decrease by 0.1846 years, holding other variables fixed.

b. The regression coefficient is -0.2358, meaning that when health expenditures increase by 1% of gross domestic product, life expectancy is expected to decrease by 0.2358 years, holding other variables fixed.

c. $H_0 : \beta_2 = 0$, $H_1 : \beta_2 \neq 0$. We cannot reject null hypothesis because the p-value is greater than the significance level, say, 0.05. Therefore, PUBLICEDUCATION is not a statistically significant variable.

d(i). The purpose of added variable plot is to explore the correlation between PUBLICEDUCATION and LIFEEXP after removing the effects of other variables.

d(ii). The partial correlation is

$$r = \frac{t(b_2)}{\sqrt{t(b_2)^2 + n - (k+1)}} = \frac{-0.6888}{\sqrt{-0.6888^2 + 152 - (3+1)}} = -0.0565.$$

Chapter 4

4.1 a. $R^2 = (Regression\ SS)/(Total\ SS)$.

b. F-ratio $= (Regression\ MS)/(Error\ MS)$.

c.

$$1 - R^2 = \frac{Total\ SS}{Total\ SS} - \frac{Regression\ SS}{Total\ SS} = \frac{Error\ SS}{Total\ SS}.$$

Now, from the right-hand side, we have

$$\begin{aligned}\frac{R^2}{1-R^2}\frac{(n-(k+1))}{k} &= \frac{(Regression\ SS)/(Total\ SS)}{(Error\ SS)/(Total\ SS)}\frac{(n-(k+1))}{k} \\ &= \frac{Regression\ SS}{Error\ SS}\frac{(n-(k+1))}{k} \\ &= \frac{(Regression\ SS)/k}{(Error\ SS)/(n-(k+1))} \\ &= \frac{Regression\ MS}{Error\ MS} = F\text{-ratio}.\end{aligned}$$

d. F-ratio $= 0.17$.

e. F-ratio $= 19.8$.

4.3 a. The third level of organizational structure is captured by the intercept term of the regression.

b. H_0 : TAXEXEMPT is not important, H_1 : TAXEXEMPT is important. $p = 0.7833 > 0.05$, so we do not reject the null hypothesis.

c. Because p-value $= 1.15e^{-12}$ is less than significance level $\alpha = 0.05$, MCERT is an important factor in determining LOGTPY.

c(i). The point estimate is 0.416.

c(ii). The 95% confidence interval is $0.416 \pm 1.963 \times \sqrt{0.243}/(\sqrt{75}) = (0.304, 0.528)$.

d. $R^2 = 0.1463$. All the variables are statistically significant.

e. $R^2 = 0.9579$. Only LOGNUMBED is statistically significant at $\alpha = 0.05$.

e(i). The partial correlation is 0.9327. The correlation between LOGTPY and LOGNUMBED is 0.9783. The partial correlation removes the effect of other variables on LOGTPY.

e(ii). The t-ratio tests whether the individual explanatory variable is statistically significant. The F-ratio tests whether the explanatory variables taken together have an significant impact on response variable. In this case, only LOGNUMBED is significant and the R^2 is high, which explains why the F-ratio is large while most of the t-ratios are small.

4.7 a. H_0 : PUBLICEDUCATION and lnHEALTH are not jointly statistically significant. That is, the coefficients of the two variables are equal to zero. H_1 : PUBLICEDUCATION and lnHEALTH are jointly statistically significant. At least one of the coefficients of the two variables is not equal to zero. To make a decision, we compare the F statistics with critical value; if F statistics are greater than the critical value, we reject the null hypothesis. Otherwise, we do not.

F-*ratio* $= (7832.5 - 6535.7)/(2 \times 44.2) = 14.67$. The 95% of F distribution with $df_1 = 2$ and $df_2 = 148$ is approximately 3.00.

Because *F-ratio* is less than the critical value, we cannot reject the null hypothesis. That is, PUBLICEDUCATION and lnHEALTH are not jointly significant.

b. We can see that the life expectancy varies across different regions.

c. H_0: All betas corresponding to the REGION Factor are zero, H_1: At least one beta is not zero. To make the decision, we compare the p-value with significance level $\alpha = 0.05$. If $p < \alpha$, we reject the null hypothesis. Otherwise, we do not. In this case, $p = 0.598 > 0.05$, so we do not reject the null hypothesis. REGION is not a statistically significant determinant of LIFEEXP.

d(i). If REGION = Arab state, $\widehat{LIFEEXP} = 83.3971 - 2.7559 \times 2 - 0.4333 \times 5 - 0.7939 \times 1 = 74.9249$. If REGION = sub-Saharan Africa, $\widehat{LIFEEXP} = 83.3971 - 2.7559 \times 2 - 0.4333 \times 5 - 0.7939 \times 1 - 14.3567 = 60.5682$.

d(ii). The 95% confidence interval is $-14.3567 \pm 1.976 \times 1.8663 = (-18.044, -10.669)$.

d(iii). The point estimate for the difference is 18.1886.

Chapter 5

5.1 a. From equation (2.9), we have

$$\begin{aligned} h_{ii} &= \mathbf{x}_i' \left(\mathbf{X}'\mathbf{X}\right)^{-1} \mathbf{x}_i \\ &= \begin{pmatrix} 1 & x_i \end{pmatrix} \frac{1}{\sum_{i=1}^n x_i^2 - n\overline{x}^2} \begin{pmatrix} n^{-1}\sum_{i=1}^n x_i^2 & -\overline{x} \\ -\overline{x} & 1 \end{pmatrix} \begin{pmatrix} 1 \\ x_i \end{pmatrix} \\ &= \frac{1}{\sum_{i=1}^n x_i^2 - n\overline{x}^2} \left(n^{-1}\Big(\sum_{i=1}^n x_i^2 - n\overline{x}^2\Big) + \overline{x}^2 - 2\overline{x}x_i + x_i^2 \right) \\ &= \frac{1}{n} + \frac{(x_i - \overline{x})^2}{(n-1)s_x^2}. \end{aligned}$$

b. The average leverage is

$$\bar{h} = \frac{1}{n}\sum_{i=1}^n h_{ii} = \frac{1}{n} + \frac{1}{n}\sum_{i=1}^n \frac{(x_i - \bar{x})^2}{(n-1)s_x^2} = \frac{1}{n} + \frac{1}{n} = \frac{2}{n}.$$

c. Let $c = (x_i - \bar{x})/s_x$. Then,

$$\frac{6}{n} = h_{ii} = \frac{1}{n} + \frac{(x_i - \bar{x})^2}{(n-1)s_x^2} = \frac{1}{n} + \frac{(cs_x)^2}{(n-1)s_x^2} = \frac{1}{n} + \frac{c^2}{n-1}.$$

For a large n, x_i is approximately $c = \sqrt{5} = 2.236$ standard deviations away from the mean.

5.3 a. The plots are not presented here. When viewing them, it is difficult to detect linear patterns from the plot of GDP versus LIFEEXP. The logarithmic transform of GDP spreads out values of GPD, allowing us to see linear patterns. Similar arguments hold for HEALTH, where the pattern in lnHEALTH is more linear.

c(ii). It is both. The standardized residual is −2.66, which exceeds the cutoff of 2, in absolute value. The leverage is 0.1529, which is greater than the cutoff of $3 \times \bar{h} = 3 \times (k+1)/n = 0.08$.

c(iii). The variable PUBLICEDUCATION is no longer statistically significant.

Chapter 6

6.1 a. The variable involact is somewhat right skewed but not drastically so. The variable involact has several zeros that may be a problem with limited dependent variables. The variable age appears to be bimodal, with six observations that are 28 or less and the others greater than or equal to 40.

	Mean	Median	Standard Deviation	Minimum	Maximum
Race	34.9	24.5	32.6	1.0	99.7
Fire	12.3	10.4	9.3	2.0	39.7
Theft	32.4	29.0	22.3	3.0	147.0
Age	60.3	65.0	22.6	2.0	90.1
Income	10,696	10,694.0	2,754	5,583	21,480
Volact	6.5	5.9	3.9	0.5	14.3
Involact	0.6	0.4	0.6	0.0	2.2

b. The scatterplot matrix (not presented here) shows a negative relation between volact and involact, a negative relation between race and volact, and a positive relation between race and involact. If there exists racial discrimination, we would expect Zip codes with more minorities to have less access to the voluntary (less expensive) market, meaning that they have to go to the involuntary market for insurance.

c. Table of correlations

	Race	Fire	Theft	Age	Income	Volact	Involact
Race	1.000						
Fire	0.593	1.000					
Theft	0.255	0.556	1.000				
Age	0.251	0.412	0.318	1.000			
Income	−0.704	−0.610	−0.173	−0.529	1.000		
Volact	−0.759	−0.686	−0.312	−0.606	0.751	1.000	
Involact	0.714	0.703	0.150	0.476	−0.665	−0.746	1.000

d(i). The coefficient associated with race is negative and statistically significant.

d(ii). The high-leverage Zip codes are numbers 7 and 24. Race remains statistically, negatively significant. Fire is no longer significant, although income becomes significant.

e. Race remains positively, statistically significant. Similarly, the role of the other variables do not change depending on the presence of the two high-leverage points.

f. Race remains positively, statistically significant. Similarly, the role of the other variables does not change depending on the presence of the two high-leverage points.

g. Leverage depends on the explanatory variables, not the dependent variables. Because the explanatory variables remained unchanged in the three analyses, the leverages remained unchanged.

h. The demand for insurance depends on the size of the loss to be insured, the ability of the applicant to pay for it and knowledge of insurance contracts. For homeowners' insurance, the size of the loss relates to house price; type of dwelling structure; available safety precautions taken; and susceptibility to catastrophes such as tornado, flood, and so on. Ability to pay is based on income, wealth, number of dependents, and other factors. Knowledge of insurance contracts depends on, for example, education. All of these omitted factors may be related to race.

i. One would expect Zip codes that are adjacent to one another (i.e., contiguous) to share similar economic experiences. We could subdivide the city into homogeneous groups, such as inner city and suburbs. We could also do a weighted least squares where the weights are given by the distance from the city center.

Chapter 7

7.1 a.

$$\begin{aligned} \mathrm{E}\ y_t &= \mathrm{E}\,(y_0 + c_1 + \cdots + c_t) = \mathrm{E}\ y_0 + \mathrm{E}\ c_1 + \cdots + \mathrm{E}\ c_t \\ &= y_0 + \mu_c + \cdots + \mu_c = y_0 + t\mu_c. \end{aligned}$$

b.

$$\begin{aligned} \mathrm{Var}\ y_t &= \mathrm{Var}(y_0 + c_1 + \cdots + c_t) = \mathrm{Var}\ c_1 + \cdots + \mathrm{Var}\ c_t \\ &= \sigma_c^2 + \cdots + \sigma_c^2 = t\sigma_t^2. \end{aligned}$$

7.3 a(ii). No. There is a clear downward trend in the series, indicating that the mean changes over time.

b(i). The t-ratios associated with the linear and quadratic trend portions are highly statistically significant. The $R^2 = 0.8733$ indicates that the model fits well.

b(ii). The sign of a residual is highly likely to be the same as preceding a subsequent residuals. This suggests a strong degree of autocorrelation in the residuals.

b(iii). $\widehat{EURO_{702}} = 0.808 + 0.0001295(702) - 4.639 \times 10^{-7}(702)^2 = 0.6703$.

c(i). This is a random walk model.

c(ii). $\widehat{EURO_{702}} = 0.6795 + 3(-0.0001374) = 0.679088$.

c(iii). An approximate 95% prediction interval for $EURO_{702}$ is $0.679088 \pm 2(0.003621979)\sqrt{3} \approx (0.66654, 0.691635)$.

Chapter 8

8.1 $r_1 = \left(\sum_{t=2}^{5}(y_{t-1} - \bar{y})(y_t - \bar{y})\right) / \left(\sum_{t=1}^{5}(y_t - \bar{y})^2\right) = -0.0036/0.0134 = -0.2686$.

$r_2 = \left(\sum_{t=3}^{5}(y_{t-2} - \bar{y})(y_t - \bar{y})\right) / \left(\sum_{t=1}^{5}(y_t - \bar{y})^2\right) = 0.0821$.

8.3 a. $b_1 = \left(\sum_{t=2}^{T}(y_{t-1} - \bar{y}_-)(y_t - \bar{y}_+)\right) / \left(\sum_{t=2}^{T}(y_{t-1} - \bar{y}_-)^2\right)$,

where $\bar{y}_+ = \left(\sum_{t=2}^{T} y_t\right)/(T-1)$ and $\bar{y}_- = \left(\sum_{t=1}^{T-1} y_t\right)/(T-1)$.

b. $b_0 = \bar{y}_+ - b_1\bar{y}_-$.

c. $b_0 \approx \bar{y}\left[1 - \left(\sum_{t=2}^{T}(y_{t-1} - \bar{y}_-)(y_t - \bar{y}_+)\right) / \left(\sum_{t=2}^{T}(y_{t-1} - \bar{y}_-)^2\right)\right] \approx \bar{y}\,[1 - r_1]$.

8.6 a. Because the mean and variance of the sequence do not vary over time, the sequence can be considered weakly stationary.

b. The summary statistics of the sequence are as follows:

Mean	Median	Std.	Minimum	Maximum
0.0004	0.0008	0.0064	−0.0182	0.0213

Under the assumption of white noise, the forecast of an observation in the future is its sample mean, that is, 0.0004. This forecast does not depend on the number of steps ahead.

c. The autocorrelations for the lags 1–10 are:

0	1	2	3	4	5	6	7	8	9	10
1.000	−0.046	−0.096	0.019	−0.002	−0.004	−0.054	−0.035	−0.034	−0.051	0.026

Because $|r_k/se(r_k)| < 2$ ($se(r_k) = 1/\sqrt{503} = 0.0446$) for $i = 1, \ldots, 10$, none of the autocorrelations is strongly statistically significant different from zero except for lag 2. For lag 2, the autocorrelation is $0.096/0.0446 = 2.15$ standard errors below zero.

Chapter 11

11.1 a. The probability density function is

$$\mathrm{f}(y) = \frac{\partial}{\partial y}\mathrm{F}(y) = (-1)(1+e^{-y})^{-2}e^{-y}(-1) = \frac{e^{y}}{(1+e^{y})^2}.$$

b.

$$\mu_y = \int_{-\infty}^{\infty} y\mathrm{f}(y)dy = \int_{-\infty}^{\infty} y\frac{e^{y}}{(1+e^{y})^2}dy = 0.$$

c.

$$\mathrm{E}\ y^2 = \int_{-\infty}^{\infty} y^2\mathrm{f}(y)dy = \pi^2/3.$$

Because $\mu_y = 0$, the standard deviation is $\sigma_y = \pi/\sqrt{3} = 1.813798$.

d. The probability density function for y^{**} is

$$\begin{aligned}\mathrm{f}^{*}(y) &= \frac{\partial}{\partial y}\Pr(y^{**} \le y) = \frac{\partial}{\partial y}\Pr(y^{*} \le y\sigma_y + \mu_y) \\ &= \sigma_y \mathrm{f}(y\sigma_y) = \sigma_y \frac{e^{y\sigma_y}}{(1+e^{y\sigma_y})^2}.\end{aligned}$$

11.3 Let $\Pr(\varepsilon_{i1} \le a) = F(a) = \exp(-e^{-a})$ and $f(a) = \frac{dF(a)}{da} = \exp(-e^{-a})e^{-a}$. Then

$$\begin{aligned}\Pr(\varepsilon_{i2}-\varepsilon_{i1} \le a) &= \int_{-\infty}^{\infty} F(a+y)f(y)\,dy = \int_{-\infty}^{\infty} \exp\left[-e^{-y}(e^{-a}+1)\right]e^{-y}\,dy \\ &= \int_{\infty}^{0} \exp(-zA)z\,d(-\ln z) = -\int_{\infty}^{0} \exp(-zA)\,dz \\ &= \frac{\exp(-zA)}{A}\Big|_{\infty}^{0} = \frac{1}{A} = \frac{1}{1+e^{-a}},\end{aligned}$$

with $A = e^{-a} + 1$ and $z = e^{-y}$. Thus,

$$\pi_i = \Pr(\epsilon_{i2}-\epsilon_{i1} < V_{i1} - V_{i2}) = \Pr(\epsilon_{i2} - \epsilon_{i1} < \mathbf{x}_i'\boldsymbol{\beta}) = \frac{1}{1+\exp(-\mathbf{x}_i'\boldsymbol{\beta})}.$$

11.5 From equation (11.5) we know that

$$\sum_{i=1}^{n}\mathbf{x}_i\left(y_i - \pi(\mathbf{x}_i'\mathbf{b}_{MLE})\right) = \sum_{i=1}^{n}(1\ x_{i1}\ \cdots\ x_{ik})'\left(y_i - \pi(\mathbf{x}_i'\mathbf{b}_{MLE})\right) = (0\ 0\ \cdots\ 0)'.$$

From the first row, we get $\sum_{i=1}^{n}\left(y_i - \pi(\mathbf{x}_i'\mathbf{b}_{MLE})\right) = 0$. Dividing by n yields $\overline{y} = n^{-1}\sum_{i=1}^{n}\widehat{y}_i$.

11.7 a. The derivative of the logit function is

$$\frac{\partial}{\partial y}\pi(y) = \pi(y)\frac{1}{(1+e^{y})} = \pi(y)(1-\pi(y)).$$

Thus, using the chain rule and equation (11.5), we have

$$\mathbf{I}(\boldsymbol{\beta}) = -\mathrm{E}\,\frac{\partial^2}{\partial \boldsymbol{\beta} \partial \boldsymbol{\beta}'} L(\boldsymbol{\beta}) = -\mathrm{E}\,\frac{\partial}{\partial \boldsymbol{\beta}'} \left(\sum_{i=1}^{n} \mathbf{x}_i \left(y_i - \pi(\mathbf{x}_i'\boldsymbol{\beta}) \right) \right)$$

$$= \sum_{i=1}^{n} \mathbf{x}_i \frac{\partial}{\partial \boldsymbol{\beta}'} \pi(\mathbf{x}_i'\boldsymbol{\beta}) = \sum_{i=1}^{n} \mathbf{x}_i \mathbf{x}_i' \pi(\mathbf{x}_i'\boldsymbol{\beta})(1 - \pi(\mathbf{x}_i'\boldsymbol{\beta})).$$

This provides the result with $\sigma_i^2 = \pi(\mathbf{x}_i'\boldsymbol{\beta})(1 - \pi(\mathbf{x}_i'\boldsymbol{\beta}))$.

b. Define $\mathbf{a}_i = \mathbf{x}_i \left(y_i - \pi(\mathbf{x}_i'\boldsymbol{\beta}) \right)$ and $\mathbf{H}_i = \frac{\partial}{\partial \boldsymbol{\beta}'} \mathbf{a}_i = -\mathbf{x}_i \mathbf{x}_i' \pi'(\mathbf{x}_i'\boldsymbol{\beta})$. Note that $\mathrm{E}(\mathbf{a}_i) = \mathbf{x}_i \mathrm{E}\left(y_i - \pi(\mathbf{x}_i'\boldsymbol{\beta}) \right) = \mathbf{0}$. Further define $b_i = \frac{\pi'(\mathbf{x}_i'\boldsymbol{\beta})}{\pi(\mathbf{x}_i'\boldsymbol{\beta})(1-\pi(\mathbf{x}_i'\boldsymbol{\beta}))}$. With this notation, the score function is $\frac{\partial}{\partial \boldsymbol{\beta}} L(\boldsymbol{\beta}) = \sum_{i=1}^{n} \mathbf{a}_i b_i$. Thus,

$$\mathbf{I}(\boldsymbol{\beta}) = -\mathrm{E}\left(\frac{\partial^2}{\partial \boldsymbol{\beta}\, \partial \boldsymbol{\beta}'} L(\boldsymbol{\beta}) \right) = -\mathrm{E}\left(\frac{\partial}{\partial \boldsymbol{\beta}'} \sum_{i=1}^{n} \mathbf{a}_i b_i \right)$$

$$= -\mathrm{E}\left(\sum_{i=1}^{n} \left(\left(\frac{\partial}{\partial \boldsymbol{\beta}'} \mathbf{a}_i \right) b_i + \mathbf{a}_i \frac{\partial}{\partial \boldsymbol{\beta}'} b_i \right) \right)$$

$$= - \sum_{i=1}^{n} \left[\mathrm{E}(\mathbf{H}_i) b_i + \mathrm{E}(\mathbf{a}_i) \frac{\partial}{\partial \boldsymbol{\beta}'} b_i \right]$$

$$= - \sum_{i=1}^{n} \mathbf{H}_i b_i = \sum_{i=1}^{n} \mathbf{x}_i \mathbf{x}_i' \frac{\left(\pi'(\mathbf{x}_i'\boldsymbol{\beta}) \right)^2}{\pi(\mathbf{x}_i'\boldsymbol{\beta})(1 - \pi(\mathbf{x}_i'\boldsymbol{\beta}))}.$$

11.8 a(i). The plots are not presented here.

	Mean	Median	Std.	Minimum	Maximum
CLMAGE	32.531	31.000	17.089	0.000	95.000
LOSS	5.954	2.331	33.136	0.005	1,067.700

a(ii). Not CLMAGE, but both versions of LOSS appear to differ by ATTORNEY.

ATTORNEY	CLMAGE	LOSS	lnLOSS
1	32.270	9.863	1.251
2	32.822	1.865	−0.169

a(iii). SEATBELT and CLMINSUR appear to be different; CLMSEX and MARITAL are less so.

	CLMSEX		MARITAL				CLMINSUR		SEATBELT	
ATTORNEY	1	2	1	2	3	4	1	2	1	2
1	325	352	320	329	6	20	76	585	643	16
2	261	390	304	321	9	15	44	594	627	6

a(iv). Number of values missing is shown as

CLMAGE	LOSS	CLMSEX	MARITAL	CLMINSUR	SEATBELT
189	12	16	41	48	N/A

b(i). The variable CLMSEX is statistically significant. The odds ratio is $\exp(-0.3218) = 0.7248$, indicating that women are 72% times as likely to use an attorney as men (or men are 1/0.72 = 1.379 times as likely to use an attorney than women).

b(ii). CLMSEX and CLMINSUR are statistically significantly and SEATBELT is somewhat significant, as given by the p-values. CLMAGE is not significant. MARITAL does not appear to be statistically significant.

b(iii). Men use attorneys more often – the odds ratio is $\exp(-0.37691) = 0.686$, indicating that women are 68.6% times more likely to use an attorney than are men.

b(iv). The logarithmic version, lnLOSS, is more important. In the final model without LOSS, the p-value associated with lnLOSS was tiny ($<2e^{-16}$), indicating strong statistical significance.

b(v). All variables remain the same, except one of the MARITAL binary variables becomes marginally statistically significant. The main difference is that we are using an additional 168 observations by not requiring that CLMAGE be in the model.

b(vi). For the systematic component, we have

$$\begin{aligned}
\mathbf{x}'\mathbf{b}_{MLE} = {} & 0.75424 - 0.51210 * (CLMSEX = 2) \\
& + 0.04613 * (MARITAL = 2) \\
& + 0.37762 * (MARITAL = 3) \\
& + 0.12099 * (MARITAL = 4) \\
& + 0.13692 * (SEATBELT = 2) \\
& - 0.52960 * (CLMINSUR = 2) \\
& - 0.01628 * CLMAGE + 0.98260 * lnLOSS = 1.3312.
\end{aligned}$$

The estimated probability of using an attorney is

$$\widehat{\pi} = \frac{\exp(1.3312)}{1+\exp(1.3312)} = 0.791.$$

c. Women are less likely to use attorneys. Those not wearing a seatbelt (SEATBELT = 2) are more likely to use an attorney (though not significant). Single (MARITAL = 2) are more likely to use an attorney. Claimants not uninsured (CLMINSUR = 2) (are insured) are less likely to use an attorney. The higher the loss, the more likely it is that an attorney will be involved.

11.11 a. The intercept and variables PLACE%, MSAT, and RANK are significant at the 5% level.

b(i). The success probability for this case is 0.482.
b(ii). The success probability for this case is 0.281.
b(iii). The success probability for this case is 0.366.
b(iv). The success probability for this case is 0.497.
b(v). The success probability for this case is 0.277.

Chapter 12

12.1 Take derivative of equation (12.2) with respect to μ and set the first order condition equal to zero. With this, we have $\partial L(\mu)/\partial\mu = \sum_{i=1}^{n}(-1 + y_i/\mu) = 0$; that is, $\hat{\mu} = \bar{y}$.

12.3 a. From the expression of the score equation (12.5),

$$\left.\frac{\partial}{\partial \boldsymbol{\beta}} L(\boldsymbol{\beta})\right|_{\beta=\mathbf{b}} = \sum_{i=1}^{n} (y_i - \widehat{\mu}_i) \begin{pmatrix} 1 \\ x_{i,1} \\ \vdots \\ x_{i,k} \end{pmatrix} = \mathbf{0}.$$

From the first row, we have that the average of residuals $e_i = y_i - \widehat{\mu}_i$ is equal to zero.

b. From the $(j+1)st$ row of the score equation (12.5), we have

$$\sum_{i=1}^{n} e_i x_{i,j} = 0.$$

Because residuals have a zero average, the sample covariance between residuals and x_j is zero, and hence the sample correlation is zero.

12.5 a. The distribution of COUNTOP has a long tail and is skewed to the right. The variance ($12.5^2 = 156.25$) is much bigger than the mean, 5.67.

Minimum	1st Quartile	Median	Mean	3rd Quartile	Maximum
0.00	0.00	2.00	5.67	6.00	167.00

b. Yes, the tables suggest that most variables have a significant impact on COUNTOP.

c. The Pearson's chi-square statistic is 55,044.

d(i). All of the variables appear to be statistically significant.

d(ii). The coefficient of GENDER is 0.4197. Roughly, we would expect women to have 42% more outpatient expenditures than men.

d(iii). The chi-square statistic is 33,214 – lower than the one in part (b) (55,044). This indicates that the covariates help with the fitting process. The statistical significance also indicates that the covariates are statistically significant but the overdispersion is suspect – see d(iv).

d(iv). Now most of the variables remain statistically significant, but the strength of statistical significance has decreased dramatically. It is not clear whether the income variable is statistically significant.

e(i). All of the variables appear to be statistically significant. The income variable is perhaps the least important.

e(ii). The chi-square statistic is 33,660 – greater than the Poisson model (33,214) but less than the one in part (b) (55,044). This suggests that the two models fit about the same, with the Poisson having the slight edge. The AIC for the basic Poisson is 22,725 – which is much higher than the AIC for the negative binomial (10,002). Thus, the negative binomial is preferred to the basic Poisson. However, the quasi-Poisson is probably as good as the negative binomial.

e(iii). From the output, the likelihood ratio test statistic is 18.7 – based on 4 degrees of freedom, the p-value is 0.000915. This indicates that income is a statistically significant factor in the model.

f. For GENDER, education, personal health status, anylimit, income, and insurance, the models report the same sign and statistically significant effects. RACE does not appear to be statistically significant in the logistic regression model. For REGION, the signs appear to be the same, although the statistical significance has changed.

Appendix 1

Basic Statistical Inference

Appendix Preview. This appendix provides definitions and facts from a course in basic statistical inference that are needed in one's study of regression analysis.

A1.1 Distributions of Functions of Random Variables

Statistics and Sampling Distributions. A *statistic* summarizes information in a sample and hence is a function of observations $y_1, \ldots, y_n$. Because observations are realizations of random variables, the study of distributions of functions of random variables is really the study of the distributions of statistics, known as *sampling distributions*. Linear combinations of the form $\sum_{i=1}^n a_i y_i$ are an important type of function. Here, $a_1, \ldots, a_n$ are known constants. To begin, we suppose that $y_1, \ldots, y_n$ are mutually independent random variables with $\mathrm{E}\ y_i = \mu_i$ and $\mathrm{Var}\ y_i = \sigma_i^2$. Then, by the *linearity of expectations*, we have

$$\mathrm{E}\left(\sum_{i=1}^n a_i y_i\right) = \sum_{i=1}^n a_i \mu_i \quad \text{and} \quad \mathrm{Var}\left(\sum_{i=1}^n a_i y_i\right) = \sum_{i=1}^n a_i^2 \sigma_i^2.$$

An important theorem in mathematical statistics is that, if each random variable is normally distributed, then linear combinations are also normally distributed.

Linearity of Normal Random Variables. Suppose that $y_1, \ldots, y_n$ are mutually independent random variables with $y_i \sim N(\mu_i, \sigma_i^2)$. (Read $\sim$ to mean "is distributed as.") Then,

$$\sum_{i=1}^n a_i y_i \sim N\left(\sum_{i=1}^n a_i \mu_i, \sum_{i=1}^n a_i^2 \sigma_i^2\right).$$

There are several applications of this important property. First, it can be checked that if $y \sim N(\mu, \sigma^2)$, then $(y - \mu)/\sigma \sim N(0, 1)$. Second, assume that $y_1, \ldots, y_n$ are identically and independently distribution (i.i.d.) as $N(\mu, \sigma^2)$ and take $a_i = n^{-1}$. Then, we have

$$\overline{y} = \frac{1}{n}\sum_{i=1}^n y_i \sim N\left(\mu, \frac{\sigma^2}{n}\right).$$

Equivalently, $\sqrt{n}\,(\overline{y} - \mu)/\sigma$ is standard normal.

Thus, the important sample statistic $\overline{y}$ has a normal distribution. Further, the distribution of the sample variance s_y^2 can also be calculated. For $y_1, \ldots, y_n$ are i.i.d. $N(\mu, \sigma^2)$, we have that $(n-1)\, s_y^2/\sigma^2 \sim \chi^2_{n-1}$, a χ^2 (chi-square) distribution with $n-1$ degrees of freedom. Further, $\overline{y}$ is independent of s_y^2. From these two results, we have that

$$\frac{\sqrt{n}}{s_y}\,(\overline{y} - \mu) \sim t_{n-1},$$

a t-distribution with $n-1$ degrees of freedom.

A1.2 Estimation and Prediction

Suppose that $y_1, \ldots, y_n$ are i.i.d. random variables from a distribution that can be summarized by an unknown parameter θ. We are interested in the quality of an estimate of θ and denote $\widehat{\theta}$ as this estimator. For example, we consider $\theta = \mu$ with $\widehat{\theta} = \overline{y}$ and $\theta = \sigma^2$ with $\widehat{\theta} = s_y^2$ as our leading examples.

Point Estimation and Unbiasedness. Because $\widehat{\theta}$ provides a (single) approximation of θ, it is referred to as a *point estimate* of θ. As a statistic, $\widehat{\theta}$ is a function of the observations $y_1, \ldots, y_n$ that varies from one sample to the next. Thus, values of $\widehat{\theta}$ vary from one sample to the next. To examine how close $\widehat{\theta}$ tends to be to θ, we examine several properties of $\widehat{\theta}$, in particular, the *bias* and *consistency*. A point estimator $\widehat{\theta}$ is said to be an *unbiased estimator* of θ if $\mathrm{E}\,\widehat{\theta} = \theta$. For example, because $\mathrm{E}\,\overline{y} = \mu$, $\overline{y}$ is an unbiased estimator of μ.

Finite Sample versus Large Sample Properties of Estimators. Biasedness is said to be a *finite sample* property because it is valid for each sample size n. A *limiting*, or *large, sample* property is *consistency*. Consistency is expressed in two ways, *weak* and *strong*, consistency. An estimator is said to be *(weakly) consistent* if

$$\lim_{n\to\infty} \Pr\left(|\widehat{\theta} - \theta| < h\right) = 1,$$

for each positive h. An estimator is said to be *strongly consistent* if $\lim_{n\to\infty} \widehat{\theta} = \theta$, with probability one.

Least Squares Estimation Principle. In this text, two main estimation principles are used, *least squares* estimation and *maximum likelihood* estimation. For the least squares procedure, consider independent random variables $y_1, \ldots, y_n$ with means $\mathrm{E}\, y_i = \mathrm{g}_i(\theta)$. Here, $\mathrm{g}_i(\cdot)$ is a known function up to θ, the unknown parameter. The least squares estimator is that value of θ that minimizes the sum of squares

$$\mathrm{SS}(\theta) = \sum_{i=1}^{n} (y_i - \mathrm{g}_i(\theta))^2 .$$

Maximum Likelihood Estimation Principle. Maximum likelihood estimates are values of the parameter that are "most likely" to have been produced by the data. Consider the independent random variables $y_1, \ldots, y_n$ with probability function $\mathrm{f}_i(a_i, \theta)$. Here, $\mathrm{f}_i(a_i, \theta)$ is interpreted to be a probability mass function for discrete y_i or a probability density function for continuous y_i, evaluated at

a_i, the realization of y_i. The function $\mathrm{f}_i(a_i, \theta)$ is assumed known up to θ, the unknown parameter. The likelihood of the random variables $y_1, \ldots, y_n$ taking on values $a_1, \ldots, a_n$ is

$$\mathrm{L}(\theta) = \prod_{i=1}^{n} \mathrm{f}_i(a_i, \theta).$$

The value of θ that maximizes $\mathrm{L}(\theta)$ is the *maximum likelihood estimator.*

Confidence Intervals. Although point estimates provide a single approximation to parameters, *interval estimates* provide a range that include parameters with a certain prespecified level of probability, or *confidence*. A pair of statistics, $\widehat{\theta}_1$ and $\widehat{\theta}_2$, provide an interval of the form $\left[\widehat{\theta}_1, \widehat{\theta}_2\right]$. This interval is a $100(1-\alpha)\%$ confidence interval for θ if

$$\Pr\left(\widehat{\theta}_1 < \theta < \widehat{\theta}_2\right) \geq 1 - \alpha.$$

For example, suppose that $y_1, \ldots, y_n$ are i.i.d. $N(\mu, \sigma^2)$ random variables. Recall that $\sqrt{n}\left(\overline{y} - \mu\right)/s_y \sim t_{n-1}$. This allows us to develop a $100(1-\alpha)\%$ confidence interval for μ of the form $\overline{y} \pm (t\text{–}value) s_y/\sqrt{n}$, where *t–value* is the $(1-\alpha/2)$*th* percentile from a t-distribution with $n-1$ degrees of freedom.

Prediction Intervals. Prediction intervals have the same form as confidence intervals. However, a confidence interval provides a range for a parameter, whereas a prediction interval provides a range for external values of the observations. From observations $y_1, \ldots, y_n$, we seek to construct statistics $\widehat{\theta}_1$ and $\widehat{\theta}_2$ such that $\Pr\left(\widehat{\theta}_1 < y^* < \widehat{\theta}_2\right) \geq 1 - \alpha$. Here, y^* is an additional observation that is not a part of the sample.

A1.3 Testing Hypotheses

Null and Alternative Hypotheses and Test Statistics. An important statistical procedure involves verifying ideas about parameters. That is, before the data are observed, certain ideas about the parameters are formulated. In this text, we consider a *null hypothesis* of the form $H_0 : \theta = \theta_0$ versus an *alternative hypothesis*. We consider both a *two-sided alternative*, $H_a : \theta \neq \theta_0$, and *one-sided alternatives*, either $H_a : \theta > \theta_0$ or $H_a : \theta < \theta_0$. To choose between these competing hypotheses, we use a test statistic T_n that is typically a point estimate of θ or a version that is rescaled to conform to a reference distribution under H_0. For example, to test $H_0 : \mu = \mu_0$, we often use $T_n = \overline{y}$ or $T_n = \sqrt{n}\left(\overline{y} - \mu_0\right)/s_y$. Note that the latter choice has a t_{n-1} distribution, under the assumptions of i.i.d. normal data.

Rejection Regions and Significance Level. With a statistic in hand, we now establish a criterion for deciding between the two competing hypotheses. This can be done by establishing a *rejection*, or *critical, region*. The critical region consists of all possible outcomes of T_n that leads us to reject H_0 in favor of H_a. To specify the critical region, we first quantify the types of errors that can be made in the decision-making procedure. A *Type I error* consists of rejecting H_0 falsely and a *Type II error* consists of rejecting H_a falsely. The probability of a Type I error is called the *significance level*. Prespecifying the significance level is often enough to determine the critical region. For example, suppose that $y_1, \ldots, y_n$ are i.i.d. $N(\mu, \sigma^2)$ and we are interested in deciding between $H_0 : \mu = \mu_0$ and $H_a : \mu > \mu_0$. Thinking of our test statistic $T_n = \overline{y}$, we know that we would like to reject H_0 if $\overline{y}$ is larger that μ_0. The question is how much larger? Specifying a significance level α, we want to find a critical region of the form $\{\overline{y} > c\}$ for some constant c. To this end, we have

$$\begin{aligned}
\alpha &= \Pr(\text{Type I error}) = \Pr(\text{reject } H_0 \text{ assuming } H_0 : \mu = \mu_0 \text{ is true}) \\
&= \Pr(\overline{y} > c) = \Pr(\sqrt{n}\,(\overline{y} - \mu_0)\,/s_y > \sqrt{n}\,(c - \mu_0)\,/s_y) \\
&= \Pr(t_{n-1} > \sqrt{n}\,(c - \mu_0)\,/s_y).
\end{aligned}$$

With $df = n - 1$ degrees of freedom, we have that $t\text{–}value = \sqrt{n}\,(c - \mu_0)\,/s_y$ where the *t–value* is the $(1 - \alpha)th$ percentile from a t-distribution. Thus, solving for c, our critical region is of the form $\{\overline{y} > \mu_0 + (t\text{–}value)/s_y/\sqrt{n}\}$.

Relationship between Confidence Intervals and Hypothesis Tests. Similar calculations show that, for testing $H_0 : \mu = \mu_0$ versus $H_a : \theta \neq \theta_0$, the critical region is of the form

$$\left\{\overline{y} > \mu_0 + (t\text{–}value) \Big/ \frac{s_y}{\sqrt{n}} \quad \text{or} \quad \overline{y} > \mu_0 + (t\text{–}value) \Big/ \frac{s_y}{\sqrt{n}}\right\}.$$

Here, the t-value is a $(1 - \alpha/2)th$ percentile from a t-distribution with $df = n - 1$ degrees of freedom. It is interesting to note that the event of falling in this two-sided critical region is equivalent to the event of μ_0 falling outside the confidence interval $\overline{y} \pm (t\text{–}value)s_y/\sqrt{n}$. This establishes that confidence intervals and hypothesis tests really report the same evidence with a different emphasis on interpretation of the statistical inference.

***p*-value.** Another useful concept in hypothesis testing is the *p-value*, which is shorthand for *probability value*. For a data set, a p-value is defined as the smallest significance level for which the null hypothesis would be rejected. The p-value is a useful summary statistic for the data analyst to report because it allows the reader to understand the strength of the deviation from the null hypothesis.

Appendix 2

Matrix Algebra

A2.1 Basic Definitions

- *Matrix*: a rectangular array of numbers arranged in rows and columns (the plural of matrix is *matrices*).
- *Dimension* of the matrix: the number of rows and columns of the matrix.
- Consider a matrix $\mathbf{A}$ that has dimension $m \times k$. Let a_{ij} be the symbol for the number in the ith row and jth column of $\mathbf{A}$. In general, we work with matrices of the form

$$\mathbf{A} = \begin{pmatrix} a_{11} & a_{12} & \cdots & a_{1k} \\ a_{21} & a_{22} & \cdots & a_{2k} \\ \vdots & \vdots & \ddots & \vdots \\ a_{m1} & a_{m2} & \cdots & a_{mk} \end{pmatrix}.$$

- *Vector*: a (column) vector is a matrix containing only one column ($k = 1$).
- *Row vector*: a matrix containing only one row ($m = 1$).
- *Transpose*: transpose of a matrix $\mathbf{A}$ is defined by interchanging the rows and columns and is denoted by $\mathbf{A}'$ (or $\mathbf{A}^{\mathrm{T}}$). Thus, if $\mathbf{A}$ has dimension $m \times k$, then $\mathbf{A}'$ has dimension $k \times m$.
- *Square matrix*: a matrix in which the number of rows equals the number of columns; that is, $m = k$.
- *Diagonal element*: the number in the rth row and column of a square matrix, $r = 1, 2, \ldots$.
- *Diagonal matrix*: a square matrix in which all nondiagonal numbers are equal to zero.
- *Identity matrix*: a diagonal matrix in which all the diagonal elements are equal to one and that is denoted by $\mathbf{I}$.
- *Symmetric matrix*: a square matrix $\mathbf{A}$ such that the matrix remains unchanged if we interchange the roles of the rows and columns; that is, if $\mathbf{A} = \mathbf{A}'$. Note that a diagonal matrix is a symmetric matrix.

A2.2 Review of Basic Operations

- *Scalar multiplication.* Let c be a real number, called a *scalar* (a 1×1 matrix). Multiplying a scalar c by a matrix $\mathbf{A}$ is denoted by $c\mathbf{A}$

and defined by

$$c\mathbf{A} = \begin{pmatrix} ca_{11} & ca_{12} & \cdots & ca_{1k} \\ ca_{21} & ca_{22} & \cdots & ca_{2k} \\ \vdots & \vdots & \ddots & \vdots \\ ca_{m1} & ca_{m2} & \cdots & ca_{mk} \end{pmatrix}.$$

- *Matrix addition and subtraction.* Let $\mathbf{A}$ and $\mathbf{B}$ be matrices, each with dimension $m \times k$. Use a_{ij} and b_{ij} to denote the numbers in the ith row and jth column of $\mathbf{A}$ and B, respectively. Then, the matrix $\mathbf{C} = \mathbf{A} + \mathbf{B}$ is defined as the matrix with the number $(a_{ij} + b_{ij})$ to denote the number in the ith row and jth column. Similarly, the matrix $\mathbf{C} = \mathbf{A} - \mathbf{B}$ is defined as the matrix with the number $(a_{ij} - b_{ij})$ to denote the numbers in the ith row and jth column.
- *Matrix multiplication.* If $\mathbf{A}$ is a matrix of dimension $m \times c$ and B is a matrix of dimension $c \times k$, then $\mathbf{C} = \mathbf{AB}$ is a matrix of dimension $m \times k$. The number in the ith row and jth column of $\mathbf{C}$ is $\sum_{s=1}^{c} a_{is} b_{sj}$.
- *Determinant.* A determinant is a function of a square matrix, denoted by det($\mathbf{A}$), or $|\mathbf{A}|$. For a 1×1 matrix, the determinant is det($\mathbf{A}$) $= a_{11}$. To define determinants for larger matrices, we need two additional concepts. Let $\mathbf{A}_{rs}$ be the $(m-1) \times (m-1)$ submatrix of $\mathbf{A}$ defined be removing the rth row and sth column. Recursively, define det($\mathbf{A}$) $= \sum_{s=1}^{m} (-1)^{r+s} a_{rs} \mathbf{A}_{rs}$, for any $r = 1, \ldots, m$. For example, for $m = 2$, we have det($\mathbf{A}$) $= a_{11}a_{22} - a_{12}a_{21}$.
- *Matrix inverse.* In matrix algebra, there is no concept of division. Instead, we extend the concept of reciprocals of real numbers. To begin, suppose that $\mathbf{A}$ is a square matrix of dimension $m \times m$ such that det($\mathbf{A}$) $\neq 0$. Further, let $\mathbf{I}$ be the $m \times m$ identity matrix. If there exists a $m \times m$ matrix $\mathbf{B}$ such that $\mathbf{AB} = \mathbf{I} = \mathbf{BA}$, then $\mathbf{B}$ is the *inverse* of $\mathbf{A}$ and is written as $\mathbf{B} = \mathbf{A}^{-1}$.

A2.3 Further Definitions

- *Linearly dependent vectors*: a set of vectors $\mathbf{c}_1, \ldots, \mathbf{c}_k$ is said to be linearly dependent if one of the vectors in the set can be written as a linear combination of the others.
- *Linearly independent vectors*: a set of vectors $\mathbf{c}_1, \ldots, \mathbf{c}_k$ is said to be linearly independent if they are not linearly dependent. Specifically, a set of vectors $\mathbf{c}_1, \ldots, \mathbf{c}_k$ is said to be linearly independent if and only if the only solution of the equation $x_1\mathbf{c}_1 + \cdots + x_k\mathbf{c}_k = 0$ is $x_1 = \cdots = x_k = 0$.
- *Rank of a matrix*: the largest number of linearly independent columns (or rows) of a matrix.
- *Singular matrix*: a square matrix $\mathbf{A}$ such that det($\mathbf{A}$) $= 0$.
- *Nonsingular matrix*: a square matrix $\mathbf{A}$ such that det($\mathbf{A}$) $\neq 0$.
- *Positive definite matrix*: a symmetric square matrix $\mathbf{A}$ such that $\mathbf{x}'\mathbf{Ax} > 0$ for $\mathbf{x} \neq \mathbf{0}$.

- *Nonnegative definite matrix*: a symmetric square matrix **A** such that $\mathbf{x}'\mathbf{A}\mathbf{x} \geq 0$ for $\mathbf{x} \neq \mathbf{0}$.
- *Orthogonal*: two matrices **A** and **B** are orthogonal if $\mathbf{A}'\mathbf{B} = \mathbf{0}$, a zero matrix.
- *Idempotent*: a square matrix such that $\mathbf{AA} = \mathbf{A}$.
- *Trace*: the sum of all diagonal elements of a square matrix.
- *Eigenvalues*: the solutions of the nth degree polynomial $\det(\mathbf{A} - \lambda\mathbf{I}) = 0$; also known as *characteristic roots* and *latent roots*.
- *Eigenvector*: a vector **x** such that $\mathbf{Ax} = \lambda\mathbf{x}$, where λ is an eigenvalue of **A**; also known as a *characteristic vector* and *latent vector*.
- *Generalized inverse*: of a matrix **A**, a matrix **B** such that $\mathbf{ABA} = \mathbf{A}$. We use the notation $\mathbf{A}^-$ to denote the generalized inverse of **A**. In the case that **A** is invertible, then $\mathbf{A}^-$ is unique and equals $\mathbf{A}^{-1}$. Although there are several definitions of generalized inverses, the foregoing definition suffices for our purposes. See Searle (1987) for further discussion of alternative definitions of generalized inverses.
- *Gradient vector*: a vector of partial derivatives. If $\mathrm{f}(\cdot)$ is a function of the vector $\mathbf{x} = (x_1, \ldots, x_m)'$, then the gradient vector is $\partial \mathrm{f}(\mathbf{x})/\partial \mathbf{x}$. The ith row of the gradient vector is is $\partial \mathrm{f}(\mathbf{x})/\partial x_i$.
- *Hessian matrix*: a matrix of second derivatives. If $\mathrm{f}(\cdot)$ is a function of the vector $\mathbf{x} = (x_1, \ldots, x_m)'$, then the Hessian matrix is $\partial^2 \mathrm{f}(\mathbf{x})/\partial \mathbf{x} \partial \mathbf{x}'$. The element in the ith row and jth column of the Hessian matrix is $\partial^2 \mathrm{f}(\mathbf{x})/\partial x_i \partial x_j$.

Appendix 3

Probability Tables

Probability tables are available at the book web site, http://research.bus.wisc.edu/RegActuaries.

A3.1 Normal Distribution

Recall from equation (1.1) that the probability density function is defined by

$$\mathrm{f}(y) = \frac{1}{\sigma\sqrt{2\pi}} \exp\left(-\frac{1}{2\sigma^2}(y-\mu)^2\right),$$

where μ and σ^2 are parameters that describe the curve. In this case, we write $y \sim N(\mu, \sigma^2)$. Straightforward calculations show that

$$\mathrm{E}\, y = \int_{-\infty}^{\infty} y\mathrm{f}(y)dy = \int_{-\infty}^{\infty} y \frac{1}{\sigma\sqrt{2\pi}} \exp\left(-\frac{1}{2\sigma^2}(y-\mu)^2\right) dy = \mu$$

and

$$\begin{aligned}\mathrm{Var}\, y &= \int_{-\infty}^{\infty} (y-\mu)^2 \mathrm{f}(y)dy \\ &= \int_{-\infty}^{\infty} (y-\mu)^2 \frac{1}{\sigma\sqrt{2\pi}} \exp\left(-\frac{1}{2\sigma^2}(y-\mu)^2\right) dy = \sigma^2.\end{aligned}$$

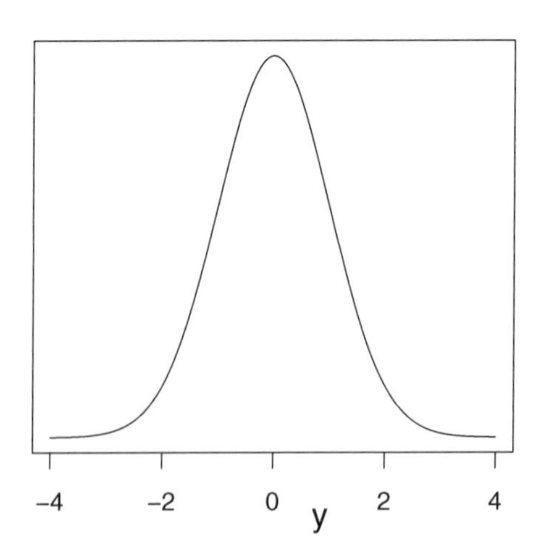

Figure A3.1 Standard normal probability density function.

Thus, the notation $y \sim N(\mu, \sigma^2)$ is interpreted to mean that the random variable is distributed normally with mean μ and variance σ^2. If $y \sim N(0, 1)$, then y is said to be *standard normal*.

Table A3.1
Standard Normal Distribution Function

x	0.0	0.1	0.2	0.3	0.4	0.5	0.6	0.7	0.8	0.9
0	0.5000	0.5398	0.5793	0.6179	0.6554	0.6915	0.7257	0.7580	0.7881	0.8159
1	0.8413	0.8643	0.8849	0.9032	0.9192	0.9332	0.9452	0.9554	0.9641	0.9713
2	0.9772	0.9821	0.9861	0.9893	0.9918	0.9938	0.9953	0.9965	0.9974	0.9981
3	0.9987	0.9990	0.9993	0.9995	0.9997	0.9998	0.9998	0.9999	0.9999	1.0000

Notes: Probabilities can be found by looking at the appropriate row for the lead digit and column for the decimal. For example, $\Pr(y \leq 0.1) = 0.5398$.

A3.2 Chi-Square Distribution

Chi-Square Distribution. Several important distributions can be linked to the normal distribution. If $y_1, \ldots, y_n$ are i.i.d. random variables such that each $y_i \sim N(0, 1)$, then $\sum_{i=1}^{n} y_i^2$ is said to have a *chi-square distribution* with parameter n. More generally, a random variable w with probability density function

$$\mathrm{f}(w) = \frac{2^{-k/2}}{\Gamma(k/2)} w^{k/2-1} \exp(-w/2), \qquad w > 0,$$

is said to have a chi-square with $df = k$ degrees of freedom, written $w \sim \chi_k^2$. Easy calculations show that for $w \sim \chi_k^2$, we have $\mathrm{E}\ w = k$ and $\mathrm{Var}\ w = 2k$. In general, the degrees-of-freedom parameter need not be an integer, though it is for the applications of this text.

Table A3.2 Percentiles from Several Chi-Square Distributions

	Probabilities									
df	0.6	0.7	0.8	0.9	0.95	0.975	0.99	0.995	0.9975	0.999
1	0.71	1.07	1.64	2.71	3.84	5.02	6.63	7.88	9.14	10.83
2	1.83	2.41	3.22	4.61	5.99	7.38	9.21	10.60	11.98	13.82
3	2.95	3.66	4.64	6.25	7.81	9.35	11.34	12.84	14.32	16.27
4	4.04	4.88	5.99	7.78	9.49	11.14	13.28	14.86	16.42	18.47
5	5.13	6.06	7.29	9.24	11.07	12.83	15.09	16.75	18.39	20.52
10	10.47	11.78	13.44	15.99	18.31	20.48	23.21	25.19	27.11	29.59
15	15.73	17.32	19.31	22.31	25.00	27.49	30.58	32.80	34.95	37.70
20	20.95	22.77	25.04	28.41	31.41	34.17	37.57	40.00	42.34	45.31
25	26.14	28.17	30.68	34.38	37.65	40.65	44.31	46.93	49.44	52.62
30	31.32	33.53	36.25	40.26	43.77	46.98	50.89	53.67	56.33	59.70
35	36.47	38.86	41.78	46.06	49.80	53.20	57.34	60.27	63.08	66.62
40	41.62	44.16	47.27	51.81	55.76	59.34	63.69	66.77	69.70	73.40
60	62.13	65.23	68.97	74.40	79.08	83.30	88.38	91.95	95.34	99.61
120	123.29	127.62	132.81	140.23	146.57	152.21	158.95	163.65	168.08	173.62

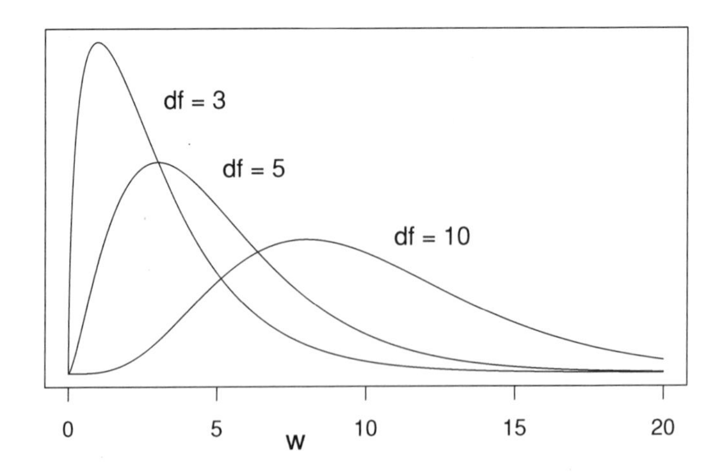

Figure A3.2 Several chi-square probability density functions. Shown are curves for $df = 3$, $df = 5$, and $df = 10$. Greater degrees of freedom lead to curves that are less skewed.

A3.3 *t*-Distribution

Suppose that y and w are independent with $y \sim N(0, 1)$ and $w \sim \chi^2_k$. Then, the random variable $t = y/\sqrt{w/k}$ is said to have a t-distribution with $df = k$ degrees of freedom. The probability density function is

$$\mathrm{f}(t) = \frac{\Gamma\left(k + \frac{1}{2}\right)}{\Gamma(k/2)} (k\pi)^{-1/2} \left(1 + \frac{t^2}{k}\right)^{-(k+1/2)}, \qquad -\infty < t < \infty.$$

This has mean 0, for $k > 1$, and variance $k/(k-2)$ for $k > 2$.

Table A3.3 Percentiles from Several t-Distributions

	Probabilities									
df	0.6	0.7	0.8	0.9	0.95	0.975	0.99	0.995	0.9975	0.999
1	0.325	0.727	1.376	3.078	6.314	12.706	31.821	63.657	127.321	318.309
2	0.289	0.617	1.061	1.886	2.920	4.303	6.965	9.925	14.089	22.327
3	0.277	0.584	0.978	1.638	2.353	3.182	4.541	5.841	7.453	10.215
4	0.271	0.569	0.941	1.533	2.132	2.776	3.747	4.604	5.598	7.173
5	0.267	0.559	0.920	1.476	2.015	2.571	3.365	4.032	4.773	5.893
10	0.260	0.542	0.879	1.372	1.812	2.228	2.764	3.169	3.581	4.144
15	0.258	0.536	0.866	1.341	1.753	2.131	2.602	2.947	3.286	3.733
20	0.257	0.533	0.860	1.325	1.725	2.086	2.528	2.845	3.153	3.552
25	0.256	0.531	0.856	1.316	1.708	2.060	2.485	2.787	3.078	3.450
30	0.256	0.530	0.854	1.310	1.697	2.042	2.457	2.750	3.030	3.385
35	0.255	0.529	0.852	1.306	1.690	2.030	2.438	2.724	2.996	3.340
40	0.255	0.529	0.851	1.303	1.684	2.021	2.423	2.704	2.971	3.307
60	0.254	0.527	0.848	1.296	1.671	2.000	2.390	2.660	2.915	3.232
120	0.254	0.526	0.845	1.289	1.658	1.980	2.358	2.617	2.860	3.160
∞	0.253	0.524	0.842	1.282	1.645	1.960	2.326	2.576	2.807	3.090

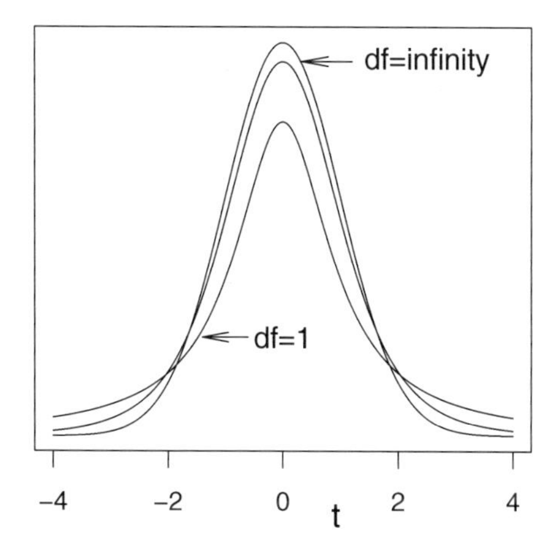

Figure A3.3 Several t-distribution probability density functions. The t-distribution with $df = \infty$ is the standard normal distribution. Shown are curves for $df = 1$, $df = 5$ (not labeled), and $df = \infty$. A lower df means fatter tails.

A3.4 F-Distribution

Suppose that w_1 and w_2 are independent with distributions $w_1 \sim \chi^2_m$ and $w_2 \sim \chi^2_n$. Then, the random variable $F = (w_1/m)/(w_2/n)$ has an F-distribution with parameters $df_1 = m$ and $df_2 = n$, respectively. The probability density function is

$$\mathrm{f}(y) = \frac{\Gamma\left(\frac{m+n}{2}\right)}{\Gamma(m/2)\Gamma(n/2)} \left(\frac{m}{n}\right)^{m/2} \frac{y^{(m-2)/2}}{\left(1+\frac{m}{n}y\right)^{m+n+2}}, \qquad y > 0.$$

This has a mean $n/(n-2)$, for $n > 2$, and a variance $2n^2(m+n-2)/[m(n-2)^2(n-4)]$ for $n > 4$.

Table A3.4 Percentiles from Several F-Distributions

df_1	df_2 = 1	3	5	10	20	30	40	60	120
1	161.45	10.13	6.61	4.96	4.35	4.17	4.08	4.00	3.92
2	199.50	9.55	5.79	4.10	3.49	3.32	3.23	3.15	3.07
3	215.71	9.28	5.41	3.71	3.10	2.92	2.84	2.76	2.68
4	224.58	9.12	5.19	3.48	2.87	2.69	2.61	2.53	2.45
5	230.16	9.01	5.05	3.33	2.71	2.53	2.45	2.37	2.29
10	241.88	8.79	4.74	2.98	2.35	2.16	2.08	1.99	1.91
15	245.95	8.70	4.62	2.85	2.20	2.01	1.92	1.84	1.75
20	248.01	8.66	4.56	2.77	2.12	1.93	1.84	1.75	1.66
25	249.26	8.63	4.52	2.73	2.07	1.88	1.78	1.69	1.60
30	250.10	8.62	4.50	2.70	2.04	1.84	1.74	1.65	1.55
35	250.69	8.60	4.48	2.68	2.01	1.81	1.72	1.62	1.52
40	251.14	8.59	4.46	2.66	1.99	1.79	1.69	1.59	1.50
60	252.20	8.57	4.43	2.62	1.95	1.74	1.64	1.53	1.43
120	253.25	8.55	4.40	2.58	1.90	1.68	1.58	1.47	1.35

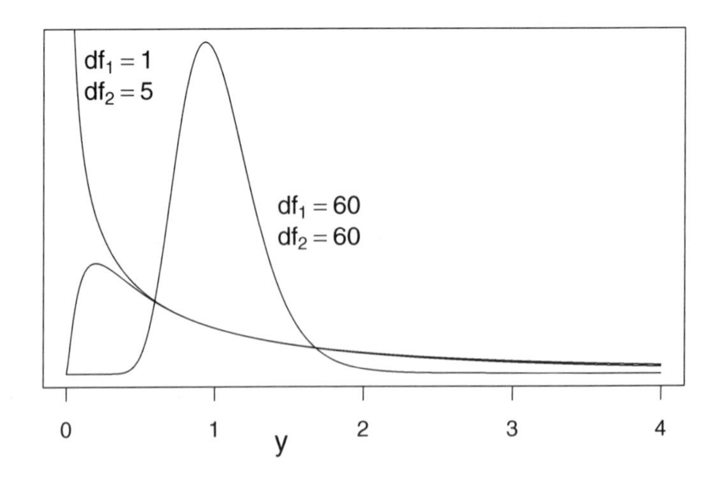

Figure A3.4 Several F-distribution probability density functions. Shown are curves for (i) $df_1 = 1, df_2 = 5$, (ii) $df_1 = 5, df_2 = 1$ (not labeled), and (iii) $df_1 = 60$, $df_2 = 60$. As df_2 tends to ∞, the F-distribution tends to a chi-square distribution.

Index